ELEVENTH EDITION

Introduction to the
Biology of
Marine Life

John F. Morrissey
Sweet Briar College

James L. Sumich
Grossmont College, Emeritus

Deanna R. Pinkard-Meier
University of San Diego

JONES & BARTLETT
LEARNING

World Headquarters
Jones & Bartlett Learning
5 Wall Street
Burlington, MA 01803
978-443-5000
info@jblearning.com
www.jblearning.com

Jones & Bartlett Learning books and products are available through most bookstores and online booksellers. To contact Jones & Bartlett Learning directly, call 800-832-0034, fax 978-443-8000, or visit our website, www.jblearning.com.

Substantial discounts on bulk quantities of Jones & Bartlett Learning publications are available to corporations, professional associations, and other qualified organizations. For details and specific discount information, contact the special sales department at Jones & Bartlett Learning via the above contact information or send an email to specialsales@jblearning.com.

09059-8

Production Credits

VP, Executive Publisher: David D. Cella
Executive Editor: Matthew Kane
Associate Editor: Audrey Schwinn
Senior Production Editor: Nancy Hitchcock
Marketing Manager: Lindsay White
Production Services Manager: Colleen Lamy
Manufacturing and Inventory Control Supervisor: Amy Bacus

Composition: Cenveo® Publisher Services
Cover Design: Michael O'Donnell
Rights & Media Specialist: Jamey O'Quinn
Media Development Editor: Shannon Sheehan
Cover Image: © R. Gino Santa Maria/Shutterstock, Inc.
Printing and Binding: LSC Communications
Cover Printing: LSC Communications

Library of Congress Cataloging-in-Publication Data
Names: Morrissey, John F. (John Francis), 1960- | Sumich, James L.
    | Pinkard-Meier, Deanna R.
Title: Introduction to the biology of marine life.
Description: Eleventh edition/John F. Morrissey, James L. Sumich, Deanna R.
    Pinkard-Meier.|Burlington, Massachusetts: Jones & Bartlett Learning,
    [2018] | Includes index.
Identifiers: LCCN 2016038190 | ISBN 9781284090505
Subjects: LCSH: Marine biology.
Classification: LCC QH91 .S95 2018 | DDC 578.77–dc23
LC record available at https://lccn.loc.gov/2016038190

6048

Printed in the United States of America
20 19 18 17 16   10 9 8 7 6 5 4 3 2 1

Dedication

I dedicate my efforts on this book to my late mother,

Diane Sue Pinkard, who taught me to appreciate

the ocean from an early age, and who supported me

through all of my endeavors. Her love for all things

ocean related made a lasting impression on me as

I pursued a career in the marine sciences.

Brief Contents

Contents

INDEX OF BOXES

Preface

As this new edition is being prepared, the longest global coral bleaching and die-off event in recorded history is occurring, and there is evidence that the oceans have become more acidic due to increased carbon dioxide emissions. As scientists worldwide attempt to deal with new marine environmental problems arising and old problems persisting, continued understanding of the sea and its inhabitants is absolutely necessary. For nonscientists, if there is a desire to preserve our world ocean, which provides so many services to us, and to continue to enjoy having fun in the sea and eating seafood, we must expand our knowledge of the sea and how our daily activities affect its health. Familiarity with the effects of a changing climate will help one understand the complexities of rising sea temperatures and sea levels. An understanding of the biology of corals and other animals will enable scientists to predict the magnitude of impacts to these organisms when the water is too warm or the chemistry changes too much. Knowledge of the chemistry of seawater will provide one with the ability to understand the many changes to the water column that are being observed as ocean acidification becomes more widespread. Perhaps most important, a study of the biology of marine life will help one appreciate the reasons why so many organisms are harmed due to these changes to our world ocean, such as corals, which support entire ecosystems. All of this insight and much more is contained within this eleventh edition.

Audience

We have written *Introduction to the Biology of Marine Life* to engage introductory, college-level students in the excitement and challenge of understanding marine organisms, the environments in which they live, and the challenges they face as the marine environment changes. We assume no previous knowledge of marine biology; however, some exposure to the basic concepts of biology is helpful. This book uses selected groups of marine organisms to develop an understanding of biological principles and processes that are basic to all forms of life in the sea. To build on these basics, we present information dealing with several aspects of taxonomy, evolution, ecology, behavior, and physiology of these selected groups. We hope that a student's venture into this exciting field provides some flavor of the mix of disciplines that constitutes modern biological science. Moreover, we hope that this text cultivates an appreciation for the need to understand marine geology (the seafloor), marine physics (waves, tides, and currents), and marine chemistry (the composition of seawater) before a complete understanding of marine biology can be achieved.

Organization

Although we intend the sequence of topics to be flexible, we have presented our material in four sections.

We begin with an introduction to the sea as a habitat (Chapter 1), highlighting the many ways that the ocean realm differs tremendously from more familiar terrestrial environments, especially in terms of the chemistry of seawater and the geology of the seafloor. We have added an entire chapter on motion within the sea (Chapter 2), highlighting the mechanisms behind ocean movements and how ocean movements affect marine life. Then we provide a brief summary of basic chemical and biological principles that are not unique to the ocean for the beginning student (Chapter 3). The next portion of the book summarizes all life in the sea, with two chapters dedicated to autotrophic producers, large and small, and marine bacteria and viruses (Chapters 4 and 5); one chapter for microbial and invertebrate consumers (Chapter 6); and two chapters covering marine vertebrates (fish, amphibians, and reptiles in Chapter 7; birds and mammals in Chapter 8). We have organized the third section of this new edition around the major marine habitats: estuaries (Chapter 9), coastal seas (Chapter 10), coral reefs (Chapter 11), the open ocean (Chapter 12), and the deep sea (Chapter 13). Chapter 14 includes descriptions of the polar seas and their inhabitants, as well as a discussion of global climate change, including suggestions for lowering one's ecological footprint. Finally, we describe the history and current status of marine fisheries and aquaculture in Chapter 15.

New to the Eleventh Edition

The tenth edition of this text was well received by a wide audience, and this was very encouraging while preparing the eleventh edition. Many wonderful suggestions were made by reviewers and readers of earlier editions, and it was these suggestions, along with creative ideas suggested by the editors at Jones & Bartlett Learning, that led to a textbook that has been greatly improved and updated. This eleventh edition represents our continuing efforts to meet the needs of our readers more completely, and to adapt to changing educational platforms. We are certain that students and instructors alike will be pleased to see the many ways that the book has been augmented with new information, new features, and refreshed dialogue.

Chapter 1 has been updated and improved by making the material more concise and moving information on ocean chemistry and oceanography to the new Chapter 2, "Physical and Chemical Oceanography." The new Chapter 2 includes examples of the linkages between ocean water movement and the fate of marine organisms and additional information on El Niño. Chapter 3 now includes a clarified discussion of photosynthesis and a new section that introduces evolution by natural selection. Chapter 4 has been expanded to cover marine bacteria, Archaea, and viruses and includes updated photosynthesis terminology and an introduction to phylogenetics.

Chapter 5 now contains the latest hypotheses on the phylogenetic relationships of marine algae. Chapter 6 includes a discussion of the abandonment of the term "protista" and more detailed information on invertebrate phyla, including new discussions of the phylum Tardigrada and meiofauna. Chapter 7 includes more detailed information on fish physiology, a discussion of *Hox* genes, and a new example of hermaphroditism in fish. Chapter 8 now includes information on echolocation; additional information on manatees, dolphins, porpoises, and sperm whales; and updated information on population statuses of several groups of marine mammals. Chapter 9 has been augmented with expanded coverage of nutrient pollution, the biology of mangroves, and suggestions for improving the health of estuaries. Chapter 10, which covers the biology of coastal seas, offers coverage on hot topics in marine science such as connectivity. Chapter 11 has been updated with a large amount of recent research examples, new hypotheses on coral spawning strategies, the effects of ocean acidification on coral, and new information on Marine Protected Areas. Chapter 12 now includes more information on the orientation of organisms in the open sea, including zooplankton behavior. An updated Chapter 13 includes new information about technological advancements in deep-sea research and new discoveries in the deep sea. Chapter 14 is a new chapter on the polar seas. It contains information moved from other chapters on animals that inhabit polar climates, new information on climate change and sea ice research, and suggestions for reducing our greenhouse gas emissions. Chapter 15 highlights the latest trends in fisheries science, updated fisheries statistics, the recent increase in aquaculture, and concludes the book with suggestions for becoming a steward of the ocean.

The Student Experience

Each of the chapters is designed to introduce key concepts, reinforce understanding, and encourage independent investigation and education.

- **Student Learning Outcomes**—Listed at the beginning of every chapter, these learning objectives prepare students for the material they will be learning.
- **Chapter Outlines**—The chapter's framework is clearly laid out to help students plan their reading and study.

CHAPTER

5

Marine Macroalgae and Plants

STUDENT LEARNING OUTCOMES

1. Compare and contrast the major groups of marine macroalgae and plants.

2. Choose one group of marine plants (Anthophyta), and describe how the group is adapted for submerged life.

3. Explain why green and red algae are more closely related to land plants than to brown algae.

4. Compare and contrast the major body parts of macroalgae to those found in land plants.

5. Identify which areas of the world contain kelp forests, and describe why kelp forests are present in those locations.

6. Develop a hypothesis to answer the following question: why are red algae found in great abundance in the tropics?

7. Create a list of conditions necessary for high net primary productivity (NPP).

8. Evaluate the NPP information provided in text, tables, and figures, and explain why it is so variable in different regions of the ocean.

CHAPTER OUTLINE

5.1 Division Anthophyta
Submerged Seagrasses
Case Study: Johnson's Seagrass (*Halophila johnsonii*): The First and Only Marine Plant Listed Under the Endangered Species Act
Emergent Flowering Plants
5.2 The Seaweeds
Structural Features of Seaweeds

5.3 Geographic Distribution
RESEARCH in Progress: Marine Plants and Algae as Nursery Habitats
5.4 Global Marine Primary Production
Study Guide
References

- **High-Quality, Carefully Rendered Illustrations and Figures**—More than 340 NEW and revised photos and illustrations are included in this edition to help support visual learners, clarify key concepts, and enhance the students' reading experience.

Figure 2.38 Graph depicting the filtering rate of a marine barnacle as a function of water temperature. The filtering rate drops dramatically as the barnacle is placed outside the range of tolerable temperatures.
Modified from Southward, A. J., *Helgol. wiss. Meersuntersuch* 10 (1964):391–401.

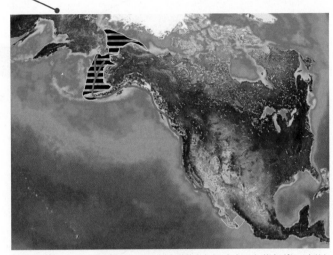

Figure 8.32 Migratory route (red line) of the North Pacific gray whale, with summer feeding (black hatching) and winter breeding areas (purple) indicated. Primary productivity is indicated by water color, with green representing the highest chlorophyll content and dark blue the lowest.
Courtesy of NASA.

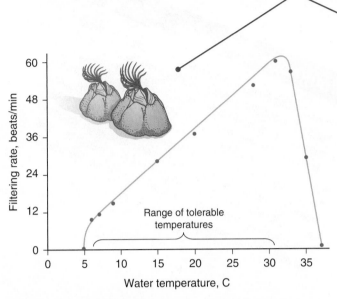

Figure 4.7 A common marine silicoflagellate slightly smaller than the coccolithophores shown in Figure 4.6.
NOAA Photo Library.

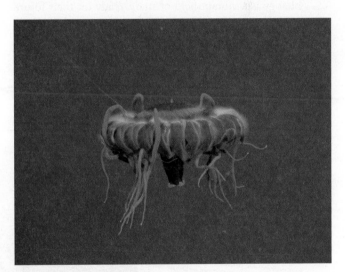

Figure 13.8 A crown jellyfish, *Atolla*

Figure 8.22 A manatee floating, surrounded by snappers in Crystal River, Florida.
Keith Ramos/USFWS.

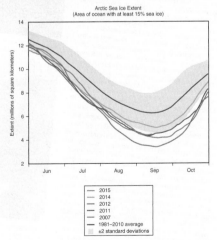

Figure 14.22 Average Arctic sea ice extent for 1981 to 2010 compared to the five lowest ice measurements recorded. Data are provided by the National Snow and Ice Data Center.
Courtesy of National Snow and Ice Data Center.

- **Did You Know?**—These NEW boxes are scattered throughout each chapter, providing interesting marine biology material for students and generating curiosity about the marine world.

- **Research in Progress**—One of our primary objectives is to show students that marine biology, like all sciences, is a dynamic and active field. Each year, recent discoveries about the sea are published in thousands of new scientific papers. A popular feature in previous editions, these boxes have been fully updated to reflect current ongoing research that will encourage students to learn more about real work being done in the field of marine biology. Through these boxes we hope to show the process of science, as well as to suggest to our readers that marine biology is a vibrant field of study, ready for their future contributions. These boxes now include Critical Thinking Questions for student engagement.

DID YOU KNOW?

Hagfish slime is unique among animal mucus because it contains protein threads that help it expand dramatically when it contacts seawater. The sticky, expanding mucus is what was thought to deter predators, and in 2011 this behavior was finally caught on video for observation by researchers. Because the slime expands, it basically chokes fish that are trying to eat the hagfish by clogging their gills. Video from 2011 also caught hagfish preying on other fish, a previously unknown behavior for this fish that was considered a scavenger, feeding on dead and decaying matter, and sometimes on small living invertebrates. These unique features are likely several of the characteristics that have allowed hagfish to persist for so long with success.

RESEARCH in Progress

A Day in the Life of Meroplankton

Orienting in the open ocean requires special adaptations for detecting the environment and slight changes to the environment. Meroplankton are particularly affected by their abilities to manipulate their positions, because successful settlement to their next habitat, the juvenile habitat, is crucial for survival. For many years the mechanisms behind larval settlement were poorly understood, and it was assumed that most meroplanktonic larvae were subject to currents and had little control over their locations in their environment. The large number of gametes released by adults was thought to make up for the low survival rates of larvae that only make it to the settlement habitat by chance.

Research efforts have been focused on larval settlement in an attempt to tease out cues for this dramatic shift in habitat use. Environmental cues such as lunar phase, tidal cycle, water temperature, scent, and salinity have now been correlated with the timing of settlement events for numerous species. Settlement often has an age and/or size requirement, and once one or both of these requirements have been met larvae use one or more of the environmental cues listed above to time their transition from the meroplankton to the benthos. Now that it has been established that a variety of environmental cues influence settlement, several questions remain: (1) how do larvae with limited swimming capabilities leave the pelagic realm and settle to the benthic realm? (2) are larvae more capable swimmers than previously thought? and (3) are larvae capable of orienting to increase their chances of settling to a suitable habitat?

Observing larvae in their natural environment is a difficult task. They are tiny, their distributions are mainly unpredictable and patchy, and their depths range drastically. Laboratory observations of reared larvae provide some interesting information, but the question of whether behaviors are similar in the laboratory setting and the natural environment introduces uncertainty. One group of scientists led by Dr. Claire Paris is working in the subtropical and tropical Atlantic, the Great Barrier Reef, the Red Sea, and the North Sea fjords to observe larvae in their environment using novel research methods. They are using a combination of a newly invented floating laboratory placed *in situ* (named the Drifting *In Situ* Chamber [DISC]; Figure A) and personal observations while freediving. Bubbles and noise from scuba equipment disturb the environment, thus Dr. Paris, a national record freediver, uses her freediving skills to place her equipment in the ocean and to deploy larvae into underwater behavioral arenas with little disturbance to the animals and the surrounding area (Figure B). The DISC allows for tracking the orientation and behavior of fish larvae in their

natural environment because it drifts with the currents and is made of transparent acrylic and mesh material, keeping it open to sight, sound, odor, and other natural environmental fluctuations that may exist. Environmental sensors and a camera that operates all hours of the day and night are mounted on the DISC to record environmental parameters and swimming behaviors (Figure C).

The results of the DISC floating laboratory experiments are intriguing thus far and appear to coincide with observations of larvae swimming freely outside of a DISC environment. In a study conducted at several sites around the Great Barrier Reef, hundreds of damselfish larvae clearly oriented themselves so that they were swimming in a southerly direction. Their swimming behavior was affected by the amount of sun available, with swimming behaviors more consistent under sunny skies and less so under cloudy skies. Time of day and angle of the sun (i.e., sun azimuth and elevation) also affected swimming behaviors. Results of observations of larvae in the DISC were very similar to those of free-swimming larvae. The DISC, however, allows cue manipulation to find the mechanisms for orientation. For example, olfactory cues propagating in the open ocean as turbulent ebb plumes help fish larvae to find their way back to isolated atolls.

DISC observations of lobster postlarvae in the Florida Straits revealed clear patterns of swimming direction in which they were keeping a significant bearing. In addition, the lobster postlarvae swam day and night, in contrast to results from a laboratory investigation that indicated swimming during the day only. Swimming orientation was generally against the prevailing northward-flowing Gulf Stream and was adjusted to veer toward the coast during ebb tides; these adjustments in swimming direction allowed them to remain on a shoreward trajectory. Lobster postlarvae also adjusted their swimming direction relative to wind direction, presumably using the wind to help them swim toward the coast.

The results of the DISC studies indicate that although larvae may not be strong swimmers given their small size, they are very capable of controlling their positions in the water to produce their desired end result: finding appropriate settlement habitat. Fish and invertebrate larvae appear to be very sensitive to small changes in their environment and use these changes as cues for orienting, not only for the final settlement act, but during the days or weeks prior to settlement to maintain a suitable location along their journey. This area of research is wide open for new discoveries, as the vast majority of marine animals have a pelagic larval phase, and additional cues for larval orientation certainly exist and have yet to be discovered.

Critical Thinking Questions

1. Do you think that use of the DISC and other similar devices allows for observations of the natural behavior of zooplankton? Why or why not?

2. It appears that some larvae orient better in sunny conditions than in cloudy conditions. Why do you think this is the case?

For Further Reading

Kough, S. A., C. B. Paris, and E. Staaterman. 2014. The *in situ* swimming and orientation behavior of spiny lobster (*Panulirus argus*) postlarvae. *Marine Ecology Progress Series* 504:207–219.

Leis, J. M., C. B. Paris, J.-O. Irisson, M. N. Yerman, and U. E. Siebeck. 2014. Orientation of fish larvae *in situ* is consistent among locations, years and methods, but varies with time of day. *Marine Ecology Progress Series* 505:193–208.

Paris, C. B., J. Atema, J.-O. Irisson, M. Kingsford, G. Gerlach, and C. M. Guigand. 2013. Reef odor: A wake up call for navigation in reef fish larvae. *PLoS ONE* 8(8):e72808.

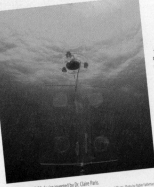

Figure A　The DISC device invented by Dr. Claire Paris.
Courtesy of Dr. Claire Paris, Rosenstiel School of Marine & Atmospheric Science, University of Miami. Photo by Robin Faillettaz.

Figure B　Dr. Claire Paris and her field assistant (and husband), Ricardo Paris, recovered an oceanographic sensor by freediving.
Courtesy of Dr. Claire Paris, Rosenstiel School of Marine & Atmospheric Science, University of Miami. Photo by William Trubridge.

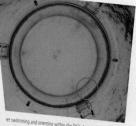

...er swimming and orienting within the DISC device, photographed by ...e red circle indicates the location of the fish.
...nstiel School of Marine & Atmospheric Science, University of Miami.

- **Case Studies**—A NEW Case Study box has been added to each chapter, taking a closer look at a particular organism. Each Case Study also includes Critical Thinking Questions.

Case Study

The Great Pacific Garbage Patch

Currents and gyres not only transport living organisms but also anything that is present in the water and light enough to be moved along within a water mass. The large gyres pictured in Figure 2.15 have been a topic of interest for researchers for decades, and just recently the North Pacific Gyre has been in the spotlight. Located within the North Pacific Gyre are several areas known as "garbage patches" (**Figure A**). Human garbage dumped into the sea from land or boats accumulates in regions of the ocean to form these patches. Most of the garbage is made of plastic.

The locations of garbage patches are somewhat predictable, because on a large scale the locations of major currents and gyres are predictable. The actual composition of the garbage patches changes daily, though, because currents change with changing weather patterns, and the water is constantly mixing up the contents of the patches. The North Pacific Gyre has gained attention in recent years, but garbage patches are found in many other areas of Earth's oceans. Floating debris accumulates within any moving water mass, and gyres are particularly good current patterns for this accumulation of junk.

An erroneous assumption that is often made concerning the garbage patches is that large pieces of trash are found floating around; however, most of it is actually small. Scientists studying these areas describe the trash as a peppery soup with larger garbage floating through it. The larger garbage may be derelict fishing gear, abandoned boating equipment, and a variety of plastic items. Birds and fish often eat the plastic items, mistaking them for food (**Figure B**). Over time, garbage made of plastic breaks down into smaller pieces, some microscopic, and floats around with the currents. Some of the worst areas of accumulated garbage are not even visible to the common person sailing right through the area, unless he or she is looking closely for plastic peppery soup.

The question now is the following: how do we clean up the oceans and get rid of these garbage patches? Cleaning up the ocean is a difficult task. Because the plastics are mostly very small, there is no feasible method to remove them. The best thing we can do now that we are aware of these garbage patches is to prevent the future addition of more debris. Until we stop adding more garbage to the ocean, all of the cleanup efforts possible will not make any difference to remedy the problem. We can do many things

Figure B A deceased albatross full of plastic trash.
NOAA Marine Debris Program.

in our everyday lives to ensure that our trash does not make it into the ocean. Reuse or recycle whenever possible. Never, ever, pollute. It sounds simple, but many well-intentioned beachgoers accidentally leave beach toys, water bottles, suntan lotion bottles, and a variety of other trash behind. Last, avoid buying plastics when there is an alternative product available. This includes limiting (or omitting altogether) purchases of plastic disposable items. Plastic never goes away but eventually breaks down into tiny potentially harmful substances that fish and other marine organisms ingest. It really is up to us to stop adding to the garbage patches that exist in our world ocean.

Critical Thinking Questions

1. Make a list of all the products you use in one day, including items used for food and drink storage (e.g., water bottle) and those used for entertainment (e.g., cell phone). Circle the products you use that are made of any kind of plastic, and add a star next to those that are single-use plastic items. Propose alternative products that can be used in place of all of the plastic in your life.

2. Propose a possible solution for reducing plastic in the sea, and describe how your idea can be implemented.

Figure A The locations of large accumulations of trash in the North Pacific Gyre.
NOAA Marine Debris Program.

- **Study Guide**—Each chapter closes with a Study Guide section containing useful study tools for instructors and students:
  - **Topics for Discussion and Review**—These questions encourage further in-depth exploration of covered topics and have been evaluated and revised, updated, or replaced as needed from the previous edition.
  - **Key Terms**—A Key Terms list, including page numbers for all terms, is included to help students learn new marine biology vocabulary.
  - **Key *Genera***—A list of Key *Genera* discussed in each chapter is a helpful study tool for students learning about new organisms.
  - **References**—Although this text includes more material than might be covered in one semester, instructors can select and mold the material to match their teaching styles and time limitations. With judicious use of outside supplementary readings, such as those suggested in the References section of the Study Guide, this text can easily provide the structure for a two-semester or upper-level course.

STUDY GUIDE

TOPICS FOR DISCUSSION AND REVIEW

1. Mesopelagic fishes differ from more familiar epipelagic and coastal fishes in many ways. Summarize these differences.

2. Summarize the proposed uses of photophores in marine animals.

3. What is the difference between holoplankton and meroplankton? List three well-known examples of each type of plankton.

4. Describe the common buoyancy structures used by pelagic marine animals.

5. Discuss the proposed advantages of vertical migration for mesopelagic species. Why do some mesopelagic species migrate to deeper waters at night?

6. Describe the various mechanisms used by zooplankton to collect diffuse food.

7. Identify the proposed cues used by marine animals to orient in space during their long migrations. Are larvae capable of orienting, or are they just subject to the current they are traveling in?

KEY TERMS

antitropical distribution 337	mesopelagic zone 340
countershading 340	neuston 348
deep sound-scattering layers (DSSLs) 342	photophores 341
diurnal 342	physoclistous swim bladder 349
epipelagic zone 336	physostomous swim bladder 349
holoplankton 332	pneumatic duct 348
isolume 344	pneumatophore 343
magnetoreception 352	rete mirabile 350
marine snow 340	vertical migration 342
meroplankton 332	

KEY GENERA

Aegisthus	Corolla
Architeuthis	Cyclothone
Argyropelecus	Euphausia
Aristostomias	Eurypharynx
Bolinichthys	Gigantactis
Bolinopsis	Glaucus
Calanus	Janthina

Loligo	Regalecus
Melanocetus	Sagitta
Oikopleura	Sapphirina
Oithona	Sebastes
Opisthoproctus	Thalassiosira
Pegea	Thysanoessa
Physalia	Velella

REFERENCES

Allan, J. D. 1976. Life history patterns in zooplankton. *American Naturalist* 110:165–180.

Alldredge, A. 1976. Appendicularians. *Scientific American* July:94–102.

Barham, E. G. 1966. Deep scattering layer migration and composition: Observations from a diving saucer. *Science* 151:1399–1403.

Bargu, S., C. L. Powell, S. L. Coale, M. Busman, G. J. Doucette, and M. W. Silver. 2002. Krill: A potential vector for domoic acid in marine food webs. *Marine Ecology Progress Series* 237:209–216.

Boden, B. P., and E. M. Kampa. 1967. The influence of natural light on the vertical migrations of an animal community in the sea. *Symposium of the Zoological Society of London* 19:15–26.

Boyd, C. M. 1976. Selection of particle sizes by filter-feeding copepods: A plea for reason. *Limnology and Oceanography* 21:175–179.

Bright, T., F. Ferrari, D. Martin, and G. A. Franceschini. 1972. Effects of a total solar eclipse on the vertical distribution of certain oceanic zooplankters. *Limnology and Oceanography* 17:296–301.

Brinton, E. 1962. The distribution of Pacific euphausiids. *Bulletin. Scripps Institution of Oceanography* 8:51–270.

Cushing, D. H. 1968. *Fisheries Biology.* Madison, WI: University of Wisconsin Press.

Denton, E. J., and J. P. Gilpin-Brown. 1973. Flotation mechanisms in modern and fossil cephalopods. *Advances in Marine Biology* 11:197–268.

Eastman, J. T., and A. L. DeVries. 1986. Antarctic fishes. *Scientific American* 255:106–114.

Gilmer, R. W. 1972. Free-floating mucus webs: A novel feeding adaptation for the open ocean. *Science* 176:1239–1240.

Giorgio, P. A., and Duarte, C. M. 2002. Respiration in the open ocean. *Nature* 420:379–384.

Hoar, W. W. 1983. *General and Comparative Physiology.* Englewood Cliffs, NJ: Prentice Hall.

Jackson, G. A. 1990. A model of the formation of marine algal flocs by physical coagulation processes. *Deep-Sea Research* 37:1197–1211.

Jensen, O. P., S. Hansson, T. Didrikas, J. D. Stockwell, T. R. Hrabik, T. Axenrot, and J. F. Kitchell. 2011. Foraging, bioenergetic, and predation constraints on diel vertical migration: Field observations and modeling of reverse migration by young-of-year herring *Clupea harengus. Journal of Fish Biology* 78:449–465.

Jumper, G. Y., Jr., and R. C. Baird. 1991. Location by olfaction: A model and application to the mating problem in the deep-sea hatchetfish *Argyropelecus hemigymnus. American Naturalist* 138:1431–1458.

Kanwisher, J., and A. Ebling. 1957. Composition of swim bladder gas in bathypelagic fishes. *Deep-Sea Research* 4:211–217.

- **Additional Online Study Tools**—Practice activities, prepopulated quizzes, and an interactive eBook with Web Links to relevant sites are available for self-study.

Teaching Tools

A variety of teaching tools are available via digital download and multiple other formats to assist instructors with preparing for and teaching their courses.

Lecture Outlines in PowerPoint Format—The Lecture Outlines in PowerPoint format provide lecture notes and images for each chapter of *An Introduction to the Biology of Marine Life, Eleventh Edition*. Instructors with Microsoft PowerPoint can customize the outlines, art, and order of presentation.

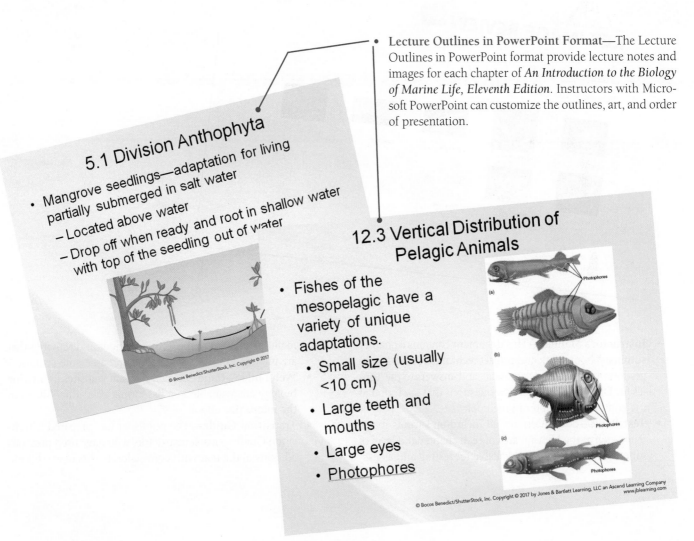

5.1 Division Anthophyta

- Mangrove seedlings—adaptation for living partially submerged in salt water
  - Located above water
  - Drop off when ready and root in shallow water with top of the seedling out of water

© Bocos Benedict/ShutterStock, Inc. Copyright © 2017

12.3 Vertical Distribution of Pelagic Animals

- Fishes of the mesopelagic have a variety of unique adaptations.
  - Small size (usually <10 cm)
  - Large teeth and mouths
  - Large eyes
  - Photophores

© Bocos Benedict/ShutterStock, Inc. Copyright © 2017 by Jones & Bartlett Learning, LLC an Ascend Learning Company
www.jblearning.com

- **Key Image Review and Unlabeled Art Bank**—The Key Image Review provides access to the illustrations, photographs, and tables that Jones & Bartlett Learning holds the copyright to or has permission to reprint digitally. These images are not for sale or distribution but may be used to enhance existing slides, tests, quizzes, or other classroom material. The Unlabeled Art Bank, consisting of 14 illustrations available on PowerPoint slides, can be used as part of a labeling exercise in an exam or can be provided to students for taking notes.

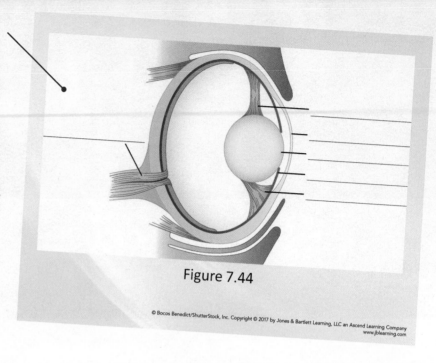

Figure 7.44

© Bocos Benedict/ShutterStock, Inc. Copyright © 2017 by Jones & Bartlett Learning, LLC an Ascend Learning Company
www.jblearning.com

- **Instructor's Manual**—This document contains a chapter summary, homework and project recommendations, suggestions for using the lab manual, and answers to the Topics for Discussion and Review questions that are found at the end of each chapter in the book.
- **Test Bank**—600 questions, all including various metadata, such as the relevant area of each chapter and level of functional taxonomy, are available for testing and assessment, in addition to the 750+ questions and activities that are included in the online study and assessment tools.
- **Web Links**—Hand-selected relevant websites for marine biology are available in a list format or as direct links in the interactive eBook.
- **Transition Guide**—The publisher has prepared a Transition Guide to assist instructors who have used previous editions of the text with conversion to this new edition.

Lab Manual

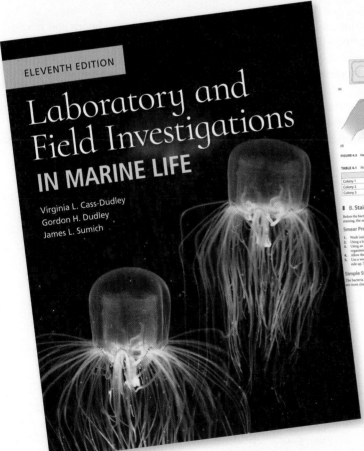

Now published for the first time in full color, *Laboratory and Field Investigations in Marine Life, Eleventh Edition* by Virginia L. Cass-Dudley, Gordon Dudley, and James L. Sumich is a unique marine biology laboratory and field manual that engages students in the excitement and challenges of understanding marine organisms and the environments in which they live. The laboratory and field activities are designed to encourage students to develop their own powers of critical observation and analysis.

Students will benefit from a thorough examination of topics such as the physical and chemical properties of seawater, marine microbes, algae, and a wide variety of invertebrate and vertebrate animals through observation and critical-thinking activities. The new and more convenient spiral binding allows the manual to lay flat while students work, and they can easily tear out pages to submit for a grade.

Laboratory and Field Investigations in Marine Life, Eleventh Edition is an ideal resource to accompany *Introduction to the Biology of Marine Life, Eleventh Edition* and the clear choice for an engaging, hands-on exploration of marine biology.

Lab Manual ISBN-13: 978-1-284-09054-3
Main Text + Lab Manual Bundle ISBN-13: 978-1-284-12406-4

Acknowledgments

Much credit for the ongoing development of this text goes to students and instructors who have used previous editions and have offered valuable comments and criticisms. We thank our instructors of the past and colleagues of the present for their contributions to this book. Special thanks also go to the many colleagues and institutions that graciously permitted use of their exceptional photographs. Finally and especially, we thank our present and former students for their interest and enthusiasm in discovering rewarding methods of communicating this information.

Many colleagues at numerous institutions reviewed drafts of various editions and collectively improved the text. We thank the following reviewers who have been generous with their time and comments:

Holly Ahern, Adirondack Community College
William G. Ambrose, Bates College
Gil Bane, Kodiak College
Paul A. Billeter, Charles County Community College
Brenda Blackwelder, Central Piedmont Community College
James L. Campbell, Los Angeles Valley College
Gregory M. Capelli, College of William & Mary
Sneed Collard, University of West Florida
Harold N. Cones, Christopher Newport University
Susan Cormier, University of Louisville
J. Nicholas Ehringer, Hillsborough Community College
Gina Erickson, Highline Community College
Paul E. Fell, Connecticut College
Susan Flanagan, Nunez Community College
Robert T. Galbraith, Crafton Hills College
Dominic Gregorio, Cypress College
Lynn Hansen, Modesto Jr. College
Marty L. Harvill, Bowling Green State University
Floyd E. Hayes, Pacific Union College
Richard Heard, Gulf Coast Research Lab
Rozalind Jester, Florida SouthWestern State College
Yan Jiao, Virginia Polytechnic Institute & State University
Mary Katherine Wicksten, Texas A&M University
Susan Keys, Springfield College

Matthew Landau, Stockton State College
Nan Ho, Las Positas College
Cynthia Lewis, San Diego State University
Vicky J. Martin, University of Notre Dame
Richard Anthony Matthews, Bainbridge State College
Jeremy Montague, Barry University
Donald Munson, Washington College
Valerie Pennington, Southwestern College
Mary K. Rapien, Bristol Community College
Richard A. Roller, University of Wisconsin–Stevens Point
Mary Beth Saffo, University of California–Santa Cruz
Allan Schoenherr, Fullerton College
L. Scott Quackenbush, Los Angeles Valley College
Robert E Shields, U.S. Coast Guard Academy
Rebecca Shipe, University of California–Los Angeles, Institute of the Environment and Sustainability
Cynthia C. Strong, Bowling Green State University
Catherine A. Teare Ketter, University of Georgia
Doug Tupper, Southwestern College
Jefferson T. Turner, Southeastern Massachusetts University
Jacqueline Webb, University of Rhode Island
John T. Weser, Scottsdale Community College
Robert Whitlatch, University of Connecticut
Richard B. Winn, Duke University Marine Laboratory

We also acknowledge the pleasant and professional editorial and production team at Jones & Bartlett Learning. Our special thanks go to Audrey Schwinn for coordinating the flow of materials; to Jamey O'Quinn for coordinating the photo research for new images and handling the permission process; to Shannon Sheehan for managing the revisions and improvements to the illustrations; to Nancy Hitchcock for guiding copyediting and the evaluation of page proofs; to Jennifer Coker for copyediting the manuscript; and to Pamela Andrada from Beyond Words Proofreading for reviewing the pages.

In closing, we encourage you, students and instructors alike, to immerse yourself in this material as much as possible and in as many ways as you can invent. Spend time at the seashore just wading about.

Walk along a beach after high tide and examine the biological treasures that the sea left behind. Sit on the edge of a rocky tide pool and watch the action before you. If you can, swim, snorkel or dive for a closer look. If you don't swim, learn. Watch how young children observe things, and mimic their enthusiasm.

Pick up the less fragile organisms for a closer look.

Take a day trip on a fishing boat. Volunteer at a local aquarium even if you think you don't yet know enough to contribute; you'll learn. Mostly, it is a matter of investing time—time in the field and time in the classroom. You will get to experience the fun stuff only if you put in the time.

Deanna R. Pinkard-Meier
University of San Diego

John F. Morrissey
Sweet Briar College

James L. Sumich
Grossmont College

About the Authors

John F. Morrissey earned his B.A. and M.A. degrees in Biology from Hofstra University. After teaching marine biology and coral reef ecology in Jamaica for 1 year, he then earned his Ph.D. in Marine Biology and Fisheries from the University of Miami's Rosenstiel School of Marine and Atmospheric Science. His dissertation research concerned the movement patterns, diel activity, and habitat selection of lemon sharks in Bimini, Bahamas. Since then, Dr. Morrissey has studied the biology of sharks, skates, and rays all over the world, including Jamaica, Japan, the Azores, and the Canaries. He has been on the board of directors of the American Elasmobranch Society since 1996.

For 16 years he taught marine biology, a field course in tropical marine biology, and comparative anatomy at Hofstra University, where he won the Distinguished Teacher of the Year Award in 2006. In 2007, Dr. Morrissey moved to Sweet Briar College in central Virginia, along with his egg-laying colony of 100+ chain catsharks, to teach marine biology, comparative vertebrate anatomy, and animal physiology. He won their Excellence in Teaching Award in 2010. He lives on a dirt road in the woods with his wife (who is also his research partner) and their four spoiled cats.

James L. Sumich received his M.S. in Biological Oceanography at Oregon State University, joined the biology faculty at Grossmont College, and then returned to Oregon State for a Ph.D. For his Ph.D. thesis, he studied the interactions between newborn gray whale calves and their mothers and the way each budgets its energy expenditures during the period of calf nursing.

He has taught marine mammal biology classes for graduates and undergraduates at San Diego State University, University of San Diego, and Oregon State University, where he continues to teach, as the requirements of retirement permit.

His retirement activities include continued research and writing on gray whale behavior and energetics. He recently marked the publication of the second edition revision of a textbook on the evolutionary biology of marine mammals, coauthored with Dr. Annalisa Berta and Dr. Kit Kovacs. He lives in a home he has built with his wife, Caren, in the woods near Corvallis, Oregon.

Deanna R. Pinkard-Meier earned her B.S. degree in Aquatic Biology from the University of California Santa Barbara and her M.S. degree in Marine Biology from the Florida Institute of Technology. Her early research efforts were concentrated on marine invertebrates near the California Channel Islands. She then moved her research focus and home to the tropics, studying the behavior, reproduction, and recruitment of tropical marine fishes, specifically snappers, damselfish, and wrasses. After 5 very productive years in the warm waters of the Caribbean and Florida, she ventured back home to California where she worked as a research fisheries biologist for the National Oceanic and Atmospheric Administration (NOAA), studying the endangered white abalone and declining rockfish populations with a remotely operated vehicle.

For the past decade Deanna has taught many biology courses at a variety of universities and community colleges in San Diego, while still publishing research with hardworking former colleagues at NOAA. She especially enjoys teaching marine biology, organismal biology, and evolution and introducing students to field survey techniques used in the intertidal or from a boat. She resides in Cardiff-by-the-Sea, California, with her husband and two sons, where they can often be found jumping in the ocean for a surf session or snorkel. She is active in her local community where she participates in Surfrider Foundation events and volunteers as a visiting scientist at local schools.

The Ocean as a Habitat

STUDENT LEARNING OUTCOMES

1. Become acquainted with terminology used in the study of marine biology and oceanography. Use flash cards or other techniques to help memorize important terms in **bold**.

2. Begin to think in an ocean-minded manner, in oceanic size and time scales, and from the perspective of an organism living in the sea.

3. Analyze different methods for mapping the seafloor. Understand how mapping the seafloor will be useful for our study of the ocean and its inhabitants.

4. Explain why we refer to the oceans of the world as "one world ocean."

5. Recall the various areas of the seafloor and water column, and describe how these classifications may affect how marine organisms are distributed.

6. Become familiar with evolution by natural selection.

CHAPTER OUTLINE

The National Oceanic and Atmospheric Administration's ship, the *Okeanos Explorer,* as seen from the water. The mission of this ship is to explore our unknown ocean for the purpose of discovery and to advance our knowledge of the sea.

© Courtesy of National Oceanic and Atmospheric Administration.

The ocean is an extremely unique and dynamic habitat. First and foremost, it is full of water, a phenomenal and yet simple substance, with properties that allow a great diversity of life forms to exist. The ocean is not just full of plain water (H_2O) though but rather the somewhat mysterious saltwater (H_2O plus *many other substances*) that you will read more about throughout this book. Another unique fact about the ocean is that it is always moving, which may be challenging or helpful to organisms living on or below the surface. Additionally, resources such as light and oxygen are in limited supply, something that land dwellers do not need to consider.

Earth's oceans are home to an extraordinary variety of living organisms adapted to the special conditions of the sea. The characteristics of these organisms and the variety of marine life itself are consequences of the many properties of the ocean habitat. This chapter provides a survey of the developmental history and present structure of the ocean basins. Adaptations to these properties and oceanographic processes have molded the ocean's inhabitants through their very long history

of evolutionary development. It is thought that life first evolved in the ocean several billion years ago, so, in essence, living conditions in the ocean helped shape all living organisms that evolved later on some level, even those currently living on land.

As students of the Earth's ocean, gaining a new perspective will help in our understanding of this nonterrestrial environment. Humans naturally tend to see the world from a human point of view, with human scales of time and distance, with land under our feet and air surrounding us. To begin to understand the marine environment of our home planet and how it and its inhabitants evolved to their present forms, we must broaden our perspective to include very different time and distance scales. Terms such as "young" and "old" or "large" and "small" have limited meaning unless placed in some useful context. **Figure 1.1** compares size scales for a few common oceanic inhabitants, and **Figure 1.2** includes time scales for the Earth, ocean, and living organisms. Throughout this book, these scales are revisited, and others are introduced to help you develop a practical sense of the time and space scales experienced by

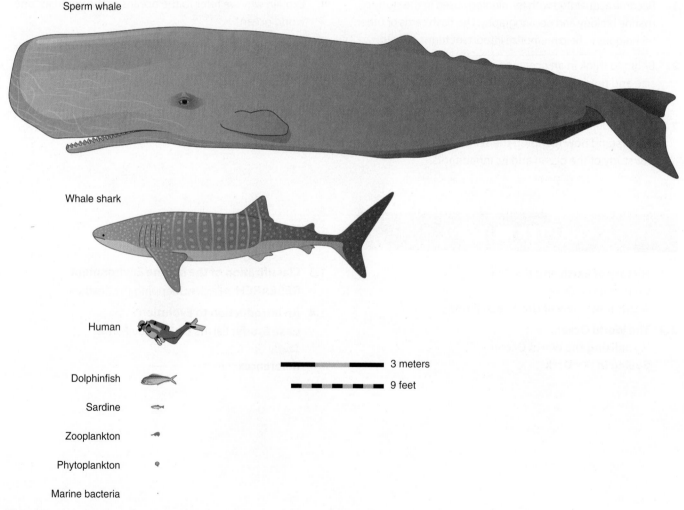

Figure 1.1 A size comparison of common marine organisms and a human. Images are to scale for the sperm whale, whale shark, human, dolphinfish, and sardine. The zooplankton, phytoplankton, and marine bacteria images are magnified approximately 10 times so that they are visible on the page.

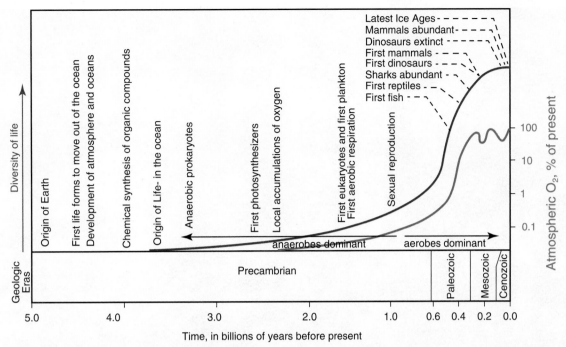

Figure 1.2 A summary of some biological and physical milestones in the early development of life on Earth. The brown curve represents the relative diversity of life; the green curve represents the O₂ concentration of the atmosphere.

marine organisms. As you read about the marine environment, attempt to think in terms of these scales.

1.1   History of Earth and the Sea

Our Earth is thought to have been formed about 4.6 billion years ago. Although some disagreement exists over the exact mechanism of the formation of the solar system and Earth, the most widely accepted explanations of the origin of our solar system indicate that the planets aggregated from a vast cloud of cold gas and dust particles into clusters of solid matter. These clumps continued to grow as gravity attracted them together. As Earth grew in this manner, pressure from the outer layers compressed and heated Earth's center. Aided by heat from decay of radioactive elements, the planet's interior melted. Iron, nickel, and other heavy metals settled to the core, whereas the lighter materials floated to the surface and cooled to form a density-layered planet with a relatively thin and rigid crust (**Figure 1.3**).

Early in Earth's history, volcanic vents poked through the crust and tapped the upper mantle for liquid material and gases that were then spewed out over the surface of the young Earth, and a primitive atmosphere developed. Thick water vapor was certainly present. As it condensed, it fell as rain, accumulated in low places on Earth's surface, and formed primitive oceans. Additional water may have arrived as "snowballs" from space in the form of comets colliding with the young Earth. Atmospheric gases dissolved into accumulating seawater, and other chemicals, dissolved from rocks and carried to the seas by rivers, added to the mixture, eventually creating that complex brew of water, ions, and molecules that we call seawater.

Since their initial formation, the ocean basins have experienced considerable change. New material derived from Earth's mantle has extended the continents so that they are now larger and stand higher than at any time in the past. The oceans have kept pace, getting deeper with accumulations of new water from volcanic gases and from the chemical breakdown of rock. Earth's early life forms (represented by marine bacterial fossils that are

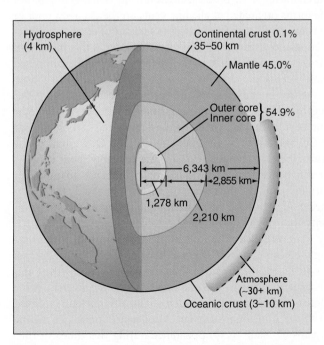

Figure 1.3 A section through the Earth representing the density-layered interior structure and the thickness of each layer. The Earth's crust is so thin that it is represented by a black line.

Figure 1.4 An artist's depiction of life forms present during the Cambrian Explosion.

around 3.5 billion years old) also had a significant impact on the character of the physical environment. Whether the earliest life forms originated at alkaline vents on the deep-sea floor or in warm pools at the sea's edges is a matter of continuing speculation and research. What is clear, however, is that life on this planet requires water; in fact, all living organisms are composed mostly of water.

Early in life's history on Earth, molecular oxygen (O_2) began to be produced in increasing amounts by microscopic photosynthetic **prokaryotes**, as they converted carbon dioxide (CO_2) and water into sugars and O_2. The O_2 content of the atmosphere 600 million years ago was probably about 1% of its present concentration. It was not much, but it was an important turning point, the time when organisms that could take advantage of O_2 in aerobic respiration became dominant and organisms not using O_2 (anaerobes) became less prevalent.

The **evolution** of more complex life forms using increasingly efficient and variable methods of energy utilization set the stage for an explosion of marine species. During the Cambrian period (542 to 488 million years ago) the diversity of life increased dramatically, so much so that this period is referred to as the *Cambrian Explosion*. Most major groups of marine organisms made their appearance during this time. Worms, sponges, corals, fishlike creatures, and the distant ancestors of terrestrial animals and plants were abundant, but life at that time could exist only in the sea, where a protective blanket of

seawater shielded it from intense solar radiation (**Figure 1.4**). How do we know about the diversity of life during the Cambrian period, or any other time in the distant past? Fossil evidence provides us with most of the information we have, and various advanced dating techniques allow for estimates of the ages of fossils. The most interesting fossils discovered from the Cambrian period are from areas containing shale material. Shale is very fine material that allows for rapid fossilization of soft or hard body parts, and an increased amount of fine details are preserved from specimens.

As O_2 became more abundant in the upper atmosphere, some of it was converted to **ozone** (O_3). The process of forming ozone absorbed much of the lethal ultraviolet radiation coming from the sun and prevented the radiation from reaching Earth's surface. The O_2 concentration of the atmosphere 400 million years ago is estimated to have reached 10% of its present level and achieved its current concentration in the Mesozoic era (about 200 million years ago). The additional ozone screened out enough ultraviolet radiation to permit a few life forms to abandon their sheltered marine home and colonize the land. Only recently have we become aware that industrialized society's increasing use of aerosols, refrigerants, and other atmospheric pollutants is gradually depleting this protective layer of ozone. Figure 1.2 provides a general timeline for some of the major events in the early development of life on Earth.

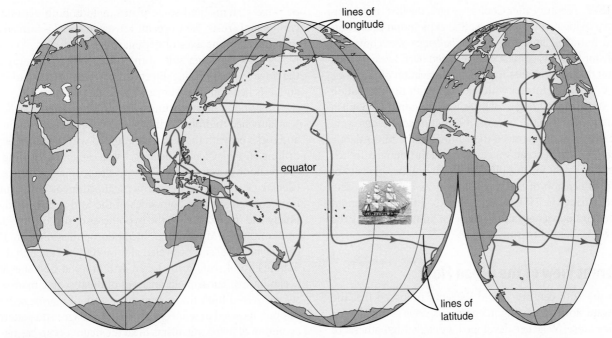

Figure 1.5   An "orange peel" projection of the Earth's surface with latitude and longitude lines at 30-degree intervals. The red line tracks the voyage of HMS *Challenger* (inset).
Courtesy of Steve Nicklas, NOS, NGS/NOAA.

DID YOU KNOW?

Geological periods are marked by large changes such as major extinction events. If even one of the major extinction events in Earth's history had not taken place, then life as we know it today would not exist. Extinction events remove some species from existence, making room for new species to evolve and take over previously occupied areas. The mass extinction of almost all dinosaurs at the end of the Cretaceous period made room (and a much safer environment) for large mammals to thrive.

Charting the Deep

As fascinated as people are with the sea, it can be assumed that people must have explored their local coastal environments very early in their history, but few early discoveries were recorded. By 325 B.C., Pytheas, a Greek explorer, had sailed to northwestern Europe and developed a method for determining **latitude** (**Figure 1.5**). About a century later, Eratosthenes of Alexandria, Egypt, provided the earliest recorded estimate of Earth's size, its first dimension. His calculated circumference of 39,690 km was only about 1% less than today's accepted value of 40,008 km. During the Middle Ages, Vikings, Arabians, Chinese, and Polynesians sailed over major portions of Earth's oceans. By the 15th century, all the major inhabitable land areas were occupied; only Antarctica remained unknown to and untouched by humans. Even so, precise charting of the ocean basins had to await several more voyages of discovery.

Between 1768 and 1779, James Cook, an English navigator, conducted three exploratory voyages, mostly in the

Southern Hemisphere. He was the first to cross the Antarctic Circle and to understand and conquer scurvy (a disease caused by a deficiency of vitamin C). He is best remembered as the first global explorer to make extensive use of the marine chronometer developed by John Harrison, a British inventor. The chronometer, a very accurate shipboard clock, was necessary to establish the **longitude** of any fixed point on the Earth's surface. Together with Pytheas's 2,000-year-old technique for fixing latitude, reasonably accurate positions of geographic features anywhere on the globe could be established for the first time, and our two-dimensional view of Earth's surface was essentially complete. Today, coastal Long Range Navigation (LORAN) stations and satellite-based global positioning systems (GPSs) enable individuals to determine their position to within a few meters anywhere on Earth. With GPS technology currently available on most cellular phones, it is difficult to imagine a time when navigation was a complicated task.

DID YOU KNOW?

Captain Cook's voyages were not for the faint of heart. Over the 25 years he led expeditions, many sailors died, food was scarce, dates of arrival back home were anything but certain, and the pay was minimal. Only the most dedicated or desperate of men dared join an expedition. In fact, during one expedition to Australia the trip turned into somewhat of a disaster as navigation near what would later be called the Great Barrier Reef led to a grounded ship. Yet Captain Cook himself was a determined man, and he did all he could to keep his men alive. His passions were exploration and navigation, and he needed brave and competent men to join him.

In 1872, a century after Cook's voyages, the first truly interdisciplinary global voyage for scientific exploration of the seas departed from England. The HMS *Challenger* was converted expressly for this voyage. The voyage lasted over 3 years, sailed almost 69,000 **nautical miles** in a circumnavigation of the globe (Figure 1.5), and returned with such a wealth of information that 10 years and 50 large volumes were required to publish the findings. During the voyage, 492 depth soundings were made. These soundings traced the outlines of the Mid-Atlantic Ridge under 2 km of ocean water, plumbed the Mariana Trench to a depth of 8,185 m, and filled in rough outlines globally of the third dimension of the world ocean, its depth. The data collected during the HMS *Challenger* expedition contributed greatly to our knowledge of the marine environment during the infancy of the field of marine science.

A Different View of the Ocean Floor

Early in the 20th century, Alfred Wegener proposed that the oceans were slowly changing in enormous ways and had been since they arose. Wegener developed a detailed hypothesis of what eventually became known as **continental drift** to explain several global geologic features, including the remarkable jigsaw-puzzle fit of some continents (especially the west coast of Africa and the east coast of South America). He proposed that our present continental masses had drifted apart after the breakup of a single supercontinent, **Pangaea**. His evidence seemed ambiguous at the time, though, and most scientists remained unconvinced. It was not until the early 1960s that new evidence compelled two geophysicists to independently, and almost simultaneously, propose the closely related concepts of **seafloor spreading** and **plate tectonics**. In hindsight, these related concepts seem completely obvious—that the Earth's crust is divided into giant irregular plates (**Figure 1.6**). These rigid crustal plates float on the denser and slightly more plastic mantle material. Each plate edge is defined by oceanic trench

or ridge systems, and some plates include both oceanic and continental crusts. New oceanic crustal material is formed continually along the axes of oceanic ridges and rises. As crustal plates grow on either side of the ridge, they move away from the ridge axis in opposite directions, carrying bottom sediments and attached continental masses with them (**Figure 1.7**).

In 1968, a new and unusual ship, the *Glomar Challenger*, was launched to probe Earth's history as recorded in sediments and rocks beneath the oceans. Equipped with a deck-mounted drilling rig, the *Glomar Challenger* was capable of drilling into the seafloor in water over 7,000 m deep. Within 2 years, the *Glomar Challenger* recovered vertical sediment core samples from enough sites on both sides of the Mid-Atlantic Ridge to finally and firmly confirm the hypotheses of seafloor spreading and continental drift. Before being decommissioned in 1983, the *Glomar Challenger* traveled almost 700,000 km and drilled 318,461 m of seafloor in 1,092 drill holes at 624 sites in all ocean basins. Subsequent analyses of microscopic marine fossils recovered from this tremendous store of marine sediment samples have led to refined estimates of the ages and patterns of evolution of living organisms in all the major ocean basins. The *JOIDES Resolution*, named after the HMS *Resolution* commanded by Captain Cook, continued the work that the *Glomar Challenger* had begun and is still in operation today (**Figure 1.8**). Data from core samples collected from this vessel add to our continued understanding of the history of Earth and our world ocean.

The changes that seafloor spreading and plate tectonics have wrought on the shapes and sizes of the oceans have been and will continue to be quite impressive. Currently, the African continent is drifting northward on a collision course with Europe, relentlessly closing the Mediterranean Sea. The Atlantic Ocean is becoming wider at the expense of the Pacific Ocean. Australia and India continue to creep northward, slowly changing the shapes of the ocean basins they border. Occasional violent earthquakes are only incidental tremors in this monumental collision of crustal plates. The rates of seafloor

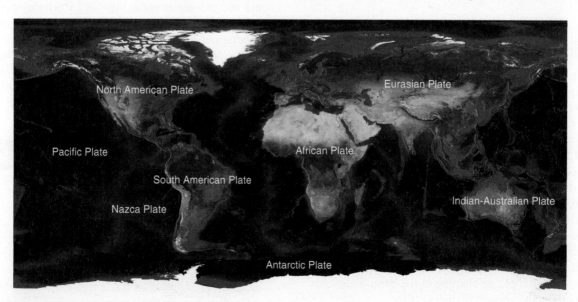

Figure 1.6 The major plates of Earth's crust. Compare the features of the map with those of Figure 1.12.
Courtesy of Reto Stockli, NASA Earth Observatory.

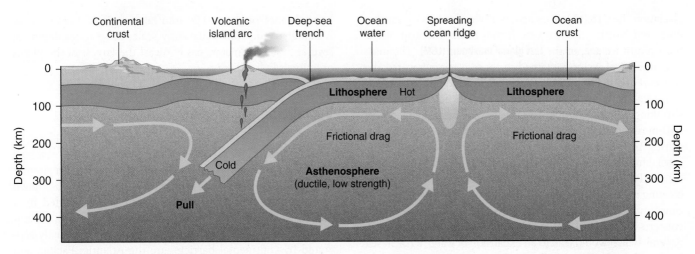

Figure 1.7   Side view of a spreading ocean floor, illustrating the relative motions of oceanic and continental crusts. New crust is created at the ridge axis, and old crust is lost in deep-sea trenches.

spreading have been determined for some oceans, and they vary widely. The South Atlantic is widening about 3 cm each year (or approximately your height in your lifetime). The Pacific Ocean is shrinking somewhat faster. The fastest seafloor spreading known, 17.2 cm/yr, was measured along the East Pacific Rise, a mid-oceanic ridge at the meeting point of five plates.

The breakup of the megacontinent Pangaea produced ocean basins where none existed before. The seas that existed 200 million years ago have changed size or have disappeared altogether. Some of the past positions of the continents and ocean basins, based on our present understanding of the processes involved, are reconstructed in **Figure 1.9**. Excess crust produced by seafloor spreading folds into mountain ranges (the Himalayas are a dramatic example) or slips down into the mantle and remelts (Figure 1.7). Consequently, most marine fossils older than about 200 million years can never be studied; they, too, have been carried to destruction by the "conveyor belt" of subduction, the sinking of seafloor crust at trench locations to be remelted in the mantle. Ironically, the only fossil evidence we can find for the first 90% of the evolutionary history of marine life is found in landforms that were once ancient seabeds.

On much shorter time scales, other processes have been at work to alter the shapes and sizes of ocean basins. During the past 200,000 years, our planet has experienced two major episodes of global cooling associated with extensive continental

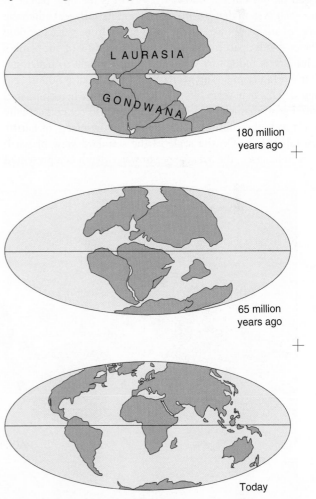

Figure 1.9   About 200 million years ago, the megacontinent, Pangaea, separated into two large continental blocks, Laurasia and Gondwana. Over time, these blocks fragmented into the smaller continents that we see today, and they continue to drift apart.

Data from Dietz and Holden 1970.

Figure 1.8   The ocean drill ship, *JOIDES Resolution*.

glaciation. Just 18,000 years ago, northern reaches of Europe, Asia, and North America were frozen under the grip of the most recent ice age, or the **last glacial maximum (LGM)**. The massive amount of water contained in those glaciers lowered the sea level about 150 m below its present (and also its preglacial) level. Between 18,000 and 10,000 years ago, melting and shrinking of these continental glaciers were accompanied by a 150-m rise in global sea level and the flooding of land exposed during the LGM. Coral reefs, estuaries, and other shallow coastal habitats were modified extensively during this flooding. Currently, warmer summer temperatures around Antarctica are creating real concern about the potential for melting glaciers to cause another rise in global sea levels of a few meters. Another concern for sea level rise is increasing ocean temperatures in general, which would lead to an expansion of water, a rise in sea level, and temperature challenges for many marine organisms. Current estimates indicate a continuous sea level rise of greater than 0.3 cm/yr worldwide.

1.2   The World Ocean

Based on our current knowledge, Earth is the only planet in our solar system that has liquid water at its surface. Unlike the faces of any planet we can see from Earth, from space our Blue Planet stands apart from all others because 70% of our planet's solid face is hidden by water too deep for light to penetrate. Our world ocean has an average depth of about 3,800 m (2.4 miles). This may seem like a lot of water, but when compared with Earth's diameter of 12,756 km, the world ocean is actually a relatively thin film of water filling the low places of Earth's crustal surface. On the scale of the standard view of Earth seen from space, the average ocean water depth is represented

by a distance of about 0.04 mm (about 1/1,000 of an inch). Although shallow on a planetary scale, these water depths of several thousand meters easily dwarf the largest of the plants or animals living there.

Visualizing the World Ocean

Being so vast in depth, volume, and area, it can be difficult to obtain a grasp of the ocean as one large entity. Yet the waters of the world's ocean are all interconnected, and one can attempt to visualize Earth's marine environment in this way, as shown in **Figure 1.10**. The Antarctic continent is surrounded by a "Southern Ocean," which has three large embayments extending northward. These three oceanic extensions, partially separated by continental barriers, are the Atlantic, Pacific, and Indian Oceans. Other smaller oceans and seas, such as the Arctic Ocean and the Mediterranean Sea, project from the margins of the larger ocean basins. These broad connections between major ocean basins permit exchange of both seawater and the organisms living in it, reducing and smoothing out differences between adjacent ocean basins.

Figure 1.11 represents a more conventional view of the world ocean, showing separations into four major ocean basins—the Atlantic, Pacific, Indian, and Arctic—and without the emphasis on the extensive southern connections apparent in Figure 1.10. The format of Figure 1.11 is often more useful because our interest in the marine environment has been focused on the temperate and tropical regions of Earth.

The equator is a very real physical boundary extending across the tropical center of the large ocean basins. The curvature of Earth's surface causes areas near the equator to receive more concentrated radiant energy from the sun than equal-sized

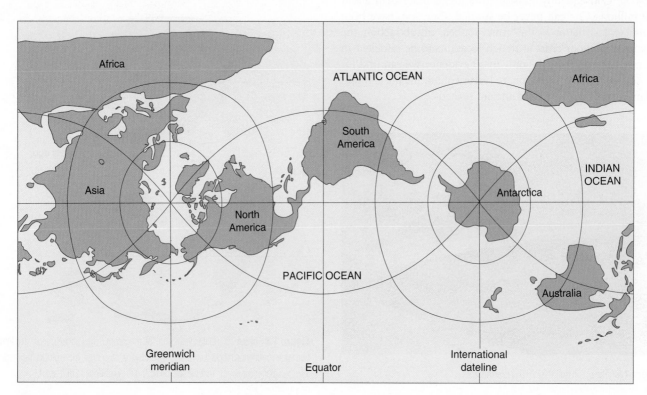

Figure 1.10 A modified polar view of the world ocean, emphasizing the extensive connections between major ocean basins in the Southern Hemisphere.

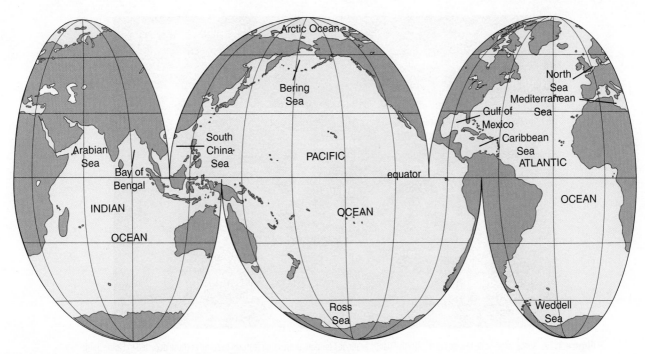

Figure 1.11   An "orange peel" equatorial view of the world ocean.

areas in polar regions where the sun's energy is spread out. The resultant heat gradient from warm tropical to cold polar regions establishes the basic patterns of atmospheric and oceanic circulation. Surface ocean current patterns display a nearly mirror-image symmetry in the northern and southern halves of the Pacific and Atlantic Oceans. This symmetry establishes it as a natural, although intangible, focus for the graphic representation of these features and of the life zones they define.

Nearly two thirds of our planet's land area is located in the Northern Hemisphere. The Southern Hemisphere is an oceanic hemisphere, with 80% of its surface covered by water. The Pacific Ocean alone accounts for nearly one half of the total ocean area. A few descriptive measurements for features of the six largest marine basins are listed in **Table 1.1**.

Maximum oceanic depths extend to over 11,000 m, but most of the ocean floor lies shallower, at depths between 3,000 and 6,000 m below the sea surface. As a terrestrial comparison, the maximum height of a mountain above land is Mount Everest, measuring 8,848 m tall. The tallest mountain is Mauna Kea in Hawaii, which reaches a total of 10,205 m high from a submarine base underwater in the Hawaiian trough. Only 4,205 m of the mountain are visible above water, which is why Mount Everest is considered the tallest peak. An image of the northern and central parts of the Atlantic Ocean (**Figure 1.12**) illustrates some of the larger scale features of the ocean floor. The **continental shelf**, which extends seaward from the shoreline and is actually a structural part of the continental landmass, would not be considered an oceanic feature if sea level were lowered by as little as 5% of its present average depth. The width of continental shelves varies, from being nearly absent off southern Florida to over 800 km wide in the Arctic Ocean north of Siberia. Continental shelves account for 8% of the ocean's surface area; this is equivalent to about one sixth of Earth's total land area.

Most continental shelves are relatively smooth and slope gently seaward. The outer edge of the shelf, called the **shelf break**, is a vaguely defined feature that usually occurs at depths between 120 and 200 m. Beyond the shelf break, the bottom steepens slightly to become the **continental slope**. Continental slopes reach depths of 3,000 to 4,000 m and form the boundaries between continental masses and the deep ocean basins.

TABLE 1.1	Some Comparative Features of the Major Ocean Basins			
Ocean or Sea	**Area (km²)**	**Volume (km³)**	**Average Depth (m)**	**Maximum Depth (m)**
Pacific	165.2	707.6	4,282	11,033
Atlantic	82.4	323.6	3,926	9,200
Indian	73.4	291.0	3,963	7,460
Arctic	14.1	17.0	1,205	4,300
Caribbean	4.3	9.6	2,216	7,200
Mediterranean	3.0	4.2	1,429	4,600
Other	18.7	17.3		
Totals (average)	*361.1*	*1,370.3*	*(3,795)*	

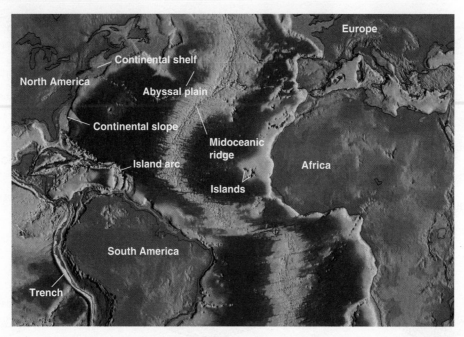

Figure 1.12 Some large-scale features of the North Atlantic seafloor. Compare these large features to some of those depicted in Figure 1.14.
Courtesy of National Geophysical Data Center/NOAA.

A large portion of deep ocean basins consists of flat sediment-covered areas called **abyssal plains**. These plains have almost imperceptible slopes, much like those of eastern Colorado. The sediment blanket over abyssal plains often completely buries smaller crustal elevations, called *abyssal hills*. Most abyssal plains are situated near the margins of the ocean basins at depths between 3,000 and 5,000 m.

Oceanic **ridge and rise systems**, such as the Mid-Atlantic Ridge and East Pacific Rise, occupy over 30% of the ocean basin area. The ridge and rise systems are rugged, more or less linear features that form a continuous underwater mountain chain encircling Earth. The Mid-Atlantic Ridge actually resembles a 20- to 30-km-wide canyon like the East African Rift Valley, sitting atop a broadly elevated seafloor aligned down the center of the Atlantic Ocean. The top of the Mid-Atlantic Ridge may extend 4 km above the surrounding abyssal hills, and isolated peaks occasionally extend above sea level to form islands such as Iceland and Ascension Island. The East Pacific Rise is much lower and broader, with a barely perceptible rift about 2 km wide.

Trenches are distinctive ocean-floor features, with depths usually extending deeper than 6,000 m. Most trenches, including the seven deepest, are located along the margins of the Pacific Ocean. The bottom of the Challenger Deep, in the Mariana Trench of the western North Pacific, is 11,033 m deep, the greatest ocean depth found anywhere. This is as far below sea level as commercial jets typically fly above the sea surface. The enormous depths of trenches impose extreme conditions of high water pressure and low temperature on their inhabitants. Although trenches account for less than 2% of the ocean bottom area, they are integral parts in the processes of seafloor spreading and plate tectonics, acting as sites where crustal plates tens or hundreds of millions of years old are finally subducted back into the mantle to be remelted.

Most oceanic islands, seamounts, and abyssal hills have been formed by volcanic action. Oceanic **islands** are volcanic

mountains that extend above sea level; **seamounts** are volcanic mountains whose tops remain below the sea surface. Most of these features are located in the Pacific Ocean where volcanic activity is common. Islands in tropical areas are often submerged and capped by coral atolls or fringed by coral reefs. These reefs form some of the most beautiful and complex animal communities found anywhere.

Seeing in the Dark

Humans are visual beings; we have relatively good vision and continually examine our surroundings through the clarity of air. In the oceans, however, the opacity of seawater has long thwarted our ability to examine its depths in detail. Fortunately, the opacity of seawater to light is compensated by its transparency to sound. The early ancestors of dolphins capitalized on this feature of water about 20 million years ago by evolving sophisticated biosonar systems capable of high-resolution target discrimination. Biosonar is based on the production of sharp sounds, detecting the echo as it bounces off a target, and then measuring the time delay between sound production and echo return. In a technologic sense, we are catching up with dolphins with the development of various types of electronic **sonar** (*so*und *n*avigating *r*anging) systems. Without going into the history of the development of sonar, it is sufficient to note that several different types of sonar systems with different resolutions have been central to obtaining detailed pictures of the deep ocean floor. Although sonar technically uses sound information, we use computer software to convert sounds into visual signals we can easily examine and relate to.

The most widespread and familiar type of sonar is operated from surface ships. You may be familiar with commercial versions of these as fish finders or depth finders on personal boats. At sea, surface sonar systems operate from a single ship-based transmitter to produce a single line or track of sequential depth measurements below the ship as it is under way. This system provides an acoustic view of the seafloor in two dimensions, with poor resolution of small structures or fine detail.

Multibeam sonar systems also are operated from surface vessels but with an array of numerous sound transmitters and receivers arranged from bow to stern. The many overlapping sound beams return much more information about the seafloor, information that would be a useless jumble of data without computers to decipher it. Multibeam sonar can map a swath of ocean bottom several kilometers wide in a single pass and with much higher resolution than single-transmitter systems. The result is the acoustic equivalent of a continuous strip of aerial photographs of the ocean floor (**Figure 1.13**).

Side-scan sonar systems are variants of the multi-beam technique. As its name implies, side-scan sonar directs sound beams to the sides of the ship's track. Images obtained with side-scan sonar show in rich detail the fine texture of the sea-floor, equivalent to close-up photographs of the ocean floor. The method of sonar used by scientists will vary depending on the specific research question. For studies with large survey areas, multibeam systems are ideal. For studies requiring finer details of the ocean floor, side-scan sonar is sometimes a better option. Regardless of the sonar method used, the resulting picture of the seafloor is clearer and provides more information when seas are calm during the survey period. Although computer software can correct for some ship movements, the corrections that can be made are limited.

1.3   Classification of the Marine Environment

The large size and enormous complexity and variability of the marine environment make it a difficult system to classify. Many systems of classification have been proposed, each reflecting the interest and bias of the classifier. A scientist interested in marine

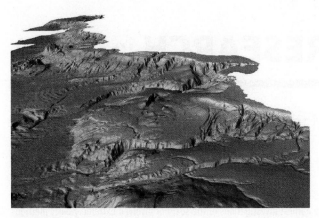

Figure 1.13 A multibeam sonar image of the coastal margin of southern California. Variegated colors depict varying depths.
Courtesy of National Oceanic and Atmospheric Administration.

worms living within the sediments will classify the environment using a much finer scale than a scientist studying the migration patterns of sea turtles across hundreds or thousands of kilometers of ocean. The system presented here is a modified version of a widely accepted scheme proposed by Hedgpeth over a half-century ago, appropriate for large-scale classifications of the marine environment. The terms used in **Figure 1.14** designate particular zones of the marine environment. These terms are easily confused with the names of groups of organisms that normally inhabit these zones. The boundaries of these zones are defined on the basis of physical characteristics such as water temperature, water depth, and available light.

Working downward from the ocean surface, the limits of **intertidal zones** are defined by tidal fluctuations of sea level along the shoreline. The splash, intertidal, and inner shelf zones occur in the **photic** (lighted) **zone**, where the light intensity is great

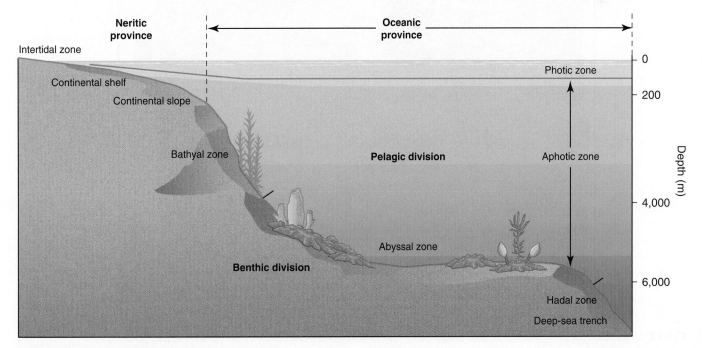

Figure 1.14 A system for classifying the marine environment.
Data from J. W. Hedgpeth, ed., *Treatise on Marine Ecology and Paleoecology* (Geological Society of America, 1966).

RESEARCH in Progress

Mapping the Seafloor

The earliest investigations of the seafloor involved extremely basic methods such as the use of ropes and lead weights dropped to the bottom to gain a sense of depth and very rough seafloor bathymetry. These methods were time-consuming and not as accurate as one would like, but at the time they offered bits and pieces of information about the seafloor that were previously completely unknown. As technology improved, oceanographers seeking finer details of the seafloor and marine biologists hoping to gain more information about the available seafloor habitat for organisms took great interest in advanced seafloor mapping methods. The development of sonar techniques opened up the seafloor to observation by scientists and mariners. Surprisingly, despite advances in sonar technology, until very recently only a small fraction of the ocean seafloor had been mapped to any extent because sonar is costly and time-consuming. Sonar requires a ship to navigate very slowly over the area to be mapped while sending and receiving sound wave signals to and from the seafloor, a process commonly known as "mowing the lawn." Because the ocean is so vast, only areas of high interest such as the coastlines, marine reserves, shipping channels, and known shipwreck areas have been mapped with sonar (**Figures A** and **B**).

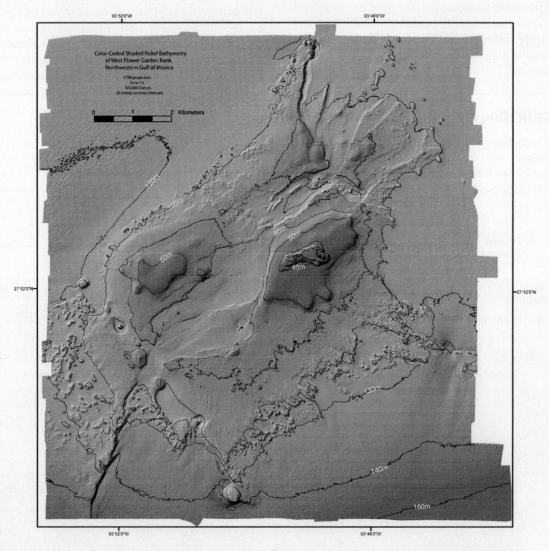

Figure A Multibeam imaging of West Flower Garden Bank, a marine reserve in the Gulf of Mexico. Colors and bathymetry lines (20-m contours) indicate depths.
Courtesy of U.S. Geological Survey.

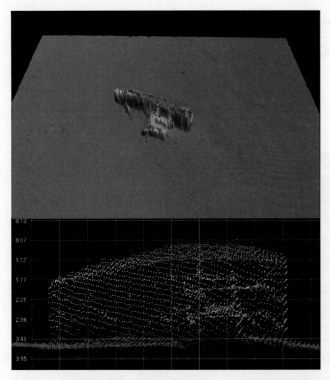

Figure B Multibeam imaging of a shipwreck in Kachemak Bay, Alaska.
Courtesy of National Oceanic and Atmospheric Administration.

Satellite data can provide some seafloor characteristic information on very large scales. Satellites cannot penetrate the sea surface, but they can detect the minor differences in sea level height caused by large features on the seafloor. These minor differences provide a signal and reveal features such as ocean trenches, seamounts, and large oceanic ridges. A recent project aimed at mapping the entire ocean bottom using satellite data provided maps at a resolution of 5 km. This means that objects 5 km or larger can be detected, so large trenches or ridges are visible, but anything smaller than 5 km is not. The results of these mapping efforts revealed around 20,000 previously unknown seamount features, which are underwater mountains that rise 1 km or more from the seafloor. Although these results are exciting to geologists and provide information about important deep-sea features, 5-km resolution is not fine enough detail to be useful to address many marine research questions.

Maps made using sonar data provide higher resolution, and these exist for only about 10% to 15% of the seafloor at a resolution of 100 m. Maps with even finer resolution that can display features several meters in size exist for less than 1% of the seafloor! These finer resolution maps are what many mariners and marine scientists are interested in, because it is at these finer resolutions that smaller features become visible to reveal a more realistic view of the seafloor. Why is it so interesting, and in some cases critical, to reveal the fine-scale features of the seafloor? From a safety perspective, mariners attempting to navigate the seas safely and to avoid potentially dangerous obstacles benefit from detailed seafloor maps. Some features are tall but narrow and pose the risk of a ship strike if their locations are unknown. From a marine science perspective, knowledge of seafloor features provides insight into the available habitat, or homes, for the many sea creatures interacting closely with their seafloor environments. Some larger marine animals appear to use seafloor features for navigation, so scientists attempting to piece together migration routes find detailed seafloor features essential for studies of migratory behaviors.

Many benthic organisms require specific seafloor features for their lifestyles, whether it be hiding between rocks or using underwater landmarks to meet up with a mate. For example, the endangered white abalone is found only on or very near rocks where it is able to blend in with the environment, feed on drift kelp, and encounter other individuals while releasing eggs or sperm into the water. Scientists studying white abalone use detailed multibeam sonar maps to determine where they should focus their research efforts to search for abalone (**Figure C**). Knowledge of seafloor features saves scientists invaluable time because searching in areas of barren sand or rocks with very high relief will yield no abalone to study.

Detailed maps of seafloor characteristics are becoming increasingly valuable to scientists as the many associations of organisms with specific bottom features are discovered. Over time a larger area of the world ocean will be mapped with higher accuracy, but presently most of the sea remains a mystery.

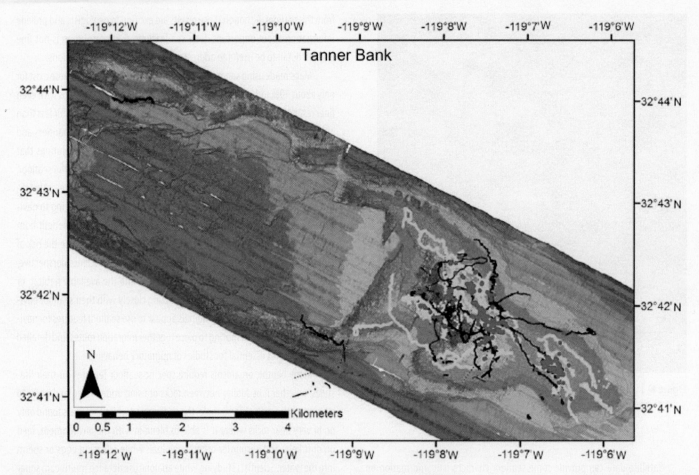

Figure C Results of white abalone population surveys (pink dots) overlaid onto multibeam bathymetry survey data (variegated colors indicate depth). Details of the seafloor are visible, as are ship tracks (horizontal lines).

Courtesy of National Oceanic and Atmospheric Administration. Adapted from Butler, J., M. Neuman, D. Pinkard, R. Kvitek, and G. Cochrane. 2006. The use of multibeam sonar mapping techniques to refine population estimates of the endangered white abalone *(Haliotis sorenseni). Fishery Bulletin* 104:521–532.

Critical Thinking Questions

1. Think of some requirements organisms have for life in the sea. Which of these requirements may include features of the seafloor that are visible with high-resolution sonar? How might revealing seafloor features aid biologists in their studies of marine benthic organisms?

2. Why are marine geologists and marine biologists interested in different levels of seafloor details? Try to think of a research topic that could make use of both broad-scale maps with less detail and fine-scale maps with many details.

For Further Reading

Afonso, P., N. McGinty, and M. Machete. 2014. Dynamics of whale shark occurrence at their fringe oceanic habitat. *PLoS ONE* 9(7):e102060. doi:10.1371/journal.pone.0102060.

Butler, J., M. Neuman, D. Pinkard, R. Kvitek, and G. Cochrane. 2006. The use of multibeam sonar mapping techniques to refine population estimates of the endangered white abalone (*Haliotis sorenseni*). *Fishery Bulletin* 104:521–532.

Sandwell, D. T., R. Dietmar Muller, W. H. F. Smith, E. Garcia, and R. Francis. 2014. New global marine gravity model from CryoSat-2 and Jason-1 reveals buried tectonic structure. *Science* 346(6205):65–67.

enough to accommodate photosynthesis. The depth of the photic zone depends on conditions that affect light penetration in water, extending much deeper (up to 200 m) in clear tropical waters than in murky coastal waters of temperate or polar seas (sometimes less than 5 m). The rest of the ocean volume is the perpetually dark **aphotic** (unlighted) **zone**, where the absence of sunlight prohibits photosynthesis.

The **benthic division** refers to the environment of the sea bottom. The inner shelf includes the seafloor from the low-tide line to the bottom of the photic zone. Beyond that, to the edge of the continental shelf, is the outer shelf. The **bathyal zone** is approximately equivalent to the continental slope areas. The abyssal zone refers to abyssal plains and other ocean-bottom areas between 3,000 and 6,000 m in depth. The upper boundary of this zone is sometimes defined as the region where the water temperature never exceeds 4°C. The hadal zone is that part of the ocean bottom below 6,000 m, primarily the trench areas. Deep-sea trenches are located within the hadal zone and sometimes extend very deep.

The **pelagic division** includes the entire water mass of the ocean. For our purposes, it is sufficient to separate the pelagic region into two provinces: the **neritic province**, which includes the water over the continental shelves, and the **oceanic province**, the water of the deep ocean basins. Each of these subdivisions of the ocean environment is inhabited by characteristic assemblages of marine organisms.

DID YOU KNOW?

The deepest parts of the ocean include deep-sea trenches that are formed by subduction, a process that includes two of Earth's plates colliding and one dropping below the other, causing the seafloor to bend. Life exists even in these deep regions of extreme conditions and no light. The deepest known part of the ocean, the Challenger Deep, is part of the Mariana Trench and is approximately 11 km deep! Only two manned expeditions have successfully explored the Challenger Deep: the first was over 50 years ago, and the second was in 2012. It looks like marine biologists have some more exploring to do in deep-sea trenches.

1.4   An Introduction to Evolution

The physical classification scheme described in the previous section is useful for us to make sense of the vast ocean realm from a physical perspective. From a biological perspective, the classifications are useful to distinguish different areas available for organisms to reside and all of the environmental conditions in these areas, known as **habitats**. Each marine organism is adapted to live in certain conditions, and as a result each life zone around the globe is represented by a suite of organisms adapted to live there. Once **eukaryotic** life evolved there was an explosion of diversity as organisms moved into the many habitat areas available in the sea. Many new species arose, some species were not well adapted for life and became extinct, and the species we see today evolved from earlier ancestral forms.

The main mechanism of evolution is **natural selection**. Organisms best suited for their environments will survive to reproduce and pass on their genes. If this continues over many generations there will be a higher proportion of genes coding for favorable features in the **population**. Eventually, the favorable features will become more common in the population, and the unfavorable features will become less common or even entirely absent. Because the environment is constantly changing there is no perfect state for an organism, and evolution is not goal oriented. The best state for survival is really a moving target as unpredictable environmental changes occur. The combination of the changing environment and the ability of species to adapt or become extinct has led to the great diversity of life that we see today. The ocean realm is no exception; marine organisms are well adapted for life in their unique environments, and changes in physical oceanographic conditions (e.g., temperature or ocean chemistry), or changes to the availability of resources (e.g., food) often lead to differential survival rates. The description of evolution presented above is a very basic introduction to the topic; we will examine the unique adaptations for survival and the evolution of marine organisms as one of the major themes discussed throughout this book.

Case Study

Earth's Earliest Life Forms

One of the most intriguing questions scientists have pondered for decades is exactly how and when the first life forms evolved on Earth. It is estimated that the Earth is 4.5 billion years old, and it was long assumed that the Earth was uninhabitable for about a billion years. Early Earth was thought to be extremely hot, with frequent volcanic activity and an abundance of chemicals that did not support life. Atmospheric oxygen levels were likely low, prohibiting life from flourishing until photosynthetic bacteria (microbes) transformed the atmosphere. Recent evidence indicates that early Earth may not have been so inhospitable, as fossils from miniature microbial communities that flourished along coasts 1 billion years after the Earth formed have been discovered.

For many years the primary evidence of early life forms was in the form of stromatolites, relatively large structures formed by cyanobacteria that fossilize well and can be dated (**Figure A**). The oldest known stromatolites are dated at approximately 3.45 billion years and are found at the Strelley Pool formation in Australia. Other fossilized microbial formations found at the same location in Australia include 3.4-billion-year-old sulfur-eating microbes.

In 2013, a scientist was exploring the geology of a study location in Australia and noticed some interesting wavy-looking formations in the rocks (**Figure B**). Although many scientists had walked over the same area and likely seen the wavy rocks, their importance had not been recognized. It turns out that the wavy formations are evidence of microbial life, and these microbes have been dated to 3.5 billion years, the oldest

(continues)

Figure A Stromatolites.
Courtesy of National Oceanic and Atmospheric Administration, Office of Oceanic and Atmospheric Research, National Undersea Research Program.

fossilized life forms known to exist. Under the microscope the wavy forms appear as thin black filaments in between sand grains, which is a characteristic formation of microbial mats. Due to their location along the ancient shoreline, it is presumed that the microbes were photosynthesizing rather than using minerals from the rocks as their energy source. This new information suggests that photosynthesis began at least 3.5 billion years ago, which is earlier than previously thought.

Figure B Microbial mat fossils that extend the fossil record of the earliest life forms by 300 million years.
© RGB Ventures/SuperStock/Alamy Stock Photo.

Is it likely that these 3.5-billion-year-old fossils provide evidence of the absolute first life forms? No, it is not likely. These microbial mats were made up of photosynthetic bacteria, and it is more likely that nonphotosynthetic bacteria evolved first. The challenge with detecting early life forms is that they must leave behind evidence in the form of fossils or in carbon isotopes. In 2015, 4.1-billion-year-old zircon was discovered that potentially once harbored life, but these results have not been verified. The search for the first life forms is ongoing and always will be, as there is no way to know with certainty when life first evolved, and any new evidence is just one more piece of the early life puzzle. What is certain is that we will continue to be amazed by new discoveries and information that will paint a clearer picture of what the young Earth and seas were like.

Critical Thinking Questions

1. Why can it be assumed that life evolved earlier than 3.5 billion years ago even though the oldest fossils are from this time?

2. What is the importance of discovering early life? What information do early life forms provide for scientists studying life today?

For Further Reading

Bell, E., and M. Harrison. 2015. Potentially biogenic carbon preserved in a 4.1 billion-year-old zircon. *Proceedings of the National Academy of Science* 112(47):14518–14521.

Nofke, N., D. Christian, D. Wacey, and R. M. Hazen. 2013. Microbially induced sedimentary structures recording an ancient ecosystem in the ca. 3.48 billion-year-old Dresser Formation, Pilbara, Western Australia. *Astrobiology* 13(12):1103–1124.

STUDY GUIDE

TOPICS FOR DISCUSSION AND REVIEW

1. Volcanic activity, earthquakes, and the distribution of continents on Earth today are all due to a single process. State what this process is, and describe how it is responsible for these phenomena.

2. Label a diagrammatic cross-section of the North Atlantic Ocean with all benthic features (such as trenches and shelf breaks) presented in this chapter. Then discuss whether this is the best way to categorize the seafloor.

3. What is sonar, and why is it perhaps the best way to determine the dimensions of ocean basins?

4. What is the photic zone, and what zones occur within the photic zone?

5. Why is classifying the marine environment useful and important?

6. What is the most basic definition of evolution by natural selection?

KEY TERMS

abyssal plains 10	nautical mile 6
aphotic zone 15	neritic province 15
bathyal zone 15	oceanic province 15
benthic division 15	ozone 4
continental drift 6	Pangaea 6
continental shelf 9	pelagic division 15
continental slope 9	photic zone 11
eukaryotic 15	plate tectonics 6
evolution 4	population 15
habitat 15	prokaryote 4
intertidal zone 11	ridge and rise system 10
island 10	seafloor spreading 6
last glacial maximum (LGM) 8	seamount 10
	shelf break 9
latitude 5	sonar 10
longitude 5	trench 10
natural selection 15	

REFERENCES

Broad, W. J. 1997. *The Universe Below: Discovering the Secrets of the Deep Sea.* New York: Simon & Schuster.

Clark, P. U., N. G. Pisins, T. F. Stocker, and A. J. Weaver. 2002. The role of the thermohaline circulation in abrupt climate change. *Nature* 415:863–869.

Cloud, P. 1989. *Oasis in Space: Earth History from the Beginning.* New York: W. W. Norton.

Detrick, R. 2004. The engine that drives the earth. *Oceanus* 42(2):6–12.

Dietz, R. S., and J. C. Holden. 1970. Reconstruction of Pangaea: Breakup and dispersion of continents, Permian to present. *Journal of Geophysical Research* 75(26):4939–4956.

Duxbury, A. C., and A. B. Duxbury. (1991). *An Introduction to the World's Oceans.* Dubuque: Wm. C. Brown Publishing.

Earle, S. 2001. *National Geographic Atlas of the Ocean: The Deep Frontier.* Washington, DC: National Geographic.

Ellis, R. 2000. *Encyclopedia of the Sea.* New York: Knopf.

Geist, E. L., V. V. Titov, and C. E. Synolakis, 2006. Tsunami: Wave of change. *Scientific American* 294:56–63.

Hogg, N. 1992. The Gulf Stream and its recirculation. *Oceanus* 35:28–37.

Janin, H., and S. A. Mandia. (2012). *Rising Sea Levels: An Introduction to Cause and Impact.* Jefferson, NC, and London: McFarland & Company, Inc. Publishers.

Jenkyns, H. C. 1994. Early history of the oceans. *Oceanus* 36:49–52.

Keleman, P. B. 2009. The origin of the land under the sea. *Scientific American* 300(2):42–47.

Kunzig, R. 1999. *The Restless Sea: Exploring the World Beneath the Waves.* New York: W. W. Norton.

Leier, M. 2001. *World Atlas of the Oceans: More Than 200 Maps and Charts of the Ocean Floor.* Richmond Hill, Ontario: Firefly Books.

Mann, M. E. 2007. Climate over the past two millennia. *Annual Review of Earth and Planetary Sciences* 35:111–136.

McDonald, K. C., and P. J. Fox. 1990. The mid-ocean ridge. *Scientific American* 262:72–79.

Melville, W. K., and P. Matusov. 2002. Distribution of breaking waves at the ocean surface. *Nature* 417:58–63.

Satake, K., and A. F. Atwater. 2007. Long-term perspectives on giant earthquakes and tsunamis at subduction zones. *Annual Review of Earth and Planetary Sciences* 35:349–374.

Tarduno, J. 2008. Hot spots unplugged. *Scientific American* 298(1):87–93.

Weller, R. A., and D. M. Farmer. 1992. Dynamics of the ocean's mixed layer. *Oceanus* 35:46–55.

CHAPTER 2

Physical and Chemical Oceanography

STUDENT LEARNING OUTCOMES

1. Describe the major interactions of the ocean and the atmosphere and how these interactions impact life on land.

2. Compare the most common types of waves, including their causes and their relative size.

3. Compare the three most common types of tides observed, and become familiar with where in the world each type is typically found.

4. Describe the Coriolis effect and how it determines the direction of surface water movement and upwelling.

5. Be able to label all of the currents and gyres pictured in Figure 2.15.

6. Explain how massive volumes of water are moved around the globe at depth.

7. Recall the many unique properties of seawater, and describe how these properties determine how marine organisms are distributed.

CHAPTER OUTLINE

2.1 The Ocean in Motion
Atmospheric Circulation
Waves
Ocean Tides
Surface Currents
Vertical Water Movements
RESEARCH in Progress: The Rotating World of Larvae Traveling in Gyres

Case Study: The Great Pacific Garbage Patch

2.2 Properties of Seawater
Pure Water
Seawater
Study Guide
References

The interactions of weather patterns and the ocean lead to surface ocean waves, such as the breaking wave displayed, and vertical mixing of water, as seen in the subsurface view in this image.
© Willyam Bradberry/Shutterstock, Inc.

Ocean water is constantly in motion, moving and mixing nutrients, oxygen, and heat within the water. Wave action, tides, currents, and density-driven vertical water movements often enhance mixing and reduce variations in **salinity** and **temperature**. Oceanic circulation processes also serve to disperse floating organisms and their eggs, spores, and larvae. For all marine organisms, movement of the sea leads to toxic body wastes being carried away and food, nutrients, and essential elements being replenished. Organisms living on land are also greatly affected by movements of ocean water; water masses with varying temperatures are moved around by currents, leading to varying temperatures on land. In many cases, the sea moderates temperatures on land, making life more pleasant for humans and other terrestrial organisms. The unique properties of water allow for all life on Earth to exist, and seawater creates an environment where the majority of organisms on Earth reside. In this chapter we will examine the physical oceanographic components of the motions of the sea and delve into the many properties of seawater. This information is crucial for our further investigations of life in the sea, because organisms are greatly impacted by water properties and movements, leading to many of the phenomenally unique adaptations observed in marine life.

2.1   The Ocean in Motion

The many movements of ocean water are a complex and dynamic mixture of interactions between weather patterns at sea and on land. Water movements are not only affected by weather patterns, but in many cases water circulation influences weather patterns, leading to a cycle of interactions between the sea and the atmosphere. Circulation of air in the form of wind drives most ocean water movements, and water movements that transfer large volumes of water around the globe affect the climate on land.

Atmospheric Circulation

The **atmosphere** is the combination of water vapor, gases, and particles in the air that surround Earth in a layer. Before beginning our discussion of water movement it is useful to discuss air movement, also known as *atmospheric circulation*. Almost all of the mechanisms of water movement that exist are driven by air movement, and, more specifically, wind. The exception to this is **tides** and tidally driven water movement, which are driven by the alignment of Earth, the sun, and the moon (discussed later in this chapter). The curvature of the sphere of Earth causes the sun to warm latitudes of Earth's surface and overlying air near the equator more than at the poles. Solar energy is not reduced as the sun moves away from the equator, but the area over which the sun shines at the equator is much smaller than the area over which it shines at the poles (**Figure 2.1**). This differential warming of the atmosphere by the sun drives patterns of winds that blow

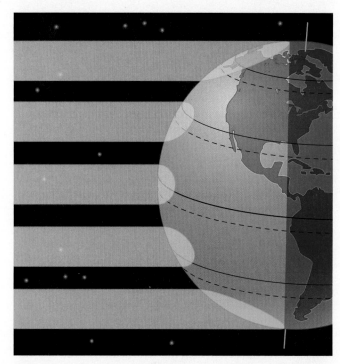

Figure 2.1 A portion of Earth with the angle of sunlight striking at several sample latitudes. Notice that the sun strikes directly, and therefore most intensely, at the equator.

DID YOU KNOW?

A hurricane forms when one huge mass of circulating air moves out over a warm part of the ocean. Hurricanes impact life on land by producing huge quantities of rain, large waves, and extreme winds that may cause damage and, unfortunately, loss of human life. The picture below is of Hurricane Katrina, a category 5 hurricane that led to massive destruction and the greatest financial cost of any natural disaster in the United States. Healthy coastal environments with well-developed vegetation will help to soak up the excess water moved toward land by hurricanes, but the demise of coastal wetlands has left many areas, like New Orleans, vulnerable. Hurricanes are yet another (albeit extreme) weather phenomenon caused by the interactions of the ocean and the atmosphere.

Hurricane Katrina, a category 5 hurricane that devastated the coast of Louisiana in 2005.
National Oceanic and Atmospheric Administration/National Climatic Data Center.

across the ocean surface, creating waves and surface currents. In addition to these net horizontal water movements, **upwelling**, a process of vertical water movement that increases productivity by photosynthesizers and will be discussed later in this chapter, is also caused by winds.

Atmospheric circulation is closely linked in a cyclical nature to ocean circulation. Oceanic currents transfer heat from the equator to the rest of the world. Currents absorb heat at the equator and transfer it to the poles. From the poles, currents send cool water toward the equator, where heat is absorbed once again. This transfer of heat between the ocean and land is substantial. The majority of the thermal energy at Earth's surface is stored in the oceans. The temperature of water masses distributed around the globe determines what species can survive in each location of the sea.

Water from the sea evaporates into the air above as **water vapor**. These water droplets in the air may remain near the sea or may rise up above the sea, become cooled with rising and cooling air, and accumulate into clouds. If enough water droplets accumulate in the clouds and the temperature is low enough, the water falls from the sky in the form of rain or snow. If the rain falls over the ocean, then it is returned to the sea immediately. If the rain falls over the land, it may still be returned to the sea by river runoff or even storm drains. Weather patterns over the sea and land are driven by these close, intermingling processes occurring daily in the ocean and in the atmosphere (**Figure 2.2**).

Waves

When most people think of an ocean wave, they think of waves that start offshore and then end up breaking on the beach (**Figure 2.3**). In reality, the term **wave** in the ocean includes water disturbances that vary in how the wave is formed and how the wave is displayed. We will begin our discussion of waves with those that are formed by wind, which includes **wind waves**, the typical waves that break on the beach. Some wind waves never reach the shore but dissipate out at sea. Wind waves start out as **capillary waves**, tiny ripples in the ocean that are the first step in transferring energy from the wind to the water. These tiny capillary waves are all over the ocean surface, and if they reach a certain **wavelength** (see below), they are considered wind waves.

Wind waves typically travel in a repeating series of alternating wave crests and troughs. The size and energy of waves are dependent on the wind's velocity, duration, and **fetch** (the distance over which the wind blows in contact with the sea surface). Wind waves range in height from the very small capillary waves a few millimeters high to monster storm waves surpassing 30 m in height (**Figure 2.4**). Waves are commonly characterized by their height, wavelength (**Figure 2.5**), and **period** (the time required for two successive wave crests to pass a fixed point). Regardless of their size, these general features of waves apply to all ocean waves.

Once created by surface winds, ocean waves travel away from their area of formation. Only the wave shape advances, however, transmitting the energy forward. The water particles

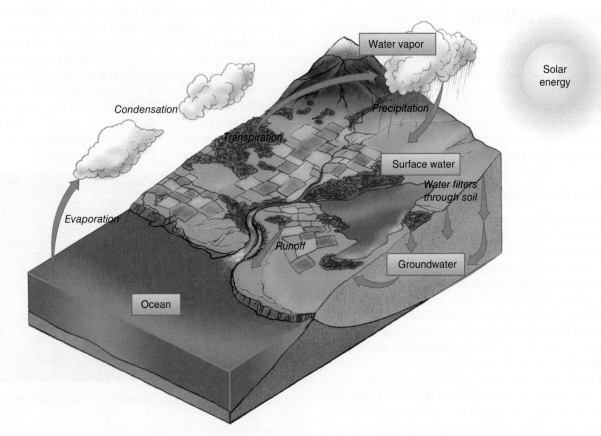

Figure 2.2 The water cycle, displaying the interactions of the atmosphere and ocean.

Figure 2.3 A wind wave breaking on the beach. Waves with nice form and size provide human entertainment value.
Courtesy of Deanna Pinkard-Meier.

themselves do not advance in the direction of the wave. Instead, their paths approximate vertical circles with little net forward motion (Figure 2.5). If waves from various directions collide they may cancel each other out, and the path of energy ends. Another possibility is that wave energy from two or more waves may actually combine, creating an even larger wave as a result. The result of wave interactions all depends on the direction of the original source of wave energy for each wave. Waves provide an important mechanism to mix the near-surface layer of the sea. The depth to which waves produce noticeable motion is about one half the wavelength. As wavelengths seldom exceed 100 m in any ocean, the effective depth of mixing by wind-driven waves is generally no greater than 50 m. Mixing moves nutrients up to the surface, providing phytoplankton with some of their necessary components for photosynthesis, and moves oxygen from the surface to deeper waters, providing organisms with oxygen.

Once waves enter shallow water they behave differently from open-ocean waves. When the water depth is less than one half the wavelength, bottom friction begins to slow the forward speed of the waves. This slowing causes the waves to become higher and steeper. At the point where the wave height becomes greater than about one seventh of the wavelength, the wave top becomes unstable as it outruns its base. It then pitches forward and breaks. The energy released on shorelines (and on the organisms living there) by breaking waves can be enormous and is a major force in shaping the physical and biological characters of most coastlines.

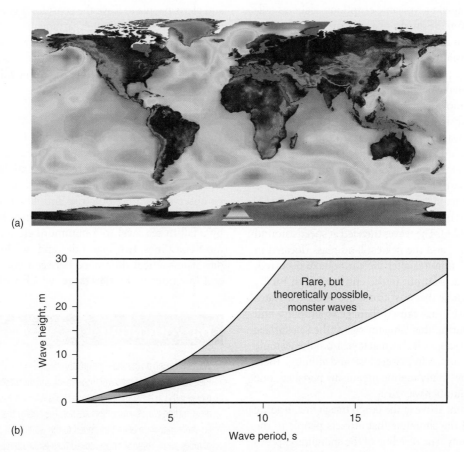

Figure 2.4 (a) Range of ocean surface wave heights, determined over a 10-day period by the TOPEX/Poseidon satellite. (b) Wave periods and their corresponding range of heights.
Courtesy of Space Science and Engineering Center, University of Wisconsin, Madison.

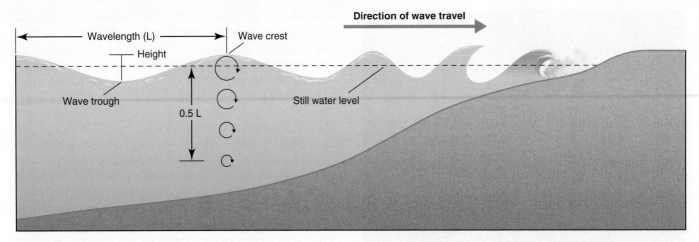

Figure 2.5 Waveform and pattern of water motion in a deep-water wave as it moves to the right toward a shoreline. Circles indicate orbits of water particles diminishing with depth. Little water motion occurs below a depth equal to one half of the wavelength.

Waves that are caused by forces other than wind include tides and **tsunami**. Tides, the longest of all ocean waves, are caused by gravity and the motion of Earth and will be discussed in detail in the next section. Tsunami is a set of waves of epic proportions, caused by large, abrupt disturbances of the sea surface. The term *tsunami* is the combination of the words *harbor* (*tsu*) and *wave* (*nami*) in Japanese. Potential disturbances that can cause tsunami include earthquakes, underwater landslides caused by earthquakes, volcanic eruptions, icebergs falling from glaciers, or even the impact of a large meteorite landing in the ocean. Tsunami is most often caused by large earthquakes underwater or in coastal regions, and the larger tsunami are caused by earthquakes that are greater than 7.0 on the **Richter scale**. When an earthquake occurs, continental plates shift. It is the resettling vertical movement of the plates displacing a large volume of water quickly that leads to the creation of the tsunami waves. Some tsunami waves dissipate out at sea and never reach land; others travel toward land at great speeds, and when they near a coastline they slow down, gain height, and move in a huge wall of water toward land.

During a large tsunami event originating in Japan in 2011, tsunami waves radiated out from the epicenter of the earthquake in all directions (**Figure 2.6**). The waves traveled at speeds around 200 m/s (470 miles/hr), and the nearest land mass (located in Japan) was hit just 15 minutes after the earthquake occurred. A tsunami warning was sent out to the entire western Pacific Ocean, and areas as far as the West Coast of the United States experienced large waves that caused damage to property and loss of human life. During this tsunami event, the sea surface was lifted 6 m (20 feet) above its normal level from the tectonic activity. Over 20,000 human lives were lost, and although there is no way to know how many marine organisms perished, one can assume that the number was great.

Sadly, some tsunami arrive at the beach **trough** first, leading to a swift receding of the shoreline that attracts people to the shoreline out of curiosity. The receding of the shoreline exposes sea creatures and is quite a sight to see, but what arrives next, the actual **crest** of the wave, is what causes death and destruction

(**Figure 2.7**). It is not only the great heights of tsunami waves that cause destruction but also the enormous lengths. Tsunami waves do not just break on the sandy beach and then end, but they continue to sweep through entire coastal towns, carrying huge volumes of water onto land. The severity of tsunami events has led to the creation of a Tsunami Warning Network, a communication system that warns coastal cities of possible tsunami events when an earthquake occurs. The network has saved many lives and alerts people to leave the coast for higher ground before the monster waves hit. Although earthquakes, which are the primary cause of tsunami, cannot be predicted, with advances in technology we are able to better monitor the oceans after an earthquake takes place to predict whether the result will be a tsunami.

Ocean Tides

Tides are ocean surface phenomena familiar to anyone who has spent time along a seashore. Tides are actually very long-period waves that are usually imperceptible in the open ocean and only become noticeable near the shoreline, where they can be observed as a periodic rise and fall of the sea surface. The maximum elevation of the tide, known as **high tide**, is followed by a fall in sea level to a minimum elevation, or **low tide**. On most coastlines, two high tides and two low tides occur each day. The vertical difference between consecutive high tides and low tides is the **tidal range**, which varies from just a few

DID YOU KNOW?

Waves breaking on the shore are greatly affected by variables such as the shape of the coastline, the steepness of the seafloor, and the presence of structures (e.g., a pier or jetty) in the surrounding area. This is why some areas are well known for great surfing waves, and others are not. The energy that sends waves breaking on shore may have been transferred to the water from hundreds or thousands of miles away. Some scientists spend their entire careers making predictions of wave heights and other variables of a swell for surfers around the world.

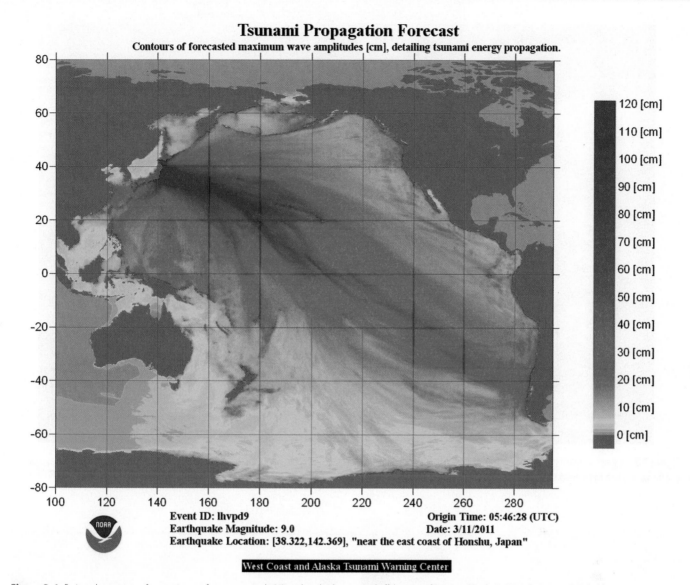

Figure 2.6 Projected movement of tsunami waves from a magnitude 9.0 earthquake that occurred off the coast of Japan on March 11, 2011. Data were compiled from gauges monitored by the National Tsunami Warning Center, and additional data were provided by several countries in the area of impact.

National Oceanic and Atmospheric Administration/West Coast and Alaska Tsunami Warning Center.

centimeters in the Mediterranean Sea to more than 15 m in the long, narrow Bay of Fundy between Nova Scotia and New Brunswick, as seen in **Figure 2.8**. The global tidal range averages about 2 m. In areas with large tidal fluctuations, the change in tide greatly impacts marine organisms inhabiting the shoreline. The changing amount of water coverage only allows organisms with adaptations to resist **desiccation** to survive. Some marine larvae (young fish and invertebrates) rely on tides to transport them into their juvenile habitat.

In 1687, in his *Principia Mathematica*, Sir Isaac Newton explained ocean tides as the consequence of the gravitational attraction of the moon and sun on the oceans of Earth. According to Newton's law of universal gravitation, our moon, because of its closeness to Earth, exerts about twice as much tide-generating force as does the more distant but much larger sun. The constantly changing position of Earth relative to the moon and sun nicely account for the timing of ocean tides, yet

Newton's equilibrium model of ocean tides is seriously deficient in its ability to explain real tidal patterns in real oceans. Because landmasses like continents and islands exist in the world ocean and the ocean varies greatly in depth and bathymetry, the models for observed tidal patterns are much more complicated. Those of you interested in Newton's model and other models of ocean tides should consult a current oceanography textbook. For our purposes, a description of the results of tide-producing forces, rather than their causes, will suffice.

The moon completes one orbit around Earth each lunar month (27.5 days). Hypothetically, if Earth were completely covered with water, two bulges of water, or lunar tides, would occur: one on the side of Earth facing the moon and the other on the opposite side of the globe (**Figure 2.9**). As Earth makes a complete rotation every 24 hours, a point on Earth's surface (indicated by the marker in Figure 2.9) would first experience a high tide (a), then a low tide (b), another high tide

Figure 2.7 Destruction from the 2004 tsunami that occurred in Banda Aceh, Sumatra, Indonesia. This photo was taken 6 weeks after the tsunami hit, and at that time it still appeared as if the sea and land had merged indefinitely.
Courtesy of Photographer's Mate 1st Class Jon Gesch/U.S. Navy.

Figure 2.8 The shoreline at Hopewell Rocks in New Brunswick, Canada, at low tide (left image) and high tide (right image).
(a) © Melissa King/Shutterstock, Inc. (b) © gvictoria/Shutterstock, Inc.

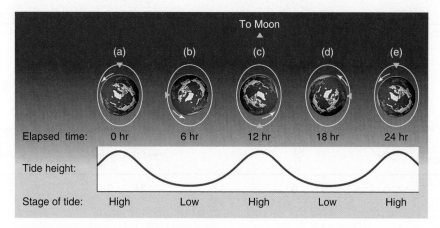

Figure 2.9 A depiction of the high tides when under tidal bulges (a, c, and e above) and low tides when at right angles to tidal bulges (b and d above) experienced each day as Earth rotates. The reference point on Earth (Florida) is indicated by a yellow marker.

(c), another low tide (d), and finally another high tide (e). During that rotation, however, the moon advances in its own orbit so that an additional 50 minutes of Earth's rotation is required to bring our reference point directly in line with the moon again. Thus, the reference point experiences only two equal high and two equal low tides every 24 hours and 50 minutes (a **lunar day**).

In a similar manner, the sun–Earth system also generates tide-producing forces that yield a solar tide about one half as large as the lunar tide. The solar tide is expressed as a variation on the basic lunar tidal pattern, not as a separate set of tides. When the sun, moon, and Earth are in approximate alignment (at the time of the new moon and full moon, **Figure 2.10**), the solar tide has an additive effect on the lunar tide, creating several days of extra-high high tides and very low low tides known as **spring tides**. One week later, when the sun and moon are at right angles to each other relative to Earth, the solar tide partially cancels the effects of the lunar tide to produce moderate tides known as **neap tides**. During each lunar month, two sets of spring tides and two sets of neap tides occur.

So far, only the effects of tide-producing forces in a not very realistic ocean covering a hypothetical planet without continents have been considered. What happens when continental landmasses are taken into consideration? The continents block the westward passage of the tidal bulges as Earth rotates under them. Unable to move freely around the globe, these tidal impulses establish complex patterns within each ocean basin that may differ markedly from the tidal patterns of adjacent ocean basins or even other regions of the same ocean basin.

Figure 2.11 shows some regional variations in the daily tidal configuration at three stations along the east and west coasts of North America. Portland, Maine, experiences two high tides and two low tides each lunar day. The two high tides are quite similar to each other, as are the two low tides. Such tidal patterns, referred to as **semidiurnal** (semidaily) **tides**, are characteristic of much of the East Coast of the United States. The tidal pattern at Pensacola, Florida, on the Gulf Coast, consists of one high tide and one low tide each lunar day. This is a **diurnal**, or daily, **tide**. Different yet is the daily tidal pattern at San Diego,

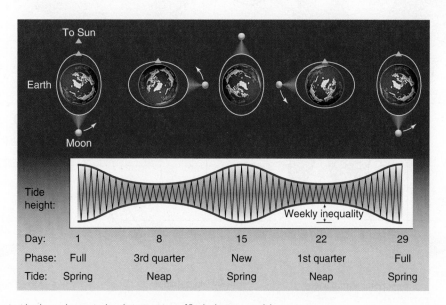

Figure 2.10 Weekly variation in tides due to changes in the relative positions of Earth, the moon, and the sun.

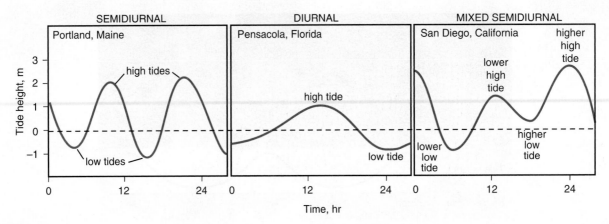

Figure 2.11 Three common tidal patterns in various areas of the United States.

California. There, two high tides and two low tides occur each day, but successive high tides are quite different from each other. This type of tidal pattern, characteristic of the west coast of North America, is a **mixed semidiurnal tide**. **Figure 2.12** outlines the geographic occurrence of diurnal, semidiurnal, and mixed semidiurnal tides for coastal areas.

Tidal conditions for any day on a selected coastline can be predicted because the periodic nature of tides is easily observed and recorded. For the most part, prediction of the timing and amplitude of future tides is based on the astronomical positions of the sun and moon relative to Earth and on historical observations of actual tidal occurrences at tide-recording stations along coastlines and in harbors around the world. The National Ocean Survey of the U.S. Department of Commerce uses information from these records to compile and publish annual "Tide Tables of High and Low Water Predictions" for principal ports along most coastlines of the world.

DID YOU KNOW?

The term **king tide** refers to the very highest high and lowest low tides experienced in an area. The term is not typically used by scientists, but conditions during this natural, normal, and cyclical phenomenon are of increasing interest as an important indicator of what coastlines might look like if sea levels rise due to climate change. King tides occur when the moon is not only aligned with Earth at a given point but is closest to Earth. It is predicted that sea-level rise will make king tides our everyday tides, so the observed coastal flooding that occurs with king tides in some areas is disconcerting. Observing king tides can help communities plan and prepare for rising sea levels.

Surface Currents

Ocean surface currents occur when winds blow over the ocean at a constant direction and velocity for a sufficient period of time

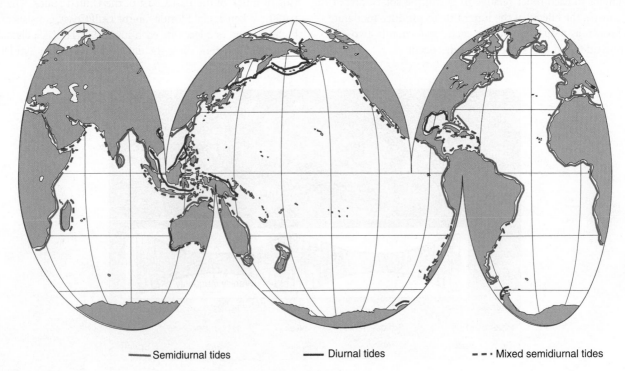

Figure 2.12 The geographic occurrence of the three major types of tides found worldwide.

to transfer momentum to the water through friction. Unlike ocean waves, surface currents do represent the actual horizontal transport of water molecules. The momentum transferred to the sea by winds drives regular patterns of broad, slow, relatively shallow ocean surface currents. Some currents transport more than 100 times the volume of water carried by all of Earth's rivers combined. Currents of such magnitude greatly affect the distribution of marine organisms and the rate of heat transported from tropical to polar regions.

These currents are driven by stable patterns of winds at the ocean surface. Three major wind belts occur in the Northern Hemisphere. The **trade winds**, near 15° N latitude, blow from northeast to southwest. The **westerlies**, in the middle latitudes, blow primarily from the west and southwest. And the **polar easterlies**, at very high latitudes, blow from east to west. Each of these wind belts has its mirror-image counterpart in the Southern Hemisphere. As the surface layer of water is moved horizontally by these broad belts of surface winds, momentum is transferred first to the sea surface and then downward. The speed of the deeper water steadily diminishes when the movement can no longer overcome the viscosity of the water. Eventually, at depths generally less than 200 m, the speed of wind-driven currents becomes negligible.

The surface water moved by the wind does not flow parallel to the wind direction but experiences an appreciable deflection, known as the **Coriolis effect**. The Coriolis effect influences moving air and water masses by causing a deflection that is affected by Earth's rotation; the deflection is to the right in the Northern Hemisphere and to the left in the Southern Hemisphere. As successively deeper water layers are set into motion by the water above them, they undergo a further Coriolis deflection away from the direction of the water just above to produce a spiral of current directions from the surface downward in a pattern known as an **Ekman spiral** (**Figure 2.13**). The magnitude of the Coriolis deflection of wind-driven currents varies from about 15 degrees in shallow coastal regions to nearly 45 degrees in the open ocean. The net Coriolis deflection from the wind headings creates a pattern of wind-forced ocean-surface currents that flows primarily in an east-to-west or west-to-east direction.

Except for the region just north of Antarctica, the continuous flow of these wind-driven east–west currents is obstructed by continents. This causes water transported by currents from one side of the ocean to pile up against continental margins on the other side. The surface of the equatorial Pacific Ocean, for example, is about 2 m higher on the west side than it is on the east. The opposite is true in the middle latitudes of both hemispheres, where the east side is higher. Eventually, the water must flow away from these areas of accumulation. Either it flows directly back against the established current, producing a **countercurrent**, or it flows as a **continental boundary current** parallel to a continental margin from areas of accumulation to areas where water has been removed. Both these current patterns exist, but they are particularly clear in the North Pacific Ocean (**Figure 2.14**). An east-flowing Equatorial Countercurrent divides the west-flowing North Pacific Equatorial Current. The north–south-flowing continental boundary currents merge into the east–west currents to produce large circular current patterns, or **gyres**. Similar current patterns are found in the other major ocean basins (**Figure 2.15**). Gyres move water from one location to another, redistributing water of varying temperatures around the globe. Gyres are responsible for the movement of relatively cool water toward the tropics and relatively warm water toward the poles.

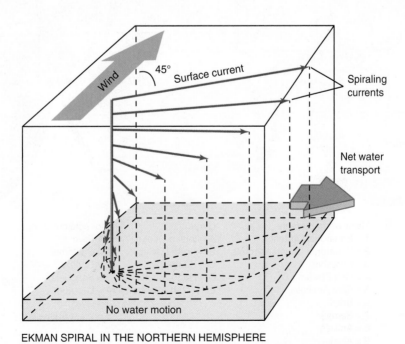

EKMAN SPIRAL IN THE NORTHERN HEMISPHERE

Figure 2.13 A spiral of current directions, indicating greater deflection to the right (in the Northern Hemisphere), which increases with depth due to the Coriolis effect. The arrow length represents relative current speed. A similar pattern, but including deflection to the left, is observed in the Southern Hemisphere.

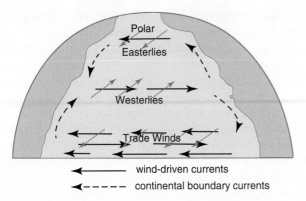

Figure 2.14 Generalized surface-current flow in the North Pacific Ocean. Blue arrows indicate general directions of ocean-surface winds.

Gyres are huge circular current patterns that are rimmed by large permanent currents. Smaller circular current patterns that are more temporary loops of swirling water are known as **eddies**. Although eddies are considered temporary, they can travel long distances before dissipating, and therefore transport water (and hitchhiking marine organisms) long distances. Gyres and eddies are either considered cyclonic (counterclockwise rotation) or anitcyclonic (clockwise rotation). They are also categorized based on whether they have a warm inner core or cold inner

core. The temperature of the water inside the core is determined by the temperature of the water where the gyre is formed. Cold and warm core gyres can easily be distinguished by examining a satellite image of the water mass, as seen in **Figure 2.16a**, and their formation (detailed in **Figure 2.16b**) can be tracked over time. Some gyres can reach diameters of up to 1,000 km!

Maps similar to Figure 2.15 are useful for describing long-term average patterns of surface ocean circulation; however, they tend to hide the subtle intricacies that exist in these currents at any moment in time. Current maps are analogous to the blurred images taken of a night freeway scene when the camera shutter is held open for hours. The pattern of traffic flow is obvious, yet the details of vehicles' slowing, accelerating, and changing lanes are completely lost. The continued development of satellite monitoring of ocean surface phenomena provides an improved approach for visualizing and understanding global-scale surface current patterns on a daily or even hourly timescale. **Figure 2.17** is a satellite image of a portion of the North Atlantic Ocean, including the Gulf Stream. This image emphasizes ocean-surface temperature differences and reveals remarkable meanders, constrictions, and nearly detached rings of Gulf Stream water as the current flows north and east along the path shown in Figure 2.15.

The major current patterns shown in Figure 2.15 and described above are those "typical" at any given time, but

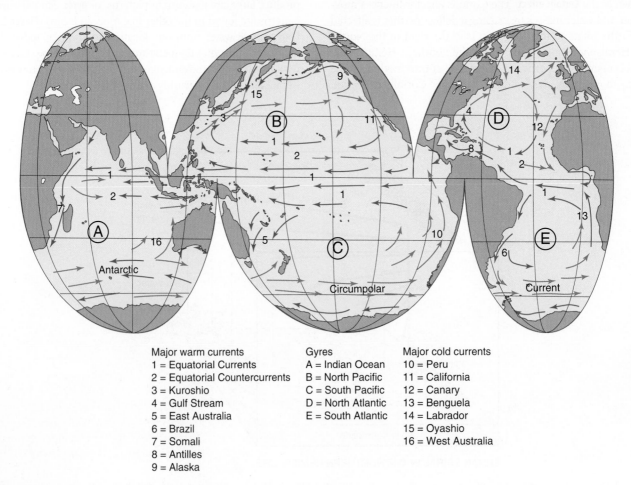

Major warm currents	Gyres	Major cold currents
1 = Equatorial Currents	A = Indian Ocean	10 = Peru
2 = Equatorial Countercurrents	B = North Pacific	11 = California
3 = Kuroshio	C = South Pacific	12 = Canary
4 = Gulf Stream	D = North Atlantic	13 = Benguela
5 = East Australia	E = South Atlantic	14 = Labrador
6 = Brazil		15 = Oyashio
7 = Somali		16 = West Australia
8 = Antilles		
9 = Alaska		

Figure 2.15 The major surface currents of the world ocean.

Modified from Pickard, G. L., and W. J. Emory, eds. *Descriptive Physical Oceanography* (Pergamon Press, 1982).

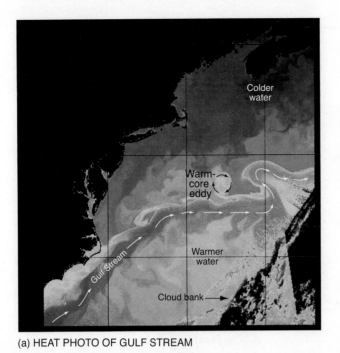

(a) HEAT PHOTO OF GULF STREAM

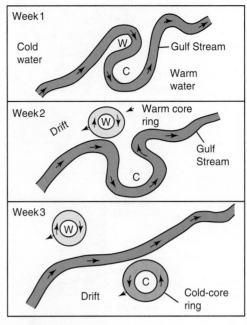

(b) FORMATION OF RINGS

Figure 2.16 (a) NOAA satellite image of the Gulf Stream with a very apparent warm-core eddy pinched off from the main current. Note that the warm-core eddy rotates clockwise. (b) The formation of eddies from a main current takes several weeks to occur. Warm or cold water is trapped in a rotating current (clockwise for warm-core eddies and counterclockwise for cold-core eddies).
Courtesy of NOAA.

Figure 2.17 NOAA satellite image of the Gulf Stream off the U.S. East Coast. Water surface temperatures are represented by a range of colors, with the coldest temperatures represented by violet (−2°C to 9°C); the warmest by red (26°C to 28°C); and blue, green, and orange in between. The Gulf Stream can be identified by an orange band of water at the southern end near Miami, Florida, and then a yellow band of water toward the northern end off the coast of Cape Hatteras. The Gulf Stream moves warm water northeast.
Courtesy of Donna Thomas/MODIS Ocean Group NASA/GSFC SST product by R. Evans et al., University of Miami.

because the ocean is dynamic in its interaction with the atmosphere there are often departures from typical patterns. **El Niño** is a phenomenon that represents a strong departure from the more typical current patterns in the central Pacific Ocean that drive coastal and equatorial upwelling events. Upwelling is a highly influential process that brings cool, deep, and nutrient-rich waters to the surface and will be discussed in detail later in this chapter. El Niño is characterized by a prominent warming of the equatorial Pacific surface waters. El Niño appears to occur irregularly every 2 to 7 years, and each occurrence lasts from several months to well over a year. The El Niño phenomenon is associated with the Southern Oscillation, a trans-Pacific linkage of atmospheric pressure systems, and the climatic anomaly has come to be known collectively as **El Niño-Southern Oscillation**, or **ENSO**. Normally, the trade winds blow around the South Pacific high-pressure center located near Easter Island and then blow westward to a large Indonesian low-pressure center. As these winds move water westward, the water is warmed, and the **thermocline** (a vertical zone of decreased temperature) is depressed from about 50 m below the surface on the east side of the Pacific to about 200 m deep on the west side. ENSOs occur, for reasons not well understood, when this pressure difference across the tropical Pacific relaxes and both surface winds and ocean currents either cease to flow westward or actually reverse (**Figure 2.18**). Although the effects of an ENSO event are somewhat variable, they are usually global in extent and occasionally severe in impact. ENSO events not only bring severe weather, but also higher than normal water temperatures in some areas lead to an expansion of geographical distributions for many marine organisms.

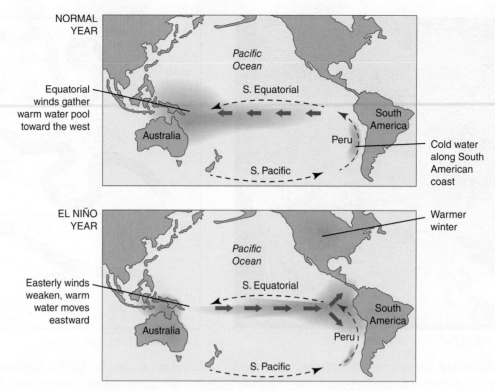

Figure 2.18 A depiction of ENSO-neutral, or "normal," conditions, and ENSO-positive, or El Niño, conditions.
© Designua/Shutterstock, Inc.

The 1982–1983 ENSO event, for example, was associated with heavy flooding on the West Coast of the United States, intensification of the drought in sub-Saharan Africa and Australia, and severe hurricane-force storms in Polynesia. Surface ocean water temperatures from Peru to California soared to as much as 8°C above normal. The 1997–1998 ENSO event caused similar disruptions but was even more severe (**Figure 2.19**). These strong El Niño events and the associated buildup of warm, less dense water block upwelling of nutrient-rich waters, and, in effect, the entire food web is altered and coastal marine populations of many organisms decline. During severe El Niño years, some fish and fish-eating seabird populations almost completely disappear. Eventually, the area of warm tropical water dissipates, and El Niño conditions are replaced by cooler eastern tropical Pacific surface temperatures, low rainfall, and well-developed coastal upwelling along Peru and northern Chile. These conditions are considered normal, or ENSO neutral.

The 2002–2004 episode is ranked in the top 10 El Niño events of the past 50 years. A similar El Niño event appeared in 2006, but it had an unusually short duration, collapsing in early 2007. When the dramatic effects of El Niño conditions were experienced starting in the early 1990s scientists started to investigate the patterns more extensively. The National Oceanic and Atmospheric Administration (NOAA) developed a team to model and attempt to predict ENSO events so that people can be more prepared. At this point in time anyone with Internet access can view the current ENSO index daily. ENSO events have been in decline in recent years, with ENSO-neutral conditions common, and even some conditions considered to be the opposite of El Niño, **La Niña**. La Niña conditions include intense upwelling events at the equator caused by strengthened trade winds, cooler water temperatures in the eastern Pacific, and drought conditions in many areas.

As this edition is being prepared, the West Coast of the United States has been experiencing severe drought conditions for several years, and many people are hoping a predicted strong El Niño event will bring much needed rain to the area. NOAA data predicted a strong El Niño for the fall and winter of 2015–2016, although whether this El Niño event brought enough rain to alleviate the drought is unlikely. What this El Niño event has resulted in is a change in the distribution of some sea life. Large fish normally inhabiting warm tropical waters moved north of their typical distribution, leading to epic fishing conditions in southern and central California. Several yellow-bellied sea

DID YOU KNOW?

There is a huge difference between currents that flow along the west coast of a continent (called *eastern boundary currents*) and those that flow along the east coast of a continent (called *western boundary currents*). Western boundary currents are the fastest and deepest currents found. They are narrow and carry large volumes of warm water from south to north. An example is the Gulf Stream Current, located along the East Coast of the United States, which has a flow rate of at least 55 million m/s! Western boundary currents are also very likely to form eddies, circular flowing currents that pinch off of a main current. Eastern boundary currents are the opposite: they are wide, shallow, and slow and carry cold water toward the equator. Eddies do not tend to form as often from eastern boundary currents. The California Current is an example of an eastern boundary current.

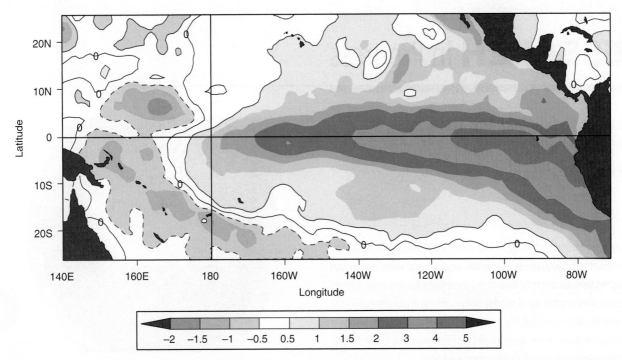

Figure 2.19   Observed sea-surface temperature anomaly, in degrees Celsius, in the Equatorial Pacific Ocean based on a 7-day average in mid-September 1997. Notice the tongue of unusually warm water extending westward from the coasts of Ecuador and Peru.

snakes that are normally found in tropical waters washed ashore in southern California, surprising beachgoers who were not expecting an encounter with a venomous aquatic snake.

Vertical Water Movements

Vertical movements of ocean water are produced by upwelling and sinking processes. These processes tend to break down the vertical stratification established by the **pycnocline**, a vertical zone of increased density. Localized areas of upwelling are created by several oceanic processes that bring deeper nutrient-rich waters to the surface. One type, coastal upwelling, is produced by winds blowing surface waters away from a coastline by Ekman transport. In the Northern Hemisphere, winds that blow parallel to the west coast of a continent from the north lead to a net transport of water to the right of the wind direction, which in this example is away from the continent. The surface waters that are taken away from shore by the net transport due to wind are replaced by deeper water rising to the surface (**Figure 2.20**). Near-shore currents, which veer away from the shoreline, produce

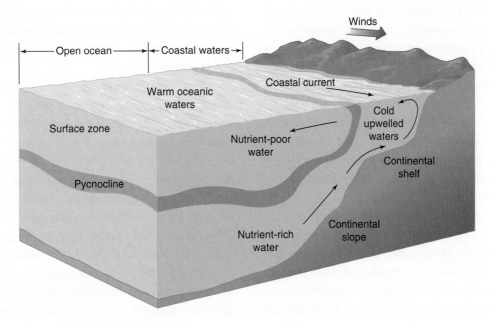

Figure 2.20   Coastal upwelling in the Northern Hemisphere.

RESEARCH in Progress

The Rotating World of Larvae Traveling in Gyres

Rotating water masses such as gyres and eddies are significant to marine life because they trap small planktonic organisms and transport them to areas that they may not have made it to on their own. Larvae (baby fish and invertebrates) may be deposited or settle out of an eddy many hundreds of kilometers from the area where they were spawned. The transport of larvae inside an eddy broadens the distribution of the species involved and increases genetic diversity. Nonliving matter may also be transported by gyres and eddies.

Research has been conducted for many years to attempt to tease out the reasons behind the distribution of marine larvae. The settlement location of marine larvae affects where the future adult population will reside. When the adults are edible fish, humans are concerned with where they will be located. Although identifying the factors that determine the location of marine larvae is a difficult research question to tackle, scientists have made some headway on this topic. The transport of larvae in eddies to some extent has been assumed; larvae live in the water column and eddies move through the water column, so some larvae must end up inside these features. What cannot be assumed and is really difficult to actually observe is the level of transport and the specific larvae involved. Scientists use a combination of research methods to answer several questions, including the following: (1) when is an eddy present? (2) what types of larvae are found inside the eddy, and in what abundance? and (3) how does the eddy affect survival and growth of the larvae and other organisms living within it?

To determine the presence of an eddy, scientists use a combination of physical oceanographic sensors to measure current direction and velocity, water temperature, and salinity underneath the sea (**Figure A**) and satellite data to visualize temperature at the sea surface (Figures 2.16 and 2.17). Measuring current directions and velocities and subsurface sea temperature provides information about the water column where the larvae are actually living. Observing temperature at the sea surface is a convenient and clear way to visibly track eddies as they move along a path, even though the surface temperature varies somewhat from the water below.

Once an eddy has been identified, the task then is to collect larvae for species or family identification and abundance estimates. Ideally, larvae collection is already taking place when an eddy passes by to compare larval abundance in the absence of an eddy to abundance during the passage of an eddy. Larval collection is challenging; larvae are tiny, mobile, and fragile. Some methods used to collect larvae include plankton nets towed behind a boat and light traps that

Figure A A mixture of larval invertebrates and vertebrates collected from waters offshore of the upper Florida Keys.
Courtesy of Dr. Evan D'Alessandro, University of Miami, Rosenstiel School of Marine & Atmospheric Science.

are left out at night to attract and capture larvae (**Figure B**). Several studies along the East Coast of the United States have successfully captured and surveyed larvae within eddies and have shown a dramatic increase in abundance and diversity of larvae during the passage of eddies. In a study conducted in the Charleston Gyre region, eddies entrained larvae from at least 91 fish families. In a similar study conducted in the Florida Keys, larvae from at least 42 fish families were collected from the water during an eddy event using light traps, and a large time series of data collections before and after the event was also collected for comparison. The significance of these findings is that these eddies can move fish and invertebrates very far from where they were released as eggs, and the duration of many eddies is as long as the larval phase of many species, 30 days or longer.

Eddies seem to form a miniature ecosystem that is very different from the surrounding water masses in temperature and other physical parameters, and, as a result, can be very high in productivity. The abundance of high-level predators has been demonstrated to be quite large within eddies, because the increased number of larvae inside the water mass provides food for small fish, large fish feed on small fish, and eventually the entire food web is affected. In a study conducted in the Straits of Florida, reef fish larvae transported within eddies had higher growth rates than those residing nearby, but outside of, eddies. Not only do eddies transport larvae, but they appear to have an impact on the entire food web and actually affect the growth rates of larvae. Although the studies highlighted here have provided some very useful information about eddies and their impacts on marine life, there is much to learn in other geographic locations and for a variety of fish and invertebrate species.

Figure B Light trap design to capture larvae in the water column.
Courtesy of Dr. Evan D'Alessandro, University of Miami, Rosenstiel School of Marine & Atmospheric Science.

Critical Thinking Questions

1. How might researchers predict where larvae that originated in a particular location may end up settling, especially if an eddy is present in the spawning location? What type of information is required to make such a prediction?

2. Provide several examples of living conditions inside an eddy that may be favorable for survival of some larvae. Provide several examples that may be detrimental to the survival of some larvae.

For Further Reading

Atwood, E., J. T. Diffy-Anderson, J. K. Horne, and C. Ladd. 2010. Influence of mesoscale eddies on ichthyoplankton assemblages in the Gulf of Alaska. *Fisheries Oceanography* 19(6):493–507.

Godø, O. R., A. Samuelsen, G. J. Macaulay, R. Patel, S. S. Hjøllo, J. Horne, S. Kaartvedt, and J. A. Johannessen. 2012. Mesoscale eddies are oases for higher trophic marine life. *PLoS ONE* 7(1):e30161. doi:10.1371/journal.pone.0030161.

Govoni, J. J., J. A. Hare, & E. D. Davenport. 2013. The distribution of larval fishes of the Charleston Gyre Region off the Southeastern United States in winter shaped by mesoscale, cyclonic eddies. *Marine and Coastal Fisheries: Dynamics, Management, and Ecosystem Science* 5:1, 246–259. doi:10.1080/19425120.2013.820245.

Shulzitski, K., S. Sponaugle, M. Hauff, K. Walter, E. K. D'Alessandro, and R. K. Cowen. 2015. Close encounters with eddies: Oceanographic features increase growth of larval reef fishes during their journey to the reef. *Biological Letters* 11(1):20140746. doi:10.1098/rsbl.2014.0746.

Sponaugle, S., T. Lee, V. Kourafalou, and D. Pinkard. 2005. Florida Current frontal eddies and the settlement of coral reef fishes. *Limnology and Oceanography* 50:1033–1048.

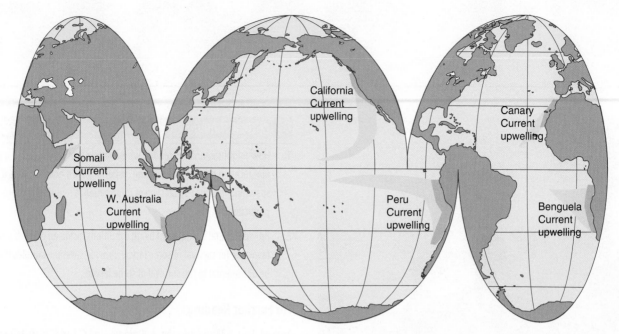

Figure 2.21 Principal regions of coastal upwelling (blue) and down-current areas of increased primary productivity (green).

the same result. Four major coastal upwelling areas occur in the California, Peru, Canary, and Benguela Currents, and lesser ones occur along the coasts of Somalia and western Australia. With the exception of the Somali Current, these currents are on eastern sides of subtropical current gyres (**Figure 2.21**) and flow toward the equator.

Another type of upwelling is more limited in extent and normally exists only in the central Pacific Ocean. In effect, this upwelling is known as *equatorial upwelling* and occurs during ENSO-neutral or La Niña periods. The Pacific Equatorial Current flows westward, straddling the equator. The Coriolis effect causes a slight displacement to the right for the portion of the current in the Northern Hemisphere and to the left for the portion of the current in the Southern Hemisphere. The resultant divergence of water away from the equator creates an upwelling of deeper water to replace the water that has moved away (**Figure 2.22**). The resulting increase in phytoplankton growth feeds a "downstream" population of zooplankton to the north and south of the equator.

In large discontinuous patches around the Antarctic continent, a different type of upwelling occurs. As a consequence of thermohaline circulation patterns outlined in **Figure 2.23** and described in detail below, massive volumes of nutrient-rich North Atlantic deep water slowly, yet continuously, drift toward the ocean surface in the Antarctic Divergence Zone between 60° S and 70° S latitude (**Figure 2.24**). For most of the year, phytoplankton production is inhibited by the absence of light. In summer, however, this liquid conveyor belt flowing under much of the Atlantic Ocean delivers massive amounts of dissolved nutrients into the **photic zone** to support one of the richest communities in the marine environment. The continuous availability of deep-water nutrients that can be used by photosynthetic organisms accounts for the high productivity so

characteristic of these upwelling regions. Several of the world's most important fisheries are based in upwelling areas.

Vertical circulation that results from varying surface water density is known as **thermohaline circulation**. The physical processes that increase seawater density and cause water to sink are strictly surface features that affect water temperature or salinity. Seawater sinking from the surface is usually highly oxygenated because it has been in contact with the atmosphere, and thus it transports dissolved oxygen to deep areas of the ocean basins that would otherwise be **anoxic** (lacking oxygen). The chief areas of sinking are located in the colder latitudes, where sea surface temperatures are low and densities are high. After sinking, this dense water continues to spread and flow horizontally as very slow and ill-defined deep ocean currents. Remarkably, time spans of a few hundred to a thousand years are required for water that sinks in the North Atlantic to reach the surface again in the Southern Hemisphere. Figure 2.23 outlines the general patterns of large-scale deep-ocean thermohaline circulation.

On a somewhat smaller scale, the Mediterranean and Black Seas provide two contrasting examples of density-driven thermohaline circulation. In the arid climate of the Mediterranean Sea, particularly at its eastern end, evaporation from the sea surface greatly exceeds precipitation and runoff, so this area of the sea is constantly losing water. The resulting high-salinity and high-density water sinks and fills the deeper parts of the Mediterranean basin. The sinking of surface water provides substantial mixing and O_2 replenishment for the deep water of the Mediterranean and is similar to the deep circulation of the open ocean. Part of this deep dense water eventually flows out of the Mediterranean over a shallow sill at Gibraltar and down into the Atlantic Ocean. To compensate for the outflow and losses due to evaporation, nearly 2 million m³ of Atlantic surface water flows into the Mediterranean each second. The currents

Figure 2.22   Upwelling along the equator. The light blue band near the equator represents areas of high nutrients and high phytoplankton biomass due to upwelling.
Courtesy of NOAA.

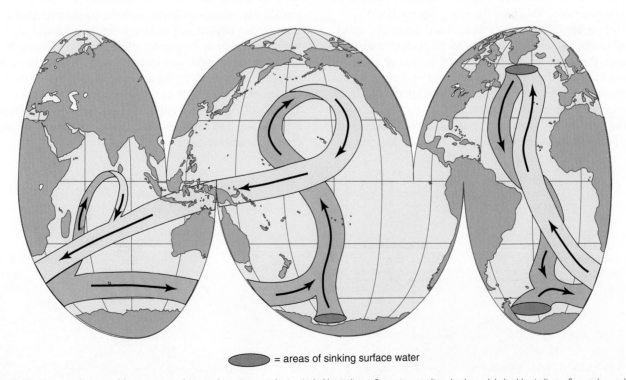

= areas of sinking surface water

Figure 2.23   The general pattern of deep-ocean circulation in the major ocean basins. Light blue indicates flow at intermediate depths, and darker blue indicates flow at deeper depths. Sinking of surface water occurs at high latitudes indicated by ovals.

Data from Broecker, W. S., et al., *Nature* 315 (1985):21–26.

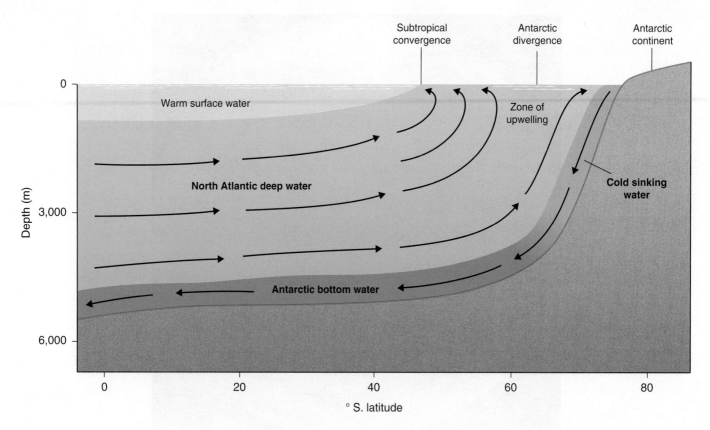

Figure 2.24 Interface of the South Atlantic Ocean and the Antarctic continent indicating the flow of water masses driving upwelling in the area.

at Gibraltar can be compared with two large rivers flowing in opposite directions, one above the other (**Figure 2.25**).

Like the Mediterranean, the Black Sea is isolated by a shallow sill (at the Bosporus). In contrast to the Mediterranean Sea, however, the Black Sea is characterized by a large excess of precipitation and river runoff, so this area of the sea is constantly gaining water. In this sense, the circulation of the Black Sea resembles that of some semienclosed fjords of Scandinavia and the west coast of Canada. The dilute surface waters of the Black Sea form a shallow low-density layer that does not mix with the higher salinity denser water below. Instead, it flows into the Mediterranean Sea through the Bosporus (Figure 2.24). Low-salinity oxygen-rich surface water does not sink, and thus the more common oxygen-dependent forms of marine life are restricted to the uppermost layer. Below 150 m, the Black Sea is stagnant and anoxic. Yet these anoxic deep waters of the Black Sea (more than 80% of its volume) are by no means lifeless. The rain of organic material from above (known as **marine snow**) accumulates and provides abundant nourishment for several types of anaerobic bacteria.

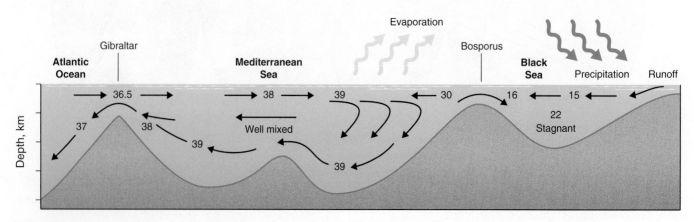

Figure 2.25 A comparison of the deep-ocean circulation patterns of two marginal seas, the Mediterranean Sea and the Black Sea. Numbers represent salinity in ‰.

Case Study

The Great Pacific Garbage Patch

Currents and gyres not only transport living organisms but also anything that is present in the water and light enough to be moved along within a water mass. The large gyres pictured in Figure 2.15 have been a topic of interest for researchers for decades, and just recently the North Pacific Gyre has been in the spotlight. Located within the North Pacific Gyre are several areas known as "garbage patches" (**Figure A**). Human garbage dumped into the sea from land or boats accumulates in regions of the ocean to form these patches. Most of the garbage is made of plastic.

The locations of garbage patches are somewhat predictable, because on a large scale the locations of major currents and gyres are predictable. The actual composition of the garbage patches changes daily, though, because currents change with changing weather patterns, and the water is constantly mixing up the contents of the patches. The North Pacific Gyre has gained attention in recent years, but garbage patches are found in many other areas of Earth's oceans. Floating debris accumulates within any moving water mass, and gyres are particularly good current patterns for this accumulation of junk.

An erroneous assumption that is often made concerning the garbage patches is that large pieces of trash are found floating around; however, most of it is actually small. Scientists studying these areas describe the trash as a peppery soup with larger garbage floating through it. The larger garbage may be derelict fishing gear, abandoned boating equipment, and a variety of plastic items. Birds and fish often eat the plastic items, mistaking them for food (**Figure B**). Over time, garbage made of plastic breaks down into smaller pieces, some microscopic, and floats around with the currents. Some of the worst areas of accumulated garbage are not even visible to the common person sailing right through the area, unless he or she is looking closely for plastic peppery soup.

The question now is the following: how do we clean up the oceans and get rid of these garbage patches? Cleaning up the ocean is a difficult task. Because the plastics are mostly very small, there is no feasible method to remove them. The best thing we can do now that we are aware of these garbage patches is to prevent the future addition of more debris. Until we stop adding more garbage to the ocean, all of the cleanup efforts possible will not make any difference to remedy the problem. We can do many things

Figure B A deceased albatross full of plastic trash.
NOAA Marine Debris Program.

in our everyday lives to ensure that our trash does not make it into the ocean. Reuse or recycle whenever possible. Never, ever, pollute. It sounds simple, but many well-intentioned beachgoers accidentally leave beach toys, water bottles, suntan lotion bottles, and a variety of other trash behind. Last, avoid buying plastics when there is an alternative product available. This includes limiting (or omitting altogether) purchases of plastic disposable items. Plastic never goes away but eventually breaks down into tiny potentially harmful substances that fish and other marine organisms ingest. It really is up to us to stop adding to the garbage patches that exist in our world ocean.

Critical Thinking Questions

1. Make a list of all the products you use in one day, including items used for food and drink storage (e.g., water bottle) and those used for entertainment (e.g., cell phone). Circle the products you use that are made of any kind of plastic, and add a star next to those that are single-use plastic items. Propose alternative products that can be used in place of all of the plastic in your life.

2. Propose a possible solution for reducing plastic in the sea, and describe how your idea can be implemented.

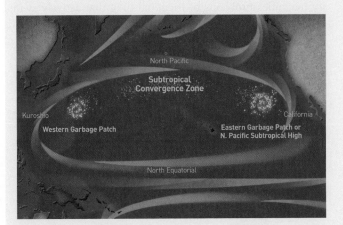

Figure A The locations of large accumulations of trash in the North Pacific Gyre.
NOAA Marine Debris Program.

2.2 Properties of Seawater

Several common properties of seawater are crucial to the survival and well-being of the ocean's inhabitants, and many water properties affect the oceanographic processes described in the previous section. Water accounts for 80% to 90% of the volume of most marine organisms. It provides buoyancy and body support for swimming and floating organisms and reduces the need for heavy skeletal structures. Minerals for building structures such as hard shells are generally abundantly available in seawater. Water is also the medium for most chemical reactions needed to sustain life. The life processes of marine organisms, in turn, alter many fundamental physical and chemical properties of seawater, including its transparency and chemical makeup, making organisms an integral part of the total marine environment. Understanding the interactions between organisms and their marine environment requires a brief examination of some of the more important physical and chemical attributes of seawater. The characteristics of pure water and seawater differ in some respects, and thus we consider first the basic properties of pure water and then examine how those properties differ in seawater.

Pure Water

Water is a common yet very remarkable substance. It is the only substance on Earth that is abundant as a liquid (mostly in oceans), with substantial quantities left over as a gas in the atmosphere and as a solid in the form of ice and snow. Individual water molecules have a simple structure, represented by the molecular formula H_2O. Yet the collective properties of many water molecules interacting with each other in a liquid are quite complex. Each water molecule has one atom of oxygen (O) and two atoms of hydrogen (H), which together form water (H_2O). Some characteristics of these and other biologically important molecules and ions are described in this chapter.

The many unusual properties of water stem from its molecular shape: a four-cornered tetrahedron with the two hydrogen atoms forming angles of about 105 degrees with the oxygen atom. This molecular shape is simplified to two dimensions in **Figure 2.26**. This atomic configuration creates an asymmetric water molecule, with the oxygen atom dominating one end of the molecule and the hydrogen atoms dominating the other end. The **covalent bond** between each hydrogen and the oxygen atom is formed by the sharing of two negatively charged electrons. The oxygen atom attracts the electron pair of each bond, causing the oxygen end of each water molecule to assume a slight negative charge. The hydrogen end of the molecule, by giving up part of its electron complement, is left with a small positive charge. The resulting electrical polarization of water molecules, one end with a positive charge and the other end with a negative charge, has profound consequences for liquid water. When water is in its liquid form, each end of one water molecule attracts the oppositely charged end of other water molecules. This attractive force creates a weak bond, a **hydrogen bond** or **H-bond**, between adjacent water molecules (**Figure 2.27**). These bonds are much weaker and less stable than the covalent bonds within a single water molecule and are continually breaking and reforming as they change partners with other water molecules millions of times per second. Take care not to infer that H-bonds are less important than other chemical bond types because they are termed *weak*. It is the H-bonds that lead to the unique and very useful features of water, discussed below.

The H-bonds between water molecules require that water must be warmed to a much higher temperature to boil than that needed for other substances with similar molecular sizes, such as O_2 or CO_2. Water's high freezing point of 0°C and boiling point at 100°C causes most water at Earth's surface to exist as a liquid, making life as we know it possible. Hydrogen bonding also accounts for several other unique and important properties of water. Some of these properties are listed in **Table 2.1** and are discussed in the following paragraphs.

Viscosity and Surface Tension

Hydrogen bonding between adjacent water molecules within a mass of liquid water creates a slight "stickiness" between these molecules. This property, known as **viscosity**, has a marked effect on all marine organisms. The viscosity of water reduces the sinking tendency of some organisms by increasing the frictional resistance between themselves and nearby water molecules. As a result, less energy is required by these organisms to maintain their positions in the water column. At the same time, viscosity magnifies problems of frictional drag that actively swimming animals must overcome. More energy is required for active movement through water because of its high viscosity.

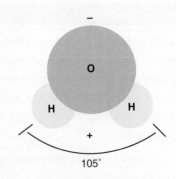

Figure 2.26 The arrangement of H and O atoms in a molecule of water (H_2O). The oxygen end of the molecule has a slight negative charge because O pulls electrons toward it more than H. For the same reason, the H end has a slight positive charge.

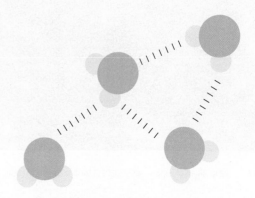

Figure 2.27 Hydrogen bonding between adjacent molecules of liquid water. The black dashed lines represent hydrogen bonds.

TABLE 2.1 Some Biologically Important Physical Properties of Water

Property	Comparison with Other Substances	Importance in Biological Processes
Boiling point	High (100°C) for molecular size	Causes most water to exist as a liquid at Earth surface temperatures
Freezing point	High (0°C) for molecular size	Causes most water to exist as a liquid at Earth surface temperatures
Surface tension	Highest of all liquids	Crucial to position maintenance of sea-surface organisms
Density of solid	Unique among common natural substances	Causes ice to float and inhibits complete freezing of large bodies of water
Latent heat of vaporization	Highest of all common natural substances (540 cal/g)	Moderates sea-surface temperatures by transferring large quantities of heat to the atmosphere through evaporation Inhibits large-scale freezing of the oceans
Latent heat of fusion	Highest of all common natural substances (80 cal/g)	Moderates daily and seasonal temperature changes
Solvent power	Dissolves more substances in greater amounts than any other liquid	Maintains a large variety of substances in solution, enhancing a variety of chemical reactions
Heat capacity	High (1 cal/g/°C) for molecular size	Stabilizes body temperatures of organisms

At the surface of a water mass (such as the air–sea boundary), the mutual attraction of water molecules creates a flexible molecular "skin" over the water surface. This, the **surface tension** of water, is sufficiently strong to support the full weight of a water strider (**Figure 2.28**). Both surface tension and viscosity are temperature dependent, increasing as the temperature decreases.

Density–Temperature Relationships

Most liquids contract and become denser as they cool. The solid form of these substances is denser than the liquid form.

Over most of the temperature range at which pure water is liquid, it behaves like other liquids. At 4°C or above, the density increases with decreasing temperature. Below 4°C, however, water behaves differently: the density–temperature pattern of pure water reverses so that the density begins to decrease as temperatures fall below 4°C. One model used to explain this unique behavior of water proposes that at near-freezing temperatures less dense icelike clusters consisting of several water molecules form and disintegrate very rapidly within the body of liquid water. As liquid water continues to cool, more clusters form, and the clusters remain intact longer. Eventually, at 0°C,

Figure 2.28 A water strider (*Halobates*) is completely supported by the surface of the water due to water's high surface tension. *Halobates* is one of the few completely marine insects.
© Vasiliy Koval/Shutterstock, Inc.

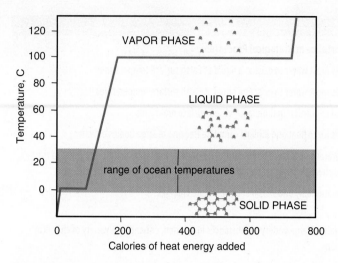

Figure 2.29 Graph depicting the amount of heat energy required to cause temperature and phase changes in water.

all the water molecules become locked into a rigid, solid crystal lattice of ice (**Figure 2.29**). The ice formed is about 8% less dense than liquid water at the same temperature, and thus ice always floats on liquid water. This is an unusual, but very fortunate, property of water. Without this unique density–temperature relationship, ice would sink as it formed, and lakes, oceans, and other bodies of water would freeze solid from the bottom up. Winter survival for organisms living in such an environment below the sea surface would be much more difficult, and life would not be possible for organisms adapted to live on top of the sea ice.[1]

Heat Capacity

Heat is a form of energy, the energy of molecular motion, that is also known as *kinetic energy*. The sun is the source of almost all energy entering Earth's surface heat budget. At the surface of the sea, some of the sun's radiant energy is converted to heat energy that is then transferred from place to place primarily by **convection** (mixing) and secondarily by **conduction** (the exchange of heat energy between adjacent molecules). Heat energy is measured in **calories**.[2]

Water has the ability to absorb or give up heat without experiencing much of a temperature change. To illustrate the high **heat capacity** of water, imagine a 1-g block of ice at −20°C on a heater that provides heat at a constant rate. Heating the ice from −20°C to 0°C requires 10 calories, or 0.5 calories per degree of temperature increase (the heat capacity of ice); however, converting 1 g of ice at 0°C to liquid at 0°C requires 80 calories. Conversely, 80 calories of heat must be removed from 1 g of liquid water at 0°C to freeze it to ice at the same temperature. This is referred to as water's **latent heat of fusion**. Continued

heating of the 1-g water sample from 0°C requires 1 calorie of heat energy for each 1°C change in temperature (the heat capacity of liquid water) until the boiling point (100°C) is reached. At this point, further temperature increase stops until all of the water is converted to water vapor. For this conversion, 540 calories of heat energy are necessary (water's **latent heat of vaporization**). Figure 2.29 summarizes the energy requirements for these changes in water temperature. The high heat capacity and the large amount of heat required for evaporation enable large bodies of water to resist extreme temperature fluctuations. Heat energy is absorbed slowly by water when the air above is warmer and is gradually given up when the air is colder, providing a crucial global-scale temperature-moderating mechanism for marine environments and adjacent land areas.

Solvent Action

The small size and polar charges of each water molecule enable it to interact with and dissolve most naturally occurring substances, especially salts, which are composed of atoms or simple molecules, called **ions**, that carry an electrical charge. Salts held together by **ionic bonds** (bonds between oppositely charged adjacent ions) are particularly susceptible to the solvent action of water. **Figure 2.30** illustrates the process of a salt crystal dissolving in water. Initially, several water molecules form weak H-bonds with each sodium (Na^+) and chloride (Cl^-) ion, and they eventually overcome the mutual attraction of those ions that previously bound them together in the crystalline structure. As more Na^+ and Cl^- ions are removed in this way, the solid crystal structure disintegrates, and the salt dissolves.

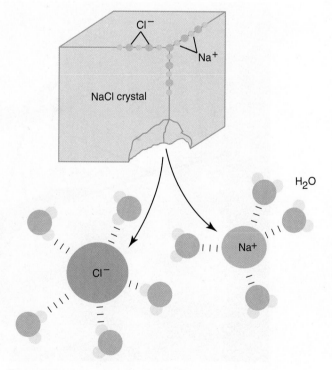

Figure 2.30 A salt crystal and the action of charged water molecules in dissolving the crystal to dissociated sodium (Na^+) and (Cl^-) ions.

---

[1] The maximum density of pure water is used to define the fundamental metric measure of mass, the gram, which is defined as the mass of pure water at 4°C contained in the volume of 1 cubic centimeter. Thus, the density, the ratio of mass to volume, of pure water is 1.000 g/cm³.

[2] A calorie is a unit of heat energy, defined as the quantity of heat needed to elevate the temperature of 1 g of pure water 1°C.

Water is not a good universal solvent, however, as it is not an effective solvent for some large organic molecules, such as waxes and oils, or for small molecules that lack electrical charges. A notable and biologically crucial example of the second category is O_2; water can only dissolve a few parts per million of O_2.

Seawater

Seawater is the accumulated product of several billion years of water condensing as rain from the atmosphere, eroding rocks and soil, and washing it all to the sea. About 3.5% of seawater is composed of dissolved compounds from these sources. The other 96.5% is pure water. Traces of all naturally occurring substances probably exist in the ocean and can be separated into three general categories: (1) inorganic substances, usually referred to as *salts*, including nutrients necessary for plant growth; (2) dissolved gases such as N_2, O_2, and CO_2; and (3) organic compounds derived from living organisms. Organic compounds dissolved in seawater include fats, oils, carbohydrates, vitamins, amino acids, proteins, and other substances. Some compounds are valuable sources of nutrition for marine bacteria and some other organisms. Current research indicates that other organic compounds, especially synthetically created ones such as polychlorinated biphenyls (PCBs) and other chlorinated hydrocarbons, have accumulated in marine food chains and have had serious negative impacts on the development and reproduction of some forms of marine life.

Dissolved Salts

Salts account for most dissolved substances in seawater. The total amount of dissolved salts in seawater is referred to as its *salinity* and is measured in parts per thousand (‰) rather than in parts per hundred (%). Salinity values range from nearly zero at river mouths to greater than 40‰ in arid areas, such as the Red Sea. Yet, in open-ocean areas away from coastal influences, salinity averages approximately 35‰ and varies only slightly over large distances (**Figure 2.31**).

Salinity is altered by processes that add or remove salts or water from the sea. The primary mechanisms of salt and water addition or removal are evaporation, precipitation, river runoff, and the freezing and thawing of sea ice. When evaporation exceeds precipitation, it removes water from the sea surface, thereby concentrating the remaining salts and increasing the salinity. Excess precipitation decreases salinity by diluting the sea salts. Freshwater runoff from rivers has the same effect. **Figure 2.32** illustrates the average annual north–south variation of sea-surface evaporation and precipitation. Areas with more evaporation than precipitation (Figure 2.32, brown-shaded areas) generally correspond to the high surface salinity regions shown in Figure 2.31 and to the latitudes with most of the great land deserts of the world. In polar regions, where seawater can freeze, only the water molecules are incorporated into the developing freshwater ice crystal. The dissolved ions are excluded from the growing ice crystal, causing the salinity of the remaining liquid seawater to increase. The process is reversed when ice melts. Freezing and thawing of seawater are usually seasonal phenomena, resulting in small long-term salinity differences.

When salts dissolve in water, they release both positively and negatively charged ions. The more common ions found in seawater are listed in **Table 2.2** and are grouped as major or minor constituents according to their abundance. The major ions account for greater than 99% of the total salt concentration in seawater. In relation to each other, concentrations of these major ions remain remarkably constant even though their total abundance may differ from place to place. Most of the more abundant ions enumerated in Table 2.2 are either important

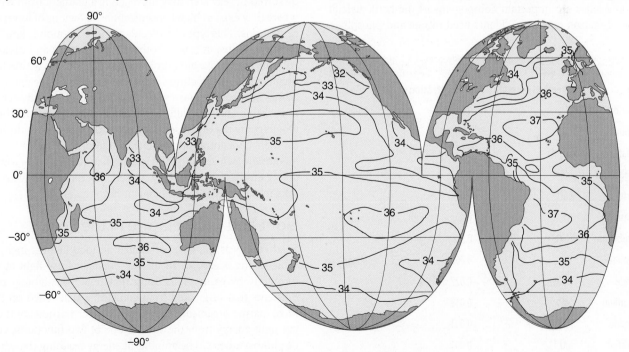

Figure 2.31 Geographic variations of sea surface salinities, expressed in parts per thousand (‰).

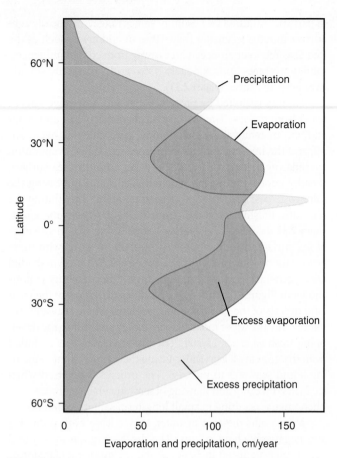

Figure 2.32 Average global north–south variation of sea surface evaporation and precipitation.
Modified from G. Dietrich, *General Oceanography* (Interscience Publishers, 1963).

components of the bodies of marine organisms or are used for crucial physiological processes. Magnesium, calcium, bicarbonate, and silica are important components of the hard skeletal parts of marine organisms. Plants need nitrate and phosphate

TABLE 2.2 Major and Minor Ions in Seawater of 35‰ Salinity

Ion	Chemical Formula	Concentration (‰)	
Chloride	Cl^-	19.3	Major
Sodium	Na^+	10.6	
Sulfate	SO_4^{2-}	2.7	
Magnesium	Mg^{2+}	1.3	
Calcium	Ca^{2+}	0.4	
Potassium	K^+	0.4	
Bicarbonate	HCO_3^-	0.1	
Bromide	Br^-	0.066	Minor
Borate	$B(OH)_4^-$	0.027	
Strontium	Sr^{2+}	0.013	
Fluoride	F^-	0.001	
Silicate	SiO_4^{4-}	0.001	

Plus traces of other naturally occurring elements.

for the synthesis of organic material. A recently discovered phenomenon called **ocean acidification** is altering the amounts of some ions in seawater very slightly, and organisms relying on these ions are suffering consequences.

Salt and Water Balance

Each living organism, from the smallest marine bacteria to an enormous blue whale, has been masterfully crafted through thousands or millions of years of evolution with very specific needs for life. The well-being and survival of all living organisms require that they maintain relatively constant internal environmental conditions, within physiological limits specific to each species. **Homeostasis** is the term used to describe the tendency of living organisms to control or regulate fluctuations of their internal environment so that they can maintain an internal state that is relatively steady, even when their external environment changes. Homeostasis is the result of coordinated biological processes that regulate conditions such as body temperature or blood ion concentrations. When working properly, these processes result in a dynamic regulation of conditions that vary within definite and tolerable limits. This section describes those processes that affect the homeostasis of salt and water exchange between the body fluids of an organism and its seawater environment.

The body fluids of marine organisms are separated from seawater by boundary membranes that participate in several vital exchange processes, including absorption of oxygen and nutrients and excretion of waste materials. Small molecules, such as water, easily pass through some of these membranes, but the passage of larger molecules and the ions abundant in seawater is blocked. Such membranes are called **selectively permeable membranes**; they allow only small molecules and ions to pass through while blocking the passage of larger molecules and ions. When substances are free to move, as they are when dissolved in seawater, they move along a gradient from regions where they exist in high concentrations to regions of lower concentrations. This type of molecular or ionic motion is known as **diffusion**. Diffusion causes both water molecules and dissolved substances to move along concentration gradients within living organisms and sometimes across selectively permeable membranes between organisms and their surrounding seawater. The remarkable aspect of diffusion is that it does not require the use of energy by the organism; molecules are moved around based on their concentrations on either side of the semipermeable membrane (whether an individual cell or the outside surface of an organism), and no energy is used. The balance of water and salts in the body is termed *osmotic balance*.

Light and Temperature in the Sea

Solar energy is the basis of almost all food chains, including those in the ocean, so all marine organisms use light at least indirectly by eating something that relied on light or using nutrients that originated from something that relied on light. Most marine organisms living in the upper portions of the sea use light energy from the sun for one of two functions, vision or photosynthesis. The amount of energy reaching the sea surface through the atmosphere depends on the presence of dust,

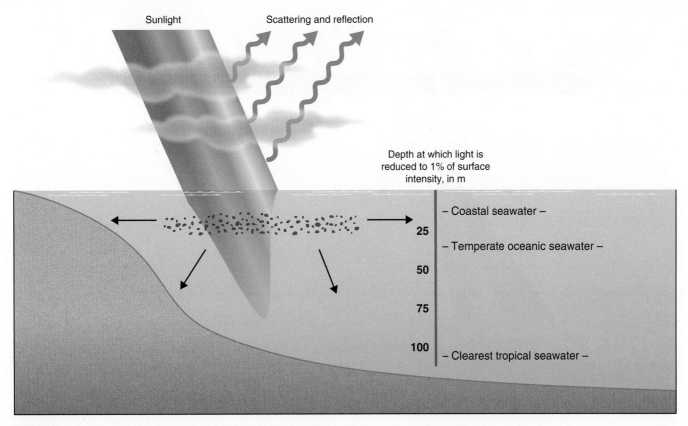

Sunlight

Scattering and reflection

Depth at which light is
reduced to 1% of surface
intensity, in m

– Coastal seawater –

25

– Temperate oceanic seawater –

50

75

100

– Clearest tropical seawater –

Figure 2.33 Sunlight striking the sea surface. The violet and red ends of the spectrum are absorbed first; the blue range penetrates the deepest.

clouds, and gases that absorb or scatter a portion of the incoming solar radiation (**Figure 2.33**). On an average day, about 65% of the sun's radiation arriving at the outer edge of our atmosphere reaches Earth's surface. The intensity of incoming solar radiation is reduced when the angle of the sun is low, as it is in winter or at high latitudes, and a portion of the light that does make it through the atmosphere is reflected back into space by the sea surface.

Of the broad spectrum of the sun's electromagnetic radiation (**Figure 2.34**, top), most marine animals can visually detect only a very narrow band near the center of the spectrum. Our eyes visually respond only to the portion labeled **visible light** in Figure 2.34 (violet through red), and most other animals with eyes, whether they see in color or not, respond visually to approximately the same portion of the electromagnetic radiation spectrum.

The band of light energy used by animals for vision broadly overlaps that used in photosynthesis. Photosynthetic organisms must remain in the upper region of the ocean (the photic zone) where solar energy is sufficient to support rates of photosynthesis that at least match their own respiratory needs. The depth of the photic zone is determined by how rapidly seawater absorbs light and converts it to heat energy. Dissolved substances, suspended sediments, and even plankton populations diminish the amount of light available for photosynthetic activity and cause the depth of light penetration to differ dramatically between coastal and oceanic water (Figure 2.33).

As sunlight travels through Earth's atmosphere and into the sea, its color characteristics are altered as seawater rapidly absorbs or scatters the violet and the orange-red portions of the visible spectrum, leaving the green and blue wavelengths to penetrate deeper. Even in the clearest tropical waters, almost all red light is absorbed in the upper 11 m. Clear seawater is most transparent to the blue and green portions of the spectrum (450 to 550 nm); 10% of the blue light penetrates to depths of 100 m or more. However, even this light is eventually absorbed or scattered (**Figure 2.35**). The deeper penetration and eventual backscattering of blue light account for the characteristic blue color of clear tropical seawater. Coastal waters are commonly more turbid, with a greater load of suspended sediments and dissolved substances derived from land runoff. Here, there is a shift

DID YOU KNOW?

The colors humans and marine critters see are actually the portion of the visible light spectrum that is reflected and how our eyes respond to the wavelengths of light. Because seawater appears blue to our eyes, we know that every color of light but blue is being absorbed by the water and blue is reflected by the water. Marine organisms have additional constraints on vision because light is absorbed or scattered much more quickly in water than in the air. As water depth increases fewer colors are visible, leading to an environment lacking some of the colors land animals are accustomed to.

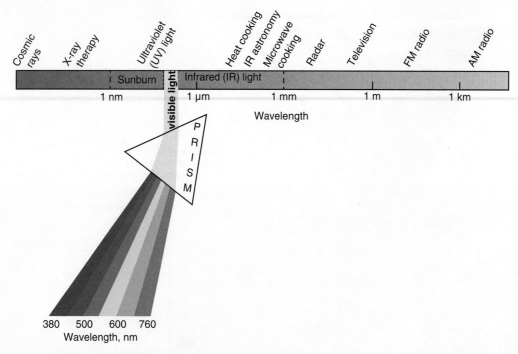

Figure 2.34 The electromagnetic radiation spectrum, highlighting the small portion known as the visible light spectrum. When passed through a prism, visible light is separated into its component colors.

in the relative penetration of light energy, with green light penetrating deepest. In many coastal regions, green light is reduced to 1% of its surface intensity in less than 30 m. Photosynthetic organisms feature adaptations to these different light regimes.

Temperature Effects

Water temperature is one of the greatest determining factors of the distributions of organisms in the ocean. When sunlight is absorbed by water molecules, it is converted to heat energy, and the motion of the water molecules increases. Temperature, commonly reported in almost all countries of the world as degrees

Celsius (°C), is the way we measure and describe that change in molecular motion. **Figure 2.36** includes the range of temperatures for living organisms in °C and degrees Fahrenheit (°F). Temperature is a universal factor governing the existence and behavior of living organisms. Life processes cease to function above the boiling point of water, when protein structures are irreversibly altered (as when you cook an egg), or at subfreezing temperatures, when the formation of ice crystals damages cellular structures, but between these absolute temperature limits, life flourishes.

The high heat capacity of water limits marine temperatures to a much narrower range than air temperatures over land (Figure 2.36). Some marine organisms survive in coastal tropical lagoons at temperatures as high as 40°C. Bacteria associated with deep-sea hydrothermal vents sometimes experience water temperatures above 60°C, and as a result are referred to as *extremophiles*. Other deep-sea animals spend their lives in water perpetually within 1 or 2 degrees of 0°C. Penguins and a few other birds and mammals are adapted to extreme cold and commonly tolerate air temperatures far below 0°C in polar regions. Emperor penguins even manage to incubate and hatch eggs under these conditions (**Figure 2.37**), but these are exceptions; most marine species typically experience water temperatures between 5°C and 30°C.

Individual activity, cell growth, oxygen consumption, and other physiological functions, collectively termed **metabolism**, proceed at temperature-regulated rates. Most animals lack mechanisms for body temperature regulation. These are **poikilotherms** (often inappropriately described as cold-blooded). These organisms are also referred to as **ectotherms**; their body temperatures vary with and are largely controlled by outside environmental

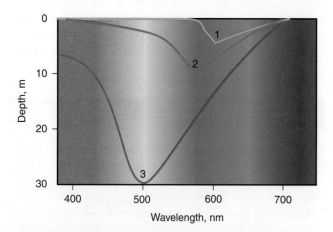

Figure 2.35 The depth of light penetration in three different water types: (1) very turbid coastal water, (2) moderately turbid coastal water, and (3) very clear tropical water. Note the shift to bluer light in clearer water.

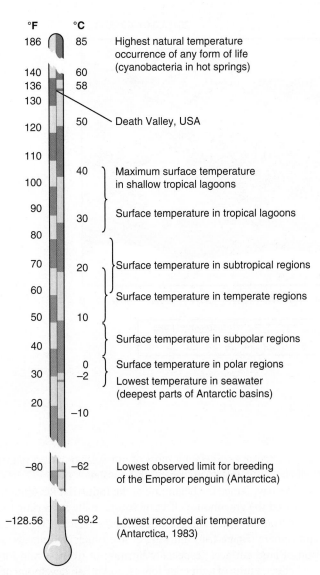

°F	°C	
186	85	Highest natural temperature occurrence of any form of life (cyanobacteria in hot springs)
140	60	
136	58	
130		
120	50	Death Valley, USA
110		
100	40	Maximum surface temperature in shallow tropical lagoons
90	30	Surface temperature in tropical lagoons
80		
70	20	Surface temperature in subtropical regions
60		Surface temperature in temperate regions
50	10	
40		Surface temperature in subpolar regions
30	0	Surface temperature in polar regions
	−2	Lowest temperature in seawater (deepest parts of Antarctic basins)
20	−10	
−80	−62	Lowest observed limit for breeding of the Emperor penguin (Antarctica)
−128.56	−89.2	Lowest recorded air temperature (Antarctica, 1983)

Figure 2.36 The temperature ranges of a variety of living organisms and for the major marine climatic regions.

temperatures. The terms *poikilotherm* and *ectotherm*, often used interchangeably, refer to distinct aspects of body temperature control. Poikilotherms experience varying body temperatures in a 24-hour day and do not regulate their body temperatures physiologically; external conditions govern the body temperatures of ectotherms. In the sea, the temperature-moderating properties of water restrict fluctuations of temperatures experienced by marine ectotherms. Most marine organisms are simultaneously ectothermic and poikilothermic.

For marine ectotherms, water temperature is a principal factor controlling metabolic rates. Marine ectotherms generally have fairly narrow optimal temperature ranges, bracketed on either side by wider and less optimal, but still tolerable, ranges. Within these tolerable temperature limits, the metabolic rate of many poikilotherms is roughly doubled by a 10°C temperature increase. This, however, is only a general rule of thumb; some processes may accelerate six-fold with a 10°C temperature increase, whereas other processes may change very little. The actual effect of water temperature on the feeding rate of a typical marine ectotherm, a submerged barnacle, is shown in **Figure 2.38**.

Only birds and mammals use physiological mechanisms to maintain nearly constant body temperatures throughout a 24-hour day. They are known as **homeotherms**. Their normal core body temperatures are maintained between 37°C and 40°C by the production of heat by internal tissues and organs. Thus, they are also considered **endotherms**. Endothermic homeotherms are less restricted by environmental temperatures than are their poikilothermic neighbors. As a result, they range widely to exploit resources over all thermal regimes available in the sea. Although they are less limited geographically than ectotherms, one trade-off is that endotherms use large amounts of energy to maintain their constant internal body temperature. Thus, endotherms have high energetic needs (they need to eat often).

A few large tunas, billfish, and sharks exhibit a thermoregulatory condition intermediate to the two just discussed. These fishes are poikilothermic, and thus their body temperatures

Figure 2.37 Emperor penguins lay eggs and rear their young on ice during the harsh Antarctic winter.

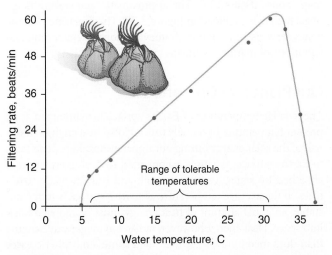

Figure 2.38 Graph depicting the filtering rate of a marine barnacle as a function of water temperature. The filtering rate drops dramatically as the barnacle is placed outside the range of tolerable temperatures.

Modified from Southward, A. J., *Helgol. wiss. Meersuntersuch* 10 (1964):391–401.

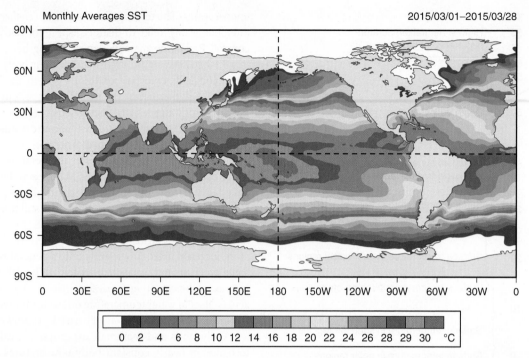

Figure 2.39 Monthly average sea surface temperatures worldwide.
Courtesy of NOAA/OAR/ESRL PSD, Boulder, Colorado, USA.

fluctuate with that of the surrounding seawater. Even so, they are unlike most other poikilotherms because they retain some of the heat produced by their swimming muscles. These animals are endothermic, yet they lack the constant body temperatures characteristic of birds and mammals.

The distribution of various forms of marine life is closely associated with geographic differences in seawater temperatures. In general, surface ocean temperatures are highest near the equator and decrease toward both poles. This temperature gradient establishes several north–south-trending marine climatic zones (**Figure 2.39**). The approximate temperature range of each zone is included in Figure 2.36. These marine climatic zones serve as an important framework for understanding distributions of marine organisms and their habitats.

Our Planetary Greenhouse

The average temperature of Earth's surface is maintained at its present temperature by a finely tuned global heat engine. About half of the solar energy hitting our upper atmosphere penetrates to Earth's surface, where it is converted to heat energy as it is absorbed by water, vegetation, soil, and human-made structures. If the average temperature of Earth's surface is to remain stable, an equal amount of heat energy must be radiated back into space. Heat energy, however, radiates at longer wavelengths than does incoming visible light, and some atmospheric gases are more transparent to visible light than they are to radiated heat. These atmospheric greenhouse gases (especially water vapor, carbon dioxide, methane, and ozone) serve as a natural part of the global heat budget system by trapping heat near

Earth's surface and keeping most of our solar-powered planet well above the freezing temperature of water.

We have, since the beginning of the Industrial Revolution, enhanced the greenhouse effect by substantially increasing the concentrations of natural greenhouse gases such as CO_2 in our atmosphere (**Figure 2.40**). Burning of fossil fuels and devegetation of land surfaces (especially burning of tropical rain forests, clear-cutting of temperate forests, and urban development) appear to be the main sources of the excess CO_2; combustion of fossil fuels adds CO_2 to our atmosphere, and devegetation removes plants that would have removed CO_2 from our atmosphere if left undisturbed.

Salinity–Temperature–Density Relationships

Seawater density is a function of both temperature and salinity, increasing with either a temperature decrease or a salinity increase. Under typical oceanic conditions, temperature fluctuations exert a greater influence on seawater density because the range of marine temperatures is much greater (–2°C to 30°C) than the range of open-ocean salinities. **Figure 2.41** graphically demonstrates the relationships between the temperature, salinity, and density of water.

Seawater sinks when its density increases. Thus, the densest seawater is naturally found near the sea bottom; however, the physical processes that create this dense water (evaporation, freezing, cooling) occur only at the ocean surface. Consequently, dense water on the sea bottom originally must have sunk from the ocean surface. This sinking process is the only

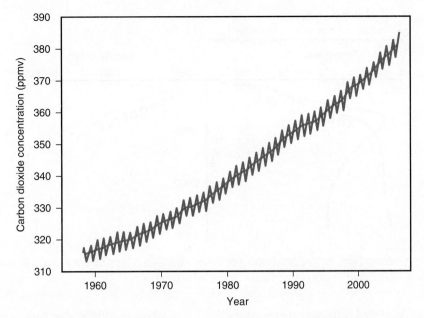

Figure 2.40 Pattern of atmospheric CO$_2$ levels over five decades. The slight annual fluctuations are due to seasonal uptake of CO$_2$ by plants.
Modified from Robert A. Rohde Atmospheric carbon dioxide, Global Warming Art, October 1, 2008, http://www.globalwarmingart.com/wiki/Image:Mauna_Loa_Carbon_Dioxide_png.

mechanism available to drive circulation of water in the deep portions of ocean basins. An obvious feature in most oceans is a thermocline, a subsurface zone of rapid temperature decrease with depth (about 1°C/m). The temperature drop that exists across the thermocline creates a zone of comparable density increase known as a *pycnocline* (**Figure 2.42**). The large density differences on either side of the thermocline effectively separate the oceans into a two-layered system: a thin well-mixed surface layer above the thermocline overlying a heavier, cold, thick, stable zone below. The thermocline and resulting pycnocline inhibit mixing and the exchange of gases, nutrients, and sometimes even organisms between the two layers. In temperate and polar regions, the thermocline is a seasonal feature. During the winter, the surface water is cooled to the same low temperature

as the deeper water. This cooling causes the thermocline to disappear and results in wintertime mixing between the two layers. Warmer marine climates of the tropics and subtropics are more

DID YOU KNOW?

The thermocline is detectable by many organisms, including humans. Scuba divers exploring the seafloor can actually feel a difference in water temperature as they pass through the thermocline and can see the dramatic difference in density (the water looks cloudy) as they pass through the pycnocline. In fact, in the cold waters off the coast of places like Maine, some divers choose to dive above the thermocline because above is cool enough, and below is just too chilly!

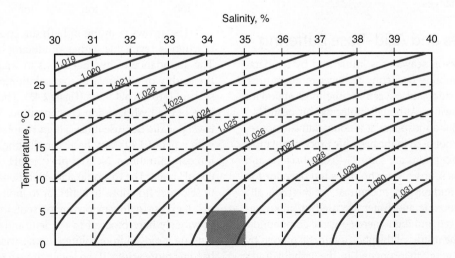

Figure 2.41 Temperature–salinity–density diagram for seawater. Blue curved lines represent density values (g/cm³) resulting from the combined effects of temperature and salinity. Approximately 75% of the ocean is quite uniform and is represented by values in the blue box.

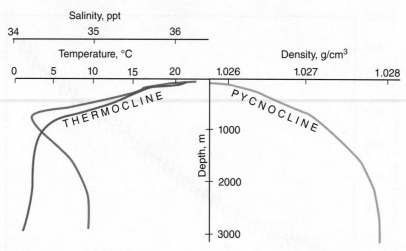

Figure 2.42 Variations in water temperature (red curve; including characteristics of the thermocline) and salinity (green curve) at a GEOSECS station in the western South Atlantic Ocean. The resulting density profile and characteristics of the pycnocline are shown at the right (yellow curve).

often characterized by well-developed permanent thermoclines and their associated pycnoclines.

Water Pressure

Organisms living below the sea surface constantly experience the pressure created by the weight of the overlying water. Although we think of water as a supportive substance that provides buoyancy, water is also heavy and weighs down on sea creatures. At sea level, pressure from the weight of Earth's envelope of air is about 1 kg/cm², or 15 lb/in², or 1 **atmosphere (atm)**. Pressure in the sea increases another 1 atm for every 10-m increase in depth to more than 1,100 atm in the deepest trenches. Most marine organisms can tolerate the pressure changes that accompany moderate changes in depth, and some even thrive in the constant high-pressure environment of the deep sea; however, organisms with gasses trapped in spaces in their bodies (like fish with swim bladders or whales with lungs) that collapse and expand with depth (and pressure) changes have evolved some sophisticated solutions to the problems associated with large and rapid pressure changes.

Dissolved Gases and Acid–Base Buffering

The solubility of gases in seawater is influenced by temperature, with greater solubility occurring at lower temperatures. Nitrogen, carbon dioxide, and oxygen are the most abundant gases dissolved in seawater. Although molecular nitrogen (N_2) accounts for 78% of our atmosphere, it is comparatively nonreactive and therefore is not used in the basic life processes of most organisms. Notable exceptions are some N_2-fixing bacteria and the occasional careless scuba diver who dives too deep for too long. Carbon dioxide and oxygen, in contrast, are metabolically very active. Carbon dioxide and water are used in photosynthesis to produce oxygen and high-energy organic compounds. Respiration reverses the results of the photosynthetic process by releasing the usable energy incorporated in the carbohydrates and fats of an organism's food. In contrast to photosynthesis, oxygen is used in respiration, and carbon dioxide is given off.

Carbon dioxide is abundant in most regions of the sea, and concentrations too low to support plant growth seldom occur. Seawater has an unusually large capacity to absorb CO_2 because most dissolved CO_2 does not remain as a gas. Rather, much of the CO_2 combines with water to produce a weak acid, carbonic acid (H_2CO_3). Typically, carbonic acid dissociates to form a hydrogen ion (H^+) and a bicarbonate ion (HCO_3^-) or two H^+ ions and a carbonate ion (CO_3^{2-}). These reactions are summarized in the following chemical equations:

$$CO_2 + H_2O \leftrightarrow H_2CO_3 \quad\quad (1)$$
carbon   water   carbonic
dioxide   acid

$$H_2CO_3 \leftrightarrow H^+ + HCO_3^- \quad\quad (2)$$
carbonic   hydrogen   bicarbonate
acid   ion   ion

$$HCO_3^- \leftrightarrow H^+ + CO_3^{2-} \quad\quad (3)$$
bicarbonate   hydrogen   carbonate
ion   ion   ion

The arrows pointing in both directions indicate that each reaction is reversible, either producing or removing H^+ ions. The abundance of hydrogen ions in water solutions controls the acidity or alkalinity of that solution and is measured on a scale of 0 to 14 pH units (**Figure 2.43**). The pH units are a measure of the hydrogen ion concentration. Water with a low pH is very acidic because it has a high H^+ ion concentration. Water with a pH of 14 is very basic (or alkaline) and denotes low H^+ ion concentrations. Neutral pH (the pH of pure water) is 7 on the pH scale. The carbonic acid–bicarbonate–carbonate system in seawater functions to **buffer**, or to limit changes in, seawater pH. If excess hydrogen ions are present, the reactions described previously here proceed to the left, and the excess hydrogen ions are removed from solution. Otherwise, the solution would become more acidic. If too few hydrogen ions are present, more are made available by the conversion of carbonic acid to bicarbonate and bicarbonate to carbonate (a shifting of the above

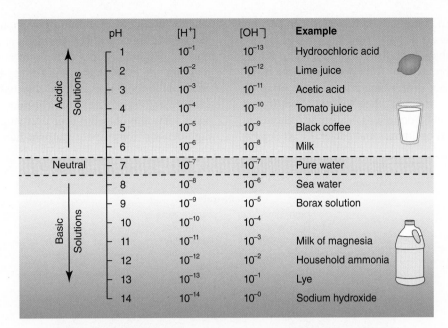

Figure 2.43 The pH scale, including concentrations of H⁺ and OH⁻ ions.

reactions to the right). Historically, in open-ocean conditions, this buffering system has been very effective, limiting ocean water pH fluctuations to a narrow range between 7.5 and 8.4. This buffering system also functions as a crucial component of our planet's ability to accommodate the increasing concentrations of atmospheric CO_2 resulting from our industrial and cultural practices on land. Recent research on the pH of the ocean has indicated that there has been a slight decrease of pH (ocean acidification). Even slight changes to normal pH values can have detrimental impacts on marine organisms. When the pH changes, the equations shown above shift, and fewer carbonate ions are in solution. Organisms that secrete calcium carbonate shells or skeletons (e.g., reef-building corals) may not be able to survive without the necessary carbonate ions. Research on this topic is ongoing and part of a heated discussion on the negative impacts humans are having on our ocean.

Oxygen in the form of O_2 is necessary for the survival of most organisms. The major exceptions are some anaerobic microorganisms. Water, however, is not a good solvent for O_2. The concentration of O_2 in seawater generally remains between 0 and 8 parts per million, not much for active, oxygen-hungry animals. Oxygen is used by organisms in all areas of the marine environment, including the deepest trenches; however, the transfer of oxygen from the atmosphere to seawater and the production of excess oxygen by photosynthetic marine organisms are the only global-scale processes available to introduce oxygen into seawater. Both of these processes occur only at or near the surface of the ocean. Oxygen consumed near the bottom can only be replaced by oxygen from the surface. If replenishment is not rapid enough, available oxygen supplies will be reduced or removed completely. Oxygen replenishment occurs by very slow diffusion processes from the oxygen-rich surface layers downward and also by density-driven sinking that carries oxygen-enriched waters to deep-ocean basins. At depths of about 1,000 m, animal respiration and bacterial decomposition use O_2 as fast as it is replaced, creating an **oxygen minimum zone**. **Figure 2.44** illustrates a typical vertical profile of dissolved oxygen concentration from the surface to the bottom of the sea.

Dissolved Nutrients

Nitrate (NO_3^-) and phosphate (PO_4^{3-}) dissolved in seawater are crucial fertilizers of the sea. Unlike terrestrial plants that can take up nutrients from the soil, marine plants and algae move nutrients from the seawater directly into their bodies. Nitrate is the most common reactive form of nitrogen in seawater. These and smaller amounts of other nutrients are used by photosynthetic organisms living in the near-surface waters. These same nutrients eventually are excreted back into the water at all depths as waste products of the organisms that consume and digest photosynthetic organisms. This once-living waste material sinks as part of a process that eventually removes

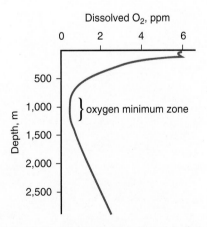

Figure 2.44 Vertical distribution of dissolved O_2 in the North Pacific Ocean during winter, including the oxygen minimum zone.

Data from Barkley, 1968. Oceanographic Atlas of the Pacific Ocean. University of Hawaii Press, Honolulu.

nutrients from near-surface waters and increases their concentrations in deeper waters.

The vertical distribution of dissolved nutrients is usually opposite that of dissolved oxygen. This opposing pattern of vertical oxygen and nutrient distribution reflects the contrasting biological processes that influence their concentrations in seawater. Oxygen is normally produced by near-surface photosynthesizers and consumed by animals and bacteria at all depths, whereas nutrients are consumed by photosynthesizers near the surface and are excreted by organisms at all depths. These contrasting vertical patterns of nutrient and oxygen abundance reinforce the temperature- and density-based stratification of much of the ocean into a two-layer system (**Figure 2.45**), a warm, low-density, oxygen-rich, nutrient-poor surface layer of ocean water a few tens of meters deep overlying a much thicker, cold, high-density, oxygen-poor, nutrient-rich layer below.

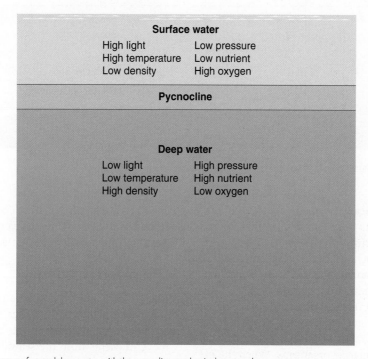

Figure 2.45 The major features of the sea surface and deep water, with the pycnocline as a barrier between the two water masses.

STUDY GUIDE

TOPICS FOR DISCUSSION AND REVIEW

1. Why is the movement of ocean water important to marine organisms as well as terrestrial organisms?

2. Which types of water movements are initiated by wind?

3. Compare and contrast upwelling and downwelling. Which process leads to the introduction of nutrient-rich waters to a coastline?

4. What are the major characteristics of a wave, and what determines the point when a wave will break?

5. What causes a tsunami? Are tsunami predictable?

6. How are tides predicted, and what are the different tide levels called?

7. How are currents formed? How are gyres formed?

8. What is the Ekman spiral, and how is it linked to the Coriolis effect? How do these interacting processes impact water movement at the surface versus farther down in the water column?

9. Describe the conditions that define El Niño and La Niña events.

10. List three properties of pure water that are highest of all common natural substances.

11. List and describe the major physical and chemical features of seawater that change markedly from the sea surface downward.

12. Compare and contrast the osmotic changes that a Portuguese man-of-war (jellyfish-like organism) will experience while being blown from the open sea into the mouth of a large river.

13. Describe how the heating of Earth is similar to the heating of a greenhouse.

KEY TERMS

anoxic 34	countercurrent 27
atmosphere 19, 48	covalent bond 38
buffer 48	crest 22
calorie 40	desiccation 23
capillary wave 20	diffusion 42
conduction 40	diurnal tide 25
continental boundary current 27	ectotherm 44
	eddy 28
convection 40	Ekman spiral 27
Coriolis effect 27	El Niño 29

El Niño-Southern Oscillation (ENSO) 29	photic zone 34
endotherm 45	poikilotherm 44
fetch 20	polar easterlies 27
gyre 27	pycnocline 31
heat capacity 40	Richter scale 22
high tide 22	salinity 19
homeostasis 42	selectively permeable membrane 42
homeotherm 45	semidiurnal tides 25
hydrogen bond (H-bond) 38	spring tide 25
ion 40	surface tension 39
ionic bond 40	temperature 19
king tide 26	thermocline 29
La Niña 30	thermohaline circulation 34
latent heat of fusion 40	tidal range 22
latent heat of vaporization 40	tide 19
low tide 22	trade winds 27
lunar day 25	trough 22
marine snow 36	tsunami 22
metabolism 44	upwelling 20
mixed semidiurnal tide 26	viscosity 38
neap tide 25	visible light 43
ocean acidification 42	water vapor 20
oxygen minimum zone 49	wave 20
period 20	wavelength 20
	westerlies 27
	wind wave 20

REFERENCES

Baker, J. J., and G. E. Allen. 1981. *Matter, Energy and Life: An Introduction to Chemical Concepts*. Reading, MA: Addison-Wesley.

Berner, R. A., and A. C. Lasaga. 1989. Modelling the geochemical carbon cycle. *Scientific American* 260:74–81.

Bignami, S., I. C. Enochs, D. P. Manzello, S. Sponaugle, and R. K. Cowen. 2013. Ocean acidification alters the otoliths of a pantropical fish species with implications for sensory function. *Proceedings of the National Academy of Sciences of the United States of America* 110:7366–7370.

Broecker, W. S., D. M. Peteet, and D. Rind. 1985. Does the ocean–atmosphere system have more than one stable mode of operation? *Nature* 315:21–26.

Garrison, T. 2013. *Oceanography: An Invitation to Marine Science*, 8th ed. Boston: Cengage Learning.

Gordon, A. L. 1986. The southern ocean and global climate. *Oceanus* 26:34–44.

Hempel, G. 1991. Life in the Antarctic sea ice zone. *Polar Record* 27:249–254.

Hua, L., Y. Yu, and D.-Z. Sun. 2015. A further study of ENSO rectification: Results from an OGCM with a seasonal cycle. *Journal of Climate* 28(4):1362–1382. doi:10.1175/JCLI-D-14-00404.1

Jones, P. D., and T. M. L. Wigley. 1990. Global warming trends. *Scientific American* 263:84–91.

Moore, III, B., and B. Bolin. 1986. The oceans, carbon cycle, and global climate change. *Oceanus* 29:16–26.

Parrilla, G., A. Lavin, H. Bryden, M. Garcia, and R. Millard. 1994. Rising temperatures in the subtropical North Atlantic Ocean over the past 35 years. *Nature* 369:48–51.

Pauly, D., and V. Christensen. 1995. Primary production required to sustain global fisheries. *Nature* 374:255–257.

Philander, G. 1989. El Niño and La Niña. *American Scientist* 77:451–459.

Pickard, G. L., and W. J. Emory., eds. 1982. *Descriptive Physical Oceanography*. New York: Pergamon Press.

Pinet, P. R. 2016. *Invitation to Oceanography*, 7th ed. Sudbury, MA: Jones & Bartlett Learning.

Post, W. M., T. H. Peng., W. R. Emanuel., A. W. King., V. H. Dale, and D. L. DeAngelis. 1990. The global carbon cycle. *American Scientist* 78:310–326.

Rasmusson, E. M. 1985. El Niño and variations in climate. *American Scientist* 73:168–177.

Richardson, P. L. 1993. Tracking ocean eddies. *American Scientist* 81:261–271.

Satake, K. (Ed.). 2006. *Tsunamis: Case Studies and Recent Developments*. New York: Springer Science.

Staaterman, E., and C. B. Paris. 2013. Modeling larval fish navigation: The way forward. *ICES Journal of Marine Science*. doi:10.1093/icesjms/fst103

Yalciner, A. C., E. N. Pelinovsky, E. Okal, and C. E. Synolakis. 2003. *Submarine Landslides and Tsunamis*. NATO Science Series, Earth and Environmental Sciences, vol. 21.

Patterns of Associations

STUDENT LEARNING OUTCOMES

1. Create a simple diagram of the seafloor and water column. Be able to label the diagram with all of the marine organism types included in Figure 3.1, and provide definitions of the organism types.

2. Describe the major differences between ecological adaptation and evolutionary adaptation. Know which of these occurs during the lifetime of an organism.

3. Compare asexual and sexual reproduction in terms of requirements, end products, and genetic diversity. Provide an example organism that uses each type of reproduction.

4. Examine a phylogenetic tree (cladogram), and evaluate the information provided by the tree. Explain why it is useful and interesting to place organisms on a phylogenetic tree instead of simply naming species.

5. Describe the path of nutrients and energy through an ecosystem. Explain why food webs are an effective method for displaying trophic relationships.

6. Compare life in the sea to life on land. Identify some of the challenges the first land-dwelling animals faced when they transitioned from the sea.

CHAPTER OUTLINE

Image displays various taxa all potentially interacting in a small area of the ocean. What we cannot see in this image are the thousands, or even millions, of microscopic organisms residing in an area like the one shown.
NMFS/SWFSC.

Upon first glance, the photo on the chapter opener page appears to simply include two colorful plant-looking organisms that are actually animals, a gorgonian (orange) and a hydrocoral (purple). When examined further, one may start to notice other creatures: a small fish hiding near the hydrocoral, dozens of small sea anemones covering the rock, and smaller fish schooling in the background. Even further inspection reveals some fuzzy-looking algae and what appears to be a small blob on the rock, but is actually a sponge, likely the first animal group to arise. The majority of organisms existing in the environment where this photo was shot are not even visible without the aid of a microscope, including tiny **plankton** and marine bacteria. Living organisms require space in which to live, material and energy from their surroundings, and an uninterrupted ancestral lineage to provide the genetic heritage that defines what they are and how they function. To satisfy these requirements, living organisms organize themselves into complex patterns of associations that are often difficult to consider in their entirety. To cope with this complexity, scientists divide these complex systems into smaller, more manageable subunits and then organize these subunits by relating them to the whole system on the basis of certain characteristics. The classification of the physical marine environment is a simple example of this approach for a nonliving system.

To be useful, any classification scheme must present the information in a manner that is generally accepted by a large majority of people. Of course, scientists often disagree, and multiple opinions of the best way to classify living things exist, but we must attempt to classify organisms to better understand them. Doing so requires an orderly framework to classify the available information so that it becomes more meaningful or useful and, most important, more understandable. In this chapter, three different but sometimes overlapping systems that we commonly use to describe the manner by which marine organisms interact with each other are introduced. These are based on the following for marine organisms: (1) the physical space they occupy in the oceans, (2) their individual evolutionary histories and **taxonomy**, and (3) the feeding relationships they have evolved to sustain life (known as **trophic relationships**).

3.1   Spatial Distribution

To better understand and begin to compartmentalize our vast Earth, including the world ocean and all of the inhabitants of the land and sea, it is useful to break things up into a hierarchy of organization. The broadest level of classifying Earth is the **biosphere**, which includes all of Earth's potential habitat areas. The biosphere is broken down into **biomes**, which on land are primarily differentiated by the dominant vegetation present. In the marine environment, the vegetation is mostly algae and

single celled, and zones deeper than the photic zone have no vegetation, so the biome level of classification in the oceans is not as relevant. The next level of classification is **ecosystem**, which includes all of the living and nonliving components of an area. An example of a marine ecosystem is a coral reef ecosystem, including all of the organisms and the oceanographic conditions present within the coral reef area. A **community** only includes the living components in a given area. Referring back to the coral reef example used for ecosystems, the community component includes all of the living organisms within the coral reef ecosystem, but it does not include nonliving components such as oceanographic conditions. A **population** includes just one species and all of the individuals of that species in a defined area, an area where individuals are likely to encounter one another and have the potential to interact and mate. The marine environment has many examples of species with numerous populations that are separated by small or large distances and are therefore separate populations. The boundaries of populations are not always clear, and genetic studies are often conducted to distinguish populations. The next level of classification includes **species**, which is the finest scale we will examine for most of this book. Individuals of a species are similar enough physically and genetically to interbreed freely in nature and produce viable, fertile offspring.

Broad classifications of the marine environment are useful to distinguish potential habitats for particular species. A simple way to classify marine organisms is according to where they live (**Figure 3.1**). Although Figure 3.1 is similar in appearance to a figure used to classify the marine environment into broad categories of habitats, you will notice that 3.1 includes living organisms. The categories indicated in Figure 3.1 refer to large and general groups of organisms based on what part of the marine environment they occupy. The **benthos** includes all organisms living on the sea bottom (the **epifauna**) or in the sediment of the seafloor (the **infauna**). This definition is often extended to include those fish and other swimming animals that spend most of their time on or near the ocean bottom. Benthic plants or algae, because of their need for sunlight, are restricted to intertidal areas and shallow inner shelves of the ocean's margins where the seafloor lies within the photic zone. Below the photic zone, plants and algae disappear and animals, microbes, and fungi survive by eating one another or by feeding on the rain of organic material sinking from the sunlit waters above.

The large actively swimming marine animals found in the pelagic division are the **nekton**. This group includes a large variety of vertebrates (e.g., fish, reptiles, birds, and mammals) and a few types of invertebrates, such as squid and some large crustaceans (e.g., shrimp). In contrast, pelagic plankton (derived from the Greek word *planktos*, which means "wandering") are defined by their inability to swim against strong currents. Carried about by water currents, plankton have little or no ability to control their precise geographic distribution, although some have reasonable abilities to swim vertically. Plankton are usually small, mostly microscopic, organisms; however, some planktonic jellyfish have tentacles greater than 15 m long and a bell measuring 2 m across. Photosynthetic members of the plankton are termed **phytoplankton**. They are nearly all microscopic,

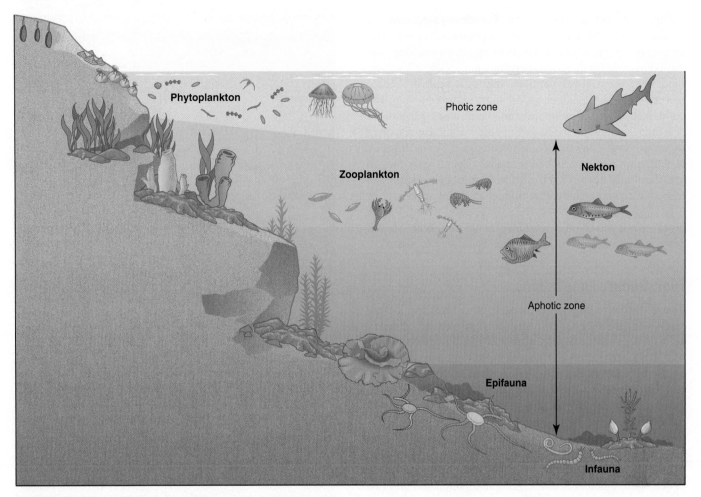

Figure 3.1 A spatial classification of marine organisms.

either a single cell or loose aggregates of a few cells, and, like plants, are restricted to the sunlit photic zone to photosynthesize. Unlike stationary plants, some phytoplankton are mobile, and are capable of migrating to deeper waters during nighttime hours or to avoid high ultraviolet (UV) light levels during the brightest period of the day. **Zooplankton** are the nonphotosynthetic plankton, ranging in size and complexity from microscopic single-celled microbes to large multicellular animals. Zooplankton are distributed throughout the pelagic division of the marine environment. Sometimes distinctions between these major groups are not clear. Many fish, for example, begin life as tiny larval zooplankton and then gradually develop into

nektonic animals as their size increases and their swimming abilities improve. Thus, many fish fit into two categories during their lifetime: zooplankton and nekton. Despite some fuzziness around the boundaries of these definitions, they serve as a convenient way of referring to major associations of marine organisms living under similar environmental conditions.

Spanning the major categories outlined in Figure 3.1, different life forms occur along gradients of latitude (North–South reference on the globe), reflecting changes in water temperature and light availability, and with distances from continental and island shorelines.

DID YOU KNOW?

Some phytoplankton can swim! They have specialized structures for moving through the water, usually a flagellum or multiple flagella (long tail-like structures). If they drop too low in the water column and need to move back up into a sunnier location, they can swim up. If they are being bombarded by too much sunlight or UV radiation, they can sink back down into a dimmer environment. Each species of phytoplankton has a specific light range where it thrives. Too much light may harm it and too little light will not provide enough solar energy for photosynthesis.

3.2 Evolutionary Relationships and Taxonomic Classification

Conditions on land and in the ocean are in a constant state of flux. Particularly in the ocean, environmental conditions are highly variable throughout a single day, and throughout the several billion–year history of life in the oceans, conditions have changed quite dramatically. All living organisms are subject to some degree of both ecological and evolutionary adaptations to these changing conditions. **Ecological adaptations** are adjustments made by individuals during their lifetimes in response to changing environmental conditions. **Evolutionary adaptations** are products of the changing response of a population of individuals over many generations. Changes in organisms resulting from evolutionary adaptations are sometimes not apparent for thousands or millions of years and, in contrast to ecological adaptations, are never apparent within an organism's lifetime. The combined results of ecological and evolutionary adaptations determine whether individuals will obtain sufficient resources to survive until they successfully reproduce. By the simplest of definitions, to reproduce successfully means only that an organism must replace itself with an offspring also capable of reproducing successfully. If reproduction is not successful or if an organism does not survive to reproduce, the organism's **genes** are not passed on to a future generation. If genes are not passed on, the characteristics of an individual are not passed on.

Evolutionary Adaptations

Natural selection in populations occurs over generations because of differential rates of reproduction and mortality in different gene lineages within a population. Over long periods of time, biological mechanisms that create genetic variations and work on those variations to create change through time have given rise to genetically isolated populations of organisms. Populations can become reproductively isolated from other closely related populations by any of several extrinsic or intrinsic factors. Common extrinsic mechanisms include geographic and climatic barriers to migration and, consequently, to **gene flow**. Intrinsic isolating mechanisms include behavioral, anatomical, or ecological differences that might split a population into two or more subgroups and prevent their interbreeding.

A clear example of intrinsic isolating mechanisms exists in two different populations of Pacific Northwest killer whales, which appear to be well into the process of becoming separate species. A "resident" population feeds almost exclusively on fish, whereas a "transient" population preys principally on marine mammals, especially harbor seals and sea lions. These two populations of killer whales use different foraging strategies, even when in close proximity to each other. These strategies reflect not only different prey selection, but also different pod traditions. Transient groups commonly encircle their prey, often using repeated tail slaps or ramming actions to kill their prey before consuming it (**Figure 3.2**). Differences in the foraging behaviors of the two overlapping populations in the Pacific Northwest persist through time, with each group expressing different vocalizations and different social group sizes, as well as different prey preferences and foraging behaviors. **Table 3.1** summarizes some important differences between these two populations.

Reproductively isolated populations experience different environmental (and therefore different selective) pressures and, given sufficient time, will likely diverge from each other as they evolve. Eventually, isolated populations evolving in different directions may achieve enough uniqueness to be accorded the status of species. It is this repetitive pattern of reproductive isolation and genetic divergence that, through more than 3 billion years of Earth's history, has been the source of the tremendous variety of species that we see today. It should be noted that

Figure 3.2 Tail slap of a transient killer whale directed at its sea lion prey.
© Francois Gohier/Photo Researchers, Inc.

much of what we know about evolution by natural selection was proposed in the 1800s by Charles Darwin, one of the earliest scientists to study evolution. Darwin's theory of evolution has not been refuted since he proposed it, and evidence continues to be gathered in support of his groundbreaking ideas. Any serious student of biology will read his famous book *On the Origin of Species by Means of Natural Selection* (1859).

Gene flow can occur when individuals from one population migrate into and interbreed with another population of the same species. The impact of gene flow is highly variable, depending on the degree of genetic difference between the two populations, as well as the number of immigrants from one population to the other. If either of these factors is large, genetic differences in subsequent generations can occur and spread rapidly through the affected population.

Natural populations of organisms are able to reproduce in numbers that exceed those needed to maintain the population size and that are in excess of the number their habitat can

TABLE 3.1 A Comparison of Foraging-Related Differences Between Transient and Resident Killer Whales of the Pacific Northwest

Character	Residents	Transients
Group size	Large (3–80)	Small (1–15)
Dive pattern	Short and consistent	Long and variable
Temporal occurrence	With salmon runs	Unpredictable
Foraging areas	Deep water	Shallow water
Vocalize when hunting?	Frequently	Less frequently
Prey type	Salmon and other fish	Marine mammals
Relative prey size	Small	Large
Prey sharing	Usually no	Usually yes

Modified from Baird, R. W., P. A. Abrams, and L. M. Dill. 1992. Possible indirect interactions between transient and resident killer whales: Implications for the evolution of foraging specializations in the genus Orcinus. Oecologia (Berl.) 89:125–132.

support, known as the **carrying capacity**. Eventually, expanding populations exceed their carrying capacity by outgrowing their necessary resources (i.e., food, water, and shelter), and competition between individual members of the population intensifies. An individual's ability to survive and reproduce depends on its physical and behavioral uniqueness, which is often controlled by genes. The majority of resources required for life are in limited supply, which causes many individuals to perish before reaching sexual maturity. Only those equipped to compete and survive in their current local environment succeed in passing on their genetic traits to future generations. This is natural selection.

The offspring inherit characteristics that, in turn, provide a similar ability to compete and survive. Superior **fitness** may result from a resistance to disease, starvation, or climatic variations; an enhanced ability to sense predators or prey; or perhaps a capacity to reproduce quickly. This competition, differential survival, and ability to reproduce and pass genes on to the next generation are summarized in the overworked phrase associated with evolution by natural selection, "survival of the fittest." However, the rules and conditions for survival change continuously and unpredictably. The primary selection factor for one generation might be a food shortage and for the next generation disease. As a result, "survival of the fitter" might be a more appropriate phrase, because organisms never evolve to fit their total environment perfectly or permanently. The environment is continuously changing.

The basic biological units of evolutionary adaptation, then, are populations. Evolutionary adaptation occurs in populations, never in individuals. Individuals perish regardless of whether their populations continue to exist, adapt, and evolve. Of the millions of species that have evolved in more than 3 billion years of life's history on Earth, only about 5% are **extant**, or exist today. These are the temporary winners.

In the sea, these current winners continually confront variations in temperature; salinity; and available oxygen, light, and food, as well as attempts by their neighbors to crowd them out or consume them. In day-to-day ecological time, these stresses mold the structures of communities and ecosystems. Over much longer periods of time, they shape the evolutionary path of populations through time. Collectively, these processes working over millions of years have resulted in an extraordinary diversity of organisms, a diversity estimated to be around 10 million distinct species of **eukaryotic** organisms currently inhabiting this planet, while only 1.2 million species have actually been identified. It is estimated that approximately 91% of marine organisms currently in existence have yet to be identified.

What is the source of all this fantastic diversity? If life originated only once on our planet, why are the seas populated with so many kinds of organisms? How do we explain the presence of nearly 17,000 species of annelid worms or 100,000 species of mollusks if each descended from a single ancestral worm or mollusk?

The diversity of life forms that we see today represents the physical expression of the organisms' genetically coded information that has been passed on from generation to generation. Every cell of each living organism has somewhere between 500 and 55,000 different genes incorporated into chromosomes,

depending on the species. The genes located on these chromosomes govern the regulatory and structural features of cells. In this way, each organism maintains its own coded repository for all of the required information needed to develop, function, and behave in its own unique manner. When it reproduces, it passes copies of its genetic information to its offspring. High genetic diversity is a key component for the long-term survival of populations of species, because it provides a suite of genes in the gene pool for natural selection to act on. If genetic diversity is low, events such as a disease or dramatic increase in predator abundance can more easily wipe out a population.

We often refer to reproduction as the process by which we replicate ourselves from one generation to the next. In fact, reproduction might better be described as the process by which sets of genes are transferred through generations of organisms. Organisms that reproduce asexually first copy their chromosomes and then themselves. The products of their reproductive efforts are genetically identical to each other and to the parent (**Figure 3.3**). The low genetic diversity of asexual organisms can be a detriment if, for example, a disease is spread through a population; if organisms with the particular genetic makeup that are negatively affected by the disease reproduce asexually, the offspring are at risk as well since they are all identical genetic copies of the parent. Although the low genetic diversity that results from asexual reproduction can be a large disadvantage, being asexual does have some advantages. Oftentimes, energetic demands are lower and reproduction is more efficient when there is no need to find a mate; reproduction occurs more quickly and more often; and for organisms with a favorable genetic makeup, the offspring carry the same favorable genes.

Sexual reproduction, however, is fundamentally different. The whole point of sexual reproduction is to provide a mechanism whereby diverse, rather than identical, offspring can be produced by the same parents. In its most basic form, sexual reproduction involves the unpairing of paired **chromosomes** (gene-carrying molecules) in a process known as **meiosis** and then recombining them at **fertilization** to form totally new chromosome pairings. When mature, sexually reproducing adults produce **gametes** (either eggs or sperm) by meiosis. Meiosis is a cell-division process in which the chromosomes of the gametes produced include one of each of the pairs of

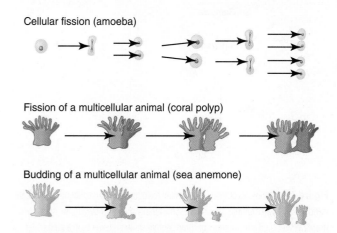

Cellular fission (amoeba)

Fission of a multicellular animal (coral polyp)

Budding of a multicellular animal (sea anemone)

Figure 3.3 Three methods used by asexually reproducing marine organisms.

chromosomes characteristic of the other cells of the adult individuals (**Figure 3.4**). Halving of chromosome numbers, the **haploid** chromosome condition, in the formation of gametes is a necessary component of sexual reproduction because gamete production is followed by fertilization, the remaining obligatory part of sexual reproduction. In fertilization, the chromosomes carried by the sperm cell are combined with those of the egg cell to form a **zygote** with double the chromosome number of either of the gametes. This double, or **diploid**, set of chromosomes is carried by all cells in the development to sexual maturity. Because the sequence of events in plant and algae sexual reproduction is so different from that of animals, these patterns are examined separately for plants/algae and animals.

The genetic diversity expressed by the offspring of sexually reproducing parents stems from several sources, including mutations, genetic recombination during meiosis, and sexual reproduction itself. Mutations are random structural alterations of the information coded in a gene. They may result from errors in the process of copying existing genetic codes, or they may be induced by external factors such as radiation. Although some mutations bestow survival advantages on those who carry them, many mutations are deleterious and sometimes fatal. Even so, it has been the slow accumulation over time of nonfatal mutations that has provided genetic diversity, the raw material on which sexual reproduction has operated. Genetic recombination during meiosis occurs when chromosomes separate and then recombine during an event termed *crossing over*. During crossing over, **homologous** chromosomes exchange small pieces, and therefore swap genes. The result of genetic recombination is new gene combinations that differ from the combinations seen in either parent. With parents of slightly different genetic

histories available, sexual reproduction (in essence, meiosis followed by fertilization) provides the mechanism to repackage the genetic diversity of parents into large numbers of different offspring. One pair of human parents, for example, each with 23 pairs of chromosomes, hypothetically could produce 2^{23}, or greater than 8 billion, genetically different children. When other chromosomal recombination maneuvers at meiosis are considered, the amount of diversity generated by the sexual reproduction of two individuals is almost limitless.

Sexual reproduction creates genetic diversity, but it is not responsible for evolutionary changes in the occurrence of particular genes in a population over time (evolutionary adaptation). Evolutionary adaptations are the expressions of the changing responses of a population of individuals to changes or variations in their environment over many generations. In natural populations, the major mechanisms for evolutionary change over time are gene flow and natural selection.

Taxonomy and Classification

Biologists estimate that around 10 million different species of organisms exist on Earth today. Of these, only about 1.2 million have been identified and formally described, and we know very little about the vast majority of those. Although most of the nonbacteria species that have been described are land-dwelling insects, the diversity of life in the sea is immense. Because of the evolutionary processes that started in the sea and have operated there for the past 3 to 4 billion years of Earth's history, each of these species exhibits some genetic relationship to all other species.

How do we organize and manipulate all the information we have accumulated regarding the evolutionary histories and relationships of millions of species? Sometimes evolutionary relationships between organisms are obvious; for example, a tuna and a mackerel are quite similar. At other times, however, such relationships are more obscure; for example, a sea star is more closely related to humans than many other simpler animal groups. The taxonomic system of classification provides a means to deal with this vast and often confusing array of diversity by reflecting these evolutionary relationships of organisms. The process of taxonomic classification consists of three basic steps. First, closely related groups of individual organisms must be recognized, named, and described. Next, these groups, called **taxa** (singular, *taxon*), are assigned Latin (or Latinized Greek) names according to procedures established by international conventions. Finally, the described and labeled groups are fitted into a hierarchy of larger, more inclusive taxa. The current system of naming species is termed *binomial nomenclature* and includes the genus and species names, typically in Latin and always italicized. For example, *Parastichopus parvimensis* is the species name for the warty sea cucumber. *Parastichopus* is the genus name and *parvimensis* is the specific species term. The words used in species names most often represent Latin terms that describe something about the animal, but the scientist who officially identifies the species for the first time is able to choose the specific term. Ideally taxonomic classification schemes accurately represent evolutionary, or **phylogenetic**, relationships, but

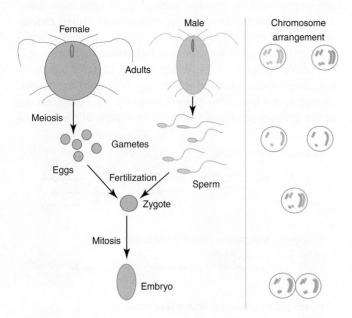

Figure 3.4 The basic components of sexual reproduction using a marine invertebrate (sand dollar) as an example. The chromosome arrangement is shown to the right. For simplicity and clarity, the number of chromosomes shown is fewer than a sand dollar actually possesses.

sometimes simply classifying a species leads to placing it in a larger grouping (like a genus, for example) that does not include the species to which it is most closely related.

The discovery and description of species and the recognition of the patterns of phylogenetic relationships among them are based on the processes of biological evolution. Patterns of relationships among species are based on changes in the features or characters of an organism. **Characters** are the varied inherited characteristics of organisms that include DNA makeup, anatomical structures, physiological features, and behavioral traits. Evolution of a character may be recognized as a change from a preexisting or **ancestral** character state to a new or **derived** character state. A fundamental underpinning of phylogeny is the concept of **homology**, the similarity of features resulting from common ancestry. Two or more features are homologous if their common ancestor possessed the same feature; for example, the flipper of a seal and the flipper of a walrus are homologous because their common ancestor had flippers.

The basic tenet of phylogenetic taxonomy, or **cladistics** (from the Greek word *klados*, meaning "branch"), is that shared derived character states provide strong evidence that two or more species possessing these features share a common ancestry. In other words, the shared derived features represent unique evolutionary events that may be used to link two or more species together in a common evolutionary history. Thus, by sequentially linking species together based on their common possession of derived shared characteristics, the evolutionary history of those taxa can be inferred.

Evolutionary relationships among taxonomic groups (such as a species) are commonly represented in the form of a **cladogram**, or **phylogenetic tree**, a branching diagram that represents our current hypothesis, or testable best estimate, of phylogeny (**Figure 3.5**). The lines of the cladogram are known as **lineages**, and the groups used for classification are known as **clades**. Lineages

represent the sequence of descendant populations through time. Branching of the lineages occurs at **nodes** on the cladogram, which represent the common ancestor of descendants. Descendants arise from **speciation events**, a splitting of a lineage resulting in the formation of two species from one common ancestor. The two Pacific Northwest killer whale populations described earlier are thought to be at such a speciation node.

The fundamental unit of taxonomic classification, then, is the species. The most widely accepted definition of a species is that it is a group of closely related individuals that are similar in appearance and that can and normally do interbreed and produce fertile offspring. The free exchange of genetic information between individuals of such groups connects each individual to a common gene pool and steers them along a common evolutionary path, with entire populations adapting to environmental influences over long periods of time.

This widely accepted definition of a species, however, poses special problems for the classification of marine organisms. Because of the environmental extremes occupied by many marine organisms, it is often quite difficult, or even impossible, to study them alive, and often little is known of their reproductive habits. Moreover, many marine species (especially planktonic microbes) are asexual, and thus the previously mentioned biological species concept is meaningless in that context. In such cases, another somewhat circuitous definition is used: a species is a group of closely related individuals classified as a species by a competent taxonomist on the basis of anatomy, physiology, and other characteristics (including genetic comparisons, where possible). Whichever definition is used, the species is regarded as a functional biological unit that can be identified and studied.

Assigning names to species or larger groups of organisms is a process more regimented than merely recognizing and describing the species. Common names are often used in localized areas, but the lack of standardization in the use of common names detracts from their widespread usefulness and acceptance. To some people, the name *dolphin* refers to an air-breathing porpoise-like marine mammal (**Figure 3.6a**). To others, a dolphin is a prized game fish (**Figure 3.6b**) that is also known as *mahi-mahi* in Hawaii and as *dorado* in Spanish-speaking countries. The confusion created by these common names is eliminated when species and other taxonomic groups are assigned names that are accepted by international agreement as standard group names.

The naming of a species does not complete the taxonomic classification process. The species is only part of a larger classification scheme that consists of a hierarchy of taxonomic categories, with a recent addition of the domain category:

> Domain
> > Kingdom
> > > Phylum (Division)
> > > > Class
> > > > > Order
> > > > > > Family
> > > > > > > Genus
> > > > > > > > Species

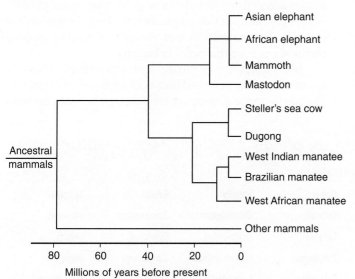

Figure 3.5 A cladogram illustrating the relationships between Sirenians (manatees, sea cows, and the dugong) and elephants and their close relatives based on differences in their mitochondrial DNA sequences. Time scale for speciation events estimated from rates of DNA change.

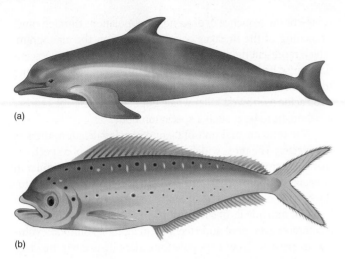

(a)

(b)

Figure 3.6 (a) Common dolphin, *Delphinus*; (b) dolphinfish, *Coryphaena*, also known as *dorado* or *mahi mahi*.

Each category is constructed so that it encompasses one or more categories from the next lower level. These groups are not completely arbitrary, however. Each group reflects the evolutionary relationships known or assumed to exist between its component taxa on the basis of its anatomy, physiology, embryology, and biochemistry. It is because much of the evolutionary history of some lineages is not known in detail that a classification based on little information may not accurately reflect the lineage's actual relationships with other closely related groups. Ideally, each genus is composed of a group of very closely related, but genetically isolated, species. Families include related genera that have many features in common. Orders include related families based on generalized characteristics. Classes, phyla (singular, phylum), kingdoms, and domains are increasingly inclusive categories based on even more general features.

Table 3.2 summarizes the taxonomy of a few organisms. Dolphins and blue whales are more closely related to each other than to the other organisms listed in Table 3.2, and thus they are placed in the same infraorder, Cetacea, which includes other whales but excludes all other species of organisms. Copepods do not resemble whales or dolphins, yet their evolutionary connections are closer to both dolphins and blue whales (organisms in the same kingdom) than they are to mangroves in the Kingdom Plantae. In this way, the taxonomic system of classification is a nested hierarchy that serves as a framework to support our understanding of the evolutionary relationships that exist between groups of organisms.

In **Figure 3.7**, the major phyla of marine organisms are arranged to illustrate the presumed evolutionary relationships of each group. Only phyla with several free-living nonparasitic marine species are included. These phyla are grouped into kingdoms, and then the three domains outlined in Figure 3.7. The classification system used in this text is slightly modified from a recently introduced and widely accepted system that groups living organisms into three domains based on their cellular structures, their modes of nutrition, and their deduced patterns of evolutionary relationships. The five-kingdom classification system used for decades is undergoing scrutiny by many scientists, as the catch-all Kingdom Protista has been abandoned, as has the Kingdom Monera that lumped together bacteria and archaea. Some organisms that were previously members of the Kingdom Protista have been placed in new kingdoms, and others still await kingdom membership. For our discussion we will recognize three major kingdoms of eukaryotes (Plantae, Animalia, Fungi), differentiate bacteria from archaea, and then discuss protists as a group, keeping in mind that the kingdom membership of many protists is unresolved.

Two domains, Bacteria and Archaea, include very simple single-celled organisms. These small organisms lack much of the complex subcellular structure found in other, more complex cells, a condition described as **prokaryotic** (**Figure 3.8**). A cell wall provides form and mechanical support for the cell. Inside the cell wall, a selectively permeable plasma membrane separates the internal fluid environment (the **cytoplasm**) from the exterior environment of the cell and regulates exchange between the cell and its external medium. In some species, limited movement is provided by a whiplike flagellum. Internally, the genetic information is coded and stored in a single circular strand of DNA. Small ribosomes use that information to direct the synthesis of enzymes. The enzymes, in turn, control and regulate all other chemical reactions that occur in bacteria.

Bacteria are the most prolific organisms on Earth. They represent most of the described species of single-celled organisms. The phylogenetic patterns of relationships between bacterial

TABLE 3.2 Taxonomic Classification of Some Marine Organisms							
Taxonomic Category							
Organism	**Kingdom**	**Phylum**	**Class**	**Order**	**Family**	**Genus**	**Species**
Blue whale	Animalia	Chordata	Mammalia	Cetacea	Balaenopteridae	*Balaenoptera*	*musculus*
Dolphin	Animalia	Chordata	Mammalia	Cetacea	Delphinidae	*Delphinus*	*delphis*
Dolphinfish	Animalia	Chordata	Osteichthyes	Perciformes	Coryphaenidae	*Coryphaena*	*hippurus*
Copepod	Animalia	Arthropoda	Crustacea	Calanoida	Calanidae	*Calanus*	*finmarchicus*
Mangrove	Plantae	Tracheophyta	Magnoliopsida	Lamiales	Avicenniaceae	*Avicennia*	*germinans*
Tintinnid	Protista	Ciliophora	Ciliata	Spirotricha	Tintinnidae	*Halteria*	*grandinella*

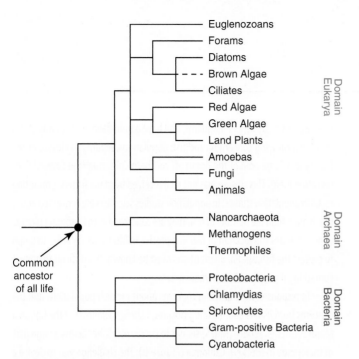

Figure 3.7 A phylogenetic tree illustrating a current hypothesis of the evolutionary relationships of the major groups of organisms based on molecular data. This tree employs the three-domain classification system, is based on phylogenetic analysis, and only includes major groups of organisms. The dashed line leading to brown algae indicates uncertainty of the position of this group within the clade.

species are not well understood, and for our convenience these important decomposers are grouped into two domains, the Bacteria and Archaea.

Despite the name, the archaea actually are not likely as old as bacteria. Some genetic comparisons indicate that archaea are sufficiently different from bacteria to warrant a separate domain status. More research on the origins of this recently discovered group is expected to clarify their taxonomic status. Marine archaea include several diverse groups of "extremophiles" with

descriptive names, including *thermophiles* ("heat lovers") that live in extremely hot water around deep-sea volcanic vents, *halophiles* ("salt lovers") in high-salinity coastal ponds and lagoons, and *barophiles* ("pressure lovers") found at extreme ocean depths.

Cyanobacteria, also known as blue-green algae, are photosynthetic members of the Domain Bacteria, and many are important in marine phytoplankton communities. In addition to their importance as photosynthesizers, cyanobacteria are a crucial source of nitrogen in some low-nutrient ecosystems. They use a process called *nitrogen fixation* to convert atmospheric nitrogen to forms that can be used by plants for photosynthesis, such as nitrate or ammonia. Fossil remains over 3 billion years old of simple cells very much like modern cyanobacteria have been reported from scattered sites in Africa and Australia. These were likely the first photosynthesizers in Earth's early oceans. Some other bacteria are also photosynthetic, but most can be found as decomposers in all marine habitats.

The prokaryotic members of Bacteria and Archaea have been eclipsed in most environments by groups of relatively large and ecologically dominant organisms, the eukaryotes. The complexity and diversity of eukaryotic cells are responsible for much of the immense variety of life forms on Earth today. These larger and structurally more complicated eukaryotic cells possess a **nucleus** and a variety of other membrane-bound structures not found in prokaryotes (**Figure 3.9**). The centrally located nucleus is formed by the chromosomes and their surrounding membrane. The enzymes involved in respiration and energy release are associated with numerous small **mitochondria**. Many of the enzyme-synthesizing **ribosomes** are free in the cytoplasm; others are arranged on a membranous **endoplasmic reticulum**. Food particles are ingested and stored in vacuoles within the

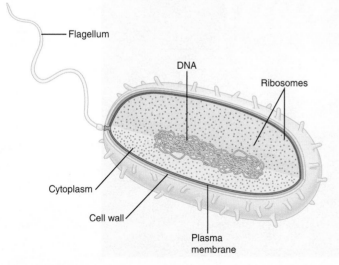

Figure 3.8 A diagram of the major components of a prokaryotic bacterial cell, *Bacillus*.

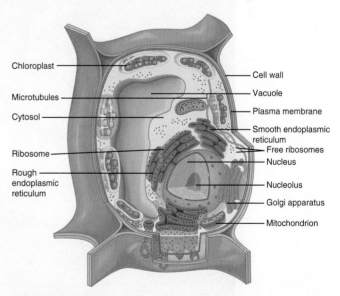

Figure 3.9 A diagram of the major components of a eukaryotic plant cell. Animal cells are similar in their components, but lack rigid cell walls, chloroplasts (used for photosynthesis), and a large central vacuole. Animal cells have centrioles, which are used in cell division, and plant cells do not.

RESEARCH in Progress

Spatial Distribution of Rockfishes off the West Coast of the United States

The spatial distribution of highly mobile marine animals is often difficult to determine, and fish are no exception. In the past, many scientists used data from fishers (men and women who fish) to estimate numbers and distributions as they reported the depth of the seafloor below their catch. These estimates include large sources of error, and depth estimates are really only valid for benthic fish living right on the seafloor. Even if the seafloor depth is accurate, depth alone does not provide information about the habitat where the fish are living. Rockfish are a large group of fish with variable lifestyles (genus *Sebastes*; **Figure A**). Despite what their name implies, they do not all live near rocks. Many do live near rocks on the seafloor, but some swim well above the seafloor along large seamounts and steep walls of rock reef. Others live on the bottom near sand or mudflats. Rockfishes are fished along the western Pacific coast by sport and commercial fishers. They are relatively easy to catch, grow

to a decent size by fishers' standards, and have a mild flavor. In the early 2000s there was growing concern for rockfish populations, leading to closure of the fishery in a large area of the ocean off southern California (the Cowcod Conservation Area). This closure was not well received by many fishers, prompting rockfish population studies. In population studies scientists are concerned about much more than the number of fish. They are concerned with the bigger picture, which includes the location of fish in a precise habitat type. In order to begin to protect the fish, spatial distributions must be known. It is difficult to protect something if you don't know where it lives.

In response to the growing concerns about rockfish populations and the concerns from fishers about fishing closures, fisheries scientists at the National Marine Fisheries Service (NMFS) set out to survey rockfishes across a large part of their range in southern California (**Figure B**). The challenge was to devise a way to study rockfishes in their habitat without removing or harming them. The technology used to observe fish and their habitat was (and still is) a remotely operated vehicle (ROV) equipped with high-resolution still and video cameras and various sensors for sampling the seawater (**Figure C**). In combination with the ROV surveys, acoustic surveys using sonar were completed to try to identify specific acoustic signatures of the fish. Popular on fishing boats, "fish finders" can also be used to classify fish by attributes such as size and location above the

Southwest Fisheries Science Center, NOAA Fisheries Service

Figure A Starry rockfish (*Sebastes constellatus*) near the seafloor.
NOAA/SWFSC.

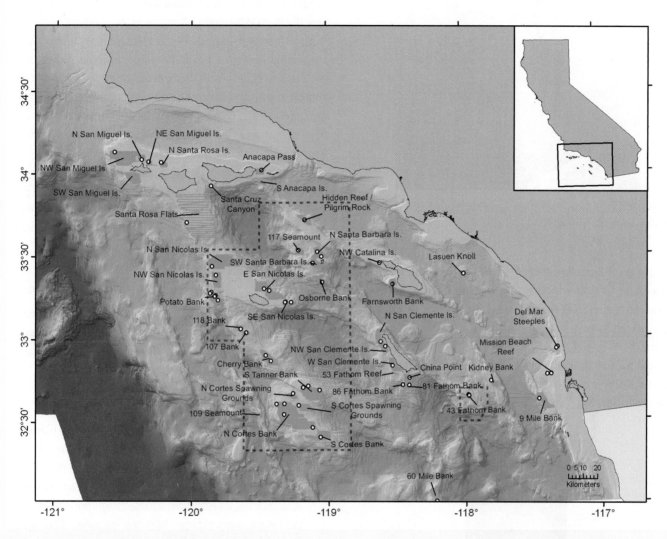

Figure B Map of collaborative acoustic and ROV survey sites. The green lines represent areas surveyed with acoustics and the ROV. The two Cowcod Conservation Areas (CCAs) are indicated by the dashed line.
NOAA/SWFSC.

seafloor. Most of the surveys by scientists at NMFS were conducted along with another type of expert: local fishers. A local fishing boat, the CPFV *Outer Limits*, was equipped with scientific equipment, and the crew set off on a long mission to observe rockfish in their habitats. Fishers provided scientists with locations of their fishing sites and local knowledge of the marine environment.

After many months of surveying sites in southern California, several facts became apparent. Each species of rockfish has a very specific spatial distribution and use of its habitat. There is no value in making generalizations about rockfishes. Bocaccio (*S. paucispinis*), one of the species that scientists had feared

was in decline, appeared to be quite abundant at most of the survey locations. Cowcod (*S. levis*), canary (*S. pinniger*), and yelloweye rockfish (*S. ruberrimus*), and several other species of concern to scientists, were not found in high numbers, indicating that protection was necessary. Since these and other surveys have been completed, these species have had their Endangered Species Act statuses changed, ranging from official designations of Species of Concern to Threatened status. In addition, many rockfish populations currently appear to be in good standing and can be fished sustainably. These population rebounds are huge victories for fisheries managers and fishers!

Figure C   ROV used to survey the seafloor, equipped with video and still cameras and various instrumentation, deployed from the CPFV *Outer Limits*.
NOAA/SWFSC.

The spatial distributions of fish were so diverse that six categories were created to attempt to place the fish in habitat use classifications: mudflat, sand, cobble, low-relief reef, high-relief reef, and vertical rock face. Along with the video data that placed rockfish in their habitat types, acoustic data were collected to observe the seafloor and fish schools on or above it (**Figure D**). Habitat types observed with the video camera on the ROV generally agreed with habitat types classified by sonar; this result is exciting, because sonar data are much less time-consuming to collect and can be used for larger survey areas. In general,

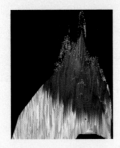

Figure D   Three-dimensional representation of the seafloor and fish schools above the bottom at survey sites.
NOAA/SWFSC.

scientists were pleased with the amount of information they collected, and they were able to add to the current information on abundance and habitat use of rockfishes in southern California. The assistance and expertise of local fishers was an integral part of the study, and very important, *because accurate results providing a real picture of what is going on underwater with fish populations is what all interested parties desire*. It is studies like these that help policy makers implement necessary restrictions on activities that may lead to unsustainable fish populations, and also assist with removal of restrictions when populations rebound and can be fished responsibly once again.

Critical Thinking Questions

1.   Think of three pieces of knowledge fishers might possess that scientists might not. How is the information provided by fishers useful in scientific studies of fish populations?

2.   Why is the precise spatial location and specific habitat use for each species of rockfish crucial knowledge when assessing population statuses?

3.   Have you ever eaten rockfish? Look up this group of fish on the Monterey Bay Aquarium Research Institute's Seafood Watch website and become familiar with what they look like, their common names, and which species are the best options for your next dinner.

For Further Reading

Pinkard, D. R., D. Kocak, and J. L. Butler. 2005. Use of a video and laser system to quantify transect area for remotely operated vehicle (ROV) rockfish and abalone surveys. *Proceedings of OCEANS 2005* 3:2824–2829.

Stierhoff, K. L., D. W. Murfin, D. A. Demer, S. A. Mau, and D. R. Pinkard-Meier. 2016. Improving the estimations of transect length and width for underwater visual surveys of targets on or near the seabed. *ICES Journal of Marine Science*. doi:10.1093/icesjms/fsw110.

Yochum, N., R. M. Starr, and D. E. Wendt. 2011. Utilizing fishermen knowledge and expertise: Keys to success for collaborative fisheries research. *Fisheries* 36(12):593–605.

cell. Other subcellular structures are involved in excretion of wastes, osmotic balance, and other cellular chores. In addition to these cellular structures common to all eukaryotes, photosynthetic eukaryotes typically possess **chloroplasts** and cell walls. Chloroplasts serve as the sites of photosynthesis, an important process described in the next section. Cell walls provide shape and support in a manner similar to those of prokaryotic cells. Cell walls also provide structural resistance to the stresses internal osmotic pressures place on fragile cell membranes. Often, cell walls alone are sufficient to deal with pressures generated by ion imbalances across cell membranes.

The eukaryotic organisms that are single celled or that consist of simple aggregations of similar cells but are not fungi are placed in the broad group Protista. Most nonparasitic protists are aquatic or live in soil. This group includes the rest of the major phytoplankton, a few terrestrial fungus-like groups, and several nonphotosynthetic phyla sometimes referred to as *protozoans*.

From the basic unit of the eukaryotic cell, multicellular organisms of even greater size and organizational complexity have evolved. Fungi form a kingdom of living organisms that includes single and multicellular representatives. Multicellular fungi are complex organisms whose bodies are composed of **hyphae**, which are fine threadlike tubes containing numerous haploid nuclei. The hyphae have cell walls made of chitin (similar to the material of arthropod exoskeletons) and often lack cross walls, and thus true cells of fungi are difficult to define. Diffuse masses of hyphae form a mycelium, the nonreproductive part of a fungus. Fungi are much more abundant and diverse on land, where their windblown spores effectively disperse them over long distances. About 1,500 species of fungi live in the sea and, along with bacteria, function as decomposers in most benthic environments, and a few, with their photosynthetic algal **symbionts**, exist as lichens on intertidal rocks.

The last two eukaryotic kingdoms, Plantae and Animalia, include groups of organisms familiar to most people. Both kingdoms consist of multicellular, and therefore usually larger, organisms. Plants are photosynthetic nonmotile organisms with cellulose cell walls and life cycles that include alternating **gametophyte** and **sporophyte** generations. Most members of the Kingdom Plantae live on land, yet several thrive in coastal marine communities.

DID YOU KNOW?

Did you know that even today, with the vast amount of technology and scientific expertise available, we are not sure which type of animal evolved first? For many years it was assumed and widely accepted that sponges, basic pore-bearing animals composed of just a handful of cell types, were the simplest and thus most closely related animals to the ancestor of all animals. Recent studies of the molecular makeup of early animal groups have led to a new hypothesis: ctenophores, small jelly-like creatures with sticky tentacles known as comb jellies, were the earliest animals, and sponges became simplified secondarily. There is debate in the scientific community about this hypothesis, but the evidence indicates that genes of the ctenophore nervous system are completely different from those found in all other animals. This unique factor alone is what has led to the idea that ctenophores are the oldest animal phylum. What do you think?

Animals lack cell walls and have some muscle-contracting and nerve-conducting capabilities. These attributes provide flexibility and mobility, two of the evolutionary hallmarks of animals.

3.3 Trophic Relationships

Relationships between different organisms residing in the same area can be described by their **trophic** associations. Organisms living within a particular environment are interconnected somewhere along a trophic system, whether they directly interact or not. For example, it is difficult to see what association a tiny marine phytoplankter (singular for *phytoplankton*) may have with a top-level predator such as a great white shark, but there are several critical connections that exist. The phytoplankter photosynthesizes, producing oxygen for all organisms to breathe and sugars for herbivores to consume. It is consumed by zooplankton, which, in turn, is food for small fish. The small fish is consumed by a larger fish, which becomes food for a sea lion. Here is where the connection finally reaches the great white shark, as the shark consumes the sea lion. This approach involves determining what each organism eats and what eats each organism within an ecosystem. Sometimes the connections are clear, but there are many examples of systems that involve complex webs of interactions, which is one topic we examine in this section. It is the trophic connections that these organisms forge with others to obtain food that weave much of the fabric of communities of organisms in all ecosystems. In this section, we examine some of the main characteristics of these trophic interactions.

Harvesting Energy

Living organisms require two fundamental things from their nourishment: matter and energy. Matter is necessary for repair, growth, and reproduction. Energy is needed to maintain the structured chemical state that distinguishes living organisms from nonliving masses of similar material. All living organisms on Earth satisfy their energy needs by using molecules of **adenosine triphosphate (ATP)** as their currency of energy exchange.

Photosynthesis is a biochemical process that uses specialized light-absorbing pigments, such as **chlorophyll**, to absorb some of the abundant energy of the sun's rays. In all photosynthetic organisms except cyanobacteria, the photosynthetic pigments and enzymes are contained in the chloroplasts of the cell. Chloroplasts are a bright green color as a result of their high chlorophyll content, and can be thought of as light-gathering antennae. The pigments in chloroplasts harness energy from particular ranges of the visible light spectrum and use this energy to fuel various chemical reactions. ATP and other high-energy substances are made and then used to synthesize important organic molecules, including sugars (e.g. glucose), amino acids, and lipids, from CO_2 and H_2O. For the present, photosynthesis can be summarized by the following general equation:

$$6CO_2 \quad + \quad 6H_2O \quad \xrightarrow{\text{chlorophyll}} \quad C_6H_{12}O_6 \quad + \quad 6O_2$$

| carbon dioxide | water | chlorophyll sunlight | sugar | oxygen |

Most nonphotosynthetic organisms on Earth rely either directly or indirectly on the energy-rich organic substances produced by photosynthetic organisms. In environments with limited amounts of free O_2 (such as **anoxic** basins or deep-ocean bottom muds) and abundant supplies of organic material, **anaerobic respiration** (respiration without O_2) provides a mechanism to obtain energy for use in cellular processes. Several variations of anaerobic respiration are exhibited by plants and animals, yet all release energy from organic substances without using O_2. In alcoholic fermentation, for example, sugar is degraded, or broken down, to alcohol and CO_2. Energy is released in the form of ATP:

$$C_6H_{12}O_6 \rightarrow 2(C_2H_5OH) + 2CO_2 + energy$$

| sugar | respiratory enzymes | alcohol | carbon dioxide | (equivalent to 2 ATP) |

In most eukaryotic organisms, respiratory enzymes housed in the mitochondria are more complex than those used for anaerobic respiration and completely break down high-energy compounds such as sugar to carbon dioxide and water and, in the process, release greater amounts of energy:

$$C_6H_{12}O_6 + 6O_2 \rightarrow 6CO_2 + 6H_2O + energy$$

| sugar | oxygen | respiratory enzymes | carbon dioxide | water | ~32 ATP |

This process uses oxygen and is therefore referred to as **aerobic respiration**. In aerobic respiration, each molecule of sugar yields about 16 times as much energy as it would if used in anaerobic respiration. Consequently, organisms that metabolize food with oxygen in this manner obtain a tremendous energetic advantage over their anaerobic competitors.

The transfer of matter and energy for use in metabolic processes has shaped the evolution of a close interdependence of three major categories of marine organisms: **autotrophic producers**, **heterotrophic consumers**, and **decomposers**. *Autotrophic* means "self-nourishing"; these organisms use photosynthesis to build high-energy organic carbohydrates or lipids and other cell components from water, carbon dioxide, and small amounts of inorganic nutrients (primarily nitrate and phosphate) found in seawater. They are the **primary producers** of marine ecosystems, providing the first level of organic nutrients in the form of sugars produced by photosynthesis. Primary producers are placed in the first **trophic level**. Some bacterial autotrophs extract energy from inorganic compounds to build high-energy organic molecules. These autotrophs are **chemosynthetic**. Heterotrophs must ingest food to meet their energetic demands.

Consumers and decomposers, which are heterotrophs, are unable to synthesize their own food from inorganic substances and depend on autotrophs for nourishment. These organisms have some specialization in terms of how they gain nutrition, yet all ultimately depend on autotrophs for their energy. Animals that feed on autotrophs are **herbivores** and occupy the second trophic level; those that prey on other animals are **carnivores** and occupy the third and higher trophic levels. The decomposers, primarily bacteria and fungi, exist on **detritus**, the waste

products and dead remains of organisms from all the other trophic levels. Whatever their specialized feeding role may be, all heterotrophs metabolize the organic compounds synthesized originally by primary producers to obtain usable energy.

Organic compounds produced by autotrophs become the vehicle for the transfer of usable energy to the other inhabitants of the ecosystem. A distinction must be made between the flow of essential nutrients and the flow of energy in an ecosystem. The movement of nutrient compounds and dissolved gases is cyclical in nature, and nutrients are recycled, going from autotrophs to consumers to decomposers, and then eventually back to the autotrophs (**Figure 3.10**). Most ecosystems function as nearly closed systems, and thus materials move from one ecosystem component to another in biogeochemical cycles. These cycles link living communities of organisms with nonliving reservoirs of important nutrients within their respective ecosystems (**Figure 3.11**).

Food Chains and Food Webs

In contrast to the cyclic flow of materials in ecosystems, the flow of energy is only in one direction, from the sun through the autotrophs to the consumers and decomposers. Living organisms are not highly efficient in their use of energy. Less than 1% of the solar energy that makes it through the atmosphere to the sea surface is absorbed by autotrophs. Then a substantial portion of the energy captured in the photosynthetic process is used for cellular maintenance, growth, and reproduction. Thus, only a small fraction of the energy from photosynthesis is actually available to consumers.

A similar decrease in available energy occurs between herbivores and carnivores (**Figure 3.12**). Laboratory and field studies of marine organisms place the efficiency of energy transfer

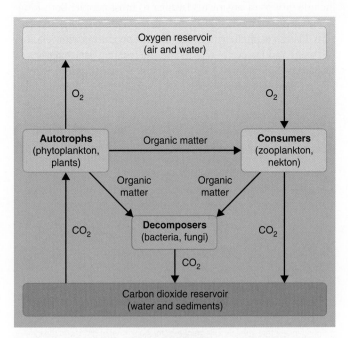

Figure 3.10 Simplified paths of the flow of oxygen and carbon in an idealized marine ecosystem.

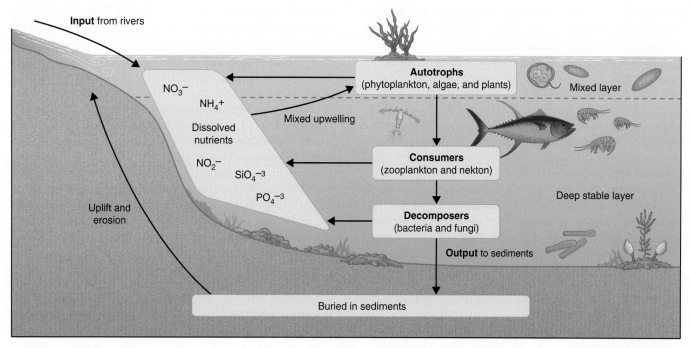

Figure 3.11 Biogeochemical cycle of nutrients, showing the major marine reservoirs. Most nutrients are available to autotrophs dissolved in deep stable waters, which then must be brought near the surface for use by autotrophs living in the photic zone. Similar cycles exist for biologically important chemicals such as carbon.

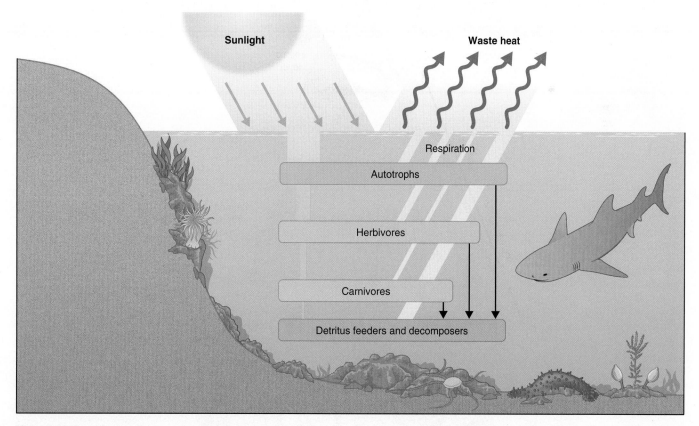

Figure 3.12 Energy flow in a marine ecosystem. Sunlight first captured by photosynthetic autotrophs is eventually degraded by their own cellular respiration or that of their consumers, and eventually is lost as waste heat.

from one trophic level to the next between 6% and 20%. In other words, only 6% to 20% of the energy available to any trophic level is usually passed on to the next level. A widely accepted average efficiency is 10%; however, recent studies of some benthic communities and fish populations provide examples of energy efficiencies that are substantially higher. These relationships can be illustrated as a food pyramid, arranged in a linear fashion to illustrate the decrease in available energy and material from lower to higher trophic levels. **Figure 3.13** illustrates such a food pyramid, proceeding from producers (phytoplankton) to herring at the third trophic level.

However, marine communities seldom exist as simple straight-line food chains. Rather, a complex interconnected food web provides a more realistic model of the paths that nutrients and energy follow through the living portion of ecosystems as its members feed on each other. They may be grazing food webs, commencing with autotrophs and progressing through a succession of grazers and predators, or they may be parallel detritus food webs, built on the waste materials and dead bodies from the grazing food webs. Open-ocean food webs with few nutrients available for photosynthesis will differ drastically from

food webs occurring in nearshore areas of frequent upwelling, where nutrients and phytoplankton are abundant.

With only a few nearshore and deep-sea exceptions, the first trophic level of marine food webs is occupied by widely dispersed microscopic phytoplankton. This microscopic character of most marine primary producers imposes a size restriction on many of the occupants of higher marine trophic levels. Because very few animals are adapted to feed on organisms much smaller than themselves, marine herbivores that feed on phytoplankton are usually quite small. Most large marine animals are carnivores and usually occupy higher levels in the food web. In contrast, the plants of most terrestrial ecosystems are generally quite large. As a result, most large terrestrial animals are herbivores, and fourth trophic–level animals are extremely rare on land.

Figure 3.14 outlines the major trophic relationships of the pelagic members of an example temperate marine community. The herring, like many of the other organisms of this food web, is an opportunistic feeder that does not specialize on only one type of food organism. Because of the complex feeding relationships of the herring, it is very difficult to place it in a particular

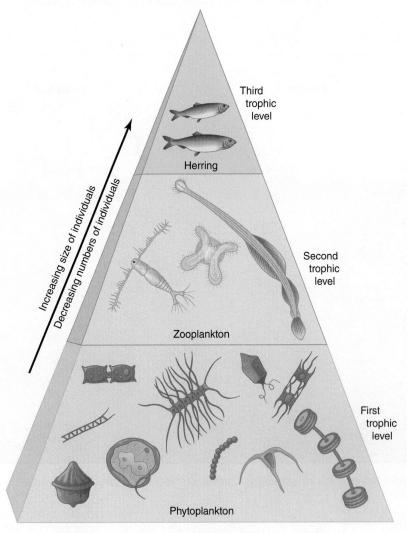

Figure 3.13 Food pyramid, beginning with phytoplankton at the base, ending with adult herring at the top.

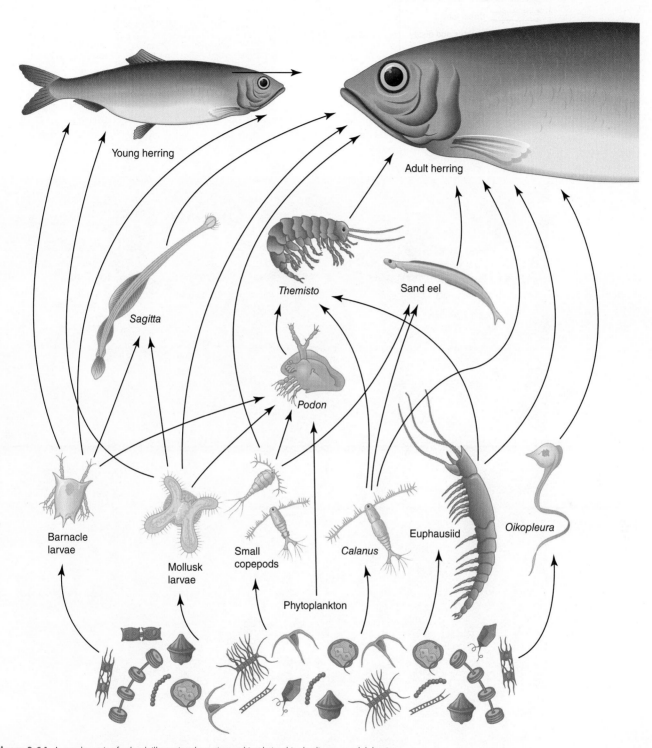

Figure 3.14   A sample marine food web illustrating the major trophic relationships leading to an adult herring.

Modified from Hardy A. C. *Fishery Investigations*. Volume 7.(3) 1924. The herring in relation to its animate environment Part 1; p. 140. (Series 2).

trophic level. The adult herring occupies the third level when feeding on *Calanus* copepods, the fourth level when feeding on sand eels, and either the fourth or fifth trophic level when feeding on the amphipod *Themisto*. Even the complex feeding relationships outlined in Figure 3.14 is an oversimplification because it ignores other marine animals that compete with the herring for the same food sources, the predators of herring,

and the detritus food webs that develop in parallel with grazing food webs. **Figure 3.15** outlines some of these more common energy and nutrient pathways of a pelagic marine ecosystem and includes humans in the top area of the complex food web.

Some marine organisms obtain their food by establishing highly specialized **symbiotic relationships**. The term *symbiosis* denotes an intimate and prolonged relationship between two

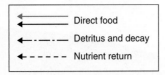

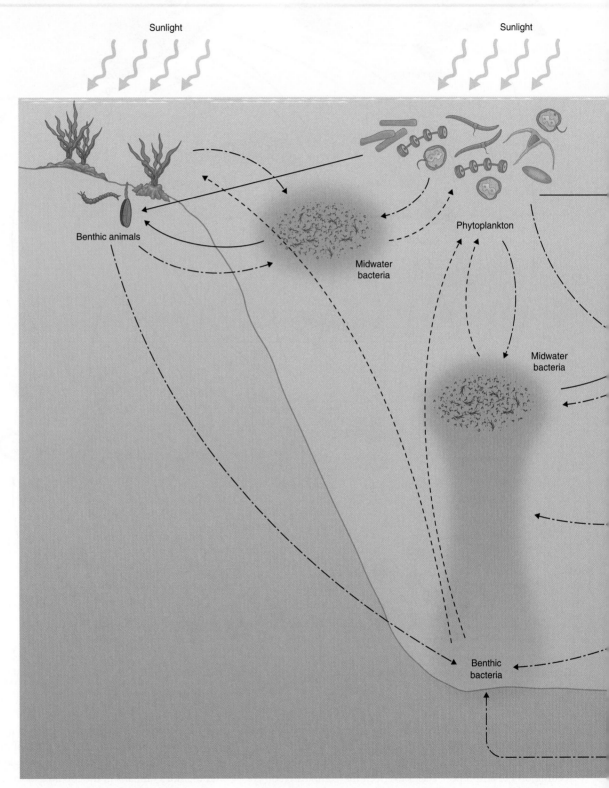

Figure 3.15 A representative food web including benthic and pelagic organisms and humans. Humans directly affect all levels of the food web with the exception of detritus, bacteria, and deep-water nekton. The red lines indicate human take of marine organisms.

Modified from W. D. Russell-Hunter, *Aquatic Productivity* (Macmillan, 1970).

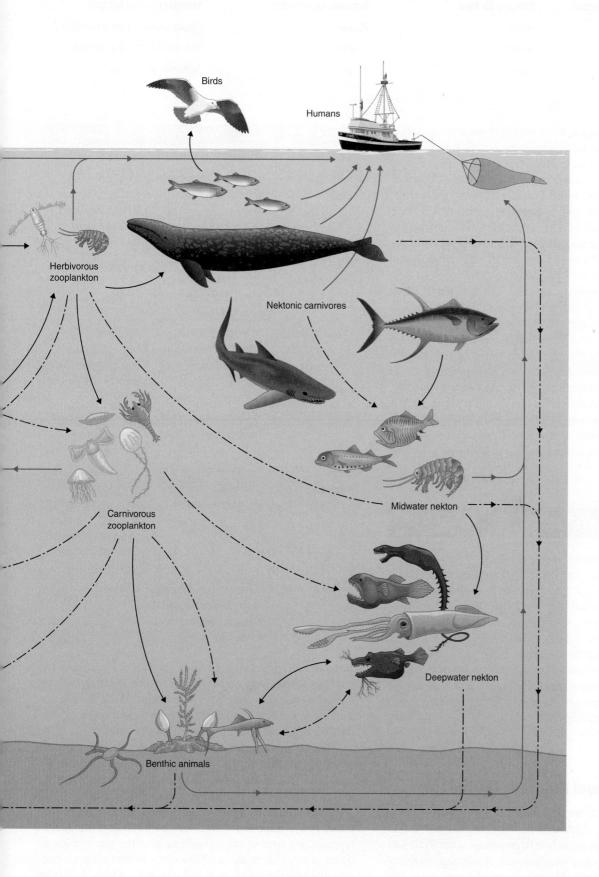

Birds

Humans

Herbivorous
zooplankton

Nektonic carnivores

Carnivorous
zooplankton

Midwater nekton

Deepwater nekton

Benthic animals

TABLE 3.3 Range of Symbiotic Interactions Between Hosts and Symbionts

Type of Symbiotic Interaction	Outcome for Host	Outcome for Symbiont	Example (+, −, or neutral)
Mutualism	Positive	Positive	Cleaner wrasse (+) and sea turtle (+)
Commensalism	Neutral	Positive	Shrimp (+) and sea slug (neutral)
Parasitism	Negative	Positive	Tapeworm (+) and tuna (−)

(or more) species living on, in, or very near one another, and in which at least one species obtains some benefit from the relationship. Commonly, that benefit is food. Symbiotic relationships can be viewed as a spectrum of interactions (**Table 3.3**), ranging from **mutualism**, which provides an obvious benefit to the symbiont and to its host, to **parasitism**, which directly benefits the symbiont at the expense of the host. Common marine examples of mutualism include several different types of cleaning symbiosis, in which small fish or shrimp maintain obvious cleaning stations that are visited by larger host fish to have damaged tissues or parasites removed. The symbiont (fish or shrimp) gains food, and the host fish is healthier once parasites or damaged tissues are removed.

Parasites live on or in their hosts and make their presence felt not by killing their hosts, but rather by reducing the host's food reserves, resistance to disease, and general vigor. The infected host then is more likely to become a casualty of infection, starvation, or predation but not of the parasite directly. Intermediate between mutualism and parasitism is a broad category of **commensalistic** interactions, in which the symbiont benefits but there is an insignificant, or at least poorly known, effect on its host.

DID YOU KNOW?

Did you know that decomposers are some of the most important organisms living in our oceans? Decomposers break down dead matter so that it can be used by other organisms. What types of organisms use the products of decay, you might ask? Well the products of decaying animals and plants are nutrients. Nutrients are required by plants and phytoplankton during photosynthesis, the process that provides oxygen in the water and eventually in the air, and sugars that are the basis of the food chain. Fungi assist with the breakdown of dead and decaying material to make it available to photosynthesizers. The role of decomposers is yet another example of how many processes in the ocean are related, often operate in cycles, and greatly affect humans. So thanks for helping plants produce oxygen for us, decomposers!

3.4 The General Nature of Marine Life

Although modern marine organisms share many basic structural and behavioral characteristics with their terrestrial relatives, marine life is unique in several important ways (**Table 3.4**). Marine organisms exist within a dense, circulating, interconnected seawater medium. The movement of waves, tides, and

TABLE 3.4 A Comparison of Life in the Sea and Life on Land

Variable	Marine	Terrestrial
Temperature (daily and seasonal)	*Daily*: Steady temperature with relatively small changes *Seasonal*: Larger changes	*Daily*: Can fluctuate greatly *Seasonal*: Very large changes in some climates
Light for photosynthesis	Limited to the photic zone; majority of the ocean does not receive enough light for photosynthesis	Not limited
Ultraviolet radiation	Only organisms in surface waters are affected; water absorbs light quickly	All organisms are exposed to some degree. Benefit to some extent by providing vitamin D
Water	Never limited, except in the intertidal zone	Limited by water for drinking; patchily available
Support of organisms' body weights	Water provides support; plants and animals need less structural support	Air provides none, and effects of gravity are felt to a greater extent because of no extra support; plants and animals need more structural support
Oxygen	Content varies with location in the sea, depth, temperature, and other components of seawater	Not limited

currents stirs and mixes organisms, their food, and their waste products so that these organisms are never completely isolated from the effects of their neighbors. Populations of even the smallest unicellular planktonic organisms can become widely distributed by currents and moving water masses.

The biology of marine organisms is, to a large extent, the biology of the very small. It is the phytoplankton that initially establish much of the structural character of marine life. Even in very productive areas of the open ocean, the concentration of phytoplankton is thousands of times more dilute than a healthy meadow. The dispersed nature, extremely small size, and rapid reproductive rates of phytoplankton limit the size and abundance of other life in the sea. Most of the heterotrophs are congregated near the photic zone and its supply of food. At greater

Case Study

Sea Otters in the Pacific Northwest

Figure A Sea otter floating on his back between foraging dives.
© Doug Meek/Shutterstock, Inc.

Common name: Sea otter

Scientific name: *Enhydra lutris* (several subspecies exist)

Distribution: The southern subspecies was previously found in California from Baja to Half Moon Bay, but it is currently restricted to areas north of Santa Barbara. One small but growing population was transplanted to San Nicolas Island, south of Santa Barbara. The northern subspecies is found from Kodiak Island through the western Aleutian Islands, which are off the coast of the Alaskan mainland.

Endangered Species Act status: Alaskan and southern populations are listed as threatened by the U.S. Fish and Wildlife Service.

Threats: Habitat destruction, pollution, food availability.

Sea otters (*Enhydra lutris*) are the smallest marine mammal species, yet they have a large impact on the kelp forest ecosystems where they reside as top predators (**Figure A**). They are voracious carnivores, able to eat around 25% of their body weight in one day in order to keep up with their high metabolic demands. Sea otters and other top predators have been declining in number in recent decades at rates higher than organisms at lower trophic levels. Declines have been due to many factors, but a very clear factor was hunting for sea otter fur.

In a trophic system, the loss of top predators often leads to an increase in abundance of herbivores that are normally preyed upon by the top predators. Increased numbers of herbivores, in turn, lead to a decrease in marine plants and algae. In the kelp forest ecosystem, sea otters eat fish and various invertebrates, including sea urchins, which are herbivorous echinoderms related to sea stars (**Figure B**). The predator–prey relationship between sea otters and sea urchins has earned sea otters the status of a **keystone species**. Keystone species have a greater impact on an ecosystem than their abundance would suggest. Oftentimes, the impacts of a keystone species are not realized until their numbers begin to decline; it is then that the whole food web is disrupted, and in the case of sea otters in the kelp forest, sea urchins take over. When sea otters are present, sea urchins remain hidden, favoring pieces of drift kelp and smaller algae. When sea otters are scarce, the sea urchins come out of hiding and begin mowing down large kelp plants. Areas where sea urchins have grazed extensively are termed *urchin barrens* because the kelp canopy is gone, leaving a barren bottom of sand with little other life.

Figure B *Strongylocentrotus franciscanus* (red sea urchin) and *S. purpuratus* (purple sea urchin) grazing on kelp.
© Greg Amptman/Shutterstock, Inc.

This predator–prey interaction and the serious impacts of lower numbers of predators is an excellent example of how important each species is within a trophic system. A decline in number of any one of the species in a food web, and certainly a keystone predator, will shift the entire ecosystem out of its normal rhythm. In the case of sea otters, not only is the ecosystem affected, but recent research has shown that a decrease in sea urchins resulting from predation by sea otters may actually lead to measurably less CO_2 entering the atmosphere. Because CO_2 contributes to climate change, keeping sea otter and other top-level predator populations at stable and healthy levels may actually decrease the rate at which our planet's climate is changing. Although sea otter populations have been in decline, there is hope that with protective measures in place sea otter populations will increase in size and inhabit their historic range. Evidence suggests that the small spread of populations in southern California has already begun to lead to healthier kelp forests.

Critical Thinking Questions

1. Why do sea otters have such high metabolic demands? If their food sources were to become less abundant, what would result for sea otter populations?

2. If this species of sea otter were to become extinct, what do you predict the new kelp forest ecosystem would look like?

For Further Reading

Hatfield, B., and T. Tinker. 2014. Spring 2014 California sea otter census results. USGS Western Ecological Research Center. http://www.werc.usgs.gov/ProjectSubWebPage .aspx?SubWebPageID=24&ProjectID=91

Wilmers, C. C., J. A. Estes, M. Edwards, K. L. Laidre, and B. Konar. 2012. Do trophic cascades affect the storage of flux of atmospheric carbon? An analysis of sea otters and kelp forests. *Frontiers in Ecology and the Environment* 10:409–415.

depths, the density of marine populations tends to decrease as the food supply diminishes. Below the photic zone, most marine life is dependent on the rain of detritus from above. The sea has few algae- or plant-dominated communities (a few notable exceptions are seagrass meadows and kelp beds). Instead, the majority of living communities in the sea are organized around coral reefs, mussel beds, and other assemblages of large and dominant animal members.

Many of the substances produced by marine primary producers are not consumed directly by herbivores but are dissolved into seawater. These substances, including lipids and amino acids, are eventually absorbed by suspended bacteria at all depths. These bacteria, in turn, become food for consumers capable of harvesting them. These microscopic phytoplankton or even smaller bacteria are food for suspension and filter feeders, small and large, that use filtering and trapping techniques and devices to collect these minute food particles suspended in seawater.

In a typical two-layered ocean system, most marine organisms and all photosynthetic ones occupy the very shallow near-surface, nutrient-rich, sunlit photic zone. The photic zone is separated vertically by the pycnocline from the colder, darker aphotic zone, which is sparsely populated by animals, fungi, and **microbes**, organisms dependent on the rain of food from the photic zone above.

Finally, the sea provides buoyancy and structural support to many strikingly beautiful organisms, but if these organisms are removed from the water their delicate and fragile forms collapse into shapeless masses. Seawater also supports some extremely large animals. Deep-sea squids longer than 15 m have been observed, and squids 20 m or even 30 m in length are not improbable. Some blue whales approached weights of 200 tons before the largest members of their populations were removed by commercial whaling. These animals are exceptional and stand out in sharp contrast to the generally tiny nature of most life forms in the sea.

DID YOU KNOW?

Did you know that life evolved in the oceans over 3 billion years ago? You may ask how we know this, because humans were not around at that time. The earliest evidence of life is from fossil stromatolites that have been dated to 3.5 billion years using radiometric dating. Stromatolites are formed by cyanobacteria as they layer and bind thin films of sediment together. Fossil evidence indicates that prokaryotes were alone on Earth for nearly 2 million years before Earth was in a state where eukaryotes could survive. All of the earliest life forms evolved in the oceans, and life did not transition to land until 1 billion years ago for prokaryotes and 500 million years ago for eukaryotes. The transition to land was a long process with many necessary adaptations to make the leap out of the water. The ocean is where life began, and it is where many species still thrive today.

STUDY GUIDE

TOPICS FOR DISCUSSION AND REVIEW

1. Consider the many ways in which one could organize and classify items commonly found in the grocery store. Is there a best criterion to use under all circumstances?

2. Label the following organisms (crab, squid, shark, jellyfish, sand worm, clam, dolphin, sea anemone) with one or more of the following adjectives (benthic, infaunal, epifaunal, pelagic, nektonic, planktonic). Consider whether these labels are consistent throughout the entire life span of each species.

3. List five differences between a plant and an animal.

4. What is the taxonomic hierarchy for our species, *Homo sapiens*?

5. Compare and contrast the following three groups of prokaryotes: bacteria, cyanobacteria, and archaea.

6. Do plants perform photosynthesis, respiration, or both?

7. Organisms, including humans, perform anaerobic respiration at specific times or specific places. Provide an example activity that leads to the use of anaerobic respiration in humans.

8. Using common names (such as *dolphin* or *phytoplankton*), provide an example of a marine organism that plays each of the following ecological roles: producer, consumer, decomposer, photosynthesizer, chemosynthesizer, heterotroph, herbivore, carnivore, detritus feeder, suspension feeder, filter feeder, predator, parasite, mutualistic symbiont, and commensal symbiont. (*Hint:* Many of these terms are not mutually exclusive.)

9. Although all living organisms are fundamentally similar, generate a list of contrasting characteristics between marine and terrestrial photosynthesizers (such as a diatom and an oak tree).

KEY TERMS

adenosine triphosphate (ATP) 65
aerobic respiration 66
anaerobic respiration 66
ancestral character 59
anoxic 66
autotrophic 66
benthos 54
biome 54
biosphere 54
carnivore 66
carrying capacity 57
character 59
chemosynthetic 66
chlorophyll 65
chloroplast 65
chromosome 57
clade 59
cladistics 59
cladogram 59
commensalistic 72
community 54
consumer 66
cytoplasm 60
decomposer 66
derived character 59
detritus 66
diploid 58
ecological adaptation 55
ecosystem 54
endoplasmic reticulum 61
epifauna 54
eukaryotic 57
evolutionary adaptation 56
extant 57
fertilization 57
fitness 57
gamete 57
gametophyte 65
gene 56
gene flow 56
haploid 58
herbivore 66
heterotrophic 66
homologous 58
homology 59
hyphae 65
infauna 54
keystone species 73
lineage 59
meiosis 57
microbe 74
mitochondria 61
mutualism 72
natural selection 56
nekton 54
node 59
nucleus 61
parasitism 72
photosynthesis 65
phylogenetic 58
phylogenetic tree 59
phytoplankton 54
plankton 54
population 54
primary producer 66
producer 66
prokaryotic 60
ribosome 61
speciation event 59
species 54
sporophyte 65
symbiont 65
symbiotic relationship 69
taxa 58
taxonomy 54
trophic 65
trophic level 66
zooplankton 55
zygote 58

LIST OF KEY *GENERA*

Avicennia
Balaenoptera
Calanus
Coryphaena
Delphinus
Enhydra

Halteria	**Strongylocentrotus**
Parastichopus	**Themisto**
Sebastes	

REFERENCES

Baird, R. W., P. A. Abrams, and L. M. Dill. 1992. Possible indirect interactions between transient and resident killer whales: Implications for the evolution of foraging specializations in the genus *Orcinus*. *Oecologia* 89:125–132.

Fuhrman, J. A. 1999. Marine viruses and their biogeochemical and ecological effects. *Nature* 399:541–548.

Gerbersdorf, S. U., and H. Schubert. 2011. Vertical migration of phytoplankton in coastal waters with different UVR transparency. *Environmental Sciences Europe 2011* 23:36.

Hardy, A. C. 1924. The herring in relation to its animate environment, pt. 1. The food and feeding habits of the herring. *Fisheries Investigations*, London, Series II, 7:1–53.

Hutchinson, G. E. 1961. The paradox of the plankton. *American Naturalist* 95:137–145.

Landry, M. R. 1976. The structure of marine ecosystems: An alternative. *Marine Biology* 35:1–7.

Levinton, J. S. 2001. *Marine Biology*. New York: Oxford University Press.

Longhurst, A. R. 2001. *Ecological Geography of the Sea*, 2nd ed. San Diego: Academic Press.

Mora, C., D. P. Tittensor, S. Adl, A. G. B. Simpson, and B. Worm. 2011. How many species are there on earth and in the ocean? *PLoS Biology* 9(8):e1001127. doi:10.1371/journal.pbio.1001127

Murphy, G. I. 1968. Pattern of life history and the environment. *American Naturalist* 102:391–403.

Reese, J. B., L. A. Urry, M. L. Cain, S. A. Wasserman, P. V. Minorsky, and R. B. Jackson. 2014. *Campbell Biology*, 10th ed. San Francisco: Pearson Education.

Roughgarden, J. 1972. Evolution of niche width. *American Naturalist* 106:683–718.

Ruggiero, M. A., D. P. Gordon, T. M. Orrell, N. Bailly, and T. Bourgoin. 2015. A higher level classification of all living organisms. *PLoS ONE* 10(4):e0119248. doi:10.1371/journal.pone.0119248

Ryan, J. F., K. Pang, C. E. Schnitzler, A.-D. Nguyen, R. T. Moreland, D. K. Simmons, B. J. Koch, W. R. Francis, P. Havlak, S. A. Smith, N. H. Putnam, S. H. D. Haddock, C. W. Dunn, T. G. Wolfsberg, J. C. Mullikin, M. Q. Martindale, and A. D. Baxevanis. 2013. The genome of the ctenophore *Mnemiopsis leidyi* and its implications for cell type evolution. *Science* 13:342(6164).

Schoener, T. W. 1974. Resource partitioning in ecological communities. *Science* 185:27–29.

Sinclair, M. 1988. *Marine Populations: An Essay on Population Regulation and Speciation*. Seattle: University of Washington Press.

Smith, R. C., D. Ainley, K. Baker, E. Domack, S. Emslie, B. Fraser, J. Kennett, A. Leventer, E. Mosley-Thompson, S. Stammerjohn, and M. Vernet. 1999. Marine ecosystem sensitivity to climate change. *BioScience* 49:393–403.

Steele, J. H. 1991. Marine functional diversity. *BioScience* 41:470–474.

Valentine, J. W. 1973. *Evolutionary Ecology of the Marine Biosphere*. Englewood Cliffs, NJ: Prentice Hall.

Van Valen, L. 1974. Predation and species diversity. *Journal of Theoretical Biology* 44:19–21.

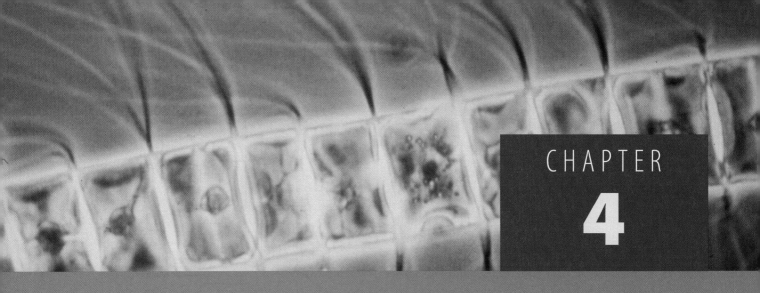

Marine Microbes

STUDENT LEARNING OUTCOMES

1. List and compare the main features of the major phyto-plankton groups.

2. Identify and describe at least one example of a symbi-otic relationship that involves phytoplankton.

3. Explain how the small size of phytoplankton is related to their ability to remain in the photic zone.

4. Describe the function of spines or other elongated structures observed in phytoplankton.

5. Explain how phytoplankton survive when conditions are less than ideal for photosynthesis.

6. Discuss the relationship between the two major phases of photosynthesis, the light reactions and the light-independent reactions.

7. Explain how marine bacteria and viruses are relevant to the lives of phytoplankton.

CHAPTER OUTLINE

A chain of marine diatoms (*Chaetoceros*) with elongated structures, lined up end to end. Chains of phytoplankton with spiny structures drift downward slowly.
NOAA Photo Library.

M uch of the special nature of life in the sea is due to the extremely small and most abundant marine primary producers, the **phytoplankton**. From providing food at the base of the food web to producing **bioluminescence** in the surf, phytoplankton are truly unique, diverse, and remarkable organisms. In this chapter, the general features of these important primary producers and other **microbes** are introduced, including features of interest for the most commonly encountered groups. We will explore the process of photosynthesis and the principal factors that affect the rates in which phytoplankton produce material to fuel the rest of the marine ecosystem. Lastly, we will discuss other marine microbes, such as bacteria and viruses, and their roles in food webs.

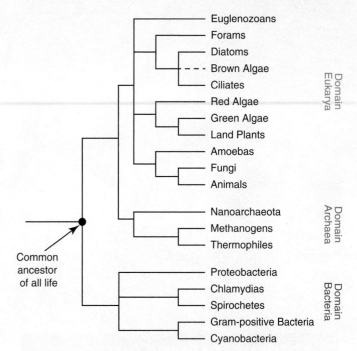

Figure 4.1 A phylogenetic tree illustrating a current hypothesis of the evolutionary relationships of the major groups of organisms based on molecular data. This tree employs the three-domain classification system, is based on phylogenetic analysis, and only includes major groups of organisms.

4.1   Phytoplankton Groups

Marine phytoplankton belong to two domains, Eukarya and Bacteria, and within those domains almost all belong to five major divisions (phyla): Cyanobacteria, Haptophyta, Ochrophyta, Bacillariophyta, and Dinophyta (**Table 4.1** and **Figure 4.1**). They are all single-celled microscopic organisms found dispersed throughout the photic zone of the oceans, where they accomplish the large majority of primary productivity in the marine environment. **Figure 4.2** compares the range of sizes of phytoplankton cells and the categories into which they are grouped. Only in the last few decades has it been possible to collect representative samples of the exceptionally small **picoplankton** and **ultraplankton**. As our knowledge of these very small phytoplankton groups improves, our understanding of their contribution to marine food webs is also increasing. Presently, it is thought that the most important primary producers in all marine environments, but especially in oceanic waters, are **nanoplankton** sized or smaller.

TABLE 4.1 Major Divisions of Marine Phytoplankton and Their General Characteristics

Division	Common Name	Approximate Number of Living Species	General Size and Structure	Photosynthetic Pigments	Storage Products	Habitat
Cyanobacteria	Blue-green algae	4,484	Unicellular, prokaryotic, nonflagellated, microscopic	Chlorophyll *a* Carotenes Phycobilins	Starch	Mostly benthic
Ochrophyta	Golden-brown algae, silicoflagellates	668,113	Unicellular, often flagellated, microscopic	Chlorophyll *a, c* Fucoxanthin	Chrysolaminarin Oils	Planktonic and benthic
Haptophyta	Coccolithophores	693	Unicellular, flagellated, microscopic	Xanthophylls Chlorophyll *a, c*	Oils	Planktonic and benthic
Bacillariophyta	Diatoms	14,008	Unicellular, nonflagellated, microscopic	Chlorophyll *a, c* Xanthophylls Fucoxanthin	Oils	Planktonic and benthic
Dinophyta (also known as Miozoa)	Dinoflagellates	3,286	Unicellular or colonial, flagellated, microscopic	Chlorophyll *a, c* Xanthophylls Carotenes	Starch Fats Oils	Planktonic

Data from algaebase.org.

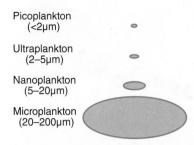

Picoplankton
(<2μm)

Ultraplankton
(2–5μm)

Nanoplankton
(5–20μm)

Microplankton
(20–200μm)

Figure 4.2 Relative sizes of phytoplankton groups. All are enlarged 1000x. At the same magnification, a human hair is as thick as this page is wide.

Cyanobacteria

Marine cyanobacteria have been the object of much recent study for their key roles in global carbon and nitrogen cycles, their potential use in "green" technologies such as biofuels, and their role as the source of chloroplasts within the earliest photosynthetic eukaryotes. Their small cell size (most are smaller than 5 μm in diameter) makes them very difficult to collect and study. Their cell structure (**Figure 4.3**) is typical of prokaryotes, with only a few of the complex membrane-bound organelles so obvious in larger eukaryotic cells. Photosynthesis in cyanobacteria is similar to that in eukaryotic autotrophs, requiring chlorophyll *a* and producing oxygen. Cyanobacteria are not newcomers to marine environments, and evidence indicates that they were the first photosynthetic organisms. Fossil stromatolites made by cyanobacteria over 3 billion years ago are remarkably similar to modern ones found at the edges of tropical lagoons in Australia (**Figure 4.4**) and the Bahamas. Marine cyanobacteria are especially abundant in intertidal and estuarine areas, with a smaller role in oceanic waters. Some species of cyanobacteria produce dense blooms in warm-water regions. The red phycobilin pigment of *Oscillatoria* is responsible for the color and name of the Red Sea.

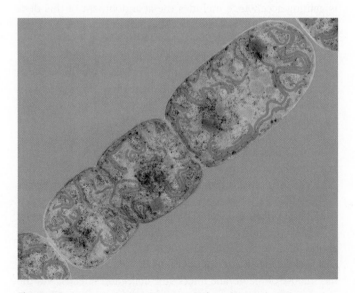

Figure 4.3 A transmission electron micrograph of a marine cyanobacterium, *Synechoccus*.
© Phototake/Alamy Images.

Figure 4.4 Stromatolites measuring over 1 m high grow in and above the water on a shallow sandy bottom of Shark Bay, Australia.
© 3Dchef/iStock/Getty Images Plus.

Benthic cyanobacteria can be found almost everywhere light and water are available. These organisms are individually microscopic and usually inconspicuous, but they may aggregate to produce large macroscopic colonies. One abundant form, *Lyngbya*, develops long strands or hollow tubes of cells nearly a meter in length. In recent years, this mermaid hair, or fire weed, has bloomed in unnatural densities in at least a dozen areas around the world, causing eye, nose, and mouth irritations and dermatitis in swimmers; harmful shading of seagrasses; and clogging of fishing nets. Fish are said to vacate areas that experience *Lyngbya* blooms. Reproduction of cyanobacteria is usually accomplished by cellular fission. Occasionally, a growing colony will fragment to disperse the cells. More complex modes of reproduction, involving motile or resistant stages, are also known.

On temperate seashores, some species of cyanobacteria develop interwoven strands (**Figure 4.5**) that appear as tarlike patches or mats encrusting rocks in intertidal or splash zones. Other species can be found in abundance on mudflats of coastal marshes, estuaries, and in association with tropical coral reefs. One well-studied cyanobacterium, *Microcoleus*, is a major component of the complex laminated microbial mats that construct modern stromatolites (Figure 4.4).

Several species of cyanobacteria are able to produce their own food by a method other than photosynthesis. This alternative method is **nitrogen fixation**, which includes the conversion of nonreactive atmospheric nitrogen (N_2) to nitrate, ammonia, or other reactive forms of nitrogen to satisfy metabolic needs. The process of nitrogen fixation is not well understood, but it is known to be limited to bacteria and archaea. The capability of some cyanobacteria to use nitrogen fixation as a metabolic process when conditions are not ideal for photosynthesis expands the potential geographic distribution of this group. Nitrogen-fixing cyanobacteria are fairly common in nearshore regions and are often associated with large marine plants, but they are rare in oceanic waters.

An important exception to this general observation of the rarity of nitrogen fixation in oceanic waters is the abundance of *Trichodesmium*, or sea sawdust, throughout low-nutrient tropical and subtropical oceans and western boundary currents

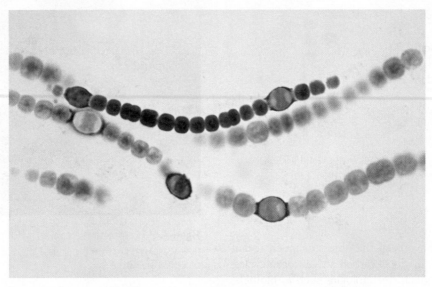

Figure 4.5 A micrograph of the colonial cyanobacterium *Anabaena*, which contains spores (akinetes) and nitrogen-fixing heterocysts along a chain of vegetative cells.
© Ed Reschke/Peter Arnold, Inc.

(such as the Gulf Stream and the Kuroshio Current). This large (0.5 × 3 mm), mat-forming, nitrogen-fixing cyanobacterium is unusual in many respects; for example, it does not have any specialized cells, such as heterocysts, for nitrogen fixation. In addition, it fixes atmospheric nitrogen under fully aerobic conditions, all the while producing molecular oxygen via photosynthesis. This should be impossible because nitrogenase, the enzyme responsible for nitrogen fixation, is inhibited by the presence of oxygen. Moreover, its rate of nitrogen fixation peaks at noon when sunlight is most intense and oxygen production is maximized. The key to the success of *Trichodesmium* is the presence of intracellular gas vesicles. These provide dynamic buoyancy that results in a **diurnal** cycle of vertical movements that may be the result of its production of relatively dense photosynthetic products during the day (so-called cell ballasting). Hence, *Trichodesmium* has a **circadian clock**, calibrated by the sun, making it only the second documented prokaryote to display an internal rhythm. These many unique characteristics help to make *Trichodesmium* the most important primary producer in open waters of the tropical North Atlantic Ocean, where it produces 165 mg C/m²/day. Additionally, it introduces 30 mg of new nitrogen per m²/day to the photic zone, an amount greater than the estimated flux of nitrate across the thermocline. When these production estimates are coupled with the fact that floating mats of *Trichodesmium* provide a unique pelagic habitat that supports a complex microcosm of bacteria, protozoans, fungi, hydrozoans, and copepods, sea sawdust is elevated to a major player in the trophic dynamics of tropical seas worldwide.

Cyanobacteria exhibit a strong tendency to form symbiotic associations with other organisms. Examples of symbiosis with animals are common, and some can even be found inhabiting marine planktonic diatoms, such as *Rhizosolenia*. Other cyanobacteria live as **epiphytes**, attached to larger plants. Some epiphytic cyanobacteria inhabit turtle grass beds along the Gulf Coast of the United States where they play an important role as nitrogen fixers in the overall fertility and productivity of these seagrass beds. These symbiotic associations are not surprising; a symbiotic relationship involving the engulfing of a cyanobacterium by a single-celled heterotrophic eukaryote is thought to be the pathway for the evolution of the chloroplast, and thus photosynthesis, in the earliest eukaryotic photosynthetic organisms.

Coccolithophores, Silicoflagellates, and Diatoms

Several phytoplankton groups have been under recent scrutiny regarding their relationships and division names. The information presented here is the most recent available (as of 2016), but it should be noted that the relationships between these groups are under review and debate. Historically, the division Chrysophyta included many freshwater algae and the marine coccolithophores, silicoflagellates, and diatoms. A new hypothesis that is gaining acceptance includes the abandonment of this division and separation of the coccolithophores, silicoflagellates, and diatoms into three divisions: Haptophyta, Ochrophyta, and Bacillariophyta, respectively. Like all other eukaryotic autotrophs, the primary photosynthetic pigment for these three divisions is chlorophyll *a* contained in their chloroplasts. In addition, organisms in these three divisions share some common features such as accessory chlorophyll *c* and golden or yellow-brown xanthophyll pigments, which lead to their golden-brown coloration. They also have mineralized cell walls or internal skeletons made of silica or calcium carbonate, providing a hard structure for members of these groups. Some species possess flagella for motility, but, like other planktonic organisms, they are not capable of swimming against ocean currents.

The coccolithophores and silicoflagellates are relatively abundant in some marine areas. Most marine coccolithophores and silicoflagellates are nanoplanktonic. Only in recent decades has the use of membrane filters, fine collection screens, and the wider application of scanning electron microscopic techniques provided us with a much better look at these very small cells (**Figure 4.6**).

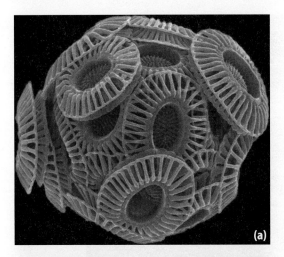

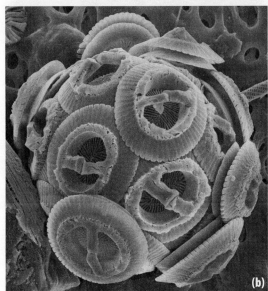

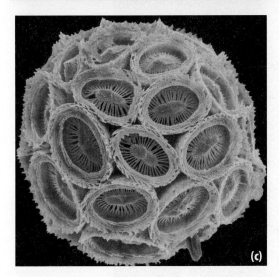

Figure 4.6 Scanning electron micrographs of three coccolithophore cells, each showing clearly their dense coverings of coccoliths. All are magnified approximately 4,500x. (a) *Emiliania huxleyi*, (b) *Gephyrocapsa oceanica*, and (c) *Coronosphaera mediterranea*.
© Steve Gschmeissner/Science Source.

Coccolithophores are unicellular, with numerous small calcareous plates, or **coccoliths**, embedded in their cell walls (Figure 4.6). Although these plates are commonly observed in marine sediments, it was not until 1898 that the photosynthetic cells (coccolithophores) producing coccolith remains from seafloor sediments were directly observed. It has been suggested that coccoliths serve as a "sunscreen" to reflect some of the abundant light in clear tropical waters, permitting these organisms to thrive in areas of very high light intensity. Coccolithophores are found in all warm and temperate seas and may account for a substantial portion of the total primary productivity of tropical and subtropical oceans. In the Sargasso Sea, for instance, a single species, *Emiliania huxleyi* (Figure 4.6a), seems to be responsible for much of the photosynthesis occurring there. However, the photosynthetic role of coccolithophores in global marine primary production is not yet well measured and is a topic of current research.

The silicoflagellates, like the coccolithophores, were first recognized and identified from fossil skeletons in marine sediments. Silicoflagellates have internal, and often beautifully ornate, silicate skeletons. They have one or two flagella and many small chloroplasts (**Figure 4.7**). The significance of silicoflagellates as marine primary producers has not been evaluated, but their contribution is thought to be small. Reproduction in coccolithophores and silicoflagellates is mostly by cellular fission.

The most obvious and often the most abundant members of the phytoplankton are the diatoms (class Bacillariophyceae). The collection of a plankton sample in any part of the world leads to a nearly guaranteed find of at least one, and potentially many, diatom species within the sample. Although diatoms are unicellular, some species occur in chains or other loose aggregates of cells. Cell sizes range from less than 15 μm to 1 mm (1,000 μm). Most diatoms are between 50 and 500 μm in size and are typically much larger than coccolithophores or silicoflagellates (**Figure 4.8**). Diatoms have a cell wall, or **frustule**, composed of pectin with large amounts (up to 95%) of silica. The frustule consists of two closely fitting halves, an **epitheca** and a thicker **hypotheca**, which fits tightly inside the epitheca (**Figure 4.9**). Planktonic diatoms usually have many small chloroplasts scattered throughout the cytoplasm, but in low light intensities the chloroplasts aggregate near the cell ends.

Diatoms exist in an immense variety of forms based on two basic cell forms. The frustules of most planktonic species appear radially symmetrical from an end view. Circular, triangular, and modified square shapes are common, and these are known as **centric** diatoms. Other diatoms, especially the benthic forms, tend to be elongated and display various types of bilateral symmetry, and are known as **pennate** diatoms. Only pennate diatoms are capable of locomotion. The mechanism for locomotion is thought to involve a wavelike motion on the cytoplasmic surface that extends through a groove (the **raphe**) in the frustule. This flowing motion is accomplished only when the diatom is in contact with another surface. Diatoms capable of locomotion are generally restricted to living near shallow water sediments or on the surfaces of larger plants and animals.

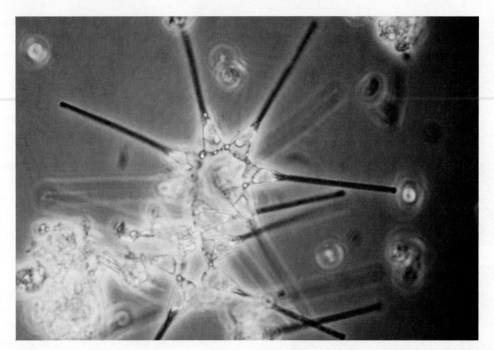

Figure 4.7 A common marine silicoflagellate slightly smaller than the coccolithophores shown in Figure 4.6.
NOAA Photo Library.

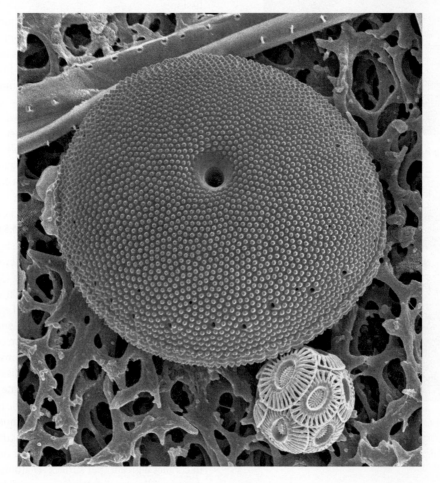

Figure 4.8 The size difference between a typical centric diatom and coccolithophore.
© Steve Gschmeissner/Science Source.

Figure 4.9 Scanning electron micrograph of *Thalassiosira*, a coastal diatom, clearly showing the epitheca, hypotheca, and a connecting girdle of cell wall material.
© Dee Breger/Photo Researchers, Inc.

The silica frustules of diatoms exhibit sculptured pits arranged irregularly or in striking geometric patterns (**Figure 4.10**). The outer pit connects with fine inner pores to facilitate exchange of water, nutrients, and waste between the diatom's cytoplasm and the external environment. The complex

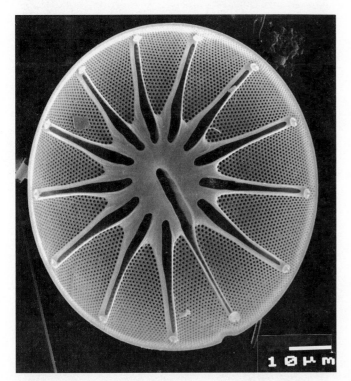

Figure 4.10 A scanning electron micrograph of a centric diatom, *Asteromphalus heptacles*.
Courtesy of Dr. José Luis Iriarte M., Universidad Austral de Chile.

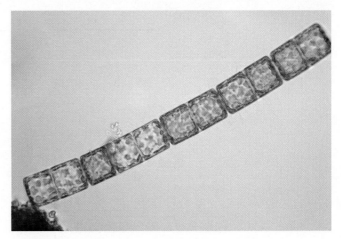

Figure 4.11 A chain of *Melosira* diatoms.
© Nancy Nehring/Getty Images.

sculpturing of diatom frustules reduces the amount of silicate required for cell wall construction while dramatically increasing its mechanical strength to resist the crushing effect of predators' jaws. The highly sculptured, yet very transparent, frustules of diatoms also act like fiber-optic light guides, possibly to direct light efficiently to the photosynthetic chloroplasts contained in the interior of the cell.

Diatoms and most other unicellular protists reproduce asexually by simple cell division. An individual parent cell divides in half to produce two daughter cells (**Figure 4.11**). This method of reproduction can yield a large number of diatoms in a short period of time. When conditions for growth are favorable, a single diatom requires less than 3 weeks to produce 1 million daughter cells. Populations of diatoms and other rapidly dividing protists thus have the capacity to respond rapidly to take advantage of improved growth conditions; however, their enormous reproductive potential usually is limited by predation or availability of light or nutrients. These limitations are addressed later in this chapter.

The sizes and shapes of diatoms with their rigid frustules create a peculiar pattern of cellular reproduction (**Figure 4.12**). During diatom cell division, two new frustule halves are formed inside the original frustule (Figure 4.12b). One is the same size as the hypotheca of the parent cell (Figure 4.12a) and is destined to become the hypotheca of the larger daughter cell (Figure 4.12c). The other newly formed frustule half is smaller, and becomes the new hypotheca for the smaller daughter cell. Each daughter cell receives its epitheca from the original frustule of the parent cell. The daughter cells grow (Figure 4.12c) and repeat the process (Figures 4.12d and 4.12e). This method of cell division efficiently recycles the old frustules, resulting in a slight decrease in the average cell size with each successive cell division. This size reduction, however, has not been observed in all natural diatom populations, suggesting that some species are able to continually adjust their cell diameter.

When cells reach a minimum of about 25% of the original cell size, these small diatoms shed their enclosing frustules (now very small), and the naked cell, known as an **auxospore**, flows out. The auxospore enlarges to the original cell size, forms a new frustule, and begins dividing again to repeat the entire

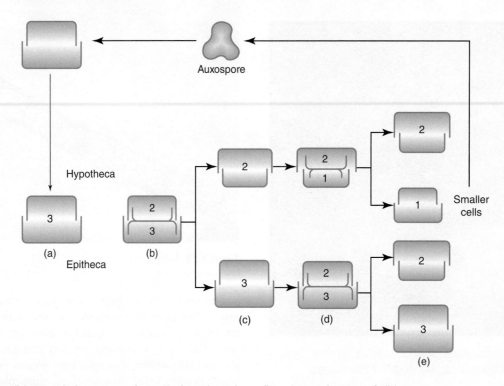

Figure 4.12 Diatom cell division and subsequent size reduction. Numbers indicate relative cell sizes. Letters indicate steps of cell division.

sequence. Occasionally, diatoms in the auxospore stage fuse with others in a form of sexual reproduction.

The variety of planktonic diatom species existing in temperate waters is impressive. **Figure 4.13** illustrates a few of the more common types. Benthic diatoms can be found on almost any solid substrate in shallow seawater: mud surfaces, rocks, larger marine plants and algae, human-made structures, and the hard shells of marine animals. One type, *Cocconeis*, even forms a thin film on the bellies of blue whales. Other benthic diatoms secrete a sticky mucilage pad to glue adjacent cells into complex chains and branching colonies a few centimeters long (**Figure 4.14**).

When compared with planktonic diatoms, the geographic distribution of benthic diatoms is very limited because of their need for light and for solid substrates. Still, benthic diatoms make a significant contribution to the total amount of primary production in estuaries, bays, and other shallow water areas. Some species of these diatoms are also key components in the **ecological succession** of species that grow on docks, boats, and other human-made structures. Although diatoms are quite small in size and one of the first groups of organisms to grow on human-made structures, studies have found that marine bacteria are usually the very first organisms to settle and grow on new or freshly denuded underwater structures. Development of a diatom film a few cells thick quickly follows bacteria and is succeeded by more complex populations of larger algae and invertebrate animals.

Dinophyta

The division Dinophyta includes dinoflagellates that are mostly marine autotrophs, but many freshwater species exist. A few species are not photosynthetic, and, like other marine heterotrophs,

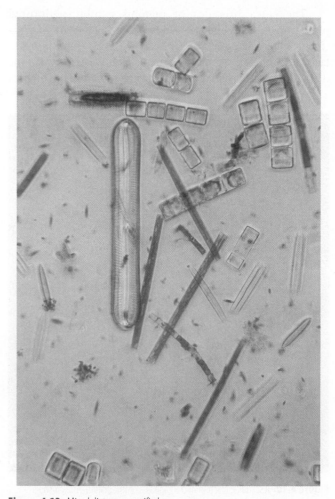

Figure 4.13 Mixed diatoms magnified.
© Comstock Images/Getty Images.

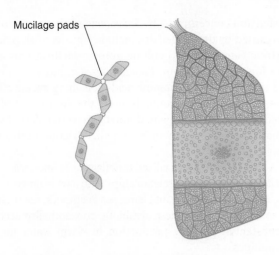

Mucilage pads

Figure 4.14 A benthic diatom, *Isthmia*, forming long complex chains of cells. A close-up of a single cell is shown to the right.

they obtain energy from organic compounds dissolved in seawater or by ingesting particulate bits of food; however, most marine dinoflagellates are photosynthetic, and their share of the total marine plant production is significant. In warm seas, dinoflagellates often contribute more to primary production than diatoms do. Dinophyta includes members that happen to possess a diversity of unique characteristics that affect the lives of humans. These characteristics will be discussed below.

Dinoflagellates are typically unicellular, with a large nucleus, two flagella, and several small chloroplasts containing photosynthetic pigments similar to those of diatoms

(**Figure 4.15**). Dinoflagellates are able to move with the assistance of two **flagella** (singular form: *flagellum*). One broad ribbon-like flagellum encircles the cell in a transverse groove and spins the cell on its axis. The other flagellum projects forward and pulls the cell, providing forward motion. Cell sizes range from 25 to 1,000 μm. In armored forms, the cell wall consists of irregular cellulose plates arranged over the cell surface. The plates may be perforated by many pores. Spines, wings, horns, or other ornamentations also may decorate the cell wall.

Dinoflagellates most commonly reproduce asexually by longitudinal cell division. Each new daughter cell retains part of the old cell wall and quickly rebuilds the missing part after cell division. Intermittent sexual reproduction has been reported in a few species; it is rapid and usually occurs in the dark, making it difficult to observe in natural conditions. In good growing conditions, the rate of cell division is extremely rapid and is similar to that of diatoms. Under optimal growth conditions, dense concentrations of dinoflagellates are produced very quickly, known as a **plankton bloom**. Plankton blooms may occur for all types of phytoplankton, although some species are known to produce blooms of great magnitude. Cell concentrations in these blooms are often so dense (up to 1 million cells/L) that they color the water red, brown, or green.

At night, dense blooms of bioluminescent forms (such as *Noctiluca* or *Ceratium*) become visible as a faint glow when disturbed by a ship's bow, a swimmer, or a wave breaking onshore (**Figure 4.16**). This bioluminescent glow is often highlighted by pinpoint flashes of light from larger crustaceans or ctenophores. The biological production of light occurs in several species of

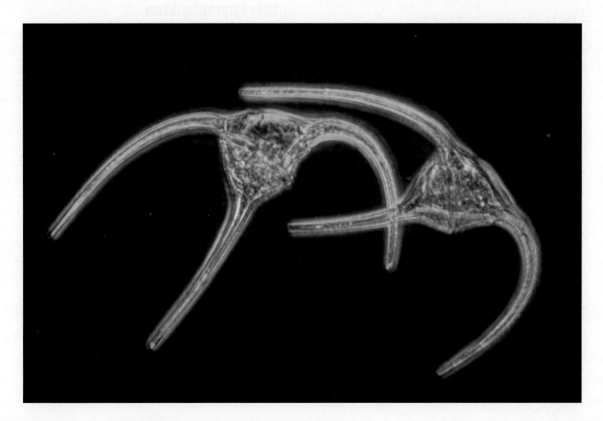

Figure 4.15 This light micrograph of a dinoflagellate, *Oxytoxum*, illustrates the major cellular features.

Figure 4.16 Bioluminescence display during a plankton bloom dominated by dinoflagellates.
© bjonesmedia/Getty Images.

dinoflagellates, some marine bacteria, and all major phyla of marine animals. Bioluminescence is produced when luciferin, a relatively simple protein, is oxidized in the presence of the enzyme luciferase. The light-producing reaction is a very energy-efficient process, producing light but almost no heat. In some species of the dinoflagellate *Gonyaulax* light production follows a daily rhythm, with maximal light output occurring just after midnight.

In the warm coastal waters of the East and Gulf Coasts of the United States, the dinoflagellate *Karenia* produces toxins that in bloom conditions are known as toxic **red tides**, or preferably, **harmful algal blooms (HABs)**. HABs can cause high mortality in fish and other marine vertebrates, and may be caused by a variety of microbes. These dinoflagellate toxins either interfere with nerve functions, resulting in paralysis, or irritate lung tissues of air-breathing vertebrates, including humans. Widespread mortality of coastal fish and marine mammals sometimes occurs after particularly intense HABs, fouling beaches and nearshore waters with their decomposing bodies.

Another form of indirect toxicity associated with dinoflagellates is created when animals (particularly shellfish) feed on dinoflagellates and accumulate toxins that make their flesh toxic. People who eat butter clams (*Saxodomus*) during the summer, for instance, occasionally experience paralytic shellfish poisoning from saxitoxin; however, this toxin is actually produced by the dinoflagellate *Alexandrium*, which is ingested and concentrated by *Saxodomus*. Other toxic conditions caused by dinoflagellates include ciguatera fish poisoning (from *Gambierdiscus*) and neurotoxic shellfish poisoning (from *Karenia*). A diatom, *Pseudo-nitzschia*, has been implicated in several recent domoic acid–poisoning events affecting marine birds and mammals along the West Coast of the United States. Like saxitoxin, domoic acid is transferred to its ultimate victims through an intermediate victim, in this case herbivorous krill.

A small group of specialized dinoflagellates known as *zooxanthellae* form symbiotic relationships with a wide range of animals, including corals, giant clams, sea anemones, sea urchins, and some flatworms. These symbiotic zooxanthellae account for substantial primary production in warm water marine communities.

DID YOU KNOW?

Dinoflagellates are absolutely crucial to the health of coral reefs! The tiny critters called *polyps* living inside the large, hard coral structure depend on photosynthesis by dinoflagellates as a main food source. You may have heard the term **coral bleaching**. Coral bleaching occurs when corals lose their dinoflagellate symbionts. It is not just the coral that benefits from this relationship though. Dinoflagellates are provided with a safe place to live, on the polyp of a coral. It's a win-win relationship that greatly affects the entire coral reef ecosystem.

Other Phytoplankton

With sampling and microscopic techniques continuing to improve, tiny phytoplankton from several other taxonomic groups are being recognized as important contributors to the trophic systems of many marine communities. Discoveries of new species also lead to reworking of **phylogenetic relationships** and division memberships for these tiny organisms. Members of two additional classes of Chrysophyta and three unicellular classes of the division Chlorophyta are sometimes found in filtered samples of coastal seawater. Chlorophytes are much more common in freshwater, and that is where most of the research on this group is concentrated. Because most of the identified marine unicellular chlorophytes have been obtained from estuaries and coastal waters, freshwater origins for many of the species found in seawater samples are likely. **Table 4.2** compares the distribution of cell sizes for the major groups of marine phytoplankton, based on the size terms used in Figure 4.2.

TABLE 4.2 Size Ranges of the Major Groups of Marine Phytoplankton

	Cyanobacteria	Diatoms	Silicoflagellates	Coccolithophores	Dinophyta	Chlorophyta
Picoplankton	+	+	+	+		+
Ultraplankton	+	+	+	+		+
Nanoplankton		+	+	+	+	+
Microplankton		+			++	

Modified from Platt, T., Li, K. W. K. (eds.) (1986). Photosynthetic picoplankton. Canadian Bulletin of Fisheries and Aquatic Science 214: 1–583.

4.2   Special Adaptations for a Planktonic Existence

Survival, and therefore evolutionary success, of all phytoplankton hinges on their ability to obtain sufficient nutrients and light energy from the marine environment for photosynthesis. Phytoplankton cells must be widely dispersed in their seawater medium to increase their ability to absorb dissolved nutrients, yet they must remain in the relatively restricted photic zone to absorb sufficient sunlight. These opposing conditions for successful planktonic existence have established some fundamental characteristics to which all phytoplankton and, indirectly due to food web dynamics, all other marine life have become adapted.

Phytoplankton have little or no ability to move horizontally under their own power and must depend on the ocean's surface currents for dispersal. Adaptive features that prolong their time spent in the horizontally moving surface currents, including eddies, or tidal waters located within the photic zone also serve to increase their geographic distribution.

Size

One of the most characteristic features of all phytoplankton is their small size. Almost without exception they are microscopic, which suggests that smallness must provide a strong selective advantage for phytoplankton. Why is this so? In contrast to land plants, phytoplankton are constantly bathed in seawater that not only provides nutrients and water, but also carries away waste products. Exchange of these materials in water is accomplished by diffusion directly across the cell membrane. A thin membrane allows for more rapid diffusion across the surface, and small size allows for easier movement of molecules once they have entered the cell. More of the body of a phytoplankton is exposed to or very close to the outside environment because the overall organism is tiny. As a result, moving molecules from the outside environment to any part of the cell, even the absolute center, is much easier, often requiring no energy. Contrast the above scenario involving phytoplankton with movement of molecules into and throughout a large shark. Moving molecules from the water to the absolute center of a shark is a much larger distance, and requires specialized processes, including specialized structures such as gills, for example. Specialized structures require energy to use, and thus are energetically costly. Phytoplankton, or any smaller organisms, are able to take advantage of their small size to save energy while easily moving vital molecules into and within their cells.

The quantity of materials required by the cell depends on factors such as the rate of photosynthesis and growth, but if these factors are held constant the basic material requirements of the cell are proportional to the size or, more precisely, to the volume of the cell; however, the ability of the cell to satisfy its material requirements is not a function of its volume but rather the extent of cell surface across which the materials can diffuse (as described above). Thus, the ratio of cell surface area to cell volume is crucial to these small cells. Smaller cells with higher surface area-to-volume ratios achieve an advantage in the competition to enhance diffusive exchange between their internal and external fluid environments (**Figure 4.17**). In the same way, a nanoplankton-sized diatom 10 μm in diameter has a 10-fold larger surface area-to-volume ratio than a microplankton-sized diatom with a cell diameter of 100 μm, and it presumably has an equally large advantage when competing for nutrients.

A reduction of cell size is an effective and widespread means of achieving high surface area-to-volume ratios, but there are

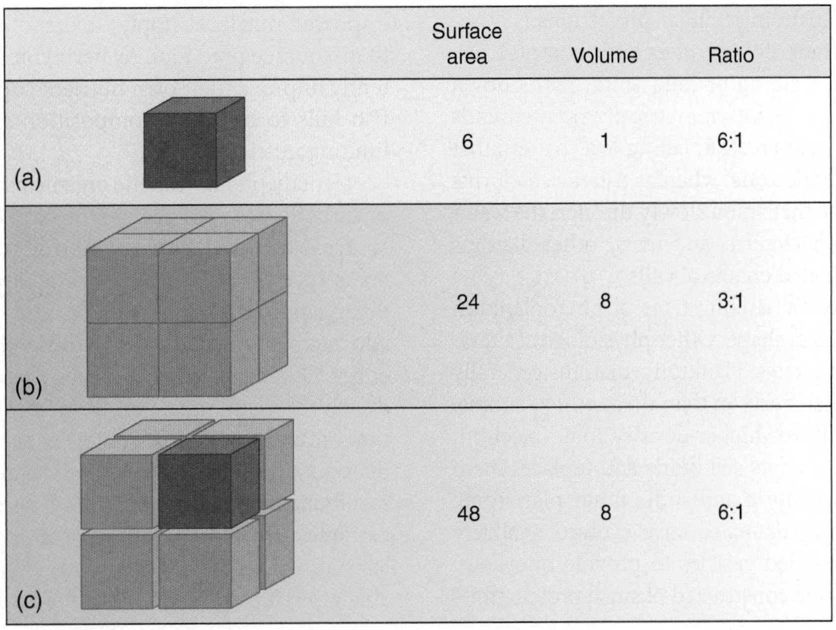

	Surface area	Volume	Ratio
(a)	6	1	6:1
(b)	24	8	3:1
(c)	48	8	6:1

Figure 4.17 With increasing size (a and b), the ratio of surface area to volume decreases, unless the larger structure (c) remains subdivided so that the interior surfaces are exposed.

other means. Many phytoplankton species have evolved complex shapes and structures that increase the surface area while adding little or nothing to the volume. Cell shapes resembling ribbons, leaves, or long bars, and cells with bristles or spines are all common mechanisms to increase the amount of surface area exposed to the outside environment without adding much to their volume. Cell vacuoles filled with seawater are common in diatoms. These vacuoles make cells larger, but the actual volume of living protoplasm requiring nutrients is only a fraction of the total volume of the cell.

Long spines and horns may also make some phytoplankton less desirable to herbivorous grazers. For example, some evidence suggests that copepods, herbivorous zooplankton, prefer nonspiny diatoms to spiny ones. Spines, cell chaining, and cell elongation all may be economical methods of increasing apparent cell size to discourage predators and reduce mortality.

Sinking

Phytoplankton cells are generally a bit denser than seawater and tend to sink away from surface waters and sunlight. In contrast to what one may think based on the known sunlight requirement for phytoplankton, the problem for phytoplankton is to not float, because floating would create intense crowding and competition at the sea surface for light, nutrients, and space. Instead, phytoplankton need to sink, but sink slowly so that some small portion of any reproducing cell line has at least a few members carried upward by turbulent mixing even as most continue their slow downward slide through the photic zone.

Phytoplankton have many adaptations that slow the sinking rate and prolong their trip through the photic zone. One very effective method already discussed is to increase frictional resistance to sinking through water by increasing surface area-to-volume ratios with small cell sizes or the production of spines or other surface-expanding cellular projections.

Other cells reduce their sinking rates with complex cell or chain shapes that trace zigzag or long spiral paths down through the water column. The asymmetrically pointed ends of individual *Rhizosolenia* cells create a "falling-leaf" pattern that prolongs its stay in the photic zone, whereas *Asterionella* forms long curved chains of cells that spiral slowly through the water (**Figure 4.18**). *Eucampia*, *Chaetoceros*, and many other diatoms form similar spiraled or coiled chains of cells.

Adaptations for reducing sinking rates of phytoplankton are not limited to variations in shape. Other physiological alterations greatly reduce sinking rates. Planktonic diatoms generally produce thinner and lighter frustules than do benthic diatoms. The diatom *Ditylum* excludes higher-density ions (calcium, magnesium, and sulfate) from its cell fluids and replaces them with less dense ions. *Oscillatoria* and some other planktonic cyanobacteria, such as *Trichodesmium*, have evolved relatively sophisticated internal gas-filled vesicles to provide buoyancy. The walls of these vesicles are constructed of small protein units

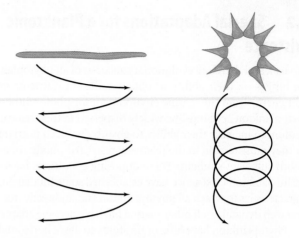

Figure 4.18 Sinking patterns of the elongate diatom, *Rhizosolenia* (left), and the spiral chain-forming diatom, *Asterionella* (right).

that can withstand outside water pressures experienced anywhere within the photic zone.

Adjustments to Unfavorable Environmental Conditions

The optimal growth period for phytoplankton in nonupwelling temperate and polar seas is limited by reduced sunlight in winter and limited nutrient supplies in summer. Faced with the prospect of weeks or months with reduced photosynthesis, phytoplankton in these regions have limited options. Some simply move, some switch to other energy sources, and others simply persist until conditions improve. The first strategy does not generally apply to diatoms, but motility, limited as it is, is extremely important to phytoplankton with flagellated cells. A swim of merely one or two cell lengths is often sufficient to place the cell away from its excreted wastes and into an improved nutrient supply. Toxins of dinoflagellates also serve to discourage predation by herbivores and sometimes inadvertently improve their own nutrient supply by causing extensive fish kills to hasten decomposition and more quickly renew limiting nutrients.

Strictly photosynthetic organisms must rely on stored lipids or carbohydrates for their short-term energy needs. When that source is depleted, some phytoplankton still have alternatives. Some species can improve their ability to harvest light by producing more chloroplasts that contain photosynthetic enzymes and pigments or by moving those chloroplasts closer to cell edges. Other species can actually absorb dilute but energy-rich dissolved organic material from surrounding seawater to tide them over. When these strategies have been exhausted, many diatoms produce dormant cysts, capsules that have reduced metabolic activity and increased resistance to environmental extremes (**Figure 4.19**). Many nearshore species of dinoflagellates also produce dormant stages during periods of unfavorable growth conditions. With the return of improved growing

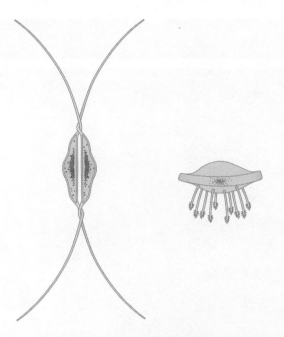

conditions, these dormant cells germinate and commence photosynthesis and growth. It is these adaptations allowing for survival during times of less than ideal conditions that contribute to the great success of phytoplankton.

Figure 4.19 Inactive resistant stages of two species of *Chaetoceros*.

DID YOU KNOW?

Some phytoplankton exhibit behaviors to evade predators. It has been known for years that phytoplankton with flagella can swim toward or away from light, but never before has swimming behavior been associated with escaping being eaten. A recent study on predator–prey interactions between a golden-brown phytoplankton and its zooplankton predator showed just that; the phytoplankton swam at greater speeds when the predator was present. No other photosynthesizing organisms are known to do this.

Case Study

Curious Coccolithophores

Coccolithophores are unique phytoplankton in several ways. First, and most obvious when observed under a microscope, they look nothing like other phytoplankton, with their round shape and coccolith armor that gives the group its name. The armor is actually a series of hard plates called *coccoliths* that are produced internally, pushed outside of the cell membrane, and then cemented in place (Figure 4.6). Second, they are primary producers with an added advantage for carbon sequestration: not only do coccolithophores use CO_2 for photosynthesis like all photosynthetic organisms, but they also trap additional carbon in their $CaCO_3$ coccoliths. Although it is well known that coccolithophores are important primary producers, this group was excluded from the news headlines until recently when they were highlighted for their strange behavior during plankton blooms; as coccolithophores produce new coccoliths, they shed the old ones, leading to ocean waters full of milky white shells.

In general, phytoplankton blooms are not uncommon and are often very visible. The blooms that gain the most attention are red tides caused by dinoflagellates, which lead to a deep red–colored sea, and often glowing bioluminescence at night (Figure 4.16). The most commonly occurring blooms are a little less extravagant and are caused by various phytoplankton that produce a green chlorophyll signature visible in satellite images (Figure 4.22). Less common, and the focus of our discussion, are blooms of coccolithophores that appear reflective, white, or milky. The reflection and white hue of the blooms in question lead to an ocean color that is lighter than normal, often a bright turquoise color, and is visible in satellite images.

It is the coccoliths on the outer layer of the coccolithophore cell that produce the reflective chalky white appearance when the layer is shed by the organism. This chalky-colored shell actually leaves behind a chalklike substance when the organism sheds the coccolith or dies, and when millions shed their coccoliths or die layers of chalk may build up on the ocean floor. Some areas where the seafloor is lifted above the ocean display a buildup of layer upon layer of coccoliths that create chalk cliffs. These cliffs take hundreds or thousands of years to form, and are only really visible when they are above water (**Figure A**).

Figure A Moens Cliff in Denmark, formed by fossilized coccoliths that built up over millions of years to create these chalk cliffs that tower over 128 m above sea level.
© Dan Bach Kristensen/Shutterstock, Inc.

(continues)

Case Study *(continued)*

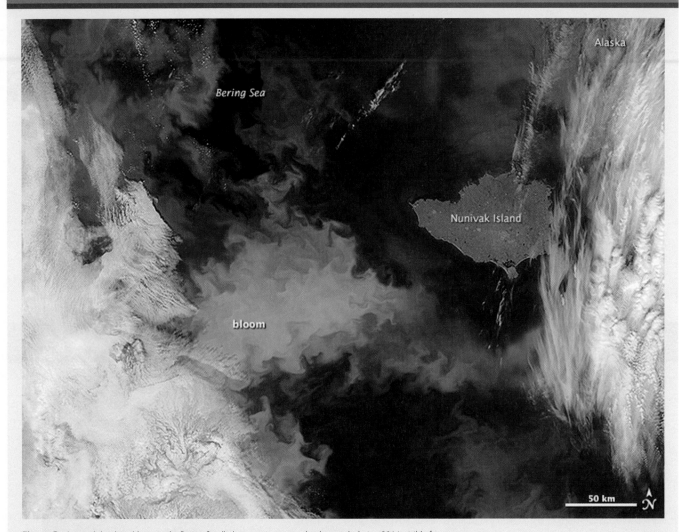

Figure B   A coccolithophore bloom in the Bering Sea (light turquoise-green cloudy water), during 2014, visible from space.
NASA Visible Earth EOS Project.

Although thought to be uncommon, these "white tides" caused by coccolithophore blooms have been observed more frequently in recent years. A strong event that took place in the Bering Sea near Alaska was identified from NASA satellite imagery in 2014 (**Figure B**) and was visible to astronauts in space for several days. Prior to the late 1990s, coccolith blooms in this area were rare, but after 1997 blooms occurred annually for the next 5 years and still remain common. Scientists hypothesize that different weather patterns and ocean mixing frequency are leading to the increased coccolithophore blooms.

A strong and rare event was identified along the coast of Santa Barbara, California, in June 2015. Coincidentally, this bloom occurred just days after an oil pipeline failed and spewed thousands of gallons of crude oil into the ocean just north of the coccolithophore bloom area. The oil company responsible for the spill quickly began to attempt to clean up the mess (**Figure C**), and scientists at the University of California–Santa Barbara quickly collected samples to identify the species of coccolithophore in the nearby plankton bloom. They also began to look for any possible connections between the oil spill and the coccolithophore bloom event. At the time of this publication no clear connection had been found between the two events, but it is possible that conditions in the ocean were altered from the oil spill, indirectly causing the coccolithophore bloom.

Although it would appear that coccolithophores may be negatively affected by ocean acidification because they use calcium carbonate ($CaCO_3$) to produce their coccoliths, the evidence supporting this assumption conflicts. A large amount of research has been dedicated to assessing the potential impacts of ocean acidification on all marine taxa, and several studies have concluded that phytoplankton with hard parts may be less susceptible to negative impacts of a reduction of available $CaCO_3$ than other organisms, such as marine snails. The results of laboratory experiments seem to be very species specific, with some coccolithophore species forming thinner shells in ocean acidification conditions, some species displaying no difference, and at least one species growing thicker shells. Laboratory studies are limited in their scope, because conditions in the lab are never exactly the same as those in nature. One field study measuring the thickness of coccoliths left behind did find that coccoliths from organisms living during an earlier global warming event had thinner or smaller shells during

Case Study *(continued)*

Figure C Cleanup efforts after a devastating oil spill off the coast of northern Santa Barbara County, California.
© Courtesy of Deanna Pinkard-Meier.

this warmer time, but the differences were not extreme, and other unidentified variables may have caused the different shell characteristics. Coccolithophores and other marine phytoplankton appear to take advantage of higher CO_2 levels for photosynthesis, and, although results are unclear at this point, it is likely that phytoplankton are negatively impacted by the shifting chemistry that occurs with higher CO_2 levels in some way, even if indirectly.

Critical Thinking Questions

1. Identify three unique features of coccolithophores and hypothesize why each unique feature may have contributed to this group's success over evolutionary history.

2. Current research suggests that marine algae are less affected by ocean acidification than shell-forming marine animals. Create a hypothesis explaining why you think this may be the case.

4.3 Primary Production in the Sea

The two major categories of autotrophs in the sea, the pelagic phytoplankton and the attached benthic plantlike macroalgae, differ in much more than their size and physical appearance. These differences reflect adaptations to the very different physical and chemical terrains of the benthic and pelagic divisions of the marine environment. The narrow sunlit benthic fringe of the ocean is home to a variety of large, relatively long-lived, attached marine algae. Yet these macroalgae account for only about 5% to 10% of the total amount of photosynthetically produced material in the ocean each year.

From our shore-based perspective, relatively large benthic macroalgae gain immediate attention because of their high **standing crop** (the amount of plant material alive at any one time), but this is a poor indicator of their share of overall primary production. Most marine primary production is accomplished by the small, dispersed, pelagic phytoplankton. On an oceanic scale, the larger nearshore macroalgae are only minor players in the process of marine photosynthesis.

The term **primary production** is more or less interchangeable with *autotrophy* or *photosynthesis*; it is the biological process of creating high-energy organic material from carbon dioxide, water, and other nutrients. The organic material synthesized by the primary producers ultimately is transferred to other trophic levels of the ecosystem. Consider this example: A neatly trimmed lawn contains an easily measured amount of living plant material, its standing crop. If the lawn is maintained throughout a summer, it will be periodically mowed to maintain the same height or, in other words, the same standing crop. During that summer, the lawn clippings will total much more than the standing crop, but the lawn clippings are not part of the lawn. They represent the primary production that occurred during the summer. In an analogous sense, it is the rapidly consumed phytoplankton production, with typically very low standing crops, that fuels the metabolic processes of most of the consumers living in the sea.

Standing crop sizes at any given moment are governed by a balance between crop increases (cell growth and division) and crop decreases (sinking and grazing). Most of the primary production of a healthy, actively growing phytoplankton population is not used in **cellular respiration** but instead contributes to the existing standing crop. Old populations or healthy cells in poor growing conditions use a larger portion of their gross production in cellular respiration, and net production declines.

As in our lawn analogy, the standing crop of a healthy phytoplankton population measured on successive days may demonstrate little or no increase, suggesting that no net production occurred from one day to the next. A more likely explanation is that significant net production did occur, but it replaced the portion of the crop lost to grazers (the "mowers") and to sinking. Thus, the relationship between standing crop and productivity depends to a large degree on the **turnover rate** of newly created cells.

The turnover rate of phytoplankton populations is typically extremely rapid. In good growing conditions, many species of large phytoplankton divide once each day, and several of the smaller species divide even faster. The coccolithophore shown in Figure 4.6a, for instance, undergoes almost two divisions per day. Its population can be completely replaced, or turned over, twice each day, and thus comparatively few cells exist in the water at any one time. Even higher turnover rates are expected for the smaller picoplankton and ultraplankton and in benthic algae.

The total amount of organic material produced in the sea by photosynthesis represents the **gross primary production** of the marine ecosystem. Gross primary production is difficult to measure in nature; nonetheless, it is useful as a base of reference for understanding the production potentials of marine communities and ecosystems. A portion of the organic material produced by photosynthesis is used in cellular respiration by photosynthesizers to sustain their own life processes. Any excess production is used for growth and reproduction and is referred to as **net primary production**. Net marine primary production represents the amount of organic material available to support the consumers and decomposers of the sea. Different types of living material contain various proportions of water, minerals, and energy-rich components. To avoid some of the problems encountered when comparing different types of primary producers, we commonly report standing crops in grams of organic carbon (g C). This unit represents approximately 50% of the dry weight and 10% of the live, or wet, weight of the standing crop. Primary production rates are listed in units of grams of organic carbon fixed by photosynthesis under a square meter of sea surface per day or per year (g C/m² per day or g C/m² per year, respectively).

The following discussion of the global aspects of marine primary production is necessarily biased toward phytoplankton, but the general concepts discussed here also apply to the attached macroalgae.

Measurement of Primary Production

Rates of primary production in the sea vary widely in time and in space, and animals that rely on the autotrophs for food must adapt to those patterns of variation. These production rates, and the ecological factors that affect them, have become clearer with the development of techniques for measuring primary production in the sea. Theoretically, the net photosynthetic rate of a phytoplankton population can be estimated by measuring the rate of change of some chemical component of the photosynthetic reaction, such as the rate of O_2 production or CO_2 consumption by phytoplankton.

The light bottle/dark bottle (LB/DB) technique was the classic approach used to study primary production in marine phytoplankton throughout much of the last century. With this method, measured changes in O_2 consumption and production were used to estimate phytoplankton respiration and photosynthetic rates. **Figure 4.20** describes an idealized version of the LB/DB concept and is used here to introduce some of the basic concepts involved in directly measuring marine primary productivity.

The extraordinary amount of time and material resources that were needed to conduct a LB/DB study meant that locations actually sampled in a single study might be several hundred kilometers and many days apart. Environmental changes that occurred as the research vessel steamed from one station to the next could not be measured nor were the details between stations examined. It simply was assumed that the data collected at the sample stations could be averaged over the vast areas between stations and between sampling periods. The complexity and richness of small- to moderate-scale spatial variations in phytoplankton abundance were missed, as were the day-to-day variations occurring at any sampling station. Even when a procedure that used radioactive carbon (^{14}C) as a tracer of CO_2 in photosynthesis was introduced in the middle of the last century, most of the logistical limitations of the LB/DB approach remained.

In contrast to ship-based sampling, satellite sampling can provide a general, and instantaneous, overview of a large portion of ocean (**Figure 4.21**). Satellites cannot directly measure marine primary productivity. Instead, subtle changes in ocean surface color, which signify fluctuations in population densities of various types and quantities of marine phytoplankton, are observed and measured. The more phytoplankton present, the greater the concentration of chlorophyll pigments and the greener the water. In the 1980s, a coastal zone color scanner (CZCS) aboard the NIMBUS 7 weather satellite was used to make the first global-scale

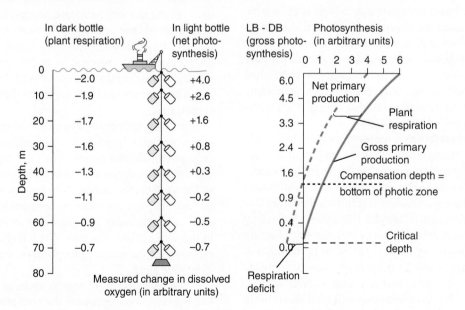

Figure 4.20 The results of a hypothetical light bottle (LB) and dark bottle (DB) experiment. Water samples from 10-m depth increments are replaced at original depths in paired LBs and DBs (left). After a period of time, the bottles are retrieved and changes in O_2 are determined. DB values indicate O_2 decreases at each depth due to respiration without photosynthesis. LB values represent O_2 changes from photosynthesis and respiration (net primary production) in the light. The difference between the two values (LB – DB) is the gross primary production. With these values, net and gross primary production curves (right) can be drawn to represent the variation in photosynthesis with depth.

Figure 4.21 Composite satellite views of the North Atlantic Ocean along the northeast coast of the United States. (Top) Phytoplankton concentrations, ranging from low (dark blue) to high (red). (Bottom) Corresponding sea surface temperature of the area shown, ranging from warm (red) to cold (dark blue). Generally, phytoplankton concentrations are highest where water is coldest.
Courtesy of NASA.

measurements of ocean surface color that, with calibration from ship-based direct measurements, were used to estimate phytoplankton standing crops and growth rates and to extrapolate shipboard productivity measurements to large oceanic areas.

Remote sensing of ocean color by satellite was the first technique to measure marine primary productivity on a global scale with enough resolution to permit analyses of phytoplankton changes over time scales of weeks or years. Since the CZCS was lofted into orbit, satellite imagery has revolutionized our view of primary productivity patterns in the ocean. As is apparent in Figure 4.21, distribution patterns of phytoplankton are complex and show some similarities with sea-surface temperature distributions. Patches and eddies of phytoplankton are common. In upwelling areas, plumes of phytoplankton-rich water extend as much as 200 km offshore. Before satellite observations, these variations in marine phytoplankton distribution, which dominate the photic zone, were completely unknown (**Figure 4.22**).

Since the CZVS ceased operation in 1986, a variety of additional satellites have been launched to provide ocean color data. For about one decade the sea-viewing wide field-of-view sensor (SeaWiFS) operated as a follow-on sensor to the CZCS. The SeaWiFS mission was a part of NASA's Earth Science Enterprise, which was designed to understand our planet better by examining it from space. The SeaWiFS sensor provided 1-km resolution at the sea surface in eight color bands. Because this orbiting sensor was able to view every square kilometer of cloud-free ocean every 48 hours, these

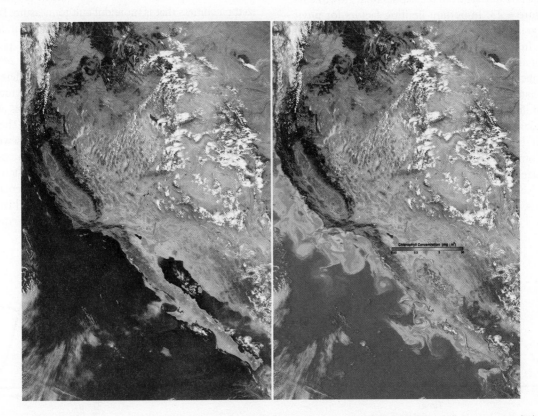

Figure 4.22 A phytoplankton bloom along the California coast imaged by SeaWiFS during the summer for true color (left) and for chlorophyll *a* concentrations (right), which are representative of the abundance of phytoplankton. The true-color image on the left is impressive, but it does not provide the information about the quantity of phytoplankton that the image on the right does.
Courtesy of GeoEye and NASA SeaWiFS Project.

satellite-acquired ocean-color data constituted a valuable tool for determining the abundance of ocean biota on a global scale and could be used to assess the dynamics of short-term changes in primary productivity patterns. The image in Figure 4.22 is from the SeaWiFS system. The newest satellite mission, Pre-Aerosol, Clouds and ocean Ecosystem (PACE), is set to launch any day. The focus of this satellite mission is to provide extended data records on ocean ecology, the carbon cycle, and clouds and aerosols.

Factors That Affect Primary Production

Continued photosynthesis by marine phytoplankton depends on a set of interacting biotic (biological) and abiotic conditions. If nutrients, sunlight, space, and other conditions necessary for growth are unlimited, phytoplankton population sizes increase in an exponential fashion (**Figure 4.23**). In nature, phytoplankton populations do not continue to grow unchecked, as the unlimited growth curve in Figure 4.23 suggests. Rather, their sizes are controlled by their tolerance limits to crucial environmental factors (including predators) or by the availability of substances for which they have a need. Any condition that exceeds the limits of tolerance or does not satisfy the basic material needs of an organism establishes a check on further population growth and is said to be a **limiting factor**.

Phytoplankton populations limited by one or a combination of these factors deviate from the exponential growth curve shown in Figure 4.23. Important limiting factors for phytoplankton are grazing by herbivores and the availability of light and nutrients. In the ocean, each major group of phytoplankton responds differently to combinations of these factors. In general, diatoms and silicoflagellates thrive in lower light intensities and colder water than do dinoflagellates and coccolithophores. Consequently, conditions that promote the growth of either group tend to exclude the other. These factors are examined below, first alone and then in concert, in an attempt to describe some sense of the complex dynamic interactions that exist between photosynthesizers and their immediate surroundings. These same factors also influence or regulate the growth of large multicellular plants.

Grazing

The trophic interrelationships of marine primary producers and their small herbivorous grazers (mostly zooplankton and small fish) can be complex. Intensive grazing can decrease the standing crop and sometimes the productivity of a phytoplankton population. Ideally, grazing rates should adjust to the magnitude of primary productivity to establish a rough balance between producer and consumer populations. Photosynthetic rates do limit the average size of the animal populations that primary producers support, and short-term fluctuations of both phytoplankton and grazer populations typically occur. The magnitude of these fluctuations tends to be moderated somewhat by stabilizing **feedback mechanisms** between all trophically related populations. An abundant food supply permits the grazers to reproduce and grow rapidly (**Figure 4.24**). Eventually, however, they consume their prey more quickly than the prey can be replaced. Overgrazing reduces the phytoplankton population and its photosynthetic capacity, causing food shortages, starvation, and consequent reductions of the enlarged herbivore populations. When grazing intensity is reduced after a herbivore population crash, the phytoplankton population may recover, increase in size, and again set the stage with an abundant food source to cause a repeat of the entire cycle. Similar fluctuations in population size may reverberate through many trophic levels of the food web.

In addition to the large-scale geographic variations in phytoplankton density observed from satellites (Figures 4.21 and 4.22), marine phytoplankton also exhibit much smaller-scale localized patchiness that is more difficult to measure. Dense patches of phytoplankton tend to alternate with concentrated patches of zooplankton. The inverse concentrations of phytoplankton and zooplankton densities displayed in Figure 4.24 stem, in part, from the effects of grazing and because of differences in their reproductive rates. Initially, a dense patch of phytoplankton provides favorable growth conditions for herbivores attracted from adjacent water into the phytoplankton patch. The grazing rate increases in the area of the patch and declines elsewhere. Production soon decreases in the original patch and increases in adjacent areas. Eventually, the original phytoplankton patch is eliminated by the increasing numbers of grazers. The adjacent areas become the new phytoplankton patches and attract herbivores from the recently overgrazed region to repeat the entire sequence.

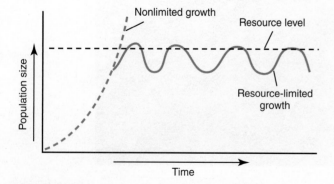

Figure 4.23 Patterns of population growth when resources are limited (solid line) and unlimited (dashed line). The dashed horizontal line represents the population level that can be supported by the amount of limited resource available.

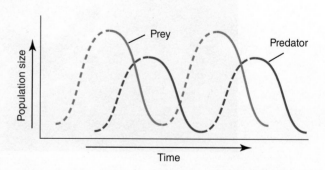

Figure 4.24 Generalized population changes of a phytoplankton prey species and its zooplankton predator, oscillating between limited (solid line) and unlimited (dashed line) phases of population growth.

Figure 4.25 Langmuir streaks on the surface of Lake Mendota, Wisconsin.
© Brad Mitchell/Alamy Stock Photo.

Similar patchy patterns of distribution may be established and maintained physically by **Langmuir cells**, named after Irving Langmuir, who first clarified their structure after he observed the macroalgae *Sargassum* in the North Atlantic floating on the sea surface in long rows parallel to the wind direction. This material is often evident at the surface as long parallel "slicks," foam lines, or rows of floating debris (**Figure 4.25**). Although Langmuir cells extend only a few meters deep, they may create particle and nutrient traps under the convergences. Phytoplankton and particulate debris that accumulate under the convergences attract grazing zooplankton in concentrations often 100 times as dense as those in adjacent areas.

Light

The requirement for light creates a fundamental limit on the distribution of all marine photosynthetic organisms and, indirectly due to food web dynamics, the distribution of many marine organisms. To live, photosynthesizers must remain in an area of the photic zone where enough light penetrates the water for them to photosynthesize. The depth of the photic zone is determined by a variety of conditions, including the atmospheric absorption of light, the angle between the sun and the sea surface, and water transparency.

Water is not very transparent to light. This low transparency causes light intensity in seawater to diminish quickly as it penetrates downward from the sea surface. At some depth, the light intensity is reduced to about 1% of its summertime surface intensity, and photosynthesis occurs at a rate that is balanced by respiration of phytoplankton. This depth, known as the **compensation depth** (Figure 4.20), a depth of zero net primary production, defines the bottom of the photic zone and varies from a few meters deep in coastal waters to more than 200 m in clear tropical seas. In clear tropical waters, the compensation depth often extends below 100 m not only during summer, but throughout the year. In higher latitudes, it may reach 30 to 50 m in midsummer, but it nearly disappears during winter. These are average compensation depths for mixed

phytoplankton communities composed of many different species; each species has its particular compensation depth. At depths above the bottom of the photic zone (i.e., above this compensation depth), the rate of photosynthesis exceeds the rate of photorespiration, and net photosynthesis occurs from that point up to the sea surface. Compensation depths, where light intensity is sufficient to enable phytoplankton cells to compensate for their own respiratory needs via photosynthetic output, are not to be confused with **critical depth** (Figure 4.20). The critical depth for any phytoplankton cell is the depth to which it can be mixed yet still spend enough time above the compensation depth such that its daily respiratory needs are able to be met by its own photosynthetic production. If a photosynthetic cell sinks or is mixed below its critical depth, it will die.

In moderate and low light intensities, photosynthesis by phytoplankton exhibits a direct relationship to light intensity (**Figure 4.26**). At higher light intensities, photosynthetic rates

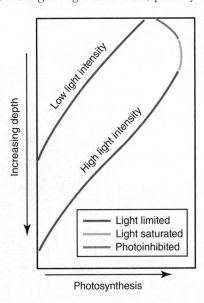

Figure 4.26 Relationship between photosynthesis and depth at low and high light intensities.

do not follow the light intensity curve; rates may stabilize or even decrease nearer the sea surface because the photosynthetic machinery of phytoplankton cells is saturated with or even inhibited by too much light. Higher light intensities near the sea surface fail to promote further increases in photosynthesis.

Remarkably, phytoplankton from different environments exhibit the ability to adjust their photosynthetic rates to varying light intensities; therefore, the saturation light intensity for any phytoplankton population changes with changing sets of environmental conditions. Variations in saturation light intensities are also found among major phytoplankton groups. Dinoflagellates and coccolithophores seem to be better adapted than diatoms to intense light. As a result, their relative contribution to the total marine primary production is greater than that of diatoms in tropical and subtropical regions, and they are able to flourish near the sea surface in temperate regions.

Photosynthetic Pigments

The photosynthetic apparatus of all marine primary producers except cyanobacteria is located in the chloroplasts of actively photosynthesizing cells. Cyanobacteria use the whole cell as their photosynthetic apparatus, so pigment molecules are stored throughout the cell. For all other groups, the chloroplasts contain the pigment systems. The pigments found in the chloroplasts include several types of chlorophyll and various amounts of other photosynthetic pigments (Table 4.1). There, these pigments absorb light energy and convert it to forms of chemical energy that can be used by the photosynthesizers and by those that consume phytoplankton.

Both cyanobacteria and eukaryotic autotrophs use an elaborate part photosynthetic process involving complex pigment systems called photosystems and two distinct sets of chemical reactions, the **light reactions** and **light-independent reactions** (also known as the **Calvin cycle** or **dark reactions**). In the light reactions of photosynthesis (**Figure 4.27**), photons of light are absorbed by chlorophyll *a* molecules located in two photosystems, and

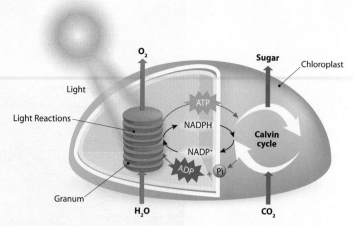

Figure 4.27 Diagrammatic representation of the photosynthetic mechanism of eukaryotic autotrophs occurring in a chloroplast. Arrows indicate the direction of molecular movement as various molecules are produced. The Calvin cycle is also known as the light-independent reactions.
Designua/Shutterstock, Inc.

energy is shuttled from photosystem II to photosystem I. The photons energize electrons and pump them through a series of other enzymes whose function is to manage some of that electron energy and transfer it to adenosine triphosphate (ATP) and another high-energy electron carrier molecule, NADPH. As the term implies, light is needed to drive the light reactions; light reactions only occur during the day when light levels are sufficient. The end products of the light reactions are used to fuel the light-independent reactions.

The pigment systems and enzymes involved in the light reactions are housed within flattened sacs called *thylakoids* that are stacked to form numerous **grana** within each chloroplast (**Figure 4.28**). The **stroma** surrounds the grana and contains the enzymes needed for the next step of photosynthesis,

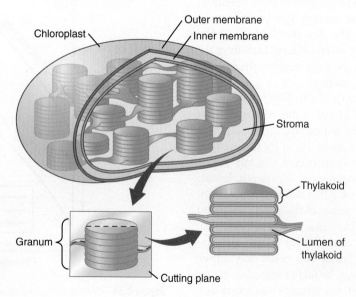

Figure 4.28 A chloroplast and its major components.

the light-independent reactions. Light energy is not necessary to maintain the light-independent reactions, but the high-energy ATP and NADPH produced by the light reactions are. Energy from these substances is used in the light-independent reactions to synthesize carbohydrates, lipids, and the other organic compounds needed by the cell. Other organisms take advantage of the energy-rich products produced by the light-independent reactions, such as glucose, when they consume algae or plant material. A by-product of the light-independent reactions is oxygen, which all organisms (including photosynthesizers) need to sustain life.

Chlorophyll appears green for the same reason coastal seawater appears green. Both absorb more of the available light energy from the violet and red ends of the visible spectrum, leaving the green light to be reflected back or to penetrate more deeply. Chlorophyll serves as the basic energy-absorbing pigment for land plants; however, within a few meters of seawater much of the red and violet portions of the visible spectrum are absorbed before they reach the chloroplasts of most marine plants. Because chlorophyll best absorbs energy from red and violet light, its effectiveness as an absorber of available light energy is greatly reduced in seawater.

All photosynthetic organisms contain the specific chlorophyll type, chlorophyll *a*, which is the pigment that allows for light energy transfer during the process of photosynthesis. The evolutionary response of most marine primary producers has been to supplement the light-absorbing ability of chlorophyll *a* with **accessory pigments** (**Figure 4.29**). These pigments absorb light energy from spectral regions where chlorophyll *a* cannot, and then transfer the energy to chlorophyll *a* for use in the light reactions. In a comparison of the process of photosynthesis to a soccer game, accessory pigments are the team members that take control of the ball and dribble it downfield toward the goal. Chlorophyll *a* is the receiver of the

last pass before the goal is scored, and the player that scores. Chlorophyll *a* receives the glory, but in most cases the player's job would not be possible without the accessory pigment teammates. Figure 4.29 illustrates the complementary effect of chlorophyll *a* and accessory pigments such as fucoxanthin, which is found in diatoms, dinoflagellates, and brown algae. Fucoxanthin absorbs light primarily from the blue and green region of the spectrum, the region where chlorophyll absorbs light least effectively. In combination, chlorophyll and fucoxanthin are capable of absorbing energy from most of the visible light spectrum. Another group of accessory pigments, the phycobilins, are found in red algae and cyanobacteria. These pigments have absorption spectra much like that of fucoxanthin. These and other accessory pigments listed in Table 4.1 have enabled various groups of phytoplankton to adapt to the limited conditions of light availability in seawater by absorbing light energy at almost any depth within the photic zone. The variety of accessory pigments found in different amounts within the bodies of algae also results in the beautiful color variation we see within marine photosynthetic organisms.

Nutrient Requirements

The nutrients required by all primary producers are a bit more complex than might be indicated by the general photosynthetic equation:

$$6CO_2 + 6H_2O \rightarrow C_6H_{12}O_6 + 6O_2$$

Proper growth and maintenance of cells depend on the availability of more than just water and carbon dioxide because plants are composed of compounds that cannot be assembled from C, H, and O alone. These nutrient requirements can be best understood by determining the basic composition of the cell itself. Chemical analysis of a hypothetical "average"

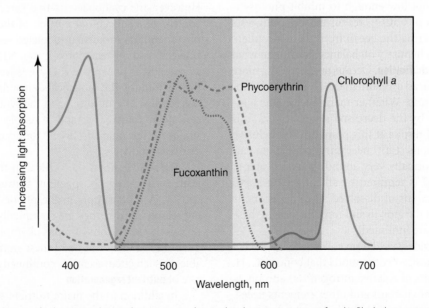

Figure 4.29 Patterns of light absorption for three major photosynthetic pigments: phycoerythrin (an accessory pigment found in Rhodophyta, cryptomonads, and Cyanobacteria); fucoxanthin (an accessory pigment found in brown algae, golden brown algae, and Dinophyta); and chlorophyll *a*.
Modified from M. B. Saffo, New Light on Seaweeds. *BioScience* 37 (1987):654–664.

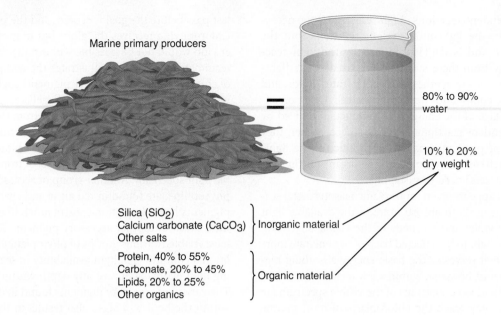

Figure 4.30 Chemical composition of typical marine primary producers.

marine primary producer might yield the results shown in **Figure 4.30**. In general, marine primary producers experience no difficulty in securing an adequate supply of water. Most are continuously and completely bathed by seawater, and few cells of any marine plant are seriously isolated from the external water environment.

Now let's discuss some nutrients that may be limited. Coccolithophores and some seaweeds are equipped with cell walls or internal skeletons of calcium carbonate ($CaCO_3$). Carbon dioxide for carbonate formation and for photosynthesis exists in seawater as carbonic acid (H_2CO_3), bicarbonate (HCO_3^-), and carbonate (CO_3^{2-}). The abundance of these ions in seawater is influenced by photosynthesis, respiration, water depth, and pH. Although the concentration of total CO_2 present in seawater is not low enough to inhibit photosynthesis or the formation of $CaCO_3$, recent evidence suggests that too much CO_2 entering the ocean from the atmosphere is throwing the ocean's chemistry off balance, leading to what has been termed **ocean acidification**. Oceans are becoming more acidic (lower pH), and as a result there is a limited supply of CO_3^{2-} in some areas. Whether marine algae are actually negatively affected by the decreased supply of CO_3^{2-} is a topic of research, and results at this point are inconclusive and variable. Calcium ions (Ca^{2+}) necessary for calcium carbonate formation are normally very abundant in seawater at all depths. Silica (SiO_2) is required by silicoflagellates and diatoms, and concentrations of dissolved silica occasionally become so depleted that the growth and reproduction of these phytoplankton groups are inhibited.

Organic matter is a widely used term collectively applied to those biologically synthesized compounds that contain C, H, usually O, lesser amounts of reactive nitrogen (N) and phosphorus (P), and traces of vitamins and other elements necessary to maintain life. Proteins, carbohydrates, and lipids are the most abundant types of organic compounds in living systems. Each contains carbon, hydrogen, and oxygen in various

ratios. **Figure 4.31** summarizes the generalized nutrient needs of photosynthetic cells.

How much of each of these elements do primary producers require? Chemical analyses of whole phytoplankton cells grown under various light conditions provide an average atomic ratio of approximately 110(C):230(H):75(O):16(N):1(P). Carbon, hydrogen, and oxygen are abundantly available from carbonate (CO_3^{2-}) or bicarbonate ions (HCO3$^-$) and water (H_2O). Reactive nitrogen is much less plentiful but is present in seawater as nitrate (NO_3^-), with lesser amounts of nitrite (NO_2^-) and ammonium (NH_4^+). High concentrations of molecular nitrogen (N_2), which constitutes 78% of the Earth's atmosphere, are also dissolved in seawater; however, most marine organisms are not metabolically equipped to use this nonreactive form of N. However, the cyanobacteria that can, such as *Trichodesmium*, contribute a substantial portion of the total N used by other phytoplankton in nutrient-depleted seas by converting nonreactive N_2 to more reactive NO_3^- and NH_4^+. Phosphorus, present principally as phosphate (PO_4^{3-}), is less abundant in seawater than is nitrate. The biological demands for phosphate are also less but just as crucial (e.g., in the synthesis of ATP, DNA, and cell membranes). The ratio of usable N and P in seawater is similar to the ratio of 16N:1P found in living cells of marine primary producers.

Figure 4.32 shows the vertical distribution patterns of silicate, nitrate, and phosphate in seawater. These nutrients are usually in short supply in the photic zone during the growing season because they are continually consumed by primary producers. In periods of rapid phytoplankton growth, needed quantities of one or more of these nutrients may not be available. In such circumstances, continued growth is limited by the rate of **nutrient regeneration**.

In addition to the major nutrient elements just described, marine autotrophs require several other elements in minute amounts. These **trace elements** include iron, manganese, cobalt, zinc, copper, and others. Depletion of iron in English Channel

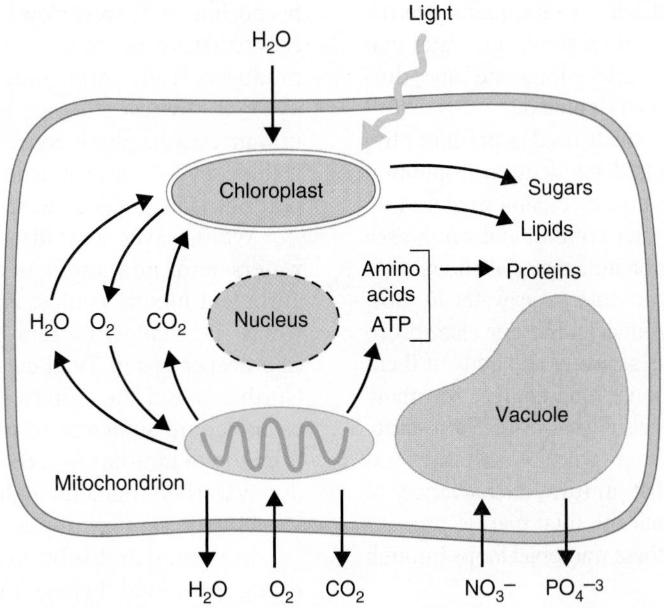

Light
H2O
Chloroplast
Sugars
Lipids
Amino acids ATP
Proteins
H2O O2 CO2
Nucleus
Vacuole
Mitochondrion
H2O O2 CO2
NO3− PO4−3

Figure 4.31 A simplified photosynthetic cell, illustrating the chemical requirements and products of several components of the cell. Note that some of the products of photosynthesis are used for cellular respiration and that some of the products of cellular respiration are used for photosynthesis.

waters has been observed during spring diatom blooms, suggesting that iron availability may limit the size or composition of phytoplankton populations. More recently, the results of both laboratory and field attempts to enrich seawater artificially with iron demonstrated marked increases in phytoplankton growth. In natural systems such as the North Atlantic Ocean, massive inputs of iron from windborne dust carried from the Sahara Desert of Africa (**Figure 4.33**) suggest that phytoplankton growth rates may be higher than in comparable areas isolated from similar inputs of iron-rich dust.

Vitamins, too, are crucial for the proper growth and reproduction of primary producers. Some species of diatoms, for example, require more vitamin B_{12} during auxospore formation than at other times. Some can synthesize their own vitamins; others must rely on free-living bacteria to provide these and other essential vitamins that they cannot synthesize for themselves.

Nutrient Regeneration

Most of the **biomass** produced by marine photosynthesis is eventually consumed by herbivores and is converted to more herbivore bodies or is formed into fecal wastes. In either case, these compact particles quickly become colonized by bacteria and sink as **marine snow** to depths well below the photic zone. It is these deep, cold portions of the world ocean that contain

Nutrient concentration, mg-atoms/L
1
10
100
0
1000
2000
3000
Depth, m

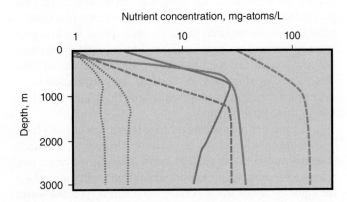

Figure 4.32 Distribution of dissolved silicate (dotted lines), nitrate (dashed lines), and phosphate (solid lines) from the surface to 3,000 m in the Atlantic (blue) and Pacific (orange) Oceans.

Modified from H. U. Sverdrup et al., *The Oceans: Their Physics, Chemistry, and Biology* (Prentice Hall, 1942).

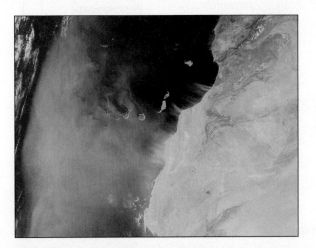

Figure 4.33 SeaWiFS image of airborne Saharan dust being carried westward over the Canary Islands and beyond into the North Atlantic Ocean.

Courtesy of GeoEye and NASA SeaWiFS Project.

the large reserves of dissolved nutrients. Consequently, nutrient-rich waters are almost always cold waters, and water that is warm has likely been at the sea surface for some time, during which its nutrient load has been depleted.

Regeneration of the nutrients initially used to produce phytoplankton cells or marine plants is dependent on respiration by consumers and on decomposition of organic material by bacteria and fungi living in the water column and on the sea floor. Bacterial action decomposes organic material and returns phosphates, nitrates, and other nutrients to seawater in inorganic form for reuse by primary producers. Bacteria also absorb dissolved organic compounds from seawater and convert them to living cells that become additional food sources for many benthic and small planktonic animals (**Figure 4.34**). These **microbial loops** divert organic material from typical planktonic food webs to populations of bacteria that, in turn, feed a variety of microzooplankton. As much as half the total marine primary productivity may be directed into these microbial loops through planktonic bacteria.

Figure 4.32 indicates that major concentrations of limiting nutrients accumulate below the photic zone, where they cannot be accessed by photosynthesizers. These two fundamental needs, light within the photic zone and nutrients from well below the photic zone on the sea floor, impose severe restrictions on the rates of primary production. For much of the ocean, the sunlit photic zone is isolated from the nutrients within deeper waters by a well-developed and permanent

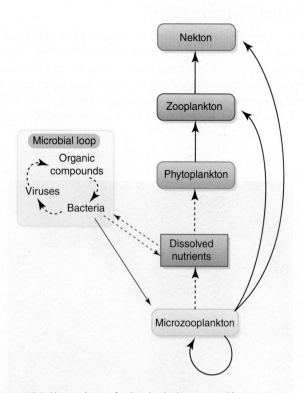

Figure 4.34 Major pathways of cycling dissolved nutrients and food particles through a microbial loop (left) and a particulate food web (right). Dissolved organic materials and inorganic nutrients are indicated with dashed lines.

pycnocline. Here, very slow molecular diffusion is the only process to return nutrients to the photic zone. Marine primary producers really thrive only in those parts of the sea where dynamic physical processes move colder nutrient-laden waters upward into the photic zone. These large-scale mixing processes include small-scale turbulence and upwelling that rapidly transport nutrient-rich deep water upward.

Wind, waves, and tides create turbulence in near-surface waters and mix nutrients from deeper water upward. Turbulent mixing is most effective over continental shelves, where the shallow bottom prevents the escape of nutrients into deeper water. Tidal currents in the southern end of the North Sea and the eastern side of the English Channel, for example, are sufficient to mix the water almost completely from top to bottom. As a result, summer phytoplankton productivity there remains high as long as sunlight is sufficient to maintain photosynthesis.

In tropical and subtropical latitudes of most oceans, the strong year-round thermocline and associated pycnocline near the base of the photic zone act as a strong barrier to inhibit upward mixing of deep nutrient-rich waters (**Figure 4.35**, top). Consequently, these low-latitude regions have very low rates of primary production, comparable with terrestrial deserts, and crystal clear surface waters. This is a bit ironic, because these areas experience the most sunlight year-round, and thus have the highest potential for photosynthesis from a sunlight perspective. The low photosynthetic rates displayed in the tropics serve to emphasize the importance of nutrients in photosynthetic processes.

Pycnoclines also develop in temperate waters to restrict the return of deep-water nutrients, but only on a seasonal basis (Figure 4.35, middle). During winter, the surface water cools and sinks. The pycnocline disappears, and deeper nutrient-rich water is mixed with the surface water. As solar radiation increases in the spring, the surface water warms, and the thermocline is reestablished. A well-developed summer pycnocline at temperate latitudes resembles the permanent pycnocline of tropical and subtropical waters and creates an effective barrier, blocking nutrient return to the photic zone. With shorter days and cooler weather in autumn, the pycnocline weakens and then disappears in winter. Without a pycnocline to interfere, **convective mixing** in temperate regions resumes, continuing from late fall to early spring, but in higher latitudes continuous heat loss from the sea to the atmosphere and low amounts of solar radiation result in year-round convective mixing (Figure 4.35, bottom). Low-light conditions rather than scarce nutrients usually limit the primary production in these polar regions.

The process of **upwelling** is crucial to move nutrients from deep, cold water areas to the surface. Upwelling is usually seasonal and is controlled by wind blowing warm surface waters away from an area (normally the coast, but also along the equator). Once warm surface water is removed, the warm water is replaced by cool, nutrient-rich water from deeper areas. **Figure 4.36** illustrates the influence of upwelling on nutrient availability in the photic zone. The nitrate concentrations at a depth of about 50 m are 5 to 10 times higher in the

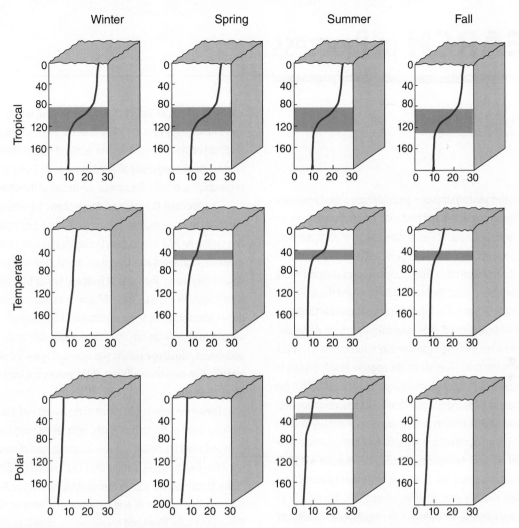

Figure 4.35 Seasonal development and destruction of thermoclines and associated pycnoclines in tropical (top), temperate (middle) and polar (bottom) oceans. The vertical axes display depth (m) and the horizontal axes display temperature (degrees Celsius).

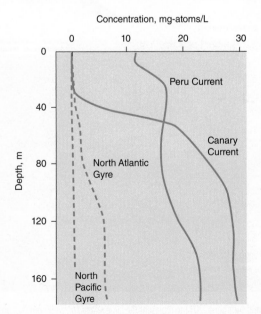

Figure 4.36 A comparison of the vertical distribution of nitrate in upwelling areas (solid curves) and adjacent nonupwelling central ocean regions (dotted curves).

Modified from J. Walsh. The Role of Ocean Biota in Accelerated Ecological Cycles: A Temporal View. *Journal of BioScience* 34 (1984):499–507.

upwelling systems than at similar depths in water masses not affected by upwelling. Collectively, the dynamic interactions of all of the processes described here that affect marine primary productivity create complex spatial and temporal patterns of phytoplankton abundance.

4.4   Small Size with a Large Impact: Marine Bacteria, Archaea, and Viruses

Our discussion of marine bacteria thus far has been general and limited to microbial loop dynamics, with the exception of one group, the cyanobacteria, because cyanobacteria are one of the major groups of photosynthetic microbes. Marine bacteria are actually the most abundant organisms in the sea and make up the bulk of marine microbes. In the past, their small size (generally several microns in length) led to difficulties studying bacteria, but recently scientists have focused research efforts on these tiny, abundant, and very impactful life forms. Many marine bacteria benefit other organisms in vital ways. Just recently it was discovered that some bacteria assist with the growth of

RESEARCH in Progress

Harmful Algal Blooms

Phytoplankton and other photosynthesizers generally have a good reputation for their roles in primary production and producing oxygen for other organisms to use in metabolic processes. However, some species of phytoplankton, when found in high enough numbers, can negatively affect other organisms, including humans. *Harmful algal bloom (HAB)* is the term used to describe such an event that may negatively affect the health of humans, and this topic is on the forefront of scientific research. This topic is of such concern that there is an entire scientific journal dedicated to it, entitled *Harmful Algae*. Several key questions are encountered when studying algae that may harm humans: (1) Which species are harmful to humans? (2) What are the negative health impacts to humans? (3) What conditions trigger a harmful algal bloom? and (4) What can we do to better prepare for HABs and minimize their health impacts on humans?

Researchers have sought answers to these questions and many more for HAB events. At this point, approximately 100 species of phytoplankton have been identified that are toxic to humans, although this number will likely increase as more research is conducted. In the past, HABs were called *red tides*. This term is no longer typically used for HABs, because toxic blooms may be green, brown, or even colorless, depending on the algal species causing the bloom. HABs are most common in warmer waters, but can take place in cool

waters, too. Areas in the South and Southeastern United States have experienced increasingly frequent HABs in recent years, leading to beach closures, shellfish bed closures, fish kills, and deaths of marine mammals and seabirds. Humans may be negatively affected by the HABs if they eat shellfish exposed to the toxins or breathe the air near the surface of the water.

The National Oceanic and Atmospheric Administration (NOAA) has responded to increasing HAB events by creating a data collection and distribution system for HAB information. During blooms, data such as cell counts and species identifications are mapped and provided for scientists and the public to observe (**Figure A**). The states of Texas and Florida have additional systems in place to monitor and track HABs. Whenever an HAB occurs near San Diego, California, scientists at the Scripps Institution of Oceanography (SIO) collect water samples to identify the species of algae responsible. In the spring of 2015, an underwater microscope camera was mounted to the SIO pier, making species identification even more efficient. If the species is a known HAB species, the public is alerted.

During the spring of 2015 one of the largest HAB events occurred along the West Coast of the United States, with record toxin levels for some locations and, most interestingly, a very long time span of occurrence. Most HABs last a few days to weeks, but the West Coast event lasted months. The culprit of this bloom was the diatom *Pseudo-nitzschia* (**Figure B**). *Pseudo-nitzschia* is responsible for domoic acid poisoning and amnesic shellfish poisoning. Researchers have developed oceanographic models to simulate the path of travel for the blooms. By predicting where the bloom will travel in advance,

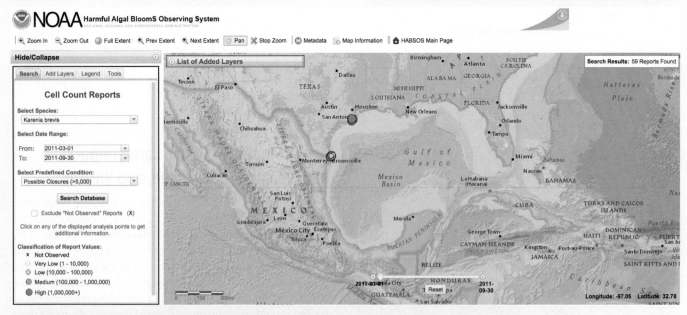

Figure A The NOAA Harmful Algal BloomS Observing System. The date range displayed is for the spring and summer of 2011 to represent the large *Karenia brevis* bloom that occurred during this time.

Courtesy of NOAA.

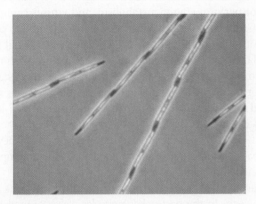

Figure B *Pseudo-nitzchia* specimens observed with a light microscope.
NOAA Northwest Fisheries Science Center.

people can be prepared and cautious about shellfish and even water exposure. The West Coast bloom is known to have tainted areas from central California to at least Vancouver Island, with some reports from Alaska. The cause of the bloom was unknown, but researchers will continue searching for information.

A record HAB took place along the coast of Texas in 2011, and the cause was linked to drought conditions. HAB species seem to flourish in warm, salty water, and a lack of rain during drought years maintains high salinity levels. The large geographic range (from Galveston to South Padre Island) and dense concentration of algae (more than 100,000 cells/mL) in this bloom made it unique, and it was the largest bloom seen in the area in over a decade. Fish kills occurred, and people could not go near the water without coughing and experiencing burning eyes. The culprit of most of the HABs in Texas and the west coast of Florida is a dinoflagellate, *Karenia brevis.* Research was conducted tracking oceanographic patterns and the abundance of *K. brevis* from 1996–2011, and the following trends were observed: (1) alongshore winds were weaker during HAB years, leading to weak downwelling that maintained the algae in the area; and (2) water from the south was present near the shore in Texas during HAB years, providing populations of HAB algae. Much of the research was conducted using drifters to see where the water would move algae, and data from oceanographic sensors were used to model water movement.

Although potentially harmful to humans, HABs are not to be feared. Most HABs are visible, and in areas where they are known to occur scientists are now monitoring waters for HAB events. The public is notified as soon as possible when an event is identified, and with simple precautions no human harm will result. The more we know about HAB species and the causes of blooms, the more we can prepare for HAB events.

Critical Thinking Questions

1. Harmful algal blooms can cause problems for other organisms living in the same area as the bloom or for humans entering the sea or eating poisoned shellfish. Think of at least one adaptive significance of HABs in the overall ecosystem where they occur. What may be an ultimate cause for this phenomenon?

2. Many HABs are quite visible in the water, producing bright colors or bioluminescence. What is the adaptive significance of the high visibility of these events? Is there an advantage to the HAB-producing algae to alert other organisms of their presence?

For Further Reading

Giddings, S. N., P. MacCready, B. M. Hickey, N. S. Banas, K. A. Davis, S. A. Siedlecki, V. L. Trainer, R. Kudela, N. Pelland, and T. P. Connolly. 2014. Hindcasts of harmful algal bloom transport on the Pacific Northwest coast. *Journal of Geophysical Research Oceans* 119:2439–2461. doi:10.1002/2013JC009622

Thyng, K. M., R. D. Hetland, M. T. Ogle, X. Zhang, F. Chen, and L. Campbell. 2013. Origins of *Karenia brevis* harmful algal blooms along the Texas coast. *Limnology and Oceanography: Fluids and Environments* 3:269–278. doi:10.1215/21573689-2417719

diatoms by secreting the growth hormone auxin, which is known to benefit plants. Many bacteria help the entire food web by converting nutrients into usable forms. An example of this is the crucial process of **nitrification**, which involves several steps. Some species of proteobacteria, a very diverse group of bacteria, are involved in the first step of nitrification, converting ammonium to nitrite. A gram-negative bacteria, *Nitrobacter*, is one of several groups of bacteria involved in the second step of nitrification, the conversion of nitrite to nitrate. The process of nitrification makes nitrogen available to photosynthesizers and is an absolutely essential process. Some marine bacteria harm other organisms by crowding or surrounding them or, in humans, causing illness. **Figure 4.37** displays a marine bacteria, *Vibrio parahaemolyticus*, adhering to a diatom. The diatom is eaten by filter-feeding shellfish, and humans that eat undercooked shellfish may be infected by the bacteria and become ill.

Members of the domain Archaea look similar to bacteria and are also single-celled prokaryotes, but molecular evidence indicates that they are different enough to warrant their own domain. Archaea were once thought to only inhabit extreme environments, such as hydrothermal vents or very low oxygen environments, but it is now known that archaea are very common in the ocean. Some *Thaumarchaea* marine archaea contribute to the first step of nitrification by converting ammonium to nitrite. It is thought that they may produce nitrous oxide, a greenhouse gas, during the conversion process.

A **virus** is a tiny (20 to 300 nm) particle of protein-coated genetic material that infects a living organism intracellularly. Most scientists consider viruses to be nonliving because they cannot regulate movement of substances into or out of themselves or perform energy metabolism, and viruses cannot replicate themselves without using the cellular machinery of their host. Although nonliving, viruses are covered in detail in any general biology textbook that is published, making their importance to living organisms and the ambiguity of their status as living or nonliving apparent. In our discussion, viruses will be referred to as *biological agents*. Viruses cause several well-known human diseases, including acquired immune deficiency syndrome (AIDS), rabies, the flu, and the common cold. They often infect bacteria and phytoplankton, and as a result are important components of microbial loops.

Marine viruses are the most abundant biological agents in the ocean. In the past, virus particles were counted in seawater samples by the expensive, tedious, and time-consuming method of electron microscopy. Today, viral particles are counted by epifluorescence microscopy after staining with fluorochrome stains, such as DAPI and SYBR Green I. Instead of transmitting light through a sectioned specimen, epifluorescence microscopy

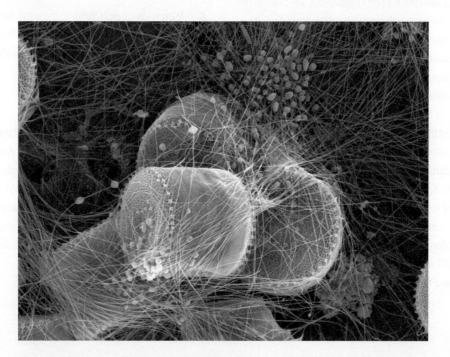

Figure 4.37 A scanning electron micrograph of the bacteria *Vibrio parahaemolyticus* adhering to diatoms. This bacteria can cause human health problems such as gastroenteritis. The bacteria are the small organisms with a long "tail" and "head," and the diatoms are the relatively large cylinder-shaped organisms.
NOAA Northwest Fisheries Science Center.

transmits light through the objective onto the specimen, which results in a much greater image intensity.

Because of their great abundance and ability to infect all organisms, including bacteria and phytoplankton, marine viruses influence many biogeochemical and ecological processes. Infection by marine viruses affects dynamics of the microbial loop and the cycling of nutrients in the ocean, total respiration in the system, production of dimethyl sulfide gas (which links marine production with climate change), distributions and sinking rates of particles, biodiversity and distribution of bacteria and phytoplankton, and diversity and transfer of genes between marine microbes.

Perhaps most significant to marine production is the effect that viral infection has on phytoplankton. Viruses may divert carbon from marine zooplankton and, by extension, the rest of the food web, by lysing bacteria and phytoplankton. Moreover, viral infections have been shown to cause behavioral and other changes in phytoplankton, sometimes destabilizing the oscillating trophic relationship between a phytoplankton species and its zooplankton grazer (Figure 4.24), leading to a collapse of phytoplankton blooms and severe impacts on marine production. Much work needs to be done before these effects are evaluated and quantified.

DID YOU KNOW?

One liter of seawater collected from near the surface of the ocean will typically contain *at least* 10 billion microbes (phytoplankton and nonphotosynthetic bacteria) and somewhere near 100 billion viruses! Most of the viruses in the ocean have yet to be identified, and for those that are known it is usually unclear which organisms they infect. Marine viruses appear to be less picky than terrestrial ones, and may infect more than one species or even individuals from multiple genera. Much remains to be discovered in the world of marine microbes.

STUDY GUIDE

TOPICS FOR DISCUSSION AND REVIEW

1. Why are most marine photosynthesizers unicellular?

2. Describe the very unique pattern of cellular reproduction observed in diatoms.

3. Summarize the antisinking strategies used by marine phytoplankton.

4. Distinguish between the terms *standing crop* and *primary production*.

5. How do gross and net primary production differ, and how are they measured?

6. How is the LB/DB technique used to measure primary production in the sea? Is this method still useful even with advances in technology?

7. Describe the population fluctuations commonly observed in phytoplankton and their herbivorous grazers.

8. What is the compensation depth? Describe its significance to phytoplankton communities. How does it differ from the critical depth?

9. List and describe five mechanisms of nutrient regeneration in the sea.

10. Describe how some marine bacteria help other organisms and how some harm other organisms.

KEY TERMS

accessory pigment 97	ecological succession 84
auxospore 83	epiphyte 80
bioluminescence 78	epitheca 81
biomass 99	feedback mechanism 94
Calvin cycle 96	flagellum 85
cellular respiration 91	frustule 81
centric 81	grana 96
circadian clock 80	gross primary production 91
coccolith 81	
compensation depth 95	harmful algal bloom (HAB) 86
convective mixing 100	hypotheca 81
coral bleaching 86	Langmuir cell 95
critical depth 95	light reactions 96
dark reactions 96	light-independent reactions 96
diurnal 80	

limiting factor 94	phytoplankton 78
marine snow 99	picoplankton 78
microbe 78	plankton bloom 85
microbial loop 100	primary production 91
nanoplankton 78	raphe 81
net primary production 92	red tide 86
nitrification 104	standing crop 91
nitrogen fixation 79	stroma 96
nutrient regeneration 98	trace element 98
ocean acidification 98	turnover rate 91
pennate 81	ultraplankton 78
phylogenetic relationship 86	upwelling 100
	virus 104

KEY *GENERA*

Alexandrium	Lyngbya
Anabaena	Microcoleus
Asterionella	Nitrobacter
Asteromphalus	Noctiluca
Ceratium	Oscillatoria
Chaetoceros	Oxytoxum
Cocconeis	Pseudo-nitzschia
Ditylum	Rhizosolenia
Emiliania	Sargassum
Eucampia	Saxodomus
Gambierdiscus	Synechoccus
Gonyaulax	Thaumarchaea
Isthmia	Trichodesmium
Karenia	Vibrio

REFERENCES

Amin, S. A., L. R. Hmelo, H. M. van Tol, B. P. Durham, L. T. Carlson, and K. R. Heal. 2015. Interaction and signalling between a cosmopolitan phytoplankton and associated bacteria. *Nature* 522: 98–101. doi:10.1038/nature14488

Anderson, D. M. 1994. Red tides. *Scientific American* August:62–68.

Anderson, D. M., and D. Wall. 1978. Potential importance of benthic cysts of *Gonyaulax tamarensis* and *G. excavata* in initiating toxic dinoflagellate blooms. *Journal of Phycology* 14:224–234.

Austin, B. 1988. *Marine Microbiology*. New York: Cambridge University Press.

Bainbridge, R. 1957. Size, shape and density of marine phytoplankton concentrations. *Biological Review* 32:91–115.

Bargu, S., C. L. Powell, S. L. Coale, M. Busman, G. J. Doucette, and M. W. Silver. 2002. Krill: A potential vector for domoic acid in marine food webs. *Marine Ecology Progress Series* 237:209–216.

Beaufort, L., I. Probert, T. de Garidel-Thoron, E. M. Bendiff, D. Ruiz-Pino, N. Metzl, C. Goyet, N. Buchet, P. Coupel, M. Grelaud, B. Rost, R. E. M. Rickaby, and C. de Vargas. 2011. Sensitivity of coccolithophores to carbonate chemistry and ocean acidification. *Nature* 476:80–83. doi:10.1038/nature10295

Blankenship, R. E. 2010. Early evolution of photosynthesis. *Plant Physiology* 154(2):434–438.

Boney, A. D. 1992. *Phytoplankton*. London: E. Arnold.

Bougis, P. 1976. *Marine Plankton Ecology*. New York: Elsevier.

Brown, O. B., R. H. Evans, J. W. Brown, H. R. Gordon, R. C. Smith, and K. S. Baker. 1985. Phytoplankton blooming off the U.S. East Coast: A satellite description. *Science* 229:163–167.

Burkholder, J. M. 1999. The lurking peril of *Pfiesteria*. *Scientific American* 281:42–49.

Carmichael, W. W. 1994. The toxins of cyanobacteria. *Scientific American* January:78–86.

Carpenter, E. J., and K. Romans. 1991. Major role of the cyanobacterium *Trichodesmium* in nutrient cycling in the North Atlantic Ocean. *Science* 254:1356–1358.

Carr, N. G., and B. A. Whitton. 1983. *The Biology of Cyanobacteria*. Berkeley, CA: University of California Press.

Chisholm, S. W. 1992. What limits phytoplankton growth? *Oceanus* 35:36–46.

Coleman, G., and Coleman, W. J. 1990. How plants make oxygen. *Scientific American* February:50–58.

Cushing, D. H., and J. J. Walsh. 1976. *The Ecology of Seas*. Philadelphia: W.B. Saunders.

Dale, B., and C. M. Yentsch. 1978. Red tide and paralytic shellfish poisoning. *Oceanus* 21:41–49.

Dawes, C. J. 1981. *Marine Botany*. New York: John Wiley & Sons.

DeLong, E. F. 2007. Microbial domains in the ocean: A lesson from the archaea. *Oceanography* 20:124–129.

Fleming, R. H. 1939. The control of diatom populations by grazing. *Journal du Conseil Permanent International pour l'Exploration de la Mer* 14:210–227.

Fryxell, G. A. 1983. New evolutionary patterns in diatoms. *BioScience* 33:92–98.

Gledhill, D. K., M. M. White, J. Salisbury, H. Thomas, I. Mlsna, M. Liebman, B. Mook, J. Grear, A. C. Candelmo, R. C. Chambers, C. J. Gobler, C. W. Hunt, A. L. King, N. N. Price, S. R. Signorini, E. Stancioff, C. Stymiest, R. A. Wahle, J. D. Waller, N. D. Rebuck, Z. A. Wang, T. L. Capson, J. R. Morrison, S. R. Cooley, and S. C. Doney. 2015. Ocean and coastal acidification off New England and Nova Scotia. *Oceanography* 28(2):182–197. http://dx.doi.org/10.5670/oceanog.2015.41

Gregg, M. C., T. B. Sanford, and D. P. Winkel. 2003. Reduced mixing from the breaking of internal waves in equatorial waters. *Nature* 422:513–515.

Griffin, D. W. 2002. The global transport of dust. *American Scientist* 90:228–235.

Guiry, M. D., and Guiry, G. M. 2015. *AlgaeBase*. World-wide electronic publication, National University of Ireland, Galway. Available from http://www.algaebase.org. Accessed on June 27, 2015.

Hamm, C. E., R. Merkel, O. Springer, P. Jurkojc, C. Maier, K. Prechtel, and V. Smetacek. 2003. Architecture and material properties of diatom shells provide effective mechanical protection. *Nature* 421:841–843.

Hardy, A. H. 1971. *The Open Sea: Its Natural History. Part I: The World of Plankton. Part II: Fish and Fisheries*. Boston: Houghton Mifflin.

Hargraves, P. E., and F. W. French. 1983. Diatom resting spores: Significance and strategies. In: *Survival Strategies of the Algae*. G. A. Fryxell, ed. New York: Cambridge University Press. pp. 49–68.

Harris, G. P. 1986. *Phytoplankton Ecology*. New York: Chapman and Hall.

Harvey, E. L., and S. Menden-Deuer. 2012. Predator-induced fleeing behaviors in phytoplankton: A new mechanism for harmful algal bloom formation? *PLoS One* 7(9):e46438. doi:10.1371/journal.pone.0046438

Honhart, D. C. 1989. *Oceanography from the Space Shuttle*. Washington, DC: Office of Naval Research.

Humm, H. J., and S. R. Wicks. 1980. *Introduction and Guide to the Marine Blue-Green Algae*. New York: Wiley.

Iida, T., K. Mizobata, and S. I. Saitoh. 2012. Interannual variability of coccolithophore *Emiliania huxleyi* blooms in response to changes in water column stability in the eastern Bering Sea. *Continental Shelf Research* 34:7–17.

Jenkins, W. J., and J. C. Goldman. 1985. Seasonal oxygen cycling and primary production in the Sargasso Sea. *Journal of Marine Research* 43:465–491.

Jumars, P. A. 1993. *Concepts in Biological Oceanography: An Interdisciplinary Primer*. New York: Oxford University Press.

Kauff, F., and B. Budel. 2011. Phylogeny of cyanobacteria. *Progress in Botany* 72:209–224.

Kaufman, P. B. 1989. *Plants: Their Biology and Importance*. Reading, MA: Addison-Wesley.

Krogmann, D. W. 1981. Cyanobacteria (bluegreen algae)—Their evolution and relation to other photosynthetic organisms. *BioScience* 31:121–124.

Lalli, C. M., and T. R. Parsons. 1997. *Biological Oceanography: An Introduction*. Oxford: Butterworth Heinemann.

Landry, M. R. 1976. The structure of marine ecosystems: An alternative. *Marine Biology* 35:1–7.

Lembi, C. A., and J. R. Waaland, eds. 1988. *Algae and Human Affairs*. New York: Cambridge University Press.

Lipps, J. H. 1970. Plankton evolution. *Evolution* 24:1–22.

Malone, T. C. 1971. The relative importance of nannoplankton and net plankton as primary producers in tropical oceanic and neritic phytoplankton communities. *Limnology and Oceanography* 16:633–639.

Margulis, L., D. Chase, and R. Guerrero. 1986. Microbial communities. *BioScience* 36:160–170.

Marshall, H. G. 1976. Phytoplankton density along the eastern coast of U.S.A. *Marine Biology* 38:81–89.

Mills, E. L. 1989. *Biological Oceanography: An Early History, 1870–1960*. Ithaca, NY: Cornell University Press.

Mincer T. J., M. J. Church, L. T. Taylor, C. M. Preston, D. M. Karl, and E. F. DeLong. 2007. Quantitative distribution of presumptive archaeal and bacterial nitrifiers in Monterey Bay and the North Pacific Subtropical Gyre. *Environmental Microbiology* 9:1162–1175.

Moll, R. A. 1977. Phytoplankton in a temperate-zone salt marsh: Net production and exchanges with coastal waters. *Marine Biology* 42:109–118.

Moore, R. E. 1977. Toxins from blue-green algae. *BioScience* 27:797–802.

Okada, H., and A. McIntyre. 1977. Modern coccolithophores of the Pacific and North Atlantic oceans. *Micropaleontology* 23:1–55.

Paasche, E. 1968. Biology and physiology of coccolithophorids. *Annual Review of Microbiology* 22:71–86.

Pace, N. R. 1997. A molecular view of microbial diversity and the biosphere. *Science* 276:734–740.

Perry, M. J. 1986. Assessing marine primary production from space. *BioScience* 36:461–466.

Peterson, M. N. A., ed. 1993. *Diversity of Oceanic Life: An Evaluative Review*. Washington, DC: Center for Strategic and International Studies.

Platt, T., and W. K. W. Li, eds. 1986. Photosynthetic picoplankton. *Canadian Bulletin of Fisheries and Aquatic Sciences* 214:583.

Platt, T., C. Fuentes-Yaco, and K. T. Frank. 2003. Spring algal bloom and larval fish survival. *Nature* 423:398–399.

Pomeroy, L. W. 1974. The ocean's food web, a changing paradigm. *BioScience* 24:499–504.

Qasim, S. Z., P. M. A. Bhattuthiri, and V. P. Devassy. 1972. The effect of intensity and quality of illumination on the photosynthesis of some tropical marine phytoplankton. *Marine Biology* 16:22–27.

Raymont, J. E. G. 1983. *Plankton and Productivity in the Oceans*. Volume 1: Phytoplankton. New York: Pergamon Press.

Roman, M. R., H. Dam, R. Le Borgne, and X. Zhang. 2002. Latitudinal comparisons of equatorial Pacific zooplankton. *Deep-Sea Research (Part II, Topical Studies in Oceanography)* 49:2695–2711.

Ronning, K., E. Beliveau, E. McCaffery, C. Omlor, and E. Rosenblum. Defining phylogenetic relationships of Ochrophyta using 18S rRNA: Existence of three major clades in which Bacillariophyta is basal. *Euglena* 1(2):52–59.

Round, F. E., R. M. Crawford, and D. G. Mann. 1990. *The Diatoms*. New York: Cambridge University Press.

Rowan, K. S. 1989. *Photosynthetic Pigments of Algae*. New York: Cambridge University Press.

Russell-Hunter, W. D. 1970. *Aquatic Productivity*. New York: Macmillan.

Ryther, J. H. 1969. Photosynthesis and fish production in the sea. *Science* 166:72–76.

Saffo, M. B. 1987. New light on seaweeds. *BioScience* 37:654–664.

Scagel, R. F., R. J. Bandoni, J. R. Maze, G. E. Rouse, W. B. Schofield, and J. R. Stein. 1980. *Nonvascular Plants*. Belmont, CA: Wadsworth.

Smayda, T. J. 1970. The suspension and sinking of phytoplankton in the sea. *Oceanography and Marine Biology (Annual Review)* 8:353–414.

Smith, D. L. 1996. *A Guide to Marine Coastal Plankton and Marine Invertebrate Larvae*. Dubuque, IA: Kendall/Hunt Publishing Company.

Smith, W. O. Jr., and D. M. Nelson. 1986. Importance of ice edge phytoplankton production in the southern ocean. *BioScience* 36:251–257.

Steeman Nielsen, E., 1975. *Marine Photosynthesis*. Amsterdam: Elsevier.

Steidinger, K. A., and K. Haddad. 1981. Biologic and hydrographic aspects of red tides. *BioScience* 31:814–819.

Sverdrup, H. U., M. W. Johnson, and R. H. Fleming. 1942. *The Oceans: Their Physics, Chemistry, and Biology*. Englewood Cliffs, NJ: Prentice-Hall.

Vinyard, W. C. 1980. *Diatoms of North America*. Eureka, CA: Mad River Press.

Von Dassow, P., G. van den Engh, M. D. Iglesias-Rodriguez, and J. R. Gittins. 2012. Calcification state of coccolithophores can be assessed by light scatter depolarization measurements with flow cytometry. *Journal of Plankton Research* 34:1011–1027.

Walsby, A. E. 1977. The gas vacuoles of blue-green algae. *Scientific American* August:90–97.

Walsh, J. J. 1984. The role of ocean biota in accelerated ecological cycles: A temporal view. *BioScience* 34:499–507.

Werner, D., ed. 1977. *The Biology of Diatoms*. Berkeley, CA: University of California Press.

CHAPTER

5

Marine Macroalgae and Plants

STUDENT LEARNING OUTCOMES

1. Compare and contrast the major groups of marine macroalgae and plants.

2. Choose one group of marine plants (Anthophyta), and describe how the group is adapted for submerged life.

3. Explain why green and red algae are more closely related to land plants than to brown algae.

4. Compare and contrast the major body parts of macroalgae to those found in land plants.

5. Identify which areas of the world contain kelp forests, and describe why kelp forests are present in those locations.

6. Develop a hypothesis to answer the following question: why are red algae found in great abundance in the tropics?

7. Create a list of conditions necessary for high net primary productivity (NPP).

8. Evaluate the NPP information provided in text, tables, and figures, and explain why it is so variable in different regions of the ocean.

Although commonly found along coasts, marine macroalgae are actually quite rare in the open ocean due to light limitations.
© Tao Jiang/Shutterstock, Inc.

Benthic marine algae are much more likely to be familiar to a typical seashore visitor than are phytoplankton because they are brightly colored, coastal, and highly visible multicellular organisms typically large enough to pick up and examine. Historically, macroalgae were considered part of the Kingdoms Plantae and Protista, but recent reworking of phylogenetic trees and a new view of plant ancestry has led to a removal of macroalgae from the Kingdom Plantae, an abandonment of the Kingdom Protista, and general confusion of phylogenies. By some accounts, brown algae and some phytoplankton are now part of the Kingdom Chromista, united by the presence of the accessory pigment fucoxanthin; by other accounts brown algae remain part of the broad, fairly generic group called protists that also includes nonphotosynthetic microbes. Although the Kingdom Protista has been abandoned, the protist group is still useful on some level when describing organisms that, by definition, are not true plants or animals (**Figure 5.1**). Green and red algae were considered by many to be part of the Kingdom Plantae, but phylogenetic studies have led to confusion of their kingdom membership. They are in the clade Archaeplastida, which includes red and green algae, charophyceans, and land plants. Green algae as a group are thought to be more closely related to and to share a common ancestor with terrestrial plants. Evidence for the evolution of plants from a green alga includes chemical, genetic, and morphological similarities to green algae. Others still lump green and red macroalgae (and all macroalgae) into the protist group. Clearly, the classification of algae is unresolved and a work in progress, and to avoid confusion for our discussion we will refer to brown and multicellular red and green algae as **macroalgae** or *seaweeds*.

Like phytoplankton and terrestrial plants, macroalgae need sunlight for photosynthesis and are confined to the photic zone, but the additional need for a hard substrate on which to attach limits the distribution of benthic macroalgae to the narrow fringe around the shorelines of the oceans where the sea bottom is within the photic zone. Some benthic macroalgae inhabit intertidal areas and must confront the many tide-induced stresses that affect their animal neighbors. Their restricted nearshore distribution limits the global importance of benthic plants as primary producers in the marine environment because they only cover a very small percentage of the ocean bottom. Yet within the nearshore communities in which they live, they play major roles as first trophic–level organisms.

The abundant plant groups so familiar on land—ferns, mosses, and seed plants—are poorly represented or totally absent from the sea. Instead, most marine macroalgae belong to two divisions (phyla): the brown algae in the division Ochrophyta (previously Phaeophyta) and the red algae in the

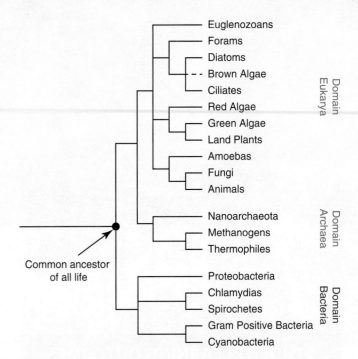

Figure 5.1 A phylogenetic tree illustrating a current hypothesis of the evolutionary relationships of the major groups of organisms based on molecular data. This tree employs the three-domain classification system, is based on phylogenetic analysis, and only includes major groups of organisms.

division Rhodophyta. These groups are almost completely limited to the sea. Two other divisions, the Chlorophyta and the flowering plants, Anthophyta, are found most commonly in freshwater and on land, yet they are important members of some shallow coastal marine communities. The characteristics of these divisions are summarized in **Table 5.1**. We begin our examination of large marine photosynthesizers with a familiar and more modern group, the Anthophyta (flowering plants), and then proceed to the seaweeds, whose ancestors preceded terrestrial plants.

5.1   Division Anthophyta

Marine flowering plants are found in great numbers and densities in localized areas along some seashores and in backwater bays and sloughs. In some areas, they help form the coastline or even small islands by supporting and accumulating sediments with their roots. Seagrasses are exposed to air only during very low tides, whereas salt marsh plants and mangroves are **emergent** and are seldom completely immersed in seawater. These coastal-dwelling plants display a secondary adaptation to the marine environment by just a few species of what is a predominantly terrestrial plant group, the flowering plants (division Anthophyta). Flowering plants are multicellular and are characterized by true leaves, stems, and roots, with water and nutrient-conducting structures running through all three of these basic structures.

TABLE 5.1 Major Divisions of Marine Algae and Plants and Their General Characteristics

Division/Phylum (common name)	Approximate Number of Living Species	General Size and Structure	Photosynthetic Pigments	Storage Products	Habitat
Ochrophyta (class Phaeophyceae) (brown algae)	2,040	Multicellular, macroscopic	Chlorophyll *a, c* Xanthophylls Carotenes	Laminarin and others	Mostly benthic
Rhodophyta (red algae)	7,101	Unicellular and multicellular, mostly macroscopic	Chlorophyll *a* Carotenes Phycobilins	Starch and others	Benthic
Chlorophyta (green algae)	6,034	Unicellular and multicellular, microscopic to macroscopic	Chlorophyll *a, b* Carotenes	Starch	Mostly benthic
Anthophyta (flowering plants)	300,000	Multicellular, macroscopic	Chlorophyll *a, b* Carotenes	Starch	Benthic

Data from algaebase.org/taxonomy.

Submerged Seagrasses

Twelve genera of seagrasses (classified in four families), including about 60 species, are dispersed around coastal waters of the world. Half of these species are restricted to the tropics and subtropics and are seldom found deeper than 10 m. The four common genera found in the United States are *Thalassia*, *Zostera*, *Phyllospadix*, and *Halodule*. *Thalassia*, or turtle grass (**Figure 5.2a**), is common in quiet waters along most of the Gulf Coast from Florida to Texas. *Zostera*, or eelgrass (**Figure 5.2b**), is widely distributed along both the Atlantic and Pacific coasts of North America. *Zostera* normally inhabits relatively quiet shallow waters but occasionally is found as deep as 50 m in clear water. Surf grass, *Phyllospadix* (**Figure 5.2c**), is found on both sides of the North Pacific and inhabits lower intertidal and shallow subtidal rocks that are subjected to considerable wave and surge action. *Halodule* prefers sandy areas with lower salinity in the Atlantic, Caribbean, southeastern Pacific, and Gulf of Mexico.

Most seagrasses produce horizontal stems, or **rhizomes**, that anchor the plants in soft sediments or attach them to rocks (Figure 5.2). From the buried rhizomes, many upright leaves develop to form thick green lawns of vegetation. These plants are a staple food for nearshore marine animals and migratory birds. Densely matted rhizomes and roots also accumulate sediments and organic debris to alter and shape the living conditions of the area.

Seagrasses reproduce either vegetatively by sprouting additional vertical leaves from the lengthening horizontal rhizomes or from seeds produced in simple flowers. The purpose of most showy flowers on land plants is to attract insects or birds so that **pollen** grains are transferred from one flower to another and cross-fertilization occurs. Pollen grains contain the plant's sperm cells. Submerged seagrasses use water currents for pollen transport; pollination occurs underwater in all seagrasses.

Some seagrasses, including *Zostera*, produce threadlike pollen grains about 3 mm long (about 500 times longer than their cargo, the microscopic chromosome-carrying sperm cells). After release, the next step is for the pollen grains of *Zostera* to

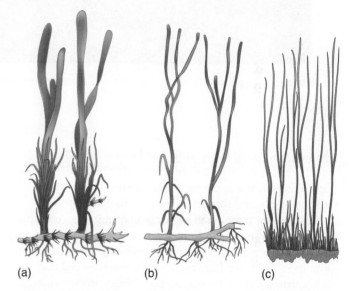

(a) (b) (c)

Figure 5.2 Three common seagrasses from different marine climatic regions: (a) turtle grass, *Thalassia*; (b) eelgrass, *Zostera*; and (c) surf grass, *Phyllospadix*.

become ensnared on the **stigma** (the pollen-receptive structure of the female flower) of another plant, and fertilization then occurs. *Thalassia* produces small round pollen grains released in a thread of sticky slime. When the slime thread lands on the appropriate stigma of another plant (also covered with a surface film of slime), the two slime layers combine to produce a firm bond between the pollen grain and the stigma, and fertilization follows. This two-component adhesive acts like epoxy glue to produce a strong bond after the separate components are mixed. It also provides a mechanism for selecting between compatible and foreign types of pollen grains. The stigma–pollen bond will only be formed on contact with pollen of the same species. Foreign pollen grains do not adhere and are washed away, possibly to try again on another plant.

Mature seeds of each type of seagrass are adapted to their preferred habitat and conditions in the habitat. Eelgrass seeds drop into the mud and take root near the parent plant, whereas the fruits of *Thalassia* may float for long distances before

Figure 5.3 A typical view of a seagrass covered seafloor. This photo was taken in the Florida Keys National Marine Sanctuary, and includes a large jack in the background. Courtesy of NOAA.

releasing their seeds in the surf. The fruits surrounding individual seeds of *Phyllospadix* are equipped with bristly projections. When shed into the surf, these bristles snag branches of small seaweeds, and the seeds germinate in place.

Like reef-forming corals and tropical mangroves, countless seagrass **blades** growing in tropical lagoons provide a large surface area on which other organisms (**epibionts**) can attach and grow. In St. Ann's Bay, Jamaica, researchers determined that, on each square meter of seafloor, seagrass blades provide an average of nearly 300 m² of surface on which epibionts can attach. This vast expanse of surface area does not go unnoticed by local organisms. About 175 species of plants, algae, and animals have been observed living attached to blades of turtle grass in the Caribbean region, including various micro- and macroalgae, sponges, hydroids, sea anemones, amphipods, ectoprocts, tunicates, annelids, and snails. These same seagrass beds support the foraging activities of a few species of marine mammals that are unusual because, unlike the large majority of marine mammals, they are herbivores.

If you were to visit a tropical seagrass meadow with a mask and snorkel, your first impression would likely be that there is little apparent life in seagrass beds besides the occasional juvenile fish passing by or cryptic seahorse attached to a blade (**Figure 5.3**). This is because, other than the epibionts, which simply appear as a whitish fuzz on older blades, you would encounter very few animals. The primary reason for the small number of animals among seagrass is that this habitat experiences a great deal of sedimentation. Waves arrive from the open sea and break on the reef flat, creating sediment-laden currents that stream into the lagoon. In the lagoon, the currents slow as they are forced to meander through millions of seagrass blades. This decreased current velocity is insufficient to transport larger sediment particles, and they begin to settle onto the seafloor among the seagrass. Many organisms cannot endure this high rate of sedimentation because it interferes with feeding, it hinders respiration by clogging gills, it easily abrades soft tissues, and it buries smaller organisms in an avalanche of particles; however, the high sedimentation rates that repel many potential seagrass residents are actually attractive to deposit-feeding sea cucumbers and mojarras, silvery fishes that make a living by straining mouthfuls of sediment through their gill rakers in search of organic morsels. One group of organisms that take advantage of seagrass beds for cover is various types of juvenile fish. Dense seagrass beds provide many excellent hiding spots for fish that are small and highly susceptible to predation.

The expansive meadow of seagrass and macroalgae that grows in most tropical lagoons is an irresistible source of food to several common herbivores. In addition to high concentrations of herbivorous parrotfishes and surgeonfishes that leave the safety of the adjacent reef at night to forage in seagrass, green sea turtles and a few species of large marine mammals rely extensively on tropical seagrass beds and are important components of seagrass communities. Manatees and dugongs consume a wide variety of tropical and subtropical seagrasses, including *Enhalus*, *Halophila*, *Halodule*, *Cymodocea*, *Thalassia*, *Thalassodendron*, *Syringodium*, and *Zostera*. Algae also are eaten, but only in limited amounts if seagrasses are abundant. **Figure 5.4** displays manatee feeding behavior.

Figure 5.4   A West Indian manatee grazing on submerged plants.
© feel4nature/Shutterstock, Inc.

Case Study

Johnson's Seagrass (*Halophila johnsonii*): The First and Only Marine Plant Listed Under the Endangered Species Act

Until 1998, no marine plants or algae were listed as threatened or endangered under the Endangered Species Act (ESA). Johnson's seagrass was the first marine plant to be listed as threatened under the ESA in 1998 due to concerns about the health of the seagrass beds throughout the very limited distribution of this species, and it remains on the threatened species list today (**Figure A**). In general, marine plants and algae are less likely to be listed under the ESA because most are not harvested, or removed from the ocean, like many marine fishes, invertebrates, or corals. Although Johnson's seagrass has never been harvested, it has faced numerous threats, including damage

Figure A   Johnson's seagrass (*Halophila johnsonii*) growing along a sandy bottom. Notice the oval-shaped blades and horizontal rhizomes.
NOAA, Julie Christian.

(*continues*)

Case Study   (*continued*)

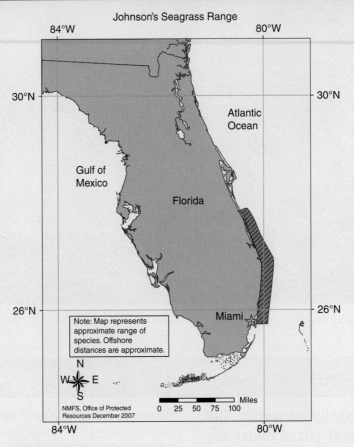

Johnson's Seagrass Range

Figure B Map of the range of Johnson's seagrass based on survey data.
Courtesy of NOAA.

from boats, anchors, moorings, and dredging. Less than optimal water quality due to urban and agricultural runoff has led to poor conditions for this species and has increased growth rates of algae that smother the seagrass. Human activities near the shore have led to increased amounts of silt (very fine bottom sediments) in the water, which cover and smother the seagrass. When the seagrass is covered by silt, photosynthetic rates decline because its leaves are not exposed to as much sunlight.

The distribution of Johnson's seagrass is very limited compared to most other seagrasses. It is only found from Sebastian Inlet in central Florida to Biscayne Bay in southern Florida (**Figure B**). One unique and defining characteristic of this species is that as far as we know, it only reproduces asexually, and it has been hypothesized that the lack of sexual reproduction has contributed to its limited distribution. Another factor that limits distribution of this species is its need for stable substrate, which can only be found in calm waters. A big physical difference between Johnson's seagrass and other seagrass species is its unique oval- or spatula-shaped leaf. The leaves also occur in pairs along a horizontal rhizome.

Some may ask why it is so important to protect and attempt to restore populations of a single marine plant species. Johnson's seagrass is a great example of a species that strongly impacts an entire ecosystem. It serves as a nursery habitat for young fishes in

the lagoon and as shelter for many other small organisms. It is food for the endangered West Indian manatee and threatened green sea turtle. It stabilizes the sediments in relatively deep areas in the coastal lagoons where it is found, areas where other seagrasses are unable to reside. It is also more tolerant of changes in temperature, salinity, and drying out than other species that reside in the same geographic area. It is unique and plays a key role in the coastal lagoon habitat of central and south Florida. Several of the basic life history characteristics of this species are unknown, and it is this information that scientists now seek to help understand how the species can be saved from future declines. Surveys of the abundance and distribution of Johnson's seagrass are limited, but in 2009 it appeared that populations were increasing. It is not too late to conserve this species if measures to reduce threats to its habitat are taken.

Critical Thinking Questions

1. Why do you think humans should be concerned with protecting a plant with such a limited distribution? Why not just let it become extinct?

2. Devise a plan for protecting this plant. Be sure to take into account the biological information presented in the case study.

Emergent Flowering Plants

Several other species of flowering plants often exist partially submerged on bottom muds of coastal salt marshes protected from strong ocean wave action. These plants are usually situated so that their roots are periodically, but not constantly, exposed to tidal flooding. They are terrestrial plants that have evolved various degrees of tolerance to excess salts from sea spray and seawater. Some even have special structural adaptations for their semimarine existence. The cordgrass, *Spartina*, for example, actively excretes excess salt through special two-celled salt glands on its leaves. Even so, several species of *Spartina* have higher experimental growth and survival rates in freshwater than in seawater. This difference strongly suggests that the salt marsh does not provide optimal growth conditions for *Spartina*, even though the salt marsh is its natural habitat. Competition with other land and freshwater plants may have forced *Spartina* and other salt-tolerant species into the more restricted areas of the salt marshes.

Salt marsh plants contribute heavily to detritus production in their protected environments as well as in nearby bays and estuaries. Some feature extensive stands containing several species of emergent grasses, especially various species of *Spartina*. At slightly higher elevations, these grasses give way to succulents (e.g., *Salicornia* and *Suaeda*), a variety of reeds and rushes, and the brush and smaller trees of the local woodland. These lush pastures are extremely productive and harbor a unique assemblage of organisms, including commercially important shellfish and finfishes. Yet as large urban centers develop near them, they have become popular sites for waste dumping, recreation, dredging and filling, and other detrimental uses. The degradation of salt marshes is a serious and worldwide problem that becomes more severe as human populations expand and place more pressure on these fragile habitats.

Several species of shrubby to treelike plants, the mangroves, create dense thickets of tidal woodlands known as **mangals** (**Figure 5.5**). Mangals dominate large expanses of muddy shores in warmer climates and are excellent examples of emergent plant-based communities. Mangroves range in size from small shrubs to 10-m-tall trees whose roots are tolerant to seawater submergence and are capable of anchoring in soft muds. Collectively, mangrove plants, the major component of mangal communities, line about two thirds of the tropical coastlines of the world (**Figure 5.6**).

Members of these mangal communities are supported on their muddy substrate by numerous prop roots that grow down from branches above the water. The pattern of mangrove development illustrates well a series of adaptations needed to exist on muddy tropical shores (**Figure 5.7**). Red mangroves (*Rhizophora*) produce seeds that germinate while still hanging from the branches of the parent tree. As the seedlings develop and grow longer, their bottom ends become heavier. When the seedlings eventually drop from the parent plant into the surrounding water, they float upright, bobbing at the water's surface; are dispersed by winds or tides; and finally implant in muddy sediments along shallow shorelines. There, the seedlings promptly develop small roots to anchor themselves and continue to mature. The resulting tangle of growing roots traps additional sediments and increases the structural complexity of mangal communities. Birds, insects, snails, and other terrestrial

Figure 5.5 Dense mangal thicket near the Florida Everglades.
© Ann Cantelow/Shutterstock, Inc.

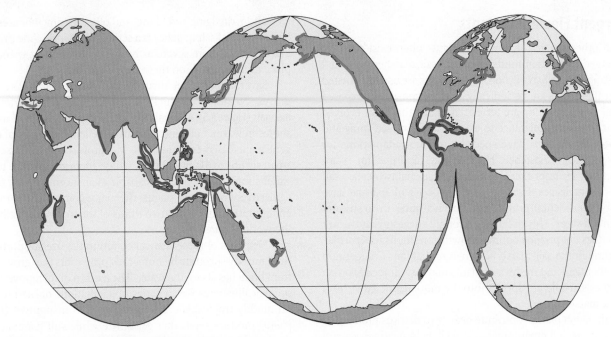

Figure 5.6 Distribution of salt marsh (orange) and mangals (maroon).

Figure 5.7 Germination cycle of a mangrove seedling.

animals occupy the upper leafy canopy of the mangroves, and a variety of fishes, crustaceans, and mollusks live on or among the root complex growing down into the mud. Mangals are another great example of a habitat that provides excellent hiding spots for juvenile fishes. Because the leafy portions of these plants are above the water level, few marine animals graze directly on mangrove plants. Instead, leaves falling from these plants into the quiet waters surrounding their roots provide an important energy source for the detritus-based food webs of these communities.

In the United States, the distribution of mangals reflects their need for warm waters protected from wave action; they

are found only along portions of the Gulf of Mexico and the Atlantic coast of Florida. The south coast of Florida is dominated by extensive interconnected shallow bays, waterways, and mangals. These mangals form a nearly continuous narrow band along the coast, with smaller fingers extending inland along creeks. Inland, toward the freshwater Everglades, the mangroves are not high, but tree height of red, black, and white mangroves (the three most common species in the Southeast United States) increases to as much as 10 m at the coast. It is these taller coastal members of mangal communities that are especially prone to hurricane damage. In 1992, Hurricane Andrew cut a swath of destruction across south Florida with sustained winds up to 242 km/hr. The accompanying storm surge lifted the sea surface more than 5 m above normal levels. Some of the more exposed coastal mangal communities experienced greater than 80% mortality, due mostly to wind effects and lingering problems of coastal erosion.

On August 29, 2005, Hurricane Katrina struck the coasts of Louisiana, Mississippi, and Alabama. This important coastline houses 15 major fishing ports, nearly 200 seafood processing plants, and nearly 15,000 state and federally permitted fishing vessels, which together produce 10% of the shrimp and 40% of the oysters consumed in the United States. Two months after Katrina made landfall, her effects on seafood production, and coastal fauna and habitats, were assessed. It is estimated that Katrina caused $1.1 billion in losses to seafood production for Louisiana and about $200 million in losses to Alabama and Mississippi, respectively. Moreover, these initial losses to seafood production may persist because benthic communities along this coastline experienced significant reductions in biodiversity as well as shifts in the composition and ranking of dominant taxa. Coastlines altered by human use and degradation are not able to withstand the large storm surge and winds that accompany a hurricane. Coastlines with healthy communities of submerged or partially submerged plants are best suited for storm conditions. The plants act like a sponge to soak up water and help to keep storm surges from flooding land.

DID YOU KNOW?

Waves breaking on shore are so strong that they can break a wooden boat to pieces, yet some seagrasses actually thrive in strong surf. Surf grass, *Phyllospadix*, has adaptations for life in the harsh conditions of the wave-swept coast. Its leaves are extremely flexible and strong, stronger than any marcroalga. Extra extensions, called *root hairs*, on its roots grow longer than on other seagrass species for anchoring onto rocks. The native people of the Pacific coast used the leaves of surf grass to weave baskets because the leaves are flexible, yet strong, when they dry out.

5.2 The Seaweeds

By far, most of the large conspicuous forms of marine photosynthesizers are seaweeds, also known as *macroalgae*. The term *seaweed* is used here in a restricted sense, referring only to macroscopic members of the divisions Chlorophyta (green algae), Ochrophyta (brown algae), and Rhodophyta (red algae) (Table 5.1). Seaweeds are multicellular photosynthetic organisms that do not produce seeds or flowers. They also do not have true roots, leaves, or stems or **vascular tissue**. Some members of Chlorophyta and Rhodophyta are unicellular; these are not considered seaweeds, but rather microalgae.

Seaweeds are abundant on hard substrates in intertidal zones and commonly extend to depths of 30 to 40 m. In clear tropical seas, some species of red algae thrive at depths as great as 200 m, and one species has been reported as deep as 268 m

in the Bahamas. Many seaweeds tolerate or even require extreme surf action on exposed rocky intertidal outcrops, where they are securely fixed to the solid substrate. Where they are abundant, seaweeds can greatly influence local environmental conditions for other types of shallow-water marine life by protecting them from waves and providing food, shade, and sometimes a substrate on which to attach and grow. Macroalgae such as *Halimeda*, *Penicillus*, *Acetabularia*, *Sargassum*, *Udotea*, and *Caulerpa* are often very common in tropical lagoons, and because some of these also precipitate calcium carbonate ($CaCO_3$) that they have extracted from seawater (like corals), they contribute additional carbonate sediment to the seafloor after they die. About 85% of the biomass of the Caribbean's *Halimeda* is $CaCO_3$, and this genus alone can contribute 3 kg of carbonate sand per square meter in just 1 year.

Structural Features of Seaweeds

Seaweeds are not as complex as the flowering plants. Seaweeds lack true roots, flowers, seeds, stems, and leaves. Nevertheless, within these structural limitations, seaweeds exhibit an unbridled diversity of shapes, sizes, and structural complexity. They also contain structures that resemble roots, stems, and leaves. Upon observing **Figure 5.8**, you will notice that a blade resembles a leaf, **haptera** resemble roots, and a **stipe** resembles a stem. The main difference between the seaweed structures and plant structures is that there is no vascular tissue, or plumbing

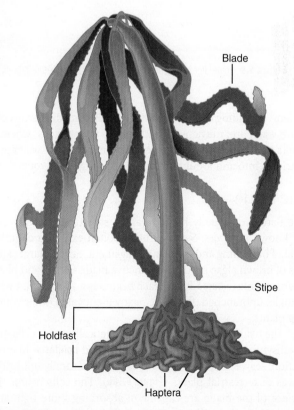

Figure 5.8 The northern sea palm *Postelsia* (brown algae) is equipped with a relatively large stipe and a massive holdfast.

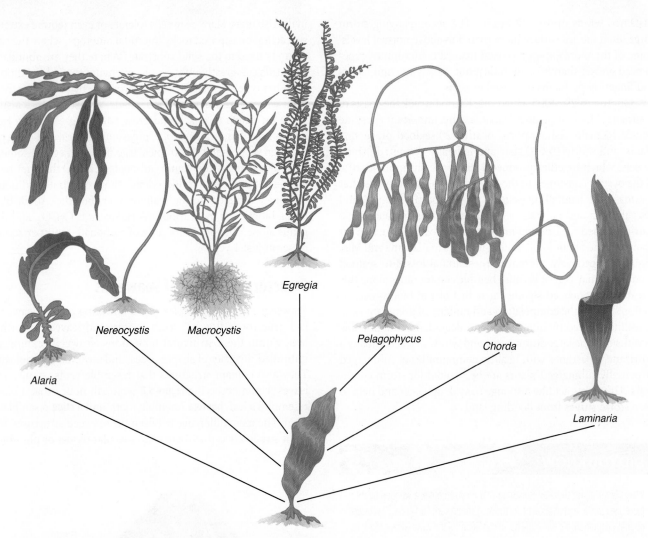

Figure 5.9 Some large kelp plants of temperate coasts. Each mature individual develops from a young individual with a single flat blade.

system, extending through seaweed structures. Most seaweeds move water and nutrients directly across their tissues, whereas plants require a complex plumbing system to move water and nutrients into and around the tissue, starting at the roots.

The Blade

The flattened, usually broad, leaflike structures of seaweeds are known as *blades*. Seaweed blades often exhibit a complex level of branching and cellular arrangement. Several larger species of brown algae produce distinctive blade shapes and blade arrangements (**Figure 5.9**), yet each begins as a young plant with a single, unbranched, flat blade nearly identical to other young **kelp** plants.

The blades house many photosynthetically active cells, but photosynthesis also occurs in the stipes and **holdfasts**. In cross section, seaweed blades (**Figure 5.10a**) are structurally unlike the leaves of terrestrial plants (**Figure 5.10b**). The cells nearer the surface of the blade are capable of absorbing more light and are photosynthetically more active than those cells near the center of the blade. "Veins" of conductive tissue and distinctions between the upper and lower surfaces are lacking in the

blades of seaweeds. Because the flexible blades either droop in the water, float erect, or are continuously tossed by turbulence, there is no defined upper or lower surface. The two surfaces of the seaweed blade are usually exposed equally to sunlight, nutrients, and water and are therefore equally capable of carrying out photosynthesis. Unlike seaweeds, flowering plants (including seagrasses) exhibit an obvious asymmetry of leaf structure, with a dense concentration of photosynthetically active cells crowded near the upper surface (Figure 5.10b). In a plant leaf, below the upper epidermis and palisade mesophyll is a spongy layer of cells separated by large spaces to enhance the exchange of carbon dioxide, which is often 100 times less concentrated in air than in seawater.

Pneumatocysts

Several large kelp species have gas-filled floats, or *pneumatocysts*, to buoy the blades toward the sunlight at the surface. Pneumatocysts are filled with the gases most abundant in air—N_2, O_2, and CO_2—although some kelp pneumatocysts also contain a small amount of carbon monoxide (CO). Again, there is a large diversity in size and structure. The largest pneumatocysts

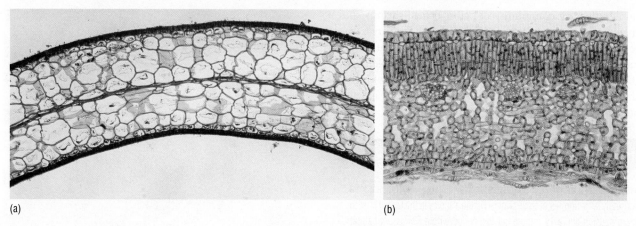

(a)

(b)

Figure 5.10 Cross sections of a blade of a typical marine alga, *Nereocystis* (a), and a typical flowering plant leaf (b). Note the complexity of the plant leaf compared to the alga.
© Biodisc/Visuals Unlimited; © Runk/Schoenberger/Alamy Images.

Figure 5.11 A portion of floating *Sargassum* with gas-filled floats becomes a mobile home for various fishes and invertebrates. Small tile fish and trigger fish call this algae home.
NOAA Ocean Explorer Gallery.

belong to *Pelagophycus*, the elkhorn kelp (Figure 5.9). Each *Pelagophycus* plant is equipped with a single pneumatocyst, sometimes as large as a basketball, to support six to eight immense drooping blades, each of which may be 1 to 2 m wide and 7 to 10 m long.

In strong contrast to *Pelagophycus*, *Sargassum* has numerous small pneumatocysts (**Figure 5.11**). A few species of *Sargassum* lead a pelagic life afloat in the middle of the North Atlantic Ocean (the "Sargasso Sea"). In the Sargasso Sea, *Sargassum* creates large patches of floating plants that are the basis of a complex floating community of crabs, fishes, shrimp, and other animals uniquely adapted to living among the *Sargassum*. Large masses of this plant community sometimes float ashore on the U.S. East and Gulf Coasts, creating odor problems for beachgoers as the dying plants decompose. In the Sea of Japan, other species of attached intertidal *Sargassum* break off and also become free-floating for extended periods of time.

Figure 5.12 The holdfast of the large kelp, *Macrocystis*, with a complex interlocking mass of haptera that help anchor it to the seafloor.
© Courtesy of Deanna Pinkard-Meier.

DID YOU KNOW?

Floating mats of a brown alga, *Sargassum*, provide a unique microhabitat for a surprisingly large number of small critters. During a survey in the Gulf of Mexico using a neuston net, a 15-minute net tow period within *Sargassum* habitat yielded over 3,000 fish in 82 kg (180 pounds) of *Sargassum*! That is 37 fish per kilogram of *Sargassum*, or 17 fish per pound. *Sargassum* is not only home to small fishes and invertebrates, but some sea turtles spend time hiding out in *Sargassum* mats to feed and drift.

The Stipe

A flexible stemlike stipe connects the wave-tossed blades of seaweeds to their securely anchored holdfasts at the bottom. An excellent example is *Postelsia*, the sea palm (Figure 5.8), which grows attached to rocks only in the most exposed surf-swept portions of the intertidal zone. Its hollow resilient stipe is remarkably well suited for yielding to the waves without breaking.

The blades of some seaweeds blend into the holdfast without forming a distinct stipe. In others, the stipe is very prominent and occasionally extremely long. The single long stipes of *Nereocystis*, *Chorda*, and *Pelagophycus* (Figure 5.9) provide a kind of slack-line anchoring system and commonly exceed 30 m in length. The complex multiple stipes of *Macrocystis* are often even longer.

Special cells within the stipes of *Macrocystis* and a limited number of other brown and red algal species form conductive tissues strikingly similar in form to those present in stems of terrestrial plants. Radioactive tracer studies have shown that these cells transport the products of photosynthesis from the blades to other parts of the plant. In smaller seaweeds, the necessity for rapid efficient transport through the stipe is minimal, and such internal transport is lacking.

The Holdfast

Holdfasts of the larger seaweeds often superficially resemble root systems of terrestrial plants; however, the basic function of the holdfast is to attach the plant to the substrate. The holdfast seldom absorbs nutrients for the plant as do true roots. Holdfasts are adapted for getting a grip on the substrate and resisting violent wave shock and the steady tug of tidal currents and wave surges. The holdfast of *Postelsia* (Figure 5.8), composed of many short, sturdy, rootlike haptera, illustrates one of several types found on solid rock.

Other holdfasts are better suited for loose substrates. Although *Macrocystis* most often anchors to rocks, the holdfast

of this genus has a large diffuse mass of haptera able to penetrate muddy or sandy bottoms and stabilize a mass of sediment for anchorage (**Figure 5.12**). Holdfasts of many smaller species do the same thing on a much smaller scale, with many fine filaments embedded in sand or mud on the sea bottom.

A variety of small red algae are epiphytes and demonstrate special adaptations for attaching themselves to other marine plants. **Figure 5.13** illustrates two common red algal epiphytes attached to a strand of surf grass. Using other marine plants as substrates for attachment is a common habit of many smaller forms of red algae.

Photosynthetic Pigments

Each seaweed division is characterized by specific combinations of photosynthetic pigments that are reflected in their color and in the common name of each division (Table 5.1). The bright grass-green color of green algae is due to the predominance of chlorophylls over accessory pigments. Green algae vary in structure from simple filaments to flat sheets (**Figure 5.14**) and diverse complex branching forms. They are usually less than 0.5 m long, but one species of *Codium* from the Gulf of California occasionally grows to 8 m in length. When compared with brown and red algae, the Chlorophyta have fewer marine species, yet in some locations their limited diversity is compensated with dense populations of individuals from one or two species.

The photosynthetic pigments of the Ochrophyta (brown algae) sometimes appear as a greenish hue, but, more often, the green of the chlorophyll is partially masked by the golden xanthophyll pigments, especially fucoxanthin, which is characteristic of this division. This blend of green and brown pigments usually results in a drab olive-green color (**Figure 5.15**). Many of the larger and more familiar algae of temperate seas belong to this division. A number of species are quite large and are sometimes collectively referred to as *kelp* (Figure 5.9). In temperate and high latitudes, these species usually dominate the marine benthic vegetation. Numerous smaller, less obvious brown algae are also common in temperate and cold waters, as well as in tropical areas.

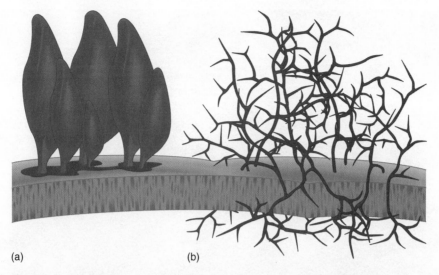

(a)   (b)

Figure 5.13   Two red algal epiphytes, (a) *Smithora* and (b) *Chondria*, attached to a leaf of *Phyllospadix*.

Figure 5.14   A healthy growth of the green algae *Ulva* lives on the sand during low tide.

Figure 5.15 The brown algae *Padina* with its intricate design and olive-green color typical of many brown algae.
NOAA Photo Library.

Red algae with red and blue phycobilin pigments, as well as chlorophyll, exhibit a wide range of colors. Some are bright green, such as *Porphyra*, the popular seaweed known as nori that is used in sushi rolls; others are sometimes confused with brown algae. However, most red algae living below low tide range in color from soft pinks to various shades of purple or red (**Figure 5.16**). Red algae are as diverse in structure and habitat as they are in coloration, and they seldom exceed a meter in length. Many red algae have very tiny and intricate patterns of branching and blade designs, making them difficult to identify without a microscope.

One might hypothesize that the green algae and seagrasses, with their large proportion of chlorophyll pigments, do not fare well at moderate depths because of their limited ability to absorb the deeper-penetrating green wavelengths of sunlight, but plants and algae can adapt to low or limited wavelength light conditions in other ways; for example, some green algae have dense concentrations of chlorophyll that appear almost black, so they are able to absorb light at essentially all visible wavelengths. In addition, most green plants have chlorophyll *b* as well as chlorophyll *a*. Chlorophyll *b* has a strong light-absorbing peak in the blue region of the visible spectrum and can collect a good fraction of the deep-penetrating blue light available in tropical waters. Still, red and brown algae, with their abundant xanthophyll and phycobilin pigments working in concert with chlorophyll, generally have a slight competitive advantage in occupying the deeper portions of the photic zone in turbid coastal waters, and because they also have chlorophylls they function at no disadvantage in shallow waters or intertidal zones.

Reproduction and Growth

Reproduction in seaweeds, as well as in most plants, is highly variable and sometimes very complex, and can be either sexual, involving the fusion of sperm and eggs, or asexual, relying on vegetative growth of new individuals. Some seaweeds reproduce both ways, but a few are limited to vegetative reproduction only. The pelagic species of *Sargassum*, for instance, maintain their populations by an irregular vegetative growth followed by fragmentation into smaller clumps. The dispersed fragments of *Sargassum* are capable of continued growth and regeneration for decades. Sexual reproduction is lacking in the pelagic species of *Sargassum* but not in the attached benthic forms of the same genus.

Much of the structural variety observed in seaweeds is derived from complex patterns of sexual reproduction, patterns that define the life cycles of seaweeds. For our purposes, these complex life cycles can be simplified to three fundamental patterns. The sexual reproduction examples of the first two types

Figure 5.16   Several species of red algae growing at approximately 75 m depth (225 feet) on McGrail Bank. Notable species include *Halymenia sp.*, composed of large, flat blades; *Gracilaria blodgettii*, a cylindrical species; *Kallyemia westii*, a pale pink species with perforated blades; and *Coelarthrum cliftonii*.
Image courtesy of FGBNMS/UNCW-NURC.

described here are not meant to cover the entire spectrum of seaweed life cycles, but are used to illustrate the basic patterns that underlie the complexity and variation involved in sexual reproduction of seaweeds.

In the life cycle of most of the larger seaweeds, an alternation of **sporophyte** and **gametophyte** generations occurs. The green alga *Ulva* represents one of the simplest patterns of alternating generations (**Figure 5.17**). This basic life cycle is a hallmark of marcroalgae and plants. The cells of the macroscopic *Ulva* sporophyte are **diploid**; that is, each cell contains two of each type of chromosome characteristic of that species. Some cells of the *Ulva* sporophyte undergo **meiosis** to produce single-celled flagellated spores. As a result of meiosis, these spores contain only one chromosome of each pair present in the diploid sporophyte and are said to be **haploid**.

The spores of *Ulva* and other green algae each have four flagella, whereas each gamete has two flagella that are equal in length and project from one end of the cell. Spores produced by *Ulva* are capable of limited swimming and then settle to the bottom. Once they reach the seafloor they immediately

germinate by a series of mitotic cell divisions to produce a large, multicellular, gametophyte generation that is still haploid. Cells of the gametophyte, in turn, produce haploid gametes, each with two flagella, that are released into the water. When two gametes from different gametophytes meet, they fuse to produce a diploid single-celled **zygote**. By repeated mitotic divisions, the zygote germinates and completes the cycle by producing a large, multicellular, diploid sporophyte once again. In *Ulva*, the sporophyte and gametophyte generations are identical in appearance. The only structural difference between the two forms is the number of chromosomes in each cell; diploid sporophyte cells have double the chromosome number of haploid gametophyte cells.

The life cycles of numerous other seaweeds are characterized by a suppression of either the gametophyte or the sporophyte generation. In the green alga *Codium* and the brown alga *Fucus*, the multicellular haploid generation is completely absent. The only haploid stages are the gametes, as is seen in animals. In other large brown algae, the gametophyte stage is reduced in size. The life cycle of *Laminaria* is similar to that of most

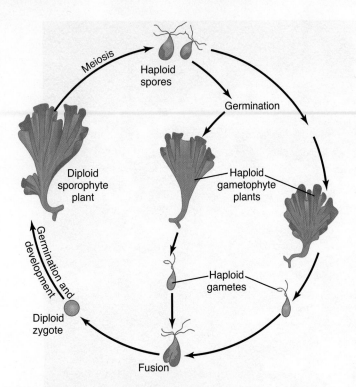

Figure 5.17 The life cycle of the green alga *Ulva*, alternating between diploid sporophyte and haploid gametophyte generations.
Data from E. Y. Dawson, *Marine Botany of Marine Plants* (Holt, Rinehart and Winston, 1966).

These haploid spores swim to the bottom and quickly attach themselves. They soon germinate into very small, yet multicellular, gametophytes. The female gametophyte produces large, nonflagellated eggs. The egg cells are fertilized in place on the female gametophyte by flagellated male gametes, the sperm cells produced by the male gametophyte. After fusion of the gametes, the resulting zygote germinates to form another large sporophyte. The flagellated reproductive cells of brown algae always have two flagella of unequal lengths, and they insert on the sides of the cells rather than at the ends.

Red algae lack flagellated reproductive cells and are dependent on water currents to transport the male gametes to the female reproductive cells. The most common life cycle of red algae has three distinct generations, somewhat reminiscent of the reproductive cycle outlined for *Ulva* (Figure 5.17). A diploid sporophyte produces haploid spores that germinate into haploid gametophytes. Instead of producing a new sporophyte, however, the gametes from the gametophytes fuse and develop into a third phase unique to red algae, the **carposporophyte**. The carposporophyte then produces **carpospores** that develop into sporophytes, and the cycle is completed.

The development of a large, multicellular seaweed from a single microscopic cell (either a haploid spore or a diploid zygote) is essentially a process of repeated mitotic cell divisions. Subsequent growth and differentiation of these cells produce a complex plant with many types of cells, each specialized for particular functions. After the plant is developed, additional cell division and growth occur to replace tissue lost to animal grazing or wave erosion; however, such cell division is commonly restricted to a few specific sites within the plant that contain **meristematic tissue** capable of further cell division. These meristems frequently occur at the upper growing tip of the plant. In kelp plants and some other seaweeds, additional meristems situated in the upper and lower portions of the stipe provide

other large kelp plants and serves as an excellent generalized example of seaweeds with a massive sporophyte that alternates with a reduced gametophyte (**Figure 5.18**). Special cells (called **sporangia**) on the blades of the diploid sporophyte undergo meiosis to produce several flagellated microscopic spores.

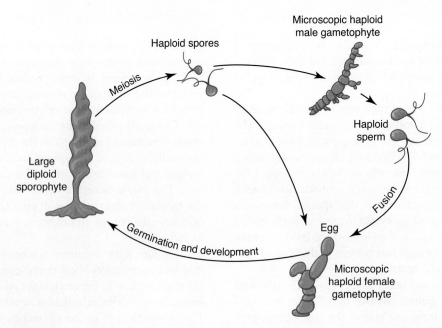

Figure 5.18 The life cycle of *Laminaria*, similar to the life cycles of other large kelps, alternates between a large diploid sporophyte and microscopic haploid gametophyte generations.

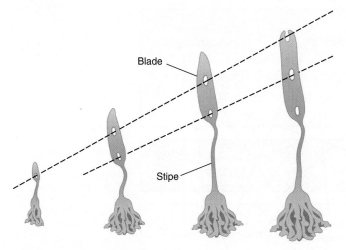

Figure 5.19 Generalized growth pattern of a kelp. Punched holes and dashed lines indicate the pattern of blade elongation.
Data from: K. H. Mann, *Marine Biology* 14 (1973):199–209.

additional cells to elongate the stipe and blades. The meristematic activity of a cell layer near the outer stipe surface of some kelp species provides lateral growth to increase the thickness of the stipe. The stipes of a few perennial species of kelp, including *Pterygophora* and *Laminaria*, retain evidence of this secondary lateral growth as concentric rings that resemble the annual growth rings of trees.

In the spring, during periods of rapid growth, the rate of stipe elongation in large *Nereocystis*, *Pelagophycus*, and *Macrocystis* plants often exceeds 30 cm/day. Many kelp species produce kelp blades resembling moving belts of plant tissue (**Figure 5.19**), growing at the base and eroding or being eaten away at the tips. At any one time, the visible plant itself (the standing crop) may represent as little as 10% of the total material it produced during a year.

Kelp Forests

Most kelp plants are **perennial**. Although they may be battered down to their holdfasts by winter waves, their stipes will regrow from the holdfast for several successive seasons. Thus, the extent of the kelp canopy and the overall three-dimensional structure of the kelp forest are quite variable over annual cycles. Occasionally, herbivore grazing or the pull of strong waves frees the holdfast and causes the plant to wash ashore. More commonly, small fragments of blades and stipes are continually eroded away to decompose into food for detritus feeders.

Along most of the North American west coast, subtidal rocky outcrops are cloaked with massive growths of several species of brown algae, dominated by either *Macrocystis* or *Nereocystis* (**Figure 5.20**). West coast kelp forests occur as an offshore band paralleling the coastline because wave action tears these plants out nearer to shore, and light does not penetrate to the seafloor farther offshore. In the dimmer light below the canopy of these large kelps exists a shorter understory of mixed brown and red algae. Together, these large and small kelp plants accomplish very high rates of primary production and support a complex community of grazers, suspension feeders, scavengers, and predators (**Figure 5.21**). From central California northward, kelp abundance varies in a predictable manner seasonally, and the fishes are dominated by several species of rockfishes in the genus *Sebastes*. Southern California kelp forest abundance varies more irregularly and is especially vulnerable to the influences of El Niño-Southern Oscillation (ENSO) events. Here, the dominant fishes include perches, damselfishes, and wrasses, reflecting the more tropical affinities of these fishes. Rockfishes are found in southern California kelp forests, but in lower abundance; rockfish densities are higher on rocky reefs farther offshore.

Compared with the richness of species observed in the western Pacific North American kelp forests, the kelp beds of the northwestern Atlantic Coast exhibit low diversity in most taxonomic groups. Unlike the U.S. West Coast, the rocky intertidal

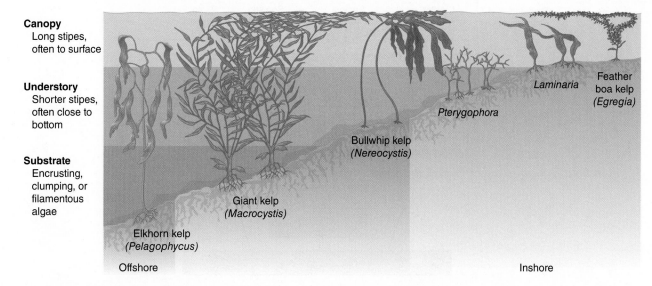

Figure 5.20 General structure of a U.S. West Coast kelp forest, with a complex understory of plants beneath the large dominant *Macrocystis* or *Nereocystis*.

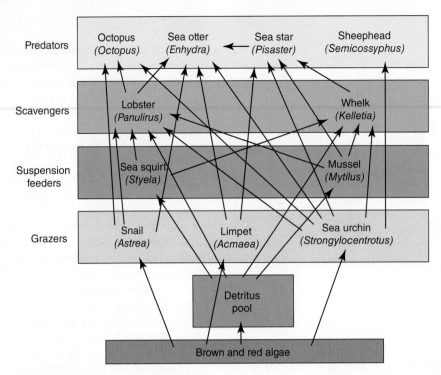

Figure 5.21 Trophic relationships of some dominant members of a southern California kelp community.

and subtidal shores of New England states and neighboring Canadian Maritime Provinces were scoured to bare rock (in places to several hundred meters below sea level) by several episodes of continental glaciation. Only since the retreat of the most recent glacial episode 8,000 to 10,000 years ago have these shores been recolonized, and that recolonization is not yet complete. In simple terms, the rocky shores of the U.S. East Coast are relatively young and over time will likely develop more complexity.

The lower species diversity of northwestern Atlantic kelp beds leads to somewhat simpler trophic interactions than those occurring in U.S. West Coast kelp forests; still, similar species occupy the same major trophic roles (**Figure 5.22**). The macroscopic primary producers are dominated by the kelp *Laminaria*, with an understory of mixed red and brown foliose algae. In clear patches below about 10 m, encrusting coralline red algae cover rock surfaces with a bright pink pavement of $CaCO_3$. These coralline crusts are maintained indirectly by the constant grazing actions of sea urchins on the larger kelp plants that would otherwise shade out the encrusting species. On this coast, the lower limit of growth for *Laminaria* and other large

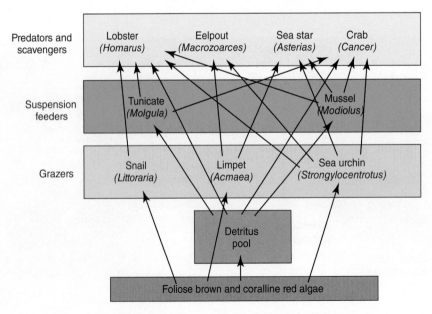

Figure 5.22 Trophic relationships of some dominant members of a New England kelp community.

kelps is controlled not by low light intensity, as it is on the West Coast, but by the presence of grazing sea urchins.

On the U.S. West Coast, too, kelp beds exist in a delicate balance with their major grazers, sea urchins. Since World War II, kelp beds on both coasts have been devastated by dense aggregations of sea urchins grazing on the holdfasts, causing the remainder of the plant to break free and wash onto the shore. These large sea urchin populations, capable of completely eliminating local kelp beds and turning them into what are termed *urchin barrens*, seem free of the usual population regulatory mechanisms—predation and starvation. A major predator of West Coast kelp bed sea urchins is the sea otter (*Enhydra*). East Coast sea urchins are similarly preyed on by the lobster *Homarus*; however, both of these predators have been subjected to intensive commercial harvesting and have experienced major population reductions in the past 200 years.

Available evidence indicates that the effects of this reduced predation have been magnified by increased concentrations of dissolved and suspended organic materials in coastal waters (mostly from urban sewage outfalls). The U.S. Office of Technology Assessment has identified more than 1,300 major industries and 600 municipal wastewater treatment plants that discharge into the coastal waters of the United States. Standard secondary treatment of sewage is intended to separate solids and to reduce the amount of organic matter (which contributes to biochemical oxygen demand), nutrients, pathogenic bacteria, toxic pollutants, detergents, oils, and grease in wastewater.

In the United States, most ocean discharges of wastewater are supposed to meet these secondary treatment standards, but many still do not, including some that discharge into southern California coastal waters. Until the mid-1980s, treated sewage containing about 250,000 tons of suspended solids was discharged from 4 large and 15 small publicly owned sewage treatment plants each day. These solids are similar to detritus from natural marine sources in their general composition and nutritional value for zooplankton and benthic detritus feeders. Measurable changes in species diversity and biomass of benthic infauna and kelp beds can be found, but these changes depend on the rate of discharge and the degree of treatment before release. Increased abundance of fishes and benthic invertebrates has been noted in the vicinity of some outfalls; at others, benthic communities have been noticeably degraded. Of the four major sewer outfalls emptying into the Southern California Bight, two had caused obvious degradation in several square kilometers around the outfall site (**Figure 5.23**). Collectively, the four outfalls significantly changed or degraded nearly 200 km² of seafloor during the 1970s and 1980s.

These energy-rich substances from treated sewage enabled sea urchin populations to evade the usual consequences that animal populations experience when they overgraze their plant food sources. These alternative sources of energy ensured that large numbers of sea urchins survived long enough after decimating one kelp bed to move to another. In central California kelp beds that sea otters have recolonized since 1950, sea urchin populations are now kept low, and kelp forests have recovered throughout most of the otters' geographic range.

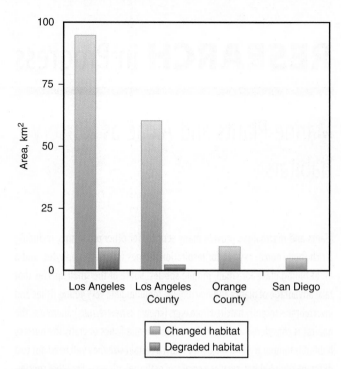

Figure 5.23 Extent of areas changed or degraded by four major sewage outfalls in the California Bight, 1978–1979, after decades of sewage with little treatment was released into the ocean.
Data from A. J. Mearns, *Marine Environmental Pollution* (Elsevier, 1981).

The kelp beds just off San Diego, however, have made a dramatic recovery since 1960 without sea otters. The recovery there was more likely due to improved urban sewage treatment, especially reducing the amounts of discharged solids (**Figure 5.24**). The activities of other predators, particularly sea stars and the California sheephead, also played important roles. In an unusual turnabout, these recovered urchin populations, so recently considered pests in need of eradication, are now themselves targets of a rapidly expanding commercial fishery to supply sea urchin roe, or "uni," to local and international sushi markets.

5.3 Geographic Distribution

The interplay of a multitude of physical, chemical, and biological variables influences and controls the distribution of marine plants and macroalgae on a local scale. For instance, on an exposed rock in the lower intertidal zone on the Oregon coast *Postelsia* may thrive, but 10 m away the conditions of light, temperature, nutrients, tides, surf action, and substrate may be such that *Postelsia* cannot survive. Nevertheless, on an ocean-wide scale, only a few factors seem to control the presence or absence of major groups of seaweeds. Significant among these are water and air temperature, tidal amplitude, and the quality and quantity of light. With these factors in mind, we can make a few generalizations concerning the geographic distribution of benthic plants.

RESEARCH in Progress

Marine Plants and Algae as Nursery Habitats

Plants and macroalgae provide many services for other organisms, including a rich food source, oxygen for respiration, homes for small epiphytes, and a great hiding place for small, mobile species. Some of the small species that take advantage of plants and macroalgae for hiding are very young fishes and invertebrates recently hatched from eggs (known as *new recruits*); therefore, the habitat is considered a nursery. For some juvenile fishes or crabs the nursery habitat is temporary, and once they reach a certain size they will head out to a different adult habitat, such as a coral reef or the pelagic zone. For other species, the nursery habitat is also the final destination, but this is less common, because many nursery areas become overcrowded with young organisms.

Nursery habitats are not a new discovery; scientists have been observing young fishes and invertebrates in these temporary habitats for decades. What researchers are exploring now are the roles that different types of nursery habitats play in the early life of each young organism. Throughout the early life history of a fish or crab, multiple habitats are typically used as a nursery, as opposed to just one type. Young fish will move between connected habitats with the tide while foraging for food or as they shift from one size or age to the next. With declines in the quality of coastal habitats, some nurseries are becoming less hospitable than they were previously, so young organisms may leave in search of a healthier area in which to reside. Research is being conducted to examine how organisms use multiple nursery habitats, how this will impact adult populations, and how regulations to protect habitat can incorporate this bigger picture to effectively protect young fishes and invertebrates during this sensitive life stage.

One area that is used by many young fish as a nursery habitat is mangrove habitat (**Figure A**). Mangrove roots provide a mazelike mass for fish to hide in, and they create a very productive miniature ecosystem. Leaves fall into the water, providing food for herbivores, and small fish hide and are sometimes eaten by larger fish that visit to forage. Many snappers, barracuda, and other reef fishes use mangrove roots as hiding areas until they are large enough to swim across seagrass out to the open ocean, an intermediate patch reef habitat, or a main reef. Mangrove nurseries have been well studied and their value is high.

Mangroves are often bordered by highly productive seagrass beds (**Figure B**). In the past, mangroves and seagrass beds were considered separate nursery habitats, but recent studies have found that many species move in between both habitats during a certain life stage. Seagrass beds are constantly moving as the blades sway with the water movement. This swaying creates an environment that predators find difficult to hunt in. Juvenile fishes and crabs

Figure A Red mangroves provide habitat for numerous small fishes and invertebrates among their roots.
© Vilainecrevette/Shutterstock, Inc.

take advantage of this confusion and dart between seagrass blades. Herbivorous organisms feed on the blades or the epiphytic algae growing on the blades. The speed of water flow is slower between seagrass blades than in the water column, so small fish require less energy to swim around and between the blades. The capability of movement between mangroves and seagrass beds allows for very effective cover for young coastal organisms. When considering the protection of coastal nursery areas, mangroves and seagrass beds should be treated equally and protection afforded to both habitats to the greatest extent possible.

Patch reefs are small reefs often located at the deep edge of a large seagrass bed or completely surrounded by seagrass beds or sand (**Figure C**). Patch reefs are permanent habitat for some organisms, but many use this habitat as an intermediate nursery area after leaving mangroves or seagrass beds but before moving out to a main reef or the pelagic zone. The reef structure provides many small holes and miniature caves for young organisms to find refuge. Once again illustrating the connectivity of many nursery areas, some fish will move from the protection of a patch reef out to a seagrass bed temporarily to feed. As organisms grow larger, seagrass beds offer less protection than a patch reef habitat, but they still offer foraging habitat.

Less known as nursery habitats are macroalgal fields. Small clumps of macroalgae have been identified as important recruitment sites for such organisms as lobsters, but large macroalgal fields were poorly studied as nursery habitat until recently. In an attempt to change this and add to our knowledge

Figure B Seagrass beds provide hiding areas in tropical and temperate seas. White and yellow fish pictured are young juvenile snappers (*Ocyurus chrysurus*) passing through.
© Rich Carey/Shutterstock, Inc.

of macroalgal beds, a study was conducted in northwest Australia to examine the importance of macroalgal fields as nursery habitat for a variety of fish species. Underwater visual surveys were conducted within macroalgal beds and at neighboring coral reefs to count the fish present. More than 20% of the total number of fish observed were found in the macroalgal fields, and more than 15% of the recruit species were observed exclusively within macroalgal fields and absent as young within coral reefs. This study highlights the importance of a variety of habitats as nurseries. Seagrass beds and mangroves are not the only important nursery habitats out there. Macroalgal beds provide hiding areas, a large abundance of fleshy algae for food, and a protective cover from the dynamic aqueous environment above.

Protecting nursery habitats seems to be the most important method for conserving fish and invertebrate adult populations. In recent years, marine reserves have gained popularity, and many have been implemented in the United States and elsewhere. Although marine reserves are effective at protecting adult organisms, if nearby nursery areas are not protected equally there will be fewer adults that make it to the reserve for protection. Based on recent research and modeling, evaluating the entire nursery habitat network

Figure C A small patch reef bordered by sand, composed mainly of dead coral, some living fire coral (orange coral in center), and soft corals, all contributing to the many layers of hiding areas for small fish.
© mychadre77/Getty Images.

of a particular fish or invertebrate and protecting each habitat type appears to be the most effective way of conserving fish and crab populations. The final question that still needs an answer is how to effectively conserve coastal habitats so that the greatest numbers of juvenile fish and invertebrates survive to adulthood. Given that nursery habitats for many species are interconnected, this question remains: how should fisheries managers focus their efforts when deciding which areas to protect?

Critical Thinking Questions

1. What do you think? How should fisheries managers focus their efforts when deciding which areas to protect?

2. How can macroalgal fields, a seemingly unstable and small habitat type, be used by young fishes and invertebrates effectively? What do you think would happen to new recruits if a large storm moves through an area and relocates a macroalgal field?

3. List two reasons nursery habitats are at greater risk of habitat destruction than other nearby habitats.

For Further Reading

Evans, R. D., S. K. Wilson, S. N. Field, and J. A. Y. Moore. 2014. Importance of macroalgal fields as coral reef fish nursery habitat in north-west Australia. *Marine Biology* 161:599–607.

Nagelkerken, I., G. G. Monique, and P. J. Mumby. 2012. Effects of marine reserves versus nursery habitat availability on structure of reef fish communities. *PLoS ONE* 7(6):e36906. doi:10.1371/journal.pone.0036906

Nagelkerken, I., M. Sheaves, R. Baker, and R. M. Connolly. 2015. The seascape nursery: A novel spatial approach to identify and manage nurseries for coastal marine fauna. *Fish and Fisheries* 16:362–371.

Parsons, D. M., C. Middleton, K. T. Spong, G. Mackay, M. D. Smith, and D. Buck-thought. 2015. Mechanisms explaining nursery habitat association: How do juvenile snapper (*Chrysophrys auratus*) benefit from their nursery habitat? *PLoS ONE* 10(3):e0122137. doi:10.1371/journal.pone.0122137

Pinkard, D. R., and J. M. Shenker. 2001. Seasonal variation in density, size, and habitat distribution of juvenile yellowtail snapper (*Ocyurus chrysurus*) in relation to spawning patterns in the Florida Keys. *Am. Zool* 41(6):1556–1557.

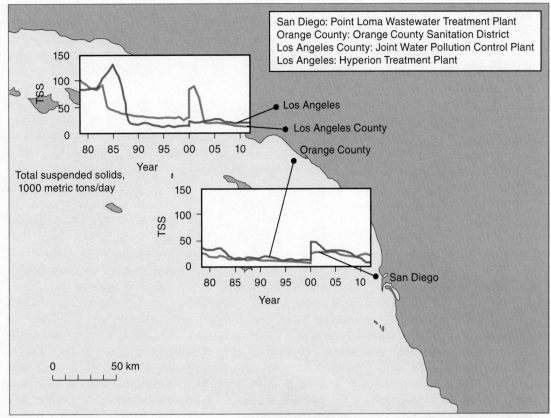

Figure 5.24 Graphs showing the reduction in discharged total suspended solids (TSS) for the past two decades at the four major sewage outfalls in the California Bight. Data from Southern California Coastal Water Research Project and Steinberger and Schiff, 2002.

Let us examine some general geographic trends in the abundance of macroalgae. The Red Sea, the tropical western coast of Africa, and the western side of Central America all have very poor representations of seaweeds. Seaweeds thrive in great amounts along the coasts of southern Australia and South Africa, on both sides of the North Pacific, and in the Mediterranean Sea. The U.S. West Coast is somewhat richer in seaweed diversity than is the East Coast. From Cape Cod northward, the East Coast is populated with subarctic seaweeds. South of Cape Cod, the effects of the warm Gulf Stream become more evident, until a completely tropical flora is encountered in southern Florida.

Some of the distribution trends of macroalgae can be explained by their **physiology**. Red algae are not necessarily rare in cold-water regions, but they are more abundant and noticeable in the tropics and subtropics. Calcareous forms of red algae (and some browns and greens as well) are characterized by extensive deposits of calcium carbonate ($CaCO_3$) within their cell walls (**Figure 5.25**). The use of calcium carbonate as a skeletal component by warm-water marine algae is apparently related to the decreased solubility of $CaCO_3$ in water at higher temperatures. In the tropics, plants expend less energy to extract $CaCO_3$ from the water, and here, coralline red algae contribute to the formation and maintenance of coral reefs. Encrusting coralline algae grow over coral rubble, cementing and binding it into larger masses that better resist the pounding of heavy surf. Some Indian Ocean "coral" reefs completely lack coral animals and are constructed and maintained entirely by coralline algae. The few calcareous forms of green algae that exist are also limited to tropical latitudes and play a large role in the production of $CaCO_3$ sediments.

Figure 5.25 Crustose coralline algae living among and on top of corals at Rose Atoll, located off the coast of American Samoa. NOAA National Ocean Service.

Figure 5.26 *Caulerpa*, one of the more invasive algal species in recent years.
NOAA Photo Library.

The small green alga *Halimeda* is one of the few green algae to also secrete a CaCO$_3$ skeleton, giving it a stony feel. *Halimeda* is a member of a remarkable group of Chlorophytes known as siphonous green algae. Although some siphonous green algae reach over a meter in length, each plant consists of one enormously long and tubular cell containing millions of nuclei, made possible by the uncoupling of the process of nuclear division from that of cell division. Two other members of this group, *Caulerpa taxifolia* and *Codium fragile*, recently have become notorious for their explosively rapid invasions as introduced exotics, *Caulerpa* in the Mediterranean Sea and even sporadically along the U.S. coasts (**Figure 5.26**) and *Codium* in shallow coastal waters of New Zealand and the U.S. Northeast. These invasions have been enhanced by the ability of these plants to fragment in storms and quickly regrow from the wave-scattered pieces.

A few of the larger species of benthic macroalgae flourish in such densities that they dominate the general biological character of their communities. Such community domination by photosynthetic organisms is common on land but is exceptional in the sea. Away from the nearshore habitats occupied by benthic macroalgae and plants, the microscopic phytoplankton prevail

as the major primary producers of the sea. In the nearshore fringe, however, mangals, salt marshes, seagrasses, and kelp beds thrive where the appropriate bottom conditions, light, and nutrients exist.

Kelp are temperate to cold-water species, with few tropical representatives. Large kelps are especially abundant in the North Pacific. Kelp beds abound with grazing herbivores that, in turn, become prey for higher trophic levels. The cool-water kelp form extensive layered forests of mixed species in both the Atlantic and Pacific Oceans. The blades of the larger *Macrocystis*, *Laminaria*, or *Nereocystis* form the upper canopy and the basic structure of these plant communities. Shorter members of other brown algal and red algal species provide secondary understory layers and create a complex three-dimensional habitat with a large variety of available niches (**Figure 5.27**). The maximum depth of these kelp beds, usually 20 to 30 m, is limited by the light available for the young growing sporophyte. The larger kelp, with their broad blades streaming at the sea surface, create substantial drag against currents and swells and are susceptible to storm damage by waves and surge. Cast on the shore, these decaying kelp are a major food source for beach scavengers.

Figure 5.27   A dense kelp forest off the California coast dominated by *Macrocystis*, a large brown alga.
© Ethan Daniels/Shutterstock, Inc.

DID YOU KNOW?

Invasive species are transported from one location to another in a variety of ways, including the aquarium trade, aquaculture, and in the ballast water of ships. In Oahu, two invasive red algae were introduced in the 1970s that have taken over the coral reefs near the introduction area. What started out as an innocent idea, to raise nonnative algae in a bay for industrial purposes, spiraled into a major problem that may never end. Any introduction of a nonnative species has major consequences on the sensitive balance within an ecosystem.

Scientists have invented a contraption called a "Super Sucker," a special vacuum, to suck up the algae from the reef in Oahu. After sucking up the algae, sea urchins are introduced to graze on any new growth of the invasive species.

5.4   Global Marine Primary Production

The large majority of marine primary production is from phytoplankton, with macroalgae only contributing a small percentage to the total in the sea. Although macroalgae are much larger in size, most are very restricted in their distribution to the shallow coast for sunlight and substrate for anchoring. Phytoplankton can live anywhere in the ocean with sufficient light and nutrients and are not restricted to the coast, and thus they are the major contributors to primary production. Obtaining accurate estimates of primary productivity from phytoplankton and macroalgae in the ocean on a global scale is no easy task, even with the help of satellite technology, as shown in upwelling regions in **Figure 5.28**. Perhaps the most difficult aspect of production estimation is the development of a model that will transpose satellite-imagery data into reliable estimates of amounts of marine productivity. Proposed algorithms typically attempt to integrate many datasets simultaneously. As if combining over a dozen datasets is not complicated enough, model-generated estimates of marine production using these datasets are greatly influenced by assumptions and data corrections made by researchers; for example, scientists at Rutgers University have estimated annual global ocean production to be between 40.6 and 50.4 billion metric tonnes of carbon per year depending on which variables they use, which is a huge range. Since 1994, the Ocean Primary Productivity Working Group, a NASA-sponsored team of oceanographers, has been comparing the performance of various productivity algorithms in an attempt to establish a NASA-resident "consensus model" for routine estimation of marine production. We view these databases, production models, and "consensus algorithms" as continuously evolving entities that become

Figure 5.28 SeaWiFS image of chlorophyll *a* concentrations in the upwelling area of the California Current (left) and the Benguela Current off South Africa and Namibia (right).
Courtesy of GeoEye and NASA SeaWiFS Project.

better refined each year, and we present current best estimates of ocean primary productivity in the following paragraphs.

Mid- and high-latitude regions, shallow coastal areas, and zones of upwelling generally support large populations of marine primary producers, but most of this production is accomplished during the warm summer months when light is not a growth-limiting factor. Open-ocean regions, especially in the tropics and subtropics, where a strong thermocline and pycnocline are permanent features, and polar seas, where light is limited through much of the year, have low rates of **net primary productivity** (NPP; 55 g C/m² per year).

Table 5.2 lists and compares the annual rates of marine NPP in several different regions (see **Figure 5.29** for a visual representation). Total NPP estimates included in syntheses such as this have been revised upward nearly 75% with the use of satellite-derived observations. About 76% of the total NPP occurs in the open ocean, spread thinly over 92% of the ocean's area. The more productive regions are very limited in geographic extent. Collectively, estuaries, coastal upwelling regions, and coral reefs produce only about 2.3 billion of the 42.5 billion tonnes of carbon produced each year.

The productivity numbers of Table 5.2 indicate that nearly 42.5 billion tonnes of carbon are synthesized each year in the world ocean, and all but 1.92 billion tonnes (95.5%) are from phytoplankton. That number is equivalent to a bit more than 90 billion tonnes of photosynthetically produced dry biomass,

TABLE 5.2	Rates of Net Primary Production for Several Ocean Regions			
Region	**Area ($\times 10^6$ km²)**	**Percentage of Ocean**	**Average (g C/m²/yr)**	**Total NPP (10^9 tonnes C/yr)**
Open ocean				
Tropics and subtropics	190	51	55	10.45
Temperate and subpolar (including Antarctic upwelling)	100	27	206	20.60
Polar	52	14	27	1.40
Continental shelf				
Nonupwelling	26.6	7.2	290	7.71
Coastal upwelling	0.4	0.1	1,050	0.42
Estuaries and salt marshes	1.8	0.05	975	1.76
Coral reefs	0.1	—	1,410	0.14
Seagrass beds	0.02	—	937	0.02

Data from Longhurst et al.,1995; Pauly and Christensen,1995; Field et al.,1998; and Gregg et al., 2003.

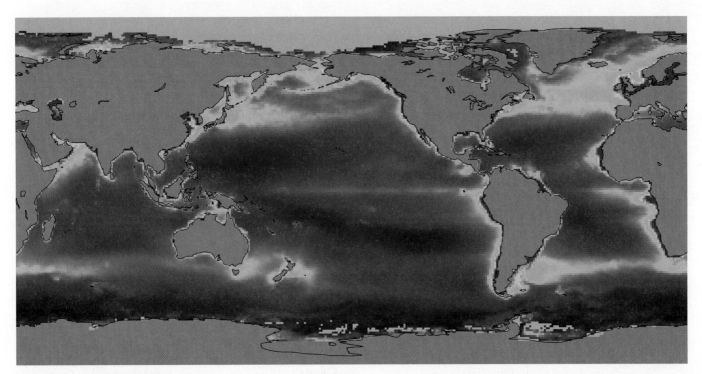

Figure 5.29 The geographic distribution of marine primary production, composed from over 3 years of observations by the satellite-borne coastal zone color scanner.
Courtesy of Robert Simmon/NASA GSFC Earth Observatory, based on data provided by Watson Gregg.

or about 15 tonnes of phytoplankton dry biomass for each person on Earth.

When compared with land-based primary production systems, NPP on land is slightly higher (about 56.4×109 tonnes C/yr), even though oceans cover more than twice as much of Earth's surface as does land. The reason for this is that terrestrial areas do not suffer a significant loss of nutrients as does the photic zone, and thus land production is 426 g C/m^2 per year (not including permanently iced areas). In contrast, marine production is 140 g C/m^2 per year. Although marine primary producers account for almost half of the total global NPP each year, at any one time phytoplankton represent only about 0.2% of the standing stock of primary producers because of their very rapid turnover rates. About 25% of ice-free land areas supports NPP rates over 500 g C/m^2 per year; in the ocean, this value is less than 2%.

The entire human population on Earth currently requires about 5 billion tonnes of food annually to sustain itself, about 12% of the total annual marine NPP. Yet for several reasons this huge amount of marine primary producers will probably never be used on a scale sufficient to alleviate the serious nutritional problems already rampant in much of the human population. Instead, this vast amount of organic material will continue to do what it has always done—fuel the metabolic requirements of the consumers occupying higher marine trophic levels.

DID YOU KNOW?

Some areas, such as the west coasts of Africa and South America, experience coastal upwelling year-round. Year-round upwelling provides cold, nutrient-rich water to support healthy macroalgae and phytoplankton populations, which contribute to global primary production. Healthy populations of photosynthesizers support herbivorous fish and zooplankton, which are food for larger carnivorous fish. The fishing grounds in these areas are considered very rich and provide an excellent source of protein for people living in the area.

STUDY GUIDE

TOPICS FOR DISCUSSION AND REVIEW

1. Recent evidence suggests that brown algae are closely related to silicoflagellates and several other phytoplankton groups. What characteristics unite these groups?

2. Terrestrial flowers are pollinated by a variety of insects, birds, and bats. How are the flowers of subtidal seagrasses pollinated?

3. The seeds of red mangroves germinate while their fruit still hangs from the parent tree. Summarize this unusual form of sexual reproduction using text and drawings.

4. *Sargassum* contains numerous small pneumatocysts to buoy the plant toward the sunlit surface and hosts a complex community of fishes and invertebrates that are uniquely adapted to living on this pelagic seaweed. What would you consider these floating mats of life?

5. A life cycle consisting of alternating gametophyte and sporophyte generations is characteristic of almost all plants and macroalgae. How do the basic features of that life cycle differ for one or more divisions of seaweeds?

6. What major characteristic of green algae (Chlorophyta) supports the hypothesis that they are ancestral to flowering plants (Anthophyta)?

7. How does the pigment content of the different macroalgae divisions influence where in the environment they may live?

8. Explain why some algae are "crusty." Which areas of the world ocean contain more crusty algae?

9. How do local assemblages of kelp, seagrasses, and mangals influence and alter the physical characteristics of the shoreline on which they live?

10. Why do phytoplankton contribute so much more to global primary production than macroalgae and marine plants even though they are so much smaller in size?

11. Where is NPP the highest? The lowest? Why?

KEY TERMS

blade   112
carpospore   124
carposporophyte   124
diploid   123
emergent   110
epibiont   112
gametophyte   123
haploid   123
haptera   117
holdfast   118
kelp   118
macroalgae   110
mangal   115
meiosis   123
meristematic tissue   124
net primary productivity (NPP)   134
perennial   125
physiology   131
pollen   111
rhizome   111
sporangia   124
sporophyte   123
stigma   111
stipe   117
vascular tissue   117
zygote   123

KEY *GENERA*

Caulerpa	Penicillus
Chondria	Phyllospadix
Chorda	Porphyra
Codium	Postelsia
Cymodaocea	Pterygophora
Egregia	Rhizophora
Enhalus	Salicornia
Enhydra	Sargassum
Fucus	Smithora
Halimeda	Spartina
Halodule	Suaeda
Halophila	Syringodium
Homarus	Thalassia
Laminaria	Thalassodendron
Macrocystis	Ulva
Nereocystis	Zostera
Pelagophycus	

REFERENCES

Baker, J. D., and W. S. Wilson. 1986. Spaceborne observations in support of earth science. *Oceanus* 29:76–85.

Bold, H. C., and M. J. Wynne. 1985. *Introduction to the Algae*. Englewood Cliffs, NJ: Prentice Hall.

Capone, D. G. 2001. Marine nitrogen fixation: What's the fuss? *Current Opinion in Microbiology* 4:341–348.

Chapman, A. R. O. 1979. *Biology of Seaweeds*. Baltimore: University Park Press.

Correll, D. L. 1978. Estuarine productivity. *BioScience* 28:646–650.

Dawson, E. Y. 1966. *Marine Botany, an Introduction*. New York: Holt, Rinehart and Winston.

Estes, J. A., and J. F. Palmisano. 1974. Sea otters: Their role in structuring nearshore communities. *Science* 185:1058–1060.

Falkowski, P. G., R. T. Barber, and V. Smetacek. 1998. Biogeochemical controls and feedbacks on ocean primary production. *Science* 281:200–206.

Goering, J. J., and P. L. Parker. 1972. Nitrogen fixation by epiphytes on sea grasses. *Limnology and Oceanography* 17:320–323.

Hagen, J. B. 2012. Five kingdoms more or less: Robert Whittaker and the broad classification of organisms. *BioScience* 62(1):67–74.

Harper, M. A., V. C. Cassie, F. Hoe Chang, W. A. Nelson, and P. A. Broady. 2012. Phylum Ochrophyta: Brown and golden-brown algae, diatoms, silicioflagellates, and kin. In: *New Zealand Inventory of Biodiversity, vol. 3. Kingdoms Bacteria, Protozoa, Chromista, Plantae, Fungi*, edited by D. P. Gordon, 114–163. Christchurch: Canterbury University Press.

Jacobs, W. P. 1994. Caulerpa. *Scientific American* 271:100–105.

Jeffries, R. L. 1981. Osmotic adjustment of the response of halophytic plants to salinity. *BioScience* 31:42–46.

Kaufman, P. B. 1989. *Plants: Their Biology and Importance*. Reading, MA: Addison-Wesley.

King, R. J., and W. Schramm. 1976. Photosynthetic rates of benthic marine algae in relation to light intensity and seasonal variations. *Marine Biology* 37:215–222.

Koehl, M. A. R., and S. A. Wainwright. 1977. Mechanical adaptations of a giant kelp. *Limnology and Oceanography* 22:1067–1071.

Lembi, C. A., and J. R. Waaland, eds. 1988. *Algae and Human Affairs*. New York: Cambridge University Press.

Longhurst, A., S. Sathyendranath, T. Platt, and C. Caverhill. 1995. An estimate of global primary production in the ocean from satellite radiometer data. *Journal of Plankton Research* 17:1245–1271.

Mann, K. H. 1973. Ecological energetics of the seaweed zone in a marine bay on the Atlantic coast of Canada. *Marine Biology* 14:199–209.

McPeak, R. H., and D. A. Glantz. 1984. Harvesting California's kelp forests. *Oceanus* 27:19–26.

Meinesz, A. 1999. *Killer Algae*. Chicago: University of Chicago Press.

Perry, M. J. 1986. Assessing marine primary productivity from space. *BioScience* 36:461–67.

Pettit, J., S. Ducker, and B. Knox. 1981. Submarine pollination. *Scientific American* 244:134–143.

Phillips, R. C. 1978. Sea grasses and the coastal environment. *Oceanus* 21:30–40.

Phleger, C. F. 1971. Effect of salinity on growth of a salt-marsh grass. *Ecology* 52:908–911.

Pimm, S. L. 2001. *The World According to Pimm: A Scientist Audits the Earth*. New York: McGraw-Hill.

Saffo, M. B. 1987. New light on seaweeds. *BioScience* 37:654–664.

Segal, R. F., R. J. Bandoni, J. R. Maze, G. E. Rouse, W. B. Schofield, and J. R. Stein. 1980. *Nonvascular Plants*. Belmont, CA: Wadsworth.

Short, F. T., T. J. R. Carruthers, B. van Tussenbroek, J. Zieman, and W. J. Kenworthy. 2010. *Halophila johnsonii*. The IUCN Red List of Threatened Species. Version 2015.2. Available from www.iucnredlist.org. Accessed July 4, 2015.

Macia, S., and M.P. Robinson. 2005. Effects of habitat heterogeneity in seagrass beds on grazing patterns of parrotfishes. *Marine Ecology Progress Series* 303:113-121.

Steele, J. H., ed. 1973. *Marine Food Chains*. Edinburgh: Oliver and Boyd.

Steele, J. H. 1974. *The Structure of Marine Ecosystems*. Cambridge, MA: Harvard University Press.

Tomlinson, P. B. 1994. *The Botany of Mangroves*. New York: Cambridge University Press.

Townsend, D. W., and M. Thomas. 2002. Springtime nutrient and phytoplankton dynamics on Georges Bank. *Marine Ecology Progress Series* 228:57–74.

Vitousek, J., P. Ehrlich, A. H. Ehrlich, and P. Matson. 1986. Human appropriation of the products of photosynthesis. *BioScience* 36:368–373.

CHAPTER 6

Microbial Heterotrophs and Invertebrates

STUDENT LEARNING OUTCOMES

1. Compare and contrast the major groups of marine protozoans and invertebrates.

2. Explain why sponges are in the Kingdom Animalia, but foraminiferans are not.

3. Describe the requirements for membership in the Kingdom Fungi.

4. Explain early development in sexually reproducing animals.

5. Distinguish between body cavity types and early developmental strategies for all of the animal groups examined in this chapter.

6. Compare and contrast all of the wormlike phyla examined in this chapter.

7. List the characteristic requirements of all chordates, and explain how invertebrate chordates fulfill these requirements.

CHAPTER OUTLINE

A unique cnidarian from the genus *Heteropolypus* is in the center of this photo, surrounded by brittle stars, smaller cnidarian polyps, and sponges. This image highlights the great diversity and prevalence of species interactions found in the invertebrate phyla.
NOAA/SWFSC La Jolla ROV team.

The story of the beginning of the Kingdom Animalia takes place in the sea and includes a large variety of small single-celled organisms with animal-like characteristics. These predecessors to animals are known as the *nonphotosynthetic microbes* or, more commonly, *nonphotosynthetic protozoans*. They were the largest organisms in the sea for several billion years, and eventually the shaping and molding of the characteristics of some nonphotosynthetic protozoans led to the evolution of true animals. In our investigation of animals we will evaluate the features that are required to be considered part of the Kingdom Animalia and what distinguishes true animals from protozoans. Some nonphotosynthetic single-celled animal-like organisms (i.e., the *protozoans*) are introduced, followed by the invertebrate phyla of the Kingdom Animalia. Our focus will be on the more abundant and obvious free-living phyla; groups that are primarily or wholly parasitic will not be discussed. The purpose of this quick romp through the more common or familiar marine protozoan and animal phyla is to provide an introduction to some of the major players making their appearance in the ocean.

Of the many phyla introduced in this chapter, only a few—in particular, various protozoans, cnidarians, nematodes, mollusks, annelids, echinoderms, and arthropods—clearly dominate the composition of most marine communities and monopolize the energy flow through those communities. These phyla are presented in an order that generally corresponds to increasing complexity and the most currently hypothesized phylogenetic relationships. The structural features listed in **Table 6.1** (and others not included here) have long served as the basis for defining patterns of relationships between animal phyla; however, some of these traditionally accepted alliances are being challenged by new molecular comparison techniques. The recent and continuing accumulation of massive amounts of molecular data from these animal phyla suggests alternative phylogenies that are at odds with traditional ones and also sometimes with each other. The use of genetic taxonomic methods is gaining popularity in the fields of evolutionary history and taxonomy, and as more of this information is compiled even newer hypotheses of animal relationships can be expected. Selecting one taxonomic arrangement from the many that have been proposed is difficult. We have taken a somewhat conservative approach here and present a traditional classification of animals, but if newer names and groupings have earned overwhelming support in the scientific community they are presented and discussed (**Figure 6.1**).

The sequence of phyla in Table 6.1 includes significant trends and major milestones in the history of eukaryotic evolution. Increased complexity and specialization of body structures are evident, especially in the systems involved in locomotion, gas exchange, excretion, feeding and digestion, circulation, and reproduction. Trends from radial to bilateral body symmetry and the evolution of body cavities are evident, as is the increased reliance on sexual reproduction. Improved sensory systems and increasingly complex brains able to integrate sensory information support expanding patterns of behavioral responses.

TABLE 6.1 Characteristics of the Major Invertebrate Phyla

Phylum/Group Name	Common Name	Number of Marine Species[*]	Tissues	Symmetry	Digestive Tract	Body Cavity	Protostome or Deuterostome
Foraminifera	Forams	8,900	—	—	—	—	—
Radiozoa	Radiolarians	429	—	—	—	—	—
Ciliophora	Ciliates	2,676	—	—	—	—	—
Labyrinthomorpha	Labyrinthulids	N/A (small)	—	—	—	—	—
Porifera	Sponges	8,421	None	None	None	—	—
Placozoa	Placozoans	1	None	None	None	—	—
Cnidaria	Sea anemones	11,296	Two layers	Radial		—	—
Ctenophora	Comb jellies	192	Two layers	Radial	Incomplete	—	—
Platyhelminthes	Flatworms	12,328	Three layers	Bilateral	Incomplete		—
Gnathostomulida	Jaw worms	98	Three layers	Bilateral		Acoelomate	—
Nemertea	Ribbon worms	1,359	Three layers	Bilateral			—
Gastrotricha	Hairy-bellied worms	494	Three layers	Bilateral	Complete	Pseudocoelom	—
Entoprocta	Entoprocts	176	Three layers	Bilateral	Complete	Pseudocoelom	—

(continues)

TABLE 6.1 Characteristics of the Major Invertebrate Phyla (*continued*)

Phylum/Group Name	Common Name	Number of Marine Species*	Tissues	Symmetry	Digestive Tract	Body Cavity	Protostome or Deuterostome
Kinorhyncha (class)	Kinorhynchs or mud dragons	188					Pseudocoelom
Nematoda	Round worms	7,145					
Bryozoa (formerly Ectoprocta)	Bryozoans or moss animals	6,146					
Brachiopoda	Lamp shells	395					
Phoronida	Horseshoe worms	17					
Mollusca	Clams	45,122		Bilateral			Protostome
Sipuncula	Peanut worms	147					
Echiurida (suborder)	Spoon worms	104	Three layers		Complete		
Annelida	Segmented worms	12,829				Coelom	
Priapulida (class)	Cactus worms	19					
Arthropoda	Crabs	57,757					
Chaetognatha	Arrow worms	131					Unresolved
Echinodermata	Sea stars	7,245		Radial-pentamerous			Deuterostome
Hemichordata	Acorn worms	130					
Chordata	Tunicates	22,238		Bilateral			

*Approximate number reported by the census of marine life in the World Register of Marine Species (WoRMS) as of July 2015. Blue shaded boxes are for characteristics that do not apply to the group.

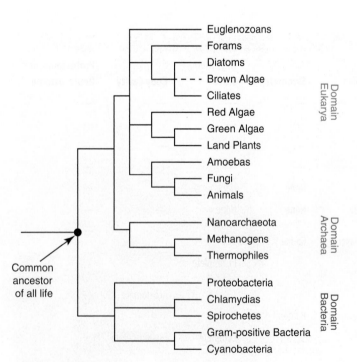

Figure 6.1 Phylogenetic tree representing a hypothesis of relationships between the major groups of organisms.

6.1 Animal Beginnings: The Protozoans

The term *protozoa* encompasses a variety of nonphotosynthetic microscopic members of various kingdoms, previously all members of the recently abandoned Kingdom Protista. They are included in this chapter because many biologists casually consider them "single-celled animals" and because a widely supported hypothesis includes the idea that the earliest animals arose from a choanoflagellate, a protozoan that forms **colonies** of individuals. Individual protozoans consist of a single cell or loose aggregates of a few cells, and they share some other features of animals described later in this chapter, including ingestion of food particles for nutrition and an absence of cell walls and photosynthetic chloroplasts.

The number of protozoan phyla varies widely from one classification system to another and will remain confusing and unresolved because the Kingdom Protista was recently abandoned; it was an artificial catch-all category for many fundamentally different phyla and divisions clearly lacking an ancestral form common to all of them. Several protozoan phyla are mostly or completely parasitic; only three groups are described here that include marine nonparasitic species that thrive in benthic and planktonic communities. Asexual reproduction by cell division is common. Sexual reproduction, when it does occur, is often quite complex, with the process of meiosis separated from that of nuclear fusion by several cell generations. These three free-living protozoan groups are most easily distinguished by their different methods of locomotion. When moving in water, these small cells encounter very high viscous forces between water molecules and, regardless of their mode of locomotion, do not swim so much as crawl through their watery environment.

DID YOU KNOW?

A marine protozoan called a choanoflagellate is a single-celled organism that looks almost exactly like a choanocyte cell found in a sponge. The similar shape and locomotion of these cell types are the major pieces of evidence for animal beginnings. Choanoflagellates have a collar-like appearance with a long flagellum sticking out the back, as do sponge choanocyte cells. The flagellum is used to move food particles around to be trapped by the collar and processed by the organism. Recent molecular evidence supports the idea that a choanoflagellate-like organism is likely the ancestor of all animals, because choanoflagellates share many genetic features with animals. These unique single-celled organisms appear to be the link between small animal-like organisms and true animals.

Phyla Foraminifera and Radiozoa

A large and widespread group, the Sarcomastigophora use either whiplike flagella or extensions of their cellular protoplasm, pseudopodia (**Figure 6.2**), or both, for locomotion. Recent evidence indicates that Sarcomastigophora should be abandoned as a phylum, because phylum membership is based primarily on means of locomotion and not evolutionary relationships. A new classification scheme for its members includes separate phyla for two of the groups of organisms that were previously considered Sarcomastigophora: the Foraminifera and Radiozoa.

About one half of all named protozoans are foraminiferans. Foraminiferans are shelled amoebas that are mostly marine. They are common in the plankton, yet more are benthic or live attached to plants and animal shells. Most foraminiferans are microscopic, although individuals of a few species grow to several millimeters in size and are visible with the naked eye. They have internal chambered shells, or *tests*, usually composed of either calcium carbonate ($CaCO_3$) or cemented sand grains. Penetrating this shell, or test, are numerous cytoplasmic filaments (Figure 6.2a) that are used for locomotion, for attachment, and for collecting food. Some planktonic foraminiferans, such as *Globigerina* (Figure 6.2b), are so widespread and abundant that their tests blanket large portions of the seafloor. After thousands of years of accumulation, this **globigerina ooze** may form deposits tens of meters thick. The famous chalk cliffs of Dover, England, are composed mainly of foraminiferan tests and coccolith plates that accumulated on the seafloor and were subsequently lifted above sea level where they were eroded into vertical seaside cliffs. This fall of carbonate skeletal material from the sea surface has and continues to be a major process to remove CO_2 from the atmosphere and sequester it away for millions of years in marine sediments.

Radiozoans, also called *radiolarians*, are entirely marine, and most members are planktonic. They are similar in size to planktonic foraminiferans and occur at all latitudes and at all depths. An internal skeleton of silica forms the beautiful symmetry often associated with radiolarians (**Figure 6.3**). This strong silica skeleton is the key to the existence of an extensive fossil record for this group, which indicates great diversity and abundance from the early Cambrian period to today. Like foraminiferans, radiozoans also use many cytoplasmic filaments to capture bacteria, detritus, and other protozoans for food. The relationship between radiozoans and foraminiferans is unclear and currently under debate.

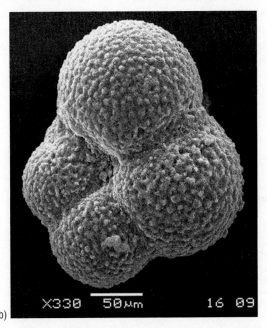

Figure 6.2 A planktonic foraminiferan, *Globigerina*. (a) Drawing of an intact individual emphasizing extended pseudopodia. (b) Scanning electron micrograph of a *Globigerina* test.
(a) Reproduced from Voyage of H.M.S. *Challenger*, 1873–1879, *Zoology*, 9:1–814. (b) Courtesy of Dr. Constantin Craciun, Babes-Bolyai University, Romania.

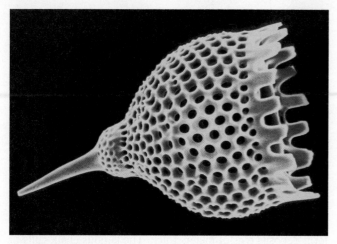

Figure 6.3 Scanning electron micrograph of the silicate skeleton of a planktonic radiolarian.
© Phototake/Alamy Images.

Figure 6.5 A marine tintinnid with a crown of cilia at one end.

Phylum Ciliophora

Members of the phylum Ciliophora are known as *ciliates* because they possess cilia as their chief means of locomotion, and it is cilia that make them one of the more complex single-celled organisms. Cilia also give them a furry or hairy appearance and explain the origin of their name, which in Latin means "eyelash." Structurally, a cilium is much like a short flagellum; however, cilia are typically much more numerous than flagella, and they move in a remarkably coordinated manner (**Figure 6.4**). Cilia work together to move water parallel to the cell surface (flagella move water perpendicular to the surface of the cell). Tintinnids are probably the most abundant of the marine ciliates. These planktonic cells live partially enclosed in a vase-shaped structure made of cemented coccoliths (stolen from their prey) or of a material secreted by the cell (**Figure 6.5**). Ciliated tentacles at one end of the cell are used for collecting bacteria and other small protozoans. A large variety of other ciliates exist in marine planktonic and benthic communities, and many other marine ciliophorans exist as parasites.

Labyrinthomorpha

The small group Labyrinthomorpha was once recognized as a phylum and includes a group of free-living organisms characterized by a network of slime through which colonies of small (about 10 μm long) spindle-shaped cells live and move. Recent evidence indicates that labyrinthomorphs do not warrant their own phylum, and may not even be closely related to the protozoans. Their position on the tree of life is very unclear, making them an intriguing group of organisms to ponder. Most species are marine, forming their colonial networks on the surfaces of seagrasses and benthic algae, and some are parasitic. Their mechanism of gliding through their slime networks is not understood and is accomplished without pseudopodia, cilia, or flagella. Reproduction is both sexual and asexual in this phylum.

6.2 Marine Fungi

Members of the Kingdom Fungi are heterotrophic organisms that acquire organic nutrients from their immediate environment. Like plants and bacteria, fungi possess a cell wall, but they fortify it with **chitin**. The ability to produce chitin (a complex polysaccharide that forms hard structures) is a **derived character** shared by specialized protozoans (the choanoflagellates), animals, and fungi. Some fungi are parasitic, absorbing nutrients from their living hosts. Most species are **saprobes** that absorb nutrients from detritus and other nonliving organic matter by secreting digestive enzymes externally and absorbing the resulting breakdown products. In addition to bacteria, saprobic fungi are the primary decomposers on Earth. Fungi recycle essential nutrients to primary producers via decay and decomposition, and fungi are the major decomposers of the cellulose and lignin from plant cell walls, substances that most bacteria cannot break down. We emphasize these free-living marine saprobes in this section.

Phyla in the Kingdom Fungi are distinguished by the methods and structures that they use during sexual reproduction

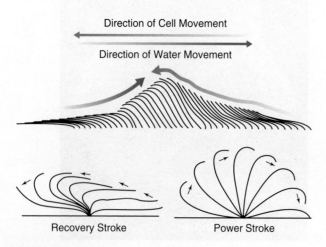

Direction of Cell Movement

Direction of Water Movement

Recovery Stroke   Power Stroke

Figure 6.4 The pattern of ciliary movement appearing as waves of alternating recovery and power strokes sweeping over the cell surface.

(although most fungi reproduce asexually as well). Some fungal species, commonly referred to as *yeasts*, are unicellular organisms that reproduce by cellular fission or sexual reproduction. Most fungi possess a multicellular body, called a **mycelium**, made of rapidly growing, tubular filaments called **hyphae**. A single fungus may produce more than 1 km of hyphal filaments per day! The many filamentous hyphae extending from a single fungal mycelium provide each fungus with an enormous surface area-to-volume ratio, a valuable adaptation for these organisms that absorb their food from the environment.

Marine fungi (**Figure 6.6**) are an ecological grouping, not a taxonomic lineage. The majority of marine fungi are considered to be obligate in that they grow and sporulate exclusively in marine or estuarine habitats. Facultative marine fungi are freshwater or terrestrial species that can grow and may be able to reproduce in the sea. Currently, about 1,300 species of marine fungi are recognized (excluding those species isolated from lichens), although little attention has been paid to them and countless species undoubtedly remain to be discovered. Discounting parasitic species, marine fungi can be found growing on wood, sediments of various particle size, algae, fallen

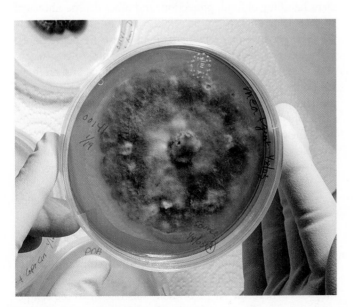

Figure 6.6 Culture of an isolated marine fungus.
Courtesy of the Pacific Northwest National Laboratory.

leaves of mangroves, seagrasses, corals, mollusks, and other living animals at all latitudes and depths.

As on land, marine fungi are major decomposers of woody or herbaceous substrata in the sea, such as mangrove litter, seagrasses, and intertidal salt marsh grasses. Specifically, marine fungi are capable of digesting lignocellulose, the ubiquitous heteropolymer of plant cell walls that consists of lignin, cellulose, and hemicellulose. Although marine borers, such as ship worms (highly derived, wormlike relatives of clams and oysters), have received a lot of credit (and blame) for degrading wood in marine habitats, marine fungi probably are much more influential because they can survive in both low oxygen conditions in marine sediments and in the extreme environmental fluctuations of intertidal regions, unlike most marine borers. In fact, more attention has been directed to mangrove-dwelling fungi than to any other type of marine fungus. About 50% of obligate marine species are known from mangrove litter, and about 150 fungal species are found only on wood, roots, and seedlings from mangroves. Moreover, marine fungi living within calcareous portions of benthic animals, such as mollusk tubes and shells, barnacle tests, and coral heads, are important producers of carbonate detritus, and they may play an important role in the decomposition of dead animals in the sea (such as the digestion of tunicin, an animal cellulose, found in the tests of tunicates).

Marine fungi interact with other living organisms in many important ways. They form three types of mutualistic symbioses with other organisms. **Lichens** are an association between a fungus and a photosynthetic organism, usually a green alga or a cyanobacterium. Until recently, only one brown alga was known to form associations with a fungus to form a lichen, but evidence for several brown algae species as symbionts now exists. Most marine lichens grow on hard substrates within the intertidal zone, and a few are permanently submerged. **Mycophycobioses** are obligate mutualisms formed between a marine fungus and a seaweed that may enable the macroalga to resist drying out during low tide. **Mycorrhizas** are a mutualistic association between a fungus and a vascular plant wherein fungal hyphae colonize a plant host's roots to gain sugars from the plant while the plant exploits the tremendous surface area of the mycelium to extract soil nutrients. Surprisingly, although mycorrhizas are well known in salt marsh grasses, no mangrove species is known to form a mycorrhiza.

Fungal diseases of marine animals and plants also are the subject of considerable attention. Most fungal infections in a marine animal are difficult to treat and often are fatal. Pathogenic fungi infect fish, bivalve mollusks, crustaceans, abalones, nudibranchs, corals, and octopuses. Of particular concern are those fungal pathogens that harm wild or aquacultured populations of commercially valuable animals. A fungal pathogen can wipe out an entire population of captive-bred animals before the pathogen has been identified by the culturist, because identification is complicated. Because fungi are differentiated by the details of their reproduction, to accurately identify many fungal pathogens observation of small details of the reproductive process is generally required. Relatively few marine fungi have been shown to harm plants, although an interesting case has occurred in marsh grasses along the southeastern coast of the United States, and a few mangrove die-offs have been caused by fungal pathogens growing on mangrove roots or leaves. Finally, fungal diseases, such as browning disease of the green alga *Chaetomorpha*, are common in marine algae, diatoms, and cyanobacteria.

6.3   Defining Animals

The boundary separating the protozoans and Kingdom Animalia is vague, and some colonial protozoans seem on the verge of crossing that boundary with their superficial appearance of multicellularity when living in colonies. Yet protozoans are not truly multicellular and do not contain contractile muscles or signal-conducting neurons as almost all animals do. In addition, there is a greater dependence in the more complex animal phyla on sexual reproduction and less on asexual reproduction more commonly seen in the protozoans. Sexual reproduction is not only more complex than asexual reproduction, it is also more energetically costly. The high costs of sexual reproduction are related to the expenses of producing and maintaining males and to the expenses associated with meiosis. Regardless of the approach taken to accomplish reproduction, all sexually reproducing organisms include the same basic elements in the process. In **hermaphroditic** animals, such as most barnacles, many snails, and some fish, some adults function in both female and male sex roles. Some species function as males and females at the same time (*simultaneous hermaphrodites*); others function as one sex and then transform to the other (*sequential hermaphrodites*).

Most other animals retain the same sex role for their entire lives, with approximately equal numbers of males and females. Either way, they still express the basic attributes of sexual reproduction, namely meiosis followed by fertilization. The diploid zygote resulting from fertilization then develops by successive cell cleavages to form a hollow ball of a few hundred cells, the **blastula**, another characteristic developmental stage of animals. This process of sexual reproduction can create enormous genetic variety in populations within short time spans. This genetic diversity is the raw material on which natural selection acts to produce adaptations in evolutionary time scales.

The first two animal phyla described, the Porifera and Placozoa, might be considered sidelines to the major trends of animal evolution. They are considered animals, yet with their individual cells functioning independently instead of within tissues they lack the coordinated patterns of cell functions seen in specialized tissues or organs of more derived animal phyla.

Phylum Porifera

Porifera is one of the few animal phyla with a widely accepted common name—the *sponges*. The sponges are among the simplest animals and likely belong closest to the base of the animal evolutionary tree. Each sponge consists of several types of loosely aggregated cells organized into a multicellular organism with a distinctive and recognizable form. Sponges lack tissues and organs, including muscles and nerves, and are not mobile once they settle to the substrate, yet they are still considered animals. Despite their structural simplicity, sponges

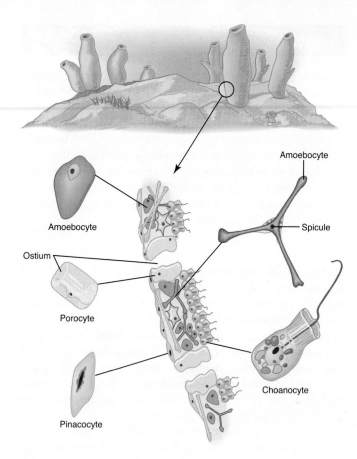

Figure 6.7   A group of marine finger sponges and several of the specialized cell types that make up the sponge wall.

share several advantages with other multicellular animals. Cells within each individual sponge can divide repeatedly to achieve larger sizes and longer life spans than are possible for individual single-celled protozoans. In addition, the specialization of cells, although it is limited in sponges, does enable more efficient handling of food, protection, and other diverse chores of survival (**Figure 6.7**).

The phylum name *Porifera* stems from the many pores, holes, and channels that perforate the bodies of sponges. In general, water is circulated through these openings into an internal cavity, the **spongocoel**, where food and oxygen are extracted by flagellated, current-producing cells, the **choanocytes**, lining the spongocoel. The water then exits through a large excurrent pore, the **osculum**. Sponges are the simplest filter feeders, and as a result must live in water with sufficient nutrients for survival. Because sponges are immobile, they cannot move to a new location if conditions become unfavorable.

Sponges are mostly marine and are usually found attached to hard substrates such as rocks, pilings, or shells of other animals. Those that attach to a mobile animal are the only sponges that are able to change their location, but they are limited to locations where the mobile animal decides to move to. Sometimes they appear radially symmetrical, but more commonly they conform to the shape of their substrate or to the sculpting influences of waves and tides. Some sponges are supported internally by a network of flexible **spongin** fibers. Commercial

Figure 6.8 A glass sponge. (a) The silicate skeleton. (b) A microscopic view of spicules.
(a) Courtesy of James Sumich. (b) Courtesy of NOAA.

natural bath sponges are actually the spongin skeleton with all living material removed. Other sponges have skeletons composed of hard and sharp mineralized **spicules**. The spicules are either calcareous ($CaCO_3$) or siliceous (SiO_2). The silicate skeleton of the deep-water glass sponge *Euplectella* is one of the most complex and beautiful skeletons of all sponges (**Figure 6.8a** and **b**). Large sponge specimens provide refuge from predators or simply a resting place for many other organisms (**Figure 6.9**).

Phylum Placozoa

The phylum Placozoa is represented by only one known species, *Trichoplax adhaerens*, and until 1971 it was misidentified as a larva from another phylum. Each animal is 2 to 3 mm long and consists of a few thousand cells shaped like a flattened plate. Like poriferans, *Trichoplax adhaerens* lacks tissues and organs. Outside aquaria, they are found throughout tropical and subtropical seas gliding on hard surfaces, although they seem to be pelagic for a part of their lives. Movement is accomplished via cilia, which cover both sides of the animal, enabling it to move in any direction. It has no obvious body symmetry. Digestion of food is accomplished by secreting enzymes externally and then absorbing the digested molecules. Like sponges, these animals exhibit limited specialization and organization of cells and can regenerate a complete animal from a single cell; asexual reproduction is common with individuals dividing in two or via budding. Although sexual reproduction is poorly known in placozoans, individuals have produced an oocyte or embryos in the laboratory. The relationship of placozoans to other animal phyla is unclear, yet it is likely that due to their simple design, like sponges, they are near the base of the animal evolutionary tree.

DID YOU KNOW?

Although filter feeding is a major characteristic of sponges, a small number of sponges are carnivorous. Instead of using choanocytes to filter feed, carnivorous sponges, which actually lack choanocytes, have cells with microscopic hooks on them for latching onto prey. They have been observed with small crustaceans inside their bodies, and because they do not have a digestive tract prey are slowly broken down by enzymes secreted by the sponge. Most of the carnivorous specimens discovered so far live near underwater volcanoes and deep-sea vents, although a few have been found in shallow water.

6.4   Radial Symmetry

Three animal phyla, the Cnidaria, Ctenophora, and Echinodermata, exhibit radially symmetrical body plans. The circular shape of radially symmetrical animals provides several different planes of symmetry to divide the animal into mirror-image halves (**Figure 6.10**). The mouth is located at the center of the body on the **oral** side; the opposite side is the **aboral** side. Radially symmetrical animals possess a relatively simple and diffuse network of nerves and lack a central brain to process sensory information or to organize complex responses. Two of these phyla, the Cnidaria and Ctenophora, are described here because they represent an early evolutionary stage of animal body structures based on the presence of tissues and a primary,

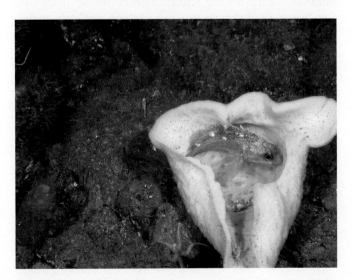

Figure 6.9 A large vase-shaped sponge housing a starry rockfish and several crustaceans.
NMFS/SWFSC.

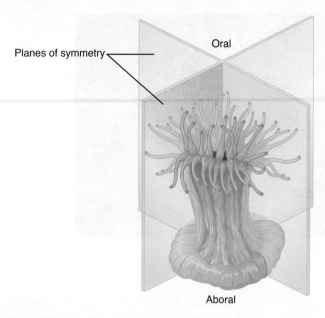

Figure 6.10 Planes of symmetry in a radially symmetrical animal.

or fundamental, radial symmetry. The Echinodermata are introduced later in this chapter because they are evolutionarily secondarily radially symmetrical, having evolved from bilaterally symmetrical ancestors, and because they exhibit bilateral symmetry as larvae and radial symmetry as adults.

Phylum Cnidaria

The phylum Cnidaria includes a large and diverse group of relatively simple yet versatile marine animals, such as jellyfish, sea anemones, corals, and hydroids. The variety of sizes, lifestyles, shapes, and body forms is truly remarkable for this relatively simple and ancient phylum. The inner and outer body walls of all cnidarians form two tissue layers and are separated by an additional gelatinous layer that is not considered tissue, called the **mesoglea**. A centrally located mouth leads to a baglike digestive tract, the **gastrovascular cavity** that is part of an incomplete gut with one opening for the entrance of food and exit of waste material. The mouth is surrounded with tentacles capable of capturing a wide variety of marine animal prey. The tentacles and, to a lesser extent, other parts of the body are armed with many microscopic harpoon-like stinging structures, the **nematocysts**. Nematocysts are produced and housed inside special cells, called **cnidocytes**, and are a characteristic of this phylum. They are discharged when stimulated by contact with other organisms. Most nematocysts pierce the prey and inject a paralyzing toxin (**Figure 6.11**), whereas others are adhesive and stick to the prey or consist of long threads that become entangled in the prey's bristles or spines.

Cnidarians exist either as free-swimming **medusa** or as attached benthic **polyps**. The two forms have essentially the same body organization. The oral side of the medusa, bearing the mouth and tentacles, is usually oriented downward. The mesoglea of most medusae is well developed and is jellylike in consistency, thus earning them the descriptive name of *jellyfish*, although they are nothing like true fish. In the polyp, the mouth

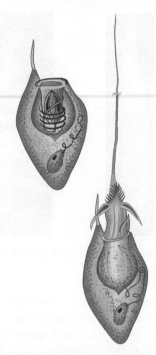

Figure 6.11 Nematocyst before discharging (top) and after discharging (bottom).

and tentacles typically are directed upward. Many species of cnidarians have life cycles that alternate between a swimming medusoid generation and an attached, benthic polypoid generation. In a generalized cnidarian life cycle (**Figure 6.12**), polyps can produce medusae or additional polyps by budding. The medusae, in turn, produce eggs and sperm that, after fertilization, develop into the polyps of the next generation.

The phylum Cnidaria consists of four classes, each characterized by its own variation of the basic cnidarian life cycle shown in Figure 6.12. The class Hydrozoa includes colonial hydroids and siphonophores, such as the Portuguese man-of-war, *Physalia*. Hydrozoans usually have well-developed medusoid and polypoid generations (**Figure 6.13a**). Different individuals of the polyp colony are specialized for particular functions, such as feeding, reproduction, or defense. In the class Scyphozoa, the polyp stage is reduced or completely absent. This class includes most of the larger and better-known medusoid jellyfish (**Figure 6.13b**). In the third class, the Anthozoa, the polyp form dominates and the medusoid generation is absent. Many anthozoans, including most corals (**Figure 6.14**) and sea fans, are colonial, but most sea anemones exist as large, solitary individuals. Unlike most cnidarians, corals and some other anthozoans (and a few hydrozoans) produce an external, often massive, calcium carbonate skeleton. The large skeletons created by corals provide habitat for a remarkable amount of marine life living on or near a coral reef. The last class, Cubozoa, includes the box jellies, and some are extremely venomous. These relatively small jellyfish are a cube-shaped medusae, and those that are toxic can kill a human being in as little as 2 minutes.

Phylum Ctenophora

The phylum Ctenophora consists of nearly 200 species. All are marine, and most are planktonic, usually preying on small

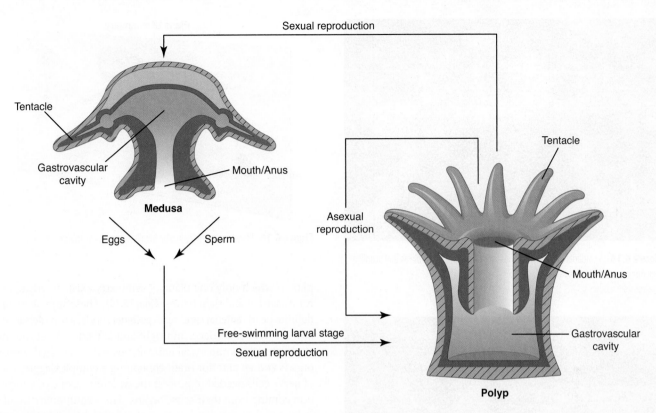

Figure 6.12 Generalized cnidarian life cycle.

zooplankton. Most individuals are smaller than a few centimeters in size, but one tropical genus (*Cestum*) may grow to exceed 2 m in length.

Ctenophores are morphologically similar to many cnidarians, with their radial symmetry, bell-shaped bodies, and streaming tentacles. Unlike cnidarians, ctenophores possess a complete digestive system (with a mouth and an anus) and colloblast cells that superficially resemble sticky cnidarian cnidocytes. One species of ctenophore, *Haeckelia rubra*, actually contains nematocysts, but is only able to sting by using the venom from the nematocysts of its prey. *Haeckelia rubra* ingests the tentacles of a particular jellyfish and then uses the venom from those tentacles to sting other organisms. Ctenophores have eight external bands of longitudinal cilia, called *comb rows*, or **ctenes** (**Figure 6.15**) that provide propulsion, whereas tentacles armed with colloblasts capture food. The comb rows led to the common name of this group, the comb jellies. Recently discovered fossil evidence indicates that some ctenophores living during the Cambrian period lacked tentacles and had hard plates on their bodies.

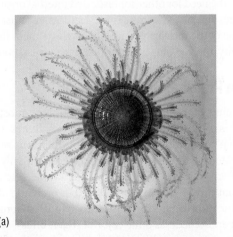

(a)

(b)

Figure 6.13 (a) What appears to be a small jellyfish, *Porpita porpita*, is one of the fascinating examples of a colonial hydrozoan. This species is only several centimeters in diameter. This specimen was photographed in the South Atlantic Bight in the southeastern United States. (b) A true jellyfish, southern California's purple-striped jellyfish, *Chrysaora colorata*, reaches 1 m in diameter.

(a) Islands in the Stream Expedition 2002. NOAA Office of Ocean Exploration and Research. (b) © Ferenc Cegledi/Shutterstock, Inc.

Figure 6.14 A cnidarian polyp with thousands of cnidocytes visible as tiny beadlike structures on the tentacles.
© Annetje/Shutterstock, Inc.

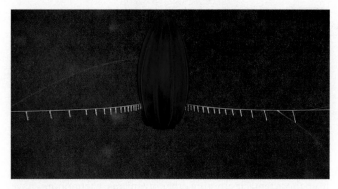

Figure 6.15 A red ctenophore that has yet to be officially named, but is commonly called the Tortugas red, with tentacles streaming out the sides and rows of visible ctenes, or combs, visible the length of the body. The deep red pigment is not visible at depth, so this species looks virtually black unless exposed to a light source.
NOAA *OKEANOS Explorer* 2014.

DID YOU KNOW?

The largest jellyfish, in the genus *Cyanea*, has a bell diameter of over 2 m (over 6 ft) and tentacles trailing behind them over 30 m (over 100 ft) in length! The common name of these jellies is the lion's mane jellyfish, as their bodies are a deep orange or brown color and resemble a lion's thick mane. It is found in cold waters, and the largest recorded specimen was observed in Massachusetts Bay in the 1800s. Although their size may seem daunting, their venom does not deliver a serious sting and is not fatal to humans. One massive stinging incident in the ocean near New Hampshire involved 150 unsuspecting swimmers and what may have been an individual lion's mane jelly, broken up into hundreds of painful stinging pieces.

6.5 Marine Acoelomates and Pseudocoelomates

With one exception (the echinoderms), the remainder of the animal phyla exhibit bilateral body symmetry and possess three tissue layers. *Bilateral symmetry* refers to a basic animal body

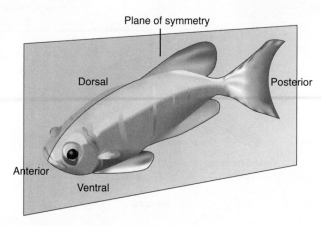

Figure 6.16 Plane of symmetry in a bilaterally symmetrical animal.

plan in which only one plane of symmetry exists to create mirror-image left and right halves (**Figure 6.16**). These animals exhibit definite head (**anterior**) and rear (**posterior**) ends, a top (**dorsal**) and bottom (**ventral**) surface, and right and left sides. Most, but not all, bilaterally symmetrical animals possess specialized sensory organs and an anterior brain containing a complex aggregation of nerve cells needed to process the widening scope of information coming from their sense organs. This evolutionary trend is culminated in the development of a defined head region containing specialized sensory receptors (for vision and the detection of chemicals and sound vibrations) and an anterior brain.

The remaining animal phyla are divided into three different groups depending on their pattern of internal body cavity development, as displayed in **Figure 6.17**. The simplest groups of bilaterally symmetrical marine animals include small, often overlooked, inhabitants of soft mud and sand. These phyla lack an internal cavity between their body wall and digestive tract (**acoelomates**) or have a poorly developed one (**pseudocoelomates**). In these phyla, a circulatory system is either absent or is a vaguely defined open system relying on contractions of the body wall or fluid under pressure rather than a heart to circulate body fluids. Small body sizes with high surface-to-volume ratios also minimize the need for other sophisticated internal organ systems such as respiratory systems. These features contrast strongly with those of the **coelomates** with their well-defined, fluid-filled body cavity, the coelom, described later in this chapter.

Phylum Platyhelminthes

Platyhelminthes, or flatworms, are acoelomates; most are parasitic. This group includes the well-known flukes and tapeworms that parasitize humans. Only some members of the class Turbellaria are free-living. Turbellarians are primarily aquatic, and most are marine. There are a few planktonic species of flatworms, but most dwell in sand or mud or on hard substrates.

Marine turbellarians are usually less than 10 cm long, flat and thin, and are sometimes quite colorful (**Figure 6.18a**). Cilia cover their outer surfaces and are best developed on the flatworms' ventral side. These cilia provide a gliding type of locomotion for moving over solid surfaces. The mouth is usually centrally located on the ventral side and leads to a baglike digestive tract.

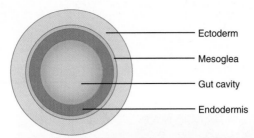

(a) 2 tissue layers, no body cavity. Example: cnidarian

— Ectoderm
— Mesoglea
— Gut cavity
— Endodermis

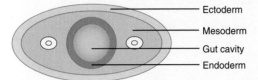

(b) 3 tissue layers, acoelomate. Example: flatworm

— Ectoderm
— Mesoderm
— Gut cavity
— Endoderm

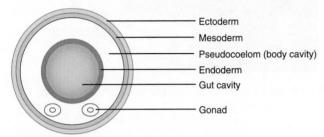

(c) 3 tissue layers, pseudocoelom. Example: nematode

— Ectoderm
— Mesoderm
— Pseudocoelom (body cavity)
— Endoderm
— Gut cavity
— Gonad

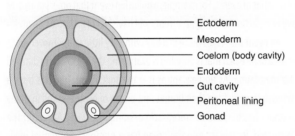

(d) 3 tissue layers, coelom. Example: arthropod

— Ectoderm
— Mesoderm
— Coelom (body cavity)
— Endoderm
— Gut cavity
— Peritoneal lining
— Gonad

Figure 6.17 Simple representations of the various body cavity and tissue layer conditions in animals with tissues.

The digestive tract is incomplete, so one opening is used for both the entrance of food and the exit of waste matter. Turbellarians are carnivorous, preying on other small invertebrates.

Phylum Gnathostomulida

Gnathostomulids, or jaw worms, are a small group of acoelomate marine worms closely related to turbellarian flatworms. Like turbellarians, jaw worms have an incomplete digestive tract. Individuals seldom exceed 2 mm in length. They live in seafloor deposits, where they scrape bacteria and algal films off sediment grains with their jaws and associated basal plate. Of the 100 or so species, many of those described live in the northwestern Atlantic. Members of this phylum are hermaphroditic, and unlike most marine invertebrates, development is direct, with no free-swimming larvae. Species studied thus far have internal fertilization, which likely explains the direct development.

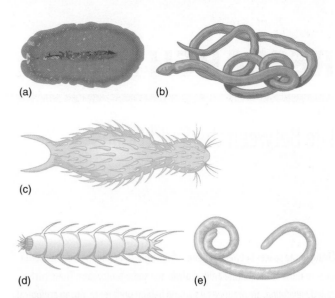

(a) (b)

(c)

(d) (e)

Figure 6.18 Simple wormlike animal phyla: (a) flatworm, (b) nemertean, (c) gastrotrich, (d) kinorhynch, and (e) nematode.

Because these organisms live in between sand grains where they are not visible to humans, it is likely that many more species exist, awaiting discovery and official species descriptions.

Phylum Nemertea

Nemerteans are benthic animals, known as ribbon worms, that are closely related to the flatworms but have a more elaborate body structure (**Figure 6.18b**). They have a simple open circulatory system, a more complex nervous system and musculature, and a complete digestive tract with mouth and anus. Individuals of one species are over 2 m long, but most are much smaller. These shallow-water animals are equipped with a remarkable **proboscis** for defense and food gathering. The proboscis can be everted rapidly from the anterior part of the body to ensnare prey. The proboscis of some nemertean worms has a piercing stylet to stab prey and inject a toxin.

Phylum Gastrotricha

Gastrotrichs include a large variety of marine species, but most are so small (usually less than 1 mm) that they go unnoticed by most observers. They are commonly known as "hairy backs" due to the cilia lining their bodies. They are cylindrical and elongated, with a mouth, feeding structures, and sensory organs at the anterior end (**Figure 6.18c**). Marine gastrotrichs inhabit sand and mud deposits in shallow water and feed on detritus, diatoms, or other very small animals. Some species have special adhesive glands to anchor to the substrate.

Phylum Nematoda

Nematodes, or roundworms (**Figure 6.18e**), are among the most common and widespread multicellular animals. Some are parasitic, such as the 9-m-long species found within the placenta of sperm whales, but many more are free-living. Most marine nematodes live in bottom sediments and are found at virtually

RESEARCH in Progress

Life Between the Sand Grains

Living in the spaces between millions of sand grains are numerous representatives of most of the marine animal phyla and some protozoans. These critters, called *meiofauna*, are grouped by size and habitat preference into an ecological unit rather than a taxonomic unit (**Figure A**). Meiofauna are generally in the range of 45 µm to 1.0 mm, and although tiny, many are visible without the aid of a microscope. They are distinguished from smaller fauna by their ability to be retained in a sieve with mesh that is larger than 45 µm and from larger fauna by their ability to be retained in a sieve with mesh that is smaller than 1 mm (**Figure B**). Meiofauna serve several key functions for ecosystems, such as their role as a food source for many benthic deposit feeders. Their role in the food web allows for nutrient exchange between the benthos and the water column, and their presence within the sediments serves to stabilize the benthos. Some examples of common and important organisms represented in coastal meiofauna include nematodes, ostracods (small arthropods), foraminiferans, and harpactacoid copepods.

Because meiofauna live within the sediments on the bottom of the ocean, they are subject to falling detritus and pollutants that settle out and accumulate in the benthos. This makes them potential indicator species of environmental health. Some meiofauna seem to be more sensitive to environmental disturbances than others; nematodes appear to be less sensitive than copepods or foraminiferans. The amount of pollutant, type of pollutant, and details of the

Figure B Rinsing a sediment sample to be preserved for species identifications.
Lophelia II 2012 Expedition, NOAA-OER/BOEM.

exposure time and habitat all factor into the potential negative effects of a pollution event or series of events. For example, a pollution event that occurs in an area with heavy wave action and strong currents will have less of an impact on meiofauna than an event that takes place in a bay with very little water movement.

One area of research that is ongoing is examining the effects of discharged sewage on meiofauna. As unpleasant as it is to think about, humans have a need to remove sewage from their homes and places of business, and in many areas the ocean has become the receptacle of human sewage. Although the sewage is treated, its release into the ocean has a number of potential environmental impacts. In most cases, pipes extend from shore out several miles where the plume is released near the seafloor. The meiofauna in these areas serve as model organisms for the effects of sewage treatment plants on the health of the ocean, because they are living in the sediments where sewage is released. Scientists collect sediment core samples at particular distances from the area of impact and identify, count, and catalog all of the organisms found in the samples (**Figure C**). Analyses of species compositions at the various sites include comparisons to before sewage release (if prerelease data are available) and comparisons between sites. Unfortunately, changes in species composition can occur quickly, and it is not clear which changes may indicate environmental stress and which are just part of the natural turnover cycle in the benthos. These facts decrease the value of simple monitoring of meiofaunal communities.

A recent study investigating the effects of polluted sediments on meiofauna was conducted using interesting methods that are arguably more valuable than basic monitoring programs. Sediments of various qualities were transplanted to a neutral location and meiofauna (namely nematodes) were assessed based on time to recovery and overall abundance. Sediment health was also determined. Assessments were done at 1, 3, and 8 weeks postdisturbance.

Figure A Typical meiofauna under a dissecting microscope. Organisms were stained using rose bengal, and those visible include ostracaods, copepods, and nematodes.
USGS, Amanda Demopoulos.

Figure C A sediment core sample extracted from the seafloor.

Image courtesy of *Lophelia II* 2009: Deepwater Coral Expedition: Reefs, Rigs and Wrecks.

Results showed that recovery time for the samples from polluted sediments was at least 3 weeks, and in some cases 8 weeks. At the end of the study the experimental (polluted) and control (natural) samples all had similar species compositions. The importance of this study stems from the findings that disturbed communities can recover if given time to do so. In the case of sewage outfalls where sewage is continuously released, it is likely that species compositions changed early on, and will never be able to recover. There is likely a "new normal" in these areas affected by sewage discharge.

Another recent study investigated the effects of the *Deepwater Horizon* oil spill on meiofauna in the Gulf of Mexico. In this case, the area experienced a huge environmental disturbance, including oil pollution and then postspill cleanup efforts, and eventually the disturbance ceased (**Figure D**). It was assumed that marine life in the entire ecosystem would be negatively impacted, but how long the effects would last and whether they would be temporary or permanent was unknown. Meiofauna were an ideal sample group of organisms to study due to their presence in the sediments where contaminants often settle out. Meiofauna were surveyed using sediment cores, and the immediate finding was a shift from animal to fungal communities. This indicated that the animals and protozoans had died, and fungi moved in to decompose the dead matter. This dramatic shift was very disturbing, so an additional study was conducted to assess the changes in meiofauna for an entire year. The study surveyed meiofaunal communities by sequencing a region of a eukaryotic rRNA gene that can help broadly distinguish taxa. Over time, the community shifted back to animals and

Figure D Cleanup efforts during the *Deepwater Horizon* Gulf oil spill included skimming the surface to try to remove oil.
NOAA.

protozoans, and fungi were less abundant. The shift back to animals took place relatively quickly, but the species compositions at the sample sites differed from pre-oil spill conditions. Whether this difference is due to natural fluctuations or the aftermath of the oil spill is still unclear, but research is ongoing to continue to assess changes and compare conditions to those seen before the oil spill.

Critical Thinking Questions

1. What do you think it is about the biology or morphology of nematode worms that seems to make them more resilient during pollution events?

2. Because of the potentially extreme negative impacts from oil spills, do you think that oil companies should be required to pay for meiofaunal monitoring within the areas where they are operating before a potential spill occurs? Why or why not?

3. Is the threat to marine life and entire ecosystems from oil transport and extraction great enough for governments to spend the billions of dollars necessary to move toward a world where oil is used sparingly, if at all? Why or why not? What ways do you reduce your oil consumption, if any?

For Further Reading

Brannock, P. M., D. S. Waits, J. Sharma, and K. M. Halanych. 2014. High-throughput sequencing characterizes intertidal meiofaunal communities in Northern Gulf of Mexico (Dauphin Island and Mobile Bay, Alabama). *Biological Bulletin* 227:161–174.

Liu, X., S. G. Cheung, and P. K. S. Shin. 2011. Response of meiofaunal and nematode communities to sewage pollution abatement: A field transplantation experiment. *Chinese Journal of Oceanology and Limnology* 29(6):1174–1185.

all water depths. In fact, nematodes are probably the most abundant multicellular animals in the marine benthic environment. Cylindrical in cross section and greatly elongated, free-living nematodes seldom exceed a few centimeters in length. Locomotion is not well developed; nematodes depend on quick bending movements of their small tubular bodies to wriggle through mud or water. Most nematodes are predators of protozoans and small animals, including other roundworms.

Phylum Entoprocta

Entoprocts are benthic and colonial, secreting thin calcareous encrustations over rocks, seaweeds, and the hard shells of other animals (**Figure 6.19**). Superficially, they resemble small colonial hydroids. Their external appearance is also quite similar to that of members of another phylum, the Bryozoa (formerly Ectoprocta).

Because entoprocts and bryozoans are superficially similar, they were formerly placed within one phylum. But recent studies of anatomy, embryonic cleavage, and their 18S rRNA genes have shown that two phyla are warranted. Individuals of both groups have U-shaped digestive tracts and a crown of tentacles projecting from the upper surface. The mouth and anus of entoprocts open within the ring of tentacles (hence the name *Entoprocta*, or "inner anus"). They feed on a wide variety of small food particles.

Phylum Tardigrada

Some animals must shed an outer layer such as a shell or **cuticle** in order to grow larger, a process called molting. Phylogenetically, these animal phyla are grouped as Ecdysozoans simply based on the need to **molt**. Ecdysozoans include the phyla Arthropoda (described below), Nematoda (described above), and several smaller phyla, including Tardigrada. Tardigrades, also known as *water bears*, are a very unique phylum with extraordinary characteristics (**Figure 6.20**). They are tiny, less than 1 mm in length, yet they resemble larger, more complex organisms. They have

Figure 6.20 A scanning electron micrograph of a tardigrade.
National Park Service, Dr. Diane Nelson, Great Smoky Mountains National Park.

four pairs of claw-bearing legs and move slowly, in contrast to most other small organisms. Most tardigrades have a complete digestive tract and strong mouth parts to suck juices from plants and a dorsal brain located along a paired ventral nervous system. The majority of tardigrades reproduce sexually, although some are known to use **parthenogenesis**, which is a process of reproduction where an egg is fertilized without the presence of a male, and so does not require a mate. They have a body cavity that is mostly open and poorly defined, but around the gonads the cavity resembles a true coelom, so they are considered to have a partial body cavity. Tardigrades are common in marine environments, but they also live in freshwater and can even live on land in moist areas.

DID YOU KNOW?

Tardigrades (also known as *water bears* or *moss piglets*) and rotifers, common tiny marine animals, typically require a moist environment to live, but can go dormant when conditions dry up. They enter a sort of suspended animation phase as their metabolism slows to near death. They can remain in this phase for days, or even years, and become revived when placed back in water. Not only can they spend time in a dormant phase, but they can also withstand tremendous heat and cold during their dormancy. Tardigrades and rotifers are tough little inhabitants of our earth and sea.

6.6 Marine Coelomates

All remaining marine animal phyla described here are coelomates, characterized by a true internal body cavity, the coelom. The coelom was a major step in the evolutionary development of more complex animal phyla. This cavity originates during embryonic development and separates the digestive tract from the body wall. The separation enables fluids in the coelom to move and promote circulation of oxygen, wastes, and nutrients. The coelom is a compartmentalized cavity, enabling isolation of various tissues, organs, or organs systems (Figure 6.17). Circulation in many coelomates is enhanced by a muscular heart pumping blood through a **closed circulatory system**, including

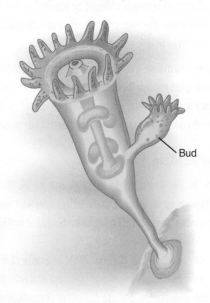

Figure 6.19 A solitary entoproct, *Loxosomella*, with its crown of feeding tentacles.

Bud

blood confined in vessels that is actively moved to and from the heart. With a coelom, the digestive tract has become more specialized and its efficiency improved. A more spacious coelom also allows larger **gonads** to increase the number of gametes for use in diverse reproductive strategies. Finally, body wall muscles function independently of the digestive tract and have a greater range of specialized actions.

Protostomes

Coelomates include most of the large, dominant, and successful marine animal phyla, as well as several smaller, lesser-known ones. Coelomates have evolved along two separate lineages, the protostomes and deuterostomes, reflecting features of their early embryonic development. This protostome–deuterostome split is characterized in modern animal groups by fundamental differences in the pattern of cellular division in zygotes after fertilization. Protostomes express spiral patterns of cleavage and a determinate pattern of embryonic development, resulting in unequal-sized cells in the preblastula stage, each fated to develop into predetermined tissues or organs of the adult. As development proceeds, protostome embryos develop an indentation, the *blastopore*, in the cells of the blastula that eventually becomes the mouth in the adult. Deuterostomes differ from protostomes in all of these features. They exhibit radial cleavage resulting in equal-sized blastomeres after fertilization, indeterminate development of preblastula cells without early embryonic determination of what adult tissues those cells will develop into, and a blastopore that develops into the anus of the eventual adult rather than the mouth. These fundamental differences are illustrated in **Figure 6.21**.

Three Marine Lophophorate Phyla

A **lophophore** is a crown of ciliated feeding tentacles found in three structurally dissimilar phyla of marine animals: Bryozoa, Phoronida, and Brachiopoda. The Bryozoa is a major animal phylum, with over 6,000 marine species commonly referred to as *bryozoans*. They are primarily members of shallow-water benthic communities, occupying the same general habitats as entoprocts (described earlier). The bryozoan mouth is located within the tentacles, but the anus is not (unlike the entoprocts). Like entoprocts, bryozoans are colonial and form encrusting or branching masses of small individuals (usually less than 1 mm in size; **Figure 6.22**). Bryozoans and entoprocts provide excellent examples of how the evolutionary pathways of quite different animal groups converge to similar adaptive forms and habits when exposed to similar environmental conditions. Entoprocts even possess a crown of feeding tentacles that resembles a lophophore, but is not considered a true lophophore. Such **convergent evolution** is a common theme in animal evolution.

The phylum Phoronida is one of the least familiar animal phyla, with fewer than 20 species of elongated burrowing animals. Although this phylum contains relatively few species, they can be very abundant in certain locations. All are marine and live in tubes in the seafloor in shallow water, and some bore into rocks or shells. When feeding on plankton, the lophophore projects out of the tube, but it can be rapidly retracted for protection. The phoronids seldom exceed 20 cm in length and have no appendages except for the lophophore.

The phylum Brachiopoda, the lamp shells, was a very successful group in the past, with more than 30,000 extinct species having been described. Fewer than 400 still survive. *Lingula*,

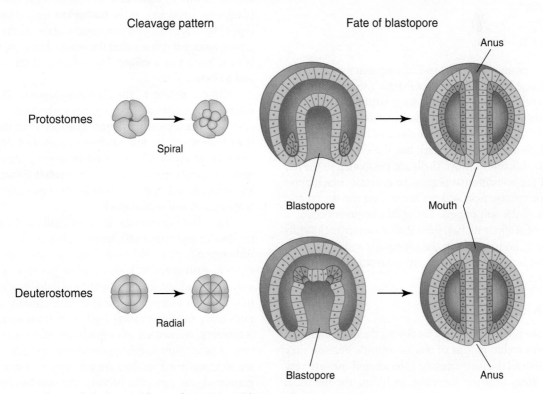

Figure 6.21   Comparison of some early developmental features of protostomes and deuterostomes.

Figure 6.22 A magnified view of a branched colony of ectoprocts with extended feathery lophophores.
© blickwinkel/Alamy Images.

Figure 6.23 A brachiopod, not to be confused with a clam, using its ciliated lophophore to feed.
© Cultura RM Exclusive/Alexander Semenov/Getty Images.

for example, has an unbroken fossil history that extends back over the past one half billion years of Earth's history. All brachiopods are benthic during the adult stage (**Figure 6.23**) and live attached to the sea bottom by a muscular stalk. The outer calcareous shells superficially resemble those of bivalve mollusks (the next phylum to be described), but the symmetry of the shells is quite different. Bivalve shells are positioned to the left and right of the soft internal organs. In contrast, brachiopod shells are not symmetrical and are located on the dorsal and ventral sides of the soft organs. Living brachiopods occupy a wide variety of seafloor niches, from shallow-water rocky cliffs to deep muddy bottoms. As in the phoronids and bryozoans, the ciliated lophophore gathers minute suspended material for nutrition from seawater.

Phylum Mollusca

Members of the phylum Mollusca are among the most abundant and easily observable groups of marine animals because they have adapted to all the major marine habitats, and they are very common in areas near the shoreline, including the intertidal zone and along human-made structures such as pier pilings.

It is difficult to characterize such a large and diverse group as the phylum Mollusca, but a few common traits are observable. Mollusks are unsegmented animals. Most mollusks have a hard external shell surrounding the internal organs and use a large muscular foot for locomotion, anchorage, and securing food. Most mollusks have an array of specialized sense organs in the anterior region of their body near the brain. This pattern of body organization, known as **cephalization**, is most apparent in squids and octopuses.

This phylum is commonly divided into seven classes, although some disagreement exists over phylogenetic groupings. Representatives of five of these classes are quite common and are shown in **Figure 6.24**. In four of these classes, the early planktonic larval form is the **trochophore** larva (**Figure 6.25**). This type of larva is also found in annelid worms. As the trochophore grows, a ciliated tissue called the velum develops, and the larva is then known as a **veliger**. The velum is used to collect food and to swim.

Chitons belong to the class Amphineura. They are characterized by eight calcareous plates embedded in their dorsal surfaces. These animals, found in rocky intertidal areas, use their large muscular foot to cling to protected depressions on wave-swept rocks. Chitons feed by grazing algae from rocks with a rasping tongue-like organ, the **radula** (**Figure 6.26**). Radulas are also found in three other classes of mollusks: gastropods, scaphopods, and cephalopods.

The class Gastropoda includes snails and slugs (marine, freshwater, and terrestrial), limpets, abalone, and nudibranchs. Although one-piece shells are characteristic of this class, several types of gastropods lack shells. Most gastropods are benthic; only a few without shells, or with very light ones, have successfully adapted to a pelagic lifestyle. Like chitons, many gastropods graze on algae; others feed on detritus and organic-rich sediments. Numerous gastropods are also successful predators of other slow-moving animal species. Some cone snails are so venomous that they are a danger to humans. Although gastropods are generally benthic, the nudibranch pictured in **Figure 6.27** can swim in the water column.

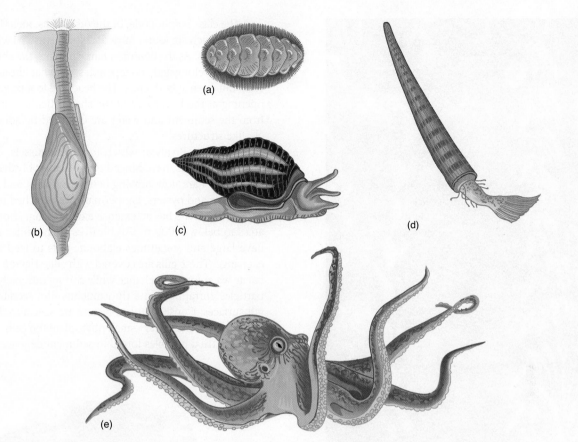

Figure 6.24 Representatives of the common classes of mollusks: (a) Amphineura, (b) Bivalvia, (c) Gastropoda, (d) Scaphopoda, and (e) Cephalopoda.

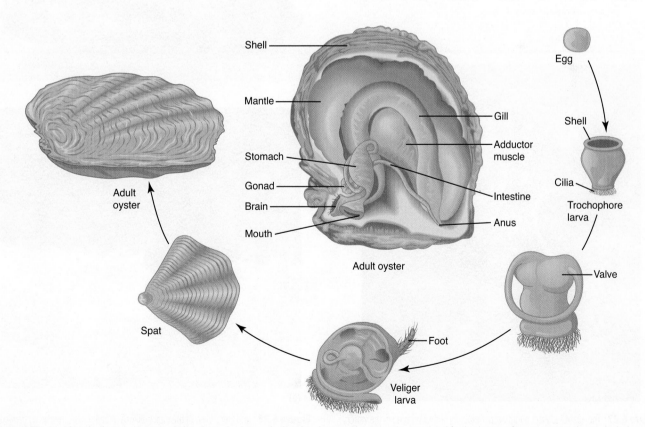

Figure 6.25 The life cycle of a typical oyster, including the trochophore and veliger larval stages. About 2 weeks are required for development from egg to spat, the attached stage prior to adulthood.

The class Scaphopoda, or the tusk shells, includes over 500 species, which are found buried in sediments in a wide range of water depths. As the common name implies, the shells of these animals are elongated and tapered, somewhat like an elephant's tusk, but open at both ends. The head and foot project from the opening at the larger end of the shell. Microscopic organisms from the sediment and water are captured by adhesive tentacle-like structures.

The class Bivalvia, which includes mussels, clams, oysters, and scallops, have hinged two-piece, or bivalve, shells. As adults, most are slow-moving benthic animals, and some, such as mussels and oysters, are permanently attached to hard substrates. This class has an extensive depth range, from intertidal areas to below 5,000 m. All bivalves lack radulas and instead have large and sometimes elaborate gills to feed and for gas exchange. These gills are covered with cilia (**Figure 6.28**) that circulate water for gas exchange while sorting extremely small food particles entrapped on a thin mucous film secreted over the gill surfaces. Consequently, bivalves are specialized to feed on suspended bacteria, very small phytoplankton cells, and microscopic detrital particles found in sediment deposits.

Figure 6.26 Scanning electron micrograph of the radula of a gastropod mollusk.
Courtesy of Dr. Carole S. Hickman, University of California Berkeley.

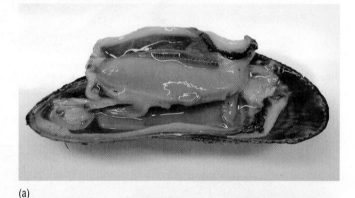

(a)

Figure 6.27 The Spanish dancer nudibranch, *Hexabranchus*, swimming in the waters of Maui. This species is exceptionally large and adapted for basic swimming.
NOAA/Jill Samzow, Pacific Islands Fisheries Science Center.

(b)

Figure 6.28 (a) Lateral view of the intact gill of a ribbed mussel, *Geukensia demissa*. (b) A micrograph of the gill edge with cilia.
(a) Courtesy of John Morrissey. (b) Courtesy of James Sumich.

Molluscan evolution has reached its zenith in the class Cephalopoda, which includes squids, octopuses, cuttlefish, and nautiluses. Members of this class are carnivorous predators with sucker-lined tentacles in most species (nautiluses lack suckers); well-developed sense organs and large brains; reduction or loss of the external shell typical of most other mollusks; and, in some species, very large body sizes. The eyes of squids and octopuses are remarkably similar to our own, with a retina, cornea, iris, and a lens-focusing system. A unique propulsion system, using high-speed jets of water, provides excellent maneuverability and swimming speeds greater than those of any other marine invertebrate. Octopuses also use their eight arms to crawl over the seafloor gracefully. Octopuses sometimes have arm spans exceeding several meters. The giant squid, *Architeuthis*, which may reach 18 m in length and weigh over 1 tonne, is by far the largest living invertebrate species.

Rigid-walled gas containers used for buoyancy are found in only a few types of cephalopods. All cephalopods are believed to have evolved from an ancestral form that had an external shell. *Nautilus* is the only living cephalopod that has retained its external shell (**Figure 6.29**). The shells of other living cephalopods are either reduced to an internal chambered structure, as in the cuttlefish (*Sepia*) and *Spirula* (a deep-water squid), or are absent entirely, as in octopuses. In squids other than *Spirula*, a thin chitinous structure (the pen) extends the length of the mantle tissue and represents the last vestige of what was once an internal shell.

Nautilus, *Spirula*, and cuttlefish all have numerous hard transverse partitions, or *septa*, that separate adjacent chambers of the shell. In *Nautilus*, only the last and largest chamber is occupied by the animal. As *Nautilus* grows, it moves forward in its shell and adds a new chamber by secreting another transverse septum across the area it just vacated. When a new chamber is formed, water is removed and is replaced by gases (mostly N_2) from tissue fluids. The gases diffuse inward, and the total pressure of the gases dissolved within the chambers never exceeds 1 atm.

These chambered cephalopods are confronted with the same depth-limiting factor that plagues submarines. Their depth ranges are limited by the resistance of their shells to increased water pressure. Each species has a crucial implosion depth at which the external water pressure becomes too great for the design and strength of its shell and the shell collapses. The implosion depth of *Nautilus* shells, for example, was previously thought to be around 500 m, but recent observations of this animal at 700 m has led to an adjustment of the implosion depth to around 800 m. This species migrates vertically for feeding and protection from prey, thus a large range of suitable depths is a benefit for survival.

Some Wormlike Protostomes

Most wormlike marine invertebrates live in soft mud or sand deposits as **infauna** or **meiofauna**. Their elongated body forms permit effective burrowing movements despite a lack of rigid internal skeletons to support the muscles of locomotion. Muscles in the body wall work against the enclosed fluid contents of the body to allow burrowing actions and other body movements. The fluids cannot escape and are essentially incompressible. As such, they provide a **hydrostatic skeleton** for the muscles of the body wall. In the more effective burrowing worms, these muscles are arranged in two sets: circular muscle bands around the body and longitudinal muscles extending the length of the body, much like the muscles of the human digestive tract. Like all other muscles, these muscles can work only by contracting. When the circular muscles of a worm's body contract, the worm's diameter decreases, squeezing its hydrostatic skeleton and forcing the body to elongate, much as squeezing a long, inflated balloon at one end causes it to expand at the other end. If the rear of the body is anchored, contracting the circular muscles pushes the anterior end forward. When the circular muscles relax, the longitudinal muscles can then contract to shorten the body and make it fatter or move it forward. These two types of muscles continue to work in opposition to each other to provide an effective sediment-burrowing motion for a large variety of marine worms. The sipunculid worm shown in **Figure 6.30** provides a visual representation of what happens when circular and longitudinal muscles contract or relax, as the worm is wide and fat in some areas and long and slender in others.

A newly defined phylum, Cephalorhyncha, includes small worms that possess a cuticle, and therefore must molt to grow. This phylum includes several worm groups that were previously considered individual phyla, two of which are discussed below, Priapulida and Kinorhyncha. The class Priapulida has fewer than 20 species. Priapulid worms (**Figure 6.31**) live buried in intertidal to abyssal sediments in warm or cold latitudes and seldom exceed 10 cm in size. Priapulid worms are detritus feeders or are predatory, feeding on soft-bodied invertebrates they

Figure 6.29 A *Nautilus* specimen swimming in the western Pacific.
NOAA Fisheries.

Figure 6.30 Sipunculid worms, *Sipunculus*.
Courtesy of Dave Cowles, Rosario Marine Invertebrates (http://rosario.wallawalla.edu/inverts).

Figure 6.31 A marine priapulid worm.

capture with their eversible introvert. Priapulids are found in hypersaline ponds and anoxic muds, as well as more moderate benthic habitats. Members of the class Kinorhyncha are exclusively marine and superficially resemble gastrotrichs (described above). They also are cylindrical and elongated but are covered with a cuticle that is segmented (Figure 6.18d). Marine kinorhynchs occur in sand and mud deposits, sometimes in densities of more than 2 million individuals per square meter, where they feed exclusively on bacteria. Approximately 188 species of kinorhynchs have been identified.

Sipuncula, another phylum of wormlike animals, are found throughout the world ocean. The 150 species of sipunculids, or peanut worms, are entirely marine. Most species are found in the intertidal zone, but their distribution extends to abyssal depths. Peanut worms are benthic. They live in burrows, crevices, or other protected niches, often in competition with other wormlike animals. Sipunculids range from 2 mm to over 50 cm in size and have a cylindrical body that is capped by a ring of ciliated tentacles surrounding the anterior mouth (Figure 6.30).

Pogonophorans are noted for their complete lack of an internal digestive tract and are thought to absorb dissolved organic material or to use symbiotic bacteria to provide their energy needs. They are restricted to the deep sea and can reach lengths of 1.5 m. These giant specimens were an exciting discovery in deep-sea waters that were once deemed uninhabitable. Pogonophorans can live and thrive in very hot, sulfur-filled water.

Phylum Annelida

The phylum Annelida is usually represented by the familiar terrestrial earthworm; however, this phylum also contains a diverse and successful group of marine forms with more than 7,800 species in the class Polychaeta alone. Polychaete worms, like other annelids, are segmented. Segmentation has evolved separately in both major lineages of coelomates, in annelids and arthropods in protostomes, and in chordates in the deuterostome line. The body cavity and internal organs contained within polychaete worms are subdivided into a linear series of structural units called **metameres**. The result of segmentation is a sequential compartmentalization of the worm's hydrostatic skeleton and surrounding muscles. This permits a greater degree of localized changes in body shape and a more controlled and efficient form of locomotion.

Some polychaetes ingest sediment to obtain nourishment, others are carnivorous, and many use a complex tentacle system (**Figure 6.32**) to filter microscopic bits of food from the water. Suspension-feeding polychaetes often occupy partially buried tubes and are common in intertidal areas; however, they are also found in deeper water. The 40 species of *Tomopteris* are planktonic

Figure 6.32 Stiff, ciliated radioles of tube-dwelling polychaete worms filter plankton and transport it down toward the mouth.
© Frank & Joyce Burek/age fotostock.

throughout their lives and can emit yellow bioluminescence from their footlike parapodia. The remarkably large tube worms discovered in deep-sea vent communities were initially assigned to the phylum Pogonophora; however, this phylum has been abandoned, and it has been determined that the large vent tube worms are actually polychaete annelids (**Figure 6.33**).

The Echiuridae is a small family of benthic polychaete marine worms that resemble peanut worms in size and general shape. Echiurids are common intertidally, but they are occasionally found at depths exceeding 6,000 m. Most echiurids live in burrows in the mud. One remarkable feature of echiurids is

Figure 6.33 *Riftia* tube worms live in extreme conditions among other unique creatures in the deep sea.
Image courtesy of NOAA *Okeanos Explorer* Program.

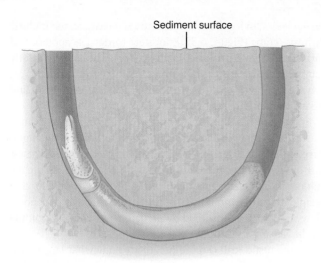

Sediment surface

Figure 6.34 The fat innkeeper worm, *Urechis*, in its burrow.

their extensible *proboscis*, a feeding and sensory organ that projects from their anterior. The proboscis of some species is longer than the remainder of the body and is quite effective for gathering food by "mopping" the sediment while the worm remains in the protected confines of its burrow. *Urechis*, an echiurid of the California coast known as the fat innkeeper, has a very short proboscis. The proboscis secretes a mucous net from the animal to the wall of its U-shaped burrow (**Figure 6.34**). The burrow is also often inhabited by small crabs, shrimps, or other casual guests. Water is pumped through the burrow by repeated waves of contractions along the worm's body wall. As water passes through the mucous net, bacteria and other extremely small food particles are trapped. When the net is clogged with food, the worm consumes it and constructs another.

Phylum Arthropoda

Like annelids, arthropods are segmented along the length of their bodies. In addition to the advantages of segmentation, marine arthropods possess a distinctive hard **exoskeleton** made of a complex blend of a long-chain polysaccharide, called *chitin*, embedded in a protein matrix. This rigid outer skeleton serves not only as an impermeable barrier against fluid loss and microbial infection, but also as a system of levers and sites for muscle attachment. Its structure resists deformations caused by contracting muscles, allowing faster responses and greater control of movements. Flexing of the body and appendages is limited to thin membranous joints located between the rigid exoskeletal plates. However, exoskeletons lack some of the shape-changing advantages of segmentation and hydrostatic skeletons found in soft-bodied annelid worms.

The exoskeleton also restricts continuous growth. Periodically, arthropods shed, or molt, their old exoskeleton, and it is replaced by a new larger one as the animal quickly expands to fill it. Each molt is followed by an extended period of time with no growth. For some species of arthropods, the number of molts is fixed, with molting and growth ceasing at the adult stage. For others, growth and molting continue as long as they live, although the frequency of molting diminishes with increasing age and body size.

Members of the phylum Arthropoda account for almost three fourths of all animal species identified so far, now exceeding 1 million species. Most belong to the class Insecta, a group abundant on land and in freshwater habitats. Only a few species of insects, including pelagic water striders, several types of sand and kelp flies, and water beetles, have evolved to thrive in seawater environments; however, three other groups within this phylum, the subphylum Crustacea (crabs, shrimp), class Merostomata (horseshoe crabs), and class Pycnogonida (sea spiders), are primarily or completely marine in distribution. Our view of their relationships to each other and to other arthropods is undergoing extensive revision. One widely accepted phylogeny, based on current genetic evidence, places Merostomata and Pycnogonida close to spiders (class Arachnida) and Crustacea as a subphylum and sister group to insects. Some insects are more closely related to aquatic crustaceans than they are to other terrestrial arthropods.

Two classes of arthropods are completely marine, but their diversity is limited. The first class, Merostomata, has an extensive fossil history that includes extinct water scorpions, or eurypterids, up to 3 m long. Only three modern genera exist, including the horseshoe crab, *Limulus*, an inhabitant of the Atlantic and Gulf Coasts of North America (**Figure 6.35**). The sea spiders of the class Pycnogonida are long-legged bottom dwellers with reduced bodies. Small pycnogonids only a few millimeters in

Figure 6.35 A pair of horseshoe crabs spawning in Chesapeake Bay, Maryland.
NOAA.

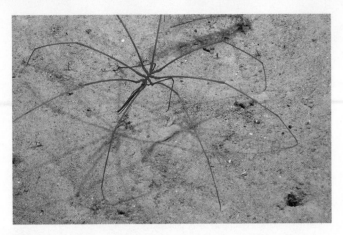

Figure 6.36 A large pycnogonid with its long proboscis extending from the head (front, thick red structure that resembles another short leg), ready to feed.
North Atlantic Stepping Stones Science Party, IFE, URI-IAO; NOAA/OAR/OER.

size are quite common in the intertidal zone. They can be collected from hydroid or bryozoan colonies or from the blades of intertidal algae. Deep-sea pycnogonids are often much larger and may have leg spans of 60 cm (**Figure 6.36**). Although pycnogonids resemble spiders, they do not create webs or possess venom, so arachnophobes need not fear the harmless sea spiders.

The third group of marine arthropods, the Crustacea, is an extremely abundant and successful group of marine invertebrates. Crustaceans are arthropods with two pairs of **antennae** and a larval stage known as a **nauplius**. Few other useful generalizations can be made concerning this class. Its members exhibit a tremendous diversity in body plans (**Figure 6.37**) and modes of feeding. The range of habitats also varies greatly, from burrowing ghost shrimp to planktonic copepods and parasitic

barnacles. Obvious and well-known crustaceans include shrimps, crabs (Figure 6.37f), lobsters (order Decapoda), and barnacles (order Cirripedia, Figure 6.37g). These large, mostly benthic, crustaceans are not representative of the entire class, however, because most marine crustaceans are very small and are major components of the zooplankton.

Copepods (subclass Copepoda, Figure 6.37d) are small crustaceans, generally smaller than a few millimeters in length. Despite their small size, their efficient filter-feeding mechanisms and overwhelming numbers in pelagic communities dictate that much of the energy available from the first trophic level of pelagic marine communities is channeled through copepods. They, in turn, are consumed by predators as diverse as minute fish larvae and huge right whales.

Euphausiid krill (Figure 6.37c) are somewhat larger than copepods, but they fill similar niches in pelagic communities. These crustaceans have a global distribution. One of the larger species of this group, *Euphausia superba*, grows to 6 to 7 cm. Found in the cold and productive waters around Antarctica, this species aggregates in large dense shoals sometimes tens of kilometers long. They are a favorite prey of fish, whales, seals, and penguins and are the basis for a potentially enormous single-species commercial fishery. Other common planktonic crustaceans include members of the class Branchiopoda (brine shrimp and cladocerans, Figure 6.37b), class Ostracoda, and the early larval stages of most other marine crustaceans.

Two groups of smaller benthic crustaceans are isopods, similar to backyard sowbugs or pillbugs, and amphipods, including beach hoppers and sand fleas. The bodies of isopods are somewhat flattened vertically, whereas amphipods are compressed laterally and exhibit more specialization of appendages, with some legs for swimming and others for jumping or

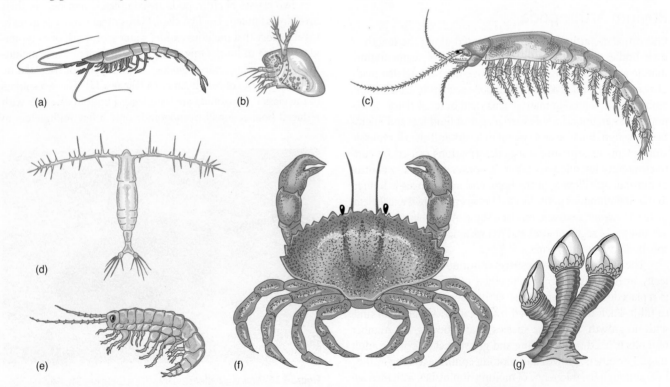

(a) (b) (c) (d) (e) (f) (g)

Figure 6.37 A variety of marine crustaceans: (a) mysid, (b) cladoceran, (c) euphausiid, (d) copepod, (e) amphipod, (f) crab, and (g) barnacle.

Case Study

American "Lobstah"

If you have ever enjoyed a lobster meal in the United States with claw meat on your plate, you have likely eaten an American lobster, *Homarus americanus*, harvested from somewhere between Labrador, Canada, and Cape Hatteras, North Carolina (**Figure A**). The American lobster is one of about 30 species of clawed lobsters, and the only clawed lobster found in the northwestern Atlantic. The largest fishery for this species is in Maine, where families of fishers have been capturing this species since colonial times. The fishery is one of the most valuable in the United States, with annual revenues in the hundreds of millions!

Figure A   A Maine lobster fisher holding an American lobster. Rubber bands are used to prevent the lobsters from harming one another or humans.
© spwidoff/Shutterstock, Inc.

American lobsters are common inhabitants of the ocean floor from shallow coastal waters out to 700 m (2,300 ft) depth. They are long-lived, and some scientists have estimated their maximum life span to be as long as 100 years. Although they live at various depths and on various substrates, wherever they live, one requirement is some kind of crevice or area to hide in. American lobsters are solitary and very territorial, defending their hiding places with their large claws. Like other crustaceans, they molt to grow, and it is estimated that this species molts around 20 to 25 times before it reaches maturity. After molting, the lobster often eats the old shell to replenish minerals, which aids in growth of the new shell. Lobster mating usually occurs when the females have soft shells, and the females store sperm for up to a year. The sperm is transferred by the female when she releases eggs to her abdomen, and the young develop for 9 to 11 months within the safety of the ventral side of their mother's tail (**Figure B**). The American lobster is carnivorous as a larva, feeding on zooplankton. As adults, they become generalist omnivores, feeding on various invertebrates, fish, and macroalgae.

Reports on the status of this species are mixed, but American lobsters have continued to exist for hundreds of years, even with heavy fishing pressure. Reports of lobster populations from the 1800s indicate that populations were much larger than what we see today, but in 2012 record catches occurred in the Gulf of Maine. For the first time in decades, the supply of lobster was higher than demand, and prices dropped. Although consumers do not really feel the effects of a price drop of $1 to $2 a pound, fishers selling thousands of pounds are greatly affected by such a price drop. As this book is being published, prices are back on the rise after a fairly slow season start in 2015. As supply increases, food suppliers are inventing new markets to sell to. American lobster is now being shipped overseas more frequently than in the past. The international interest in this

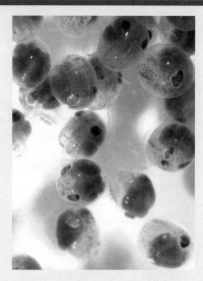

Figure B   Late stage eggs that will soon hatch, so the larvae can swim and find a suitable area to settle to.
Alicia Miller, NEFSC/NOAA.

species makes regulations more important than ever. Although populations may seem steady now, overfishing can occur very quickly. Maine populations appear to be steady or even on the rise, but populations in southern New England are experiencing low abundance, reproduction, and survival rates. The cause of the decline in New England populations is uncertain, but fishing pressure is likely contributing.

The American lobster fishery is an example of a fishery with strict regulations and historic and ongoing population data. Fishers whose target species is the American lobster are serious about their jobs, and serious about maintaining lobster populations for their own future and the futures of their children, if their children should choose to pursue lobster fishing as a career. Any time a gravid female lobster is caught a notch is stamped out of her bottom shell before she is returned to the ocean to alert future fishers of her contribution to the population (**Figure C**). It is illegal to capture females with

Figure C   A gravid female with a notched tail (left side of lobster's tail with the "V" notch).
NOAA Fishwatch.

(continues)

Case Study (*continued*)

notches. Strict size requirements have also been implemented, and traps are required to have escape vents for smaller-sized lobsters that wander in. Regulations are also in place for trap design to reduce bycatch and potential harm to marine mammals. Strict regulations and decreased permit availability have made lobster fishing a difficult career path. Many fishers have moved on to fishing other species or out of the fishing industry entirely. The American lobster is a species that appears to be able to handle fishing pressure to some degree and, with continued fisheries management, will likely provide delectable main courses for people for many years to come.

Critical Thinking Questions

1. What factors other than fishing may be contributing to declining lobster populations in New England? (*Hint*: Think of the environmental factors that tend to affect species distributions.)

2. Do you think the success of the American lobster fishery for several centuries is due to effective fisheries management or just pure luck? What factors have helped you to formulate an opinion on this topic?

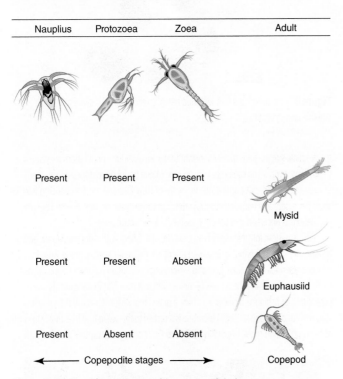

Figure 6.38 Developmental stages of three groups of planktonic crustaceans.

digging (Figure 6.37e). Both groups are found in a wide variety of marine habitats, with amphipods more commonly associated with soft bottom sediments.

Life cycles of crustaceans involve several definable stages: nauplius, protozoea, zoea, and adult (**Figure 6.38**). Each stage is punctuated by one or more molts of the exoskeleton and some accompanying structural metamorphosis. For example, the larvae of copepods enter the copepodite stage, a sexually immature form resembling adults after six molts. Five more molts lead to the adult stage; thereafter, no more molts occur. The yearly growth and reproductive cycles for a planktonic copepod are shown in **Figure 6.39**.

Deuterostomes

Only four phyla of deuterostomes are found in the sea, yet they exhibit a wide range of body plans, from relatively simple worms to radially symmetrical echinoderms to bilaterally symmetrical and segmented chordates. The evolutionary affinities of these four phyla are not resolved. The developmental characteristics they share place them closer to each other than to any of the protostomes described to this point. In particular, phylum Chaetognatha, the arrow worms, has left scientists baffled. Members of this phylum have deuterostome-like development, yet chaetognaths are placed at the base of the protostome tree by most molecular studies. We will include them below in our discussion of deuterostomes, but it should be noted that studies are ongoing and the phylogenetic position of this group is likely to change in the near future.

Two Deuterostome Wormlike Phyla

In contrast to the general benthic habitat of most marine wormlike animals, arrow worms (phylum Chaetognatha) are streamlined planktonic carnivores (**Figure 6.40**). Although they seldom exceed 3 cm in length, they are voracious predators of other zooplankton, especially copepods. Arrow worms swim with rapid darting motions and capture prey with the bristles that surround their mouth. Only about 130 species of arrow worms exist, but they are frequently very abundant in the zooplankton. Certain species of arrow worms apparently respond to and associate with subtle chemical or physical characteristics of seawater. Consequently, they are useful biological indicators of particular oceanic water types. It is thought that some arrow worms have become secondarily simplified over evolutionary time to meet the requirements of their environments.

The phylum Hemichordata (acorn worms) is a small group of benthic marine worms resembling pogonophorans and is truly part of the deuterostome lineage. Acorn worms have an anterior proboscis and a soft, flaccid body that is up to 50 cm long. These worms are generally found in shallow water and live in protected areas under rocks or in tubes or burrows.

Radial Symmetry Revisited: Phylum Echinodermata

An exclusively marine phylum, the Echinodermata are widely distributed throughout the sea. They are common intertidally and are also abundant at great depths. Almost all forms are

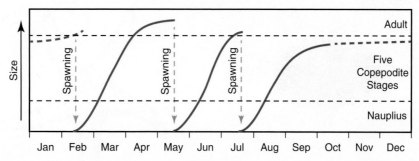

Figure 6.39 Growth and reproductive cycles of a North Atlantic copepod (*Calanus finmarchicus*). The adults of each brood produce eggs for the next brood (arrows). Dashed lines indicate overwintering of copepods in deep water that experience little growth. Data from R. S. Russel. *Journal of the Marine Biological Association, U.K.* 20(1935):309–332.

benthic as adults. Most are characterized by a calcareous skeleton, external spines or knobs, and a five-sided, or **pentamerous**, radial body symmetry (obvious in **Figure 6.41b**, **e**, and **f**). Because echinoderms develop from bilaterally symmetrical larval stages, radial body symmetry is a secondary condition in this phylum. This and other aspects of their evolutionary history separate them on the phylogenetic tree of animal evolution (Figure 6.1) from Cnidaria and Ctenophora, the two phyla characterized by primary radial body symmetry. A unique internal water-vascular system functions as a simple circulatory system and hydraulically operates numerous tube feet. The tube feet extend through the skeleton to the outside and act as respiratory, excretory, sensory, and locomotive organs.

Five classes of echinoderms are usually recognized. Representatives of each are shown in Figure 6.41. The Echinoidea are spiny herbivores or sediment ingesters variously known as sea urchins, heart urchins, and sand dollars (Figure 6.41a). The Asteroidea, or sea stars (Figure 6.41b), are usually five armed, but the number of arms may vary. Sea stars with 6, 10, and 21 arms are known. Most sea stars are carnivorous, but a few use cilia and mucus to collect fine food particles. Feather stars and

sea lilies (class Crinoidea, Figures 6.41c and **6.42**) usually attach themselves to the sea bottom with their mouths oriented upward to trap plankton and detritus with their arms and with mucous secretions. Sea cucumbers of the class Holothuroidea are sausage shaped and have a mouth located at one end of their body (Figure 6.41d). The body wall is muscular, with reduced skeletal plates and spines. A few sea cucumbers feed on plankton, but most ingest sediment and detritus. The brittle stars (class Ophiuroidea, Figure 6.41e) are smaller than most other echinoderms but are very common in soft muds, rocky bottoms, and coral reefs. Brittle stars are often found blanketing the muddy seafloor in great abundance, reaching their arms up into the water to capture fine food particles. A recently described group, the Concentricycloidea was initially considered a class, but has now been changed to infraclass status within the class Asteroidea. This group includes a few species of echinoderms known as sea daisies (Figure 6.41f). These animals are essentially flattened, armless sea stars and have been collected only at depths greater than 1,000 m.

The Invertebrate Chordates: Phylum Chordata

Body segmentation evolved twice in the history of animal evolution, once in the protostome line, leading to annelids and arthropods, and again in the deuterostome line with the phylum Chordata. Like mollusks and arthropods, chordates represent a pinnacle in animal evolution. Chordates exhibit a remarkable variety of body forms, from small gelatinous zooplankton to large fish and whales. Yet all members of this phylum possess the following structural features at some time during their development: (1) a supportive **notochord** made of cartilage extends along the midline of the body; (2) a hollow dorsal nerve cord, located above the notochord; (3) a postanal tail that extends well beyond the posterior opening of the complete digestive tract; and (4) gill pouches or slits that develop as openings on each side of the pharynx.

Two subphyla, the Tunicata (previously Urochordata) and Cephalochordata, are relatively small and are completely marine in distribution. The roughly 3,000 species of tunicates are usually divided into three classes: benthic sea squirts (**Figure 6.43**), gelatinous pelagic salps, and small planktonic larvaceans. The other subphylum, Cephalochordata, contains about 30 species, including the lancelet, *Branchiostoma*. These animals are small

Figure 6.40 *Sagitta*, a chaetognath arrow worm.

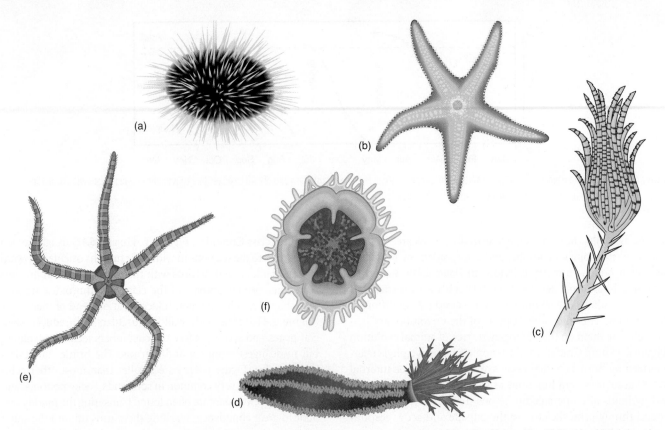

Figure 6.41 Representatives of the five living echinoderm classes, and one newly demoted class, Concentricycloidea: (a) Echinoidea, (b) Asteroidea, (c) Crinoidea, (d) Holothuroidea, (e) Ophiuroidea, and (f) Concentricycloidea (now part of Asteroidea).

and tadpole shaped and live partially buried tail first in near-shore sediments.

A unique feature of the third subphylum, the Vertebrata, is the vertebral column that replaces the notochord as the central skeletal structure of body support. Members of this subphylum include three classes of fishes and four classes of tetrapods.

Figure 6.42 This image displays members of three different echinoderm classes living closely together on the seafloor at an offshore bank in southern California. The white seastar is in the class Asteroidea, the many small brittlestars with striped arms are in the class Ophiuroidea, and the orange feathery crinoid (bottom right and top right) is in the class Crinoidea.

NOAA/SWFSC ROV team.

Figure 6.43 A nearly transparent sea squirt with a small incurrent and large excurrent opening for circulating water through its body cavity.

© Dirscherl Reinhard/age fotostock.

STUDY GUIDE

TOPICS FOR DISCUSSION AND REVIEW

1. Why was the Kingdom Protista abandoned?

2. Placozoans and labyrinthomorphs are both seemingly multicellular, yet only the placozoans are classified as animals. Why is this?

3. Many common marine animals have wormlike body forms. Why might this body shape be advantageous for mud or sand dwellers?

4. Why do you think many crucial sensory organs are concentrated in the head region of "higher" animals rather than in other parts of their bodies?

5. Prepare a table of the following mollusks: snail, clam, chiton, abalone, nudibranch, scaphopod, squid, and octopus. Complete the table by listing the following characteristics of their shells: present or absent, shape, number present, internal or external.

6. Compare and contrast the methods of locomotion used by the mollusks listed in the previous question. (*Hint*: Some mollusks are capable of more than one type of movement.)

7. List and discuss the advantages and disadvantages of the rigid arthropod exoskeleton in comparison with the fluid hydrostatic skeleton of annelid worms.

8. List the genus names of two common local intertidal animals that exhibit radial body symmetry, one that is primarily radial and one that is secondarily radial.

9. To be classified as chordates, humans must possess a dorsal hollow nerve cord, a longitudinal notochord, pharyngeal gill slits, and a postanal tail at some point during their life. When did you, or do you, possess each of these chordate characteristics?

KEY TERMS

aboral   145

acoelomate   148

antennae   160

anterior   148

blastula   144

cephalization   154

chitin   142

choanocyte   144

closed circulatory system   152

cnidocyte   146

coelomate   148

colony   140

convergent evolution   153

ctene   147

cuticle   152

derived character   142

dorsal   148

exoskeleton   159

gastrovascular cavity   146

globigerina ooze   141

gonads   153

hermaphrodite   144

hydrostatic skeleton   157

hyphae   143

infauna   157

lichen   143

lophophore   153

medusa   146

meiofauna   157

mesoglea   146

metamere   158

molt   152

mycelium   143

mycophycobiosis   143

mycorrhiza   143

nauplius   160

nematocyst   146

notochord   163

oral   145

osculum   144

parthenogenesis   152

pentamerous   163

polyp   146

posterior   148

proboscis   149

pseudocoelomate   148

radula   154

saprobe   142

spicules   145

spongin   144

spongocoel   144

trochophore   154

veliger   154

ventral   148

KEY *GENERA*

Architeuthis

Branchiostoma

Calanus

Cestum

Chaetomorpha

Chrysaora

Cyanea

Euphausia

Euplectella

Geukensia

Globigerina

Haeckelia

Heteropolypus

Hexabranchus

Homarus

Limulus

Lingula

Loxosomella

Nautilus

Physalia

Sagitta

Sepia

Sipunculus

Spirula

Tomopteris

Trichoplax

Urechis

REFERENCES

Alexander, R. M. 1979. *The Invertebrates*. New York: Cambridge University Press.

Atlantic States Marine Fisheries Commission. 2009. *American Lobster Stock Assessment Report for Peer Review*. Stock Assessment Report No. 09-01. Prepared by the ASMFC American Lobster Stock Assessment Subcommittee. A publication of the Atlantic

States Marine Fisheries Commission pursuant to NOAA Award No. NA05NMF4741025.

Barnes, R. D. 1980. *Invertebrate Zoology*. Philadelphia: Saunders College/Holt, Rinehart and Winston.

Bayne, C. J. 1990. Phagocytosis and non-self-recognition in invertebrates. *BioScience* 40:723–731.

Biddle, J. F., C. H. House, and J. E. Brenchley. 2005. Microbial stratification in deeply buried marine sediment reflects changes in sulfate/methane profiles. *Geobiology* 3:287–295. doi:10.1111/j.1472-4669.2006.00062.x

Brusca, R. C., and G. J. Brusca. 1990. *Invertebrates*. Sunderland, MA: Sinauer Associates.

Buzas, M. A., and S. J. Culver. 1991. Species diversity and dispersal of benthic foraminifera. *BioScience* 41:483–489.

Dunstan, A. J., P. D. Ward, and N. J. Marshall. 2011. Vertical distribution and migration patterns of *Nautilus pompilius*. *PLoS ONE* 6(2):e16311. doi:10.1371/journal.pone.0016311

Gasmi, S., G. Nève, N. Pech, S. Tekaya, A. Gilles, and Y. Perez. 2014. Evolutionary history of Chaetognatha inferred from molecular and morphological data: A case study for body plan simplification. *Frontiers in Zoology* 11(1):84.

Grahame, J., and G. M. Branch. 1985. Reproductive patterns of marine invertebrates. *Oceanography and Marine Biology Annual Review* 23:373–398.

Hatia, K. 2012. Diseases of fish and shellfish caused by marine fungi. In: *Biology of Marine Fungi*. C. Raghukumar, ed. New York: Springer. pp. 15–39.

Hickman, C. P., and L. S. Roberts. 1994. *Biology of Animals*, 6th ed. Dubuque, IA: Wm. C. Brown.

Kier, W. 2012. The diversity of hydrostatic skeletons. *Journal of Experimental Biology* 215:1247–1257.

Kozloff, E. N. 1987. *Marine Invertebrates of the Pacific Northwest*. Seattle: University of Washington Press.

Lundsten, L., M. H. Reiswig, and W. C. Austin. 2014. Four new species of Cladorhizidae (Porifera, Demospongiae, Poecilosclerida) from the Northeast Pacific. *Zootaxa* 3786(2):101–123.

Margulis, L., and M. Chapman. 2010. *Kingdoms and Domains: An Illustrated Guide to the Phyla of Life on Earth*, 4th ed., reprinted with corrections. Philadelphia, PA: Academic Press.

Mills, C. E., and R. L. Miller, 1984. Ingestion of a medusa (*Aegina citrea*) by the nematocyst-containing ctenophore (*Haeckelia rubra*, formerly *Euchlora rubra*): Phylogenetic implications. *Marine Biology* 78:215–221.

Ou, Q., S. Xiao, J. Han, G. Sun, F. Zhang, Z. Zhang, and D. Shu. 2015. A vanished history of skeletonization in Cambrian comb jellies. *Science Advances* 1(6):E1500092.

Ovtcharova, M., N. Goudemand, Ø. Hammer, K. Guodun, F. Cordey, T. Galfetti, U. Schaltegger, and H. Bucher. 2015. Developing a strategy for accurate definition of a geological boundary through radio-isotopic and biochronological dating: The Early-Middle Triassic boundary (South China). *Earth-Science Reviews*. doi:10.1016/j.earscirev.2015.03.006

Pilling, E. D., R. J. G Leakey, and P. H. Burkill. 1992. Marine pelagic ciliates and their productivity during summer in Plymouth coastal waters. *Journal of the Marine Biology Association* 72:265–268.

Richardson, J. 1986. Brachiopods. *Scientific American* September: 100–106.

Rouse, G. W. (2001). A cladistic analysis of Siboglinidae Caullery, 1914 (Polychaeta, Annelida): Formerly the phyla Pogonophora and Vestimentifera. *Zoological Journal of the Linnean Society* 132:55–80.

Russell-Hunter, W. D. 1979. *A Life of Invertebrates*. New York: Macmillan.

Sanders, W. B., R. L. Moe, and C. Ascaso. 2004. The intertidal marine lichen formed by the pyrenomycete fungus *Verrucaria tavaresiae* (Ascomycotina) and the brown alga *Petroderma maculiforme* (Phaeophyceae): Thallus organization and symbiont interaction. *American Journal of Botany* 91(4):511–522.

Sebens, K. P. 1977. Habitat suitability, reproductive ecology and the plasticity of body size in two sea anemone populations (*Anthopleura elegantissima* and *A. xanthogrammica*). Ph.D. Dissertation. Seattle: University of Washington.

Stanley, S. M. 1975. A theory of evolution above the species level. *Proceedings of the National Academy of Science U.S.A.* 72:646–650.

Tsuia, C. K., W. Marshall, R. Yokoyama, D. Honda, J. C. Lippmeier, K. D. Craven, P. D. Peterson, and M. L. Berbee. 2009. Labyrinthulomycetes phylogeny and its implications for the evolutionary loss of chloroplasts and gain of ectoplasmic gliding. *Molecular Phylogenetics and Evolution* 50(1):129–140.

Valentine, W. 1978. The evolution of multicellular plants and animals. *Scientific American* September:140–158.

Villee, C. A. 1984. *Biology*. Boston: McGraw-Hill.

WoRMS Editorial Board. 2015. World Register of Marine Species. Available from http://www.marinespecies.org at VLIZ. Accessed July 14, 2015.

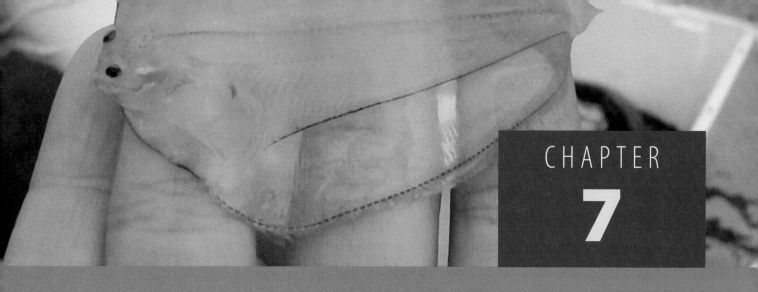

Marine Vertebrates I: Fishes and Reptiles

STUDENT LEARNING OUTCOMES

1. Define the term *vertebrate*, and list the features that all vertebrates share.
2. Compare agnathans and fish with jaws, listing the features that agnathans are missing besides jaws.
3. Discuss the main features of chondrichthyans, and identify the features that limit their range of maneuverability during swimming.
4. Compare the swimming maneuverability of chondrichthyans to that of bony fish.
5. Compare and contrast three Osteichthyes orders.
6. Describe the characteristics that enable marine amphibians, marine reptiles, and marine mammals to survive in a salty, fluid medium.
7. Explain how the following marine organisms respire and osmoregulate: bony fish, cartilaginous fish, and marine birds.
8. Summarize the vast array of behaviors marine vertebrates use for daily life and survival.
9. Explain the unique and elaborate sensory capabilities of vertebrates in the sea.

CHAPTER OUTLINE

A larval flatfish about midway through metamorphosis to a juvenile, with its left eye on top of its head.
NOAA Fisheries West Coast.

Vertebrates occupy all major marine habitats and are especially dominant in the pelagic realm of the sea. As a group, adult marine vertebrates express an enormous range of body sizes, from the stout infant fish of the Great Barrier Reef that is just under 1 cm in length to whales and sharks greater than 15 m long. Consequently, marine vertebrates exploit food sources of almost all sizes found in the sea and are important in almost all marine food webs, including those involving humans. Because of their high visibility, value as a human food resource, and occurrence throughout the ocean, marine vertebrates have been studied more than any other group of organisms in the sea. The fact that marine vertebrates are more closely related to humans than all other living beings, and therefore more relatable to humans, has also contributed to the extensive efforts to study this group. Many of the favorite marine vertebrates among students of marine biology are described as "cute" (e.g., a sea turtle) or "fuzzy" (a sea otter), and who would not want to learn more about something that is cute and/or fuzzy? We start our survey of marine vertebrates with the major groups of fish and marine reptiles, and also include an introduction to some of the basic physiology of all vertebrates. Although fish and reptiles are described most often as "slimy," "scaly," or "spiny," these groups display many advanced features that carved the way for other vertebrates, specifically mammals.

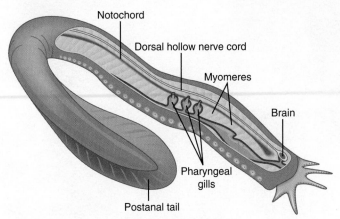

Figure 7.1 Generalized structure of a vertebrate chordate.

DID YOU KNOW?

Muscle segments, or myomeres, are not only used for movement. They have the very important function of connecting the vertebral column to the skin via tendons. When you buy a fish fillet at the store you can very easily and clearly observe myomeres, the W-shaped layers of muscle in the fillet. The middle of the "W" connects to the anterior region of the fish, and the side parts of the "W" connect toward the posterior region. The fact that vertebrate muscles are divided into myomere segments allows for much finer controlled movements. Other, more advanced vertebrates, including mammals, also have myomeres, but mammalian myomeres cannot be observed as clearly as those found in a fish due to secondary modifications.

7.1 Vertebrate Features

All chordates with a vertebral column are members of the subphylum Vertebrata. In addition to the basic chordate features, vertebrates also have specialized sensory organs and a brain consisting of a concentration of nerves positioned at the enlarged anterior end of the nerve cord. During development the notochord becomes ossified into a linear series of articulated skeletal units, the **vertebrae**. Skeletal muscles are segmented into **myomeres** that permit controlled and efficient movements. On the simpler end of the spectrum, vertebrates respire primitively with pharyngeal gills, and they eliminate digestive wastes via a cloaca that is located anterior to their tail. These minimum vertebrate structural features are illustrated in **Figure 7.1**, although many vertebrates have evolved further adaptations for more efficient gas exchange and more selective waste disposal.

Vertebrate circulatory systems are closed loops, known as a **closed circulatory system**, each consisting of a multichambered heart, arteries, capillaries, and veins. Fish have two-chambered hearts with a single circulation loop, amphibians and most reptiles have three-chambered hearts with a double circulation loop, and mammals have four-chambered hearts with a double circulation loop (**Figure 7.2**). Double circulation loops are more efficient for transporting oxygen from the lungs to the entire body. Vertebrate blood itself is unique in the animal kingdom, with all its oxygen-transporting hemoglobin completely contained in red blood cells that can circulate but cannot leave the blood vessels.

The details of the origin and early evolution of vertebrates have been blurred by the passage of nearly one half billion years. Most biologists agree that early vertebrates evolved from filter-feeding chordate ancestors that had characteristics resembling the larval stages of invertebrate sea squirts or lancelets. A major distinction between invertebrate chordates and vertebrate chordates is the number of clusters of **Hox genes** they possess; invertebrate chordates possess one cluster, and vertebrate chordates possess more than one cluster. *Hox* genes control early development, and changes in *Hox* genes correlate strongly with changes in the overall body plans of animals. Unfortunately, genetic comparisons are only possible for **extant**, or living, species, because fossils rarely retain genetic material. The earliest vertebrates may be conodonts, which are extinct soft-bodied marine animals from the late Cambrian that were discovered in 1983. Conodonts had many tiny mineralized toothlike structures that fossilized well and are commonly found by geologists. Conodonts were small in size, had a skeleton made of cartilage, and lacked jaws.

Conodonts possessed a notochord, myomeres, fin rays, and two eyes (**Figure 7.3**). Although a close relationship to other vertebrates was not completely accepted at first, many biologists

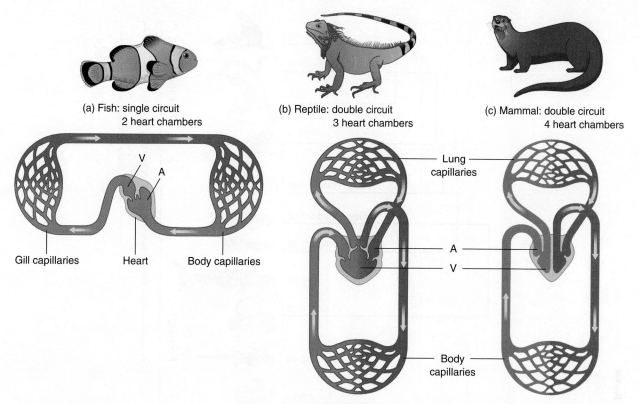

(a) Fish: single circuit
2 heart chambers

(b) Reptile: double circuit
3 heart chambers

(c) Mammal: double circuit
4 heart chambers

Gill capillaries   Heart   Body capillaries

Lung capillaries

Body capillaries

Figure 7.2 Comparison of circulatory patterns of (a) typical fish, (b) reptiles and amphibians, and (c) mammals. Note that crocodilians have four-chambered hearts.
Note: A = atrium. V = ventricle.

now agree that conodonts should be included within vertebrate ancestry. Hagfish, the most primitive surviving vertebrate (and therefore most likely to be representative of the primitive vertebrate osmoregulatory condition), have body fluids isotonic to seawater, supporting arguments for a marine origin for early vertebrates; however, essentially all other vertebrate groups have body fluids with ionic concentrations less than half that found in seawater. When living in low-salinity estuarine or freshwater conditions, reduced ion concentrations in the body fluids of vertebrates lessen the osmotic gradient and reduce the amount of metabolic energy expended on **osmoregulation**; however, normal nerve and muscle function requires that sodium ion concentrations of vertebrate body fluids are maintained at a minimum of 30% to 50% of their concentration in seawater.

One reasonable scenario proposes that early vertebrates moved between marine and freshwater habitats, possibly for spawning in freshwater, as modern lampreys and salmon do now. Such migrations would place the young stages in nutrient-rich environments, where they could feed and grow with perhaps fewer large active predators than would be encountered in the sea. In these brackish and freshwater environments, the evolution of reduced-ion body fluids would be expected to conserve substantial amounts of energy that could then be used for rapid growth. Only after gaining in body size would these fish move into marine habitats to compete for larger prey items. This is precisely what salmon and lampreys do now. From this scenario for early vertebrates, it is suggested that each major group has evolved its own solution to the problem of maintaining

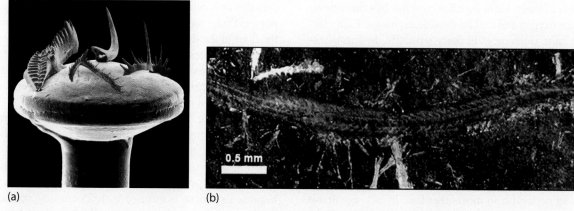

(a)   (b)

Figure 7.3 The conodont animal revealed: (a) fossilized conodont teethlike structures on the head of a pin and (b) a fossil conodont.
(a) Courtesy of Mark Purnell, University of Leicester, UK; (b) R. J. Aldridge, D. E. G. Briggs, M. P. Smith, E. N. K. Clarkson, and N. D. L. Clark, 1993. *The Anatomy of Conodonts*.

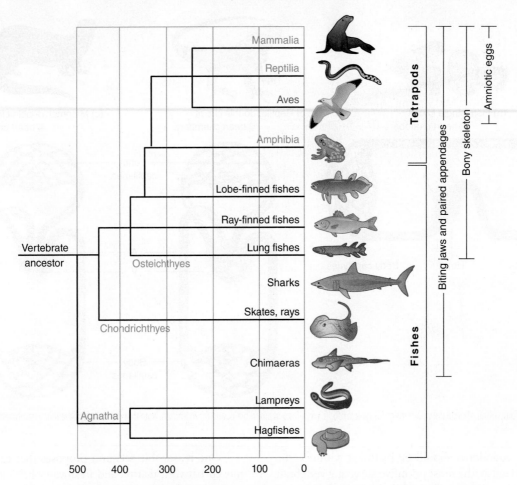

Figure 7.4 Hypothesized phylogenetic relationships of extant vertebrates, with emphases on modern fish groups. Class names are in red font, with the exception of Aves. Aves, or birds, are considered reptiles due to their apparent close relationships to some dinosaurs.

osmoregulatory balance. These adaptations are described later in this chapter.

According to the most widely accepted taxonomic groupings, marine fish include three of the seven classes of vertebrates (**Figure 7.4**); the other four classes are known collectively as *tetrapods*. The number and members of recognized vertebrate classes will likely change soon, as birds are now considered a subgroup within Reptilia because they have a theropod dinosaur ancestor. Fish are difficult to characterize precisely, but typically they live and grow in water, swim with fins, and use gills throughout their lives for oxygen and carbon dioxide exchange. Each class exhibits definite differences in body structure, in specialization of sense organs, in solutions to osmoregulatory changes, in reproductive and life history strategies, and ultimately in species diversity.

7.2 Agnatha: The Jawless Fishes

Agnathans are vertebrates that are often characterized in negatives because they lack the paired fins, bone, biting jaws, and skin scales so noticeable in most other fish. In fact, agnathans even lack vertebrae, and instead retain a cartilaginous notochord throughout their lives. Although abundant and highly diverse in Paleozoic seas, fewer than 100 species of these primitive fishes exist in the world ocean today in two very different classes (**Table 7.1**), the hagfish and the lampreys.

The 70 to 80 extant species of hagfish are entirely marine fish (**Figure 7.5**) that differ greatly from all other species of vertebrates, including lampreys. Like most invertebrates, hagfish are isotonic with seawater. In addition, they possess just one type of granular white blood cell (instead of the usual three types), they have degenerate eyes that lack eye muscles, and their inner ear contains only one semicircular canal (whereas most vertebrates possess three; see Figure 7.46). Hagfish live in mud burrows in the floor of the deep sea, eating polychaetes and scavenging on dying vertebrates by burrowing into their body cavities. Unlike lampreys and most other multicellular marine organisms, hagfish lay eggs that hatch directly into miniature versions of the adults; no larva exists.

Hagfish are best known for their highly unique slime glands that are distributed in two **ventrolateral** rows along their body. These glands contain both mucous cells and thread cells that function in synchrony to produce a thick, viscous slime that seems to deter potential predators. Hagfish slime is very viscous because its mucus is fortified with 25-cm-long molecules, the products of their unique thread cells that are jettisoned intact during slime production. Once in seawater, the thread cells

TABLE 7.1 Classes and Orders of Marine Fishes and Reptiles

Class	Order	Common Examples	Approximate Number of Marine Species
Agnatha	Myxiniformes	Hagfish	79
	Petromyzontiformes	Lampreys	14
Chondrichthyes	Chimaeriformes	Chimaeras, ratfish	52
	Rajiformes	Skates	358
	Torpediniformes	Electric rays	65
	Myliobatiformes	Sting rays, eagle rays	192
	Pristiformes	Sawsharks	7
	Squaliformes	Dogfish sharks	133
	Orectolobiformes	Whale sharks	44
	Lamniformes	Thresher sharks	16
	Carcharhiniformes	Tiger, swell, and whitetip sharks	285
Osteichthyes	Elopiformes	Tarpons, bonefish	9
	Anguilliformes	Moray eels	903
	Salmoniformes	Salmon	47
	Clupeiformes	Sardines, anchovies	326
	Gadiformes	Cods, hakes	624
	Atheriniformes	Flying fish, silversides	130
	Lampriformes	Oarfish, opah	24
	Gasterosteiformes	Seahorses, pipefish	16
	Scorpaeniformes	Rockfish, sculpins	1,554
	Perciformes	Basses, groupers, snappers, wrasses, tunas	8,473
	Pleuronectiformes	Flatfish	756
	Tetraodontiformes	Triggerfish, boxfish, puffers	412
Reptilia	Squamata	Sea snakes, marine iguana	95
	Testudines	Sea turtles	9
	Crocodilia	Crocodiles	2

rupture to release the molecular thread that will unfurl to fortify the mucus into viscous slime. Hagfish periodically tie their flexible bodies into sliding knots to clean themselves of excessive slime buildup (**Figure 7.6**). This knot-tying behavior also provides stability and leverage while tearing apart large prey items.

Although jawless, the 14 marine species of lampreys are strikingly different than hagfish, possessing many features commonly seen in more advanced, jawed vertebrates. Lampreys have large, mobile eyes; pineal and pituitary glands; a lateral line with hair cells; and neural and haemal arches associated with their notochord. Unlike hagfish, they maintain an internal osmotic pressure that is less than that of seawater, and they produce eggs that hatch into larvae. Most lampreys are **anadromous**, and thus leave the sea to spawn in freshwater, where their larvae spend several years as filter-feeding, infaunal planktivores before metamorphosing and returning to the sea to mature. In both the freshwater and marine portions of their life cycles, adults of some species of lampreys parasitize various species of bony fishes (**Figure 7.7**). Parasitic lampreys use numerous conical horny teeth within their oral disc to attach to their unfortunate hosts and rasp through their skin. Anticlotting salivary enzymes ensure continuous flow of blood from the parasitized host in a manner reminiscent of vampire bats and mosquitoes. The parasitic lifestyle of lampreys is likely a specialized form of feeding not representative of the way early vertebrates obtained their food.

Figure 7.5 An Atlantic hagfish captured in a trap during a research cruise. Notice the slime dripping off the scientist's gloved hand.
NOAA.

Figure 7.7 Two sea lampreys, *Petromyzon*, attached to a host fish.
Courtesy of USGS.

Neither hagfish nor lampreys play major roles in present-day marine communities. However, both groups of jawless fish negatively impact important fisheries around the world. Lampreys are notorious for their destruction of fisheries within the Great Lakes of North America, and hagfish routinely find and consume dying and dead fish caught on longlines, such as cod, before commercial fishers have time to retrieve their gear and its valuable catch.

DID YOU KNOW?

Hagfish slime is unique among animal mucus because it contains protein threads that help it expand dramatically when it contacts seawater. The sticky, expanding mucus is what was thought to deter predators, and in 2011 this behavior was finally caught on video for observation by researchers. Because the slime expands, it basically chokes fish that are trying to eat the hagfish by clogging their gills. Video from 2011 also caught hagfish preying on other fish, a previously unknown behavior for this fish that was considered a scavenger, feeding on dead and decaying matter, and sometimes on small living invertebrates. These unique features are likely several of the characteristics that have allowed hagfish to persist for so long with success.

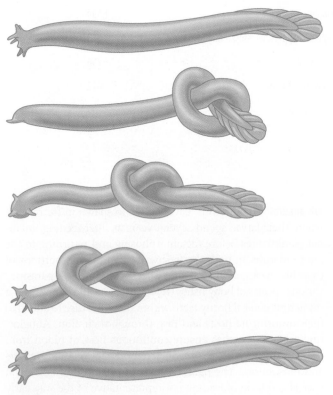

Figure 7.6 Slime removal behavior of a hagfish.

7.3 Chondrichthyes: Sharks, Rays, and Chimaeras

Chondrichthyans exhibit three basic body plans: streamlined sharks, dorsoventrally flattened rays, and the unusual chimaeras (**Figure 7.8**). The evolution of paired fins and biting jaws armed with teeth (features found in all groups of marine fishes except agnathans) provides sharks, rays, and their allies with the structures needed for better maneuverability, faster swimming speeds, and ultimately far more effective predation.

Members of this class are often referred to as the *cartilaginous fishes*, because although their skeletons may be strong, rigid, and highly mineralized, they use only cartilage for their skeletons, and the bone tissue characteristic of other vertebrates is absent in all members of this class. The difference between bone and cartilage is that cartilage is **avascular**, containing no blood vessels, whereas bone contains a highly organized system of blood vessels that deliver oxygen and nutrients to bone-producing osteocytes and remove wastes. Cartilage is a hard

Figure 7.8 A chimaera gliding along the seafloor near Ryan Canyon off the U.S. East Coast.

Image courtesy of NOAA Okeanos Explorer Program, Our Deepwater Backyard: Exploring Atlantic Canyons and Seamounts.

Figure 7.9 (a) Left side view of a typical shark, a black-tipped reef shark. (b) Dorsal view of a typical batoid, a stingray.

(a) © cbpix/Shutterstock, Inc.; (b) © Matt9122/Shutterstock, Inc.

tissue, thus cartilaginous sharks and rays have no difficulty biting through or crushing their bony prey.

Chondrichthyans tend to be larger in body size than members of the other classes of fish. Adult body lengths range from less than 20 cm for some deep-sea sharks to over 15 m for whale sharks and basking sharks. With the exception of some whales, these are among the largest living animals. Interestingly, the largest species of this group have tiny teeth and filter feed on plankton in a similar manner to the feeding habits of large baleen whales.

The 1,140 species of elasmobranchs (sharks, rays, skates, guitarfish, sawfish, and their allies) differ from the more than 50 species of chimaeras (Figure 7.8) in a number of ways. Sharks and rays typically possess rugged placoid scales on their integument, whereas the skin of chimaeras is smooth and scaleless. Elasmobranchs possess five to seven pairs of gill slits, whereas chimaeras have just one pair of lateral gill slits. Chimaeras also differ from sharks and rays in that their upper jaw is fused to their skull, or *chondrocranium*, as is that of mammals. This anatomical attribute is responsible for the name of their class, the Holocephali (or "whole heads"). Male holocephalans possess a sexually dimorphic, macelike appendage on their foreheads called a *cephalic clasper* that is used to hold females during copulation; this structure is not found in male sharks and rays. Finally, chimaeras have separate anal and urogenital openings, and their teeth are fused into a set of two upper and one lower crushing plates.

Chimaeras are relatively poorly known egg-laying inhabitants of the deep sea that ascend into shallower waters only in high latitudes and in regions of cold-water upwelling events. Three families of these interesting cartilaginous fishes are recognized today: one with a blunt snout, one with a soft plow-shaped nose, and one with a greatly flattened and extended snout that presumably provides enhanced surface area for its electrosensory organs. Chimaeras rarely leave the proximity of the seafloor, where they forage on mollusks, fish, crustaceans, worms, and echinoderms.

Sharks and rays (**Figure 7.9**) differ from one another in several important ways, one of which is not their body profile. And thus, although sharks tend to be tubular and rays tend to be depressed, there are very flattened raylike sharks, such as angel sharks (*Squatina*), and very robust, cylindrical sharklike rays, such as sawfish (*Pristis*) and the shark ray (*Rhina*) of the Indo-Pacific. Sharks and rays are easily distinguished based on the location of their gill slits; those of sharks are on the lateral surfaces of their heads, whereas those of rays are always found on their ventral surface. Also, because most rays have a caudal fin that is reduced or absent, they tend to swim using their greatly expanded pectoral fins (and thus gain the name *batoid fishes*), whereas all sharks propel themselves through the water via powerful lateral undulations of their caudal fin.

Although most sharks consume fish, as a group they select an amazing variety of prey. The largest species, whale, basking, and megamouth sharks, all use highly modified gill rakers to strain zooplankton and small fish from the water column. Horn sharks use flattened molars to crush mollusks and echinoderms. Many species of deep-sea dogfish and catsharks seem to target cephalopods and crustaceans. Large white sharks often emphasize marine mammals in their diets, whereas tiger sharks are considered to be nearly indiscriminate feeders that prey on fish, crustaceans, large gastropods, cephalopods, sea turtles, sea snakes, seabirds, and human refuse. Perhaps most unusual among the sharks are two species of midwater dogfish known as cookiecutter sharks. They are parasitic on large marine vertebrates, including other sharks, using their enormous lower teeth to carve shot glass–sized hunks of flesh from the flanks of their much larger victims. Skates, rays, and other batoids are equally diverse in terms of their prey selection. Manta and devil rays are planktivorous like the largest sharks, sawfish use their elongated and toothed rostrum to rake bivalves from the sediment, eagle rays use jets of water to excavate snails from inside the sediment, and electric rays use powerful electric organs located in their expanded pectoral fins to incapacitate their fish prey.

DID YOU KNOW?

Whale sharks spend around 7.5 hours per day actively feeding! Due to their large size and filter-feeding habits, they must filter water almost constantly to ingest enough plankton for their metabolic demands. This is also why whale sharks are known to follow plankton blooms and fish or coral spawning events. They will swim through a cloud of plankton or newly spawned eggs, swallowing a nutrient-rich meal. Plankton blooms are analogous to a healthy fast-food drive-thru for whale sharks and other filter feeders.

Cartilaginous fishes use a wide variety of reproductive strategies based on a general pattern of internal reproduction, leading to fairly small numbers of large eggs or offspring. All chimaeras, as well as some rays and benthic sharks, are oviparous, producing only a few large eggs each reproductive cycle. Some sharks, such as catsharks, produce pairs of egg cases throughout the year, and thus lack an identifiable reproductive season. No chondrichthyan fish builds a nest or guards its egg cases. Some females simply stick their egg cases into soft sediments (chimaeras and skates) or piles of rocks (horn sharks), whereas others may employ meter-long tendrils extending from the corners of the egg case to hang their egg cases from soft corals (**Figure 7.10**). Most oviparous chondrichthyans seem to scatter their egg cases on the seafloor, and thus the developing embryos are protected only by the durable egg case while being nourished by the abundant yolk inside.

Chondrichthyan egg cases are very different from the eggs of birds in two ways. First, the protective case resembles human fingernails rather than the fragile design of a bird's egg. Also, unlike the eggs of birds, chondrichthyan egg cases do not contain a unique developmental environment because they crack open at the corners soon after deposition, such that seawater flows through the egg case with every movement of the developing pup. Most egg cases contain a single pup, but those of the big skate can measure 30 cm in length and contain three to seven embryos. When the eggs hatch 3 to 13 months after laying, the young cartilaginous fish are well developed and quite capable of surviving on their own.

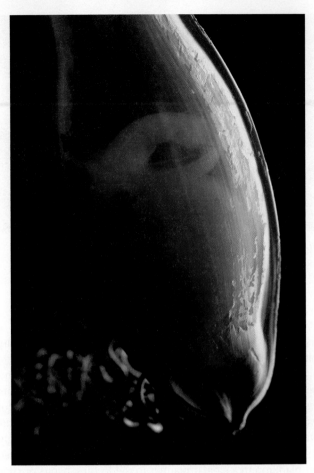

Figure 7.10 Developing swell shark embryo, *Cephaloscyllium*, enclosed in a tough protective egg case.
©Nature Picture Library/Alamy Stock Photo.

Other sharks produce eggs that are maintained within the reproductive tract of the female until they hatch. This intermediate condition between **viviparity** (live birth) and **oviparity** (egg laying) is known as **ovoviviparity**. Ovoviviparity is an adaptation for incubating developing embryos inside the mother's reproductive tract where they obtain nourishment from the yolk of their own eggs (**Figure 7.11**). Several species of ovoviviparous

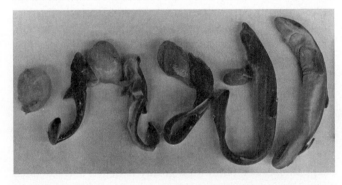

Figure 7.11 A developmental series of the dogfish shark, *Squalus*, from an egg (far left) to a completely formed embryo ready for birth (far right). Note the twins (second from left) joined to a common yolk.
Courtesy of James Sumich.

fishes provide embryonic nutrition in addition to the nutrition contained in the yolk. Only a few examples, some of which are rather exotic, are described here. Several pelagic sharks and rays, for instance, have uteri lined with numerous small projections called **trophonemata** that secrete a nutritious uterine milk for the embryos. Other sharks, such as white tips and hammerheads, absorb nutrients through a placenta-like connection between the yolk sac and the uterine wall. The embryos of the porbeagle shark, *Lamna*, lack structures with which to absorb nutrients from the reproductive tract of the female. Instead, when the embryos within a female *Lamna* have used their own yolk, they simply turn on the other eggs within the uterus and consume their would-be future siblings. With the nutrition gained from their potential siblings, these large embryos are developmentally much better prepared for a pelagic existence before leaving the protective confines of their mother. A variation of that strategy is used by the sand tiger shark, *Carcharias*. After the two surviving embryos, one in each of the two uteri, have consumed their developing siblings, they remain in the uterus and consume thousands of additional pea-sized trophic eggs released by the female's single enormous ovary. This process may continue for a year, producing two shark pups each about 1 m long! This is a notable feat for a mother only 2.5 m in length.

7.4   Osteichthyes: The Bony Fishes

As the class name implies, a key characteristic of class Osteichthyes is a skeleton of bone. Bone is stronger and lighter than cartilage and has permitted the evolution of smaller body sizes in this class, with most species growing to only a few centimeters in length. The longest bony fish, the oar fish (*Regalecus*), grows to over 16 m in length, and the heaviest bony fish, the ocean sunfish (*Mola mola*), can weigh up to 1,000 kg. Of the two subclasses of Osteichthyes illustrated in Figure 7.4, the ray-finned fishes (Actinopterygii) reach their peak diversity in marine habitats, with more than 13,000 species. In contrast, the lobe-finned fishes (Sarcopterygii) have but one living marine member, the coelacanth of the Indian Ocean and Indonesia (the few other species of living Sarcopterygii are freshwater lungfishes). The coelacanth (**Figure 7.12**) shares some common characteristics with sharks and rays, including urea accumulation for osmoregulation. The coelacanth first appeared in the fossil record some 400 million years ago and was assumed to have become extinct about 65 million years ago. Then, in 1938, a fresh (but dead) specimen was found in a South African fish market. Other individuals of this rare, but definitely not extinct, species have since been collected throughout the western Indian Ocean and Indonesia, where they live in rocky caves at depths below 100 m. Lungfishes, the freshwater lobe-finned fishes, have been proposed as possible ancestors of the land-dwelling tetrapod vertebrates (Figure 7.4).

The ray-finned fishes are divided into three major groups, of which the Teleostei is by far the largest. Within the many orders of teleost fishes, hundreds of families and thousands of species occupy marine habitats (Table 7.1). Nearly 60% of the species of living teleost fishes live in marine habitats and

Figure 7.12 A coelacanth from the eastern Indian Ocean. The fish is about 1 m long.
© Peter Scoones/Photo Researchers, Inc.

dominate the flow of energy in their communities. Several of the groups listed in Table 7.1 are familiar to most of us, either as a restaurant menu item or from encounters in the sea; others are not commonly seen except by professional biologists. A few representative species of nine common marine orders of teleost fishes are shown in **Figure 7.13**.

Regardless of where they live, nearly all teleosts share several fundamental features in common, including skeletons of bone; thin and flexible skin scales; a gas-filled buoyancy organ, the **swim bladder**; and maintenance of body fluids **hypoosmotic** to seawater. Buoyant swim bladders free the pectoral fins from the need to provide lift, as they do in sharks, to maintain or change depth. Two features in particular, swim bladders and bony skeletons, complement the membranous fins supported by rays for fine control of swimming movements. Freed of some of the structural constraints seen in the remarkably uniform body plans of cartilaginous fishes, teleosts have adapted to enormously varied aquatic habitats and exhibit a diversity of body styles and of species unmatched in the phylum Chordata.

DID YOU KNOW?

Some fish can breathe air, even without lungs. Fish living in water with low oxygen content will jump out of the water and gulp air to supplement their gill breathing in water. A marine example is a tarpon, a large game fish found in the subtropics. A tarpon can use its swim bladder to store air, and the bladder has a direct connection to the esophagus for bringing in the air through the mouth. Tarpon swim bladders are specialized to allow them to extract the oxygen from the bladder when necessary. Not only is air gulping helpful to the fish, but it provides humans with entertainment when they are lucky enough to observe a huge fish rolling or jumping out of the water.

A common, although not exclusive, theme in sexually reproducing animals is to produce approximately equal numbers of female and male offspring. The maleness or femaleness

Figure 7.13 Some representative body types of nine common orders of marine fish: (a) Gasterosteiformes: seahorse; (b) Tetraodontiformes: triggerfish and trunkfish; (c) Pleuronectiformes: halibut; (d) Atheriniformes: flying fish; (e) Clupeiformes: anchovy; (*continues*)

(a) NOAA; (b1) NOAA, Flower Garden Banks; (b2) NOAA, Flower Garden Banks; (c) Adam Obaza/NOAA; (d) Image courtesy of Bermuda: Search for Deep Water Caves 2009 Exploration; (e) NOAA, Flower Garden Banks.

Figure 7.13 (*continued*)   (f) Anguilliformes: green moray eel; (g) Perciformes: bicolor damselfish; (h) Perciformes: yellowtail snapper; (i) Perciformes: grunt; (j) Perciformes: tiger grouper; (k) Salmoniformes: salmon; (l) Scorpaeniformes: scorpionfish.

(f) NOAA, Flower Garden Banks; (g) NOAA, Flower Garden Banks; (h) NOAA, Flower Garden Banks; (i) NOAA, Flower Garden Banks; (j) NOAA, Flower Garden Banks; (k) NOAA/NEFSC; (l) NOAA, Flower Garden Banks.

of mammals, birds, and many kinds of invertebrates is determined by their complement of sex chromosomes. In these animals, the nucleus of each cell contains one pair of **sex chromosomes** and several other pairs (22 pairs in humans) of **autosomes** not directly involved in sex determination. The influence of sex chromosomes on the sex of bony fish is less straightforward and, in fact, is quite variable. Some guppies, for instance, reflect the mammalian pattern of sex chromosomes: XX is female, and XY is male. Occasionally, these sexes are reversed. The sex chromosomes of bony fish lack the absolute control over sex determination found in birds and mammals because some of the genes involved in sex determination, unlike those of mammals and birds, are also carried on the autosomes. In some fish, these autosomal sex genes apparently influence or regulate the production of sex hormones, especially **androgen**, a male hormone, and **estrogen**, a female hormone. These hormones, in turn, influence the expression of several sexual characteristics and the determination of sex.

The fluid and unfixed nature of sex determination in bony fish has been effectively exploited through the evolution of a broad range of sex ratios and reproductive habits not common in other vertebrate groups. Part of this sexual diversity is due to the separation of sexes. Separate sexes housed in different individuals eliminate the possibility of self-fertilization and its accompanying reduction in genetic variation. Even in hermaphroditic fish such as sea basses, *Serranus*, which function simultaneously as both males and females, specific behavioral interactions with mates ensure that cross-fertilization will occur.

Other species of bony fish produce offspring that all clearly show the functional characteristics of one sex as they mature; then, at some point in their lives, some or all of them undergo a complete and functional transformation to the opposite sex. These fish are hermaphroditic, but unlike *Serranus*, a **simultaneous hermaphrodite**, they are **sequential hermaphrodites**. The entire gonad functions as one sex when the fish first matures and then changes to the other sex. California sheephead, for example, become sexually mature as females at about 4 years of age. Those that survive to 7 or 8 years of age undergo sexual transformation, become functional males, and mate with the younger females. The actual ratio of females to males depends on the survival curve of the population and on the age at which sexual transformation occurs. For the sheephead, it is approximately five females to one male.

An extreme example of manipulating the sex ratio for increased reproductive fitness is found in the tropical bluehead wrasse, *Thalassoma bifasciatum* (belonging to the same family, Labridae, as the sheephead). Sexual transformation is socially driven and quite complicated. In each group of fish, one dominant male has a harem of females that he mates with. If the dominant male dies, a female or small initial phase male will take over. There are three possibilities for the life phases of this species: dominant male (terminal phase), female (initial phase), and small male (initial phase). Some individuals mature as females and remain females their entire lives. Others are initially male, but small in size, and may shift into a dominant, terminal phase male role if the terminal phase male is removed from a group. Some mature as females, but change sex

Figure 7.14 Bluehead wrasse with representatives of all life stages. The largest individuals with vertical black bands and blue bodies are males. The smaller individuals with horizontal yellow, black, and white stripes are females and/or juveniles.
NOAA, Flower Garden Banks National Marine Sanctuary.

if a terminal phase male is removed from the group. **Figure 7.14** includes the different morphologies and color schemes for different sexes and life phases in the bluehead wrasse. Note that the initial phase males and females look very similar. In populations of the bluehead wrasse, terminal phase males are produced only as they are needed, and then only from the most dominant of the remaining members of the population. A few species of sea basses do the reverse; they begin life as males and then change to females.

The physical characteristics associated with these sex changes seem to be controlled by the relative amounts of androgen and estrogen produced by the gonads as fish grow and mature. Young female sheepheads, when artificially injected with the male hormone androgen, change to males at a younger age than normal. Injections of estrogen delay sex transformation and maintain the individual in a prolonged state of femaleness. Similar roles of estrogen have been observed for bluehead wrasse females, where decreased estrogen levels lead to more aggressive behaviors and courtship of females, behaviors that will increase in intensity as a female becomes male.

7.5 Marine Amphibians

Marine tetrapods are four-limbed, air-breathing vertebrates that evolved characteristics allowing them to share the pelagic realm with numerous fishes and had a terrestrial ancestor somewhere in their distant evolutionary past. All living classes of air-breathing tetrapods (the amphibians, the reptiles, the birds, and the mammals) contain various groups that have adapted to a marine existence independently of each other (Figure 7.4). Each of these groups depends on the sea for food and may spend a good portion of time in the sea. Despite the obvious specializations of each of the four classes, these groups still share several important adaptations. They all use lungs to breathe air in an

Case Study

The Opah, a Unique Fully Endothermic Bony Fish

Opah, also known as moonfish (*Lampris guttatus*), is an oval-shaped fish found most often between 45 and 400 m (150 to 1,300 ft.), rarely surfaces, and has some physiological features beyond its odd shape that make it very unique. Opah are a brilliant, bright silver, lined with deep red pectoral, caudal, anal, and dorsal fins (**Figure A**). Upon initial observation, an opah may not appear to be a very effective swimmer or predator. Its flattened oval shape looks a bit awkward. But upon closer inspection one may notice very elongated pectoral fins that are capable of providing robust movements. It is indeed these pectoral fins that are used for rapid swimming motions through the water as the fish uses the fins like wings to swim rapidly for long periods of time and for seeking out prey.

The odd shape of this fish isn't the only feature that would indicate slow, limited swimming capabilities. The water at the depths opah are found in is extremely cold, and other fish living at a similar depth swim slowly to conserve energy. Opah do not need to swim slowly to conserve energy because they are able to maintain a constant entire body temperature regardless of the surrounding water temperature. Full-body endothermy is a feature that was previously thought to exist only in mammals and birds. Some fish, such as tuna and some sharks, can maintain portions of their bodies, like their muscles, at warmer temperatures to increase their swimming performance. After some time at depth, though, their hearts cool too much, and they must return to shallower, warmer waters to warm up. Until 2015 nobody knew that the opah was so unique and a truly endothermic animal.

So how do opah maintain a constant body temperature while swimming in cold, deep water? The key element is in their gills. Fish pass water over their gills to take up oxygen in the blood, and in cool water this action normally cools the blood passing through the gills. The opah, however, has what is known as a countercurrent heat exchange, where warm blood traveling from the body to the gills heats up the blood

that has left the gills that would otherwise become cold. In addition, the major organs and tissues, such as the gills, heart, and muscle tissue, are surrounded by a thick layer of fat for insulation. The countercurrent heating mechanism is only known for the opah, and it has completely shifted scientists' ideas about this species (**Figure B**). Opah are now thought to be fast swimmers capable of maintaining their swimming speed at depth. They rarely move into surface waters because they don't need to warm up like other fish, and they are active predators that can chase and catch very mobile prey. In addition to allowing for more rapid and prolonged swimming abilities, warm blood is thought to increase brain and eye function and help maintain overall body functions that would be hindered by cold water.

Opah are found around the world, and although they are not currently a popular sport or commercial fish species, they are sometimes caught by fishers targeting other species, and their flesh is enjoyed by people worldwide. In the United States, opah is caught in Hawaii and sold to tourists or exported to the mainland. Local legend once deemed the opah good luck, and fishers would give the fish away as a gift as a kind gesture rather than sell it. In recent years opah have been encountered by people more frequently than in the past, and scientists are not sure why. Now that we know that opah possess highly unusual thermoregulatory capabilities, they will likely be studied extensively so we can further understand their unique biology.

Figure B An opah being brought onboard a fishing vessel to insert a temperature sensor into the muscles to record body temperature data.
NOAA/SWFSC.

Figure A An opah, collected on a research vessel in the Pacific Ocean.
NOAA/SWFSC.

(continues)

Case Study (*continued*)

Critical Thinking Questions

1. The discovery of endothermy in a fish was a huge surprise to the scientific community. Can you think of any other animals in the sea that would be more likely than others to exhibit similar thermoregulatory capabilities?

2. Considering the large energetic costs associated with maintaining a constant body temperature in the sea, what are some advantages you can think of to being endothermic in this environment?

For Further Reading

Wegner, N. C., O. E. Snodgrass, H. Dewar, and J. R. Hyde. 2015. Whole-body endothermy in a mesopelagic fish, the opah, *Lampris guttatus*. *Science* 348(6236):786–789.

environment where air is only available at the sea surface. Many successfully prey on other active animals even though they cannot use their sense of smell underwater and have only limited vision. Like teleost fishes, tetrapods have body fluids that are hypoosmotic to seawater—they, too, lose water to the environment by osmosis, and with the exception of some marine reptiles and birds whose salt-secreting glands enable them to drink seawater, marine tetrapods must satisfy all of their water needs from their food. Two of these tetrapod classes, the birds and the mammals, are **homeothermic** in an environment perpetually colder than their bodies. Despite these limitations imposed by their terrestrial ancestry, several groups of tetrapods have reinvaded the sea and have done so very successfully.

Amphibians are rarely associated with the sea, perhaps because their highly permeable skin prevents them from combating the osmotic stresses and dehydration that result from immersion in seawater. However, one candidate exists for the designation of "world's only marine amphibian." *Fejervarya cancrivora* is a crab-eating frog found in the mangrove estuaries of Southeast Asia. This remarkable frog utilizes the sea only as a source of food, like polar bears, and yet has a number of physiological adaptations that enable its unique lifestyle. *Fejervarya* feeds on intertidal crabs during low tide and is exposed to 80% seawater during high tide. To prevent water loss, it stores high concentrations of urea in its plasma, as do cartilaginous fishes and the coelacanth. In fact, the concentration of urea in its blood increases 50% during high tide to compensate for potential immersion in seawater. Unlike the adult frogs, *Fejervarya* tadpoles do not store urea. However, they do possess salt-excreting chloride cells in their gills, just like fishes.

7.6 Marine Reptiles

Marine reptiles (**Figure 7.15**) such as sea turtles, sea snakes, marine crocodiles, and the marine iguana are, like other reptiles, ectothermic. Therefore, they are largely restricted to tropical latitudes. Sea snakes move up the east coast of Asia by staying within the warm, northward-flowing Kuroshio Current, and some sea turtles occur off Long Island because of the influence of the warm Gulf Stream in the region or off California during El Niño years; however, cold shock is a common cause of strandings in sea turtles that stray into higher latitudes, and most marine reptiles prefer to remain near the equator.

Marine Iguana and Crocodiles

A single species of lizard, the marine iguana of the Galapagos Islands, is marine (Figure 7.15a). Molecular data demonstrate that it is closely related to the terrestrial iguana of the Galapagos Islands, and both species share a common ancestor with continental iguanas in Ecuador and Peru. Remarkably, these same molecular data suggest that the marine iguana diverged from its terrestrial ancestors about 15 to 20 million years ago, long before the current islands of the Galapagos were formed less than 5 million years ago. Presumably, the Galapagos iguanas inhabited older islands in the chain that have since subducted below the sea's surface and remain as sea mounts on the Nazca Plate.

Marine iguanas rarely venture inland, preferring to remain within 15 m of the shoreline in very dense aggregations that can reach 75 individuals in an area that is only 3 m². Like all iguanas, the marine iguana is an herbivore; unlike all other lizards, the marine iguana dives below the surface to feed on sea lettuce (*Ulva*) and other species of macroalgae that grow on subtidal rocks. Grazing pressure is so intense that the sea lettuce remains closely cropped throughout the year, appearing as a short green fuzz on the rocks that is scraped off by the marine iguanas using their lateral teeth. Because marine iguanas are **ectotherms**, their dive time is limited to 1 hour or less by decreasing body temperature rather than their inability to hold their breath for longer periods of time. Among the adaptations that they possess that enable a marine existence are a flattened tail for underwater propulsion, long claws for grasping intertidal rocks, a dark coloration to facilitate rapid warming after feeding dives, and specialized salt glands to expel unwanted ions obtained directly from seawater and from their diet (i.e., isotonic seaweeds).

In contrast to the herbivorous habits of marine iguanas, the 23 species of crocodilians existing today are all carnivorous. Most, including alligators, caimans, and gharials, inhabit freshwater river and lake systems. Crocodiles (the family Crocodylidae, **Figure 7.16a**) are more euryhaline than their relatives, and two species routinely enter the sea. The American crocodile, *Crocodylus acutus*, is quite at home in the sea, occurring from the southern tip of Florida through the Caribbean to northern South America, and the saltwater crocodile, *Crocodylus porosus*, inhabits tropical estuaries and mangrove swamps around islands of the Indo-Australian archipelago from Asia to Australia. Although it is not uncommon to see this largest living

Figure 7.15 Representatives from the three groups of marine reptiles: (a) marine iguana, *Amblyrhynchus*, of the Galapagos Islands; (b) banded sea snake (*Laticauda colubrina*); and (c) young loggerhead sea turtle, *Caretta caretta*.

(a) © Photos.com; (b) NOAA's Coral Kingdom Collection; (c) NOAA.

(a)

(b)

Figure 7.16 (a) The saltwater crocodile, *Crocodylus*, in a north Australia coastal river, and (b) beach warning sign on Australia's northeast coast.

(a) © susan flashman/ShutterStock, Inc.; (b) © David Franklin/ShutterStock, Inc.

crocodilian at sea hundreds of kilometers from land, it is more frequently encountered near the shore, where it poses some real danger to swimmers and divers (**Figure 7.16b**).

Sea Snakes

Sea snakes are derived from the highly venomous elapids, a terrestrial family that includes cobras, mambas, coral snakes, and kraits. Sea snakes are grouped into the family Elapidae with the cobras and coral snakes, but recently two subfamilies have been recognized based on lifestyle, the Laticaudinae, which lay 10-cm-long sticky eggs above the high-tide line, and the more highly derived Hydrophiinae, which are helpless on land. Hydrophiid sea snakes never leave the sea and retain their eggs to give birth to live young there, much like the ovoviviparous sharks and rays described earlier.

RESEARCH in Progress

The Green Sea Turtles of Ascension Island

The green sea turtle is found worldwide, is known to nest in over 80 countries, and is the largest of the hard-shelled sea turtles (**Figure A**). A large number of green sea turtles feed along the coast of Brazil and nest on Ascension Island, a tiny volcanic peak only 20 km in diameter located about 2,200 km offshore. These turtles lay their eggs in the warm sandy beaches along the north and west coasts of Ascension Island. Immediately after hatching, the young turtles instinctively dig themselves out of the sand, scurry into the water, and head directly out to sea (**Figure B**). During this very short period, they are heavily preyed on by seabirds and large fish. After they are beyond the hazards of shoreline and surf, they presumably are picked up by the South Atlantic Equatorial Current and are carried toward Brazil at speeds of 1 to 2 km/h. Less than 2 months are needed to drift to Brazil passively, yet nothing is known of the young turtles' whereabouts or activities during their first year.

As Atlantic green sea turtles mature, they congregate along the mainland coast of Brazil, where they graze on turtle grass, *Thalassia,* and other seagrasses on shallow flats. Features of the nesting migration back to Ascension Island are a topic of current research, as the adult turtles do show up there in great numbers during the nesting season. Mating apparently occurs only near the nesting ground. Either the males accompany the females on their migration from Brazil to Ascension or they make a precisely timed, but independent, trip on their own.

Figure B   Newly hatched green sea turtles feverishly crawling toward the water.
Mark Sullivan, NOAA.

Figure A   A swimming green sea turtle.
Andy Bruckner, NOAA.

Either way, the males get to the nesting area and can be seen just outside the surf zone competing for the attention of females.

Females go ashore around six times during the nesting season and deposit 120 to 150 eggs each time (**Figure C** and Figure 7.18). These egg-laying episodes provide the only opportunity researchers have to capture and tag large numbers of green turtles at their nesting sites. Until recently it was not known where the females spend time between egg deposits. Tagging efforts have revealed that they do not go far; they appear to rest close to shore, waiting for the next round of egg laying. The green sea turtles' food source (seagrasses and macroalgae) does not occur near Ascension Island, so females must conserve their energy because they do not feed during the nesting period. Because males do not leave the water, almost nothing is known about their migratory behavior. Tagging results indicate that the females leave Ascension Island after laying

Figure C A female green sea turtle digging her nest.
Mark Sullivan, NOAA.

their eggs and then return to the Brazilian coast. Recent research on and around Ascension Island using satellite-tagging methods has revealed that females return to Ascension Island every 3 to 4 years to nest.

It has been hypothesized that adult green turtles use a straightforward navigation system, using the sun on their spawning migration back to their Ascension Island nesting sites. Ascension Island lies due east of Brazil at 8° S latitude. Adult turtles may use the height of the noonday sun to judge latitude, swim to the east at 8° S latitude, and eventually make landfall on Ascension Island. Island-finding by the turtles might be improved if once they were close to Ascension Island on the down-current side they detected a characteristic chemical given off by the island. No one is sure how well green turtles can use celestial cues (if at all) or how well they can detect chemicals dissolved in water. Until these aspects of turtle biology are studied further, the guidance system of green turtles must remain hypothetical. Earlier tagging studies led to the following informative conclusions about green sea turtle navigation based on the tracks recorded by the satellite tags: (1) straight courses can be maintained at sea over long distances; (2) some individuals may perform exploratory movements in different directions, but still end up at the island; (3) the turtles are capable of correcting course using external stimuli; and (4) they initially maintain the same direction as the west–south-westerly current.

Green sea turtles used to be incredibly abundant. It has been estimated that about 33 million green sea turtles lived in the Caribbean region in the early 1800s. The harvesting of turtles for meat and raiding of their nests for eggs by humans devastated the Caribbean population. Formerly, before they were hunted to near extinction, sirenians (manatees in the Atlantic and dugongs in the Indo-Pacific) also used to be very abundant in these lagoonal systems, grazing like "sea cows" in an aquatic meadow. Upon the decline in numbers of sirenians and green sea turtles, herbivorous fishes and sea urchins became the dominant consumers of seagrasses.

The story of Caribbean green sea turtle populations has changed for the better according to a recent analysis. Monitoring of Ascension Island beaches began in 1977 and continues today. Analysis of 36 years of data has revealed excellent news—green sea turtles in the Caribbean are on the rebound, and quickly. In 1977, an estimated 3,700 clutches (groups of eggs) were deposited, and in 2013 the number of clutches deposited was 23,700. This is a six-fold increase and wonderful news for conservationists and the local people on Ascension Island. The focus now is on conservation and tourism. Tourists can visit sea turtles on the beaches during nesting season and watch the nesting process. To keep the turtles safe, strict guidelines are in place to minimize disturbances during these sea turtle nesting observations. This is one story of recovery that provides evidence that under strict protection species that were once seriously depleted can experience population rebounds.

Critical Thinking Questions

1. Most animals mate often, some even daily, but most mate at the least one time per year. Satellite tagging has revealed that green sea turtles may only nest every 2 to 3 years. Provide a hypothesis for why this species might mate so infrequently.

2. Caribbean green sea turtles provide a great example of how populations can rebound. How can scientists use this story to motivate the general population of people to conserve marine life?

For Further Reading

Luschi, G., C. Hays, C. Del Seppia, R. Marsh, and F. Papi. 1998. The navigational feats of green sea turtles migrating from Ascension Island investigated by satellite telemetry. *Proceedings of the Royal Society of London* 265:2279–2284.

Weber, S. B., N. Weber, J. Ellick, A. Avery, R. Frauenstein, B. J. Godley, J. Sim, N. Williams, and A. C. Broderick. 2014. Recovery of the South Atlantic's largest green turtle nesting population. *Biodiversity and Conservation* 23(12):3005–3018.

Like their terrestrial relatives, sea snakes are highly venomous, although they are difficult to excite, and bites on human divers are quite rare (Figure 7.15b). A common misconception is that the mouths of sea snakes are not large enough to bite a human, but this is simply not true. Like many wild animals, sea snakes are not likely to bite unless they feel threatened or are provoked. The 70 or so extant species all inhabit Indo-Pacific waters, with the exception of *Pelamis platurus* in the eastern tropical Pacific. Sea snakes are not found in the Atlantic Ocean, and all reports of sightings there stem from sightings of moray and snake eels.

Sea snakes are extremely specialized for life in the ocean: the distal end of their tail is laterally flattened into an oarlike structure for swimming; the large ventral scales common in terrestrial snakes are reduced or absent in most species; their valved nostrils are dorsally located on their snouts to make taking a breath at the surface just a bit easier; and uptake of oxygen through their skin while underwater has been demonstrated. Like all advanced snakes, sea snakes possess only a right lung, but unlike terrestrial species the single lung of sea snakes extends back to the cloaca and is not highly vascularized, suggesting that it serves a hydrostatic function similar to the swim bladders of bony fish.

Sea Turtles

The eight species of sea turtles that survive today are classified into two families, seven hard-shelled species in the family Chelonidae and the soft-shelled leatherback sea turtle, *Dermochelys coriacea*, in the family Dermochelyidae (**Figure 7.17**). Leatherbacks are adapted to colder water than other sea turtles (and marine reptiles in general) and are the most widely distributed of all sea turtles as a result. They routinely travel far from the tropics, easily reaching the North Sea, Barents Sea, and Newfoundland in the Atlantic and ranging from New Zealand to Alaska in the Pacific. Their ability to thrive in very cold water is poorly understood, yet several hypothetical attributes that may enable leatherbacks to enter cold waters have been proposed. These include protection gained from their thick and oily dermis, behavioral thermoregulation by basking in the sun at the

surface and absorbing solar energy through their black backs, and heat retention via circulatory countercurrent heat exchangers in their flippers. In addition, many biologists credit their cold-water tolerance to **gigantothermy**, a phenomenon wherein large organisms with relatively little surface area are able to stay warm for very long periods of time (once they successfully heat up their large volume via surface basking). Leatherbacks are the largest living sea turtles and are one of the largest living reptiles. The largest specimen ever measured, an individual that stranded on the coast of Wales in 1988, had carapace and flipper lengths of nearly 260 cm and a body weight that exceeded 900 kg. Nevertheless, most species of marine turtle are tropical and subtropical and are often common in reef areas.

All sea turtles except the closely related green and black sea turtles are carnivores, and most consume a variety of reef-dwelling benthic invertebrates. Loggerheads seem to prefer large mollusks, Kemp's ridley and olive ridley turtles target crabs, and Indo-Pacific flatback sea turtles emphasize sea cucumbers in their diets. Hawksbill sea turtles have been described as having a "diet of glass" in that they frequently consume sponges with siliceous spicules. Because leatherback sea turtles are highly oceanic, only approaching land during their breeding season, they feed mainly on gelatinous epipelagic invertebrates such as jellyfish, siphonophores, and tunicates. Unfortunately, common items in their diet closely resemble plastic debris in the sea, and leatherbacks are highly prone to ingesting discarded plastic trash that ultimately clogs their digestive tract and kills them. Only the green sea turtle, *Chelonia*, is herbivorous, preferring to graze on seagrasses, although it does consume macroalgae and kelp.

Sea turtles are well known for their need to return to land to lay their eggs in nests dug in sandy beaches above the high-tide lines of tropical and subtropical shores (**Figure 7.18**). Temperature-dependent sex determination, a phenomenon wherein the sex of some reptiles is determined not by genetics but rather by the nest temperature that the eggs experience during development, occurs in all crocodilians and many turtles, including sea turtles. In sea turtles, warmer eggs yield more female hatchlings; the opposite is true among crocodilians. The ability of sea turtles to navigate over thousands of kilometers of

Figure 7.17 The leatherback sea turtle, *Dermochelys coriacea*, on shore for nesting.
Courtesy of Canaveral National Seashore, NPS/NOAA.

Figure 7.18 The green sea turtle, *Chelonia mydas*, laying eggs in a sandy beach nest.
© Lynsey Allan/ShutterStock, Inc.

open ocean to return to nesting beaches year after year is legendary. The most striking example of this ability is displayed by green sea turtles in the Atlantic Ocean, as described in the Research in Progress box.

DID YOU KNOW?

Leatherback sea turtles are extremely deep divers. They can dive to at least 1,200 m (3,900 ft.) and hold their breaths while swimming for 85 minutes! While resting, they can hold their breath for hours. Why do they dive so deep, you may ask? It is hypothesized that they are chasing prey, such as jellyfish, which concentrate in deep waters during the day.

7.7 Physiology and Behavior of Marine Vertebrates

The marine habitat presents challenges that are very different from the challenges of terrestrial life. Air contains much more oxygen than water, so terrestrial animals have little trouble obtaining sufficient oxygen for their metabolic needs, unlike marine species. Interestingly, because they are bathed in water, many marine animals are always in jeopardy of dehydration and must deal with a chronic influx of ions and loss of water. Terrestrial organisms face similar osmotic challenges, though their mechanisms of coping are very different because they are not bathed in water. Finally, movements on land are designed to offset gravity, whereas marine animals are designed to compensate for the frictional resistance to movement through water, and some take advantage of the buoyancy and support provided by water. The elegant and novel solutions developed by marine vertebrates to respiratory, osmotic, and movement difficulties are described in the following paragraphs.

Respiration

Marine vertebrates are characteristically active animals. Their activity is fueled by the breakdown of lipids and other energy-rich foods. The oxygen used in the mitochondria of active cells either comes from the atmosphere (for tetrapods) or is dissolved in seawater (for fish). For air-breathing tetrapods that swim below the sea surface, air is rich in O_2 (21%) but is available only when they are at the surface to breathe. Fish, in contrast, use their gills to extract constantly accessible O_2 from seawater, but seawater is composed of just a tiny fraction of O_2.

However they acquire their O_2, nearly all vertebrates use hemoglobin to store and transport it in the blood. Hemoglobin is found in other animal phyla, including several inhabitants of deep-sea hot vent communities, but only vertebrates package their hemoglobin inside red blood cells. Hemoglobin is a pH-sensitive protein that has a high chemical affinity for O_2. Each hemoglobin molecule can bind with up to four O_2 molecules. In marine vertebrates, it matters little whether O_2 is obtained

with gills or with lungs; the function of hemoglobin is essentially the same.

When oxygen is transported in vertebrate blood, little exists dissolved in the plasma of the blood because a liter of blood plasma can only dissolve 2 to 3 mL of O_2. In chemical combination with hemoglobin, however, a liter of blood with red blood cells can hold about 200 mL of O_2. As an additional bonus, O_2 remains osmotically invisible when it is bound with hemoglobin, and the tendency for O_2 to continue diffusing into the red blood cells is enhanced.

All cartilaginous and bony fishes take water and dissolved gases into their mouths and pump them over their gills. Each **gill arch** supports a double row of bladelike gill filaments (**Figure 7.19**). Each flat filament bears numerous smaller secondary **lamellae** to increase further the gill surface available for gas exchange. Active fish such as mackerel have up to 10 times as much gill surface as body surface. The O_2 requirements of sedentary bottom fish are not as great, and thus their gill surfaces are not as extensive.

Microscopic capillaries circulate blood very near the inner surface of the secondary lamellae. As long as the O_2 concentration of the blood is less than that of the water passing over the gills, O_2 continues to diffuse passively across the very thin walls of the lamellae and into the bloodstream. In fish gills, the efficiency of O_2 absorption into the blood is enhanced by the direction of water flow over the gill lamellae (Figure 7.19, purple arrows), a direction reverse that of the blood flow within the lamellae. Oxygen-rich water moving opposite to the flow of oxygen-depleted blood creates an effective countercurrent system for gas exchange (**Figure 7.20**). Blood returning from the body with a low concentration of O_2 enters the lamellae adjacent to water that has already given up much of its O_2 to blood in other parts of the lamellae. As the blood moves across the lamellae, it continually encounters water with greater O_2 concentrations, and thus a favorable diffusion gradient is maintained along the entire length of the capillary bed within the lamellae, and O_2 continues to diffuse from the water into the blood over most of the gill surface. With such an efficient countercurrent O_2 exchanger, some fish are capable of extracting up to 85% of the dissolved O_2 present in the water passing over the gills. In contrast, typical air-breathing vertebrates, such as humans, generally extract and use less than 25% of the O_2 that enters their lungs.

Osmoregulation

To illustrate the basic process of salt and water balance in marine organisms, let us examine some somewhat idealized animal examples from various taxa: a sea cucumber, a salmon, and a shark. A sea cucumber avoids problems of salt and water imbalance by maintaining an internal fluid medium chemically similar to seawater (about 35‰ dissolved substances). It can easily maintain this balance as long as the salt concentrations of the fluids on either side of its boundary membranes are equal and no concentration gradient exists. This is known as an **isosmotic** condition. A state of equilibrium is maintained as long as water diffuses out of the sea cucumber as rapidly as it enters,

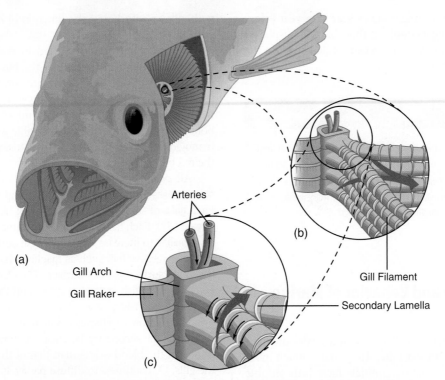

Figure 7.19 Cutaway drawing of a fish showing the position of the gills (a). Broad purple arrows in (b) and (c) indicate the direction of blood flow through the capillaries of the gill filament in a direction opposite that of incoming water. In the arteries, blue blood indicates low oxygen content, and red blood indicates higher oxygen content.

and the salt content of the internal fluids remains equal to that of the seawater outside (**Figure 7.21**, top).

If the sea cucumber is removed from the sea and placed in a freshwater lake, however, the salt concentration is then greater inside the animal (still 35‰) than outside (the body fluids are now **hyperosmotic** to the lake water), and the internal water concentration (965‰) is correspondingly less than the concentration of the lake water (1,000‰). Water molecules, following their concentration gradient, rapidly diffuse across the selectively permeable boundary membranes into the sea cucumber. The movement of water across such a membrane is a special type of diffusion known as **osmosis**. The dissolved ions, now more concentrated within the animal than outside, cannot diffuse out of the sea cucumber because this movement is blocked by the impermeability of the membranes to the ions. The net result is an increase in the amount of water inside the sea cucumber. The additional water creates an internal

osmotic pressure that is potentially damaging because the animal is incapable of expelling the excess water, and so it swells. Many other marine animals are similarly incapable of countering such osmotic stresses. As a consequence, these organisms are restricted to regions of the ocean where salinity fluctuations are small.

In contrast to the lack of control that sea cucumbers have over their osmotic situation, marine vertebrates, some marine invertebrates, and most marine plants possess well-developed osmoregulatory mechanisms. As a result, some of these organisms are free to occupy estuaries and other regions of varying salinities without osmotic stress. Salmon, which spend their early life in freshwater rivers and then grow to maturity at sea, are an example of how some organisms maintain a **homeostatic** internal medium regardless of external environmental conditions (Figure 7.21, middle).

The salt concentration of a salmon's body fluids, like those of most other bony fishes, is midway between the concentrations found in freshwater and in seawater (about 18‰). As such, the body fluids are hyperosmotic to freshwater and hypoosmotic to seawater. So these fish never achieve an osmotic balance with their external environment. Instead, they constantly expend energy to maintain a stable internal osmotic condition different from either river or ocean water. In seawater, salmon lose body water by osmosis and are constantly fighting dehydration. To counter these losses, they drink large amounts of seawater that are absorbed through their digestive tracts and into their bloodstreams. The water is retained in the body tissues, and excess salts taken in with the water are actively excreted by special

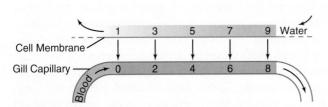

Figure 7.20 A countercurrent gas exchange system of fish gills. Nearly all the oxygen from water flowing right to left diffuses passively across the gill membrane into the blood flowing in the opposite direction. Numbers represent arbitrary oxygen units. Blue blood indicates low oxygen content, and red blood indicates higher oxygen content.

In Seawater
(35‰)

In Freshwater
(0‰)

Sea cucumber (Body fluids = 35‰)

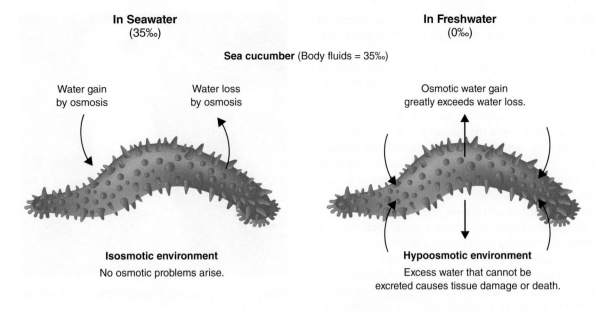

Water gain
by osmosis

Water loss
by osmosis

Osmotic water gain
greatly exceeds water loss.

Isosmotic environment
No osmotic problems arise.

Hypoosmotic environment
Excess water that cannot be
excreted causes tissue damage or death.

Salmon (Body fluids = 18‰)

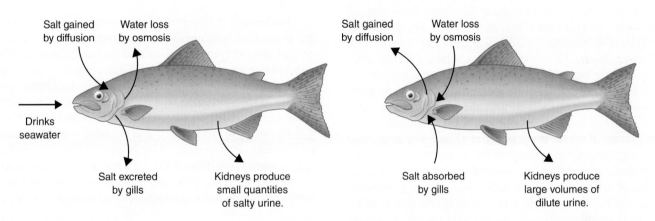

Salt gained
by diffusion

Water loss
by osmosis

Salt gained
by diffusion

Water loss
by osmosis

Drinks
seawater

Salt excreted
by gills

Kidneys produce
small quantities
of salty urine.

Salt absorbed
by gills

Kidneys produce
large volumes of
dilute urine.

Shark (Body fluids = 18–35‰)

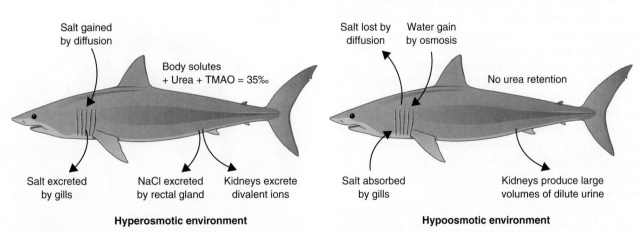

Salt gained
by diffusion

Body solutes
+ Urea + TMAO = 35‰

Salt lost by
diffusion

Water gain
by osmosis

No urea retention

Salt excreted
by gills

NaCl excreted
by rectal gland

Kidneys excrete
divalent ions

Salt absorbed
by gills

Kidneys produce large
volumes of dilute urine

Hyperosmotic environment

Hypoosmotic environment

Figure 7.21   A comparison of the osmotic conditions of a sea cucumber, a salmon, and a shark in seawater and in freshwater. TMAO = trimethylamine oxide; NaCl = sodium chloride.

chloride cells located in their gills. The kidneys produce urine as concentrated as possible to avoid losing too much water during waste disposal.

The osmotic challenges of salmon are completely reversed when, as adults, they return from the sea to spawn in freshwater. Now the problems are excessive osmotic water gain across their gills and digestive membranes and a steady loss of salts to the surrounding water. Here, the salmon drink very little fresh water. To balance the osmotic water gain, the kidneys produce large amounts of dilute urine after effectively recovering most of the salts from that urine. Needed salts are obtained from food and are also actively scavenged from the surrounding water through other specialized cells in their gills. Thus, at a considerable expense of energy, salmon maintain a homeostatic internal fluid environment in either fresh or ocean water.

Nearly all species of sharks and their relatives are marine, although one family of stingrays is restricted to freshwater rivers of South America; some Indo-Pacific sharks live in rivers; and a number of sharks, rays, and sawfish are euryhaline, perhaps even anadromous like salmon. Like most other vertebrate groups, the concentration of salts in their body fluids is much less (about 50%) than that of seawater, and thus, like the salmon, are vulnerable to osmotic water loss and salt gain through their gill tissue (Figure 7.21, bottom). Most sharks and their allies prevent water loss in their marine home by achieving osmotic equilibrium with seawater via the retention of two waste products from protein catabolism: urea and trimethylamine oxide (TMAO). Urea is a toxic nitrogenous waste that would denature proteins if not for the presence of TMAO. The inactivated urea, together with the protective TMAO and the shark's other body solutes, provides a total internal ion concentration equal to that of seawater outside the gills, and thus it loses no water to its hyperosmotic environment. To combat the steady influx of ions through its gills and into its blood plasma, it excretes them using three mechanisms. Like the salmon, sharks excrete monovalent ions out of their gills and pump divalent ions out with their urine. Unlike salmon, they do not excrete NaCl with their gills. Instead, they perform this task with a unique organ called the *rectal gland*, a finger-sized papilla located adjacent to the intestine in the posterior portion of the body cavity. This salt-excreting gland pumps NaCl into the shark's intestine so that it can be eliminated with feces. Freshwater sharks and rays, which are hyperosmotic to their environment, do not retain urea and TMAO. Instead, like salmon in freshwater, they use their gills to pump ions into their blood and they produce large volumes of very dilute urine. As described previously in this chapter, the marine frog and the coelacanth also retain urea to combat osmotic water loss to the sea.

Marine reptiles and birds have evolved a double-barreled solution to deal with the shortage of freshwater and with the extra salt loads associated with feeding at sea: special **salt glands** and kidneys. Both reptiles and birds have complex salt-excreting glands (**Figure 7.22**), one above each eye (in birds, turtles, and the marine iguana) or associated with the tongue (snakes and crocodiles), that concentrate NaCl to twice the amount of seawater. After concentration, this salty solution drips down from the eye, is blown out the nasal passages, or is washed from the mouth.

Figure 7.22 A lesser albatross with prominent nasal openings for salt excretion.
© Clara Natoli/ShutterStock, Inc.

The kidneys of both birds and reptiles convert toxic nitrogen wastes from body metabolism to **uric acid** rather than urea, as do the kidneys of some fish. Uric acid is a nearly nontoxic substance that requires only about 2 g of water to excrete per gram. The white uric acid paste is mixed with feces for elimination, and thus urine production (and its associated water loss) as we know it in mammals does not occur in either birds or reptiles.

The ability to produce uric acid is related to another major evolutionary advancement of birds and reptiles, the shelled egg, or **amniotic egg** (**Figure 7.23**). Although most sea snakes are ovoviviparous, most reptiles and all birds are oviparous, laying large, shelled eggs that require some period of incubation before

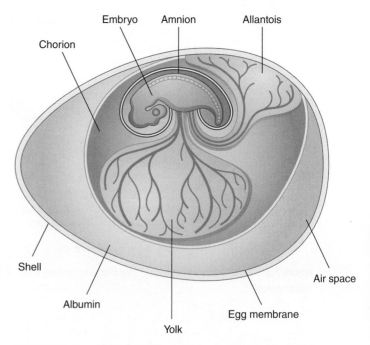

Figure 7.23 Cross section of a typical amniotic egg, showing the major internal membranes. The amnion surrounds and protects the developing embryo.

Figure 7.24 Marine turtle hatchling emerging from its egg.
© Aqua Image/age fotostock.

Figure 7.25 A pelagic great white shark, *Carcharodon charcharias*. Lift is obtained from its heterocercal tail shape and the large pectoral fins extending from the flattened underside of the body.
© Stephen Frink Collection/Alamy Images.

hatching. During this period, the enclosed developing embryo floats in its own water-filled sac, the amnion, whereas its nitrogen wastes are stored as uric acid in the allantois because these wastes cannot be eliminated across the protective outer eggshell.

Until hatching occurs, gas exchange occurs across a third membrane, the chorion. When the egg hatches, its shell, inner membranes, and accumulated uric acid are abandoned (**Figure 7.24**). The accumulation of relatively solid uric acid waste is an advantage for the developing young that is trapped inside an egg for long periods. If the waste material were liquid the developing young would be swimming in its own waste.

Fertilization of shelled eggs must occur before the protective shell is in place, and thus all birds and reptiles fertilize their eggs internally. Shelled eggs, uric acid excretion, and even internal fertilization were early adaptations for life out of water. Thus, for birds, crocodiles, turtles, and iguanas that have returned to the sea, laying and incubating their eggs ashore is still the norm.

Locomotion in the Sea

Locomotion in the sea varies greatly among taxa with different habitat preferences, body sizes and shapes, and feeding preferences, among many other variables. Most rays are lie-and-wait ambush specialists adapted to living on the sea bottom, whereas most sharks are streamlined, fast-moving, pelagic predators. When swimming, rays gracefully undulate the edges of their flattened pectoral fins or, in the cases of manta and eagle rays, flap their pectorals like large wings. Swimming sharks (and most pelagic bony fishes as well) develop thrust with their flattened caudal fin; however, the asymmetrical **heterocercal** tail (**Figure 7.25**), so characteristic of sharks, has a shape very different from that of most bony fishes (which are **homocercal**, or symmetrical with the long axis of the body). When a typical shark's caudal fin is moved from side to side, a forward thrust develops, and because of the angle of the trailing edge of the

tail, it produces some lift as well. The paired pectoral fins of sharks are flat and large and extend horizontally from the body like stubby aircraft wings (Figure 7.25). The front part of the shark's underside is nearly flat and, with the flat extended pectoral fins, meets the water at an angle and produces lift for the front part of the body to balance the lift produced by the tail. Pelagic sharks and other cartilaginous fishes lack swim bladders and need this lift to maintain their position in the water column. This mechanism for achieving lift, however, does have its disadvantages. These fish cannot stop or hover in midwater. To do so would cause them to settle to the bottom, a bottom that may be some distance away in the open ocean. Maneuverability is also reduced; the large and rigid paired fins that function as hydrofoils are not well suited for making fine position adjustments.

The variety of ecological niches occupied by teleost fishes are reflected in their diversity of specialized approaches to swimming, with associated adaptations of body shape, fins, and muscle tissues. Most teleosts, such as the surfperch in **Figure 7.26**, are generalists. From this generalist body plan, other more specialized modes of swimming can be derived: sprinting (barracuda), fine maneuvering (butterflyfish), or nearly continuous high-speed cruising (tuna). Rapidly accelerating fish, such as barracudas, tend to have thinner, more elongated bodies, possibly to reduce their chances of being seen and recognized as they rush their prey. Butterflyfish and other fine maneuverers are tall and elliptical in cross section, with large fins extending even greater distances from the body. The increased amount of body surface, while adding to the overall drag, creates large control surfaces for making fine position adjustments. Tuna have large bulky bodies packed with swimming muscles and come closest to the ideal shape for a high-speed underwater swimmer (this topic is expanded later in this chapter).

In fish species with any of the body shapes shown in Figure 7.26, the push, or thrust, needed for swimming is developed almost entirely by the backward component of the pressure of the animal's body and fins against the water. The bending motion of the anterior part of the body, initiated by

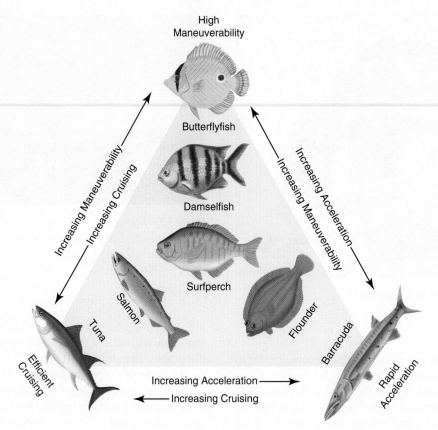

Figure 7.26 Examples of body shape specialization for three different swimming modes. Data from P. W. Webb, *Scientific American* 25(1984):74–82.

Figure 7.27 Dorsal view of the progression of a body wave as an eel swims from left to right.

the contraction of a few muscle segments (myomeres) on one side, throws the body into a curve (**Figure 7.27**). This curve, or wave, passes backward over the body by sequential contraction and relaxation of the myomeres (**Figure 7.28**). The contraction

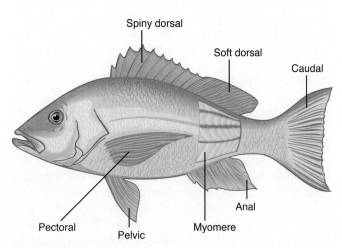

Figure 7.28 Names of fins and their positions on a derived teleost fish. A portion of the integument has been removed to reveal the arrangement of myomeres.

of each myomere in succession reinforces the wave form as it passes toward the tail. Immediately after one wave has passed, another starts near the head on the opposite side of the body, and the entire sequence is repeated in rapid succession.

The amount of forward thrust developed is magnified by the flared and flattened caudal fin at the posterior end of the body. Caudal fins typically flare dorsally and ventrally to provide additional surface area to develop thrust. One index of the propulsive efficiency of the caudal fin, based on its shape, is its **aspect ratio**, which is the square of the fin height divided by the total fin area. The caudal fins of fish exhibit a range of profiles, illustrated with increasing aspect ratios in **Figure 7.29**: round, truncate, forked, and lunate. At the low end of the range of aspect ratios is the butterflyfish, with a round caudal fin that is soft and flexible. This flexibility permits the caudal fin to be used for accelerating and maneuvering. Truncate and forked fins have intermediate aspect ratios, produce less drag, and are generally found on faster fish. These fins are also flexible for maneuverability. The lunate caudal fin characteristic of tuna, sailfish, marlin, and swordfish have high aspect ratios (up to 10 in swordfish) for reduced drag at high speeds. The shape closely resembles the swept-wing design of high-speed aircraft.

Caudal fins

Fin:	Round	Truncate	Forked	Lunate	Heterocercal
Shape:					
Aspect ratio:	1	~3	~5	7+	Variable
Fishes:	flounder, butterflyfish	salmon, barracuda	herring, perch	tuna, mackerel	shark

Figure 7.29 Examples of shapes and aspect ratios for caudal fins.

These fish are among the fastest swimming marine animals. The caudal fin is quite rigid for high propulsive efficiency but is poorly adapted for slow speeds and maneuvering. Fish with high aspect–ratio caudal fins (especially forked and lunate types) are capable of long-distance continuous swimming.

Bony fishes equipped with swim bladders have their pectoral and pelvic fins free for other uses. In most bony fishes, the paired fins are used solely for turning, braking, balancing, or other fine maneuvers. When the fish are swimming rapidly, these fins are folded back against their bodies. Yet several groups of bony fishes develop all their thrust without using their caudal fins. Wrasses, parrotfish, and surgeonfish, for example, swim with a jerky fanning motion of their pectorals and hold the remainder of their bodies straight. The greatly enlarged pectoral fins of the flyingfish in **Figure 7.30** enable this animal to glide in air for long distances, apparently as a means of escaping predators. Flyingfish build up considerable speed while just under the sea surface and then leap upward with their pectorals extended. The length of the glide is dependent on wind conditions and the initial speed of the fish as it leaves the water, and glides up to 400 m have been reported.

Triggerfish and ocean sunfish swim by undulating their anal and dorsal fins only (**Figure 7.31**). For the triggerfish, these fins extend along much of the body. The large sunfish, which reaches lengths of nearly 3 m and attains weights up to a ton, is a sluggish fish and is often seen "sunning" at the surface.

The little swimming it does is accomplished by its long dorsal and anal fins. Seahorses and the closely related pipefish rapidly vibrate their dorsal and pectoral fins to achieve propulsion. Seahorses usually swim vertically with their heads at right angles to the rest of the body (**Figure 7.32**). The prehensile tail tapers to a point and is used to cling to coral branches and similar objects.

In the pelagic environment, vertebrates often must move long distances to find food or mates or to otherwise improve

(a)

(b)

Figure 7.31 Two fish that use their dorsal and anal fins for propulsion: (a) triggerfish and (b) ocean sunfish, *Mola mola*.

(a) NOAA Flower Garden Banks National Marine Sanctuary; (b) NOAA.

Figure 7.30 A flyingfish uses enlarged pectoral fins for gliding through the air.

Shannon Rankin, NMFS, SWFSC.

Figure 7.32 The seahorse, *Hippocampus*, swims vertically, using its dorsal fin for propulsion.

conditions for their survival. Structural or behavioral adaptations that enable animals to swim with reduced energy expenditures enable them to divert more energy to growth and reproduction and contribute to the potential success of an individual.

Vertebrates are large and fast animals. Consequently, their energetic costs of locomotion are expensive and represent a major expenditure of their available resources. The cost of swimming is affected by swimming speed, the flow patterns of water around the swimmer, and two physical properties of water—density and viscosity. Water is greater than 800 times more dense than air and at least 30 times more viscous. Thus, the frictional resistance to moving through water is considerably greater than in air. As a consequence, movement through water imposes severe limitations on speed and energetic performance for vertebrates and other nektonic species, such as squid. However, moving in water does have its advantages. Propulsive forces are easier to generate in water than in air, and because the body densities of most nekton are similar to that of water, most marine nekton are approximately neutrally buoyant. Thus, swimming animals can maintain their vertical position in the water with little energy expenditure because they do not need to support their weight during locomotion as do terrestrial animals.

Most pelagic fish use side-to-side motions of their caudal fins as their chief source of propulsion, as do tetrapods as diverse as seals and sea snakes. Whales move their caudal flukes in vertical motions to achieve their swimming power. Other tetrapods use paddling (turtles) or underwater flying motions (penguins, sea lions, and many pelagic rays), using their pectoral flippers to do most of the work (**Figure 7.33**). A few invertebrate nekton are also excellent swimmers. Squids and other cephalopods take water into their mantle cavities and then expel it at high speeds through a nozzle-like siphon. The **siphon** can be aimed in any direction for rapid course corrections and for maneuvering purposes. Squids and cuttlefish also use their undulating lateral fins in much the same manner as benthic skates and rays.

One useful approach to gaining a general understanding of the energy costs associated with locomotion is to evaluate an animal's **cost of transport (COT)**. COT comparisons can be made between different modes of locomotion and for animals of different sizes. All species in all modes of locomotion have some preferred speed at which their COT is minimized. When this COT_{min} is calculated and plotted as a function of body mass, it is clear that both flying and swimming impose lower COT_{min} than does walking or running (**Figure 7.34**). Flying is energetically demanding but covers long distances in a short time, and thus the COT_{min} is low. Swimmers, regardless of size, do not need to support the weight of their bodies, and thus their COT_{min} is lower still. In contrast to fliers, however, which cannot be larger than about 40 kg, swimmers can be extremely large and still move efficiently.

Body Shape and Coloration

Whether a large whale or a swimming invertebrate, to maintain high swimming velocities or to swim in an energetically efficient manner a swimmer must overcome (or at least reduce) several different components of the total hydrodynamic drag on the

Power stroke

Glide stroke

Figure 7.33 Power and glide strokes of three pectoral-swimming tetrapods.

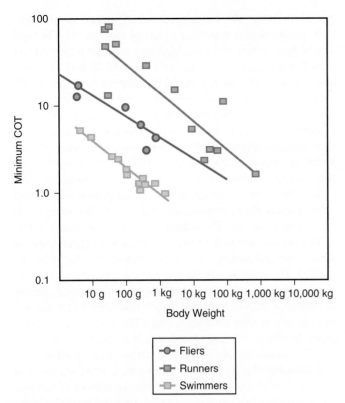

Figure 7.34 Relationship between COT$_{min}$ and body size for different modes of locomotion.
Data from D. W. Tucker, *Nature, London* 183(1959):495–501.

body. Each of these components is influenced by the swimmer's body size and shape. The body shape of a fast swimmer, such as a tuna or dolphin, is a compromise between different hypothetical body forms, each of which reduces some component of the total drag and enables the animal to slip through the water with as little resistance as possible. **Frictional drag** is a function of the extent of wetted surface an animal has in contact with the water and the density and viscosity of the water. Frictional drag is lower if the flow of water over the body surface is smooth, or laminar, and high if it is turbulent. **Pressure drag**, sometimes called *form drag*, is the consequence of displacing an amount of water equal to the swimmer's largest cross-sectional area (from a head-on view). Another drag component is **induced drag** created by the fins, flukes, or flippers that swimmers use to produce their thrust. Finally, an additional **wave drag** is created by the production of surface waves when swimming at or near the sea surface. Because all marine tetrapods must surface frequently to breathe, wave drag can contribute substantially to the total drag that they must overcome to swim.

Frictional and form drag of a fast swimmer are reduced with a streamlined body form that is roundly blunt at the front end, tapered to a point in the rear, and round in cross section. The **fineness ratio** is the ratio of an animal's body length to its maximum body diameter. For efficient swimmers, fineness ratios range from 3 to 7, with the ideal near 4.5. Most cetaceans exhibit fineness ratios near 6 or 7, with only killer whales and right whales having nearly ideal ratios of 4.0 to 4.5. Most tunas also have nearly ideal fineness ratios. This streamlined shape of

most fast nekton, excluding their fins, is the best possible body shape to reduce the several components of drag and to slip through the water with as little resistance as possible.

Unlike fish, marine birds and mammals must necessarily surface to breathe more frequently when swimming at higher speeds. Because small dolphins or penguins must surface to breathe more frequently than large whales, it is more difficult for them to escape beneath the high drag associated with the sea surface. Only by remaining at least 2.5 body diameters below the sea surface can they minimize wave drag. At high speeds, the high wave drag associated with surfacing can be partly avoided by leaping above the water–air interface and gliding airborne for a few body lengths. Although you may have assumed that dolphins leap through the air for entertainment, entertainment value is likely an added benefit, because this behavior is actually an energy saver. The aerial phase of this type of porpoising or leaping locomotion sends the animal above the high-drag environment of the water surface while simultaneously providing an opportunity to breathe. The velocity at which it becomes more efficient to leap rather than to remain submerged is known as the crossover speed. The **crossover speed** is estimated to be about 5 m/s for small spotted dolphins and increases with increasing body size until leaping becomes a prohibitively expensive mode of locomotion in cetaceans longer than about 10 m.

Fish coloration schemes vary dramatically in the sea. A variety of factors determine the best color scheme for any given species, including their habitat, typical predators, behavior, swimming speed, and whether they live solitarily or in schools. Fish living a pelagic existence are often **countershaded** with dark coloration on the dorsal surface and light coloration on the ventral surface. This coloration pattern makes the fish less visible to predators above or below it. Many benthic fish benefit from a color scheme that allows them to blend into their environment, providing camouflage, which decreases their visibility to predators. The tropics are home to many brightly colored fish, and it is this bright coloration that seems to stand out to humans when exploring a reef that may actually help these fish blend in. A coral reef ecosystem is very bright in general, so by being brightly colored a fish blends into the maze of brightly colored objects around it.

Speed

Several species of nekton are noted for their amazingly fast swimming speeds. The oceanic dolphin *Stenella* has been clocked in controlled tank situations at better than 40 km/h (approximately 25 miles/h). Top speeds of killer whales are estimated to be 40 to 55 km/h. For comparison, human Olympic-class swimmers achieve sprint speeds of only 4 to 5 km/h. Using specially designed fishing poles to measure speeds at which line is stripped from a reel, a 1-m-long barracuda has been clocked at 40 km/h. Brief sprints of similarly sized yellowfin tunas and wahoos have been clocked at more than 70 km/h, and large bluefin tunas may be capable of speeds in excess of 110 km/h. This estimate may not be as farfetched as it seems because some bluefins reach lengths of 4 m and presumably would be much faster than a fish only 1 m long, but it has yet to be confirmed.

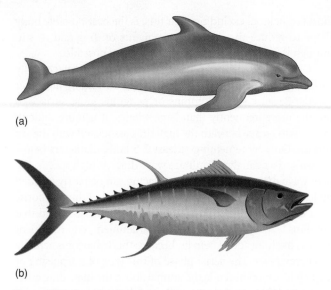

(a)

(b)

Figure 7.35 Streamlined body forms of two swift pelagic animals: (a) bottlenose dolphin, *Tursiops*, and (b) tuna, *Thunnus*.

What enables dolphins, tunas, and similar fish to swim so fast? In addition to having nearly optimal streamlined body shapes (**Figure 7.35a**), many cetaceans are faster than most fish because they are endothermic and thus can maintain high and continuous rates of power output. But what about rapidly swimming fish (**Figure 7.35b**)? The exceptional swimming abilities of tuna and tuna-like fish go beyond simply having a streamlined body form and an efficient caudal fin. Their streamlined body form is complemented by other friction-reducing features, including small and smooth scales, nonbulging eyes, fins that can be retracted into slots and out of the path of water flow when not needed for maneuvering, and numerous small median **finlets** on the dorsal and ventral surfaces of the rear part of the body function to reduce turbulence in that region.

Most of the caudal flexing of a tuna is localized in the region of the **caudal peduncle**, the region where the caudal fin joins the

rest of the body. The caudal peduncle, flattened in cross section, produces little resistance to lateral movements. The rigid caudal fin is lunate and has a high aspect ratio (usually greater than 7; see Figure 7.29). The tail beats rapidly with relatively short strokes. This type of caudal fin creates a large thrust with little drag but also provides very little maneuverability.

Nearly 75% of the total body weight of a tuna is composed of swimming muscles. In tuna, each myomere (see Figure 7.28) overlaps several body segments and is anchored securely to the vertebral column. Tendons extend from the myomeres across the caudal peduncle and attach directly to the caudal fin. Tuna swimming muscles consist of segregated masses of red and white muscle fibers. Structurally, red muscle fibers are much smaller in diameter (25 to 45 µm) than white muscle fibers (135 µm) and are rich in myoglobin, a red pigment with a strong chemical affinity for O_2 (even greater than that of hemoglobin). The small size of the red muscle cells provides extensive surface area that, in conjunction with myoglobin, greatly facilitates O_2 transfer to the red muscle cells. Physiologically, red muscle cells respire aerobically and white muscle cells respire anaerobically, converting glycogen to lactic acid.

The metabolic rate (and power output) of tuna red muscle, and probably of red muscle in other fish, is about six times as great as that of white muscle. The relative amount of red and white muscle a fish has is related to the general level of activity the fish experiences. An ambush predator such as a grouper has almost no red muscle, but its large mass of white muscle fibers can power short, fast lunges to capture prey or elude predators. At the other extreme are tunas, with over 50% of their swimming muscles composed of red muscle fibers. Electrodes monitoring activity of muscle tissues indicate that at slow normal cruising speeds only the red muscles of tunas contract. White muscles come into play only during sprints. Top speeds of about 10 body lengths per second can be maintained for about 1 second, but cruising speeds of 2 to 4 body lengths per second can be maintained indefinitely (**Figure 7.36**). The power for continuous swimming comes from the red muscle masses, with white

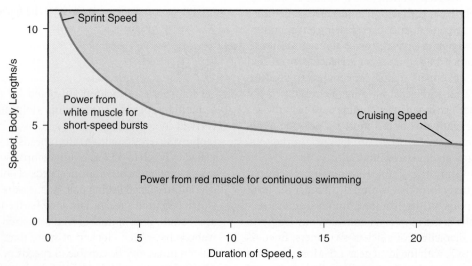

Figure 7.36 Duration of swimming speeds for white and red muscles. White muscle is powered anaerobically and thus is used for short bursts at sprint speeds and fatigues rapidly; red muscle is powered aerobically and thus maintains continuous cruising speeds.
Data from R. Bainbridge, *Journal of Experimental Biology* 37(1960):129–153.

TABLE 7.2 Elevation of Red Muscle Temperatures Above Seawater Temperatures for Some Marine Fishes			
Fish with Slightly Elevated Temperatures		**Fish with Dramatically Elevated Temperatures**	
Yellowtail (*Seriola*)	+1.4°C	Porbeagle shark (*Lamna*)	+7.8°C
Mackerel (*Scomber*)	+1.3°C	Mako shark (*Isurus*)	+4.5°C
Bonito (*Sarda*)	+1.8°C	Tuna (*Thunnus*)	+5 to +13°C occasionally to +23°C

muscle being held in reserve for peak power demands. White muscle does not require an immediate O_2 supply; it can operate anaerobically and accumulate lactic acid during stress situations. The lactic acid can be converted back into glycogen or some other substance when the demand for O_2 has diminished. Tunas, with a greater proportion of red muscle, are able to maintain a faster cruising speed than most other fish indefinitely.

Fish are generally considered to be ectothermic animals. The heat generated by metabolic processes within the body may elevate blood temperature slightly above the ambient water temperature, but the heat gain is quickly lost to the surrounding seawater because their warm venous blood is juxtaposed with cold water in the fish's gills (**Table 7.2**, left column). A few exceptionally fast fish, however, have red muscle masses that are much warmer than the surrounding water (Table 7.2, right column). The magnitude of muscle temperature elevation above the water temperature is usually consistent for each species. The one well-studied exception is the bluefin tuna (*Thunnus thynnus*), which has a consistently high red muscle temperature regardless of water temperature. In water of 25°C, for example, the core muscle temperature of the bluefin tuna is near 32°C and declines only slightly to 30°C when the animal moves to seawater with a temperature of 7°C. Another exception is the opah (*Lampris*), or moonfish, which was recently determined to be fully endothermic. This fish is described in detail in the Case Study for this chapter.

Within certain limits, metabolic processes, including muscle contractions, occur more rapidly at higher temperatures. Consequently, the power output of a warm muscle is greater than that of a cold muscle. Tuna, manta ray, the swordfish and other billfish, and lamnoid sharks, including porbeagles, makos, threshers, and white sharks, all possess fascinating heat-conserving circulatory features. The swimming muscles of most bony fishes receive blood from the dorsal aorta just under the vertebral column. The major blood source for the red muscle masses of lamnoid sharks and most tuna is a cutaneous artery under the skin on either side of the body (**Figure 7.37**). The blood flows from the cutaneous artery to the red muscle and then returns to the cutaneous vein. Between the cutaneous vessels and the red muscles are extensive countercurrent heat exchangers that facilitate heat retention within the red muscle. Cold blood returning from the gills enters the countercurrent system and is warmed by the blood leaving the warm red muscle and heading toward the gills. As a result, only cooled blood returns to the gills, and little of the heat generated in the red muscles is lost to the sea.

All of the previously described features collectively function to provide some sharks, tunas, billfish, and other similar fish with the ability to cruise continually at moderate speeds

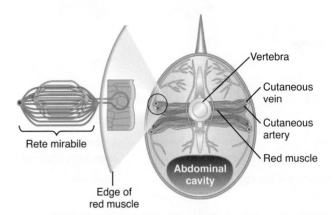

Figure 7.37 Cross section of a tuna showing the position of the red muscles (shaded) and the countercurrent system of small arteries and veins serving the red muscles.
Data from F. G. Carey, *Scientific American* 36 (1973):79–86.

and with the opportunity to be the efficient pelagic predators they are. **Table 7.3** summarizes these features and compares a tuna with a normally noncruising fish (a rockfish) that typically lies in wait for its prey.

Schooling

Successful feeding techniques used by large whales, numerous fish, and even a few birds and seals are dependent on the presence of abundant and dense aggregations of smaller animals. Hundreds of species of smaller fishes and several species of squids, sharks, and dolphins exist in well-defined social organizations called **schools**. Fish schools vary in size from a few individuals to enormous populations extending over several square kilometers (**Figure 7.38**). Schools usually consist of a single species, with all members similar in size or age. Larger fish swim faster than smaller ones, and mixed populations quickly sort themselves out according to their size. The spatial organization of individuals within a school remains remarkably constant as the school moves or changes direction. Individuals typically line up parallel to each other, swim in the same direction, and maintain fixed spacings between individuals. When the school turns, it turns abruptly, and the animals on one flank assume the lead. The spatial arrangement within schools seems to be maintained with the use of visual or vibrational cues.

Why do some species band together in schools to be so conveniently eaten by larger predators (as indicated in **Figure 7.39**)? A part of the answer seems to be that for small animals with no other means of individual defense, schooling behavior provides a degree of protection, commonly referred to as "safety in numbers." Most of our present understanding of the survival

TABLE 7.3 Functional Comparison of Some Features That Influence the Swimming Speeds of a Noncruising Fish (Rockfish) and a Specialized Cruiser (Tuna)

Characteristic	Rockfish	Tuna
Body feature		
Shape		
Front view		
Rigidity	Flexible body	Rigid body
Scales	Abundant large scales	Small scales
Eyes	Bulging eyes	Nonprotruding eyes covered with adipose lid
Percentage of thrust by body	50%	Almost none
Dorsal fin	Broad based and high	Small, fits into slot
Caudal peduncle		
Cross-sectional shape		
Keels	Absent	Present
Finlets	Absent	Present
Caudal fin		
Aspect ratio	Low, 3	High, 7–10
Rigidity	Flexible	Rigid
Maneuverability	Good	Poor
Tail-beat frequency	Low	High
Tail-beat amplitude	Large	Small
Swimming muscles		
Percentage of body weight	50% to 65%	75%
Percentage red muscle	20%	50% or more
Body temperature	Ambient	Elevated

Modified from Fierstine and Walters. *Memoirs of the Southern California Academy of Sciences* 6(1968):1–31.

they might even discourage hungry predators with the illusion of an impressively large and formidable opponent.

Schooling also can serve as a drag-reducing behavior as individuals draft behind leading individuals, much as race car drivers do. Laboratory studies with fish that instinctively school also indicate that if these fish are isolated at an early age and prevented from schooling, they learn more slowly, begin feeding later, grow more slowly, and are more prone to predation than their siblings who are allowed to school. Schooling behavior also serves as a mechanism to keep reproductively active members of a population together. Many schooling species reproduce by broadcast spawning, and dense concentrations of mature individuals spawning simultaneously ensure a high proportion of egg fertilization and probably greater larval survival.

Several species of pelagic tunas, especially bigeye and yellowfin, tend to congregate under near-surface schools of spinner or common dolphins. The behavioral reasons for the large mixed schools of pelagic tunas and small dolphins are not well understood, but one hypothesis is that the presence of dolphins may indicate the presence of food, like smaller fish, that the tunas are also interested in hunting.

Migration

In the sea, only the larger and faster swimming nekton are capable of accomplishing regular long-distance migrations, and a few examples of migratory fish are described later here. These migrations commonly serve to integrate the reproductive cycles of adults into local and seasonal variations in the patterns of primary productivity, leading to steady food sources. Many species of nekton participate in regular and directed migratory movements that are several orders of magnitude larger in both time and space scales than are the patterns of vertical migration. Some migrators require several months of each year to accomplish their oceanic treks. In general, these migrations are adaptations to better exploit a greater range of resources for feeding or reproduction. For example, the food available in spawning areas may be appropriate for larval and juvenile stages, but it might not support the mature members of the population. Thus, the adults congregate for part of the year in productive feeding areas elsewhere that may be unsuitable for the survival of the younger stages.

Migratory patterns of marine animals often exhibit a strong similarity to patterns of ocean surface currents. Juvenile stages of some species may be carried long distances from spawning and hatching areas by ocean currents. Although adults may use currents for a free ride, many types of larvae and juvenile fish are absolutely dependent on current drift for their migratory movements. The down-current drift of these young may require the adults to make an active compensatory return migration against the current flow to return to the spawning grounds.

Migrating teleosts typically move below the sea surface and well away from the coast, making it difficult or even impossible for us to observe their migratory behavior directly. Most of our understanding of oceanic migrations has been inferred from studies using visual or electronic tags and from distributional patterns of eggs, larvae, young individuals, and adults of a species. When a general progression of developmental stages from

value of schooling behavior is based on conjecture because experiments with natural populations are exceedingly difficult to conduct and evaluate. Predatory fish have less chance of encountering prey if the prey are members of a school because the individuals of the prey species are concentrated in compact units rather than dispersed over a much larger area. Moreover, once a predator encounters a school, satiation of the predator enables most members of the school to escape unharmed. Unfortunately, because commercial fishers are immune to satiation, the predatory activities of humans often result in rapid overexploitation of schooling species. Large numbers of fish in a school may achieve additional survival advantages by confusing predators with continually shifting and changing positions;

Figure 7.38 A large school of thousands of yellowback fusiliers, *Caesio*, extending along and beyond a reef.
© Zeamonkey/Getty Images.

Figure 7.39 A large school of fish visited by a blacktip reef shark, *Carcharinus*. The fish school for protection and safety in numbers.
© cbpix/Shutterstock, Inc.

egg to adult can be found extending from one oceanic area to another, a migratory route between those areas may be inferred.

Animals marked with visual tags can yield valuable information about their migratory routes and speeds, but only if the tags are repeatedly observed during migration or are recovered after the migration has been completed. The application of tagging programs is thus limited to animals that can be recaptured in large numbers (usually commercially harvested species) or to animals whose tags can be observed frequently at the sea surface. Newer techniques, such as continuous tracking of individual animals fitted with radio or **ultrasonic** transmitters, have added considerably to our knowledge of oceanic migration patterns. Radio tags attached to tetrapods that must surface for air periodically can be monitored by orbiting satellites.

These efforts are revealing more of the fine details of oceanic movements by obtaining nearly continuous global coverage of swimming, diving, and migratory behaviors of reptiles, birds, and mammals, even when they migrate into extremely remote parts of the world ocean. More recently, pop-up archival transmitting (PAT) tags have been used successfully to track fish that do not need to come to the surface to breathe. By collecting and archiving data on depth, temperature, and ambient light, PAT tags can determine a fish's position in the world ocean. Then, at a preprogrammed time, the PAT tag releases, pops up to the surface, and transmits stored data to orbiting satellites.

A classic migration example includes salmonids. Four salmon species in the genus *Oncorhynchus* live only in the North Pacific. All are anadromous; they spend much of their lives at sea and then return to freshwater streams and lakes to spawn. They deposit their eggs in beds of gravel, and the eggs remain there through the winter. After spawning, the adult salmon die. Because the migratory patterns of the various types of salmon are similar, only the patterns of the sockeye salmon are described here. After hatching in the spring, the young sockeye remain in freshwater streams and lakes for about 2 to 3 years as they develop to a stage known as **smolts**. The smolts, which

are anatomically and physiologically modified for life in deep seawater, then migrate downstream and into the sea and enter a period of heavy feeding and rapid growth.

Sockeye salmon, as well as other salmon species, follow well-defined migratory routes, usually 10 to 20 m deep, during the oceanic phase of their migrations. These migrations closely follow the surface current patterns in the North Pacific Subarctic Current Gyre (**Figure 7.40**, top), but the sockeye move faster than the currents. After several years at sea and several complete circuits of the gyre, the sockeye approach sexual maturity, move toward the coast, and seek out freshwater streams. Strong evidence supports a home-stream hypothesis that each salmon returns to precisely the same stream and tributary in which it was spawned. It spawns once there and then dies.

Tunas also have extensive migrations. Skipjack tuna are widely distributed in the warm latitudes of the world ocean. Several genetically distinct populations probably exist, but we examine only the eastern Pacific population. These tuna spawn during the summer in surface equatorial waters west of 130° W longitude (Figure 7.40, bottom). For several months, the young tunas remain in the central Pacific spawning grounds. After reaching lengths of approximately 30 cm, they either actively

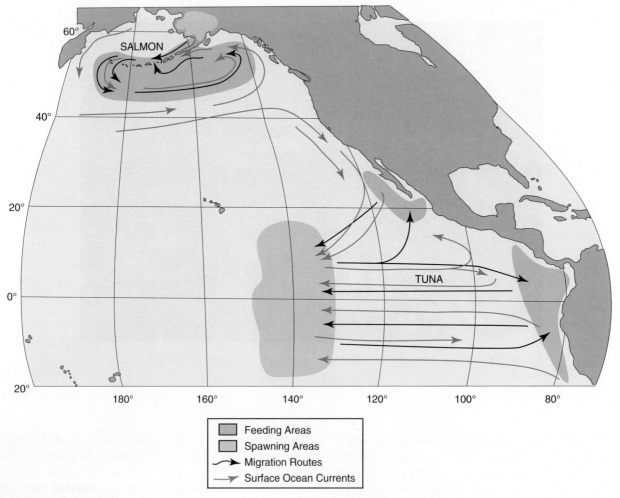

Figure 7.40 The general oceanic migratory patterns of the Bristol Bay sockeye salmon (above) and the east Pacific skipjack tuna (below). Note the apparent relationship between these migratory patterns and surface ocean currents.

Data from W. Royce et al., *Fishery Bulletin* 66(1968):441–462.

migrate or are passively carried to the east in the Pacific Equatorial Countercurrent. These adolescent fish remain in the eastern Pacific for about 1 year as they mature. Two feeding grounds, one off Mexico and another off Central America and Ecuador, are the major centers of skipjack concentrations in the eastern Pacific.

As the skipjack approach sexual maturity, they leave the Mexican and Central and South American feeding grounds and follow the west-flowing equatorial currents back to the spawning area. After spawning, the adults retrace the Equatorial Countercurrent they followed as adolescents; however, the feeding adults are seldom found as far to the east. Subsequent returns to the spawning area follow the general pattern established by the first spawning migration.

The Atlantic eel, *Anguilla,* exhibits a migratory pattern just the reverse of the Pacific salmon. This eel also migrates between freshwater and saltwater, but, in complete contrast to salmon, Atlantic eels are **catadromous**. They hatch at sea and then migrate into lakes and streams, where they grow to maturity. Two species of Atlantic eels exist: the European eel and the American eel. The distinction between the species is based on their geographic distribution and anatomical and genetic differences of the adults (**Figure 7.41a**). Both species spawn deep beneath the Sargasso Sea of the central North Atlantic. Their eggs hatch in the spring to produce a leaf-shaped, transparent **leptocephalus larva** that is about 5 cm long (**Figure 7.41b**). The leptocephalus larvae, drifting near the surface, float out of the Sargasso Sea and move to the north and east in the Gulf Stream. After 1 year of drifting, American eel larvae metamorphose into **elvers** (young eels) that move into rivers along the eastern coast of North America. The European eel larvae continue to drift for another year across the North Atlantic to the European coast (**Figure 7.42**). There, most enter rivers and move upstream. The remainder of the European population requires a third year

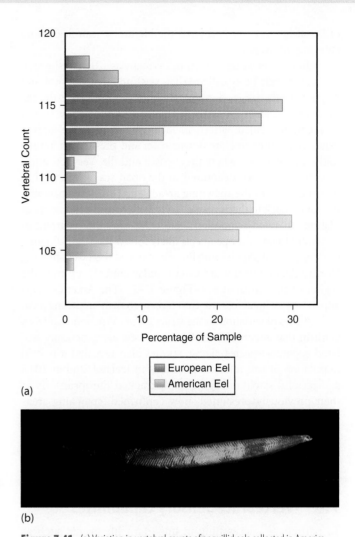

(a)

(b)

Figure 7.41 (a) Variation in vertebral counts of anguillid eels collected in America and Europe. (b) A leptocephalus larva of the eel *Anguilla*.

(a) Redrawn from Cushing, 1968. Fisheries Biology. University of Wisconsin Press, Madison. 200 pp. (b) Courtesy of E. Widder/NOAA.

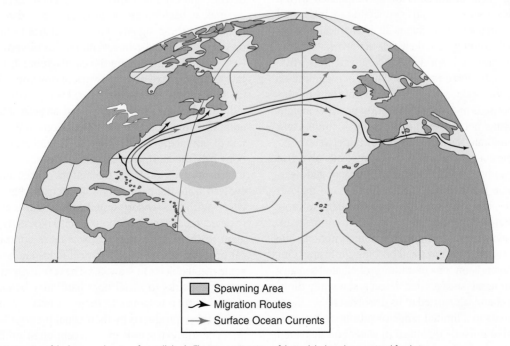

Figure 7.42 Migratory routes of the larvae and young of anguillid eels. The return migrations of their adults have been omitted for clarity.

of larval life to enter and cross the Mediterranean Sea before entering freshwater.

After several years (sometimes as many as 10) in freshwater, the mature eels (now called *yellow eels*) undergo physical and physiological changes in preparation for their return to the sea as silver eels. Their eyes enlarge, and they assume a silvery and dark countershaded pattern characteristic of midwater marine fishes. Then they migrate downstream and presumably return to the Sargasso Sea, where they spawn and die. Very few adult silver eels have been captured in the open sea, and none has been taken from the spawning area itself. Thus, the spawning migration back to the Sargasso Sea is still a matter of some speculation. To avoid swimming against the substantial current of the Gulf Stream, European eels likely follow the south-flowing Canary Current after leaving European rivers, then west-flowing North Atlantic Equatorial Current, and eventually return to the region of the Sargasso Sea (Figure 7.42). The American eels apparently swim across the Gulf Stream to their spawning area.

Studies of variation in the structure of mitochondrial DNA confirm that American and European eels are genetically isolated separate species. These studies also revealed a hybrid population of eels inhabiting streams in Iceland, and in 2013 a successful satellite tagging study tracked European eels to their previously presumed, now confirmed, spawning areas of the Sargasso Sea. Another tagging study conducted by the same group of scientists also led to the discovery that this eel is preyed upon by marine mammals at depth.

7.8   Vertebrate Sensory Capabilities

To participate successfully in important life events such as migration, feeding, mating, and simply surviving, all animals must be able to evaluate their immediate surroundings more or less continuously. These evaluations are accomplished with a variety of specialized sensory organs connecting their internal nervous systems with the various chemical, mechanical, or electromagnetic stimuli coming from their external world. Sensory organs contain receptor cells specialized to convert these environmental stimuli into nerve impulses. These nerve impulses are conducted to the brain, where perception of the stimulus occurs and a response is initiated. This section provides insight into the mechanisms behind the most obvious or important adaptations of vertebrate sensory abilities for underwater use.

Most of us have a general understanding of our own five basic sensory capabilities: taste, smell, touch, vision, and hearing; however, some fish exhibit electroreceptive and magnetoreceptive abilities that have no known counterparts in most terrestrial vertebrates, including humans. When animals are submerged in seawater, even comfortable human notions like the differences between taste and smell become confusing. Our ability to smell depends on tens of millions of ciliated sensory cells located in our nasal passages that detect and identify thousands of different chemicals carried to us dissolved in air. Taste, in contrast, responds to a limited range of substances (sugars, acids, and salts) that must be dissolved in water and delivered to a few hundred taste buds on the tongue, mouth, and lips.

These convenient distinctions between taste and smell become less clear when applied to fish that live constantly underwater and never breathe air. Thus, fish use a combination of several unique versions of the major senses to provide similar information.

Chemoreception

Both taste and smell (**olfaction**) are chemoreceptive senses. For marine fish swimming in an aquatic medium of near-uniform salinity, with a buffered pH, and a nearly complete absence of sugar, an ability comparable with our sense of taste has little use. Olfaction, however, the detection with olfactory sensory cells of chemicals dissolved in water, is highly evolved in fish. Salmon and some species of large predatory sharks respond to very low concentrations of odor molecules. Just a few parts per billion of chemicals in the water of their olfactory sacs are sufficient for recognition. With such olfactory capabilities, it is possible for predatory sharks to locate odor sources using very dilute chemical trails left by injured prey or for a migrating salmon to locate and identify the stream of its birth.

For air-breathing marine tetrapods, the olfactory situation is reversed. Regardless of how acute their olfactory sense is in air, when a reptile, bird, or mammal submerges it closes its nostrils and leaves its olfactory sense at the sea surface. Cetaceans, because they spend most of their lives below the sea surface, have completely lost all traces of their ancestral nasal structures used for olfaction.

Electroreception and Magnetoreception

Humans are totally oblivious to the weak electrical and electromagnetic energy fields generated by contractions of muscles in swimming animals, by water currents moving past inanimate objects, and even by Earth's own magnetic field. Yet organisms as small as bacteria and as large as sharks detect and respond to some of these signals. These specialized senses are known or suspected to exist in several classes of vertebrates, but the best-studied examples are the cartilaginous fishes. Sharks and rays exhibit an extensive network of tiny pores or pits arranged on their snouts and pectoral fins. Each pit connects via a short jelly-filled canal to a flask-shaped **ampulla of Lorenzini**. These ampullae are associated with extensions of the lateral line system of cartilaginous fishes (**Figure 7.43**) and of at least one marine bony fish, the marine catfish. Electroreception is accomplished by sensory cells located at the bottom of each ampulla, possibly evolved from the basic lateral line sensory hair cell (described later). With this sensory system, some sharks and rays are able to detect (at distances of about 18 cm) bioelectrical fields equivalent to those generated by the muscle contractions of typical prey species. Similar electrical fields are also produced by some metal objects in seawater. The seemingly erratic responses by some sharks to metal boat parts may be explained as the sharks' normal response to these artificial electric fields mimicking those produced by their usual prey.

The study of geomagnetic reception in animals is still in its infancy. It has been confirmed in some birds and cartilaginous

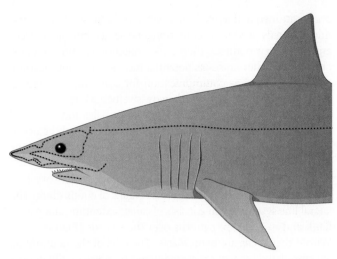

Figure 7.43 The major branches of the left lateral line system (blue) of a shark. Pores scattered in clusters over the snout are the openings to the ampullae of Lorenzini.

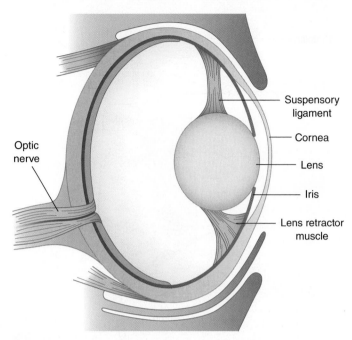

Figure 7.44 Cross section of a fish eye. Note the solid round lens that is focused by being moved toward or away from the retina by the retractor muscle.

fishes and is suspected and deemed plausible in some bony fishes such as tuna and salmon, sea turtles, and possibly some whales. The ampullae of Lorenzini are thought to be the organs of detection in sharks and rays; in other vertebrate groups, the possible organs of detection have not yet been clearly identified.

Vision

Most animals rely on ambient light from the sun, moon, or stars to illuminate what they see and also to provide the energy needed to stimulate their photoreceptor cells. A few animals, especially mid- and deep-water fishes, have light-producing photophores to illuminate their own very small visual fields. Although sunlight travels several kilometers through our atmosphere with little loss in intensity, an additional few hundred meters through the clearest ocean water so reduces the intensity that photosynthesis is impossible and vision is very limited. As light intensity is reduced, visual fields shrink to a few meters, and the range of colors available narrows to the green and blue portions of the visible light spectrum.

All marine vertebrates except hagfish obtain visual images of their surroundings with a remarkable organ, the camera eye (**Figure 7.44**). A similar type of eye has evolved independently in octopuses and other cephalopod mollusks. In vertebrate eyes, light is focused by a round lens through a light-tight and nearly spherical eye cavity to the light-sensitive receptor cells of the **retina** at the back of the eye. In contrast to eyes adapted for vision in air, fish and cetacean eyes must accommodate the higher refractive power of water. To do this, the eyes of fish and cetaceans are strongly flattened in front, with a round lens that focuses by moving nearer to or away from the retina rather than by changing shape as ours do.

The retina contains the light-sensitive rod and cone cells specialized for light detection. Typically, cones serve as high-intensity and color receptors, and rods serve as low-intensity receptors. Fish living below the photic zone usually have fewer cones than rods, and deep-sea fishes often lack cones entirely.

For all fish with only one type of cone (or no cone cells at all), vision is limited to detecting variations in light intensity; they see their world in varying shades, not varying colors. Many fish species nearer the surface and in better-illuminated marine environments such as coral reefs are capable of varying degrees of color vision. The retinas of these fish contain either two or three different types of cone cells, each type sensitive to a different range of wavelengths. In bright light and clear water, fish with three types of cones apparently possess a visual color acuity that may exceed that of humans.

Equilibrium

A crucial aspect of orienting oneself in space is knowing which way is up or down. For human divers away from surface light or diving at night, the absence of a visual horizon or other clear notions of up and down can be very disorienting. It is also needed by actively moving animals to monitor their swimming speeds and changes in direction. With this information, body positions and orientations can be updated continuously. To do this, almost all vertebrates have two types of receptors within their organs of equilibrium: one responds to the pull of gravity, and the other detects acceleration forces.

Vertebrates use a basic **sensory hair cell** design (**Figure 7.45**) to serve a variety of mechanoreceptive purposes, including equilibrium. Their organ of equilibrium, the **labyrinth organ**, is located on each side of the head (**Figure 7.46**). Each labyrinth organ consists of three semicircular canals and three smaller sac-shaped chambers. These sac-shaped chambers are the gravity detectors, with small stony secretions suspended by sensory hair cells known as **neuromasts**. The canals are the acceleration detectors. Each canal is filled with fluid and has an enlarged ampulla that

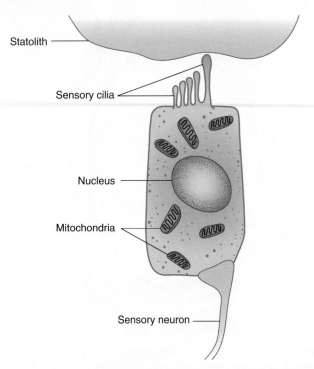

Statolith

Sensory cilia

Nucleus

Mitochondria

Sensory neuron

Figure 7.45 General structure of a mechanosensory hair cell. When the sensory cilia of the hair cell are bent, a nerve impulse is initiated and passed to the associated sensory neuron.

is lined with sensory hair cells supporting a cupula. Acceleration in any direction moves the fluid against the cupula and stimulates the neuromasts of at least one canal.

Sound Reception

Sound is transmitted through air or water as spreading patterns of vibrational energy. This energy travels at 1,500 m/s in water, about five times faster than in air. Although the sea may at times seem surprisingly silent to human divers, animals with sensory organs attuned to that medium must find it a noisy place. Swimming animals produce unintentional noises as they move, feed, and bend their bodies. Others make intentional grunts, groans, chirps, and other noises as potential means of communication. Sounds are used as communication for courtship, spawning, defending territory, or general aggressive behaviors. Add those to surface wave noises, vocalizations from whales and seals, and the mechanical noises we introduce into the sea, and the ocean becomes a cacophony of sounds within, below, and well above the range of human hearing.

Fish detect sounds with mechanoreceptive sensory hair cells nearly identical to those in their organs of equilibrium. The lateral line system of fish consists of canals extending along each flank and in complex patterns over their heads (Figure 7.43). Within the canals are neuromasts. The cilia of the neuromasts are stimulated by water movement and by pressure differences at the fish's body surface. These are communicated to the lateral line canals through pores in the skin surface. In this way, lateral line systems function to detect disturbances caused by prey or by predators, by swimming movements of nearby schooling companions, and by sound vibrations.

Another structure associated with sound detection in bony fishes is the otolith (Figure 7.46). Otoliths are small calcareous stones embedded in and associated with part of the labyrinth organ. Together, they constitute the inner ear. On each side of the head, two or three otoliths are suspended in fluid-filled sacs where they contact neuromasts. Arriving sound waves move the fish very slightly, and the denser otoliths lag behind, bending the neuromast cilia and stimulating a nerve impulse. Hair cells with different orientations relative to the otoliths may provide some directional information about the sound source.

Tetrapods have evolved very different hearing systems for detecting sounds transmitted in air. Airborne sound vibrations are transmitted from the eardrum to the cochlea of the inner ear. The cochlea is a coiled tubular structure filled with fluid and lined with yet another type of sensory hair cells. In toothed whales, the sound-processing structures of the middle ear are

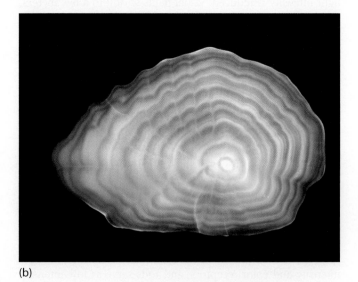

(a)   (b)

Figure 7.46 (a) Removal of an otolith from a fish, and (b) a magnified image of a sagittal otolith.

(a) NOAA/NEFSC; (b) NOAA.

The small ear bones in fish, called *otoliths*, can be used to determine the age of a fish. As a fish grows it lays down new material onto the otolith. Many fish larvae lay down one ring each day, and thus can be aged to the day as young. As fish age, they often lay down one ring each year, so adult fish can be aged to the year. To determine the age of a fish, its otoliths are sanded to reveal the rings, which are then counted. Using growth rings to age an organism may sound familiar, because a similar process is used to estimate the ages of many trees.

enclosed in a bony case, the **tympanic bulla**. The bulla is supported only by a few wisps of connective tissue to isolate it from adjacent bones of the skull by air sinuses filled with an insulating emulsion of mucus, oil, and air.

The **external auditory canal** is the usual mammalian sound channel connecting the external and middle ears; however, the auditory canal of mysticetes is completely blocked by a plug of earwax; in toothed whales, the canal is further reduced to a tiny pore or is completely covered by skin. Mapping of acoustically sensitive areas of dolphins' heads has shown the external auditory canal to be about six times less sensitive to sound than the lower jaw. The unique sound reception system in toothed whales begins with the bones of the lower jaw, which flare toward the rear and are extremely thin. Within each half of the lower jaw is a fat body (or, in some cases, liquid oil) that directly connects with the wall of the bulla of the middle ear. These fat bodies act as a sound channel to transfer sounds from the flared portions of the lower jaw directly to the middle ear. An area on either side of the forehead is nearly as sensitive as the lower jaw, providing four very sensitive separate channels for sound reception.

STUDY GUIDE

TOPICS FOR DISCUSSION AND REVIEW

1. List and describe the major evolutionary structural advances exhibited in bony fishes, cartilaginous fishes, and agnathans. Focus on the increasing complexities found in these groups.

2. Compare the osmoregulatory adaptations of marine teleosts, sharks, reptiles, and mammals. Which of these groups includes an osmoregulatory adaptation that affects the developing offspring?

3. Summarize the great variety of developmental methods observed in living cartilaginous fishes, including viviparity, oviparity, and ovoviviparity.

4. Describe the adaptive significance of salt glands and uric acid secretion for marine reptiles and birds.

5. Why is olfaction more important to marine fish as opposed to marine tetrapods?

6. Describe how sharks are able to detect weak electric fields in the sea. What do they use this adaptation for?

7. Summarize the differences between terrestrial and marine snakes. Suggest additional adaptations that would benefit sea snakes in their ability to live in the sea.

8. Generate a list of potential cues that sea turtles could use while navigating over great distances to return to their preferred nesting beaches.

KEY TERMS

amniotic egg   188

ampulla of Lorenzini   200

anadromous   171

androgen   178

aspect ratio   190

autosome   178

avascular   172

catadromous   199

caudal peduncle   194

chloride cells   188

closed circulatory
  system   168

cost of transport (COT)   192

countershading   193

crossover speed   193

ectotherm   180

elver   199

estrogen   178

extant   168

external auditory
  canal   203

fineness ratio   193

finlet   194

frictional drag   193

gigantothermy   184

gill arch   185

heterocercal   189

homeostatic   186

homeothermic   180

homocercal   189

Hox genes   168

hyperosmotic   186

hypoosmotic   175

induced drag   193

isosmotic   185

labyrinth organ   201

lamella   185

leptocephalus larva   199

myomere   168

neuromast   201

olfaction   200

osmoregulation   169

osmosis   186

osmotic pressure   186

oviparity   174

ovoviviparity   174

pressure drag   193

retina   201

salt gland   188

school   195

sensory hair cell   201

sequential
  hermaphrodite   178

sex chromosome   178

simultaneous
  hermaphrodite   178

siphon   192

smolt   198

swim bladder   175

trophonemata   175

tympanic bulla   203

ultrasonic   197

uric acid   188

ventrolateral   170

vertebrae   168

viviparity   174

wave drag   193

KEY *GENERA*

Amblyrhynchus

Anguilla

Caesio

Carcharias

Carcharodon

Caretta

Cephaloscyllium

Chelonia

Crocodylus

Dermochelys

Fejervarya

Hippocampus

Lamna

Lampris

Laticauda

Mola

Oncorhynchus

Pelamis

Petromyzon

Pristis

Regalecus

Rhina

Sarda

Scomber

Seriola

Serranus

Squalus

Squatina

Stenella

Thalassia

Thalassoma

Thunnus

Tursiops

Ulva

REFERENCES

Alderton, D. 1988. *Turtles and Tortoises of the World*. New York: Facts on File.

Alexander, R. McNeill. 1970. *Functional Design in Fishes*. London: Hutchinson.

Alexander, R. McNeill. 1988. *Elastic Mechanisms in Animal Movement*. New York: Cambridge University Press.

Blake, R. W. 1983. *Fish Locomotion*. New York: Cambridge University Press.

Bracis, C., and J. J. Anderson. 2012. An investigation of the geomagnetic imprinting hypothesis for salmon. *Fisheries Oceanography* 21(2–3):170–181.

Carey, F. G. 1973. Fishes with warm bodies. *Scientific American* February:36–44.

Carr, A. 1965. The navigation of the green turtle. *Scientific American* May:79–86.

Compagno, L., M. Dando, and S. Fowler. 2005. *Sharks of the World*. Princeton, NJ: Princeton University Press.

Demski, L. S. 1987. Diversity in reproductive patterns and behavior in teleost fishes. In: D. Crew, ed. *Psychobiology of Reproductive Behavior. An Evolutionary Perspective*. Englewood Cliffs, NJ: Prentice Hall. pp. 1–27.

Donoghue, P. C. J., and M. A. Purnell. 2005. Genome duplication, extinction and vertebrate evolution. *Trends in Ecology and Evolution* 20(6):312–319.

Erdmann, M. V., and R. L. Caldwell. 2000. How new technology put a coelacanth among the heirs of Piltdown Man. *Nature* 406:343.

Fish, F. E. 1999. Energetics of swimming and flying in formation. *Comments on Theoretical Biology* 5:283–304.

Gilmore, R. G., Jr. 2003. Sound production and communication in the spotted seatrout. In: S. Bortone, ed. *Biology of the Spotted Seatrout*. Boca Raton, FL: CRC Press. pp. 177–195.

Griffith, R. W. 1994. The life of the first vertebrates. *BioScience* 44:408–416.

Hasler, A. D., A. T. Sholz, and R. M. Horrall. 1978. Olfactory imprinting and homing in salmon. *American Scientist* 66:347–354.

Hawryshyn, C. W. 1992. Polarization vision in fish. *American Scientist* 80:164–175.

Helfman, G. S., B. B. Collette, D. E. Facey, and B. W. Bowen. 2009. *The Diversity of Fishes*, 2nd ed. London: Wiley-Blackwell.

Kalmijin, A. J. 1977. The electric and magnetic sense of sharks, skates, and rays. *Oceanus* 20:45–52.

Lohmann, K. J., N. F. Putman, and C. M. F. Lohmann. 2008. Geomagnetic imprinting: A unifying hypothesis of long-distance natal homing in salmon and sea turtles. *Proceedings of the National Academy of Sciences of the U.S.A.* 105:19096–19101.

Marsh-Hunkin, K. E., H. M. Heinz, M. B. Hawkins, and J. Goodwin. 2013. Estrogenic control of behavioral sex change in the bluehead wrasse, *Thalassoma bifasciatum*. *Integrative and Comparative Biology* 53(6):951–959.

Martin, A. P., G. J. P. Naylor, and S. R. Palumbi. 1992. Rates of mitochondrial DNA evolution in sharks are slow compared with mammals. *Nature* 357:153.

Martini, F. H. 1998. Secrets of the slime hagfish. *Scientific American* 228:70–75.

Motta, P., M. Maslanka, R. Hueter, R. Davis, R. de la Parra, S. Mulvany, M. Habegger, J. Strother, K. Mara, & J. Gardiner. 2010. Feeding anatomy, filter-feeding rate, and diet of whale sharks *Rhincodon typus* during surface ram filter feeding off the Yucatan Peninsula, Mexico. *Zoology* 113(4):199–212. doi:10.1016/j.zool.2009.12.001

Nicol, J. A. C. 1989. *The Eyes of Fishes*. New York: Oxford University Press.

Noble, R. W., L. D. Kwiatkowski, A. De Young, B. J. Davis, R. L. Haedrich, L. T. Tam, and F. A. Riggs. 1986. Functional properties of hemoglobins from deep-sea fish: Correlations with depth distribution and presence of a swim bladder. *Biochimica et Biophysica Acta* 870:552–563.

Popper, A. N., and S. Coombs. 1980. Auditory mechanisms in teleost fishes. *American Scientist* 68:429–440.

Rasmussen, A. R., J. Elmberg, P. Gravlund, and I. Ineich. 2011. Sea snakes (Serpentes: subfamilies Hydrophiinae and Laticaudinae) in Vietnam: A comprehensive checklist and an updated identification key. *Zootaxa* 2894:1–20.

Triantafyllou, M. S., and G. S. Triantafyllou. 1995. An efficient swimming machine. *Scientific American* 272:64–70.

Wahlberg, M., H. Westerberg, K. Aarestrup, E. Feunteun, P. Gargan, and D. Righton. 2014. Evidence of marine mammal predation of the European eel (*Anguilla anguilla* L.) on its marine migration. *Deep Sea Research Part I: Oceanographic Research Papers* 86:32. doi:10.1016/j.dsr.2014.01.003

Warner, R., and S. Swearer. 1991. Social control of sex change in the bluehead wrasse, *Thalassoma bifasciatum* (Pisces: Labridae). *Biological Bulletin* 181(2):199–204.

Webb, P. W. 1984. Form and function in fish swimming. *Scientific American* 251:72–82.

Webb, P. W. 1988. Simple physical principles and vertebrate aquatic locomotion. *American Zoologist* 28:709–725.

Wrootton, R. J. 1990. *Ecology of Teleost Fishes*. New York: Chapman and Hall.

CHAPTER 8

Marine Vertebrates II: Seabirds and Marine Mammals

STUDENT LEARNING OUTCOMES

1. Analyze the thermoregulatory mechanisms of seabirds and marine mammals, and compare them to those used by other marine organisms, such as fish.

2. Describe the diversity of seabird body forms, behaviors, and habitats.

3. Develop an understanding of the major groups of marine mammals, including the most useful distinguishing factors for each group and their relationships to one another and their environments.

4. Describe the characteristics of representative seabird and marine mammal species.

5. Explain the basic physiological adaptations many marine animals possess for diving below the surface, ranging from just below the surface to thousands of meters down.

CHAPTER OUTLINE

A humpback whale breaching in front of researchers in the Stellwagen Bank National Marine Sanctuary. Whales are photographed for identification and cataloging.
NOAA.

Seabirds and marine mammals are the groups of marine organisms that are most familiar to humans. They are considered "cute and fuzzy" by many, and they are constantly warm blooded, or *homeothermic*, as are humans. They are also generally more visible than fish and other marine organisms, spending time at the sea surface or even on land. Humans tend to feel connections with these organisms and take a special interest in their well-being, as evidenced by the dozens of organizations whose main goals are to protect these life forms. On the opposite end of the spectrum, historically marine mammals were the targets of hunting, and in some areas of the world they are still hunted today, driving many populations to dangerously low numbers. Regardless of the nature of their interactions with humans, seabirds and marine mammals play instrumental ecological roles in the many areas of the ocean they occupy. Some marine mammals are top predators, greatly influencing population sizes of their prey species. The largest animal to ever live, even larger than the largest dinosaurs, is a marine mammal, the blue whale. Penguins, marine birds that many would characterize as living in extreme conditions, have adaptations for survival in freezing cold temperatures with long periods of no food or shelter of any kind. These organisms, whose relatives and ancestors are terrestrial, have moved into a saline, fluid environment, and along with this lifestyle have evolved unique adaptations for survival and success. We will examine the major groups of marine birds and mammals and explore the challenges associated with maintaining a constant body temperature, finding food, maneuvering in the environment, diving below the surface, and simply surviving in the sea.

8.1 Thermoregulation

Muscle contraction is an **exothermic**, or heat-producing, phenomenon. This is immediately obvious when we become hot and sweaty during exercise, and when we use a blanket to stay warm while motionlessly watching television. This internal source of metabolic heat is just one aspect of an animal's overall heat budget.

All organisms exchange heat with their ambient environment via four different mechanisms. **Evaporation** results in a loss of body heat as body water changes its physical state from liquid to gas and leaves respiratory surfaces and skin. Most marine organisms do not lose heat this way because they are never exposed to air, but evaporation does play a significant role in the heat budget of marine tetrapods, such as seabirds and marine mammals. A second mechanism of heat transfer between an organism and its environment is via radiation. Like the sun, all objects emit **electromagnetic radiation**, usually at long infrared wavelengths, and thus an animal may exchange heat via radiation with an object in its environment even though it is not in contact with the object. Once again, heat transfer via radiation

is not significant in the sea because water blocks infrared wavelengths, but it does enable marine tetrapods to heat up while basking in the sun.

Conduction and **convection** are the dominant forms of heat transfer in the sea. Conduction is the transfer of heat through a macroscopically motionless object, such as when the metal handle of a pot gets hot because the base of the pot is touching the coil of an electric stove. Convection requires macroscopic motion of a substance to carry heat from place to place, such as when the wind chill makes you feel colder than ambient temperature alone. The primary difference between conduction and convection is that convection transfers heat energy much more quickly than conduction (for a given temperature). Thus, marine organisms lose most of their metabolic heat via convection, as currents of water flow past them, carrying away their metabolic heat. Heat transfer via conduction and convection is also greatly influenced by density, and thus contact with more dense objects and fluids will result in more rapid heat transfer. This explains two common experiences for humans: air at 65°F (18°C) feels cool, whereas water at 65°F feels very cold, and the air within a 450°F (232°C) oven feels very warm to your skin whereas the 450°F oven rack will cause a severe burn if touched. In both cases, the denser substance (water vs. air, and metal vs. air) was able to transfer heat into or out of your body much more quickly.

These two facts (i.e., that convection is more efficient than conduction and that both mechanisms are directly proportional to density) explain why humans use a variety of strategies to trap air when trying to stay warm, such as via the use of insulation in the walls of our homes, a wool sweater or down comforter, and the numerous tiny air bubbles present in a neoprene wetsuit. In all cases, trapping low-density air in a motionless chamber greatly decreases heat transfer via conduction and convection. Seabirds and marine mammals use an identical strategy by trapping motionless air within their dense feathers or **pelage**.

Fish are much more vulnerable to convective heat loss than other marine vertebrates because they use gills to exchange respiratory gases with high-density water rather than using lungs to exchange gases with low-density air. Thus, most of their metabolic heat is lost to high-density water as the colder water flows past the enormous surface area of their blood-warmed gills. Only a tiny fraction of fish species (the lamnid sharks, tunas, swordfish, and opah) have developed a way to retain metabolic heat by removing the heat energy from their warm venous blood before it returns to the gills. Nevertheless, even most of these specialized fish are not homeothermic; they are simply warmer than the water in which they swim. The opah is the one fish exception; this species has specialized gills with vessels running in such a manner that the blood is not cooled by water flowing over them.

Birds and mammals are the only true homeotherms in the animal kingdom, with the exception of the opah (an odd shaped pelagic fish), whose abilities to remain warm were discovered in 2015. Although some large fish and reptiles such as tunas and white sharks and the leatherback sea turtle can maintain body temperatures somewhat higher than the waters in which they swim, all but the opah lack the integrated physiological adaptations of true homeotherms. While at sea, birds and mammals live in direct contact with seawater much colder than their body temperatures. Many live in high-latitude, food-rich waters where water temperatures always hover near the freezing point, but even in more temperate latitudes the high heat capacity of water (about 25 times higher than air of the same temperature) results in a major heat sink and greatly affects the heat budgets of these homeotherms.

Seabirds and marine mammals exhibit several adaptations that reduce their body heat losses to tolerable levels. The streamlined bodies of seabirds and marine mammals enable them to move through air and water with minimal resistance and also to reduce the amount of body surface in contact with seawater and the amount of heat transferred to the water. The major muscles of propulsion, which generate considerable heat, are located within the animal's trunk rather than on the exposed parts of the flippers, wings, or feet. For many species, a surface layer of dense feathers, fur, or **blubber** also insulates and streamlines the body. Blubber is a sheath of fat and connective tissue of varying thickness that shields the deeper body musculature from the cold of the surrounding water. The insulation value of blubber is a function of both its thickness and its lipid content, and thus it is not effective as an insulator for flying birds, but works well for large swimming animals where a thick blubber layer does not seriously distort an animal's body shape or proportions. Body heat losses are further limited by restricting the flow of warm blood from the core of the body to the cooler skin. Vascular countercurrent heat exchangers found in the feet of birds and flukes (**Figure 8.1**), fins, and even the tongues of cetaceans also conserve heat. Arteries penetrating these appendages are surrounded by several veins carrying blood in the opposite direction. Heat from the warm blood of the central artery is absorbed by the cooler blood in the surrounding veins and carried back to the warm core of the body before much of it can be lost through the skin of the appendages. Collectively, these structures enable the relatively large-bodied birds and mammals to survive and thrive in the cold, allowing them to take advantage of the very productive parts of marine ecosystems.

DID YOU KNOW?

There are no adult marine mammals any lighter than the weight of a medium-sized dog (about 16 kg or 35 lbs.). Marine mammals are constantly fighting to maintain their body temperature as they lose heat to the cool seawater. The smallest marine mammals, sea otters, spend the majority of their time resting to conserve energy or feeding to maintain energy stores. They lose heat very quickly across their relatively large exposed surface area. Larger marine mammals have smaller surface areas to lose heat across when you consider their overall volume. A hypothetical rat-sized marine mammal would need to eat constantly, and even then would not be able to eat enough to maintain its required body temperature. This is likely why no rat-sized marine mammals exist.

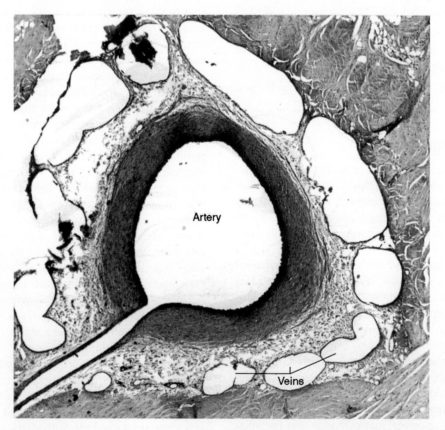

Figure 8.1 A cross section of a small artery from the tail fluke of a bottlenose dolphin. The muscular artery in the center is surrounded by several thin-walled veins carrying blood in the opposite direction.
Ridgway, S. H., *Mammals of the Sea, Biology and Medicine*, 1972. Courtesy of Dr. Sam H. Ridgway.

8.2 Seabirds

Birds (class Aves) are tetrapods with feathers and front appendages adapted for flight. Several of the basic avian adaptations for flight, including streamlined and insulated bodies, are also useful for life in marine habitats. Birds exhibit greater diversity (**Table 8.1**) than either reptiles or mammals, and their role in marine communities is substantial. The term *marine* carries a variety of meanings for birds, and many birds inhabit both the land and sea. A few birds, such as penguins (**Figure 8.2a**), spend most of their lives at sea, going ashore only to breed and raise their young. At the other extreme are some ducks, geese, and coots that are common on inland ponds and lakes, and only sometimes move into coastal waters to feed. Herons, stilts, sandpipers, turnstones, and other shorebirds (**Figure 8.2b**) venture into shallow coastal waters only to feed on benthic animals. Others, including albatrosses, petrels, gannets, pelicans, gulls, terns, and murres, are more pelagic. These birds forage extensively at sea and often rest on the sea surface rather than returning to land to roost.

Birds from penguins to pelicans prey extensively on animals living in coastal waters. Their features, such as diving styles, patterns of pursuit, and even bill shapes, differ greatly (**Figure 8.3**), depending on their prey preferences. Seabirds also show a great deal of variation in bill shapes, again depending on their food preferences (**Figure 8.4**). The shape of a bird's bill provides insight into what type of food items it consumes and also what food it is not likely able to consume. For example, the curlew featured in Figure 8.4 very clearly uses its long beak to penetrate mud to capture small benthic organisms and very clearly is not capable of crushing and consuming large seeds.

Seabird Diversity

The orders of modern birds that depend upon the sea most heavily, and thus can truly be considered to be seabirds, are the penguins (Sphenisciformes); the albatrosses, petrels, shearwaters,

(a)

(b)

Figure 8.2 (a) Emperor penguins on the Ross Sea in Antarctica. (b) A group of mixed seabird species, including willets and marbled godwits.
(a) Courtesy of Michael Van Woert/ORA/NESDIS/NOAA; (b) U.S. Fish and Wildlife Service.

TABLE 8.1 Classes and Orders of Seabirds and Marine Mammals			
Class	**Order**	**Common Examples**	**Approximate Number of Marine Species**
Aves			**645**
	Sphenisciformes	Penguins	18
	Procellariiformes	Albatrosses, petrels, shearwaters, fulmars	128
	Pelecaniformes	Cormorants, boobies, pelicans, frigatebirds	59
	Charadriiformes	Plovers, sandpipers, gulls, terns, puffins	269
	Anseriformes	Ducks, geese	111
	Ciconiiformes	Herons, egrets, grebes	20
	Falconiformes	Ospreys, eagles	5
	Gruiformes	Rails, coots	10
Mammalia			**140**
	Carnivora	Seals, sea lions, walruses, sea otters, polar bears	45
	Cetacea	Whales, dolphins, porpoises	90
	Sirenia	Manatees, dugongs	4

Pursuit diving with wings (penguins)
Pursuit diving with feet (cormorants)
Surface plunging (boobies)
Dipping (gulls)
Skimming (skimmers)

Figure 8.3 Bill shapes and feeding pursuit patterns of birds that feed at sea.

and fulmars (Procellariiformes); the pelicans, boobies, gannets, and their allies (Pelecaniformes); and the gulls, terns, auks, puffins, plovers, sandpipers, and their allies (Charadriiformes).

The 18 species of penguins, the most "marine" of the seabirds, are widely distributed in the Southern Hemisphere, from

Godwit

Ringed Plover

Curlew

Redshank

Turnstone

Oystercatcher

Figure 8.4 Bill shapes of some common wading seabirds.

the Antarctic fast ice in the south to the rocky shores of the Galapagos Islands at the equator; no penguin species inhabits the Northern Hemisphere. Penguins are highly modified, both anatomically and physiologically, to endure life in the ocean. They have lost the ability to fly, but instead use their stout, narrow wings to propel their streamlined bodies through water rapidly enough to overtake and capture fish prey. On land, their short legs are clumsy for walking, and they commonly flop down on their bellies and "toboggan" on ice or snow while pushing with their wings and feet. Their feathers are short and stout, resembling the scales of fish more than the feathers of songbirds. Penguins are accomplished divers, capable of holding their breath for long periods of time while searching for prey with eyes that are highly modified to provide excellent underwater vision.

Like penguins, the albatrosses, petrels, shearwaters, and fulmars of order Procellariiformes also spend a great deal of time away from land; however, they do not spend as much time in the water. Instead, they glide above the water's surface for hours at a time (albatrosses) or even seem to walk along the water's surface, using stiffly outstretched wings to hover as they tiptoe along in search of food (petrels). Both albatrosses and petrels can be described as being "pluck" feeders that use their curved upper bills to snatch passing baitfish from the water while floating on (albatrosses) or tiptoeing along (petrels) the surface.

Fulmars are closely related to petrels and albatrosses (**Figure 8.5**). All three are pelagic predators, flying long distances, often hundreds of kilometers from nest sites, to forage on near-surface prey. Their ability to exploit long-distance food resources is enhanced by several adaptations. They are adept fliers with long wings to help conserve energy by gliding and slope soaring the backs of ocean waves. Core body temperatures 2°C to 3°C lower than those of most other birds also conserve energy. Subdermal fat and stomach oil reserves enable them to

Figure 8.5   A northern fulmar, *Fulmarus glacialus*, in search of prey.
© David Woods/Shutterstock, Inc.

(a)

(b)

Figure 8.6   (a) Broad-billed prion gliding over the water in the Southern Ocean.
(b) Broad-billed prion beak showing the bill lamellae used for filtering krill from
the water.
(a) Lieutenant Elizabeth Crapo, NOAA Corps; (b) Courtesy of Janet Hinshaw, University of Michigan, Museum of Zoology.

endure long periods without food. Foraging at night when vertically migrating prey are nearer the sea surface is facilitated by
their well-developed olfactory sense. And finally, their eggs are
less sensitive to chilling when the adults are away.

Another Antarctic nesting bird that is restricted to the
Southern Hemisphere is the Antarctic prion, or whale-bird
(**Figure 8.6a**). It is the largest of the prions, with a wing span of
55 to 60 cm. Of the several very similar species of prions, this
is the only one that nests on the Antarctic continent as well as
on Antarctic islands to the north. In their breeding colonies,
Antarctic prions construct nests on exposed rocky cliff faces, in
cavities under boulders, or in short burrows on grassy slopes
and then lay a single egg. After 45 days of incubation, the hatchling is fed by both parents at night. Nighttime feeding of young
is also a characteristic of closely related shearwaters.

Prion chicks and adults depart the nests in March, about
50 days after hatching, and move north into subantarctic waters
for the winter months. At sea, they are very gregarious, con-
gregating in flocks containing thousands of individuals. Antarctic prions feed on krill and other crustaceans by running/
flying along the surface of the water with outstretched wings
and bill submerged in the water to scoop their food. Although
they occasionally make shallow dives to capture prey, their bills
are adapted to skim and strain water with comblike lamellae
on either side of the bill (**Figure 8.6b**). The lamellae function in
much the same manner as whale baleen does.

The pelicans, boobies, gannets, and their allies (order
Pelecaniformes) can best be described as plunge divers (see
Figure 8.3) that capture prey immediately after spearing into
the water (pelicans) or soon after a brief pursuit powered by
their rigid wings (boobies and gannets). Plunge diving can be
quite a spectacle to observe, as a bird quickly transitions from
gliding over the water to diving straight down, hitting the water
head on, and plunging below for a (hopefully) tasty fish meal.
Cormorants and frigatebirds seem to be the exception to the

rule for this group of seabirds. Cormorants differ from their
relatives in that they swim underwater for many minutes in
search or pursuit of prey, often ducking under the water after
resting on its surface. In addition, unlike other seabirds, the skin
of cormorants lacks oil glands that keep feathers dry. Therefore, after emerging from the sea they must stand in the sun
with their wings extended while their feathers dry off. Even
the flightless cormorant dries its wings after a swim (**Figure 8.7**).
The five species of frigatebirds are unique among the Pelecaniformes in that they are not able to take off from the water. They
either plunk baitfish, juvenile sea turtles, or crustaceans from
the surface like petrels and albatrosses or use their superior
speed and maneuverability to steal captured prey from other
seabirds while still on the wing. This **kleptoparasitic** behavior
results in several of their other common names, such as *pirate
birds* and *man of war birds*. Male frigatebirds also are well known
for their bright red gular pouch that they inflate to attract potential mates (**Figure 8.8**).

The final lineage of birds that commonly is considered
to contain true seabirds is the Charadriiformes (the gulls,

Figure 8.7 A flightless cormorant in the Galapagos Islands drying its wings.
Lieutenant Elizabeth Crapo, NOAA Corps.

Figure 8.9 A puffin with a prize catch.
© Randy Rimland/Shutterstock, Inc.

terns, auks, puffins, plovers, sandpipers, and their allies). Species within this large order vary considerably in terms of their reliance on the sea for food. Many charadriiform birds that frequent the shoreline at low tide (e.g., gulls, terns, plovers, sandpipers) also feed inland during periods of high water, whereas auks, murres, and puffins are considered to be the Arctic's ecological equivalent to penguins, obtaining all of their food from the sea while diving under the surface for long periods of time. Seabirds commonly employ legs and bills of various lengths (Figure 8.4) to subdivide intertidal resources along soft-sediment shorelines.

Auks and puffins are north polar versions of penguins in the Southern Hemisphere. Puffins are found throughout the North Atlantic Ocean, as far south as Newfoundland, Canada, in the west, and down to the Canary Islands, Spain, in the east. Auks are generally distributed a bit further south than puffins. Both are poor fliers and never make long flights comparable with those of fulmars or petrels, although unlike penguins, they can actually fly. Similar to penguins, they are excellent

swimmers and divers. When below the sea surface, their short wings provide thrust in a flying motion similar to that of sea lions or penguins. Prey is captured with this sort of subsurface pursuit swimming rather than by plunge diving from the surface (**Figure 8.9**).

The Arctic tern resembles a small gull. In summer, it ranges north to within 8° of the North Pole, but it nests in coastal areas all around the Arctic basin. As the northern autumn approaches, Arctic terns begin their transequatorial migration to similar latitudes in the Antarctic. This, the longest known annual migration of any animal, maintains these terns in endless summer conditions. It also causes them to overlap the distribution of a Southern Hemisphere sister species, the Antarctic tern. This is an example of the bipolar geographic distribution of a species or of two closely related species. The northern fulmar also has a sister species, the southern fulmar, that occupies a similar habitat in the Southern Hemisphere.

Seabird Life History

More than half of all bird species are classified within the order Passeriformes, the familiar songbirds that have feet and are adapted for perching. These familiar birds, such as sparrows, finches, and blue jays, differ in many ways from seabirds. For example, seabirds are usually quite large, with emperor penguins standing nearly 1.2 m (4 ft.) tall; in contrast, most songbirds are tiny. Passerine songbirds often are quite colorful, and it is not uncommon for species to be sexually dimorphic such that males can be easily distinguished from females. Seabirds are generally more muted in coloration, with many species just one or two shades of gray, black, or simply white.

Songbirds also have a life-history strategy that differs markedly from that of seabirds (**Table 8.2**). For example, the passerine

Figure 8.8 Male frigate bird with his gular pouch inflated in an attempt to attract a mate.
James P. McVey, NOAA Sea Grant Program, Hawaii, Laysan Island, Pacific Ocean.

TABLE 8.2 Life-History Characteristics of Seabirds and Songbirds

Characteristic	Seabirds	Songbirds
Age at maturity	2–9 years	1–2 years
Clutch size	1–5	4–8
Incubation period	20–69 days	12–18 days
Time of fledging	30–280 days	20–35 days
Maximum life span	12–60 years	5–15 years

DID YOU KNOW?

The bird with the largest wingspan is a marine bird, the albatross. The wingspan of the largest albatross species can be as long as 3.4 m (11 ft.)! This bird has a wingspan that is twice the length of the average human's height, yet it weighs a mere 10 kg (22 lbs.) or less. Such large wings and a relatively low body weight provides these birds with the tools they need to glide for hours and to spend almost all of their time out at sea. They take advantage of oceanic winds to reduce the energy needed to fly, gliding as much as possible.

songbirds mature quickly, at an age of just 1 to 2 years, whereas seabirds can take up to 9 years to reach sexual maturity. Songbirds also lay more eggs (four to eight) than do seabirds (just one to five). Once laid, the eggs of seabirds can be incubated for more than 2 months, whereas passerines incubate their eggs for less than 3 weeks at the most. After hatching, songbirds fledge quickly (in 1 month or less), whereas some seabirds care for their young for 9 months. Finally, as compared to seabirds, passerines have a short life span of just 5 to 15 years, whereas some seabirds live as long as 60 years.

Although passerine songbirds are closely related to one another, "seabird" is an ecological label that reflects a trophic role in the marine realm and not a similar ancestry. Therefore, it is tempting to speculate on the selective forces that resulted in somewhat dissimilar lineages of birds converging on a very similar life-history strategy once they became dependent on the sea for most or all of their food. Simply stated, seabirds mature late, have small clutches, care for developing embryos and young for long periods of time, and live much longer than most songbirds. Some biologists suggest that this particular life-history strategy reflects energetic limitations that result from adapting to a diet of seafood. They argue that the inconsistent production in temperate regions and year-round low production of tropical areas prevents seabirds from obtaining sufficient energy to fuel rapid growth and early maturity, or the production of large clutches or eggs with large yolks, or rapid growth and fledging of young. These biologists contend that seabirds have long lives to compensate for the relatively slow growth and lower reproductive potentials imposed upon them by energy limitations of the marine environment.

Other biologists hypothesize that the life history of seabirds differs from that of songbirds because of reduced predation pressure around their nesting sites. Many seabirds nest on uninhabited offshore islands or on very steep coastal cliffs that prevent access to most predators of eggs and young. By selecting predator-free nesting sites, seabirds can "take their time" when creating the next generation and thus produce fewer, larger young that are cared for over many months prior to fledging. In short, the predation-free nesting locations of many seabirds may enable them to devote more time to raising larger, healthier offspring. Caring for young for longer time periods can lead to higher survival rates of the young. From an evolutionary perspective, this benefits adults by increasing the chances that their genetic material is passed on in future generations.

8.3 Marine Mammals

Of all the tetrapod classes, only the modern mammals (class Mammalia, subclass Eutheria) are characterized by viviparity, the internal nourishment and development of a fetus. Mammals also have body hair, milk-secreting mammary glands, specialized teeth, and an external opening for the reproductive tract separate from that of the digestive system. Thus, eutherian mammals lack the cloaca that is characteristic of primitive mammals and all nonmammalian vertebrates. Marine mammals are eutherians, so they possess all of the characteristics listed above.

The three orders of marine mammals listed in Table 8.1 have experienced varying degrees of adaptation in their evolutionary transition from life on land to life in the sea. Seals, sea lions, walruses, sea otters, and polar bears (order Carnivora) are quite agile in the sea, yet all except the otter must leave the ocean to give birth. Sirenians (dugong and manatees) and cetaceans (whales and dolphins) complete their entire life cycle at sea and never leave the water, with the exception of spectacular acrobatics performed by some species that propel them above the water for very brief periods.

At birth, the young of cetaceans are capable swimmers and instinctively surface to take their first breath. Polar bear and pinniped pups are unable to swim at birth, and thus their birth must occur on land or on ice floes. The newborn of marine mammals are large; blue whale calves weigh about 3 tons at birth. Still, these newborn mammals are smaller than their parents, and thus they have higher surface area-to-volume ratios, and their insulating layers of blubber or fur are not usually well developed. Several features compensate for the potentially serious problem of heat loss and body temperature maintenance in newborn marine mammals. Terrestrial pupping in pinnipeds provides some time in the sun for growth before the pups must face their first winter at sea. The larger cetaceans spend their summers feeding in cold polar and subpolar waters and then undertake long migrations to their calving grounds in tropical and subtropical seas (**Figure 8.10**). In these warm waters, their calves have an opportunity to gain considerable weight before migrating back to their frigid summer feeding grounds.

The growth rates of the young of some marine mammal species are truly astounding. Nursing blue whales grow from 3 tons at birth to 23 tons when weaned a scant 7 months later—an average weight gain of almost 100 kg per day! The average weight globally of a human being is 62 kg. Hooded seal pups

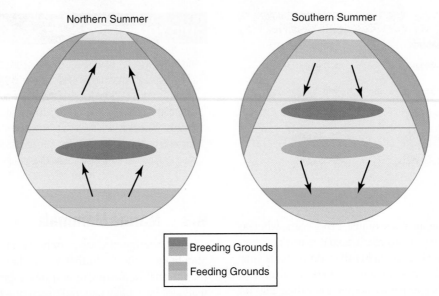

Figure 8.10 Generalized migratory patterns of large whales between summer feeding and winter breeding grounds. Northern and southern populations follow the same migratory pattern but do so 6 months out of phase with each other. Consequently, northern and southern populations of the same species remain isolated from each other, even though both populations approach equatorial latitudes.
Data from N. A. Mackintosh,. *Whales, Dolphins and Porpoises* (university of California Press, 1966).

gain more than 7 kg each day of their extremely short 4-day nursing period before weaning (**Figure 8.11**). These prodigious growth rates are supported by an abundant supply of high-fat milk. Cetacean milk is 25% to 50% fat (cow's milk ranges from 3% to 5% fat), and several species of seals produce milk that is 50% to 60% fat. The daily milk yield of a blue whale has been estimated at nearly 600 L. In pinnipeds, 2 to 5 L are more typical. In both cetaceans and pinnipeds, species occupying colder waters consistently produce milk with a higher fat content.

The energy demands made on females to produce a large offspring and then to supply it with large quantities of fatty milk until it is weaned (usually a few weeks to several months) are exceedingly high. Even the water that goes into the milk imposes additional osmotic costs that must be paid with further energy expenditures. Because of these high energetic demands,

marine mammals with long gestation periods (several months to more than a year) tend to reproduce only once per year and, with the exception of polar bears, generally produce only a single offspring each time they reproduce. In this section, several different species are compared to provide a sense of how marine mammals solve specific challenges of life in the sea.

Marine Otters

Marine otters have retained a strong resemblance to their fish-eating relatives of freshwater lakes and streams. Two species have been described, the marine otter or sea cat of South America and the sea otter of the north Pacific. The sea cat, *Lutra felina,* is very similar to its freshwater cousins, and its biology is very poorly known. This species is thought to give birth to two pups in December and January after a gestation period of 60 to 70 days. It inhabits the kelp beds of Chile and Peru and may venture into Argentine waters. Its distribution is becoming patchy, and population numbers have declined due to habitat destruction and illegal hunting practices, among other threats. Sea cats are timid and secretive loners that are rarely observed more than one or two at a time. They forage on crustaceans, mollusks, and fish in the sea and enter rivers to feed on *Macrobrachium*, a freshwater prawn. Sea cats are targeted and killed by fishers who blame them for damaging local fish, bivalve, and shrimp populations.

The sea otter, *Enhydra lutris*, of the Pacific rim (**Figure 8.12**) is the most derived of the otters and is much better known than the sea cat because of its range. Historically, sea otters occupied the entire Pacific rim, from Russia to the Baja Peninsula. Today, after more than a century of exploitation for their furs, the species is reduced to fewer than 15 populations that are scattered along its formerly continuous range. One small population now exists at San Nicolas Island in southern California as a result of

Figure 8.11 Hooded seal pups (right) are weaned only 4 days after birth, the shortest nursing period of any mammal.
© Steve Bly/Alamy Stock Photo.

Figure 8.12 A southern sea otter with her pup.
NOAA.

translocation efforts to expand southern sea otter distributions. Although reduced greatly, sea otter populations are showing steady increases and may have the potential to recover.

Sea otters have a dorsoventrally flattened tail that seems to provide additional propulsion during swimming. The tails of other otters, including the sea cat, are tapered at the ends. Sea otters are by far the largest otters, being known to reach nearly 1.5 m in length and 50 kg in mass. Although this larger size enables them to retain more heat than smaller otter species by providing them with a slightly smaller surface area relative to their body volume, sea otters still are much smaller than other marine mammals, and thus they lose a lot of heat to seawater. To combat this heat loss, sea otters have developed one of the most luxurious coats among all mammals, possessing upwards of 125,000 hairs per square centimeter of skin, which they preen incessantly throughout the day. This dense coat nearly spelled their doom during the 1800s, because the fur was highly prized by humans.

Even with their dense coat, sea otters still lose a great deal of metabolic heat to the cold seawater of the North Pacific. To combat this loss, sea otters have a very high basal metabolism that is fueled at an amazing rate: sea otters consume about 35% of their body weight each day. Sea otters prefer to eat benthic invertebrates they find along the shallow edges of their range. At some point in their evolutionary past, sea otters entered a tool-using stone age of their own. Using rocks carried to the surface with their food, they float on their backs and crack open the hard shells of sea urchins, crabs, abalones, and mussels to get at the soft tissue inside. In fact, unlike other otters, sea otters even swim on their backs, perhaps in an effort to keep their high surface–area limbs and feet out of the water.

Unlike sea cats, sea otters are very gregarious and often are seen floating on their backs with scores of other otters in aggregations called *rafts*. Rafts of otters off California often contain 50 or more individuals, whereas in Alaskan waters rafts of 2,000 otters have been seen. Sea otters are polygynous, with males defending large coastal territories that include the ranges of several adult females. Like sea cats, female sea otters give birth to one to two pups once per year.

Pinnipeds

Studies comparing molecular structures and DNA from seals, walruses, and sea lions indicate that all pinnipeds share a common evolutionary ancestry and are placed in the infraorder Pinnipedia. Pinnipeds evolved from terrestrial carnivores and have maintained their predatory habits in the sea. Only one species of walrus survives today, inhabiting shallow Arctic waters where it feeds on benthic mollusks. Seals and sea lions (**Figure 8.13**) are not as easily distinguished from each other as they are from walruses. In the open water, sea lions swim using a slow underwater "flying" motion of their front flippers (**Figure 8.14**). Seals propel themselves underwater with side-to-side movements of their rear flippers. Sea lions have visible external ear flaps, and seals have tiny holes for ears that are only visible upon close inspection. Another distinguishing feature is that, in general, sea lions are much noisier than seals, who use soft grunts for vocalizations.

The 14 species of sea lions, fur seals, and eared seals of the family Otariidae are found throughout the world ocean with the exception of the North Atlantic. Sea lions are thought to be the most primitive of modern pinnipeds in that they retain a

(a)

(b)

Figure 8.13 Two types of pinnipeds: (a) a harbor seal, *Phoca*, and (b) Steller sea lions, *Eumetopias*.
(a) © Steffen Foerster Photography/Shutterstock, Inc.; (b) © Tim Zurowski/Shutterstock, Inc.

Figure 8.14 California sea lion, *Zalophus*, swimming with typical "flying" motion of flippers.
© Eric Prine/age fotostock.

Figure 8.15 A walrus with prominent tusks.
USGS.

number of characteristics of terrestrial carnivores that are variably absent in the walrus and true seals. For example, they possess unfused hind limbs that enable them to walk on land on all fours, although in a slightly clumsy manner due to the size of their enormous flippers. They rely on fur instead of blubber for heat retention and possess several small bits of anatomy through which a large amount of heat can be lost, such as pinnae and scrotal testes.

Perhaps the best known of all sea lions is the star attraction of many zoological parks and public aquaria, the California sea lion, *Zalophus californianus*. Like all otariids, California sea lions are polygynous with strong sexual dimorphism. Males are much larger than females, reaching nearly 400 kg in mass (more than three times the weight of adult females), and possess both an enlarged sagittal crest on the top of their heads and a thickened mane of fur around their necks This latter feature gives rise to their common name, sea lion.

Because otariids routinely target the same prey species as humans, they often are considered a nuisance by commercial fishers. California sea lions prey on many valuable species of fish as well as on octopods and squid. Steller sea lions (*Eumetopias jubatus*) aggregate in areas of high prey abundance, such as offshore of river mouths during salmon runs. Some sea lions and fur seals even go so far as to interfere with fishing operations by feeding on fish that are trapped in trawls and drift nets.

Female sea lions give birth to one pup that is less than 1 m in length and less than 10 kg in weight. The pups are often born with a temporary coat, or **lanugo**, which is shed within a month of birth. Thereafter, they produce a juvenile coat, with juvenile coloration, that is lost during molting into their adult appearance.

The walruses of the Arctic Ocean seem to be an anatomical intermediate between more primitive sea lions and more derived seals (**Figure 8.15**). Like sea lions and fur seals, they have separate rear limbs and can walk on all fours, although their short tail is webbed to their thighs. Like more primitive otariids, they have four teats for nursing their pups and flippers that are mostly naked; sparse hairs are found on the tops of their flippers only. However, like the more derived true seals, walruses use their rear flippers for underwater propulsion, use blubber instead of fur for heat retention, sink tail first to dive, and lack pinnae. They use their large tusks for a variety of functions, including to haul themselves out of the water and to break holes in the ice, and males use them to protect territory or mates.

Walruses are polygynous breeders, with adult males typically growing to about twice the size of adult females. Walruses mate in water, giving females a selective advantage that does not occur on land—directly selecting the male with whom they will copulate. Pacific walruses have been described as having a lek or leklike mating system. During the breeding season, female walruses haul out on ice or rest in the water while one or more adult males station themselves in the water nearby to perform underwater visual displays while producing a series of amazingly bell-like or gonglike sounds to attract females. These ritual courtship displays are similar to the well-known lekking behaviors of some African antelopes. In all lekking species, including walruses, males only aggregate and display; females make the choice of their mating partners. The elaborate display behaviors of males allow females an opportunity to make the best genetic deal possible by getting males to advertise their own fitness before selecting a mate.

The gestation period for walruses is 15 months, and thus the annual reproductive schedules characteristic of other pinnipeds cannot be maintained. Walrus calves are nursed for about 1 year and are weaned gradually during their second year as they improve their diving abilities and foraging skills. Adult females, consequently, mate and produce a calf only every 3 years.

The 18 species of earless or true seals of the family Phocidae are scattered around the entire planet, with at least one species being found in every ocean and at every latitude, from the Arctic to Hawaii and the Antarctic to the Mediterranean Sea; two species of seals are even landlocked, one in the brackish Caspian Sea and one in the totally fresh Lake Baikal in southeast Russia. Phocids are the most derived pinnipeds, so much so that they have difficulty moving around on land. No seal can

Figure 8.16 Two bull elephant seals hauled out on a beach, vocalizing.
NOAA Fisheries.

walk; instead, they shimmy along on their fat bellies using their sternum as a fulcrum, looking more like enormous inchworms than mammals.

Tropical latitudes are dominated by monk seals of the genus *Monachus*. Historically, three species of monk seals existed, in Hawaii, the Mediterranean, and the Caribbean. Sadly, the Caribbean monk seal has not been seen since the early 1950s and is now believed to be extinct. The Mediterranean monk seal is highly endangered, with an estimated 500 individuals remaining, and its chances for survival are thought to be poor. The population of Hawaiian monk seals seems to be stable, or even increasing slightly since the 1980s.

One of the best-studied foraging and migratory patterns among pinnipeds is that of the northern elephant seal (**Figure 8.16**), one of two species of elephant seals. The following discussion will include a large amount of information on elephant seals with comparisons to other pinnipeds and other marine mammal groups. Adult northern elephant seals range over much of the North Pacific Ocean, making them nearly permanent members of the pelagic realm. They stay at sea for 8 to 9 months each year to forage, interrupted by two round-trip migrations to nearshore island breeding rookeries. With the development of microprocessor-based dive time and depth recorders, details of the migratory and foraging behavior of these deep-diving seals have been fairly well documented. By attaching time, depth, and environmental data recorders to adults as they leave their island rookeries, dive behaviors of individual animals have been recorded continuously for several months until their return to their rookeries with their time, depth, and environmental recorded data.

When ashore for breeding or for molting, adult elephant seals fast, and each fasting period is followed by prolonged periods of foraging in offshore waters of the North Pacific Ocean. The postbreeding migration averages 2.5 months for females and 4 months for males. After returning to their breeding rookery for a short molting period of 3 to 4 weeks, during which they shed their old guard hairs, they again migrate north and west into deeper water. During this postmolt migration, females forage for about 7 months and males for about

4 months (**Figure 8.17**). Both males and females then return again to their rookeries for breeding. Individuals return to the same foraging area as during postbreeding and postmolt migrations.

While at sea, both males and females dive almost continuously, remaining submerged for about 90% of the total time. Dives for both sexes average 23 minutes, with males generally having slightly longer maximum dive durations. These dive durations probably are near the aerobic dive limit for this species. Elephant seals of both sexes spend about 35% of their dive time near maximum depth, between 200 and 800 m, although some individuals have been observed diving down to 1,500 m. In general, daytime dives are 100 to 200 m deeper than nighttime dives. The principal prey of elephant seals are mesopelagic squid and fish that typically spend daylight hours below 400 m and ascend nearer the surface at night. This upward nighttime movement reflects the pattern of diurnal vertical migration of these prey species and is likely the basis for the difference in maximum depths between day and night dives of elephant seals.

Tracking of southern elephant seal dive profiles has revealed deeper dives when they are in warmer than usual temperatures. These deeper dives seemingly based on water temperature are likely due to prey species moving to deeper waters to seek out cooler temperatures. As ocean temperatures rise, many species may be forced to shift their usual areas of occurrence to remain in an area that has ample food available. Elephant seals are already thought to dive to the maximum depth that is physiologically feasible, so if prey move deeper their food sources may become unattainable.

Foraging northern elephant seals are widely dispersed over the northeast Pacific Ocean and exhibit strong gender differences for preferred foraging locations. Females tend to remain south of 50° N latitude, particularly in the subarctic frontal zone (Figure 8.17). Adult males transit farther north, through the feeding areas of females to the subarctic waters of the Gulf of Alaska and the offshore boundary of the Alaska Current flowing along the south side of the Aleutian Islands. These areas of aggregation are regions of high primary productivity and contain abundant fish and squid communities.

Twice each year, adult elephant seals return from foraging at sea to the same breeding beach on which they were born, once to breed and again several months later to molt. Although many vertebrates migrate long distances between breeding seasons, elephant seals are the only animals known to make a double migration each year. Their individual annual movements of 18,000 to 21,000 km rival those of gray and humpback whales in terms of the greatest distance traveled.

From an evolutionary perspective, mammals are ideal candidates for polygyny for several reasons. Males produce millions of sperm for each mating event, whereas females produce only a handful of eggs per breeding season. Males can mate with many females during a breeding season, but females, although able to mate numerous times, can achieve only one pregnancy each breeding season. The role males play in the successful rearing of offspring is relatively minor compared to females. Caring for offspring until they are nutritionally independent is the most

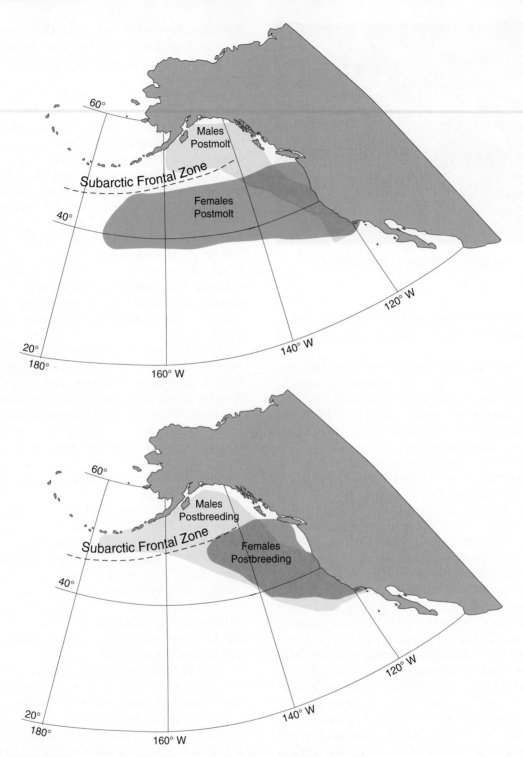

Figure 8.17 Geographic distribution of male and female elephant seals during postmolt (top) and postbreeding migrations (bottom).
Data from B. S. Steward and R. L. DeLong, *Symposium of the Zoological Society of London* 66(1993):179–194.

energetically expensive component of reproduction. With very few exceptions, care and feeding of marine mammal offspring are the sole responsibility of the offspring's mother. In many species, this disparity between the reproductive best interests of males and females of the same species is associated with **sexual dimorphism**, an obvious and often dramatic difference in the size, appearance, and behavior of adult males and females.

Males and females look and act different to attract mates and for their differing roles in the rearing of young.

Most pinniped species breed and pup on land, whereas the remainder mate and give birth on fast ice or pack ice. Species such as elephant seals that pup on land select their pupping and mating rookeries on islands free of large terrestrial predators or on isolated mainland beaches and sandbars not easily accessible

Figure 8.18 Male and female elephant seals display obvious sexual dimorphism. The females (left) lack the elongated nose, enlarged canines, thickened neck, and large size characteristic of sexually mature males.
© Bryan & Cherry Alexander Photography/Alamy Images.

to such predators. Consequently, available rookery space may become limiting, and females become densely aggregated in large breeding colonies. These dense aggregations establish conditions favorable to a polygynous mating system. Almost all pinniped species that breed on land, including elephant seals, are extremely polygynous and strongly sexually dimorphic, with males showing obvious secondary sex characteristics. Breeding male northern elephant seals are five to six times larger than adult females and have an elongated proboscis; enlarged canine teeth; and thick cornified, or thickly calloused, skin on the neck and chest as secondary sexual characteristics (**Figure 8.18**).

Adult male elephant seals arrive at rookery sites in early December, and successful breeders remain while fasting until late February or early March. Arrival of these males displaces the juveniles who used the rookeries as haul-out sites during the nonbreeding season. Pregnant females arrive in late December near the end of their pregnancy and aggregate in preferred rookery sites already occupied by dominant males. Single pups are born and begin nursing about 1 week after females arrive.

Adult females fast through their 4-week lactation, remaining on the beach near their pups. Elephant seal milk is low in fat (about 10%) in the early stages of lactation but increases to about 40% through the last half of lactation. The gradual substitution of fat for water is probably an adaptation of the mother to conserve water late in her fast; it also contributes to the developing blubber store of the pup before weaning. Pups weigh about 40 kg at birth and gain an average of 4 kg each day before weaning. Overall, nearly 60% of the total energy expenditure of a lactating elephant seal mother is in the form of milk for her pup.

At weaning, most elephant seal pups have quadrupled their birth weights; however, a few may have increased their birth weights as much as sevenfold by sneaking milk from other lactating females after their own mothers have departed to feed at sea. Weaned elephant seal pups do not immediately commence foraging for their own food. Rather, they remain on the pupping beaches for a postweaning fast lasting an average of 8 to 10 weeks. Only when pups have lost about 30% of their weaned body mass and are at least 3 months old do they enter shallow waters to feed for the first time.

Lactating elephant seal females enter estrus and mate about 2 weeks after pupping. Adult male elephant seals compete for reproductive control of these females by establishing dominance hierarchies to exclude other males from breeding activities. This type of mating system often is referred to as *female* (or harem) *defense polygyny*. Elephant seal males fast for the duration of the 3-month breeding season, and the increased lipid stores associated with large male body mass confer clear advantages in successfully withstanding these extended periods without food. The most dominant (or alpha) male in a dominance hierarchy defends nearby females against incursions by subordinate males. In areas crowded with females, an alpha male can dominate the breeding of about 50 females.

Subdominant, usually younger, males use other strategies to gain access to at least a few estrus females. Some male elephant seals occupy less preferred breeding beaches that attract fewer males with whom they must compete; however, these beaches also attract few females. In more crowded rookeries, subdominant males may accomplish occasional matings by sneaking into large harems in which alpha males are more likely to be distracted or by mobbing a female as she departs from the rookery at the end of the breeding season. Almost all of these departing females are already pregnant by a dominant male, and thus these late copulations by subdominant males are unlikely to lead to paternity.

Mating strategies of female pinnipeds have been less studied than those of males. This is, in part, due to the crowded rookery conditions of this highly polygynous species. It is sometimes difficult to distinguish a female's behavior directed at defending her pup from strictly mating-related behavior. Females may move away from or vocally protest mounting attempts by subdominant males, thus attracting the notice of higher-ranking males. A female's position within a group, especially a large group, also may influence with whom she will mate. Older or more-dominant females near the center of a harem will be more likely to mate with the dominant male, leaving less-dominant females to be pushed to the edge of the harem where they are more exposed to mating forays by subdominant males.

After the last of the females have departed the rookery, emaciated adult males begin to leave in February or March. Adult females and juveniles return to the rookery for a 1-month molting fast between mid-March and mid-May. Adult males return to the rookery in June and July for their molt. Then the juvenile animals return to use the beach as a haul-out site until the beginning of the next breeding season.

The duration of gestation in placental mammals ranges from 18 days in jumping mice to 22 to 24 months in elephants. In terrestrial species, gestation duration is roughly related to the size of the fetus: larger fetuses require longer gestation periods to develop. For elephant seals, both birth and fertilization occur in the same location and at the same time of year, but in successive years, and this gestation period must be about 1 year in duration regardless of body size. Consequently, these and all other species of pinnipeds must stretch it out to prolong gestation for a year.

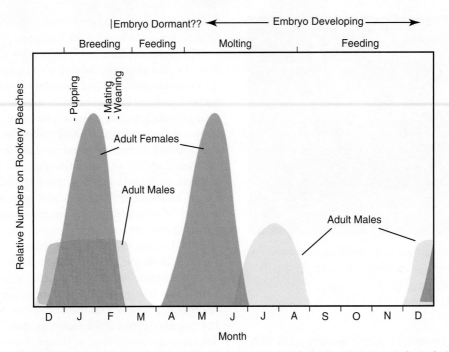

Figure 8.19 Seasonal patterns of rookery use by adult elephant seals. Periods of embryo dormancy and of active fetal growth in pregnant females are indicated.

Adjustments to fit gestation periods of less than 1 year into an annual time frame are accomplished by a remarkable reproductive phenomenon known as **seasonal delayed implantation** (**Figure 8.19**). In seasonal delayed implantation, the zygote undergoes several cell divisions to form a hollow ball of a few hundred cells, the **blastocyst**, which then remains inactive in the female's uterus for several months. After this delay, the blastocyst becomes implanted into the inner wall of the uterus, a placental connection develops between the embryo and uterine wall, and normal embryonic growth and development resume through the remainder of the gestation period. In effect, delayed implantation substantially extends the normal developmental gestation period to enable mating when adults are aggregated in rookeries rather than when dispersed at sea. By providing flexibility in the length of gestation, this reproductive strategy confines birth and mating to a relatively brief period of time ashore and enables the young to be born when conditions are optimal for their survival. Because female elephant seals are at sea and inaccessible when blastocyst implantation occurs, the physiological and hormonal mechanisms controlling delayed implantation are not well understood.

The geographic segregation of foraging areas by male and female northern elephant seals may reflect the preferences of males for larger and more oil-rich species of squid found in higher latitude, subarctic waters, even though they must travel farther through the foraging areas of the females to get to these high-latitude foraging grounds. The smaller females of this species also may have different energetic requirements that encourage them to undertake shorter migratory distances while giving up access to energy-rich prey at higher latitudes. Strikingly similar patterns of latitudinal segregation by gender in North Pacific foraging areas are exhibited by sperm whales. Their prey are also primarily mesopelagic squid, and their seasonal migrations

and diving patterns may also reflect the geographic and vertical distribution of their preferred squid prey.

The differential migrations of male and female elephant seals appear to develop during puberty, when growth rates of males are substantially greater than those of females. These patterns are well established by the time males are 4 to 5 years old. Although body mass will vary substantially on a seasonal basis in those species that experience prolonged postweaning or seasonal fasts, body length tends to increase regularly until physical maturity is reached. For marine mammals, growth to physical maturity continues for several years after reaching sexual maturity, and thus old, sexually mature individuals are often much larger than younger but still sexually mature individuals of the same gender. Second, in polygynous and sexually dimorphic species of pinnipeds and odontocetes (such as killer and sperm whales) patterns of male growth exhibit a delay in the age of sexual maturity to accommodate a period of accelerated growth into body sizes much larger than those of females (**Figure 8.20**). Before polygynous males can compete successfully for breeding territories or establish a high dominance rank, they must achieve a body size substantially larger than that of females. Males of all polygynous species delay both sexual and physical maturity as compared with females to allow for growth.

The longer wait to sexual maturity comes at a substantial cost reflected in overall mortality. For example, in northern elephant seals using central California rookeries, life histories of individual males collected between 1967 and 1986 were similar: only 7% to 14% of males survived to 8 to 9 years of age, when they could first successfully compete for females. Females, however, must only survive to about 4 years of age before they can successfully breed. The period of accelerated growth in males often corresponds to the age at which females of the same species achieve sexual maturity. Although the life

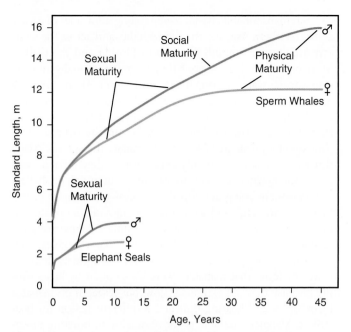

Figure 8.20 Growth curves for male (blue line) and female (pink line) northern elephant seals (bottom) and sperm whales (top).

Data from Clinton, 1994 (Clinton, W.L. 1994. Sexual selection and growth in male Northern Elephant Seals. *In: Elephant Seals, Population Ecology, Behavior, and Physiology*. Ed. by B.J. LeBoeuf and R.M. Laws. Berkeley: University of California Press), and Lockyer, 1981 (Lockyer, C. 1981. Growth and energy budgets of large baleen whales from the southern hemisphere. *Food and Agriculture Organization of the United Nations Fisheries Series* 5:379–487).

history of male northern elephant seals is geared toward high mating success late in life, the chance of living to that age of high mating success is small. Fewer than 10% of a study group of males managed to mate at all during their lifetimes, whereas the most successful male of that group mated with 121 females.

The mortality experienced by elephant seals during their first year is caused by several agents, ranging from starvation to parasitic lungworm infections to predation by white sharks. White sharks are not the capricious or mindless killers portrayed by the "Jaws" movies. Rather, they are skilled predators of pinnipeds. About 1 m long at birth, young white sharks initially feed on small teleost fish. By the time they are 3 m long, they begin to shift to the adult preference for pinnipeds, especially young seals. This shift in preferred prey coincides with movements to higher latitudes (in both hemispheres) where aggregations of seals are abundant.

The predatory behavior of white sharks on seals involves a stealthy approach along the seafloor in shallow areas where seals enter the water. White sharks exhibit strong countershading patterns with a dark gray dorsal surface, and their approach must be very difficult for seals to detect while looking down from the sea surface. When the victim is located, likely by silhouetting it against the lighter background of the sea surface, the shark lunges upward from below and bites the seal to cause profuse bleeding (**Figure 8.21**). The bleeding seal is then either carried underwater to drown in the shark's jaws or is left at the surface until it dies. This attack behavior is why many white

Figure 8.21 Typical attack behavior of a white shark on a seal. The shark approaches the seal from below to attack (left) and then may carry the seal underwater or let the seal float to bleed. White sharks in South Africa leap even higher out of the water during the strike.

shark attacks on humans are not lethal, and if they are lethal, death occurs from blood loss, not the shark actually eating a human. The white shark takes a bite to incapacitate the misidentified human prey, but humans are relatively small and thin with very little to offer a white shark energetically. Consequently, the shark does not return to finish the kill. In the summer of 2015 an Australian professional surfer, Mick Fanning, was attacked by a white shark during a World Tour surfing competition in South Africa. Fortunately for Mick, the shark did not take a bite out of him, but attacked his surfboard. Once the shark realized the surfboard was not a pinniped, it retreated, and the surfer was able to swim to the safety of a jet ski.

From analyses of stomach contents, large white sharks appear to prefer seals and whales over other kinds of prey, such as birds, sea otters, or fish. This selective preference for marine mammals that use fat-rich blubber for insulation may be related to the high metabolic demands on the shark to maintain elevated muscle temperatures and high growth rates in the temperate waters occupied by their seal prey.

Sirenians

Manatees, dugongs, and sea cows (order Sirenia, **Figure 8.22**) are large, ungainly creatures with paddle-like tails (manatees) or horizontal flukes (dugongs and sea cows) and no pelvic limbs. They are docile herbivorous animals and are now completely restricted to shallow tropical and subtropical coastal waters where they can secure an abundance of macroscopic marine and freshwater vegetation. They are the only extant marine mammal that is mostly herbivorous, and their herbivorous diet has led to several unique adaptations in their physiology and lifestyles. Plant and algal material contains less energy than meat, so sirenians must eat often. Their metabolisms are slower, and because they are endothermic they use a large amount of energy to maintain their body temperature above the ambient temperature. Like many herbivores, sirenians have adaptations for digesting plant material, including extremely long intestines and hindguts specialized for fermenting plant material. They also have specialized teeth for processing plant material.

Figure 8.22 A manatee floating, surrounded by snappers in Crystal River, Florida.
Keith Ramos/USFWS.

Sirenians inhabit coastal regions along both sides of Africa, across southern Asia and the Indo-Pacific, and across the western Atlantic from South America to Florida and the Gulf of Mexico. Within these regions they migrate seasonally, moving into warmer waters during the fall and winter. At one time, the Steller's sea cow occupied parts of the Bering Sea and the Aleutian Islands. It took hunters and whalers less than 30 years, from the time the explorer Vitus Bering discovered these slow quiet animals in 1741, to exterminate the species. Many threats to manatees currently exist because they inhabit nearshore waters where humans frequently occur and their shallow waterways are being developed for access to the open ocean. Manatees are large and generally slow moving, so boaters in narrow waterways often strike these animals. Many manatees exhibit one or more patches of long scratchlike wounds on their backs from boat propellers running over the animals. Initially, it was believed that manatees were hit so often by boats simply because they are extremely slow, but research on manatee locomotion and hearing indicates a different reason for boat strikes. Manatees are able to move quickly in bursting movements when startled, so they technically should be able to flee from an approaching boat. Manatees do not appear to hear well in the frequency range associated with boat motor sounds, particularly the frequency associated with boats moving slowly, so they do not hear the boats approaching. By the time a manatee may see an approaching boat, it is often too late, even with a quick bursting movement. In areas where manatees frequent, slow boating speeds are part of the regulations, but it appears that slow speeds are not helping to decrease manatee mortality.

Cetaceans

The evolution of members of the order Cetacea from terrestrial ancestors has led to a remarkable assemblage of structural, physiological, and behavioral adaptations to a totally marine existence. In contrast to typical mammals, cetaceans lack pelvic appendages and body hair, breathe through a single or a pair of dorsal blowholes, are streamlined, and propel themselves with broad horizontal tail flukes. Body lengths vary greatly from small dolphins a bit more than a meter long to blue whales exceeding 30 m in length and weighing over 100 tons (**Figure 8.23**). Convincing evidence of the tetrapod ancestry of whales can be seen during embryonic development. Limb buds develop (**Figure 8.24**) and then disappear before birth, leaving only a vestigial remnant of pelvic appendages.

The modern whales (cetaceans) are of two distinct types. The filter-feeding baleen whales (mysticetes) lack teeth; in their place, rows of comblike **baleen** project from the outer edges of their upper jaws (**Figure 8.25**). All other living whales (including porpoises and dolphins) are toothed whales (odontocetes) and are described later in this section. They are generally smaller than mysticetes, and most are equipped with numerous teeth to catch fish, squid, and other slippery morsels of food. The smaller odontocetes, especially, are very social and are thought by many to be highly intelligent animals. Ongoing studies evaluating cetacean intelligence and communication capabilities remain highly visible aspects of current marine mammal

Figure 8.23 A few species of cetaceans, showing the immense range of body sizes at maturity. Mysticete baleen whales are on the left (a–e), odontocete toothed whales on the right (f–j).

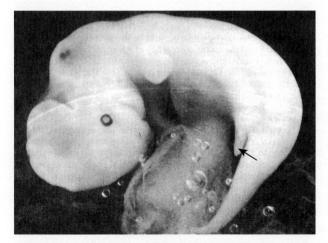

Figure 8.24 A 70-day embryo of a gray whale. Note the definite rear limb buds (arrow).

Rice and Wolman, "Life History and Ecology of the Gray Whale (Eschrichtius robustus)." Spec. Publ. No. 3, (1971). American Society of Mammalogists, Allen Press, Inc.; Courtesy Dale W. Rice.

Figure 8.25 Left-side baleen plates of a rehabilitating year-old gray whale, "JJ," with Jim Sumich (right).

Courtesy of James Sumich.

Figure 8.26 Scientists examine stranded pilot whales off the coast of southern California.
NOAA/SWFSC.

research. These studies are necessarily biased toward smaller species that are easily maintained in captivity and are complemented by information derived from examination of singly or mass-stranded animals (**Figure 8.26**) and from research conducted in the animals' natural habitat.

The largest group of baleen whales is known as the rorquals (family Balaenopteridae). Their distribution is worldwide, and blue, fin, humpback, and minke whales migrate to Antarctic waters to feed during summer (see Figure 8.23). All are fast, streamlined swimmers equipped with baleen that is intermediate in length between bowhead and gray whale baleen. The mouths of all rorqual species are enormous, extending posteriorly nearly half the total length of the body (**Figure 8.27**). All

Figure 8.27 Engulfment feeding behavior of a Bryde's whale.
© Doug Perrine/SeaPics.

members of this family have 70 to 80 external grooves in the floor of the mouth and throat. During feeding, this grooved mouth floor expands like pleats to four times its resting size. Alternating longitudinal strips of muscle and blubber interspersed with an elastic protein facilitate this extension. As the mouth floor is extended, small blood vessels in this tissue dilate to give the throat a reddish color (hence the name *rorqual*, or "red whale," for balaenopterids). As the mouth inflates, it fills with an amount of water equivalent to two thirds the animal's body weight; a mature blue whale might engulf as much as 70 tons of water in a single mouthful.

Water and the small prey it contains enter the open mouth by negative pressure produced by the backward and downward movement of the tongue and by the forward swimming motion of the feeding animal. This method of prey capture, in which large volumes of water and prey are taken in, is referred to as *engulfment feeding* (Figure 8.27). After engulfing entire shoals of small prey in this manner, the lower jaw is slowly raised to close around the mass of water and prey. Then the muscular tongue acts in concert with contraction of the muscles of the mouth floor (and sometimes with vertical surfacing behavior) to force the water out through the baleen and to assist in swallowing the trapped prey.

In addition to the engulfment behavior of other rorquals, humpback whales often produce a large 10-m-diameter curtain or net of ascending grapefruit-sized bubbles of exhaled air, referred to as **bubble-net feeding** (**Figure 8.28**). This net of bubbles is produced by one member of a group of several cooperating whales that form long-term foraging associations. They seem to confuse prey or cause them to clump into tight balls for easier capture by the whales. While one whale produces the bubble curtain, other members of the group dive below the target prey school and force it into the curtain and then lunge through the confused school from below. Low-frequency calls that are sometimes produced as the whales lunge through the school of prey may aid in orienting whales during this complex maneuver.

In 1992, the U.S. Navy began to make available to marine mammal scientists the listening capabilities of its Integrated Undersea Surveillance System (IUSS). IUSS was part of the U.S. submarine defense system developed over 30 years ago to detect and track Soviet submarines acoustically. The IUSS study has provided a wealth of acoustical information on vocalizations of large baleen whales, especially blue, fin, and minke whales. The vocalizations of blue whales are very loud low-frequency pulses between 15 and 20 Hz, mostly below the range of human hearing, whereas those of fin whales are only slightly higher at 20 to 30 Hz. Their function is not known, but two plausible explanations have been put forward. It is reasonable to conclude that if we can detect these sounds at long distances, other whales should be able to as well. They may therefore function in long-distance communication across hundreds or thousands of kilometers of open ocean. These loud, low-frequency, patterned sequences of tones propagate through water with less energy loss than do the higher-frequency whistles or echolocation clicks of small toothed whales. In addition to identity calls, these low-frequency pulses may serve an echolocation function, although a very different one than that described for

Figure 8.28 Final phase of humpback whale bubble-net feeding.
Courtesy of Stellwagen Bank National Marine Sanctuary/NOAA.

small toothed whales. The low frequency of blue and fin whale tones have very long wavelengths, ranging from 50 m for 30-Hz fin whale calls to 100 m for 15-Hz blue whale calls. If they are used for echolocation, these sound frequencies cannot resolve target features smaller than their respective wavelengths of 50 to 100 m. Some researchers have speculated that these tone pulses might be used by large mysticetes to locate very large-scale oceanic features, such as continental shelves or islands, sharp differences in water density associated with upwelling of cold water, and possibly even large swarms of krill.

Although polar bears, emperor penguins, and a few species of pinnipeds remain in summer-intensive polar and sub-polar production systems year-round, mysticete whales provide examples of the more common approach to exploiting these high-latitude production systems with intensive summer feeding followed by long-distance migrations to low latitudes in winter months.

Mysticete whales are among the largest animals on Earth, and all are filter feeders. All except the gray whale feed on planktonic crustaceans or small shoaling fishes. The character of the baleen, as well as the size and shape of the head, mouth, and body, differ markedly between species of baleen whales (see Figure 8.23). All mysticetes show reverse sexual size dimorphism, with females growing to larger maximum sizes than males. This may provide females with larger lipid stores to help offset the energetic costs associated with rapid fetal growth and lactation.

Several distinctive types of feeding behaviors have been described for baleen whales, depending on the type of whale as well as on its prey. Bowhead whales are large bulky animals with very long and fine baleen plates well adapted to collect copepods, euphausiids, and other small planktonic crustaceans using a surface feeding behavior known as skimming (**Figure 8.29**). Swimming slowly with their mouths slightly agape,

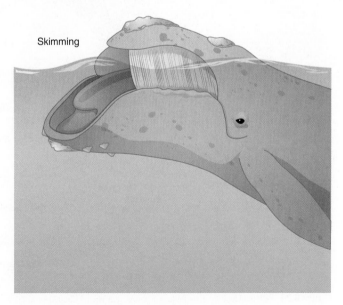

Skimming

Figure 8.29 Skimming feeding behavior of bowhead whales.

Figure 8.30 An adult gray whale surfacing. The white baleen plates are evident in its open mouth.
NOAA.

water and prey items flow into the mouth and then through the baleen where the small prey are trapped. They also use this feeding behavior well below the sea surface and sometimes even near the seafloor. Bowhead whales live all year along the edge and in leads of pack ice around much of the Arctic basin, migrating with the growth and retreat of the pack ice.

Gray whales exhibit the most distinctive feeding behavior of all mysticetes and also have the coarsest and shortest baleen (**Figure 8.30**). In their shallow summer feeding grounds of the Bering and Arctic Seas, these medium-sized whales feed on bottom invertebrates, especially amphipod crustaceans. Direct observations of gray whale feeding behavior demonstrate that these animals roll to one side and suck their prey into the side of the mouth and expel water out the other side, flushing the mud out through the filter basket formed by the coarse baleen while trapping their infaunal invertebrate prey inside the baleen.

Long-distance fasting migrations to low latitudes in winter months are typical of large mysticetes in both hemispheres, with both mating and calving occurring in warm and often protected waters. A general picture of the relationship between these migrations and reproductive timing and behavior is well illustrated by gray whales. The annual gray whale migration has been extensively studied and is the best known of the large whale migrations. These whales migrate an impressive 18,000 km round-trip each year. Most gray whales spend the summer in the Bering Sea and adjacent areas of the Arctic Ocean as far north as the edge of the pack ice. Their habit of feeding on bottom invertebrates and their limited capacity to hold their breath (4 to 5 minutes) restrict their feeding activities to the shallow portions of these seas (usually less than 70 m).

The annual migration of gray whales exists as two superimposed patterns related to the reproductive states of adult females (**Figure 8.31**). Each year approximately one half of the adult females are pregnant. These near-term pregnant females depart the Bering Sea at the start of the southbound migration 2 weeks before other gray whales (Figure 8.31, red solid line). Nonpregnant adult females generally start south a little later accompanied by adult males.

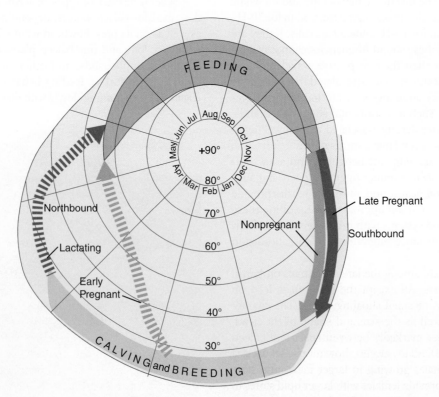

Figure 8.31 Pattern of latitudinal migrations of female gray whales through a complete 2-year reproductive cycle.
Data from Sumich, J. L., Growth in young grey whales (*Eschrichtius robustus*). *Marine Mammal Science* 2(1986):145–152.

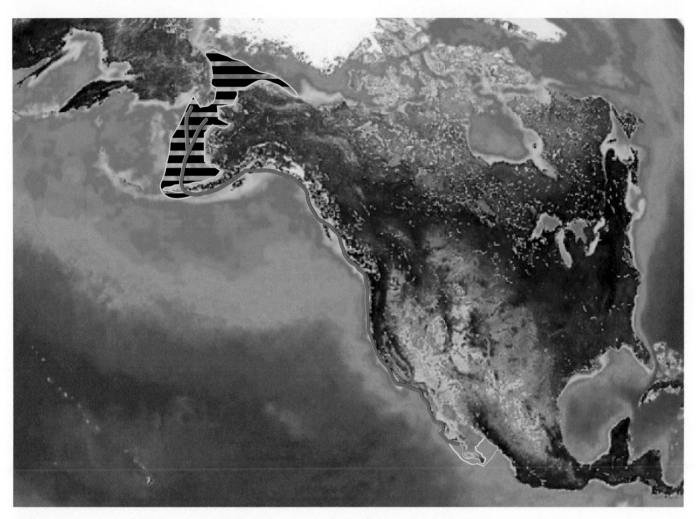

Figure 8.32 Migratory route (red line) of the North Pacific gray whale, with summer feeding (black hatching) and winter breeding areas (purple) indicated. Primary productivity is indicated by water color, with green representing the highest chlorophyll content and dark blue the lowest.
Courtesy of NASA.

The southward migration is initiated in autumn, possibly in response to shortening days or to the formation of sea ice in Arctic waters. The migration is a procession of gray whales segregated according to age and sex. After they pass through the Aleutian Islands, gray whales follow the long, curving shoreline of Alaska (**Figure 8.32**). South of British Columbia, gray whales travel reasonably close to the shoreline, usually in water less than 200 m deep, within sight of land and likely within hearing of breaking waves on shore. The average speed of southbound gray whales is about 7 km/h. At that speed, most of the whales reach the warm protected coastal lagoons of Baja California by late January.

It is in these lagoons that the pregnant females give birth to calves weighing approximately 1 ton and measuring 4 to 5 m long. The new mothers remain with their calves in the Mexican lagoons for about 2 months. During that time, the nursing calves more than double their birth weight in preparation for the rigors of a long migration back to the chilly waters of the Bering Sea. While in the lagoons and adjacent coastal areas, most lactating cows with new calves maintain a spatial separation from other age and sex groups by occupying the inner, more protected reaches of the lagoons until the other animals

depart. These females, accompanied by their calves, are the last to leave calving lagoons the following spring.

Accompanied by adult males, nonpregnant adult females arrive later than pregnant females and are also the first to leave the lagoon after approximately 30 days. They mate during the southern portion of their coastal migration or after they arrive in their winter lagoons. Both are environments of limited underwater visibility, and thus direct observations of mating behavior have not been made. Based on above-surface observations of courting activities, mating appears to be promiscuous as it is in most other baleen whale species, although actual paternity is impossible to establish from direct observations of their very active and vigorous courting encounters (**Figure 8.33**). Courting/mating groups including as many as 17 individuals have been observed; however, the intensity of the physical interactions during these courting bouts makes it impossible to determine the gender of all group participants, their interwhale contact patterns, or even accurate assessments of group sizes.

In the absence of direct underwater observations of gray whale mating behavior, arguments for promiscuous mating must be based on other evidence. In comparison with most other baleen whales, gray, bowhead, and right whale males

Figure 8.33 Two gray whales courting in their winter breeding lagoon.
© Francois Gohier/Science Source.

have substantially larger testes weight-to-body weight ratios (**Figure 8.34**) and relatively larger penises. Larger testes are presumed to produce greater quantities of sperm. This has been interpreted as evidence of **sperm competition** (a behavior well studied in several species of birds), with copulating males competing with previous copulators by attempting to displace or dilute their sperm within a female's reproductive tract, thus increasing the probability of being the male to fertilize the female's single oocyte.

much the reverse of the previous southbound trip, with nursing females and their calves the last to leave the lagoons. While still nursing, calves and their mothers migrate back to their polar feeding grounds. Traveling at a more leisurely pace than when going south, the whales reach their Arctic feeding grounds in the late spring or early summer. With food abundant there, calves are weaned, and females begin to replenish their fat and blubber reserves before migrating back to the winter breeding grounds to mate and begin the cycle again.

DID YOU KNOW?

Male humpback whales display some of the longest and far-reaching vocalizations of all marine animals. Their songs are complex, lasting up to 20 minutes, and can be heard underwater for distances over 30 km (20 miles)! Scientists still do not know exactly why humpbacks sing, but the songs are unique for each population. Humpbacks sing their songs while overwintering in the warm waters of the Hawaiian Islands in the Pacific or the Dominican Republic in the Atlantic. After winter, they head to Alaska in the Pacific or the Gulf of Maine in the Atlantic to rich feeding grounds.

The reproductive cycle of gray whales and other large baleen whales consists of three parts (**Figure 8.35**): a 12- to 13-month gestation period followed by 6 months of lactation, and then another 6 months of rest to prepare for the next pregnancy. In early spring, the northward migration begins and is

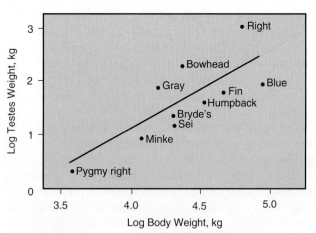

Figure 8.34 General relationship between testis size and body size for 10 species of mysticete whales. Gray, bowhead, and right whales (above the diagonal line) have relatively larger testes and are suspected of participating in sperm competition.

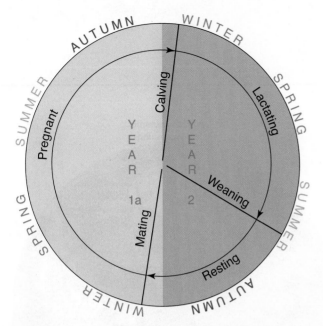

Figure 8.35 The reproductive cycle of female gray whales (*Eschrichtius robustus*), including the accompanying growth of fetus and calf during its period of dependency on the female.

Data from A. Berta and J. Sumich, *Marine Mammals: Evolutionary Biology* (Academic Press, 1999).

Odontocete Whales

Odontocete, or toothed whales, are the larger group of cetaceans, including over 70 species. Diversity of body form varies widely, from relatively small and very streamlined porpoises to one of the largest whale species, the sperm whale. Toothed whales are found worldwide, and some thrive in the extremely cold waters of the polar seas. Here we explore some representatives of odontocete whales and their unique characteristics.

The striking white and black color patterns of killer whales make them the most easily recognized of all the whales (**Figure 8.36**), and they are the most widely distributed mammal besides humans. Their distribution is cosmopolitan, from tropical

Figure 8.36 A killer whale, *Orcinus orca*, breeching in the Monterey Bay National Marine Sanctuary.

NOAA/SWFSC/Michael Richlen.

waters to ice edges of both polar regions. Captive animal displays, hit movies, and worldwide boat- and shore-based whale-watching activities have made this species a common sight to large numbers of people, and long-term research programs using photographic identification of individual whales in some populations have been crucial to improving our understanding of their biology.

Within their broad geographic range, smaller social groups of killer whales, known as **pods**, occupy smaller individual ranges. Individual killer whale pods consist of a mature female, her offspring, and her daughters' offspring. Thus, pods are tight social groups of very closely related individuals, and are **matrilineal**, with each pod member a direct descendant of the oldest female in the pod. In contrast to many other social mammals, killer whale pods include both male and female siblings that remain together all of their lives. The group behavior of this species appears to increase survival rates of adults and young.

The broad ecological and latitudinal range of killer whales reflects their wide variety of prey species, including fish, cetaceans, pinnipeds, birds, cephalopods, sea turtles, and sea otters. Killer whales are the only cetacean known to attack and consume other marine mammals consistently. Although killer whales have been described as feeding opportunists or generalists, they also can behave as feeding specialists, with the necessary flexibility to respond to variations in the type and abundance of their preferred prey.

In Icelandic coastal waters, for example, some pods of killer whales feed on marine mammals and then switch to fish prey at other times of the year. In both Antarctic and Pacific Northwest waters, prey specialization between killer whale populations occurs. In Washington and British Columbia coastal waters, a resident population feeds almost exclusively on fish, whereas a nonresident or transient population preys principally on marine mammals, especially harbor seals and sea lions. Sightings of foraging resident pods are strongly associated with seasonal variations in their most common prey, salmon. During months when salmon are unavailable, residents switch prey to herring or bottom fish and may disperse northward several hundred kilometers. In contrast, the prey of transient groups is available year-round, and despite their label, these whales tend to remain in a small area for extended periods of time.

Different populations use different foraging strategies, even when in close proximity to each other. These strategies reflect different prey types and possibly different pod traditions. Resident whales typically swim in a flank formation when hunting, possibly to maximize the likelihood of detecting prey, although individual hunting is sometimes seen at the periphery of the main pod group. Resident pods appear to prefer foraging at slack tidal periods when salmon tend to aggregate.

Echolocation (described in Section 8.5) and other vocalizations are common when resident pods feed on fish. These vocalizations are less frequent in transient groups preying on other marine mammals, possibly because mammals are more likely to detect vocalizations and exhibit avoidance responses. Mammal-eating killer whales are cooperative foragers and take advantage of an ability, probably unique among marine mammals, to capture prey larger than themselves. The degree of foraging cooperation is variable, depending on the type of prey;

RESEARCH in Progress

Global Wanderers: Gray Whales Found in Surprising Locations

Throughout much of their known fossil history, gray whales ranged widely in coastal waters of both the North Pacific and North Atlantic Oceans and likely moved between these ocean basins as well. Subfossil remains of gray whales along both sides of the North Atlantic Ocean provide ample evidence that gray whales existed there throughout much of the last several millennia. We will never know if North Atlantic gray whale populations were declining and would have disappeared naturally, but there is little question that human predation, however small, hastened their decline so that this species disappeared completely from the North Atlantic Ocean by the late 1700s and has not been observed anywhere in the Atlantic Ocean basin for over 300 hundred years.

So imagine everyone's astonishment when a gray whale, seemingly in good body condition, appeared in shallow water near the coast of Israel in May 2010 and was then resighted near Barcelona, Spain, 22 days later. The appearance of this solitary animal in the eastern Mediterranean prompted a host of questions: Was it a juvenile or adult; male or female? Was it a wanderer from one of the North Pacific populations or a rare surviving individual from what has been assumed to be an extinct North Atlantic population? None of these questions has been answered definitely, because the whale was not seen again after its brief sighting off the Spanish coast. As there have been no other verified sightings of gray whales in the North Atlantic for three centuries, it seems most reasonable to assume that this whale, likely a subadult, had wandered from a high-latitude summer foraging area in the Bering or Chukchi Sea, across the Arctic Ocean basin, then south to Europe and the Mediterranean (**Figure A**). This feat required sufficient ice-free open water along the Arctic coastline, at least in late summer, a conclusion supported by the presence of two gray whales sighted a year later in the Laptev Sea near the Russian Arctic coast.

An even more unexpected sighting was made 3 years after the sighting of the gray whale near Israel. Another wandering gray whale was observed near the Namibian coast of southeast Africa (latitude 23° S; indicated on the map in Figure A) where it remained for over a month. This is the first confirmed gray whale, fossil or living, in any ocean of the Southern Hemisphere. We have no idea of the route this whale followed for its epic journey. However, assuming it originated from the more populous eastern population, it, too, must have traveled across the Arctic Ocean basin from somewhere along the west coast of North America, a trip of at least 20,000 km (half the distance around Earth) to arrive at this location.

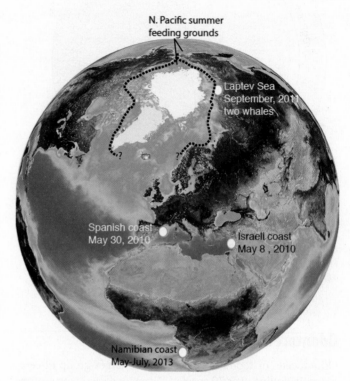

Figure A Recent sighting dates and locations of several gray whales outside their usual geographic range. Dotted lines indicate alternative probable routes across the Arctic Ocean during minimum ice coverage in late summer.
Images and data from National Snow and Ice Data Center.

Long-term climate changes have the potential to profoundly affect the foraging activities and food sources of many migratory Arctic marine animals, including gray whales. The continuing trend over the last several decades of decreasing sea ice extent in the Arctic Ocean is a consequence of global climate change, with complicated potential outcomes predicted for gray whales and other Arctic marine mammals. Like polar bears and walruses, gray whales serve as important Arctic ecosystem sentinels as they respond to changes in the physical features of their habitat. We can already observe some effects of climate change throughout the geographic range of gray whales. These effects are especially dramatic in the Arctic where increasing air and water temperatures have resulted in reductions of sea ice coverage, thickness, and seasonal duration (**Figure B**). What impacts these changes will ultimately impose on gray whales is uncertain, but some patterns are becoming evident. In response to changing conditions in the Arctic in the past several decades, gray whales have delayed their south migration and extended their foraging ranges westward across the north Siberian coast and east along the north slopes of Alaska and Canada. This current expansion of foraging areas may be countered in the near future as the retreating ice-edge habitat becomes uncoupled from the shallow

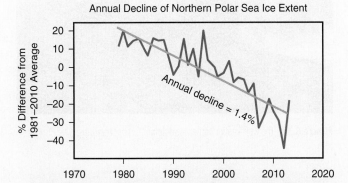

Annual Decline of Northern Polar Sea Ice Extent

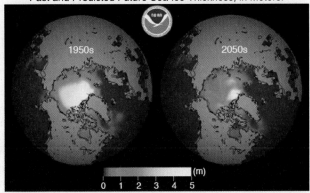

Past and Predicted Future Sea Ice Thickness, in Meters.

Figure B Areal extent of northern polar sea ice for September, 1979–2013, as percent of the 1981–2010 average (top). Projected reduction in its average thickness between 1950 and 2050 (bottom).

Images and data from National Snow and Ice Data Center.

accessible depths that foraging gray whales now depend on, reducing their accessible prey even as their foraging ranges expand. The recent detection of gray whale calls throughout the winter near Barrow, Alaska, indicates that some gray whales now completely forego the south migration, instead remaining in high latitudes throughout the year.

The ongoing shrinkage of summer ice coverage in the Arctic Ocean basin and the occurrence of these two wandering gray whales in the Atlantic Ocean raises the likelihood that this species, without any assistance or interference from us, will use the Arctic Ocean in ice-free summer months as an avenue for range expansion into the North Atlantic and possibly beyond. It may take some time, but this species just might rectify the long-ago actions of Yankee whalers that eliminated gray whales from the North Atlantic Ocean by reestablishing itself naturally in Atlantic waters after a very long absence.

Critical Thinking Questions

1. Propose two alternate hypotheses for why gray whales have been sighted in areas of the world where they were not previously known to inhabit.

2. If climate change is negatively affecting a marine mammal that is a protected species, should activities that cause climate change be made illegal? Obviously this would be difficult to enforce because the causes are indirect, but it is illegal to harm or hinder species protected under the Endangered Species Act. Ponder this fact and make an argument for or against legal repercussions for causing climate change.

For Further Reading

Sumich, J. L. 2014. *E. robustus*: *The Biology and Human History of Gray Whales*. Corvallis, OR: Whale Cove Marine Education.

Figure 8.37 Killer whale ramming a gray whale near Unimak Pass, Alaska.
John Durban/NOAA Alaska Fisheries Science Center/North Gulf Oceanic Society.

Figure 8.38 The well-known bottlenose dolphin, *Tursiops truncatus*.
NOAA.

it can involve cooperative prey encirclement and capture, division of labor during an attack, or sharing of prey after capture.

Relative to residents, transients use short and irregular echolocation trains composed of clicks that appear structurally variable and low in intensity, more closely resembling random noise. The contrasting black-and-white coloration patterns of killer whales may facilitate visual coordination and communication within feeding groups of transients. Transient groups commonly encircle their prey, taking turns hitting it with their flippers and flukes. This extended handling of prey may continue for 10 to 20 minutes before the prey is killed and consumed. When hunting alone, transient whales often use repeated percussive tail slaps or ramming actions to kill their prey before consuming it (**Figure 8.37**). Foraging by transient groups occurs at high tides when more of their pinniped prey are in the water and vulnerable to predation.

Differences in the foraging behaviors of the two overlapping populations in the Pacific Northwest persist through time, with each group expressing different vocalizations and different social group sizes as well as different prey preferences and foraging behaviors. Some researchers have suggested that the two populations in the Pacific Northwest are in the process of speciation.

Dolphins and porpoises are the smallest toothed whales, and they are also some of the most conspicuous. They are often sighted near the shoreline while surfing the waves and swimming near human swimmers or surfers. Out at sea they engage in acrobatics, riding the wake of the bows of ships and making impressive spinning jumps into the air. The most well-known dolphin is the common bottlenose, *Tursiops truncatus*. This species has been studied extensively in captivity, is popular as a show animal at major aquaria worldwide, and even had a representative star in the hit television show, *Flipper*, during the 1970s (**Figure 8.38**). There are more species of dolphins than porpoises, with 32 marine dolphins and only 6 porpoises. In comparison to porpoises, dolphins have longer snouts and bodies, dorsal fins that are more curved, and more streamlined bodies overall. Dolphins are also apparently more communicative, using whistling sounds through their blowholes to communicate with each other underwater.

Both dolphins and porpoises are extremely intelligent animals, likely due in part to their large brains. They also have a large melon on their forehead, used to generate sonar underwater (known as *echolocation*; described in Section 8.5) to help them find prey, navigate, and provide an overall picture of their environment. The use of echolocation by dolphins and other whales makes them more vulnerable to anthropogenic noise disturbances. Echolocation involves transmitting and receiving sound waves, and sound waves of extreme frequencies made by shipping noise, icebreaking, or submarines may harm whales. Some whale strandings have been attributed to Navy sonar tests off the coast, and as a result extensive studies have been conducted to determine the frequency ranges of detection for various whales. When Navy testing requires the use of sonar frequencies that may harm marine mammals, attempts are made to survey the testing area for whales to avoid harm to them.

Another group of small-sized odontocetes includes the beaked whales. Although small in size, beaked whales comprise about 25% of all cetaceans, and many species have wide distributions in most of the oceans. Little is known about this group of cetaceans because they spend most of their time in deep waters, and when they do surface they only do so briefly. Their behavior upon surfacing is very cryptic; they blow air out relatively gently, creating a small blow signature in 20- to 30-second intervals that is barely visible to observers. They are also very skittish and appear to avoid research vessels. Closely related species are difficult to differentiate from one another, even upon close examination when stranded. Most of what we know about beaked whales is from studies of stranded animals, but researchers are making valiant efforts to observe individuals in their natural environment. Recent research has revealed that beaked whales tend to favor oceanographic features such as major currents, current boundaries, and eddies, which is likely due to the availability of rich food sources within these features. Cuvier's beaked whales are one of the most frequently sighted species and are also thought to be the deepest diving marine mammals. Because beaked whales are so elusive and difficult to identify, most population estimates are uncertain, and their statuses worldwide are unknown.

By far the largest odontocete species is the sperm whale (Figure 8.23), which is found worldwide from the equator to the ice-filled poles in both hemispheres. This species displays strong sexual dimorphism, primarily in size. Males are much large than females, reaching over 15 m in length, whereas females are only around 11 m long. Both sexes display features unique among whale species, including a very large, squared off head and blowhole that is asymmetrically placed on one side of their large head. Their flippers are paddle shaped and small compared to their body size. Their brains are also small compared to body size, even though they have the largest brains of any animal. They are deep divers, and as a result spend most of their time over very deep waters. Sperm whales consume prey items found at depth that other predators are unable to consume, such as large squid, sharks, and bony and cartilaginous fishes. The famous novel *Moby Dick* by Herman Melville, written in the 1800s, includes a sperm whale as one of the main characters (**Figure 8.39**). This novel highlights the early whaling industry by providing detailed and realistic descriptions of whale hunting and the act of extracting whale oil from sperm whales.

Sperm whales display interesting social behaviors, and individuals form lasting relationships with one another, despite the fact that most of their time is spent diving individually. Females form bonds with other females in their family unit that span over several generations, and up to 12 adult females and their young will travel and live together. Strong bonds exist between females and their own young, and the young of others. Within groups individuals appear to have strong preferences for certain other individuals, and bonds last for long periods. Males leave the family by the age of 21, although some have been documented leaving much earlier at the age of 4. After leaving the family, young males initially join bachelor schools for a period of time, but eventually the largest males break away from the group to travel alone and return to breed with females in the tropics. Sperm whales are well adapted for life in a variety of environmental conditions during long-distance migrations. Although scientists have studied sperm whales extensively, population estimates are highly variable and uncertain in most areas of the world.

"Both jaws, like enormous shears, bit the craft completely in twain."

—*Page 510.*

Figure 8.39 An illustration of a sperm whale depicted on the cover of the novel *Moby Dick*, or *The Whale*.

Case Study

The Vanishing Vaquita

Common names: Gulf of California Harbor Porpoise, vaquita, cochito

Species name: *Phocoena sinus*

ESA status: Endangered

The Gulf of California harbor porpoise, also known as the *cochito* or *vaquita*, is a rare and unique marine mammal in several aspects. It is found in alarmingly low numbers in a very small geographic area and has the smallest distribution of any marine mammal. It is also the smallest cetacean, weighing between 30 to 55 kg (65 to 120 lbs.)

and measuring 1.2 to 1.5 m (4 to 5 ft.) in length. Unlike other porpoises, vaquitas have a rounded head with almost no beak, a feature that leads to an appearance more similar to, but not exactly like, a pilot whale. Also, unlike other porpoises vaquitas are shy and elusive creatures, a characteristic that helps to explain the limited ability to capture photographs of this species (**Figure A**). Currently the distribution of vaquitas appears to be restricted to the very upper portion of the northern Gulf of California, primarily within the Colorado River delta (**Figure B**). It is thought that their distribution was once much more expansive and extended to mainland Mexico, but survey data are limited.

An extensive survey was conducted in the fall of 2015 with participating scientists from the United States and Mexico. The goal of the survey was to provide the Mexican

(continues)

Case Study *(Continued)*

Figure A Two vaquitas surfacing in their shy fashion, making decent photographs rare.
NOAA/Paula Olson.

government with information about current vaquita abundance. An earlier survey in 2008 was conducted to assess whether the reduction of gill netting, a major threat to vaquitas and many other marine animals, was allowing for a recovery of the species. Initial 2015 survey data indicated that vaquita abundance was approximately the same as it was in 2008, although more data will be collected and analyzed. A steady population size may seem promising, but the estimate from 2008 was 500 to 600 individuals, which is very small and potentially unsustainable. Other sources estimate the abundance of vaquitas at a mere 97 individuals. The fall 2015 surveys comprehensively estimated abundance using passive acoustics and visual surveys. Passive acoustics are useful because data can be collected in many locations simultaneously by recording and interpreting vocalizations, and sightings of actual individuals are not necessary. No matter which abundance estimate is the most accurate, the message is clear: vaquita numbers are extremely low, and this species is in grave danger of extinction in the near future.

What can humans do to attempt to save the vaquita from extinction? The first step is to examine the reasons for their decline. The largest threats to this species are from fishers, including their use of gillnets, trawl nets, and other fishing gear that lead to vaquitas becoming bycatch. Gillnets are used to illegally capture an endangered fish, the large-bodied totoaba (*Totoaba macdonaldi*), to harvest the swim bladder for sale on the Chinese market. Like most threatened or endangered marine species, vaquitas are also threatened by pollution and habitat degradation. An additional threat is low genetic diversity; as numbers decline, the gene pool shrinks, and inbreeding becomes very likely, which can lead to offspring that do not survive.

The largest challenge is finding a balance between maintaining the livelihood of fishers and saving the remaining vaquitas. Income from fishing is sustaining many families in the northern Gulf of California, so, understandably, the idea of restricting fisheries is met with great resistance. In April 2015 a ban on gillnets was announced by Mexican president Enrique Peña Nieto. The plan includes compensating fishers who lose income from the ban and its strict enforcement. Just days after the ban was in place numerous boats were reported using gillnets and several fishers were arrested. Researchers are currently developing fishing gear that is safe for vaquitas and other animals that become bycatch. It will take some time and convincing before fishers may agree to use the new gear, but many do want to conserve their local environment. Not

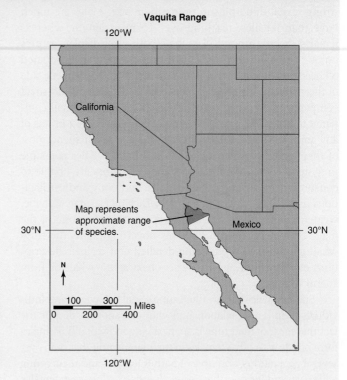

Figure B The geographic distribution of vaquitas.
NOAA.

only is a healthy ocean a more pleasant place to live near, but clean water and well-managed fisheries with little bycatch lead to a greater abundance of fish available for fishers to catch. For the sake of the vaquitas, existence, many are hopeful that the necessary changes to the fishing culture will become accepted and that more fishers will begin using less harmful fishing practices.

Critical Thinking Questions

1. Fishing activities appear to be the greatest threat to the vaquita, and a threat that can be removed. Think of a plan that would remove the threat of vaquitas becoming bycatch while still enabling fishers to make a living.

2. Do you think that the small size of the vaquita has anything to do with its high susceptibility to being caught in fishing nets? Why or why not? If not, what do you propose might be making them more susceptible?

3. Which population numbers do you trust to be closer to the actual population size, the values obtained during sonar surveys or visual surveys? Explain.

8.4 Breath-Hold Diving in Marine Tetrapods

All marine tetrapods (reptiles, birds, and mammals) breathe air to obtain O_2. The length of time they can spend underwater is controlled by an intricate balance of their capacity for storing O_2; their metabolic demands placed on that store of O_2; and their tolerance of low internal levels of O_2, or **hypoxia**. Marine reptiles are **poikilothermic**, and their relatively low body temperatures drive correspondingly low metabolic rates. **Homeothermic** birds and mammals have much higher metabolic rates and activity levels that require more O_2 to support. The mechanisms for prolonged breath-hold diving in birds and mammals are similar, and only marine mammals are emphasized in the following discussion. It should be noted that when discussing marine mammal maximum depths and times of dives that we only know what information has been recorded by scientists to date. Scientists are repeatedly astounded by new dive records of various species, and there will undoubtedly be many more surprises as we continue to track dives of marine mammals.

Aristotle recognized more than 20 centuries ago that dolphins were air-breathing mammals. Yet it was not until the classic studies conducted by Irving and Scholander nearly halfway into the 20th century that the physiological basis for the deep and prolonged breath-holding dives by marine birds and mammals was defined. Their diving capabilities vary considerably. Some are little better than the human Ama pearl divers of Japan who, without the aid of supplementary air supplies, repeatedly dive to 30 m and remain under water for 30 to 60 seconds. The maximal free-diving depth for humans is about 214 m; breath-holds lasting as long as 22 minutes have been independently achieved, although not while actively swimming. Free divers who dive the deepest use weighted sleds to submerge and drop down quickly, then an inflatable bag to return to the surface. Although impressive by human standards, even the best efforts of humans pale in comparison with the spectacular dives of some whales, pinnipeds, and penguins (**Table 8.3**). With dive times often exceeding 30 minutes, these exceptional divers are no longer closely tied to the surface by their need for air.

When diving, submerged marine tetrapods experience a triad of worsening physiological conditions: activity levels (and O_2 demands) are increasing at the very time their stored O_2 is diminishing, and CO_2 and lactic acid are accumulating in working tissues. Prolonged dives such as those listed in Table 8.3 are achieved with several respiratory adjustments in addition to just holding one's breath. As indicated in Table 8.3, breathing rates of marine mammals are decidedly lower than those of humans and other terrestrial mammals. The pattern of breathing is also quite different. In general, marine mammals exhale and inhale rapidly, even when resting at the sea surface, and then hold their breaths for prolonged periods before exhaling again. Even the largest baleen whales can empty their lungs of 1,500 L of air and refill them in as little as 2 seconds. In the larger species of whales, dives of several minutes' duration are commonly followed by several blows 20 to 30 seconds apart before another prolonged dive is attempted. This **apneustic breathing** pattern (**Figure 8.40**) is also common in diving penguins and pinnipeds.

TABLE 8.3 Diving and Breath-Holding Capabilities of a Few Mammals

Animal (genus)	Maximum Depth (m)	Maximum Duration of Breath-Hold (minutes)
Human (*Homo*)	281	11
Dolphin (*Tursiops*)	535	12
Cuvier's beaked whale (*Ziphius*)	2,992	137.5
Sperm whale (*Physeter*)	1,200	138
Fin whale (*Balaenoptera*)	500	30
Sea lion (*Zalophus*)	482	15
Weddell seal (*Leptonychotes*)	626	82
Elephant seal (*Mirounga*)	2,388	120
Walrus (*Odobenus*)	100	13
Manatee (*Trichechus*)	600	6
Sea otter (*Enhydra*)	23	4
Leatherback sea turtle (*Dermochelys*)	124	67
Hawksbill sea turtle (*Eretmochelys*)	91	53
Adélie penguin (*Pygoscelis*)	33	1.5
Gentoo Penguin (*Pygoscelis*)	127	3.5
Emperor penguin (*Aptenodytes*)	564	22

Extensive elastic tissue in the lungs and diaphragms of these animals stretches during inspiration and recoils during expiration to empty the lungs rapidly and nearly completely. Apneustic breathing provides time for the lungs to extract additional O_2 from the air held in the lungs. Dolphins can remove nearly 90% of the O_2 contained in each breath. Oxygen uptake within the **alveoli** (air sacs) of the lungs may be enhanced as lung air is moved into contact with the walls of the alveoli by the kneading action of small muscles scattered throughout the lungs. In some species, an extra capillary bed surrounds each alveolus and may also contribute to the exceptionally high uptake of O_2. Taken together, these features represent a style of breathing that permits marine mammals increased freedom to explore and exploit their environment some distance from the sea surface. Still, apneustic breathing alone cannot explain how some seals and whales are capable of achieving extremely long dive times.

Cetaceans typically dive with full lungs, whereas pinnipeds often exhale before diving. Whether they dive with full or empty lungs is of little importance because the lungs and their protective rib cage smoothly collapse as the water pressure increases with increasing depth (**Figure 8.41**). For a dive from the sea surface to 10 m, the external pressure is doubled, causing the air volume of the lungs to be compressed by half and the air pressure within the lungs to double. Complete lung collapse for most diving tetrapods probably occurs in the upper 100 m; any air remaining in the lungs below that depth is squeezed by increasing water pressure out of the alveoli and into the larger air passages, the **bronchi** and **trachea**, of the lungs.

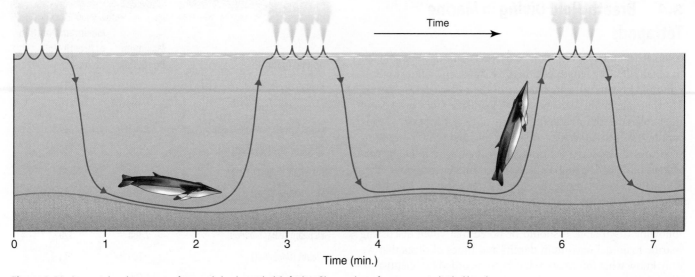

Time

Figure 8.40 Apneustic breathing pattern of a gray whale, observed while feeding. Blows at the surface represent individual breaths.

Even the trachea is flexible and undergoes partial collapse during deep dives.

By tolerating complete lung collapse during dives, these animals sidestep the need for respiratory structures capable of resisting the extreme water pressures experienced during deep dives (over 300 atm for a sperm whale at 3,000 m), and they receive an additional bonus: as the air is forced out of their collapsing alveoli during a dive, the compressed air still within the larger air passages is blocked from contact with the walls of the alveoli. Consequently, little of these compressed gases is absorbed by the blood, and marine mammals avoid the serious diving problems (decompression sickness and nitrogen narcosis) sometimes experienced by humans when they breathe compressed air at moderate depths while underwater. In humans, after prolonged breathing of air under pressure (with hard hat,

Figure 8.41 A self-portrait of Tuffy, a bottlenose dolphin, taken at a depth of 300 m. The water pressure at that depth caused the thoracic collapse apparent behind the left flipper.

Ridgway, S. H., *Mammals of the Sea, Biology and Medicine*, 1972. Courtesy of Dr. Sam H. Ridgway

hooka, or scuba gear), large quantities of compressed lung gases (particularly N_2) are absorbed by the blood and distributed to the body. As the external water pressure decreases during rapid ascents to the surface, these excess gases sometimes are not eliminated quickly enough by the lungs. Instead, they form bubbles in the body tissues and blood vessels, causing excruciating pain, paralysis, and occasionally even death. The excess N_2 absorbed from scuba gear also has a mildly narcotic effect on human divers, sometimes leading to unusual behaviors during the deepest parts of deep dives. Deep-diving marine mammals avoid both of these problems simply because the air within their lungs is forced away from the walls of the alveoli as the lungs collapse during a dive, thereby preventing excess N_2 from diffusing into the blood.

Because the collapsed lungs of deep-diving marine mammals are not effective stores for O_2, it must be stored elsewhere in the body or its use must be seriously curtailed during a prolonged dive (**Figure 8.42**). Both options are exercised by diving mammals. Additional stores of O_2 are maintained in chemical combination with hemoglobin of the blood or with myoglobin in muscle cells. Deep-diving birds and mammals have more hemoglobin-containing red blood cells, and their blood volume is also significantly higher than that of nondiving mammals. About 20% of the total body weight of elephant seals and sperm whales, for instance, is blood. Much of the additional blood volume is accommodated in an extensive network of capillaries (or *rete mirabile*) located along the dorsal side of the thoracic cavity (**Figure 8.43**). The vena cava (the major vein returning blood to the heart) in some species is baglike and elastic. In elephant seals, it alone can accommodate 20% of the animal's total blood volume. These features all contribute to large reserves of stored O_2 for use during a dive.

The swimming muscles of marine mammals are highly tolerant to the accumulation of lactate, a metabolic product of hypoxic conditions during a breath-hold dive. Therefore, muscles and other noncrucial organs (such as the kidneys and digestive tract) can be temporarily deprived of access to the

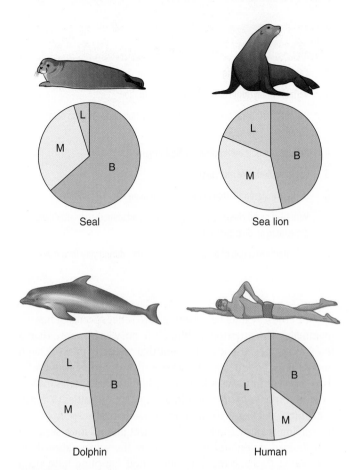

Figure 8.42 Comparison of oxygen stores in blood (B), muscles (M), and lungs (L) for several different mammals.

reserve O_2 stored in the blood. To regulate the distribution of blood, smooth muscles in the walls of arteries leading to these peripheral muscles and organs contract to reduce the flow of blood. The general term for this process is **vasoconstriction**. Most

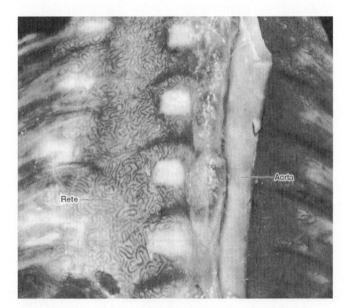

Figure 8.43 The right thoracic rete mirabile of a small porpoise, *Stenella*.
Courtesy of James Sumich.

of the circulating blood is then shunted to other vital organs, primarily the heart and brain. Simultaneously, the heart rate slows dramatically to accommodate pressure changes in a much reduced circulatory system comprising the heart and lungs, the brain, and connecting blood vessels. Other circulatory structures also help to smooth out and moderate fluctuations in the pressure of blood going to the brain. An elastic bulbous "natural aneurysm" in the **aorta** (the large artery leaving the heart) and another rete in the smaller arteries at the base of the brain both help to dampen blood pressure surges each time the heart beats.

Bradycardia is a term used to describe the marked slowing of the heart rate that accompanies vasoconstriction and probably occurs in all vertebrates experiencing reduced access to their usual supply of O_2. These two responses, peripheral vasoconstriction and bradycardia, probably occur in all diving, air-breathing vertebrates, including birds, reptiles, and mammals. The intensity of bradycardia varies widely between marine mammal groups. During experimental dives under laboratory conditions, heartbeat rates of restrained cetaceans are reduced to 20% to 50% of their predive rates. The combination of bradycardia, peripheral vasoconstriction, and other circulatory adjustments to diving has been commonly referred to as "the mammalian diving reflex." The major components of this response are summarized in **Table 8.4**.

Despite variable and even conflicting data for some cetaceans and semivoluntary diving pinnipeds, this set of characteristics was widely applied to explain long-duration breath-holds by marine mammals. Subsequent studies of free-diving Weddell seals in Antarctic waters support a substantially different picture of diving responses in unrestrained divers in their natural habitat (**Figure 8.44**). These researchers attached instrument packages to numerous free-diving seals to monitor and record dive time, depth, heart rate, core body temperature, and blood chemistry. Weddell seals were ideal subjects for this type of study because they breathe by surfacing at holes maintained in the fast sea ice. Each seal returning to its own hole to breathe after a dive offered convenient opportunities to attach and retrieve recording instrument packs.

Weddell seals perform breath-holding dives up to about 20 minutes in length without using the mammalian diving reflex presented in Table 8.4. This suggests that Weddell seals have sufficient stored oxygen at the beginning of a dive to last about 20 minutes. Only during dives lasting longer than about 20 minutes (their apparent **aerobic dive limit [ADL]**) are circulatory

TABLE 8.4	A Summary of the Mammalian Diving Reflex
1	Cessation of breathing
2	Extreme bradycardia regardless of dive duration
3	Strong peripheral and some central vasoconstriction
4	Reduced aerobic metabolism in most organs
5	Rapid depletion of muscle O_2
6	Lactate accumulation in muscles
7	Variety of blood chemistry changes during and immediately after dive

Figure 8.44 Weddell seal at a breathing hole in Antarctica.
NOAA/Michael Cameron.

TABLE 8.5 A Summary of Dive Responses in Weddell Seals (Compare with Those Listed in Table 8.4)	
1	Cessation of breathing
2	Variable bradycardia depending on dive duration
3	Variable peripheral and central vasoconstriction depending on dive duration
4	Reduced aerobic metabolism in most organs
5	Rapid depletion of muscle O_2
6	Lactic acid accumulation in muscles beginning after 20 minutes
7	Variety of blood chemistry changes during and immediately after dive, depending on dive duration
8	Voluntary reduction of body core temperatures during very long dives

responses observed that resemble the expected diving reflex described in Table 8.4. An animal's ADL is defined as the longest dive that does not lead to an increase in blood lactate concentration during the dive; therefore, if an animal dives within its ADL, there is no lactate accumulated to metabolize after the dive, and a subsequent dive can be made as soon as the depleted blood oxygen is replenished.

For dives longer than the ADL, the magnitude of physiological responses is generally related to the length of the dive, yet the picture is not a simple one. Peripheral vasoconstriction and bradycardia do occur, with the magnitude of those responses generally proportional to the duration of the part of the dive that exceeded the animal's ADL. The more a dive exceeds the ADL, the greater the accumulation of lactate and the longer it takes for the level of lactate to return to resting levels. After very long dives (1 hour or more), Weddell seals are exhausted and sleep for several hours. Although Weddell seals are capable of remaining submerged for more than an hour, they seldom do, because about 85% of their dives are within their projected 20-minute ADL.

It now appears that Weddell seals anticipate before a dive begins how long it will last and then consciously make the appropriate circulatory adjustments before leaving the sea surface. This is why it should not be called a diving reflex, but rather an integrated set of related responses. During short dives (less than 20 minutes), no adjustments are necessary. For anticipated dives of long duration, both peripheral vasoconstriction and bradycardia occur maximally at the beginning of the dive and remain that way throughout the dive. On extended dives approaching 1 hour in duration, core body temperature can be voluntarily depressed to 35°C, kept depressed between dives, and then rapidly elevated after the last dive of a dive series. Together, these responses (**Table 8.5**) enable Weddell seals to accomplish some of the longest breath-holds known for mammals. They also indicate that, at least in this species, peripheral vasoconstriction, heart rate, and core body temperature are under conscious control and argue strongly against the earlier concept of a "diving reflex" for marine mammals.

Can these results be generalized to other species of marine mammals? The question is difficult to answer for more than a few species, because it is apparent that researchers' work with unrestrained Weddell seals, whose dive duration was under their own control, was a key to elucidating this species' physiological response to breath-hold diving. Opportunities to attach instrument packages to monitor physiological responses of a free-diving mammal, release the subject animal for a series of unrestrained dives, and then recapture the subject animal for instrument package recovery and blood sampling have been limited in the past; however, such opportunities are becoming more frequent as improvements in electronic monitoring and radio telemetry techniques expand our studies of animals diving without restraint.

Although not as well studied, elephant seals (see Figure 8.18) may surpass Weddell seals in their breath-holding ability. Elephant seals spend months at sea foraging for squids and fish at depths between 300 and 1,500 m (Table 8.3). Elephant seals exhibit diving patterns that suggest they also may play a role in the avoidance of predators: diving deeply with no swimming at the surface, short surface intervals, and long-duration dives may help elephant seals to minimize encounters with white sharks. Their feeding dives are typically 20 to 25 minutes long, with females usually going to depths of about 400 m and males to depths of 750 to 800 m. Both sexes dive night and day for weeks on end without sleeping and usually spend only around 4 minutes at the surface between dives. These short surface times between long deep dives suggest that these are not unusual dives but are the norm for this species. Further studies may show that the dive responses of Weddell seals, as outlined in Table 8.5, are essentially what all marine tetrapods do to varying degrees.

Deep-diving whales, such as sperm and beaked whales, are not accessible to researchers in the way Weddell and elephant seals are, yet recent surveys reveal that a beaked whale species surpasses either seal in both maximum dive depth and maximum dive duration, and that sperm whales dive for longer durations than seal species (Table 8.3). They exhibit many of the features described here for deep and prolonged divers.

8.5   Echolocation

Seawater is not very transparent to light, but it is an excellent transmitter of sound energy. Marine mammals have very good hearing, and they have taken advantage of the sound-conducting properties of water to compensate partially for the generally poor visibility found below the sea surface. Many marine mammals can obtain some information about their surroundings simply by listening to the environmental sounds that surround them. Others, however, have evolved systems for actively producing sounds to illuminate targets acoustically for detailed examination.

Soon after the first hydrophone was lowered into the sea, it became apparent that whales and pinnipeds could generate a tremendous repertoire of underwater vocalizations. Many of the moans, squeals, and wails that we can hear are evidently for communication. Bottlenose dolphins, *Tursiops*, produce a large variety of whistle-like sounds, and captive individuals have been shown to understand complex linguistic subtleties (**Figure 8.45**). Other sounds, especially those of the humpback

whale, *Megaptera*, have a fascinating musical quality. The songs of each humpback whale population are identifiably different from the songs of other populations, are probably produced exclusively by adult males advertising to females, and are culturally transmitted from one individual to another within each population. Each song is composed of numerous phrases, some of which are repeated several times. During each breeding season the songs evolve; some phrases are modified, and others are added or deleted.

About 20% of all mammal (and even a few bird) species have overcome the problems of orienting themselves and locating objects in the dark or underwater with echolocation. **Echolocation**, also referred to as *biosonar*, consists of animals producing sharp sounds and listening for reflected echoes as the sounds bounce off target objects. Bats are well-known echolocators, but so, too, are many marine mammals, particularly toothed whales. To echolocate successfully, an animal must be able to produce an appropriate sound signal, detect its echo, and then mentally process that signal to extract meaningful information about its immediate environment.

The sounds most useful for echolocation are neither squeals nor songs but trains of broad-frequency clicks of very short duration. Much more is known about the echolocating capabilities of the smaller whales, such as *Tursiops*, because they are frequently maintained in captivity for convenient study. *Tursiops* use clicks consisting of sound frequencies audible to humans as well as higher-frequency clicks, often exceeding 150 kHz (**Figure 8.46a**). Because our human ears are sensitive to airborne sound frequencies between about 18 vibrations per second (or hertz) and 18,000 Hz (18 kHz), most of the energy in dolphin echolocation clicks is well beyond the upper range of human hearing. These frequencies are referred to as being *ultrasonic*. Each click lasts only a fraction of a millisecond and is repeated as often as 600 times each second (**Figure 8.46b**). As each click strikes a target, a portion of its sound energy is reflected back to the source. Click repetition rates are adjusted to allow the click echo to return to the animal during an extremely short silent period between outgoing clicks. The time required for a click to travel from an animal to the reflecting target and back again is a measure of the distance to the target. As that distance varies, so will the time necessary for the echo to return. Continued reevaluation of returning echoes from a moving target can indicate the target's speed and direction of travel.

Relying solely on their echolocating abilities, captive blindfolded bottlenose dolphins have repeatedly demonstrated aptitudes for discriminating between objects of a similar nature: two fish of the same general size and shape, equal-sized plates of different metals, and pieces of metal differing only slightly in thickness. In the wild, these animals must acoustically survey their surroundings, while simultaneously distinguishing their own echolocation clicks from the many other sounds so frequently present in large herds of wild dolphins.

How do whales produce the sounds involved in echolocation? The larynx of toothed whales is well muscled and complicated in structure, yet it lacks vocal cords and is not used in sound production. The elongated tip of the larynx extends across the esophagus into a common tube leading to the

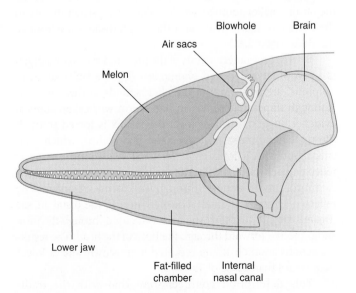

Figure 8.45 Midsection of a dolphin head, showing the bones of the head, the air passages, and the structures associated with sound production and reception.

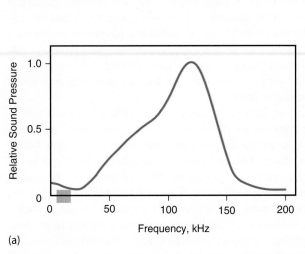

(a)

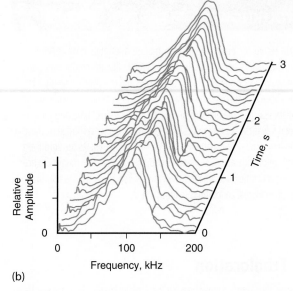

(b)

Figure 8.46 (a) Power spectrum of a typical echolocation click and (b) display of the power spectra of a single series of echolocation clicks of a bottlenose dolphin. The green band in (a) spans the frequency range of human hearing.

blowhole to separate the pathways for food and air completely. Just inside the blowhole is a pair of heavily muscled valves, the nasal plugs. Associated with the nasal plugs and a complex of air sacs branching from the nasal passage are the dorsal bursae, which drive a pair of **phonic lips** that vibrate to produce the clicks. Clicks produced here are directed forward by the concave front of the skull and then focused by the fatty lens-shaped **melon**, the rounded forehead structure so characteristic of toothed whales (**Figure 8.47**), to concentrate the clicks into narrow directional beams. Recent research indicates that some species of toothed whales may also stun fish prey with intense blasts of sound energy, presumably using the same sound production system used for echolocation.

To make this acoustic picture even more complicated, at the same time a dolphin is producing a train of rapidly repeated echolocation clicks, it can simultaneously produce frequency-modulated tonal whistle signals that vary in pitch from 2 to 30 kHz and direct those emitted sound signals forward from the melon in different directional beam patterns. It can do this while continually varying the frequency content of the clicks to adjust to the changing background noise or to the acoustic characteristics of the target.

Sperm whales are notable for their massive and very distinctive foreheads and extremely powerful and structurally complex echolocation signals. Inside their forehead is a highly specialized organ, the **spermaceti organ**, which may occupy 40% of the whale's total length and 20% of its body weight. This organ is filled with a fine-quality liquid, or spermaceti oil, once prized by whalers for candle making and for burning in lanterns. The spermaceti organ is encased within a wall of extremely tough connective tissue just above another fat-filled organ of similar size, the **junk**. The junk is thought to be homologous with the melon of smaller toothed whales. The entire structure sits in the hollow of the rostrum and the amphitheater-like front of the skull (**Figure 8.48**).

Between the anterior ends of the junk and spermaceti organ is a pair of large opposed phonic lips (**Figure 8.49a**) that, as in dolphins, are the origin of the click sounds of these whales. Although impossible to test in live sperm whales, anatomical evidence suggests that air from the larynx is forced through the right nasal passage to open the phonic lips, which then snap shut with a loud clap. Most of that sound energy is reflected backward by the frontal air sac (acting as an acoustical mirror) and channeled through the spermaceti organ to be reflected again by another acoustical mirror, the distal air sac (**Figure 8.49b**). From there, the click is focused through the junk and projected forward through the front of the head. The resulting echolocation clicks are repeated more slowly and at lower frequencies than those of *Tursiops*.

This description conflicts somewhat with the traditional view of sperm whale echolocation clicks, which have been described as reverberant pulses lasting about 24 ms and

Figure 8.47 A bottlenose dolphin (*Tursiops*) with a prominent melon.
Courtesy of James Sumich.

Figure 8.48 Sperm whale skeleton on display in the Museum of the World Ocean, Kalinigrad.
© Courtesy of Fastboy/Creative Commons.

consisting of a series of several rapidly reverberating individual clicks. The reverberant click sounds long reported in the literature appear to represent "leakage" of some of the click energy from the sides of the whale's head as the click signal bounces repeatedly between the two acoustical mirrors at either end of the spermaceti organ. Researchers recently demonstrated that when a recording hydrophone is directly in the frontal sound beam of an echolocating sperm whale, there is no click reverberation, and the sound intensity level of the single focused click is the loudest sound yet recorded from any natural biological source.

These intense clicks of sperm whales travel for several kilometers in the sea. The powerful long-range echolocation systems of sperm whales may partially explain their success as efficient predators of larger mesopelagic squid. Try to visualize these whales cruising along at the sea surface with a constant supply of air, periodically scanning the unseen depths below with a short burst of echolocation click pulses. Only when a

(a)

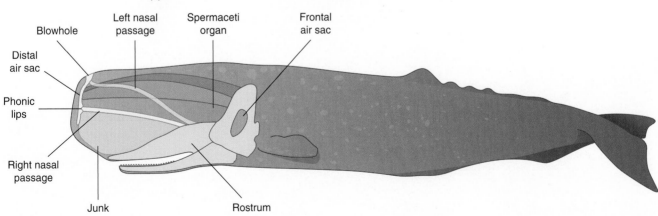

(b)

Figure 8.49 (a) Phonic lips of a sperm whale, and (b) a cutaway view of the complex structure of a sperm whale head.
(a) Courtesy of Dr. Ted W. Cranford, San Diego State University. (b) Data from K. S. Norris and G. W. Harvey, *Animal Orientation and Navigation* (NASA, 397–417, 1972).

target worthy of pursuit is detected and its location pinpointed does the whale depart from its air supply and go after its meal (**Figure 8.50**).

How common is echolocation in marine mammals? Presently, it is uncertain because it is difficult to establish whether wild populations are indeed using the echolocation-like clicks for the purposes of orientation and location. If judgments can be made from the types of sounds produced, then echolocation is assumed to occur in all toothed whales, some pinnipeds, and possibly a few baleen whales. Click series with echolocation-like qualities have been recorded in the presence of gray whales in the North Pacific and blue and minke whales in the North Atlantic. It is not unreasonable to assume that these animals use these sounds, as well as any other sensory means they possess, to find food, locate the bottom, and evaluate the nonvisible portion of their surroundings.

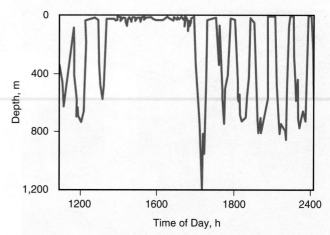

Figure 8.50 Continuous 13-hour trace of nine deep dives by a single adult sperm whale. Depths during the dives were recorded by a digital time–depth recorder.

STUDY GUIDE

TOPICS FOR DISCUSSION AND REVIEW

1. Compare and contrast the thermoregulatory strategies of marine birds and mammals.

2. List two specific structural features that distinguish each of the following marine mammal groups from the others: baleen whales, toothed whales, sea lions, seals, manatees.

3. Why do the growth rates of young marine mammals need to be truly astounding? What drives the need to grow very large, very quickly?

4. Why do all mysticetes show reverse sexual dimorphism?

5. How do killer whale pods differ from the groups or herds of many other social mammals?

6. List some unique features of sperm whales that make them extremely social animals.

7. Describe how some marine tetrapods are able to dive to depths greater than 1,000 m.

8. Describe how some marine tetrapods are able to hold their breath for more than 1 hour.

KEY TERMS

aerobic dive limit (ADL) 237	hypoxia 235
alveoli 235	junk 240
aorta 237	kleptoparasitic 211
apneustic breathing 235	lanugo 216
baleen 222	matrilineal 229
blastocyst 220	melon 240
blubber 208	pelage 207
bradycardia 237	phonic lips 240
bronchi 235	pod 229
bubble-net feeding 224	poikilothermic 235
conduction 207	seasonal delayed implantation 220
convection 207	sexual dimorphism 218
echolocation 239	spermaceti organ 240
electromagnetic radiation 207	sperm competition 228
evaporation 207	trachea 235
exothermic 207	vasoconstriction 237
homeothermic 235	

LIST OF KEY *GENERA*

Aptenodytes	Enhydra
Dermochelys	Eretmochelys
Eschrichtius	Phoca
Eumetopias	Pygoscelis
Fulmarus	Physeter
Leptonychotes	Stenella
Lutra	Trichechus
Megaptera	Tursiops
Mirounga	Zalophus
Monachus	Ziphius
Odobenus	

REFERENCES

Ainley, D. G., E. F. O'Connor, and R. J. Boekelheide. 1984. *The Marine Ecology of Birds in the Ross Sea, Antarctica.* Washington, DC: American Ornithologists Union.

Amundin, M. 1998. Sound production and hearing in marine animals. *Bioacoustics* 9(3):213–214.

Barlow, J., L. Rojas-Bracho, C. Muñoz-Piña, and S. Mesnick. 2010. Conservation of the vaquita (*Phocoena sinus*) in the northern Gulf of California, Mexico. In: R. Q. Grafton, R. Hilborn, D. Squires, M. Tait, and M. Williams, eds. *Handbook of Marine Fisheries Conservation and Management.* New York: Oxford University Press. Chapter 15.

Bartholomew, G. A. 1970. A model for the evolution of pinniped polygyny. *Evolution* 24:546–559.

Berta, A., C. E. Ray, and A. R. Wyss. 1989. Skeleton of the oldest known pinniped, *Enaliarctos mealsi. Science* 244:60–62.

Berta, A., J. L. Sumich, and K. M. Kovacs. 2005. *Marine Mammals: Evolutionary Biology*, 2nd ed. San Diego: Academic Press.

Bonnell, M. L., and R. K. Selander. 1974. Elephant seals: Genetic variation and near extinction. *Science* 184:908–909.

Bonner, W. N. 1982. *Seals and Man: A Study of Interactions.* Seattle: University of Washington Press.

Bonner, W. N. 1989. *Whales of the World.* New York: Facts on File.

Cantu-Guzman, J. C., Oliviera-Bonavilla, A. and Sanchez-Saldana, M. E. 2015. A history (1990–2015) of mismanaging the vaquita into extinction—A Mexican NGO's perspective. *Journal of Marine Animals and Their Ecology* 8:15–25.

Clarke, A., and C. M. Harris. 2003. Polar marine ecosystems: Major threats and future change. *Review Environmental Conservation* 30:1–25.

Clinton, W. L. 1994. Sexual selection and growth in male northern elephant seals. In: B. J. Le Boeuf and R.M. Laws, eds. *Elephant Seals.* Berkeley: University of California Press. pp. 154–168.

Croxall, J. P. 1987. *Seabirds: Feeding Ecology and Role in Marine Ecosystems.* New York: Cambridge University Press.

DeLong, R. L., and B. S. Stewart. 1991. Diving patterns of northern elephant seal bulls. *Marine Mammal Science* 7:369–384.

Diamond, A. W., and C. M. Devlin. 2003. Seabirds as indicators of changes in marine ecosystems: Ecological monitoring on Machias Seal Island. *Environmental Monitoring and Assessment* 88:153–181.

Elsner, R., and B. Gooden. 1983. *Diving and Asphyxia: A Comparative Study of Animals and Men.* New York: Cambridge University Press.

Fish, J. F., J. L. Sumich, and G. L. Lingle. 1974. Sounds produced by the gray whale, *Eschrichtius robustus*. *Marine Fisheries Review* 36:38–45.

Folkens, P. A, R. R. Reeves, B. S. Stewart, P. J. Clapham, and J. A. Powell. 2002. *Guide to Marine Mammals of the World*. New York: Alfred A. Knopf.

Geraci, J. R. 1978. The enigma of marine mammal strandings. *Oceanus* 21:38–47.

Gero, S., J. Gordon, and H. Whitehead. 2015. Individualized social preferences and long-term social fidelity between social units of sperm whales. *Animal Behavior* 102:15–23.

Gerstein, E. 2002. Manatees, bioacoustics and boats. *American Scientist* 90(2):154.

Herman, L. M., ed. 1980. *Cetacean Behavior: Mechanisms and Functions*. New York: John Wiley & Sons.

Hunt, G. L. J. 1991. Marine ecology of seabirds in polar oceans. *American Zoologist* 31:131–142.

Jefferson, T. A, M. A. Webber, and R. L. Pitman. (2008). *Marine Mammals of the World: A Comprehensive Guide to Their Identification*. Amsterdam: Elsevier.

Kooyman, G. 1989. *Diverse Divers: Physiology and Behavior*. New York: Springer-Verlag.

Kooyman, G. L., M. A. Castellini, and R. W. Davis. 1981. Physiology of diving in marine mammals. *Annual Review of Physiology* 43:343–357.

Laws, R. M. 1961. Reproduction, age and growth of southern fin whales. *Discovery Reports* 31:327–486.

Laws, R. M. 1985. The ecology of the southern Ocean. *American Scientist* 73:26–40.

Lockyer, C. 1981. Estimates of growth and energy budget for the sperm whale, *Physeter catodon*. *Mammals in the Sea* 3:489–504.

Mackintosh, N. A. 1966. The distribution of southern blue and fin whales. In: K. S. Norris, ed. *Whales, Dolphins and Porpoises*. Berkeley, CA: University of California Press. pp. 125–144.

McIntyre, T., H. Bornemann, P. N. De Bruyn, R. R. Reisinger, D. Steinhage, M. E. Márquez, M. N. Bester, and J. Plötz. 2014. Environmental influences on the at-sea behaviour of a major consumer, *Mirounga leonina*, in a rapidly changing environment. *Polar Research*, 33:23808. doi:10.3402/polar.v33.23808.

Nelson, C., and K. Johnson. 1987. Whales and walruses as tillers of the seafloor. *Scientific American* 256:112–117.

Nemoto, T., M. Okiyama, N. Iwasaki, and T. Kikuchi. 1988. Squid as predators on krill (*Euphausia superba*) and prey for sperm whales in the Southern Ocean. In: D. Shahrage, ed. *Antarctic Ocean and Resources Variability*. New York: Springer-Verlag. pp. 292–296.

Nerini, M. 1984. A review of gray whale feeding ecology. In: M. L. Jones, S. L. Swartz, and S. Leatherwood, eds. *The Gray Whale*, Eschrichtius robustus (Lilljeborg, 1861). New York: Academic Press. pp. 423–450.

Nettleship, D. N., G. A. Sanger, and P. F. Springer, eds. 1985. *Marine Birds: Their Ecology and Commercial Fisheries Relationships*. Ottawa: Canadian Wildlife Services.

Nevitt, G. 1999. Foraging by seabirds on an olfactory landscape. *American Scientist* 87:46–53.

Norris, K. S. 1968. Evolution of acoustic mechanisms in odontocete cetaceans. *Evolution and Environment* 297–324.

Norris, K. S., and G. W. Harvey. 1972. A theory for the function of the spermaceti organ of the sperm whale (*Physeter catodon*). In: K. S.

Norris, ed. *Animal Orientation and Navigation*. Washington, DC: NASA. pp. 397–417.

Perrin, W. F., B. Wursig, and J. G. M. Thewissen, eds. 2002. *Encyclopedia of Marine Mammals*. New York: Academic Press.

Pierotti, R., and C. A. Annett. 1990. Diet and reproductive output in seabirds. *BioScience* 40:568–574.

Pivorunas, A. 1979. The feeding mechanisms of baleen whales. *American Scientist* 67:432–440.

Quetin, L. B., and R. M. Ross. 1991. Behavioral and physiological characteristics of the Antarctic krill, *Euphausia superba*. *American Zoologist* 31:49–63.

Rice, D. W., and A. A. Wolman. 1971. *The Life History and Ecology of the Gray Whale* (Eschrichtius robustus). Stillwater, OK: American Society of Mammalogists.

Ridgway, S. H. 1972. *Mammals of the Sea, Biology and Medicine*. Springfield, IL: Charles C. Thomas.

Schorr, G. S., E. A. Falcone, D. J. Moretti, and R. D. Andrews. 2014. First long-term behavioral records from Cuvier's beaked whales (*Ziphius cavirostris*) reveal record-breaking dives. *PLoS ONE* 9(3):e92633. doi:10.1371/journal.pone.0092633.

Schreer, J. F., and K. M. Kovacs. 1997. Allometry of diving capacity in air-breathing vertebrates. *Canadian Journal of Zoology* 75:339–358.

Shirihai, H., and B. Jarrett. 2006. *Whales, Dolphins and Other Marine Mammals of the World*. Princeton, NJ: Princeton University Press.

Stevens, J. E. 1995. The Antarctic pack-ice ecosystem. *BioScience* 45:128–221.

Stewart, B. S., and R. L. Delong. 1993. Seasonal dispersion and habitat use of foraging northern elephant seals. *Symposium of the Zoological Society of London* 66:179–194.

Tovar, H., V. Guillen, and M. E. Nakama. Monthly population size of three guano bird species off Peru, 1953 to 1982. In: D. Pauly and I. Tsukayama, eds. *The Peruvian Anchoveta and Its Upwelling System: Three Decades of Change*. Callso, Peru: ICLARM Studies and Reviews. pp. 208–218.

VanBlaricom, G. R., and J. A. Estes. 1988. *The Community Ecology of Sea Otters*. New York: Springer-Verlag.

Watkins, W. A., K. E. Moore, and P. Tyack. 1985. Investigations of sperm whale acoustic behaviors in the southeast Caribbean. *Cetology* 49:1–15.

Whitehead, A. L., P. Lyver, G. Ballard, K. Barton, B. J. Karl, K. M. Dugger, S. Jennings, A. Lescroël, P. R. Wilson, and D. G. Ainley. 2015. Factors driving Adélie penguin chick size, mass and condition at colonies of differing size in the southern Ross Sea. *Marine Ecology Progress Series* 523:199–213.

Wikelski, M., and C. Thom. 2000. Marine iguanas shrink to survive El Nino. *Nature* 403:37.

Wilson, R. P., B. Culik, D. Adelung, N. Ruben Coria, and H. J. Spairani. 1991. To slide or stride: When should Adélie penguins (*Pygoscelis adeliae*) toboggan? *Canadian Journal of Zoology* 69:221–225.

Würsig, B. 1988. The behavior of baleen whales. *Scientific American* 258:102–107.

Yeates, L. C., T. M. Williams, and T. L. Fink. 2007. Diving and foraging energetics of the smallest marine mammal, the sea otter (*Enhydra lutris*). *Journal of Experimental Biology* 210(11):1960–1970.

Zopal, W. 1987. Diving adaptations of the Weddell seal. *Scientific American* 256:100–105.

CHAPTER

9

Estuaries

A common inhabitant of estuaries, a great blue heron.
Image Courtesy of National Oceanic and Atmospheric Administration.

Estuaries are semienclosed coastal embayments where at least one freshwater source, such as a river or stream, meets the sea. Freshwater and seawater mix within these embayments, creating unique and complex ecosystems with salinities that are not quite as high as the neighboring ocean water, but not as low as the rivers or streams that feed freshwater into the system. Familiar places such as the Chesapeake Bay, San Francisco Bay, Great South Bay, Tampa Bay, Puget Sound, and the Mississippi River Delta are among the 100 or so bodies of water officially designated as estuaries in the United States. Over one third of the U.S. population lives within the drainage basins of estuaries. Because the estuarine ecosystem is located at the boundary of the land and sea where humans live, drive, work, and play, these transitional coastal habitats demonstrate the problems and challenges created by human intervention into the workings and the very structures of marine ecosystems. Many estuaries are located adjacent to or below a major highway, train tracks, other roads, or even all of these human-made structures.

Estuaries are highly variable ecosystems that continually change in response to local physical, geological, chemical, and biological factors. The transition from freshwater to saltwater in estuaries of large rivers such as the Columbia River may extend over 100 km inland, whereas the estuaries of small streams may be only a few hundred meters in extent, and the water in them may be well-mixed, highly stratified, or any combination in between. The size and shape of an estuary are influenced by its depth and geological history. Tectonic movement of Earth's crust has elevated and lowered coastal areas, and the formation and melting of ice age glaciers have alternately removed and returned water from the ocean basins. The resulting changes in sea level alter the size and shape of estuaries by altering the water depth and the extent of submerged coastal features.

Physical forces at work also influence the chemistry of an estuary. When freshwater draining from a coastal watershed mixes with the seawater pushing upstream during high tides, suspended river sediments settle to the bottom and become part of the accumulating sediment blanket of the estuary. Because many pollutants are transported downstream in the river water or are adsorbed onto sediment particles, individual estuarine conditions also influence the fate and availability of pollutants to the inhabitants of the ecosystem.

In their natural states, estuaries are among the most biologically productive ecosystems on Earth. Their rates of primary productivity rival and often exceed those of coral reefs, rain forests, and even intensively cultivated corn fields. These special habitats are created by the combination of turbulent mixing, daily fluctuating tidal cycles, and the downstream flow of freshwater that usually changes seasonally in velocity and volume.

When these forces meet in an estuary, they exert considerable and complicated effects on the system, creating diverse aquatic habitats that are not quite typical of either the river or the sea. More than two thirds of the species of fish and shellfish harvested by commercial and sport fishers depend on estuaries for feeding or as nursery areas. Estuaries also provide crucial habitat for terrestrial and freshwater organisms, including many threatened, endangered, and rare species.

Estuaries and their surrounding wetlands are recognized as fragile environments that have been heavily used for and disturbed by human activities. Dredging of navigation channels in estuarine ports, filling estuarine wetlands for development, disposing of wastewaters from coastal communities, diverting rivers for irrigation purposes, and allowing pesticide- and fertilizer-contaminated runoff to flow into coastal watersheds have changed the character of estuaries and threatened their ecological integrity. Many estuaries that were once rich sources of fish, game, and shellfish have become stagnant and unproductive as a result of unregulated or poorly regulated economic exploitation and pollution. Many irreversible alterations have been made to estuarine habitats, and, as a result, an overall theme in estuarine science is restoration. Now that humans realize the many services healthy estuaries can provide, we are attempting to undo damage that has altered these coastal habitats and their species compositions. Unfortunately, runaway human population growth coupled with a rapid increase in per capita human consumption of resources in many countries makes these efforts a truly daunting challenge. Intensive and ongoing efforts requiring large amounts of money and manpower will be required to potentially restore the health of most estuaries around the world.

9.1 Types of Estuaries

Most estuaries are in their present physical state due to ancient patterns of river or glacial erosion that occurred during the last glacial maximum (LGM), when sea level worldwide was about 150 m lower than at present. These scoured river or glacial channels slowly moved into the configuration we see today as the great continental ice sheets melted and gradually flooded them. Some estuaries remain sensitive to slight changes in sea level; increases of only a few meters could dramatically increase the size of small estuaries (**Figure 9.1**), and comparable decreases could shrink others or even cause them to disappear.

Estuaries are found in some form along most coastlines of the world, but most are evident in wetter climates of temperate and tropical latitudes. In such areas, drainage of inland watersheds provides the necessary freshwater input at the head of the estuary to keep salinities below those of adjacent open-ocean waters. In North America, excellent examples of all major types of estuaries exist.

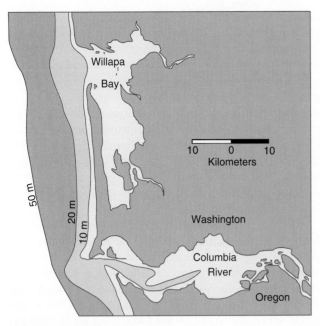

Figure 9.1 The Columbia River estuary and Willapa Bay with the present shoreline (green) and depth contours at 10, 20, and 50 m below.

Estuaries are classified by both their mode of formation and by their pattern of water circulation. First we will discuss classification by mode of formation. **Coastal plain estuaries**, also known as *drowned river valleys* (**Figure 9.2a**), lie along the north and central Atlantic Coast, the Canadian Maritime region, and many areas of the west coast of North America; this type of estuary includes the Columbia River and Chesapeake and Delaware Bays. These estuaries are broad, shallow embayments formed from deeper V-shaped channels as the sea level rose and flooded river mouths after the last episode of continental glaciation. The extent to which the sea has invaded these coastal river valleys since the LGM is determined by the steepness and size of the valley, its rate of river discharge, and the range and force of the tides of the adjacent sea. This type of estuary continues to be gradually modified as wave erosion cuts away some existing shorelines and creates others by building mudflats.

Bar-built estuaries are common along the south Atlantic Coast, the Gulf of Mexico (**Figure 9.2b**) in North America, and the coastal lowlands of northwestern Europe. These estuaries are formed as nearshore deposits of sand and mud transported by coastal wave action to build an obstruction, or **barrier island**, in front of a coastal area fed by one or more coastal streams or rivers. Often, these small coastal rivers and streams have little freshwater flow so the estuary may be partially or completely blocked by sand deposited by ocean waves. During rainy seasons, however, the increased runoff often temporarily reopens the estuary mouth. **Coastal lagoons** (Figure 9.2b) are similar to bar-built estuaries, but without a river or other strong source of freshwater input they lack the strong salinity gradients and mixing patterns characteristic of estuaries. Lagoons may receive freshwater from small seasonal streams and street runoff in urbanized areas, but the amount of freshwater is small.

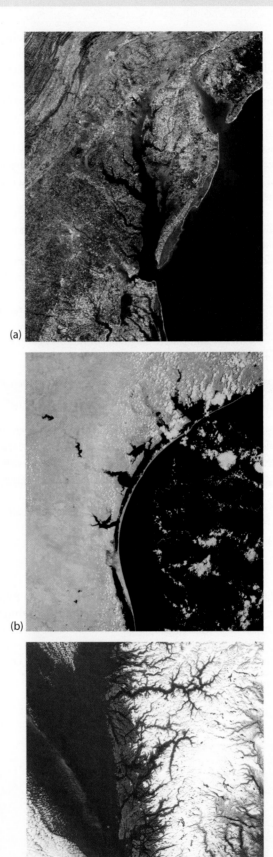

(a)

(b)

(c)

Figure 9.2 Satellite images of three types of estuaries. (a) Chesapeake and Delaware Bays, two coastal plain estuaries on the U.S. East Coast. (b) Several bar-built estuaries of the Texas coast. (c) Steep-sided fjords on the southwestern coast of Norway.
(a,b,c) Courtesy of NASA/Johnson Space Center.

Some estuaries have broad, poorly defined, fan-shaped mouths called **deltas**. The Mississippi, Mekong, Colorado, and Nile River deltas and other similar delta estuaries are created as heavy loads of sediments eroded from the upstream watersheds are deposited at the river mouth. Different still are **tectonic estuaries**, such as San Francisco Bay, created when the underlying land sank because of crustal movements of Earth. As coastal depressions created by these movements sank below sea level, they filled with water from the sea and also became natural land drainage channels, directing the flow of land runoff into the new estuary basin.

From the central North American west coast northward, estuaries become more deeply incised into coastal landforms and gradually merge into the deep glacially carved **fjords** characteristic of British Columbia and southeastern Alaska (**Figure 9.2c**). Fjords are also common on the coasts of Norway, southern New Zealand, and southern Chile. In cross section, fjords resemble the Black Sea, with deeper regions at the upstream reaches of fjords. The shallow sills at their mouths partially block the inflow of seawater and lead to stagnant conditions near the bottoms of deeper fjords.

DID YOU KNOW?

Of the 32 cities with the highest populations in the world, 22 are located on estuaries. This fact highlights the importance of estuaries and the coastal environment in general to humans, because humans choose to build cities within these ecosystems. It also reinforces the idea that estuarine life is tightly linked to human presence. With large cities built on or very near estuaries, there is no doubt that human activities will impact the health and longevity of estuaries.

9.2   Estuarine Circulation

In addition to their structural differences, estuaries also exhibit differing patterns of freshwater and seawater mixing within the basin and are also classified by their water circulation patterns. The upstream-to-downstream variations in salinity, water temperature, turbidity, and current action are complex and change markedly during a single tidal cycle and in response to seasonal changes in the volume of freshwater discharge from streams and rivers. Salinity values typically increase from the surface downward and from the estuary head downstream toward the mouth. As tides change sea level in a typical estuary, higher-density seawater moves in and out along the estuary bottom and is gradually mixed upward into the outflowing low-salinity surface water (**Figure 9.3**). Because of this mixing, there is generally an inward flow of nutrient-rich seawater along the bottom of the estuary and a net outward flow at the surface. On a localized scale, this upward mixing creates a process of estuarine upwelling that replenishes nutrients and promotes the growth of estuarine primary producers.

The shape of an estuary's basin is a major factor in determining its mixing pattern, and in effect its classification by mixing. A triangular estuary with a wide, deep mouth enables seawater to move farther upstream. The currents may be strong and the water is well mixed, and thus salinity and water density are nearly the same from the surface downward at any location within the estuary (**Figure 9.4a**). These are classified as **vertically mixed estuaries**. In narrow-mouthed estuaries, circulation is decreased, creating more pronounced vertical salinity gradients. These narrow-mouthed estuaries are often highly stratified. They have a well-defined seawater wedge under the less dense freshwater on the surface (**Figure 9.4b**), with an abrupt salinity change, or **halocline**, where seawater and freshwater meet. These are classified as **salt-wedge estuaries**. Estuaries that experience mixing at all depths, but water in the lower layers remains salty, are termed **slightly stratified**. Salinity is highest at the mouth of the estuary and decreases upstream. Fjords, described above, have restricted water circulation with the open ocean, and seawater rarely enters the fjord. Freshwater does flow out of the fjord and mix with the ocean. Overall, fjords have very little tidal mixing and remain highly stratified. By analyzing water samples taken at fixed depths throughout a tidal cycle and connecting points with the same salinity values, lines of equal salinity, called **isohalines**, can be plotted. The shape of these isohalines is useful in classifying types of estuaries by mixing pattern and in understanding the distribution of estuarine organisms.

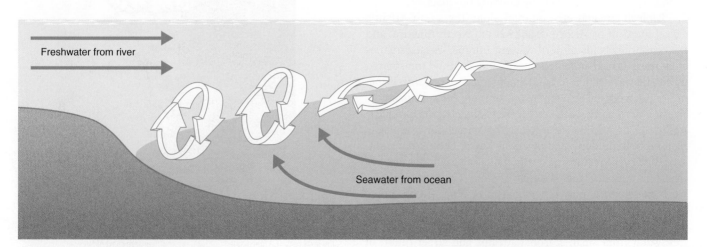

Freshwater from river

Seawater from ocean

Figure 9.3 The general pattern of freshwater and seawater mixing in an estuary.

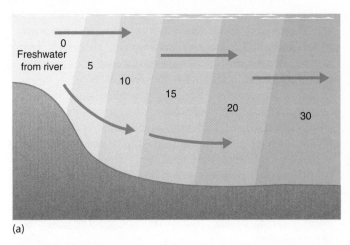

(a)

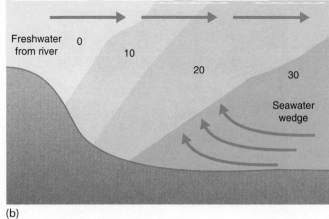

(b)

Figure 9.4 Cross sections of well-mixed (a) and stratified (b) estuaries. Numbers indicate salinities in parts per thousand (‰).

In addition to the more predictable effects of tides and river discharge, circulation in estuaries often changes rapidly and less predictably in response to short-term influences of heavy rainfall or even changing winds. The Coriolis effect also exercises its influence on circulation patterns within estuaries by forcing seawater farther upstream on the left sides (when facing seaward) of estuaries in the Northern Hemisphere and on the right sides of estuaries in the Southern Hemisphere.

The time necessary for the total volume of water in an estuary to be completely replaced is called the **flushing time**. Flushing times range from days to years depending on the estuary's combination of geography, tides, river flow, wind, and salinity gradients. The flushing time in an estuary strongly influences

the transport of nutrients and is also a crucial factor in determining the fates and impacts of pollutants in estuaries.

DID YOU KNOW?

One of the largest fjords in the world, Sognefjorden, is found in Norway and extends more than 160 km (nearly 100 miles) from the sea. At the bottom of fjords in Norway live some of the world's largest and most recently discovered coral reefs, although these reefs are not like those found in warm, tropical waters. These reefs exist in the dark, cold water under great water pressure, and scientists know very little about them. Created by glaciers, fjords are one of the very unique estuarine habitats found along just several coasts of Earth.

Case Study

A Small and Marvelously Strange Estuarine Fish, the Mangrove Rivulus

Common name: Mangrove rivulus, mangrove killifish

Scientific name: *Kryptolebias marmoratus* (formerly known as *Rivulus marmoratus*)

Distribution: From south to central Florida at the northern end of the range, south to the West Indies, South America, Cuba, the Bahamas, Jamaica, and the Mexican Yucatan Peninsula

Endangered Species Act status: Species of Concern

Threats: Habitat destruction or alteration, mosquito control, urban development

The mangrove rivulus went unnoticed for many years due to its small size, cryptic coloration, and shy nature (**Figure A**). The species reaches a maximum size of 7.6 cm (3 in.), is generally brown in coloration, mottled with black dots, and sometimes has a bit of orange on the body and fins. This species is found in the warm climates of Florida and the Caribbean, exclusively in and around estuaries lined with mangroves (**Figure B**).

The mangrove rivulus is capable of life in a wide range of conditions. Although it prefers water temperatures in the range of 18°C to 24°C, it can survive temperatures in the range of 5°C to 38°C. In the wild, it has been found in salinities ranging from 0 to 69 parts

Figure A A mangrove rivulus.
NOAA.

per thousand (‰), and in a laboratory setting can survive salinities as high as 70 to 80‰. It is also one of very few fish species that can survive out of water for extended periods—up to 30 days! It was discovered living inside tree logs, taking advantage of the hiding spot, and fully capable of performing life's functions in a dry environment. It appears that a favorite hiding spot for this species is inside land crab burrows. The mangrove rivulus has specialized gills that retain water and nutrients for extended periods

(*continues*)

Figure B Typical habitat for a mangrove rivulus, mangrove roots.
NOAA.

when out of water. When on land, the fish breathe and excrete nitrogenous wastes across their skin. When the fish return to water their gills once again function similarly to those of other water-breathing animals. These adaptations for survival in a wide range of habitats and environmental conditions make the mangrove rivulus a relatively hardy species, but, like other species, it has no defense against habitat destruction.

Not only is this species unique in its tolerance for a wide variety of living conditions, but a recent investigation led to the discovery of unique and variable reproductive methods. All mangrove rivulus are internal fertilizers that lay eggs, which is not very common in fish. Some populations are also capable of self-fertilization, producing genetic clones, which is rare in the world of hermaphrodites. In these populations it appears that no females exist. The fish are either male or simultaneous hermaphrodites. This reproductive method that does not require females is the only known example for vertebrates. Populations in Belize appear to be nonhermaphroditic, reproducing in a more traditional vertebrate manner. Eggs only require a moist environment, but do not need to be submerged in water, and they hatch in 2 to 4 weeks.

Because the mangrove rivulus is so tolerant of a wide range of conditions, it can be used as an indicator species for extremely poor conditions. This species can live in a range of salinities that most cannot endure, can survive in polluted waters, requires very little oxygen, and can move onto land if necessary. A mangrove rivulus die-off would indicate conditions that are toxic to most other life forms. The fragmentation of habitat in Florida and the Caribbean threatens this species, as mangroves are removed or thinned out and high marshes are impounded for mosquito control. Because of their various habitats, this species is very difficult to study. Currently, no estimates of population sizes exist, and studies are ongoing to unravel more of the puzzle of the life of this unique fish.

Critical Thinking Questions

1. Given that the mangrove rivulus can live in areas with very little water or even out of water, do you think this species is more closely related to terrestrial vertebrates than other fish species are? Why or why not?

2. With as much creativity as you can come up with, propose a research plan to study the mangrove rivulus, keeping in mind their patchy distribution and the challenges that scientists have had thus far while attempting to study them.

9.3   Salinity Adaptations

To survive in fluctuating estuarine conditions, benthic organisms must be able to tolerate frequent changes in salinity, which often lead to internal osmotic stresses. A small fraction of animal species that live in estuaries, especially insect larvae, a few snails, and some polychaete worms, have their closest relatives in freshwater; however, most are derived from marine forms (**Figure 9.5**) and include some of the same species found on nearby beaches. Some animal species have poorly developed osmoregulatory capabilities, so to survive, they must avoid osmotic problems by not venturing too far into the variable or low-salinity portions of estuaries. Others use a variety of adaptive strategies to overcome the osmotic problems of recurring exposure to low and variable salinities of estuarine waters. Some of these adaptations are modifications of structural or physical systems already used for survival on exposed intertidal shorelines. Oysters and other bivalve mollusks, for instance, simply stop feeding and close their shells when subjected to the osmotic stresses of low-salinity water. Isolated within their shells, they switch to anaerobic respiration and await high tide, when water higher in salinity and O_2 returns. Other animal species retreat into mud burrows that act to trap water, and as a result salinity fluctuations caused by tidal cycles are usually much less severe (**Figure 9.6**).

A few species of tunicates, sea anemones, and other soft-bodied estuarine epifauna are **osmotic conformers**. Osmotic conformers are unable to control the osmotic flooding of their tissues when subjected to low salinities, so their body

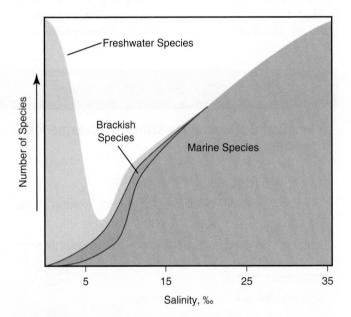

Figure 9.5 Relative contributions of freshwater, brackish, and marine species to estuarine fauna.
Data from Rename, A, *Zoologischer Anzeiger Supplementband* 7 (1934):34–74.

fluids fluctuate to remain isotonic with the water around them (**Figure 9.7**). These organisms must be able to tolerate salinity changes if they are to live in estuarine waters because they have no mechanisms in place to maintain an osmotic state that is different from the water in which they live.

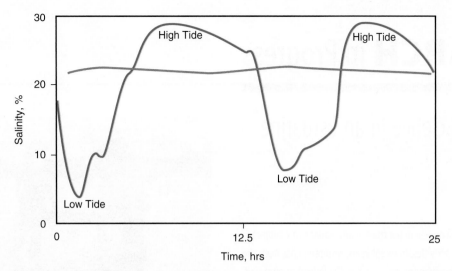

Figure 9.6 Comparison of salinity variations through a typical tidal cycle of interstitial water (green) with that of the overlying water (red) in Pocasset Estuary, Massachusetts. Data from Mangelsdorf, P. C., *Estuaries* 83 (1967): 71–79.

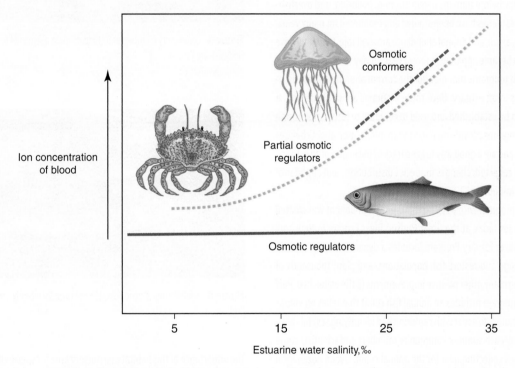

Figure 9.7 Variations in ion concentrations of body fluids or blood with changing external water salinities for osmotic conformers (cnidarians and ctenophores), partial osmotic regulators (crustaceans and squid), and osmotic regulators (vertebrates).

The most successful and abundant groups of estuarine animals have evolved mechanisms to stabilize the concentrations of ions in their body fluids despite external variations. These mechanisms are as varied as the organisms themselves, yet all involve systems that acquire essential ions from the external medium and excrete excess water as it diffuses into their bodies. The body fluids of estuarine crabs remain nearly isotonic with their external medium when in seawater but become progressively hypertonic as the seawater becomes more dilute. When

these partial osmotic regulators are subjected to reduced salinities, additional ions are actively absorbed by their gills to compensate for the ions lost in their urine (Figure 9.7). Thus, these and most other estuarine crustaceans are osmotic conformers at or near normal seawater salinities and **osmoregulators** in dilute seawater.

Most estuarine animals are **stenohaline**; they can tolerate exposure only to limited salinity ranges, and therefore occupy only a limited portion of the entire range of salinity regimes

RESEARCH in Progress

Community Science in an Estuarine System

Because of their locations in or near major cities, many estuaries are subject to great stress due to human activities. In recent years, restoration has been the goal of many city and state governments, as well as the federal government, in an attempt to return proper flow, health, and vitality to estuaries. One common challenge with restoration efforts is obtaining knowledge of the estuarine system's original state. Which species—plants, algae, and animals—were found in each estuary before humans began altering, polluting, and overfishing the area? Some estuaries have been studied and monitored for many years, and records of plant, algae, and animal abundance are available, but many have not. As this problem became apparent in recent years, many resource managers began to implement programs that monitor the current abundance and distribution of organisms in an estuary. Once these programs are implemented, a current baseline can be established and used to compare to future monitoring data. Monitoring programs provide managers with information about changes occurring in an area and are a good way to keep track of the health of an estuary. As years of data are collected, changes in species distributions and abundance are tracked and examined.

A monitoring program with a strong community component was created in 2012 and continues today at the Great Hudson River Estuary in New York State. The Hudson River Estuary Program includes a dynamic plan to improve water quality, manage and restore fish populations, and plant thousands of native plants, among many other positive improvements to the watershed. Part of the monitoring program includes an annual fish count that relies on volunteers from the community to assist with capturing and identifying several thousand fish. On one day each summer community members and scientists come together with various collecting gear for the annual count. Over a dozen local organizations participate and act as partner organizations for this large monitoring effort. Seine nets, minnow traps, and rods and reels are used to collect fish in a substantial effort to create a snapshot of fish populations for the year (**Figure A**). Fish data are collected, and soon after the fish are returned to the water, dazed but mostly unharmed.

The results of this effort thus far have provided several key pieces of information. First, they reveal a diversity of freshwater, marine, and anadromous species in various parts of the estuary. Some of the species occurrences were surprising based on previous studies on salinity ranges. Since the study's inception, 37 fish species have been identified. Well over 50% of the catch was young of the year, or juvenile, striped bass (**Figure B**) and herring (**Figure C**), emphasizing

Figure A Community members and students using a seine net to capture fish in the Hudson River Estuary.
USDA Forest Service.

Figure B Adult Atlantic striped bass, *Morone saxatilis*, recently captured.
NOAA.

the importance of this habitat as a nursery area for species humans like to consume. The results have also shown some variability at sites during different years, with fluctuating numbers of many species.

The unique and very impactful aspect of the Great Hudson River Estuary annual fish count is the ability to collect a great amount of data from a large geographic area in just 1 day. With so many people involved, the data collection that would take one scientist with his or her crew several months can take place all at once. This is important, because weather patterns are dramatic in New York State, so data collection on a single day at many sites versus a span of several months allows for more realistic comparisons between sites. Although the fish count only provides a snapshot of the fish found in the estuary during the summer, these annual counts are useful for monitoring the health of the

Figure C A bucket full of Atlantic herring, *Clupea harengus*.
NOAA Fisheries.

local fisheries and the ecosystem as a whole. As restoration efforts continue, scientists hope to continue to see steady and rising fish populations. Monitoring efforts that include the involvement of community members will likely lead to more investment in the health and restoration of the estuary, as volunteers literally end up wet and dirty while directly assisting with fish collections. The

Great Hudson River Estuary fish count and other community programs serve as models for estuary programs in other areas to implement similar strategies for public involvement. The health of our estuaries impacts all of us.

Critical Thinking Questions

1. Given that citizens are collecting a large amount of the data on the annual fish count day, do you feel that the data will be as robust as if scientists were collecting it? What are some potential inconsistencies you would be concerned with, and what could be (and may already be) done to reduce bias or inconsistent data?

2. What are three advantages to involving the local community in environmental research activities?

3. Perform an Internet search to try to find a local aquatic organization that citizens in your area can participate in. Describe the results of your search and consider joining the efforts.

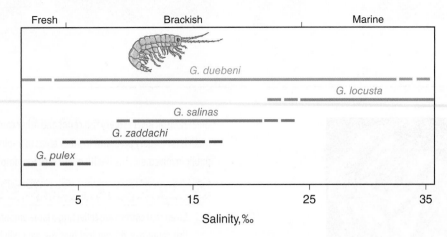

Figure 9.8 Differing salinity tolerances of five species of amphipods (*Gammarus*). Of these, only *G. duebeni* (pictured) is euryhaline.
Data from J. A. C. Nicol., *The Biology of Marine Animals* (Pitman and Sons, 1967).

available within an estuary (**Figure 9.8**). A few opportunistic species of estuarine organisms are **euryhaline**, capable of withstanding a wide range of salinities. These species can be found throughout the range of estuarine salinities, with a limited number of euryhaline species also found in high-salinity lagoons that fringe some of the world's arid coastlines. Lagoons such as those along the coast of Texas and both sides of northern Mexico have shallow bottoms, high summer temperatures, excessive evaporation, and high salinities. The osmotic problems experienced by animal species in these high-salinity lagoon populations are similar to those encountered by bony fishes in seawater and are so severe that reproduction is seldom successful. Continued immigration of euryhaline species from nearby estuaries is necessary to sustain these lagoon populations.

Osmotic adaptations of species are continually surprising scientists, as more and more marine species are discovered inhabiting estuaries, even if only on occasion. Until recently the importance of estuarine habitat as intermittent feeding or temporary nursery areas was not quite realized for many species. It might surprise you to learn that many species of sharks, often viewed as large, open-ocean predators, rely on shallow inshore areas, including wetlands and their associated tidal creeks, as their nursery grounds. For example, at least nine species of large sharks are known to use Bulls Bay, South Carolina, as a nursery area. Pregnant female blacktip, sandbar, dusky, smooth hammerhead, and spinner sharks, among others, enter the bay in the spring of each year to give live birth to their pups. Soon afterward, the large adults depart, but the juvenile sharks often remain in the shallow protected waters of the nursery for periods of up to 5 years and are osmotically capable of doing so. Our understanding of the importance of wetlands to a great variety of marine species, including even large, pelagic predators such as sharks, has been acquired in just the past 20 years and has resulted in a fundamental shift in our efforts to conserve many of these species by recognizing and protecting their estuarine habitat.

Species diversity and numbers of individuals usually decline considerably from a maximum near the ocean to a minimum near the headwaters of an estuary. The distributional patterns of estuarine animals are governed by salinity variations, patterns of food and sediment preferences, current action, water temperature variations, and competition between species. It is the collective interaction of these factors and others that establishes and maintains the distribution limits of estuarine organisms.

DID YOU KNOW?

Weather patterns can determine species compositions in an estuary. During drought periods, less freshwater will enter an estuary because river or stream flow is reduced from the lack of rain. Most organisms within estuaries have specific salinity ranges they can tolerate (stenohaline), so species distributions for some species will differ between drought periods and periods with average rainfall. A clear example of this relationship is observed in the St. Lucia estuarine system in South Africa, where during drought periods a razor clam is the dominant bivalve mollusk. When weather patterns shift and rain returns, a mussel takes over as the dominant bivalve mollusk. Estuaries are sensitive to many environmental changes, weather being no exception.

9.4 A Survey of Habitat and Community Types

Creation of Habitats with Sediments

Sediments are transported into estuaries from rivers that drain coastal watersheds and from coastal areas outside the estuary mouth. River sediment particles range in size from gravels and coarse sands to fine silts, clays, and organic detritus They are derived from erosion of river banks stripped of their natural plant cover and from the scouring of meandering river channels. As fast-moving rivers widen and slow when they enter coastal flood plains, they begin to meander, and their loads of suspended sediments settle to the bottom. As a result, estuaries serve as very effective catch basins for much of the fine suspended sediments washed off the land. Current speeds

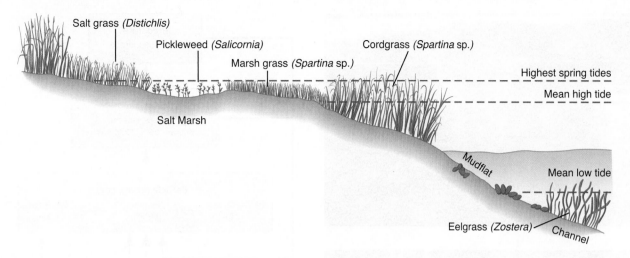

Figure 9.9 Plant-dominated salt marsh, mudflat, and channel habitats of Chesapeake Bay and other East Coast estuaries, with their vertical position relative to high tide indicated.

necessary to keep the sediment load suspended diminish in the protected and quiet waters of estuaries to a point where only the finest silts and clays remain suspended in the water.

Storms and nearshore currents of the open ocean can also move coastal sand and detritus materials into the mouth of an estuary and add to the complex mix of estuarine sediments. Typically, these deposits show a characteristic distribution of different sediment types, with coarse particles deposited at the heads of estuaries and in shallow water and finer particles settling nearer the mouth and in deeper water. These graded and sorted sediment deposits transported down from rivers and in from the sea provide a rich and varying substrate to support the estuarine communities.

Estuarine Habitats and Communities

Estuaries on both coasts of North America and in other areas of the world are ecologically crucial areas that support a wide variety of biological communities of various taxa and serve as vital resting and feeding stops within the migratory flyways of ducks, geese, bald eagles, and many species of shorebirds. All of the estuarine communities have specific functions for the overall ecosystem, but **salt marshes**, in particular, are important natural filters that trap pollutants, some of which are converted by resident bacteria to less harmful substances. Salt marshes also play a role in moderating flooding and sedimentation processes, roles that are often only recognized by humans when a severely disturbed and altered salt marsh stops providing these services.

Estuarine communities include salt marshes, **mudflats**, and **channels**, and variations of these three major community types based on salinity (**Figure 9.9**). For example, the broad category of mudflat can be furthered categorized as either fresh/brackish flat or saline/brackish flat depending on where the community occurs within an estuary and, consequently, the amount of exposure to freshwater. The areas of highest elevation are the salt marshes; they are periodically covered by estuarine water at high tides and consist of dense plant communities

that tolerate contact with seawater. Mudflats, or tidal flats, are lower in elevation than the salt marshes and are alternately submerged and exposed by changing tides. Channels, those areas that are underwater even at the lowest tides, are prevented from filling with sediments by the scouring action of tides or river flow. Many estuarine areas are surrounded by mangrove forests, which provide many valuable services to the shoreline.

Salt Marshes

Salt marshes are essentially wet grasslands, or wetlands, that grow along estuarine shores. The dominant members of these marshes are **halophytes**, a few species of plants that require seawater or at least are tolerant to it and serve as excellent hiding spots for various animals (**Figure 9.10**). Salt marshes develop in the muddy deposits around the edges of temperate and subpolar estuaries, creating a transition zone between land and other estuarine communities. Salt marshes are inhabited by several plant species, each with its own specific set of sediment, water, and exposure requirements (Figure 9.9). The lowest parts of a typical salt marsh, submerged for longer periods of time, are dominated by pickleweed, *Salicornia*, which stores excess salt in its fleshy leaves, and by marshgrass and cordgrass, *Spartina*, which have special glands that excrete excess salt. In higher marsh zones, grasses, rushes, and sedges that cannot tolerate prolonged submersion by the tides dominate the landscape.

DID YOU KNOW?

Pickleweed, or *Salicornia*, is a halophylic (salt-loving) plant that is eaten by humans. The common name, *pickleweed*, describes the plant in its younger phase when it is a gray-greenish color and resembles small pickles. The Latin genus name, *Salicornia*, means "salt horns," which describes the tips of this plant that are long, thin, filled with salt, and sometimes a reddish color. The numerous tips of the plant are edible and are harvested and eaten pickled, cooked like green beans, or added to salads. Figure 9.10c shows pickleweed.

(a)

(b)

(c)

Figure 9.10 Two types of emergent salt marsh plants. (a) A stand of cordgrass, *Spartina*, with taller mangroves behind. (b) Pickleweed or glasswort, *Salicornia*. (c) A fiddler crab hiding within a pickleweed plant.
(a) Courtesy of John Morrissey; (b) Courtesy of John Bruno/NOAA; (c) NOAA.

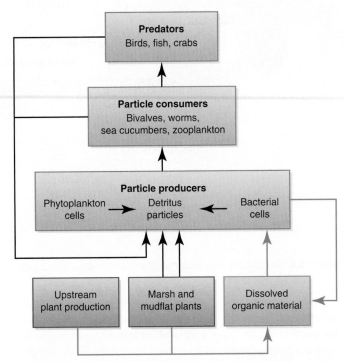

Figure 9.11 Food particle production and utilization in a typical estuary.
Data from Correll, D. L., *BioScience* 28 (1978): 646–650.

tidal creeks, where the detritus sinks and decomposes further. Each winter, salt marsh plants die back and tides (or, in cold climates, the shearing action of rising and falling tidal ice) harvest this grass and put it into the detrital food chain. It then becomes the target of decomposing bacteria, protists, and fungi. The activities of these microbes further break down the plant matter, especially the cellulose cell walls, which are undigestible by most estuarine animals, convert some of it to additional microbial biomass, and release dissolved organic materials and inorganic nutrients to be reused by other plants into the estuary.

In estuaries, bacteria and the phytoplankton of the overlying waters contribute heavily to the production of small, energy-rich, detrital food particles. These microorganisms then become a major source of food for large populations of estuarine particle consumers (**Figure 9.11**). These detritus feeders produce still more food particles in the form of feces and rejected food items. These particles are eventually recolonized by microbes and recycled into the particle pool of the estuary. It is these activities that highlight the importance of estuarine bacteria and other microorganisms to the overall ecosystem, because they play a central role in transforming the productivity of estuarine margins into small detrital food particles available to numerous other species of estuarine animals.

Mudflats

Mudflats are estuarine expanses composed primarily of rich, thick muds that are exposed to air at low tide. Only in these protected coastal environments can significant amounts of finer silt and clay particles settle out. Mudflats contain some sand,

Salt marshes form an important part of the base of estuarine food webs. Some of the plants in the estuary are eaten directly by marsh herbivores, but most of the vegetation decays and enters estuarine food webs as detritus, eventually adding nutrients to the system. The flooding and ebbing of the tides wash detritus from the marsh into the estuary and surrounding

but the sand is mixed with varying amounts of much finer silt and clay particles to produce mud. Where marine waters and rivers mix and salinity gradients are large, dissolved ions interact with the sediments and bind together to form larger particles and add to the accumulating richness of the bottom muds. These unstable, soft mud deposits serve as the principal structural foundation of soft-bottom communities that thrive in estuaries.

Three groups of primary producers are found on mudflats: diatoms, multicellular seaweeds, and seagrasses. Microscopic benthic diatoms coat the mud surfaces with a golden brown film because they contain a large amount of the accessory pigments fucoxanthin and carotene. These photosynthesizers are a rich and important food source for benthic invertebrates. Green mats of macroscopic algae, such as sea lettuce (*Ulva*), commonly cover rocks, shells, and pieces of wood debris on mudflats. These are important food sources for herbivores, especially certain worms, amphipods, and crabs.

Seagrasses are found in a variety of habitats, including at lower levels of mudflats, submerged along wave-swept sandy beaches and in the lower intertidal zone of rocky shores. Eelgrass and a few other seagrasses comprise one of the few types of flowering plants that can survive completely submerged in saltwater (**Figure 9.12**). Eelgrass gets its name from the long (up to 2 m), thin, straplike leaves that weave back and forth in the currents. Eelgrass production contributes greatly to the pool of detrital particles within an estuary. In its detrital form, it is consumed by ducks and geese, invertebrates, fish, and larval stages of insects. Algae and diatoms grow on its leaves, as do many types of hydroids, clam larvae, tunicates, bryozoans, and crustaceans. In addition to their roles in detritus food webs, seagrasses are usually the initial plants to stabilize shallow mudflats, with their roots and long leaves trapping even more fine particulate materials to add to this food-rich protected habitat. Finally, seagrasses also serve as nutrient pumps by taking nutrients from the sediment for growth and later releasing them to the water when they die at the end of the growing season. The health of seagrass beds is essential to the overall health of the estuary they occur in.

In fine-grained muds, sediment particles pack together so tightly that little water can percolate through. Organisms living in the mud face challenges, because oxygen used by mud dwellers is not rapidly replenished and their wastes are not quickly removed. Fine-grained muds with small interstitial spaces between sediment particles are effective traps for particles of organic debris. Much of the accumulated organic material is found in a thin, brownish, oxygenated surface layer about 1 cm thick. Aerobic decomposers dominate the surface oxygenated layer of mud, but their numbers decline rapidly with depth as oxygen levels decline. Beneath the oxygenated layer, anaerobes are active down to 40 to 60 cm, where their numbers also dwindle rapidly. The overwhelming abundance of both aerobic and anaerobic decomposers is responsible for most of the chemical changes that occur in estuarine sediments. The results of these chemical reactions include the decomposition of organic material, consumption of dissolved oxygen near the bottom, and the recycling of crucial plant nutrients back to the water.

Figure 9.12 A bed of eelgrass, *Zostera*, at low tide.
Courtesy of James Sumich.

Below the thin, oxygenated surface layer, the organic content of the muds usually decreases as animals and decomposing bacteria and fungi consume it. Respiration by the inhabitants of mudflats further reduces the available dissolved O_2 supply of the interstitial waters. The lower limit of O_2 penetration in organic-rich sediments is usually apparent as a color change, from light-colored sediments in the oxygenated surface layer to a dark or even black sediment in the anaerobic zone below. The anaerobic conditions of deeper muds decrease, but do not completely halt decomposition rates of organic material.

Bacteria and fungi are the major groups of marine organisms capable of using the rich organic accumulations in the anaerobic portion of muddy sediments. Without O_2, these anaerobic decomposers must use other available elements for their respiratory processes. Sulfate, the third most abundant ion in seawater, is commonly used and reduced to hydrogen sulfide (H_2S), the gas responsible for the memorable rotten-egg odor and black color so characteristic of anaerobic estuarine muds.

Other types of sediment dwellers, such as clams and mud shrimp, burrow in the muds to seek protection from predators, to be sheltered from the drying effects of the sun at low tide, or to occupy an environment where the salinity is more constant than that of the overlying water (Figure 9.6 and **Figure 9.13**). The many holes in Figure 9.13 are indications of organisms like clams living beneath the mud, but with access to the air above the mud at low tide and water above the mud during high tide.

Animals living on or in mudflats have developed a variety of feeding habits. Many of the near-surface infauna are filter or

Figure 9.13 Barren surface of a mudflat, with tubes, openings, burrows, and other evidence of abundant animal life beneath the surface.
Courtesy of John Morrissey.

Figure 9.14 A tidal channel lined by marsh plants in the National Park of Butrint, Albania, just north of Greece.
© ollirg/Shutterstock, Inc.

suspension feeders, gleaning small food particles from water currents above their burrows during higher tides. Others, such as lugworms, are deposit feeders, digesting the bacterial and organic coatings of sediment particles passing through their digestive system. Other types of burrowing animals, such as the arrow goby, leave their burrows at high tide and forage for food over the mudflat.

The epifauna of mudflats are dominated by mobile species of gastropod mollusks, crustaceans, and polychaete worms. These organisms sometimes range over a wide area of the mudflat and demonstrate only blurred, weakly established patterns of lateral zonation. Mud-dwelling infaunal organisms, however, do occupy vertically arranged zones in the sediment due to oxygen availability. A few centimeters below the mud surface the interstitial water is generally devoid of available oxygen, and the infauna must obtain their oxygen from the water just above the mud or do without. The numerous openings of tubes and burrows on the surfaces of estuarine mudflats (Figure 9.13) attest to an unseen wealth of animal life underneath. Bivalve mollusks extend tubular siphons through the anaerobic mud to the oxygenated water above. The depth to which these animals can seek protection in the mud is limited largely by the lengths of their siphons, and thus indirectly by their ages. Other infauna use the sticky consistency of fine-grained, organically rich muds to construct permanent burrows with connections to the surface for breathing.

When the tide is out, the infauna of muddy shores that normally breathe in the water must also cope with an absence of available O_2 and changing air temperatures. Some switch from aerobic to anaerobic respiration. In doing so, many encounter a dilemma in trying to match the relative inefficiency of energy-yielding anaerobic respiration with the increased energy demands forced by increasing tissue temperatures and higher metabolic rates. Larger infauna exist anaerobically only temporarily and revert to aerobic respiration as soon as they are covered by the tides, but for many of their smaller burrowing

neighbors with no direct access to the O_2-laden waters above, anaerobic respiration is a permanent feature of their infaunal existence in intertidal muds.

At high tide, submerged mudflats are visited by shore crabs, shrimps, fish, and other transients from deeper water. Some come to forage for food; others find the protected waters ideal for spawning. Their forays are only temporary, however, because they leave with the ebbing tide and surrender the mudflats to shorebirds and a few species of foraging coastal mammals. During low water, long-billed curlews and whimbrels probe for deeper infauna, whereas sandpipers and other short-billed shorebirds concentrate on the shallow infauna and small epifauna.

Channels

A channel is the part of an estuary that is filled with water under all tidal conditions (extreme right side of Figure 9.9 and **Figure 9.14**). A channel may be as broad as the entire estuary, or it may be restricted to a narrow creeklike feature between mudflats. Numerous species of planktonic organisms inhabit channels, relying on the action of currents to move them around. Crabs, oysters, starry flounders, sculpins, anchovies, and killifish are also abundant in channels during certain times of the year.

Channel areas are also used as spawning and nursery areas by many animals. Herring and sole are ocean fish that move into the protected areas of the estuary to spawn so that their offspring can feed on food available there. Crabs also use the estuaries as nursery areas. Anadromous fishes such as salmon and shad may linger for a time in estuarine channels to feed when migrating between the ocean and their spawning areas in freshwater streams. In some areas, plankton samples collected with nets in channels during a flood tide often yield large numbers of larvae from various taxa, especially during new or full moons. Many larvae are transported into the protection of an

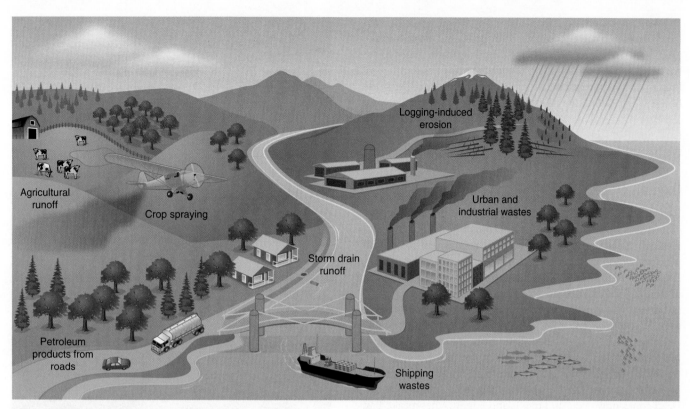

Figure 9.15 Common sources of human-made pollutants entering estuaries.

estuarine nursery habitat by tides and remain there until they are juveniles.

9.5   Environmental Pollutants

The assumption is often made that pristine ocean waters are essential for maintaining healthy marine communities. This assumption is being put to a global test as our growing human population generates an enormous and increasing burden of domestic and industrial wastes. Initially, these wastes may be dumped down sewers or sent up smokestacks, but ultimately many make their way into estuaries on their way to the ocean (**Figure 9.15**). As far as we know, the world ocean has a large but finite capacity to assimilate these waste materials without apparent degradation of water quality; however, that capacity is often exceeded in the semienclosed conditions of estuaries where mixing processes are not sufficient to dilute or disperse wastes, creating localized water-quality problems and subsequent biological disturbances. Contrary to what people believed when the world population was relatively small, dilution is *not* the solution to pollution, in estuaries or any other natural areas.

Waste materials discharged into estuarine waters are considered pollutants if they have measurable adverse effects on natural populations. Persistent contaminants such as heavy metals, pesticides, radioactive wastes, and petroleum products head the list of substances (**Table 9.1**) that even at low concentrations can adversely affect the health of marine organisms and the integrity of their natural ecological relationships because they accumulate in marine organisms, and their concentrations can be magnified as they are transferred up food chains.

TABLE 9.1   A Summary of the Sources and Effects of Some Marine Pollutants

Pollutant	Sources	Effects
Particulate material	Dredged material, sewage, erosion	Smothers benthic organisms, clogs gills and filters, reduces underwater light
Dissolved nutrients	Sewage, agricultural runoff	Increases phytoplankton blooms, decreases dissolved oxygen
Toxins	Pesticides, industrial wastes, oil spills, antifouling paint	Increases incidence of disease, contaminates seafood, suppresses immune systems, contributes to reproductive failure
Petroleum products	Tankers, drill sites, urban and industrial wastes, runoff from roads and parking lots	Smothers organisms, clogs gills, mats fur or feathers, causes anatomical and physiological abnormalities
Marine debris	Garbage, ship wastes, fishing gear	Causes physical injuries and mutilations, increases mortality

Municipal and industrial wastewater discharges (including storm drain and sewer overflows) are considered **point sources** of pollutants. Urban runoff and land-based agriculture and forest harvest activities contribute to **nonpoint sources** of estuarine pollutants. It is paradoxical and somewhat frightening that some of the worst pollutants in estuaries are the very pesticide and fertilizer products initially used to increase food production on land, and that some of the most obvious disturbances appear in marine species harvested for human consumption.

Oxygen-Depleting Pollutants

Excessive amounts of organic materials or fertilizers from agricultural runoff and sewage outfalls contain large quantities of nitrogen compounds, a common limiting nutrient for phytoplankton. Discharged into semienclosed estuaries, these nutrients promote increased phytoplankton production and blooms of dinoflagellates, such as toxic *Pfiesteria*. These high concentrations of phytoplankton eventually die and sink to the bottom. Decomposition of the excess phytoplankton biomass by microbes creates a high **biochemical oxygen demand (BOD)** and can reduce an estuary's reservoir of life-supporting oxygen.

When an estuary's BOD is high and multicellular organisms cannot survive, seasonal or permanent "dead zones" occur. Dead zones now occur in Long Island Sound, Chesapeake Bay, and nearly 150 other estuarine or very nearshore locations around the world, mostly in the Northern Hemisphere. One of the largest dead zones covers nearly 20,000 km^2 just outside the mouth of the Mississippi River (**Figure 9.16**), an area about equal to that of the state of Massachusetts. The extent of this dead zone approximately doubled during the 1990s and reaches its greatest extent in spring and summer months. From a few meters below the surface to a depth of about 60 m, fish and mobile invertebrates move away from the anoxic water, whereas attached and burrowing species are killed. As this dead zone continues to grow in size, it imposes an increasing threat to the commercially valuable fish and shrimp populations living there.

Figure 9.16 Sea-viewing wide field-of-view sensor (SeaWiFS) satellite view of the U.S. Gulf Coast, with the dead zone at the mouth of the Mississippi River.
Courtesy of GeoEye and NASA SeaWiFS Project.

DID YOU KNOW?

Although many dead zones are caused by pollution, some occur naturally in deep waters with little or no mixing. The largest dead zone in the world occurs naturally in the depths of the Black Sea, between southeastern Europe and western Asia. The upper portion of the sea contains oxygenated water because waters from the Mediterranean mix with waters from the Black Sea, but no mixing occurs at depth, creating a zone of almost no oxygen. Very little life occurs in this large, unique area.

Toxic Pollutants

We do not yet know how to determine the extent or fate of many toxic substances in the marine environment or precisely how to evaluate their effects on marine life. The effects are difficult to trace in a natural setting, and the results of laboratory tests are not always indicative of what is occurring in nature. Some of the better-known trace metals and toxic chemicals include mercury, copper, lead, and chlorinated hydrocarbons. The most common chlorinated hydrocarbons are synthetic chlorine-containing compounds created for use as pesticides or generated as by-products of the manufacture of plastics. They are among the most persistent and harmful of all toxic substances and include well-known products such as chlordane, lindane, heptachlor, dichlorodiphenyltrichloroethane (DDT), dioxins, and polychlorinated biphenyls (PCBs).

DDT

By virtue of its long and widespread use, dichlorodiphenyltrichloroethane (DDT) and its effects on marine life can serve as a model for the behavior of other persistent toxic substances in estuarine and marine systems. DDT was the first of a new class of synthetic chlorinated hydrocarbons. It became available for public use in 1945 and quickly gained international acceptance as an effective killer of most serious insect threats: houseflies, lice, mosquitoes, and several crop pests.

DDT is a persistent environmental pollutant; it does not break down or lose its toxicity rapidly. Once in seawater, DDT is rapidly adsorbed by suspended particles. Because it is nearly

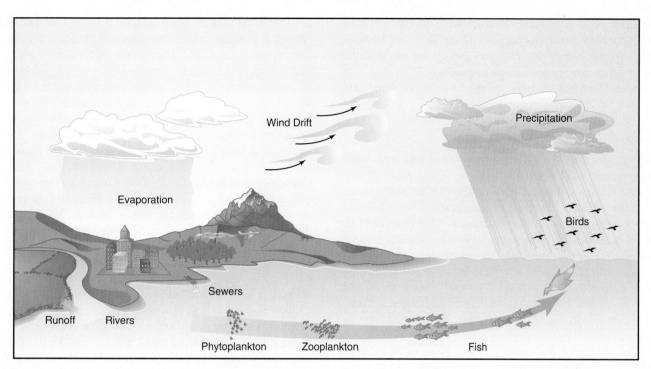

Figure 9.17 Transfer of DDT to and within marine food webs (arrows). DDT is absorbed by phytoplankton and then concentrated at each step in the food web.
Data from Epel, D. and W. L. Lee., *The American Biology Teacher* 32 (1970): 207–211.

insoluble in water, measurable levels of DDT in seawater practically never occur. Yet DDT contamination from land and air has been so pervasive that it can be found in nearly all parts of the world ocean. Antarctic penguins, Arctic seals, Bermuda petrels, and fish everywhere have measurable accumulations of DDT in their fatty tissues.

Although DDT is insoluble in water, it is quite soluble in lipids. Fatty tissues and oil droplets concentrate the DDT adsorbed on suspended particles in seawater. Phytoplankton and, to a much lesser extent, zooplankton are the initial steps in DDT's entry into marine food webs. Fish, birds, and other predators eventually consume this plankton and its load of DDT. At each step in the food web, further concentration, or **bioaccumulation**, occurs (**Figure 9.17**), eventually reaching the top carnivores, including humans.

Marine birds suffer the most devastating effects of DDT poisoning. As fish-eating predators, bald eagles, ospreys, pelicans, and other species are sometimes four or five trophic levels removed from the phytoplankton that initially absorb the DDT. The bioaccumulation of DDT that occurs at each trophic level assures these predatory birds of high DDT loads in their food. DDT and its residues block normal nerve function in vertebrates. DDT also interferes with calcium deposition during the formation of eggshells. The eggshells of birds with high DDT loads are very thin and fragile. They frequently break when laid or fail to support the weight of adult birds during incubation. The broken eggs lay in abandoned nests, mute testimony to the insidious effects of DDT.

The United States banned the general use of DDT in 1972, yet DDT is still being used in other parts of the world, each year adding to the load already existing in estuarine and marine organisms. Animals from walruses in the Arctic to penguins in the Antarctic show elevated concentrations of DDT, particularly in their blubber and other fatty tissues. By itself, DDT is a threat to the continued existence of several species of marine animals, but it is only symptomatic of the greater danger posed by the many other persistent toxins that we know even less about. Unfortunately, negative effects of contaminants often take years to show signs in organisms.

Dioxins

Dioxins are another group of about 75 chlorinated compounds gaining international notoriety. The most potent dioxin, 2,3,7,8-tetrachlorodibenzo-*p*-dioxin (2,3,7,8-TCDD) is toxic to birds and aquatic life at concentrations as low as a few parts per quadrillion and is considered by the World Health Organization to be a class I, or known, human carcinogen. Dioxins belong to a group of pollutants called the "dirty dozen," which is a group of dangerous persistent toxins, including DDT. Like DDT, dioxins are fat soluble and stable, they bioaccumulate readily, and they are suspected of causing developmental malformations; cancer; and immune, nervous, and reproductive difficulties.

Dioxins enter estuaries from many sources, but most arrive in the effluent from pulp and paper manufacturing plants that chlorinate wood pulp to produce bleached paper. Trace levels of dioxins have been found in tissues of fish collected near pulp and paper mills. Although the concentrations do not affect fish, the bioaccumulation of dioxin may occur in the consumers of these fish, including humans.

European paper-producing countries are working to reduce the amounts of dioxin discharged into the Baltic Sea by finding alternative bleaching techniques that do not use chlorine and by producing unbleached paper products. More studies to determine the extent and effect of dioxin in marine waters are currently under way.

PCBs

Polychlorinated biphenyls (PCBs) are a class of industrial chemicals originally produced in the 1930s as a cooling and insulating fluid for capacitors and transformers that are now banned in the United States, but they are still present at toxic levels in many estuarine and coastal environments. Like dioxins, PCBs are carcinogenic and immunotoxic to many animals (including humans). Studies of stranded beluga whales from a population isolated in the St. Lawrence River estuary of Canada found high tissue levels of PCBs as well as DDT, heavy metals, mirex, and other pesticides. These animals exhibited high rates of bacterial and protozoan infections, suggesting compromised immune systems.

The North Sea, surrounded as it is by many of Europe's most industrialized nations, experiences some of the highest levels of marine pollution anywhere. High on the list of these pollutants are dioxins and PCBs. During the summer of 1988, more than 17,000 common and gray seals died of a viral infection that swept the Baltic, Wadden, and North Seas. By the time it ran its course, this epidemic killed as many as 80% of some North Atlantic populations of common and gray seals. Stressed seals exhibited a variety of symptoms, including lesions, encephalitis, peritonitis, osteomyelitis, and premature abortions.

The source of this epidemic has not been established with any certainty, nor has an absolute link between this viral outbreak and any specific pollutant been demonstrated; however, PCBs in the seals' food is strongly suspected because strong experimental evidence indicates that even low concentrations of PCBs (and possibly dioxins as well) lead to suppression of the immune systems of harbor seals. PCBs also interfere with embryo implantation in seals, leading to fewer births and lower birth weights. Seal pups born of PCB-contaminated mothers may still be confronted with higher mortality rates and suppressed immune systems as they grow and mature.

To examine the extent of the pollution in the Baltic Sea, the signatory nations to the Helsinki Convention have developed pollution monitoring and control strategies. In the Pacific Rim nations, similar efforts are taking place through the Pacific Basin Consortium for Environment and Health to identify and study pollutants present in the Pacific Ocean and to develop joint international partnerships to reduce existing and manage future marine pollutants.

Organotin Compounds in Antifouling Paints

For centuries, boat hulls have been covered with paint containing high levels of copper or other metals to attempt to retard the attachment and growth of barnacles and other hull-fouling organisms. More recently, metal-based antifouling paints using biocides made from these metals have been applied. The latest entrant in this race to find a substance to block fouling organisms is tributyltin (TBT). Paints with TBT are more effective and last longer than older copper-based antifouling paints. TBT leaches from boat hulls into harbor waters but in amounts difficult to detect with current analytical techniques. Yet even in these very low concentrations (a few parts per trillion), TBT harms nontarget organisms such as oysters and clams. TBT is known to deform oyster shells and to cause chronic reproductive failure in a variety of shellfish species. In 1987, the U.S. Congress passed the Antifouling Paint Control Act that classified TBT as a restricted pesticide and severely limited its use and the amount that could be used in paints. Because TBT eventually degrades to less toxic forms in the marine environment, these new restrictions have already led to decreased concentrations of tin in estuaries and bays.

Even so, since 1987 large numbers of dolphins, seals, and sea turtles have been killed by disease in the Atlantic Ocean, Gulf of Mexico, North Sea, and Mediterranean. Bottlenose dolphins found dead on Atlantic and Gulf Coast beaches in Florida between 1989 and 1994 had elevated levels of TBT, presumably derived from boats in coastal marinas. Concentrations of tin compounds (most likely degradation products of TBT) found in the Florida-stranded bottlenose dolphins were much higher than concentrations of tin compounds found in offshore whales, presumably because bottlenose dolphins spend their lives close to shore, where antifouling paint from boats and ships has contaminated bottom sediments and local food chains. Accumulated tin compounds, combined with PCBs and DDT (which were also found at high levels), may have damaged their immune systems and left them vulnerable to the bacterial and viral infections that were their eventual cause of death.

More recently, a study of sea otters that stranded along the Pacific Rim (from Kamchatka, Russia, to Alaska, Washington, and California) from 1992 to 2002 revealed a variety of organotin compounds (including mono- to tributyltins) in their livers and estimated that the half-life of TBT in sea otters is about 3 years. Livers assayed from sea otters that stranded in California contained 34 to 4,100 ng of total organotin compounds per gram of liver tissue, and thus some individuals were found to contain enough organotins to experience adverse health effects. Nevertheless, the concentration of total organotins in the livers of sea otters stranded in California during this study period decreased significantly.

Finally, TBT has now been demonstrated to be transferred from female fish to their offspring. In 2006, a team of Japanese biologists examined surfperch (*Ditrema temmincki*) that inhabit seagrass meadows and rocky reefs off temperate coasts of Japan. This viviparous fish mates from early September through early December and then gives birth to live young in May and June of the following year. During this 6-month gestation period, nutrients and pollutants are provided to the developing young by gravid females. Surprisingly, newborn surfperches had 10 to 16 times more TBT in their tissues than did parental females, presumably because the offspring have a lesser ability to metabolize TBT. Juvenile surfperches in Japan thus are at a greater risk of TBT exposure than their parents and seem to be exposed to TBT throughout their entire life history.

Some U.S. states are taking actions to decrease the amount of harmful chemicals leeched by paint from boat hulls by limiting the types of paints that can be used. In 2012, Washington State was the first U.S. state to ban the use of copper-based paints for recreational boats under 19.8 m (65 ft.) in length. Alternative paints are being used, some that take advantage of a photoactive technology, such as the biocide Econea, or even some completely metal-free options. The effectiveness of these paints varies, and the risk to marine organisms has not been evaluated for all of these paint types. Regardless, many consumers are happy with the results, and the waters and coastal organisms of Washington will likely benefit from the reduction of metal-based pollution. Other states, including California, are following suit by creating deadlines for a reduction in the use of known dangerous bottom paints and a switch to alternatives and by being proactive by conducting studies on the effectiveness of other paint types.

Introduction of Nutrients from Human Sources

One very preventable source of pollution is from nutrients released into the water from human sources. Estuaries are particularly vulnerable to this type of pollution because they are often located between human-made structures and the ocean, and they have less flushing than marine coastal waters. Many storm drains empty directly into estuaries, and runoff from household and agricultural irrigation makes its way into estuaries and the sea. Fertilizers used on lawns and other plants near the coast run off into storm drains during rainstorms, making estuarine and nearshore marine waters unsafe for swimmers and loaded with artificial sources of nutrients that affect local aquatic life. Of particular concern are large areas of fertilized lawns such as golf courses and agricultural fields.

Increased nutrients can lead to algal blooms and **eutrophication**. When nutrients such as nitrogen and phosphorus are added to estuarine or marine habitats, it causes algal blooms, which reduce water clarity and can harm water quality. A reduction of water clarity leads to shading for other algal species and reduced visibility for animals foraging in the area. When the blooms die off, decomposers use up a large amount of oxygen to break down the millions of cells, reducing available oxygen for other organisms. This can lead to an anoxic region with reduced abundance of many organisms. Another potential outcome of artificial nutrients is a harmful algal bloom event, which can kill fish and lead to toxic filter-feeding mollusks that are dangerous for human consumption.

Preventing nutrients from entering the ocean is one environmental task that everyone can easily partake in by making informed decisions about their homes. Limiting the amount of fertilizer used on lawns, or choosing not to use any fertilizers, is an easy way to help solve this problem. Anytime you pour a chemical onto your garden, whether it is a fertilizer or weed killer, just think about the fact that traces of the chemical will very likely eventually make it into your local bodies of water. Using one or more rain barrels to capture rainwater from rain gutters is another easy and relatively inexpensive solution. The captured water can then be used to water plants during the dry season. The nonprofit organization Surfrider Foundation created a program called Ocean Friendly Gardens that encourages homeowners to design their gardens to soak up irrigation and rainwater. If homeowners can retain more rainwater on their property they will benefit from well-irrigated plants, and there will be less runoff into the ocean. These suggestions are just a few ways we can help keep our estuarine and marine waters healthier.

9.6 The Largest Estuary in the United States: The Chesapeake Bay System

Although the use of shorelines for various human activities is an integral part of all coastal economies, heavy human use contributes contaminants to the coastal river and estuary systems that can cause habitat loss and a change in the ecological integrity that cumulatively affects public health, fish and wildlife habitat, and recreational resources. Despite past regulatory and management efforts, the water quality and habitat within most estuaries around the world have been seriously degraded, and the health of these ecosystems continues to decline.

To understand some of the conflicts that surround the use of estuarine resources better, we examine in more detail one major and well-known estuary that is also the largest in the United States, the Chesapeake Bay system on the East Coast (**Figure 9.18**). The Chesapeake Bay system exemplifies the physical, chemical, and biological features of estuaries and the substantial social and political problems generated by conflicts between natural estuarine processes and the many additional uses imposed on these coastal habitats by humans.

The Chesapeake Bay system is a cascading series of five major and smaller estuaries (**Figure 9.19**). These estuaries were linked together when the ancestral Susquehanna River valley flooded after the LGM. The bed of the ancient Susquehanna River is now a deep channel, but most of the Chesapeake Bay is sufficiently shallow to allow sunlight to penetrate to the bottom.

The Chesapeake Bay drains a very large (166,000 km²), heavily populated, and agriculturally rich land area of the U.S. central Atlantic coastal plain. Approximately 17 million people live in the Chesapeake Bay watershed. Human activities have imposed some serious stresses on the Chesapeake Bay in the form of increasing loads of heavy metals, fertilizers, pesticides, and incompletely treated sewage. Increased BOD in Chesapeake Bay waters is only one of several complications resulting from this input. Yet the Chesapeake Bay continues to be used as a protein factory. Millions of blue crabs, oysters, striped bass, and other finfishes are harvested from the Chesapeake Bay each year, contributing several hundred million dollars annually to the economies of Maryland and Virginia. When the recreational, military, shipping, and other uses of the Chesapeake Bay are added to the mix of conflicting uses, we have in microcosm a picture of some of the same problems confronting other estuarine systems and the larger world ocean.

Figure 9.18 Aerial view of the Chesapeake Bay estuarine system.
Courtesy of the National Estuarine Research Reserve System/National Oceanic and Atmospheric Administration.

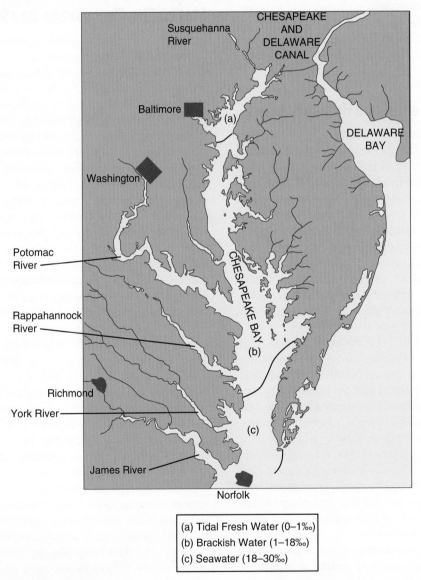

(a) Tidal Fresh Water (0–1‰)
(b) Brackish Water (1–18‰)
(c) Seawater (18–30‰)

Figure 9.19 Chesapeake Bay and its numerous smaller side estuaries, showing mean surface salinity zones.

In the Chesapeake Bay, the existing salinity gradient creates an upper-bay low-salinity zone, a mid-bay brackish zone, and a lower-bay marine zone (Figure 9.19). Although the tides in the Chesapeake Bay have an average vertical range of only 1 to 2 m, they are the major mixing influence. On longer time scales, seasonal flooding and storms have additional effects on the salinity distribution patterns.

For animals capable of tolerating the dynamic fluctuations of the Chesapeake Bay and other estuaries, these habitats offer nearly ideal nursery conditions for their young.

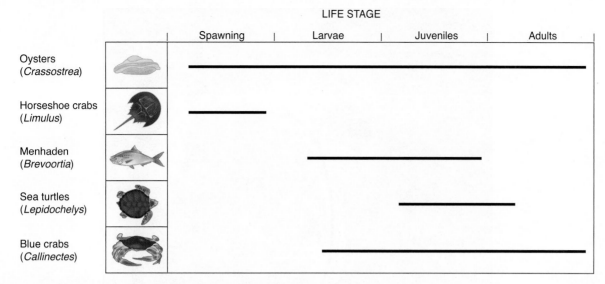

Figure 9.20   Utilization of estuaries by different life stages of five common inhabitants of Chesapeake Bay.

Estuaries provide some protection against the physical stresses of nearby open coasts, and there is an abundance of food available in a large range of particle sizes. Most fish species commercially exploited along the Atlantic and Gulf Coasts of the United States use estuaries such as the Chesapeake Bay as spawning or juvenile feeding areas. Some species occupy estuaries throughout their lives; others occupy estuaries for only a particularly crucial stage of their development. **Figure 9.20** illustrates this range of utilization patterns for a few commercially important or otherwise notable Chesapeake Bay species.

At one extreme are oysters, which typically spawn, mature, and die within the confines of the bay (although some larvae occasionally may drift to other nearby estuaries). Oysters are broadcast spawners, with fertilization and a 2-week larval development period occurring in the moving water above the benthic habitat of the adults. The double-layered circulation pattern of the Chesapeake Bay is used by oyster larvae to avoid being washed out of the estuary. During ebb tides, when most of the tidal outflow is in the surface layers, the larvae remain in the deeper inflowing seawater. During slack or incoming tides, the larvae venture into the shallower portions of the larger bay or its smaller side estuaries, where high levels of larval settling and retention occur.

Horseshoe crabs and menhaden are marine species that use the Chesapeake Bay for early life stages only. Adult horseshoe crabs move into the bay only to spawn. During high tides in the spring, these animals crawl into salt marshes at the water's edge. There the female digs a depression to deposit her eggs. The smaller male, who hitches a ride on the female's back, releases sperm to fertilize the newly deposited eggs. The eggs are then covered by sand to await hatching 2 weeks later, when they are again flooded by the next series of spring tides. After hatching, the larvae swim and feed near the surface while currents carry them out to sea where development continues through as many as 13 successive larval stages.

Menhaden is a commercially valuable fish species of the Atlantic coast, ranging from Nova Scotia to Florida. Although adult menhaden live and spawn in coastal waters, their larvae drift into a number of estuaries, including Chesapeake Bay, to continue their development. Four species of sea turtles frequent the Chesapeake Bay, mostly as juveniles. Loggerheads are by far the most common, accounting for about 90% of juvenile sea turtles observed in the bay during summer months. Juvenile Kemp's ridley, leatherback, hawksbill, and green sea turtles also enter the lower bay and its rivers in the spring, feed within the bay throughout the summer months, and then depart when fall water temperatures begin to drop. All species of sea turtles are critically threatened or endangered. The fact that Chesapeake Bay serves as an important feeding or nursery area for five of nine recognized species of sea turtles, as well as with menhaden and many other species, makes it a region of primary concern along the U.S. East Coast.

For blue crabs, the pattern of bay utilization is completely different. Adults live in estuaries along most of the U.S. Atlantic and Gulf Coasts. After mating, females seek higher salinities in the open sea before releasing their larvae (**Figure 9.21**). Larval development continues in coastal waters outside the Chesapeake Bay, where winds and coastal currents combine to keep blue crab larvae close to shore until they return to the bay as young crabs. It is at this time that exchange of individuals between neighboring estuaries sometimes occurs, preventing genetic isolation of the crabs that occupy any one estuary. Within the estuary, young blue crabs seek eelgrass beds and salt marshes as winter nursery areas for food and protection until they grow sufficiently to exploit other estuarine habitats.

Through their roles as nursery areas and feeding grounds, the value of estuaries such as the Chesapeake Bay extends far beyond the bounds of the estuary itself. Yet the very health of the Chesapeake Bay, as well as that of most of Earth's other major estuaries, has gradually but seriously deteriorated in the past few decades. Since 1960, submerged vegetation in the bay, especially eelgrass beds, the preferred habitat of overwintering blue crabs, has declined in size and abundance. Eelgrass does not do well in extremely warm temperatures, and the hot summers of 2005 and 2010 produced dramatic

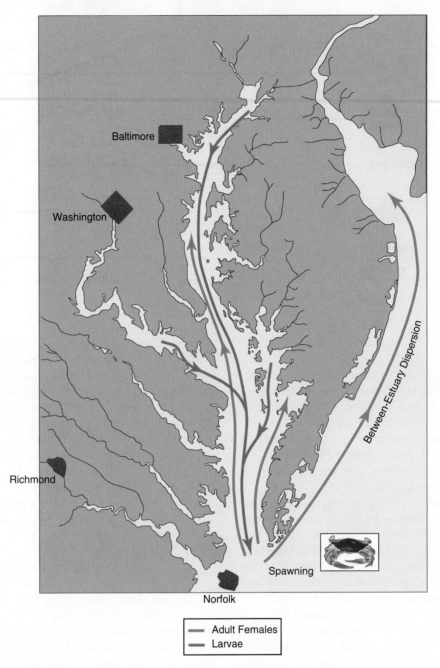

Figure 9.21 Spawning migration of adult female blue crabs and return routes of planktonic larval stages.

die-offs of eelgrass beds. Surveys conducted in 2014 indicated a slight recovery of eelgrass, and a strong recovery of widgeon-grass. Any large decline in eelgrass results in a collapse of the blue crab population. A 2013–2014 winter survey suggested that the stock of blue crabs in the Chesapeake Bay is once again near the lowest point ever recorded. Dissolved nutrient loads draining into the bay have increased, causing changes in the species composition of the estuarine waters. In the upper reaches of the Chesapeake Bay, concentrations of cyanobacteria and dinoflagellates have increased 250-fold since 1950 at the expense of diatom species. During the same time period, submerged eelgrass and cordgrass beds have declined dramatically. Presently, most of the bay water deeper than 13 m

(from near the mouth to the Rappahannock River north to a point just south of Baltimore) is a dead zone, with little or no dissolved O_2. As a consequence of these biological changes, recent harvests of commercially important species have also dropped substantially.

Serious and sustained efforts on the part of the Chesapeake Bay Program, a regional partnership of federal and state agencies, local governments, nonprofit organizations, and universities, are starting to stall these adverse changes and reduce the nutrient and toxic substance loads presently carried into the bay. It will not be easy, yet only when these efforts are successful can we ensure the continuing health of this hardy and beautiful estuary system that is the Chesapeake Bay.

STUDY GUIDE

TOPICS FOR DISCUSSION AND REVIEW

1. List the different types of estuaries and their major characteristics.

2. Describe the pattern of water circulation in a typical estuary as the tide changes.

3. Describe the distribution of salinity values in a typical estuary, from surface to sediment and from head to mouth.

4. Compare and contrast the type of sediment found at the head of an estuary as opposed to its mouth. Describe the factors that result in this differential deposition.

5. What types of plants dominate low, frequently submerged portions of wetlands? Is the flora different in higher marsh zones that do not experience prolonged submergence?

6. If you were to walk out onto a mudflat at low tide and dig a hole that is 2 feet deep, what types of organisms would you expect to find while digging this hole?

7. Why are estuaries important to nonestuarine species, such as migrating geese and anadromous shad?

8. Summarize the series of events leading from runoff of agricultural fertilizers to dangerously low depletion of dissolved oxygen within a nearby estuary.

9. What is the source of DDT? What is its residence time in the environment? What effect does it have on wildlife? Is it still being produced or used in the United States?

KEY TERMS

bar-built estuary 247	halophyte 255
barrier island 247	isohaline 248
bioaccumulation 261	mudflat 255
biochemical oxygen demand (BOD) 260	nonpoint source 260
	osmoregulator 251
channel 255	osmotic conformer 250
coastal lagoon 247	point source 260
coastal plain estuary 247	salt marsh 255
delta 248	salt-wedge estuary 248
euryhaline 254	slightly stratified estuary 248
eutrophication 263	
fjord 248	stenohaline 251
flushing time 249	tectonic estuary 248
halocline 248	vertically mixed estuary 248

KEY *GENERA*

Clupea	**Salicornia**
Gammarus	**Spartina**
Kryptolebias	**Ulva**
Morone	**Zostera**
Pfiesteria	

REFERENCES

Alexander, M. 1981. Biodegradation of chemicals of environmental concern. *Science* 211:132.

Anderson, T. H., and G. T. Taylor. 2001. Nutrient pulses, plankton-blooms, and seasonal hypoxia in western Long Island Sound. *Estuaries* 24:228–243.

Avise, J. C., and A. Tatarenkov. 2015. Population genetics and evolution of the mangrove rivulus *Kryptolebias marmoratus*, the world's only self-fertilizing hermaphroditic vertebrate. *Journal of Fish Biology* 87:519–538. doi:10.1111/jfb.12741.

Barnes, R. S. K. 1974. *Estuarine Biology. Studies in Biology*, no. 49. Baltimore: University Park Press.

Bjorndal, K. A., A. B. Bolten, and C. J. Lagueux. 1994. Ingestion of marine debris by juvenile sea turtles in coastal Florida habitats. *Marine Pollution Bulletin* 28:154–158.

Blus, L., R. G. Heath, C. D. Gish, A. A. Belisle, and R. M. Prouty. 1971. Eggshell thinning in the brown pelican: Implications of DDE. *BioScience* 21:1213–1215.

Botton, M. L., and H. H. Haskin. 1984. Distribution and feeding of the horseshoe crab, *Limulus polyphemus*, on the continental shelf off New Jersey. *Fishery Bulletin* 82:383–389.

Britton, J. C. 1989. *Shore Ecology of the Gulf of Mexico*. Dallas: Texas Press.

Broecker, W. S., and G. H. Denton. 1990. What drives glacial cycles? *Scientific American* January:49–56.

Champ, M. A., and F. L. Lowenstein. 1987. The dilemma of high technology antifouling paints. *Oceanus* 30:69–77.

Charney, J. I., ed. 1982. *The New Nationalism and the Use of Common Spaces: Issues in Marine Pollution and the Exploitation of Antarctica*. Lanham, MD: Rowman and Littlefield.

Chesapeake Bay Stock Assessment Committee. 2014. Annual report published by the Chesapeake Bay Stock Assessment Committee (CBSAC) on the status of the blue crab population in the Bay and management advice for Bay jurisdictions. Annapolis, MD: CBSAC.

Chislock, M. F., E. Doster, R. A. Zitomer, and A. E. Wilson. 2013. Eutrophication: Causes, consequences, and controls in aquatic ecosystems. *Nature Education Knowledge* 4(4):10.

Clark, R. B. 1992. *Marine Pollution*, 3rd ed. Oxford, UK: Oxford University Press.

Cloern, J., and F. Nichols. 1985. *Temporal Dynamics of an Estuary*. Boston: Kluwer Academic.

Correll, D. L. 1978. Estuarine productivity. *BioScience* 28:646–650.

Costlow, J. D., and C. G. Bookhout. 1959. The larval development of *Callinectes sapidus* Rathbun reared in the laboratory. *Biological Bulletin* 116:373–396.

Cox, J. L. 1972. DDT in marine plankton and fish in the California Current. *CalCOFI Reports* 16:103–111.

Epel, D., and W. L. Lee. 1970. Persistent chemicals in the marine ecosystem. *The American Biology Teacher* 32:207–211.

Epifanio, C. E., and R. W. Garvine. 2001. Larval transport on the Atlantic continental shelf of North America: A review. *Estuarine, Coastal and Shelf Science* 52:51–77.

Frankel, E. G. 1995. *Ocean Environmental Management: A Primer on the Role of the Oceans and How to Maintain Their Contributions to Life on Earth*. Englewood Cliffs, NJ: Prentice Hall.

Green, E. P. 2003. *World Atlas of Seagrasses*. Berkeley, CA: University of California Press.

Heinle, D. R., R. P. Harris, J. F. Ustach, and D. A. Flemer. 1977. Detritus as food for estuarine copepods. *Marine Biology* 40:341–353.

Hickie, B. E., M. C. S. Kingsley, P. V. Hodson, D. C. G. Muir, P. Béland, and D. Mackay. 2000. A modelling-based perspective on the past, present, and future polychlorinated biphenyl contamination of the St. Lawrence beluga whale (*Delphinapterus leucas*) population. *Canadian Journal of Fisheries and Aquatic Sciences* 57:101–112.

Jefferies, R. L. 1981. Osmotic adjustment and the response of halophytic plants to salinity. *BioScience* 31:42–46.

Johnson, D. R. 1985. Wind-forced dispersion of blue crab larvae in the Middle Atlantic Bight. *Continental Shelf Research* 4:733–745.

Johnson, D. R., B. S. Hester, and J. R. McConaugha. 1984. Studies of a wind mechanism influencing the recruitment of blue crabs in the Middle Atlantic Bight. *Continental Shelf Research* 3:425–437.

Laws, E. A. 1993. *Aquatic Pollution: An Introductory Text*, 2nd ed. New York: John Wiley & Sons.

Lippson, J. A., and R. L. Lippson. 1984. *Life in the Chesapeake Bay*. Baltimore: The Johns Hopkins University Press.

Mangelsdorf, P. C., Jr. 1967. Salinity measurements in estuaries. In: G. H. Lauff, ed. *Estuaries*. Washington, DC: American Association for the Advancement of Science. Publication no. 83. pp. 71–79.

Marshall, H. G. 1980. Seasonal phytoplankton composition in the lower Chesapeake Bay and Old Plantation Creek, Cape Charles, Virginia. *Estuaries* 3:207–216.

McRoy, C. P., and C. Helfferich. 1977. *Seagrass Ecosystems*. New York: Marcel Dekker.

Miller, J. M., and M. L. Dunn. 1980. Feeding strategies and patterns of movement in juvenile estuarine fishes. In: V. S. Kennedy, ed. *Estuarine Perspectives*. New York: Academic Press. pp. 437–448.

Moriarity, F. 1983. *Ecotoxicology: The Study of Pollutants in Ecosystems*. New York: Academic Press.

National Wildlife Foundation. 1989. *A Citizen's Guide to Protecting Wetlands*. Washington, DC: National Wildlife Foundation.

Nichols, F. H., J. E. Cloern, S. N. Luoma, and D. H. Peterson. 1986. The modification of an estuary. *Science* 231(4738):567–573.

Peakall, D. B. 1970. Pesticides and the reproduction of birds. *Scientific American* 222:72–78.

Peterson, D., D. Cayan, J. DiLeo, M. Noble, and M. Dettinger. 1995. The role of climate in estuarine variability. *American Scientist* 83:58–67.

Rabalais, N. N. 2002. Nitrogen in aquatic ecosystems. *Ambio* 31:102–112.

Risebrough, R. W., D. B. Menzel, D. J. Martin Jun, and H. S. Olcott. 1967. DDT residues in Pacific seabirds: A persistent insecticide in marine food chains. *Nature* 216:389–391.

San Diego Unified Port District. 2011. Safer Alternatives to Copper Antifouling Paints for Marine Vessels. USEPA Project, NP00946501-4. Final Report. January 2011.

Selander, R., S. Yang, R. Lewontin, and W. Johnson. 1970. Genetic variation in the horseshoe crab (*Limulus polyphemus*), a phylogenetic "relic." *Evolution* 24:402–414.

Siry, J. V. 1984. *Marshes of the Ocean Shore*. Austin: Texas A&M University Press.

Tyler, M. A., and H. H. Seliger. 1978. Annual subsurface transport of a red tide dinoflagellate to its bloom area: Water circulation patterns and organism distributions in the Chesapeake Bay. *Limnology and Oceanography* 23:227–246.

Valiela, I., and J. Teal. 1979. The nitrogen budget of a salt marsh ecosystem. *Nature* 280:652–656.

Walsh, J. P. 1981. U.S. policy on marine pollution. *Oceanus* 24:18–24.

Warner, W. W. 1976. *Beautiful Swimmers*. Boston: Little, Brown.

Williams, A. B. 1974. The swimming crabs of the genus *Callinectes* (Decapoda: Portunidae). *Fishery Bulletin* 72:685–798.

Wright, P. A. 2012. Environmental physiology of the mangrove rivulus, *Kryptolebias marmoratus*, a cutaneously breathing fish that survives for weeks out of water. *Integrative and Comparative Biology* 52(6):792–800.

Wurster, C. E. 1968. DDT reduces photosynthesis by marine phytoplankton. *Science* 159:1474.

Zedler, J., T. Winfield, and D. Mauriello. 1978. Primary productivity in a southern California estuary. *Coastal Zone* 3:649–662.

CHAPTER
10

Coastal Seas

A rocky shoreline with a sandy beach in the background. Scientists use transects to survey algae and animals in the intertidal zone during low tides.
Carliane Johnson/SeaJay Environmental.

More than 90% of the animal species found in the ocean and nearly all of the larger marine plants are benthic, living in close association with the sea bottom, and a large proportion of benthic organisms live along or near the coast. Due to their reliance on solar energy, benthic primary producers exist only in the shallow nearshore fringe in the photic zone, where the sea bottom is illuminated by the sun. Benthic animals range much more widely than primary producers, from high intertidal zones to cold, perpetually dark trenches more than 10,000 m deep, although species diversity is much greater in shallower areas of the sea where light is available and overall productivity is generally high.

Regardless of depth or geographic location, all benthic organisms live at the interface between the sea bottom and the overlying water. The environmental conditions they experience are defined by the characteristics of the bottom materials and the overlying water, the exchange of substances between the sediments and the overlying water, and conditions established by the other members of their communities. In this chapter, we examine the general conditions of life on temperate and subtropical coasts, from intertidal shorelines to continental shelf areas just below the low-tide line. We will explore creatures living on top of the seafloor and those living nestled in between the sand grains, along with the unique mechanisms benthic organisms have evolved for reproduction and survival.

10.1 Seafloor Characteristics

The sea bottom supports the weight of many benthic organisms, from microscopic bacteria to large animals such as the giant clam, weighing 227 kg (500 lbs.). Some animals alter the seafloor by excavating burrows or constructing tubes of soft sediments for protection. On hard rock bottoms, animals and plants secure a firm attachment to the rocks so that they can resist the tug of waves and currents. Benthic organisms are adapted for life on or in particular bottom types, and the character of life in any particular area, to a large extent, is dependent on the properties of the bottom material in the area, which varies from solid rock surfaces to very soft, loose deposits.

The sea bottom also acts as a large receptacle, accumulating plankton, waste material, and other detritus that sink from the sunlit waters above. In some regions, fallout of organic detritus from the photic zone is the only source of food for the inhabitants on the dark bottom. A variety of worms, mollusks, echinoderms, and crustaceans obtain their nourishment by ingesting accumulated detritus and digesting its organic material.

The composition of the sea bottom is determined by its constituent materials and, in shallow water, by the amount of energy available in the wind-driven waves and currents at the sea surface. The energy of waves (and their ability to move particles) decreases from the sea surface downward and disappears at depths equal to about one half the wavelength of surface

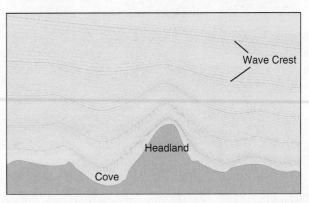

Figure 10.1 In coves and bays, refraction of advancing ocean waves spreads out the wave crests (and their energies) and concentrates them on headlands and other projecting coastal features.

waves. The coastline shape and nearshore subtidal structures also strongly influence the amount of energy expended on the shore when waves break. Wave fronts approaching shore start to slow just as they begin interacting with shallow reefs and bars. As a result, the wave fronts lag behind the rest of the wave as it approaches shore and changes shape to approximately match the curvature of an irregular coastline (**Figure 10.1**, and compare with Figure 10.14). This process redistributes the energy in the breaking waves, concentrating wave energy on headlands while spreading out and diminishing the energy of waves entering bays, coves, and other coastal indentations.

Taken together, the overall behavior of wind waves causes large amounts of energy to be expended on headlands in shallow waters (shallower than one half the wavelength of the waves) and lesser amounts of energy to be expended in coastal indentations and in deeper waters. On exposed headlands, erosion is the main result of breaking waves. These high-energy environments are continuously swept clean of fine sediment particles, detritus, and anything else not securely attached to the bottom. This debris is washed offshore into deeper water or along shore into bays or other calm-water coastal features, where it settles to the bottom and adds to the accumulating deposits already there.

Marine sediments near the shore and on the continental shelves are largely the products of erosion on land and subsequent transport by rivers (and, to a lesser extent, winds) to the sea (**Figure 10.2**). Once in the ocean, suspended sediment particles are carried and sorted by current and wave action according to their size and density. Large, dense sand grains quickly settle to the bottom near the shore. Very fine clay particles are often carried several hundred kilometers out to sea before settling.

On an oceanic scale, rocky outcrops occur in association with deep-sea ridges, rises, and volcanoes, but these outcrops are actually comparatively small and geographically isolated. Other rocky substrates are scattered around the very edges of ocean basins where wave-driven erosional processes dominate. Although relatively rare, these rocky shorelines have been more intensively studied than any other part of the ocean because of their easy access to land-based researchers and their students. Rocky intertidal and shallow subtidal areas are much easier to access than many other areas of the ocean.

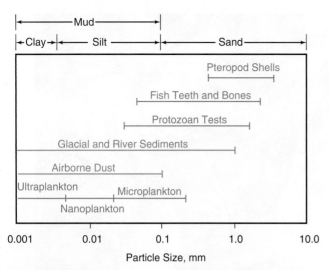

Figure 10.2 Particle size ranges for some common sources of marine sediments. Biogenic particles are shown in blue and terrigenous particles in tan.

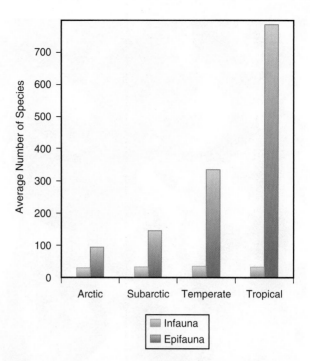

Figure 10.3 Variations in the average number of species of several bottom invertebrate groups from equal-sized coastal areas in different latitudes.

Data from G. Thorson, *Treatise on Marine Ecology and Paleoecology*. Vol I., Ecology. (Geological Society of America, 1957).

DID YOU KNOW?

To date, humans have explored less than 5% of the vast seafloor! Most of what has been explored is relatively shallow and nearshore. The number of deep-sea expeditions has increased in recent years, but the limitations of surveying the deep sea have led to a very slow process with very little area actually observed. Many mysteries of the sea are left yet undiscovered, and scientists have a lot of work to do to increase our understanding of this fascinating part of our Earth.

10.2   Animal–Sediment Relationships

Benthic animals that crawl about on the surface of the sea bottom or sit firmly attached to it are referred to as **epifauna**. Epifauna are associated with rocky outcrops or the surface of firm sediment deposits. Other benthic animals, the typically small-sized **infauna**, live within the substrate forming the bottom, where they feed and find protection from predators. **Figure 10.3** shows the relative abundance of epifauna and infauna at different marine climatic zones. Epifauna seem to be more sensitive than infauna to global-scale climatic differences, as the number of infaunal species is quite similar around the globe.

Infaunal clams, worms, and crabs are macroscopic and are familiar to anyone who has spent a few moments digging in a sandy beach or mudflat. These **macrofauna** either swallow or displace the sediment particles around them as they move. Less obvious, but no less important, are **microfauna**, microscopic infauna less than 100 μm in size that live on the sediment particles. Intermediate in size between the macrofauna and microfauna are the **meiofauna**, a very interesting and abundant group of animals. The meiofauna are also referred to as **interstitial** animals because they occupy the spaces (the *interstices*) between sediment particles. Surprisingly, many well-known types of large invertebrates have meiofaunal relatives (**Figure 10.4**).

Benthic animals mix and sort layers of sediments through their burrowing and feeding activities. Oxygen and water from the sediment surface circulate down into the sediment through their burrows. Sedimentary characteristics are further modified when particles are cemented together to form tubes and when sediments are compacted in the form of fecal pellets and castings. On rocky bottoms, the grazing activities of chitons, gastropods, and sea urchins aid the erosive processes of waves by scraping away rock particles as well as food (**Figure 10.5**) A few benthic animals, such as boring clams, are especially adapted for boring into solid rock.

The distributional patterns of benthic organisms are strongly influenced by the firmness, texture, and stability of their substrate. These features control the effectiveness of locomotion or, for nonmotile species, the persistence of their attachment to the bottom. Epifauna are most frequently associated with firm or solid bottom material. Moving over or attaching to a bottom composed of firm material, such as rock, is much easier than attempting to do the same on a bottom that is loose or soft with a relatively temporary configuration, such as fine sand.

The particle size and organic content of the bottom material limit the versatility and distribution of specialized feeding habits. Suspension feeders depend on small plankton or detritus for nutrition and use some kind of structure to capture their food. Sticky mucous nets or sheets collect minute suspended food particles from the water (**Figure 10.6**). Suspension feeders generally require clean water to avoid clogging their filters with indigestible particles; therefore, they are usually found on rocks or are associated with coarse sediments. A similar but distinct feeding strategy is filter feeding. Filter feeders move water through gills or another specialized structure and extract food particles.

Benthic environments are prime feeding grounds for predators and scavengers that feast on the residents of the bottom or on their remains. Most bottom predators and scavengers are

Figure 10.4 Microscopic view of an assortment of interstitial infauna in a benthic sample.
NOAA.

permanent members of the benthos and are eventually eaten by other benthic consumers. Fish, however, often feast heavily on intertidal animals during high tides, and shore birds replace them as predators at low tide. Sea stars are also usually carnivorous (**Figure 10.7**), and their feeding methods are quite unusual. Many sea stars extrude their stomachs outside their bodies and then begin to digest and absorb their food items externally. Predatory sea stars eat a variety of prey species, including

shellfish such as mussels, spiny sea urchins, or even other sea stars. Scavenger species such as the bat star consume benthic microbes and detritus from the sediments. Some sea star species are both predators and scavengers.

Regardless of the feeding habit used by benthic animals, the ultimate source of food is the primary producers of the photic zone. Intertidal and shallow-water benthic plants provide direct sources of nutrition for the abundant herbivorous algal grazers. Some **algal grazers** nibble away bits of the larger seaweeds (**Figure 10.8**). Most, however, scrape filmy growths of diatoms, cyanobacteria, and small encrusting plants from rocky substrates. Sea urchins use their five-toothed Aristotle's lantern mouthparts to remove algal growths. Herbivorous gastropods and chitons accomplish similar results with their filelike radula.

Figure 10.5 Sandstone erosion pits created by the rasping actions of small chitons.
Courtesy of James Sumich.

DID YOU KNOW?

Organisms living in between the sand grains, the infauna, can live in what would be considered very crowded conditions by many. In 10 cm² (1.5 in.²) of sediment, over a thousand tiny individual worms, crustaceans, and a variety of other organisms can survive and thrive. These dense communities may be hidden under the sediments, but they are feeding, sifting the sediments, and creating a lasting impact on the characteristics of the seafloor.

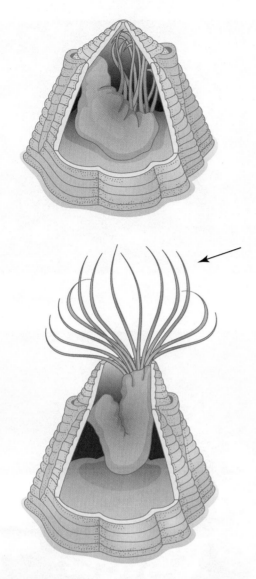

Figure 10.6   Barnacle tucked up in its shell (top) and with its feathery filtering appendages extended (bottom).
© Dorling Kindersley/Getty Images.

Figure 10.7   A group of sea stars, *Pisaster*, feeding on a bed of mussels.
Courtesy of Dave Cowles, Rosario Marine Invertebrates/http://rosario.wallawalla.edu/inverts.

Figure 10.8   A large snail grazing on seaweed.
© Christopher Poliquin/Shutterstock, Inc.

10.3   Larval Dispersal

Populations of slow-moving or sedentary benthic animals are not necessarily limited to narrow geographic ranges. Although as adults the distances they travel are quite short, reproductive efforts often lead to dispersal of future young great distances from the home range of the parents. One fifth of the common shallow-water animal species found at San Diego, California, for example, can also be found along the entire West Coast of the United States and in British Columbia. Other benthic species are even more widely dispersed, often in similar ecological conditions on opposite sides of the same ocean basin. *Mytilus edulis*, known as the bay mussel, blue mussel, or edible mussel, is common to temperate coasts on both sides of the Pacific and Atlantic Oceans.

A few animals, including a small percentage of barnacles, hitch rides as juveniles or adults on floating debris, on the hulls of ships, or in the ballast water of ships to travel transoceanic distances. An Australian barnacle, *Elminius modestus*, was apparently introduced to England by supply ships during World War II. It has since colonized most of the north and west coasts of continental Europe. In many sheltered reaches of these coastlines, *E. modestus* is competing with and replacing native barnacle populations.

A far more common and reliable adaptation for extending the geographic range of temperate- and warm-water benthic species involves the production of temporary planktonic larval stages. Planktonic larvae are part of the group of plankton known as **meroplankton** and account for the vast majority of this group. These small, feeble swimmers, bearing little resemblance to their parents (**Figure 10.9**), drift with the ocean's surface currents for some time before they metamorphose and assume their juvenile benthic lifestyles.

About 75% of shallow-water benthic invertebrate species produce larvae that remain planktonic for 2 to 4 weeks. More than 5% of the species examined in one study had planktonic larval stages exceeding 3 months, with a few as long as 6 months (**Figure 10.10**). Our understanding of ocean currents suggests that only larvae with the most prolonged larval stages

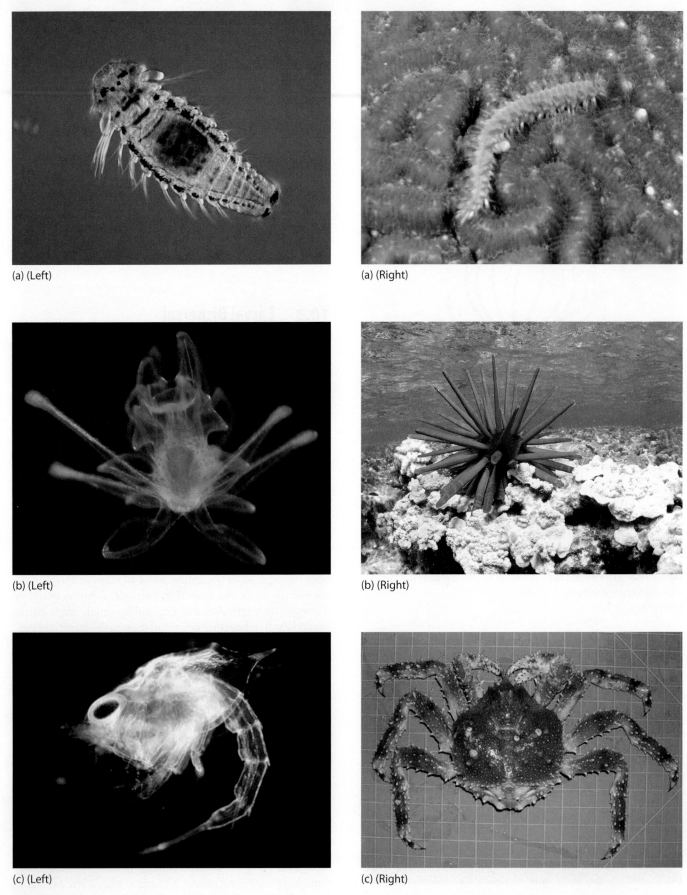

(a) (Left)

(a) (Right)

(b) (Left)

(b) (Right)

(c) (Left)

(c) (Right)

Figure 10.9 Meroplanktonic larval forms (left) and adult forms (right) of some common benthic animals: (a) polychaete worm, (b) sea urchin, and (c) crab.
(a) NOAA; (b) NOAA; (c) NOAA/AFSC.

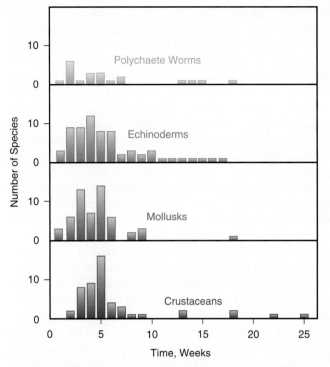

Figure 10.10 Typical duration of planktonic existence for four common groups of marine benthic invertebrates.
Data from G. Thorson. *Oceanography* (1961):455–474.

can make direct transoceanic trips before settling to the bottom. For each extra day the larvae remain in the plankton, they are exposed to additional threats of predation, increased pressures of finding food, and greater possibilities of being carried by the currents to areas where survival is unlikely. Even so, a temporary planktonic larval existence can provide several advantages to offset the enormous mortality experienced by these larvae.

Even in very slow ocean currents, drifting planktonic forms may spread far beyond the geographic limits of their adult population. Many are swept into unfavorable areas and perish, but survivors may expand their parents' original range or settle into and mix with other populations and reduce their genetic isolation. The exchange of individuals among marine populations is termed **connectivity**, and it has been a hot topic in the world of marine ecological research over the past decade. Ocean-current modeling and genetic analysis are some of the techniques that are being used to determine where collected larvae or young, recently settled juveniles originate. During their planktonic existence, many types of larvae react positively to sunlight and remain near the sea surface and their food supply, the phytoplankton. As their planktonic life draws to a close and they seek their permanent homes on the bottom, some larvae remain near the sea surface and ride into intertidal shores on waves and tides. Other larvae shun the light and swim near the bottom. Yet others seem to settle to their juvenile habitat during particular lunar phases, presumably to either take advantage of a bright full moon so they can see while settling during night-time hours or to take advantage of the darkness during a new moon to settle at night with less of a chance of being eaten by a visual predator. Most enter a swimming–crawling phase and

settle to the bottom and investigate it, and, if it is not suitable, they swim up to be carried elsewhere.

Just how larvae know when a suitable substrate is encountered is an important, but as yet unanswered, question. Chemical attractants, current speeds, types and textures of bottom material, lunar phase, and the effects of light are only partial answers to the question. Specific bottom types, such as sand or hard rock, do not attract larvae from a distance. Only after the appropriate bottom type is actually encountered may larvae be induced to remain and quickly metamorphose into a bottom-living stage. Alternatively, chemical substances diffusing from established populations of some attached animals, including oysters and barnacles, attract larvae of their own species. This attraction may be beneficial for oyster and barnacle larvae, because the presence of adults in the settling site ensures that physical conditions have been appropriate for survival. Also, the larvae's eventual reproductive success may be enhanced if they are in the vicinity of other members of their own species. For many other larvae, however, settling among their adults can be disastrous. Older established individuals generally have relatively lower demands for food and oxygen; have more stored energy; and, in general, can compete more effectively for resources with newly settled young. In times of shortages, younger or smaller individuals are usually the first to suffer. **Figure 10.11** illustrates the more obvious environmental features that may guide and influence larval settling.

Until they settle to the bottom and undergo metamorphosis to their juvenile benthic form, meroplanktonic larvae are not in competition for food or space with the adults of their species. Even so, competition for food among plankton is often rigorous. A few species with long planktonic larval phases produce large yolky eggs that provide larvae with most or all their nutritional supply; however, most species hatch from small eggs with little stored food and must begin feeding and competing with each other almost immediately.

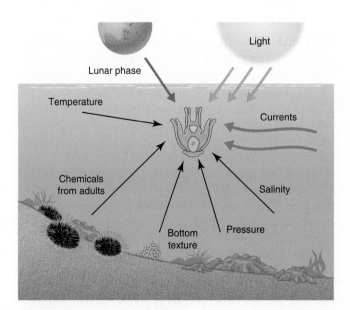

Figure 10.11 Several major environmental factors that influence the selection of suitable bottom types by planktonic larvae.

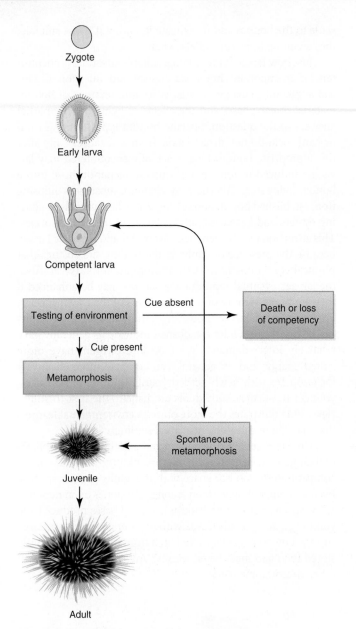

Figure 10.12 Generalized developmental pattern for planktonic larvae, illustrating available options in response to food, substrate, or other environmental cues.

It is becoming apparent that for some species the quality of larval feeding experiences strongly influences their juvenile growth and survival rates after metamorphosis. In laboratory studies, short-term food shortages of only a day or two during the larval stages of snails and barnacles led to delay of metamorphosis, increased larval mortality, and reduced growth or survival of juveniles after metamorphosis (**Figure 10.12**). It would be interesting, but is not yet possible, to follow the fates of individuals to determine whether the detrimental effects of food shortages in their early larval history persisted to adulthood and influenced their individual fitness by reducing fecundity or by delaying the age of sexual maturation.

Some benthic species, especially those in the tropics and the deep sea, spawn all year long; others, particularly those in temperate waters, have short and well-defined spawning seasons. In the latter group, the timing of reproduction or spawning is geared to produce young at times most advantageous to their survival. The spawning periods of many species of benthic animals are timed to place their larvae in the plankton community when their phytoplankton food source is abundant and readily accessible during spring and fall diatom blooms. In shallow waters, temperature and day length provide two of the more obvious cues for timing reproduction. The gonads of spring and summer spawners develop in response to rising water temperatures and lengthening days. Oysters, for instance, refrain from spawning until a particular water temperature is reached.

Regardless of how many eggs are produced, the reproductive success or fitness of an individual requires that its **fecundity** (the production of eggs or offspring) exceeds the **mortality** (the rate at which individuals are lost) of its offspring. Any population whose mortality consistently exceeds its fecundity will shrink and eventually disappear. Adaptations that increase fecundity or reduce mortality improve the chances for successful reproduction. The fecundity of some shallow-water benthic animals is truly amazing, with females of many species producing several million eggs annually. An individual sea hare of the genus *Aplysia* weighing a few kilograms produced an estimated 478 million eggs during 5 months of laboratory observations. Such excessive reproductive enthusiasm would quickly place any shoreline knee deep in sea hares if all of the spawning efforts of only a few adults survived.

Obviously, the egg and larval mortality of these species is extremely high. Of the millions of potential offspring produced, very few attain sexual maturity. The eggs of some benthic animals are fertilized internally before they are released into the water. Some species of snails, crabs, sea stars, sea cucumbers, and other invertebrates are **brooders**, retaining their larvae internally or in special brood pouches until the larvae are at reasonably advanced stages of development. For some of these invertebrates, brooding may be an adaptive consequence of small adult size. Smaller species of benthic invertebrates, with correspondingly smaller gonads, are less likely to produce sufficient planktonic larvae to equal their larger competitors. Consequently, they may opt for internal fertilization and larval incubation to reduce offspring mortality to some extent. An example of this pattern is seen in sea cucumbers in southern California. The larger species of sea cucumber releases its eggs and sperm into the water, leading to pelagic larvae that must compete and survive in the plankton. One small species of sea cucumber found at the California Channel Islands, *Pacythyone rubra*, is a brooder. *P. rubra* lives in groups of hundreds of individuals living closely clumped together. It uses internal fertilization and brooding, with young released as miniature versions of the adults within the same area as the parents.

Many of the more abundant and familiar seashore animals are **broadcast spawners**; they spew great quantities of eggs and even more remarkable quantities of tiny sperm into the surrounding water where fertilization occurs. These numerous eggs are necessarily small, and once fertilized hatch quickly into planktonic larval forms. Each larva, then, is a relatively low-cost genetic insurance policy for spawning adults, there to ensure that the genes of the parents survive another generation.

It seems that fertilization in these broadcast spawners is neither a casual nor haphazard process. Chemical substances (known as **pheromones**) are present in the egg or sperm secretions of sea urchins, oysters, corals, seaweeds, and other broadcast spawners. When shed into seawater, these pheromones induce other nearby members of the same population to spawn, and they, in turn, stimulate still others to spawn until much of the population is spawning simultaneously. As you might guess, spawning pheromones are typically species specific; pheromones of one species of broadcast spawner will induce only other members of the same species to spawn without influencing the spawning of any other species.

After spawning has occurred, sperm must still make contact with the correct type of egg, possibly in an ocean of other eggs, for fertilization to occur. A structural protein contained in the head of sperm cells of broadcast spawners binds only with other proteins on the surface coat of eggs from the same species (**Figure 10.13**), which limits the potential for the formation of **hybrid** individuals. Together with pheromones, these substances regulate the timing of spawning, the specificity of sperm for eggs of the same species, and ultimately the overall prospects for successful fertilization by broadcast spawners. All of these factors increase the chance of success for many species that use the very basic reproductive method of spewing gametes out into the water.

DID YOU KNOW?

Because many larvae are attracted to light, scientists use lighted traps to collect marine larvae. The tiny creatures swim toward the light source and end up trapped inside a fine mesh net and ready for a scientist to identify them. Light traps are a very effective way to capture many tiny marine invertebrates and larvae for study.

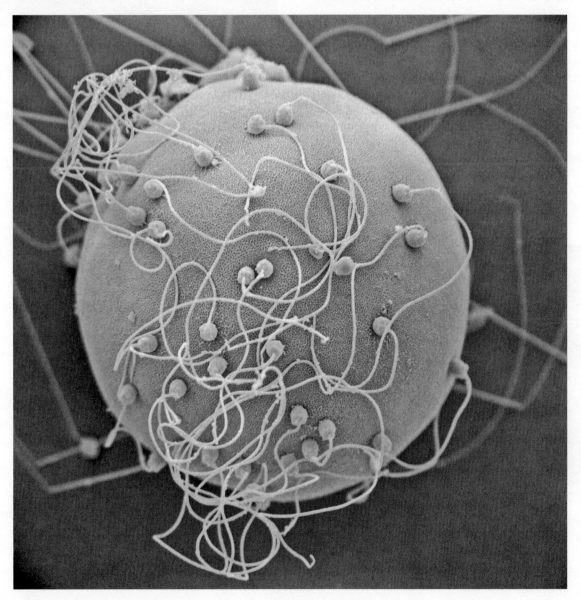

Figure 10.13 A scanning electron micrograph of a sea urchin egg with numerous sperm cells.
© Dr. David Phillips/Visuals Unlimited.

Case Study

Is the White Abalone on the Path to Extinction?

Common name: White abalone

Species name: *Haliotis sorenseni*

Endangered Species Act status: Endangered

The white abalone is a rare mollusk that was once found in great numbers in coastal waters from Morro Bay in southern California to Punta Rompiente, Baja California, Mexico (**Figure A**). It is the deepest-living abalone species in the North Pacific, inhabiting depths down to 60 m (197 ft) and rarely occurring in waters shallower than 30 m (98 ft). It is because of these great depths that white abalone populations remained stable for many years; fishing for white abalone requires more effort and involves more risks for divers than fishing for shallower abalone species does. Once shallow-water abalone species were fished severely and their numbers declined, many fishers then decided to make the switch to white abalone.

The fishery seemed stable until the 1990s when numbers began to plummet. At this time, many white abalone were found with a shriveled foot, the large muscle that enables them to anchor to rocks, and the same large muscle that is considered a delicacy by humans. Unfortunately, this condition, called *withering foot syndrome*, had been spreading through populations for at least a decade. However, this did not stop fishers from harvesting healthy specimens, and scientists were not immediately made aware of the condition so they could determine the cause and monitor populations. Withering foot syndrome afflicted other abalone species in the North Pacific, leading to declines in those populations as well. The combination of overfishing and withering foot syndrome has been catastrophic for the white abalone. In 2001, the white abalone was the first marine invertebrate to be added to the endangered species list.

Figure B   A captive-bred juvenile white abalone reared at the Bodega Marine Lab.
Kristin Aquilino/UC Davis.

White abalone are broadcast spawners, so each reproductive attempt includes females releasing thousands of eggs and males releasing millions of sperm into the water with very low fertilization and survival rates. Males and females must be located near enough for gametes to meet up during spawning. It has been determined that at the current estimated densities white abalone are unlikely to be capable of producing offspring in the wild. Their numbers are just too low, and individuals are located too far apart for successful fertilization to take place. In response to the predicted lack of new offspring, researchers have been feverishly attempting to culture white abalone in captivity. Early efforts were hindered by regulations placed on working with an endangered species, and once initiated were unsuccessful, in part because white abalone with withering foot syndrome tainted a breeding facility. Recent efforts have been much more successful, and now there are several facilities in California with captive-bred white abalone (**Figure B**). The goal of the captive breeding program is to reseed wild populations, which will include scuba divers placing white abalone in suitable habitat in the wild in hopes that they will survive and reproduce. The fate of this species seems to be in the hands of humans. Whether our efforts are sufficient or not will be determined very soon, as scientists estimate the demise of the species in the next decade if human intervention does not lead to a serious rebound.

Critical Thinking Questions

1. Do you think it is worth the time and cost associated with protecting an endangered species like the white abalone? Why or why not?

2. What potential problems are there with releasing captive-bred abalone into the wild?

3. Why are low densities so detrimental to populations of species that broadcast spawn, such as the white abalone?

Figure A   A white abalone blends into the rocks. Respiratory openings are visible along the dorsal edge of the shell.
NOAA/SWFSC.

10.4 Intertidal Communities

The coastal strip where land meets the sea is home to some of the richest and best-studied marine communities found anywhere. Although this coastal strip is narrow, its influence is enhanced by the wealth of marine organisms present and its accessibility to humans. Typically, the total biomass in a square meter at the low-tide line is at least 10 times as high as that of a comparable area on the bottom at 200 m and is several thousand times higher than that found in most abyssal areas.

The periodic rise and fall of the tides has a dramatic effect on a portion of the coastal zone known as the intertidal, or **littoral**, zone. In the littoral zone, the sea, the land, and the air all play important roles in establishing the complex physical and chemical conditions to which all intertidal organisms must adapt. Tidal fluctuations of sea level often expose intertidal organisms to severe environmental extremes, alternating between complete submergence in seawater and nearly dry, often very warm terrestrial conditions. Local characteristics of the tides, including their vertical range and frequency, determine the amount of time intertidal organisms are out of water and exposed to air. Still, regardless of their location, most intertidal regions have exposure curves that resemble those shown in **Figure 10.14**.

DID YOU KNOW?

Some areas have extremely large tidal fluctuations. The Bay of Fundy in Canada experiences a difference of over 16 m (52 ft.) between low and high tide. That is the height of a five-story building. Organisms living in areas with huge tidal ranges spend hours exposed to the air and sun. Life is harsh for these critters, but those that are adapted for life in these conditions fill a unique environmental niche.

Most intertidal organisms are marine in origin and prefer to remain in seawater, and thus exposure at low tide is a time of serious physiological stress for them. Some of the most impressive adaptations for survival are displayed by marine organisms living in the intertidal zone. When the tide is out, exposed organisms are subjected to wide variations of atmospheric conditions. The air may dry and overheat their tissues in hot weather or freeze them during cold weather. Rainfall and freshwater runoff create osmotic problems as well. Predatory land animals, such as birds, rats, and raccoons, also make their presence felt in the intertidal zone at low tide. Only at high tide are truly marine conditions restored to the intertidal zone. The returning waters moderate the temperature and salinity fluctuations brought on by the previous low tide. Requirements for life such as food, nutrients, and dissolved oxygen are replenished, and accumulated wastes are washed away.

The return of seawater during high tides refreshes and revives marine organism, but it also brings the physical assault of waves and surf. The influence of wave shock on the distribution of intertidal organisms is apparent on all exposed coastlines of the world. Surf, storm waves, and surface ocean currents shift and sort sediments, transport suspended food, and disperse reproductive products. Much of the wave energy expended on

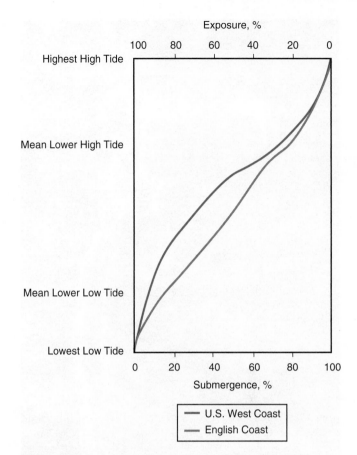

Figure 10.14 Exposure curves for the Pacific Coast of the United States and the Atlantic Coast of England.
Data from Ricketts, E. F., and J. Calvin, *Between Pacific Tides* (Stanford University Press, 1968) and from J. H. Lewis, *The Ecology of Rocky Shores* (English Universities Press, 1964).

the shore and the organisms living there eventually serves to shape and alter the essential character of the shoreline itself. Continually modified by the power of ocean waves, shorelines assume a variety of forms (**Figure 10.15**). Rocky shorelines are constantly swept clean of finer sediments by heavy surf or strong currents. Waves on beaches remove fine silt and clay particles but leave well-sorted, larger sand grains behind. The finer materials are washed out to sea or are deposited in the quiet protected waters of bays and lagoons.

The variety of tidal conditions, bottom types, and wave intensities along the shore creates a boundless assortment of living conditions for coastal organisms. It is difficult to characterize the prevailing conditions on long stretches of shoreline without risking overgeneralization. The west coast of North America, for instance, has many rugged rocky cliffs exposed to the full force of wave action, yet interspersed between these cliffs and headlands are numerous sandy beaches and quiet mud-bottom bays and estuaries. On the east coast, conditions vary from the spectacular rugged coastline of northern New England and Canada's Maritime Provinces to the extensive sandy beaches of the mid-Atlantic states. From Chesapeake Bay south to Florida, numerous coastal marshes are protected from extensive wave action by long, low barrier islands that parallel the mainland. Similar conditions with smaller tidal ranges exist along much of the Gulf Coast.

Figure 10.15 An infrared aerial photograph of a portion of the Oregon coast with protected coves, exposed headlands, sandy beaches, and offshore rocky reefs. Note the complex refraction of surface waves around the offshore reefs, apparent just behind the shore break.
Courtesy of U.S. Geological Survey.

(a)

(b)

Figure 10.16 (a) Young red mangroves colonize a tropical shoreline. (b) The network of mangrove prop roots traps sediment and detritus.
(a) © FloridaStock/Shutterstock, Inc. (b) © Seaphotoart/Shutterstock, Inc.

The southern tip of Florida is the only shoreline on the continental United States to experience tropical conditions. Tropical shorelines are typically marked by large coral reefs bordered by seagrass beds or by extensive swampy woodlands of mangroves. The coral reef system off the Florida Keys, fairly typical for the Caribbean, is unlike most other parts of the continental United States. Mangroves grow along Florida's southern coast and recently have become established along the extreme southern Texas coast (**Figure 10.16**). Included among these mangroves are several types of shrubby and treelike plants that grow together to form impenetrable thickets. Their branching prop roots trap sediments and detritus to extend existing shorelines or to form new low-lying islands.

The vertical distribution of intertidal organisms within the intertidal zone is governed by a complex set of environmental conditions that vary along gradients above and below the sea surface. Temperature, wave shock, light intensity, and wetness are some of the more important physical factors that vary along such gradients. Breaking waves can impose large forces on intertidal organisms. These organisms, in turn, demonstrate a remarkable variety of adaptations to deal with the forces associated with large waves. Several biological factors, including predation and competition for food and space, are superimposed on the physical gradients to delineate further the life zones of the shoreline. The combination of physical and biological factors frames and limits the range of an organism's existence. The range, as well as the biological role the organism plays in its habitat, defines the organism's niche. The complex interplay of physical and biological conditions and the variety of shore life itself create an abundance of niches for intertidal organisms.

Within a particular type of coastal environment, the interrelated influences of tidal exposure, bottom type, and intensity of wave shock produce an infinitely varied set of vertically arranged habitats. The vertical distribution of coastal organisms reflects the vertical changes in the shoreline's environmental conditions. Different species tend to occupy different levels or zones within the intertidal shoreline. Quite often, each zone is sharply demarcated from adjacent zones immediately above or below by the color, texture, and general appearance of the species living there. The result is a well-defined vertical series of horizontal life zones, sometimes extending for substantial distances along a coastline.

As the tides race in and out over the intertidal portions of the shoreline, vertically graded changes in temperature, light intensity, degree of predation, and other environmental factors are evident. Ocean tides are not the sole cause of vertical zonation within the intertidal shoreline. They do, however, modify and compress the pattern of zonation to make it more pronounced. These unifying and quite visible patterns of vertical zonation on rocky and sandy shorelines provide an appropriate framework within which to compare and describe the marine life of a few selected shores.

Rocky Shores

Ecosystems have two complementary pathways for energy transfer: a grazing food web and a detritus food web. The grazing food web routes energy and nutrient material from the primary producers through the grazers and predators. Detritus feeders use the bits and pieces of dead and decaying matter available from all trophic levels within the ecosystem. Within the coastal zone, neither rocky shores nor sandy beaches appear to be complete ecosystems by themselves. The trophic relations within rocky shore communities exhibit well-developed and complicated grazing food webs but very little in the way of detritus food webs. The constant and extreme water movements at rocky shores simply prohibit the accumulation of detritus and the existence of those animals dependent on it for food.

In rocky intertidal communities, patterns of distribution and abundance, as well as trophic relationships, are complex and sometimes change dramatically over short distances and from season to season. A shaded northern exposure may harbor several species absent from nearby sunny slopes. Tide pools contain an assemblage of organisms quite different from well-drained platforms 1 m away. The variety of life on one side of a boulder may differ markedly from life on the other side, and if you look under the boulder still other species may be found.

With such a bewildering array of niches available on a small stretch of shoreline, it might seem unlikely to find recurring themes of vertical zonation on widely separated shorelines. Yet similar patterns of zonation do exist on temperate rocky shores, whether in New England, Australia, British Columbia, or South Africa. Vertical zonation is such a compelling feature of life on rocky shores that considerable effort has been expended in devising schemes to identify and describe distinct intertidal subzones and their inhabitants.

The following sections examine some of the more conspicuous intertidal organisms and the adaptations that permit them to remain conspicuous. Three rather ill-defined zones—the upper, middle, and lower intertidal zones—are used for general reference. Because the boundaries separating these zones are artificial, they are frequently violated by their residents. You are likely to see species that dominate one zone scattered throughout other zones as well. This intertidal overview begins at the top, the upper intertidal zone, and proceeds downward to the shallow subtidal portions of temperate coastal shorelines.

The Upper Intertidal Zone

In the upper intertidal zone, living conditions are sometimes nearly as terrestrial as they are marine. The area is exposed to seawater only infrequently by extremely high tides and splash from breaking waves and is sparsely inhabited by marine organisms. Scattered dark mats of the cyanobacterium *Calothrix* or the lichen *Verrucaria* frequently form bands or series of tarlike patches to mark the uppermost part of the rocky intertidal. Small tufts of *Ulothrix*, a filamentous green alga, may also extend into the highest parts of the intertidal zone. These plants are tolerant to large temperature changes and are adapted to resist desiccation. The small tangled filaments of *Calothrix* are embedded in a gelatinous mass to maintain their store of water and to reduce evaporation. Lichens such as *Verrucaria* are symbiotic associations of a fungus and a unicellular alga (**Figure 10.17**). In the case of *Verrucaria*, the fungal part absorbs and holds several times its weight of water, water used by the fungus as well as by the photosynthetic algal cells that produce food for the entire lichen complex.

Only a few species of snails, limpets, and occasional crustaceans graze on the sparse and scattered vegetation of the upper intertidal zone (**Figure 10.18**). Unlike its other close marine relatives, the small snail *Littoraria* is an air breather. *Littoraria* uses a highly vascularized mantle cavity in much the same manner as land snails do for gas exchange. Some species of littorine snails are so well adapted to an air-breathing existence that they drown if forced to remain underwater. Like littorines, limpets

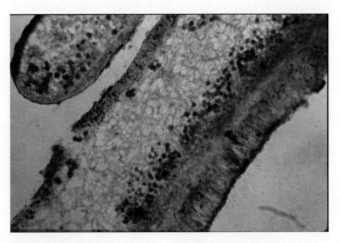

Figure 10.17 Magnified cross section of a lichen with algae cells (dark spots) embedded in fungal filaments.
Courtesy of James Sumich.

Figure 10.18 Snails and limpets grazing on sparsely distributed algae growing along the edge of a tidal pool.
© Wildlife Pictures/age fotostock.

Figure 10.19 Stunted acorn barnacles, *Chthamalus*, survive in the shallow depression of carved letters.
Courtesy of James Sumich.

of the upper intertidal (especially *Acmaea*) are amazingly tolerant of temperature changes. Both littorines and limpets can seal the edges of their shell openings against rock surfaces to anchor themselves and to retain moisture. They are algal grazers and use their filelike radulae to scrape the small algae and lichens from the rocks.

A conspicuous zone of small barnacles (**Figure 10.19**) frequently appears just below the lichens and cyanobacteria. Barnacles are filter feeders, but in the high intertidal zone they are able to feed only when wetted by high spring tides a few hours each month. While submerged, their feathery feeding appendages extend from their volcano-shaped shell and sweep the water for minute plankton (Figure 10.6). Between high tides, a set of hinged calcareous plates block the entrance to the shell and seal in the remainder of the animal, making it appear to be nonliving.

Most barnacles are hermaphroditic; they contain gonads of both sexes, yet most generally refrain from fertilizing their own eggs. During mating, a long tubular penis is extended into a neighboring barnacle, and the sperm are transferred to fertilize the neighbor's eggs. The eggs develop and hatch within the barnacle's shell and are released as microscopic free-swimming planktonic organisms known as **nauplii** (**Figure 10.20a**). After several molts of its exoskeleton, the nauplius develops into a **cypris** larva (**Figure 10.20b**). The cypris eventually settles to the bottom, selects a permanent settling site, and then cements itself to the bottom with a secretion from its antennae. Cypris larvae are attracted by the presence of other barnacles, thus ensuring settlement in areas suitable for barnacle survival and for obtaining future mates. Soon after settling, the cypris turns over, loses its larval appearance, and begins to surround itself with a wall of calcareous plates (**Figure 10.20c**) characteristic of the adult it will become.

Two species of intertidal barnacles provide a clear example of how the interplay of physical and biological factors influences the eventual vertical distribution of adult barnacles. In England, the larvae of one barnacle, *Chthamalus*, settle principally in the upper half of the intertidal zone, whereas the larvae of the other barnacle, *Balanus*, settle throughout the entire intertidal range (**Figure 10.21**). Desiccation in the upper extremes of the intertidal zone rapidly eliminates a good number of the settled *Balanus* but has little effect on *Chthamalus* at the same levels. Below the level of significant desiccation effects, however, *Balanus* is clearly the better competitor for space, overgrowing and undercutting *Chthamalus* wherever the two species overlap. The resulting distribution of adult forms of both species provides a useful generalization applicable to other attached animals living in space-limited intertidal situations: the upper limit of a species' distribution is restricted by the species' ability to cope with environmental stresses and other physical factors, such as temperature or desiccation, whereas a species' lower vertical range is limited by biological factors, especially competition with other species.

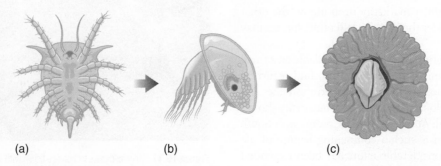

(a)   (b)   (c)

Figure 10.20 Planktonic and early benthic stages of the barnacle *Balanus*: (a) nauplius stage, (b) cypris stage, and (c) early benthic stage.

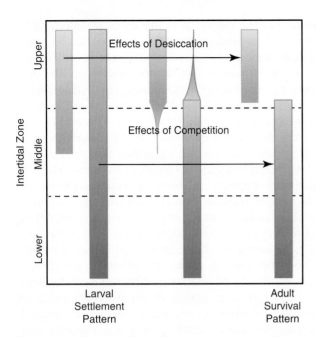

Figure 10.21 The limiting effects of desiccation and competition on the vertical distribution of two species of intertidal barnacles, *Chthamalus* (yellow bars) and *Balanus* (green bars).
Data from J. H. Connell, *Ecology* 42(1961):710–723.

The Middle Intertidal Zone

The middle intertidal zone is occupied by greater numbers of individuals and species than is the upper intertidal zone. This zone, sufficiently inundated by tides and waves, provides an abundance of plant nutrients, oxygen, and plankton food for filter-feeding animals. The lush growths of green, red, and brown algae also furnish a bountiful supply of locally produced food for herbivorous grazers.

Occasional small, water-filled tide pools protect hermit crabs, snails, nudibranchs, anemones, and a few small fish species from exposure and the physical assault of the surf. Aggregate anemones of the Pacific Coast (*Anthopleura elegantissima*) are known to withstand internal temperatures as great as 13°C above the surrounding air temperature, yet serious water loss will destroy them. These anemones combat extreme desiccation and temperature fluctuations by retracting their tentacles and attaching bits of light-colored stone and shell to themselves, presumably to reflect light and heat. Until recently *A. elegantissima* included two forms: one small, clumping form and another large, singular form. Genetic analysis revealed that the large, singular form is actually another species, *A. sola*. Another singular species exists in this genus, *A. xanthogrammica*, and is quite brilliantly green colored and large (**Figure 10.22**).

The clumped mats characteristic of the aggregate anemone are the result of a peculiar mode of asexual reproduction. To divide, these anemones pull themselves apart by simultaneously creeping in opposite directions. Each half quickly regenerates its missing portion, producing two new individuals to replace the original. All of the members of a clump resulting from this asexual fission are **clones**, or genetically identical individuals, with the same sex and color patterns. The clonal clumps of anemones are uniformly spaced and are separated from adjacent clones by

Figure 10.22 Photo containing all three of the common intertidal sea anemones in the genus *Anthopleura* found along the West Coast of the United States. From left to right: sunburst anemone (*A. sola*), giant green anemone (*A. xanthogrammica*), and several dozen aggregating anemones (*A. elegantissima*).
Dr. John Pearse/UC Santa Cruz.

bare zones about the width of a single anemone. These anemones also have separate sexes and can reproduce sexually by releasing eggs and sperm into the water.

The bare zones between anemone clumps result from a subtle type of competition between dissimilar individuals from adjacent clones. They are armed with special tentacles, or **acrorhagi**, that can inflict serious damage to anemones of opposing clones but that have no effect on individuals of the same clump. In these border wars, anemone clumps rely on a mechanism of self-recognition so that the aggressive response is directed only to members of genetically dissimilar clones.

The dominant and conspicuous members of the middle intertidal zone are mussels (*Mytilus*; **Figure 10.23**), barnacles (usually *Balanus*), some chitons and limpets, and several

Figure 10.23 Close-up view of mussels, *Mytilus*, attached to rocks in the middle intertidal zone.
Steve Lonhart / NOAA MBNMS.

Figure 10.24 Algal species exposed during low tides use thickened cell walls to prevent water loss.
Chad King / NOAA MBNMS.

species of brown algae (especially *Fucus* or *Pelvetia*). The animals securely anchor themselves to the substrate and generally present low, rounded profiles to minimize resistance to breaking waves (Figure 10.19). Macroalgae living here are secured by strong holdfasts and usually have sturdy but flexible stipes to absorb much of the wave shock. Both *Fucus* and *Pelvetia* have thickened cell walls to resist water loss during low tide (**Figure 10.24**).

In the densely populated middle intertidal zone, mussels, barnacles, brown algae, and other sessile creatures are limited by two commonly shared resources: the solid substrate on which they live and the water that provides their dissolved nutrients and suspended food. Mussels, barnacles, and algae compete for these crucial resources in different ways. In regions where physical factors are suitable for their survival, these three competing groups interact by dominating the available attachment space or by overgrowing their competitors and monopolizing the resources available from the water (**Figure 10.25**).

Available space on the rock surfaces of the middle intertidal zone is crucial for survival, yet it is seldom fully used. A number of interacting biological and physical processes occasionally create patches of open space. Sea stars and predatory snails continually remove patches of barnacles and mussels. Seasonal die offs of algae, and even battering by heavy surf, ice, and drifting logs, also clear patches for future settlement and competition.

Chance events and their relationship to seasonal patterns of reproduction have an appreciable influence on settlement patterns. Most of the middle intertidal animals have planktonic larval stages capable of settling almost anywhere within the intertidal zone. Algal spores and barnacle larvae simultaneously settling on bare rock eventually grow and compete for available space. Because of the space limitations that may exist on exposed coasts, the algae are usually squeezed out as the barnacles increase in diameter and dislodge or overgrow them. On some sheltered rocky coasts, recruitment of barnacle larvae is prevented by sweeping action of wave-tossed algal blades. Only in this manner can *Fucus* achieve and maintain spatial dominance over barnacles. If the barnacles are not removed before they are securely cemented into place, they escape the adverse effects of algal blades because of growth and may eventually force *Fucus* off the rocks.

Figure 10.25 Aggregation of tightly packed gooseneck barnacles (*Pollicipes polymerus*) compete for space in the intertidal zone.
Josh Pederson / NOAA MBNMS.

Barnacles are not necessarily safe after they have outgrown or overgrown their algal competitor. They are consumed in great numbers by sea stars, carnivorous snails, and certain fish. Even herbivorous limpets have a detrimental, and sometimes severe, impact on barnacle populations. Young barnacles are eaten or dislodged by limpets bulldozing along during their grazing activities. Limpets seem to have less effect on the small, crack-inhabiting *Chthamalus* than on the larger, more exposed *Balanus*. Thus, in the presence of limpet disturbance, *Chthamalus* gains a slight competitive advantage over the otherwise dominant *Balanus*.

The larval stages of mussels do not require bare rock exposures; they will settle on algae and barnacles and among aggregates of adult mussels. After settling, the young mussels crawl over the bottom, seeking better locations before they permanently attach themselves to the substrate with several strong elastic **byssal threads** (see Figure 10.23). Byssal threads are formed from a fluid secreted by an internal byssal gland. The fluid flows down a groove in the small tongue-shaped foot and onto the substrate. On contact with seawater, the fluid quickly toughens to form an attachment plate and a thread. Then the foot is moved slightly, and additional threads and plates are formed.

If left undisturbed, mussels eventually overgrow barnacles and algae. Seldom, however, do rocky intertidal conditions remain undisturbed for long. Mussels are extensively preyed on by sea stars (such as *Pisaster* on the Pacific Coast and *Asterias* on the Atlantic Coast). These sea stars are quite sensitive to desiccation and are limited to sites that remain submerged most of the time (**Figure 10.26**). Consequently, their impact on mussel populations is much more severe in the lower portions of the mussels' intertidal range. Young mussels are also preyed upon by *Nucella* and other predatory snails. Some mussels survive these predatory onslaughts by numerically swamping an area with more individuals than the local predators can consume. In time, the mussels may also escape through growth, becoming so large that sea stars cannot open them and predatory snails are incapable of drilling through their shells to consume the soft flesh within.

As patches of mussels are cleared out by predators or broken off by waves, they are temporarily replaced by algae or barnacles, but gradually the mussels regain their space on the rocks.

Figure 10.26   Sea stars, *Pisaster*, aggregating near the low-tide line to avoid desiccation.
Steve Lonhart / NOAA MBNMS.

In this way, diversity of species is maintained. It is ultimately the dynamic interplay resulting from competition between these dominant organisms and the patterns of disturbance that affect their survival that shapes the biological character of the middle intertidal zone. These organisms, in turn, influence the distribution and abundance of other plant and animal species.

Living on the mussel shells or among the thick masses of byssal threads beneath is a complex community of more fragile and often unseen animals. This submussel habitat protects several common species of clams, worms, shrimps, crabs, hydroids, and many types of algae. Many of these species use mussel shells as available solid substrate for attachment. Others exist because they are unable to survive in the same area without the protection afforded by the canopy of mussels overhead.

This complex association of organisms is wholly dependent on the existence of thick masses of well-anchored mussels. When natural disturbances remove the mussels and disrupt the stability of the community, they are quickly replaced by a predictable succession of nonmussel populations (**Figure 10.27**). Eventually, this process of biological succession may reach the

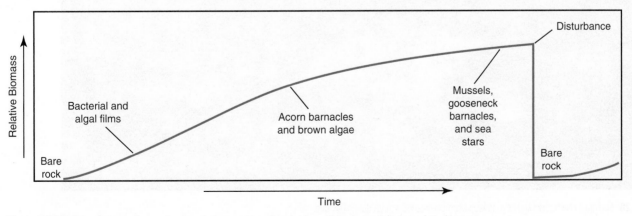

Figure 10.27   General pattern of succession through time on temperate rocky shores. The red curve indicates a relative biomass.

climax stage, in this case the mussel bed–gooseneck barnacle–sea star community. However, the structure of these communities is seldom stable for long; rather, they achieve a state of dynamic equilibrium between the stabilizing effect of succession and the many disruptive factors that reduce that stability.

The Lower Intertidal Zone

The biological character of lower intertidal rocky coasts differs markedly from the zones above it. It is difficult for some species and impossible for others to tolerate the more exposed conditions found in the upper and middle intertidal zones. The few species that do are often present in vast numbers. In the lower intertidal zone, the emphasis changes to a community with a high diversity of species, often without the conspicuous dominant types so characteristic of the middle and upper intertidal.

The lower intertidal zone of rocky shores abounds with seaweeds. Brown, red, and even a few species of green algae of moderate size spread a protective canopy of wet blades over much of the zone. In other places, extensive beds of seagrasses achieve a similar effect (**Figure 10.28**). Tufts of small filamentous brown and red algae carpet many of the rocks. Calcareous red algae become especially prolific at these levels. The pinkish hue of *Lithothamnion* encrusting rocks and lining the sides of tide pools is a common sight.

The animals of the lower intertidal zone include species from several animal phyla. It is here that the diversity,

complexity, and sheer beauty of intertidal marine life abound. On the East Coast of the United States, a large white anemone, *Metridium*, occurs in tide pools and on exposed portions of the lower intertidal zone. *Anthopleura*, a beautiful green anemone (**Figure 10.29**), occupies a similar habitat on the West Coast. Securely anchored by discs at their bases, anemones are predators of planktonic animals and small fish. They capture prey by discharging many microscopic nematocysts from special cells in their tentacles. When touched with a finger, nematocysts of sea anemones produce a slight tingling, sticky sensation.

The batteries of nematocysts found on anemone tentacles effectively discourage the hostile intentions of most predators, but they do not guarantee complete immunity against predation. A few snails and sea spiders penetrate the sides of anemones and feed on the unprotected tissues. Aeolid nudibranchs also commonly graze on anemones and the closely related hydroids. These nudibranchs possess mechanisms, not yet completely understood, that block the discharge of the toxic nematocysts. During digestion, the undischarged nematocyst-containing cnidocytes are preserved and passed to special storage sacs in the rows of finger-like **cerata** along the back of the nudibranch (**Figure 10.30**). There the cnidocytes serve as secondhand defensive stinging cells to be used against potential predators of nudibranchs.

The echinoderms are another familiar group of animals in the lower intertidal zone. Sea stars, sea urchins, brittle stars, and sea cucumbers are all quite sensitive to desiccation and salinity

Figure 10.28 Surf grass covers rocks and helps to keep intertidal organisms moist during low tide.
Chad King / NOAA MBNMS.

Figure 10.29 The green anemone, *Anthopleura xanthogrammica*.
© Weldon Schloneger/Shutterstock, Inc.

Figure 10.31 A scallop flaps its valves (shells) vigorously to jet away from a predatory sea star.
© Marevision/age fotostock.

changes and are seldom seen in abundance above the lower intertidal zone. Although slow moving, sea stars are voracious predators of mussels, barnacles, snails, an occasional anemone, and even other echinoderms (see Figure 10.7).

Sea stars continuously emit chemical substances that initiate alarm reactions in their prey; actual contact usually leads to even more vigorous escape movements. Prey species apparently recognize and identify their sea star predators by the substances they release. They react violently to the touch or presence of sea stars that usually prey on them, but they seldom respond to those not encountered in their normal habitat. When approached by some species of sea stars, many normally sessile species execute remarkable escape responses. Scallops swim jerkily away (**Figure 10.31**), clams and cockles leap clear of the sea star, and sea urchins and limpets crawl away relatively rapidly. Some sea anemones detach themselves and somersault or roll aside when touched by certain sea stars.

In summary, rocky intertidal zones of temperate shores have a dynamic pattern of organization dominated by physical forces at the upper extreme that diminish in influence downward and are gradually replaced by competition, predation, and other biological interactions. **Figure 10.32** shows the general pattern of vertical zonation of intertidal organisms on temperate rocky shores.

Figure 10.32 Vertical zonation patterns on a 3-m-high rock on the coast of Oregon.
Courtesy of James Sumich.

Figure 10.30 An aeolid nudibranch with long finger-like cerata projecting from its dorsal surface.
Douglas Mason / California High School.

RESEARCH in Progress

Mass Mortality of Millions of Sea Stars Linked to a Virus

During what was identified as the largest known mass mortality event to occur in the marine environment, populations of 20 species of sea stars began dying off and wasting away at alarming rates along the West Coast of the United States beginning in 2013. Sea stars with drooping arms and clearly decaying bodies were first found in central California and Washington populations and soon after in southern California and Oregon populations. Eventually, the condition spread as far south as Baja California, Mexico, and north into Alaska. Scientists and citizens roaming the intertidal zone were greatly concerned about these massive die-offs of millions of sea stars.

The dramatic symptoms were recognized by scientists immediately, as die-offs of "melting" sea stars have been observed since the 1970s in West Coast populations of various species. The difference this time was the magnitude of the affliction, known as *sea star wasting disease* (SSWD), and the large number of sea star species affected. Early symptoms of SSWD include behavioral changes, resulting in lethargic sea stars that curl up their arms. Soon after these early signs appear, lesions start to form, and eventually tissue death is apparent before the animal dies (**Figures A** and **B**).

Concerns over disappearing sea star populations and potential alterations of the entire food web in the shallow coastal areas led to the launch of extensive research to identify the cause of the disease. Several research questions were crucial to understanding the threat presented by this outbreak: (1) was

Figure B A diseased sea star, *Pisaster*, slowly wasting away and losing tissue. Kevin Lafferty/USGS.

the disease contagious? (2) how was the disease contracted? (3) what was the exact cause of the disease? and (4) were other echinoderms at risk?

Researchers surveyed areas up and down the West Coast to attempt to quantify the number of afflicted individuals and to assess the extent of the die-offs. They also collected specimens to analyze and try to solve the mystery of what the exact cause of the disease was. Early hypotheses included bacteria or a virus as the potential culprit, and extensive research studies were required to tease out the best possible answer. Pathogenic agents such as viruses are so abundant and so small that it is difficult to identify a specific individual agent as the cause for a particular disease. After examining sea star tissues, the researchers concluded that it was unlikely that a prokaryotic or eukaryotic pathogen was the cause, so they turned to viruses. Experiments revealed several key pieces of information: first, the pathogenic agent is the size of a virus, and second, the pathogen is transmissible from diseased sea stars to healthy sea stars. The apparent culprit is a densovirus called *sea star–associated densovirus*. This virus was present in afflicted individuals, as well as in sediments and the water column, so the disease could have spread between sea stars or from the environment to sea stars.

What is really interesting is that the sea star–associated densovirus is present in sea star samples from as far back as the 1940s and is also found in other echinoderms. One question that stems from this research is why the densovirus is killing sea stars now, but did not appear to do so in the past. A second question is how the densovirus may affect other echinoderms. In the spring of 2015, signs of a potential wasting disease event emerged in southern California sea urchin populations. At the same time, sea urchins in northern California experienced increased numbers in response to the die-offs of one of their top predators, the sea stars. Whether the West Coast is experiencing a wasting event

Figure A Healthy sea stars, *Pisaster*, clumped together in the intertidal zone.

in another echinoderm group is still unclear, but ecosystem shifts due to sea star die-offs are already occurring. The resulting communities will likely look quite different from those that existed before the large sea star die-offs. Researchers will continue to track these diseases and try to understand why echinoderms now seem more sensitive to viruses that have existed for decades. Has water chemistry due to ocean acidification or pollution changed enough to make them more vulnerable? Are they more sensitive to the virus in warmer waters, like we are now experiencing? These are just a few of the questions that remain to be answered and this massive die-off is a good example of how interconnected the marine environment and marine organisms are.

Critical Thinking Questions

1. Consider the intertidal zone without sea stars. What other organisms would increase in number?

2. The densovirus wiped out millions of sea stars from 2013 to 2015. Devise a hypothesis for why the sea stars died from the disease during this outbreak, when the disease appears to have been around for many years with no impacts on the same sea star species.

For Further Reading

Hewsona, I., J. B. Button, B. M. Gudenkauf, B. Miner, A. L. Newton, J. K. Gaydos, J. Wynne, C. L. Groves, G. Hendler, M. Murray, S. Fradkin, M. Breitbart, E. Fahsbender, K. D. Lafferty, A. M. Kilpatrick, C. M. Miner, P. Raimondi, L. Lahner, C. S. Friedman, S. Daniels, M. Haulena, J. Marliaveo, C. A. Burgem, M. E. Eisenlord, and C. D. Harvell. 2014. Densovirus associated with sea-star wasting disease and mass mortality. *Proceedings of the National Academy of Sciences USA* 111(48):17278–17283.

Sandy Beaches

Beaches and mudflats, the ecological complement to rocky shores, are depositional features of the coastal zone and are best developed along sinking (or subsiding) coastlines. Beaches and mudflats are unstable and tend to shift and conform to conditions imposed by waves and currents. Large plants find the shifting nature of soft sediments difficult to cope with, and few exist there. The few plants that have managed to adapt support even fewer grazers. Detritus food webs dominate on these depositional shores. Bits of organic material washed off adjacent rocky shores and the surrounding land or drifted in from kelp beds farther offshore sustain the detritus feeders of sandy and muddy shores. The washed up seaweed that lines the sandy shores is referred to as the **wrack line**. The wrack line provides a rich food source for terrestrial organisms and small organisms such as marine amphipods that burrow in the sand and surface to feed. It also provides nutrients for terrestrial plant growth. The wrack lines of some coastlines are quite impressive, with tall mounds of decaying seaweeds that remain for many months.

Beaches are made of whatever loose material is available. Quartz grains, black volcanic sand, or pulverized carbonate plant and animal skeletons are most common. Beaches occur where waves are sufficiently gentle to enable sand to accumulate but still strong enough to wash the finer silts and clays away. A good portion of the sand on many beaches is eroded away by large winter waves and deposited as underwater sandbars offshore. Smaller waves the following summer move the sand back on shore. Longshore currents also slowly move beach sands parallel to the shore. In response, populations of beach inhabitants may fluctuate widely from season to season and from one year to the next.

Several properties of marine sediments are defined by the size and shape of sediment particles. The size of spaces between sediment particles (the interstitial spaces) decreases with finer sediments. Interstitial-space size, in turn, regulates the porosity and permeability of sediments to water. In coarse sands, water flows freely between sand grains, recharging the supply of dissolved oxygen and flushing away wastes. Sand beaches also drain and dry out quickly during low tide.

When compared with the teeming populations of the rocky intertidal area [?], beaches appear to be biologically boring. Macroalgae and large obvious epifauna are rare. Shifting unstable sands are unsuitable platforms for surface anchorage, and nearly all of the permanent residents of beach communities dwell below the sand surface. Patterns of zonation are more difficult to demonstrate, yet under the sand there are distinguishable life zones comparable with those on rocky shores (**Figure 10.33**).

The upper portions of sandy beaches along temperate coasts are occupied by a few species of amphipods, particularly *Talorchestia*. Its common name, "beachhopper," reflects the unusual bounding mode of locomotion these small crustaceans use. Beachhoppers prefer to burrow a few centimeters into the sand during the day and are most active at night. Occasionally, they make excursions down the beach face as the tide recedes. They often emerge to feed on decaying, washed up seaweeds on the sand.

The middle beach is frequently populated by a variety of other amphipods; lugworms (*Arenicola*); dense concentrations

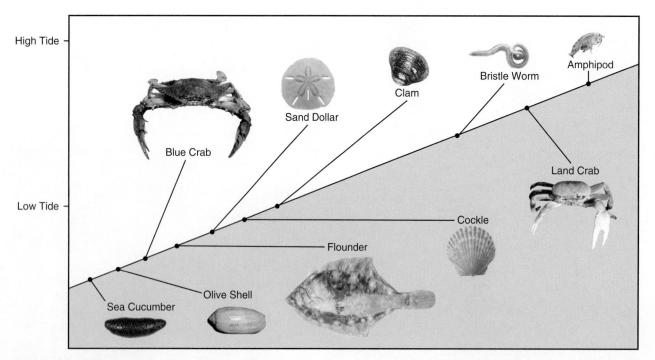

Figure 10.33 Sandy beach zonation along the East Coast of the United States. The species change rapidly from the portion permanently under water at left to the dry part of the beach above high tide at right.

Figure 10.34 Coiled fecal casting of the lugworm, *Arenicola*.
© Ismael Montero Verdu/Shutterstock, Inc.

of isopod crustaceans; and the sand crab, *Emerita*. *Arenicola* occupies a U-shaped burrow, with its head usually buried just below a sand-filled surface depression. The burrows are more or less permanent because waves stir up and move sediment and detritus to the head region where they are consumed. Mounds of coiled castings indicate the location of the other end of this sediment ingester (**Figure 10.34**).

The sand crab illustrates another feeding mode common to many beach macrofauna. When feeding, *Emerita* burrows tail first into the sand and faces down the beach (**Figure 10.35**). Only its eyes and a pair of large feathery antennae protrude above the sand. When a wave breaks over the crab and begins to recede, the antennae are extended against the rush of water known as **swash**. Entrapped phytoplankton (and possibly even large bacteria) are swept into the filtering antennae and then moved to the mouth by other feeding appendages.

Small isopods, usually less than 1 cm long, actively prey on even smaller interstitial animals that inhabit the interstitial spaces between sand grains. Many animal phyla are represented in the interstitial fauna of beaches, and a few groups

such as harpacticoid copepods and gastrotrichs are practically confined to the interstices of beach sands. Despite their diverse backgrounds, most interstitial animals exhibit a few basic adaptations needed for life between sand grains. They tend to be elongated, small (no more than a few millimeters), and move with a sliding motion between sand grains without displacing them. Examples of interstitial animals from different phyla are shown in **Figure 10.36**. Some interstitial animals are carnivorous; others feed on detritus deposits and material in suspension. A specialized feeding habit, unique to interstitial animals, is sand licking. Individual sand grains are manipulated by the animals' mouthparts to remove minute bacterial growths and thin films of diatoms.

In the lower portion of intertidal beaches, the diversity of life increases. Polychaete worms, still other amphipods, and an assortment of clams and cockles appear. Many of these lower-beach inhabitants, such as soft-shelled clams and cockles of the Atlantic Coast, represent the upper fringes of much larger subtidal populations. The small wedge-shaped bean clam, *Donax*, of the Atlantic and Gulf Coasts (but not the Pacific Coast species) migrates up and down the beach with the tides, yet it is usually considered an inhabitant of the lower beach. *Donax* responds to the agitation of incoming waves of rising tides by emerging from the sand to feed. After the wave carries the clam up the beach, it digs in to await the next wave and another ride. During ebb tides, the behavior is reversed. *Donax* emerges only after a wave breaks and begins to wash back down the beach. Thus, with little energy expenditure of its own, this small clam capitalizes on the abundance of available wave energy to carry it up and down the beach face such that it always is perfectly situated to feed in the water currents of appropriate velocities.

Donax is one of many sandy beach inhabitants to exhibit a rhythmic behavior that corresponds to the tidal cycle. Fiddler crabs quietly sit submerged at high tides and then emerge from their burrows at low tide to feed or engage in social activities. This tide-related cycle of activity exhibited by fiddler crabs is known as a **circalunadian rhythm**; it is synchronized to moon-related tidal cycles that repeat every lunar day (24.8 hours). When fiddler crabs are removed to the laboratory, their activity rhythms remain in concert with the changing tidal cycle for some time despite the absence of tidal cues. The same species of fiddler crab is light colored at night but darkens during the day; this cycle of color change depends on **circadian rhythms** based on a solar day length of precisely 24 hours.

Grunion (*Leuresthes*; **Figure 10.37**) are small fish inhabiting coastal waters of southern California. They, too, exhibit a very tightly timed spawning behavior related to ocean tide cycles. On the second, third, and fourth nights after each full or new moon of the spring and summer spawning season, grunion move up on the beach by the thousands to deposit their eggs in the sand and away from water. Their precise timing is remarkable; they spawn only during the first 3 hours immediately after the highest part of the highest spring tides. During the spring and summer, these tides occur only at night, so those who desire to observe or catch them spend late nights roaming the beaches with flashlights.

Figure 10.35 A sand crab, *Emerita*, backing into the sand in preparation for feeding.
© Stan Elems/Visuals Unlimited.

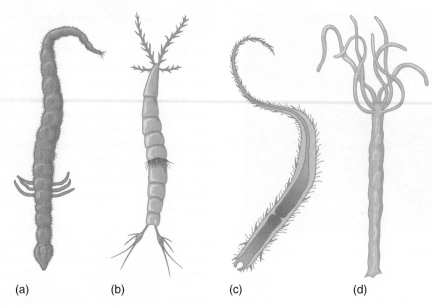

Figure 10.36 A few examples of the interstitial fauna of sandy beaches. Each is of a different phylum, yet all exhibit the small size and worm-shaped body characteristic of meiofauna: (a) a polychaete, *Psammodrilus*; (b) a copepod, *Cylindropsyllis*; (c) a gastrotrich, *Urodasys*; and (d) a hydra, *Halammohydra*.
Data from S. K. Eltringham *Life in the Mud and Sand* (Crane, Russak, 1972).

Figure 10.37 Grunion, *Leuresthes*, spawning in the sands of a southern California beach. Males coil around females that dig themselves into the sand to deposit their eggs.
USFWS.

Because the highest spring tides occur at the time of full and new moons, the grunion spawn immediately after high tides, but they spawn on successively lower tides each night (**Figure 10.38**). Thus, the eggs are buried by sand tossed up on the beach by the succeeding lower tides, and they are not washed out of the sand until the next series of spring tides. Nine or 10 days after the last spawning, tides of increasing height reach the area where the lowest eggs were buried (Figure 10.38). Wave action erodes the sand away and bathes the eggs with seawater. Almost immediately after being agitated and wetted by

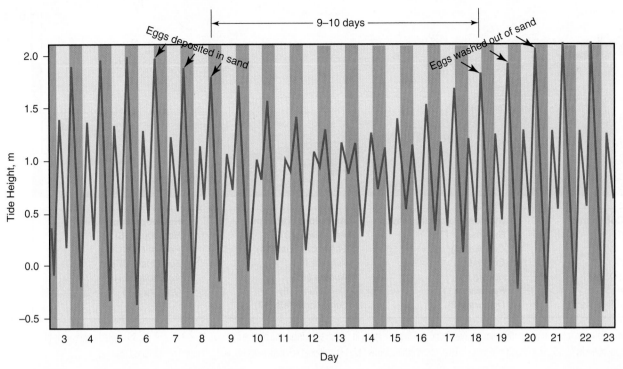

Figure 10.38 Predicted tide heights for a 3-week period at San Diego, California. Spring tides appropriate for grunion spawning occur on days 6, 7, and 8 (pointers at left). Nine to 10 days later, the next set of spring tides (pointers at right) wash the eggs from the sand, and they hatch. Shaded portions indicate night hours.

the waves, the eggs hatch, and the young grunion swim away. They remain in shallow coastal waters to feed and grow, reaching sexual maturity and their first spawning about 1 year later.

Like fiddler crabs and grunion, most, and probably even all, organisms have an innate time sense, a "biological clock." Rhythmic cycles of body temperatures, activity levels, oxygen consumption, and a host of other physiological variations occur independently of changes in or signals from the external environment. The internal "mechanism" of the clock is not known, but its existence has been demonstrated in a wide variety of organisms ranging from diatoms to humans.

10.5   Shallow Subtidal Communities

At the low-tide line, the lower intertidal zone merges with the uppermost part of the inner shelf zone. Where rocky bottoms extend below the low-tide line and are not covered by sediments the transition from intertidal to subtidal zone is gradual. Many organisms common to the lower intertidal zone are also abundant in neighboring shallow subtidal regions; however, rocky substrates eventually give way to soft sediments. In protected stretches of coastlines or below the sea surface, wave action diminishes, and loose sediments and detritus begin to accumulate. Organisms characteristic of the rocky shore are replaced by those typical of sand or mud bottoms. In some areas, seagrass beds begin in the lower intertidal zone and extend into the shallow subtidal zone. An example of this is in coral reef areas where large seagrass beds flourish around shallow reefs and near the shoreline.

Extensive studies of the life in shallow-water soft sea bottoms were initiated by the Danish biologist C. G. J. Petersen in the early part of the 20th century. His intent was to evaluate the quantity of food available for flounders and other commercially useful bottom fishes. After sorting and analyzing thousands of bottom samples from Danish seas, Petersen concluded that large areas of the level sea bottom are inhabited by recurring associations, or communities, of infaunal species (**Figure 10.39**). Each community has a few very conspicuous or abundant macrofauna as well as several less obvious forms. On other bottom types, different distinct communities of other species can be found. When exposed to similar combinations of environmental conditions, widely geographically separated shallow-bottom communities in temperate waters closely resemble each other in structure and the major types of organisms found (**Figure 10.40**). Although the actual species compositions will most often differ over large geographic areas, the types of organisms (to the class, order, family, or genus level) and the proportions at which they are found within the sediments will remain similar.

Modern benthic ecologists have found parallel shallow-water communities in much of the cold and temperate regions of the world ocean. This parallel community concept has been extended beyond obvious animal associations to include bottom type, depth, and water temperature as additional key factors in shaping benthic community structures. These infaunal communities exist only in soft, muddy marine sediments that dominate the continental shelves of the world. Because infaunal community compositions are relatively consistent across large areas with similar environmental conditions, benthic

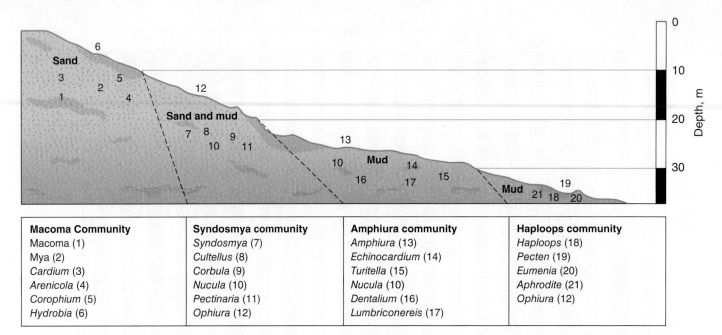

Macoma Community	Syndosmya community	Amphiura community	Haploops community
Macoma (1)	Syndosmya (7)	Amphiura (13)	Haploops (18)
Mya (2)	Cultellus (8)	Echinocardium (14)	Pecten (19)
Cardium (3)	Corbula (9)	Turitella (15)	Eumenia (20)
Arenicola (4)	Nucula (10)	Nucula (10)	Aphrodite (21)
Corophium (5)	Pectinaria (11)	Dentalium (16)	Ophiura (12)
Hydrobia (6)	Ophiura (12)	Lumbriconereis (17)	

Figure 10.39 A series of soft-bottom benthic communities found at different depths in Danish seas, including bivalve mollusks (1, 2, 3, 7, 8, 9, 10, 19), polychaete worms (4, 5, 11, 17, 20, 21), gastropod mollusks (6, 15), scaphopod mollusks (16), ophiuroid echinoderms (12, 13), echinoid echinoderms (14), and arthropod crustaceans (18).

samples are useful in assessing potential changes to an area due to pollution or other anthropogenic disturbances. Examples of anthropogenic activities that may change the benthos include fishing gear, such as trawl nets that drag along the bottom, and sewage outfalls that release treated sewage from a pipeline directly onto the benthos. Samples from infaunal communities in areas of anthropogenic disturbance are often studied and compared to samples from areas with no known human disturbance.

Infaunal communities require soft substrate and thus cannot establish themselves on rocky sea bottoms. Rocky outcrops on the shallow ocean floor in temperate areas are instead dominated by assemblages of brown algae, which anchor to rocks and provide large areas for other organisms to live on or swim

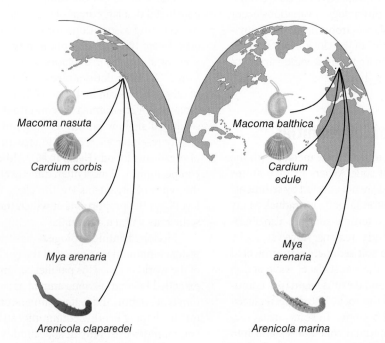

Figure 10.40 Diagram showing the close similarity in composition of soft-bottom communities in the northeast Pacific and the northeast Atlantic.

Data from G. Thorson *Treatise on Marine Ecology and Paleoecology*. Vol I., Ecology. (Geological Society of America, 1957).

Figure 10.41   Several species of kelp-community fishes sheltering near giant kelp, *Macrocystis*.
Chad King / NOAA MBNMS.

among. The large size (up to 30 m) of some kelp plants adds an important three-dimensional structure to kelp "forests" analogous to the canopy structure of terrestrial forests. Consequently, in the more complex ecological terrain of kelp forests, more niches exist than do on nearby soft sediments, and these niches are occupied by a rich diversity of invertebrates on the seafloor and numerous fish species in the kelp canopy (**Figure 10.41**). The kelp forest ecosystem represents a delicate balancing act between the kelp plants, herbivores that graze on the kelp, and carnivores that keep the herbivores in check. If the average abundance of any one of these participants in the food web is drastically altered, the whole system is thrown out of balance quickly.

STUDY GUIDE

TOPICS FOR DISCUSSION AND REVIEW

1. Why do you think most marine animals are benthic? What advantages are there to a benthic existence (as opposed to a pelagic life in the water column)?

2. Distinguish between the terms *epifauna* and *infauna*, and give an example of each type of animal. Then do the same for the terms *macrofauna*, *meiofauna*, and *microfauna*.

3. List and discuss the selective advantages of meroplanktonic larval stages for benthic animals living in shallow water.

4. Compare the species diversity of the middle and lower intertidal zones on rocky shores and on sandy beaches. What factors influence or create the differences observed?

5. Compare and contrast detritus, filter, and suspension feeding. Name a representative animal that performs each type of feeding.

6. Summarize the major factors that influence the vertical distribution of intertidal plant, algae, and animal species, and describe how these factors are influenced by tidal fluctuations.

7. Why are benthic epifauna and attached algae and plants seldom found on exposed sandy beaches?

8. Discuss the ecological relationships among intertidal mussels, barnacles, macroalgae, and sea stars on a temperate rocky coast.

KEY TERMS

acrorhagi 283	hybrid 277
algal grazer 272	infauna 271
broadcast spawner 276	interstitial 271
brooder 276	littoral 279
byssal thread 285	macrofauna 271
cerata 286	meiofauna 271
circadian rhythm 291	meroplankton 273
circalunadian rhythm 291	microfauna 271
clone 283	mortality 276
connectivity 275	nauplius 282
cypris 282	pheromone 277
epifauna 271	swash 291
fecundity 276	wrack line 290

KEY *GENERA*

Acmaea	Leuresthes
Anthopleura	Lithothamnion
Aplysia	Littoraria
Arenicola	Macrocystis
Asterias	Metridium
Balanus	Mytilus
Calothrix	Pacythyone
Chthamalus	Pelvetia
Cylindropsyllis	Pisaster
Donax	Pollicipes
Elminius	Psammodrilus
Emerita	Talorchestia
Fucus	Ulothrix
Halammohydra	Urodasys
Haliotis	Verrucaria

REFERENCES

Alongi, D. M., and P. Christoffersen. 1992. Benthic infauna and organism-sediment relations in a shallow, tropical coastal area: Influence of outwelled mangrove detritus and physical disturbance. *Marine Ecology Progress Series* 81:229–245.

Armstrong, R. A., and R. McGehee. 1980. Competitive exclusion. *American Naturalist* 115:151–170.

Brafield, A. E. 1978. *Life in Sandy Shores*. Studies in Biology no. 89. London: Edward Arnold.

Butler, J. B., M. Neuman, D. R. Pinkard, R. Kvited, and G. Cochrane. 2006. The use of multibeam sonar mapping techniques to refine population estimates of the endangered white abalone (*Haliotis sorenseni*). *Fishery Bulletin* 104:521–532.

Carlton, J. T. 1985. Transoceanic and interoceanic dispersal of coastal marine organisms: The biology of ballast water. *Oceanography and Marine Biology Annual Review* 23:313–371.

Connell, J. H. 1961. The influence of interspecific competition and other factors on the distribution of the barnacle, *Chthamalus stellatus. Ecology* 42:710–723.

Constable, A. J. 1999. Ecology of benthic macroinvertebrates in soft-sediment environments: A review of progress towards quantitative models and predictions. *Austral Ecology* 24:452–476.

Dayton, P. K. 1971. Competition, disturbance, and community organization: The provision and subsequent utilization of space in a rocky intertidal community. *Ecological Monographs* 41:351–389.

Denny, M. 1995. Survival in the surf zone. *American Scientist* 83:166–173.

Eltringham, S. K. 1972. *Life in Mud and Sand*. New York: Crane, Russak.

Epel, D. 1977. The program of fertilization. *Scientific American* November:129–138.

Farmingon, J. 1985. Oil pollution: A decade of monitoring. *Oceanus* 28:2–12.

Grahame, J., and G. M. Branch. 1985. *Reproductive Patterns of Marine Invertebrates*. Scotland: Aberdeen University Press.

Hadfield, M., and V. J. Paul. 2001. Natural chemical cues for settlement and metamorphosis of marine-invertebrate larvae. *In*: J. B. McClintock and B. J. Baker, eds. *Marine Chemical Ecology*. Boca Raton, FL: CRC Press. pp. 431–461.

Harger, J. R. E. 1972. Competitive coexistence among intertidal invertebrates. *American Scientist* 60:600–607.

Hayes, F. R. 1964. The mud–water interface. *Oceanography and Marine Biology Annual Review* 2:122–145.

Hoar, W. S. 1983. *General and Comparative Physiology*. Englewood Cliffs, NJ: Prentice Hall.

Johnson, C. H., and Hastings, J. W. 1986. The elusive mechanism of the circadian clock. *American Scientist* 74:29–36.

Kline, D. 1991. Activation of the egg by the sperm. *BioScience* 41:89–95.

Lee, J. J. 1995. Living sands. *BioScience* 45:252–261.

Lewis, J. H. 1964. *The Ecology of Rocky Shores*. London: English Universities Press.

Loughlin, T. R., ed. 1994. *Impacts of the Exxon Valdez Oil Spill on Marine Mammals*. San Diego: Academic Press.

Lubchenco, J. 1978. Plant species diversity in a marine intertidal community: Importance of herbivore food preference and algal competitive abilities. *American Naturalist* 112:23–39.

McIntyre, A. D. 1969. Ecology of marine meibenthos. *Biological Review* 44:245–290.

Mearns, A. J. 1981. Effects of municipal discharges on open coastal ecosystems. *In*: R. A. Geyer, ed. *Marine Environmental Pollution*. Amsterdam: Elsevier.

Menge, B. A., and J. Lubchenco. 1981. Community organization in temperate and tropical rocky intertidal habitats: Prey refuges in relation to consumer pressure gradients. *Ecological Monographs* 51:429–450.

Moore, P. G., and R. Seed. 1986. *The Ecology of Rock Coasts*. New York: Columbia University Press.

Newell, R. C. 1979. *Biology of Intertidal Animals*. Faversham, Kent, UK: Ecological Surveys.

Palmer, J. D. 1974. *Biological Clocks in Marine Organisms: The Control of Physiological and Behavioral Tidal Rhythms*. New York: Interscience Publishers.

Pechenik, J. A., D. E. Wendt, and J. N. Jarrett. 1998. Metamorphosis is not a new beginning. *BioScience* 48:901–910.

Ray, G. C. 1991. Coastal-zone biodiversity patterns. *BioScience* 41:490–498.

Ray, G. C., and W. P. J. Gregg. 1991. Establishing biosphere reserves for coastal barrier ecosystems. *BioScience* 41:301–309.

Reise, K. 1985. *Tidal Flat Ecology*. New York: Springer-Verlag.

Ricketts, C., J. Calvin, and J. W. Hedgepeth. 1986. *Between Pacific Tides*. Revised by D. W. Phillips. Palo Alto, CA: Stanford University Press.

Ruppert, E., and R. Fox. 1988. *Seashore Animals of the Southeast*. Columbia, SC: University of South Carolina Press.

Sanders, H. L. 1968. Marine benthic diversity: A comparative study. *American Naturalist* 102:243–282.

Scheltema, R. S. 1971. Larval dispersal as a means of genetic exchange between geographically separated populations of shallow-water benthic marine gastropods. *Biological Bulletin* 140:284–322.

Sebens, K. P. 1983. The ecology of the rocky subtidal zone. *American Scientist* 73:548–557.

Selkoe, S. A., and R. J. Toonen. 2011. Marine connectivity: A new look at pelagic larval duration and genetic metrics of dispersal. *Marine Ecology Progress Series* 436:291–305.

Steinberger, A., and K. Schiff. 2002. Characteristics of Effluents from Large Municipal Wastewater Treatment Facilities Between 1988 and 2000. S. Calif. Coastal Water Resources Project Special Publication.

Stierhoff, K. L., M. Neuman, and J. L. Butler. 2012. On the road to extinction? Population declines of the endangered white abalone, *Haliotis sorenseni*. *Biological Conservation* 152:46–52.

Strathman, R. 1974. The spread of sibling larvae of sedentary marine invertebrates. *American Naturalist* 108:29–44.

Trager, G. C., J. S. Hwang, and J. R. Strickler. 1990. Barnacle suspension-feeding in variable flow. *Marine Biology* 105:117–127.

Turner, M. H. 1990. Oil spill: Legal strategies block ecology communications. *BioScience* 40:238–242.

Underwood, A. J. 1974. On models for reproductive strategy in marine benthic invertebrates. *American Naturalist* 108:874–878.

Whitlatch, R. B. 1981. Patterns of resource utilization and coexistence in marine intertidal deposit-feeding communities. *Journal of Marine Research* 38:743–765.

CHAPTER 11

The Coral Reef Ecosystem

A healthy elkhorn coral (*Acropora palmata*) colony in St. Croix, U.S. Virgin Islands. This species has declined dramatically throughout its range and was the first coral species to be listed as a threatened species.
Courtesy of National Oceanic and Atmospheric Administration.

The tropical coral reef ecosystem is one of the most diverse and visually stunning ecosystems on Earth, with a species diversity that rivals that of rain forests. This diversity is immediately apparent when observing a coral reef underwater, with an amazing myriad of colors and unique body forms exhibited by the reef's inhabitants and a high number of life forms darting around the reef. Reef-building corals provide a three-dimensional living structure for other organisms to live near, on, or within. This structure is invaluable for the many inhabitants of the reef for survival, feeding, and reproduction, among other important life processes. The hard substrate that corals provide is in high demand and is a limited resource that is competed for. The evolutionary pathways of coral reef inhabitants have certainly included behavioral and physiological adaptations for obtaining space on the reef, and some of the behaviors exhibited by coral reef organisms are unique among the animal kingdom. In this chapter, we summarize the biology of coral reefs, a tropical ecosystem that provides, by virtue of its presence alone, millions of hectares of firm substrate and vertical structure on which countless plants and animals live, many of which are found nowhere else on Earth. The species that compose coral reefs are inherently interesting on their own, but the fact that they also create the physical structures for unique biological communities makes them especially fascinating and important subjects of study. We will also examine the biology and diversity of coral reef fishes and the health of coral reefs and their inhabitants worldwide, as coral reef mortality is a topic of grave concern.

11.1   Coral Reefs

For many people, thoughts of tropical islands conjure up images of a special type of marine ecosystem, coral reefs. Unlike the rocky substrate of intertidal communities, coral reefs are actually produced by some of the organisms that live on them. The entire reef, which may extend for hundreds of kilometers, is primarily composed of a veneer of tiny sea anemone–like creatures called *coral polyps*. These small colonial animals slowly produce the massive carbonate infrastructure of the reef itself, which a vast array of other organisms live on, around, and within.

Therein lies a wonderful biological paradox. Think of any common terrestrial ecosystem—a temperate forest, a tropical jungle, a Midwestern plain, or the field adjacent to your house. These areas are dominated by a great variety of plants (the producers) and contain just a handful of animal species, both herbivores and carnivores (the consumers). Conversely, a typical coral reef contains an impressive assemblage of consumers and just a few producers. The coral animals that create the reef feed by removing plankton from the water column, as do the many sponge species that decorate the reef and represent the second most important component of the benthic fauna on coral

reefs. Yet tropical seas are virtually devoid of plankton. That is why azure tropical waters are so transparent. A coral reef can be viewed as one giant animal that is inhabited by hundreds of other animals, such as sponges, snails and clams, squids and octopuses, sea anemones and jellyfish, shrimps and crabs, worms, and fish. This raises a number of questions: Where are the primary producers on a coral reef? Can an ecosystem violate the second law of thermodynamics by containing more consumers than producers? Why do planktivorous reef creatures, such as corals and sponges, not starve to death in the nearly plankton-free waters that surround them? In this section, we attempt to answer these fascinating biological riddles.

Coral Anatomy and Growth

Coral is a general term used to describe a variety of cnidarian species. Some grow as individual colonies; hence, not all corals produce reefs, and not all reefs are formed by corals; some reefs are formed by oysters, annelid worm tubes, red algae, or even cyanobacteria.

Reef-forming corals, the primary species that secrete the calcium carbonate ($CaCO_3$) matrix of coral reefs, are members of the class Anthozoa. All anthozoans are radially symmetrical, a morphology that is adaptive for sessile organisms, such as corals and sea anemones. Anthozoans are subdivided into two subclasses. The subclass Octocorallia, comprising soft corals, sea fans, sea whips, sea pansies, and sea pens, are characterized by the presence of polyps with eight pinnate, or feather-like, tentacles. Members of the subclass Hexacorallia have polyps possessing multiples of six smooth tentacles and include four orders of sea anemones (some exist as individuals, some in colonies, and others are tube dwellers) and three orders of corals (stony corals, false corals, and black corals). One group, the stony corals (order Scleractinia), is responsible for creating coral reefs. Stony corals and most of their cnidarian relatives are carnivores that use tentacles armed with cnidocytes that ring the mouth (**Figure 11.1**) to capture prey and push it into their gastrovascular cavity where it is digested.

Figure 11.1 Extended polyps of a coral colony. The numerous light-colored spots on the tentacles are batteries of cnidocytes.
Courtesy of George Schmahl, Flower Garden Banks National Marine Sanctuary, National Oceanic and Atmospheric Administration.

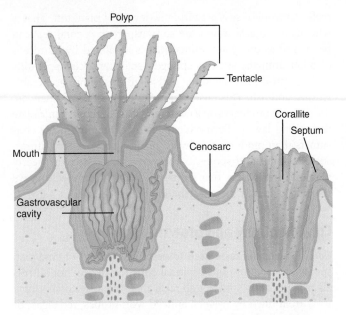

Figure 11.2 Cross section of a coral polyp and a calcareous corallite skeleton. The living coral tissue forms a thin interconnection, the cenosarc, over the surface of the reef.

to infection. Coral reefs are living entities, and despite their stony appearance they are easily harmed by disturbances to the cenosarc.

The growth rate of corals is affected by light intensity (which is affected by water motion, depth, and turbidity), day length, water temperature, plankton concentrations, predation, and competition with other corals. Stony corals exhibit a large variety of growth forms that are typically described as encrusting, massive, branching, or foliaceous (**Figure 11.3**). In addition, many species are rather **polymorphic**, expressing different growth forms in response to differences in wave exposure or depth; therefore, growth may seem like a simple parameter to measure, but for corals it is not. Techniques for monitoring the growth of corals include measuring an increase in weight, diameter, surface area, branch length, number, or a combination of these factors. Individual coral colonies may grow continually for centuries or even longer. Some species exceed several meters in size. In general, species with lighter, more porous skeletons grow more rapidly than species with denser skeletons, and branching species grow more quickly than massive species. In general, an entire reef will grow upward as much as 1 mm a year and spread horizontally 8 mm a year.

Most corals are colonial, built of numerous basic structural units, or *polyps* (**Figure 11.2**), each usually just a few millimeters in diameter. Coral polyps sit in calcareous cups, or **corallites**, an exoskeleton secreted by their basal epithelium. Several wall-like **septa** radiate from the sides of each corallite, and a stalagmite-like **columella** extends upward from its floor. Periodically, the coral polyp grows upward by withdrawing itself up and secreting a new basal plate, a partition that provides a new elevated bottom in the corallite. In addition, the coral colony also increases in diameter by adding new asexually cloned polyps to its periphery. These new polyps secrete their own $CaCO_3$ corallites that share a wall with neighboring polyps. All polyps that comprise the colony are interconnected over the lips of their corallites via a thin sheet of tissue called a **cenosarc**; therefore, touching a living coral colony in any way can easily crush the cenosarc against its own $CaCO_3$ skeleton, thus compromising the colony by leaving it open

The growth of an individual coral or an entire reef is not simply a function of the local rate of calcification for that species. The persistence of a coral colony or reef depends on a balance between the deposition and removal of $CaCO_3$ throughout the entire reef. In addition to calcium carbonate deposition by corals, several other types of organisms contribute their hard parts to the structure of coral reefs, including encrusting and segmented calcareous red and green algae; calcareous colonial hydrozoans; skeletons of crustaceans, bryozoans, and single-celled foraminiferans; mollusk shells; tests and spines of echinoderms; sponge spicules; and serpulid polychaete tubes. The loss or erosion of calcium carbonate is caused by grazers or scrapers, such as sea urchins and fish (**Figure 11.4**), and etchers, such as bacteria, fungi, and algae, that penetrate coral substrates. Infaunal organisms, such as sponges, bivalves, sipunculans, and polychaetes, also drill or bore into coral skeletons. From this encrusted, integrated base of living and dead

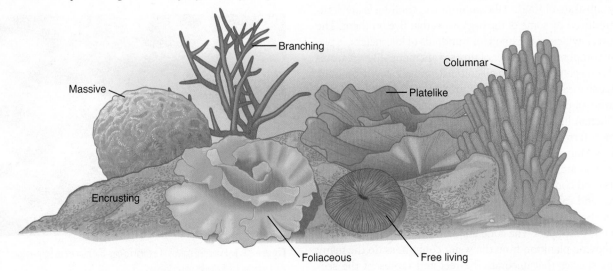

Figure 11.3 Corals exhibit a large variety of growth forms.

Figure 11.4 Parrotfish, major grazers of coral skeletal material and macroalgae, use their powerful jaws to produce large amounts of carbonate sand on the reef. These stoplight parrotfish are feeding on a star coral, *Orbicella faveolata*.
Courtesy of George Schmahl, Flower Garden Banks National Marine Sanctuary, National Oceanic and Atmospheric Administration.

skeletal remains, coral reef ecosystems have evolved as the most complex of all benthic associations.

DID YOU KNOW?

Although coral polyps are relatively tiny marine organisms, they combine to create the largest living structures on the entire Earth! Some reefs are so large that they are visible from space in satellite images. The Great Barrier Reef in Australia is the largest coral reef, stretching over a distance of 2,300 km (1,429 miles). It is made up of nearly 3,000 individual small reefs. The size and age of corals makes them truly remarkable organisms.

Coral Distribution

Like sea anemones, corals are ubiquitous. Non-reef-forming corals can be found in the deep sea (e.g., black corals) and in temperate zones (such as *Astrangia* on shipwrecks off New England), as well as in the tropics; however, there are several restrictions to the distribution of reef-forming corals, which are more abundant and diverse in the Indo-Pacific (about 700 species) than in the Atlantic Ocean (about 145 species; **Figure 11.5**). First, coral reefs generally are restricted to tropical and subtropical regions (usually below 30° latitude) where the annual sea-surface temperature averages at least 20°C. Second, coral reefs generally are better developed on the eastern margins of continents where shallow submarine platforms provide suitable habitat. Third, coral reefs generally thrive only in normal-salinity seawater; hence, reefs are rare on the eastern coast of South America because of the enormous outflow of freshwater from the Amazon River system. Fourth, reef-forming corals are usually found within 50 m of the surface in clear water on exposed surfaces.

These first two biogeographic restrictions suggest that reef-forming corals generally thrive only in warmer water, probably because only in warm waters can the high rates of $CaCO_3$ deposition needed for reef building be achieved. Hence, they are found in low latitudes and on eastern shorelines where coastal upwelling of cold water is less common and where the major ocean gyres direct warm tropical currents. These latitudinal limits of coral reef development also are often influenced by competition with macroalgae, with macroalgae being favored in higher latitudes because of increased nutrient concentrations, decreased water temperatures, and perhaps decreased grazing pressure.

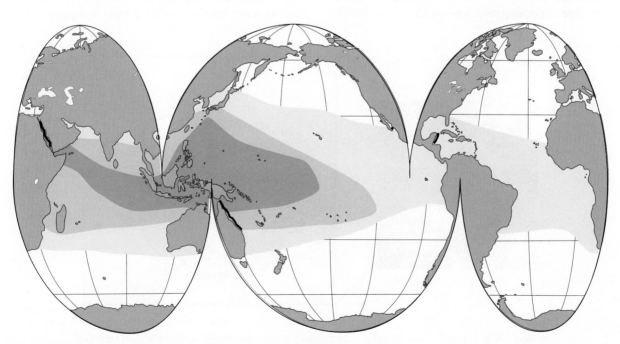

Figure 11.5 Distribution of reef-forming corals, by approximate number of genera: light blue, < 20 genera; medium blue, 20–40 genera; dark blue, > 40 genera. Heavy black lines indicate continental barrier reefs.

The third generalized limit to the global distribution of coral reefs suggests that coral animals cannot thrive in low-salinity seawater or in the sedimentation and high concentration of nutrients associated with rivers and freshwater runoff. The final biogeographic limitation—that coral reefs typically grow within 50 m of the surface in clear water on exposed surfaces—suggests that they need sunlight for their survival and growth. This limitation seems puzzling. Why would an animal (i.e., a coral colony) require sunlight, and why would their growth rates be affected by light intensity and day length, as described above? The answer to this question is also the answer to the apparent paradox posed at the beginning of this chapter.

Coral Ecology

Living intracellularly within the endodermal tissues of all reef-building, or **hermatypic**, corals are masses of symbiotic **zooxanthellae**, unicellular algae that, like all other photosynthetic organisms, require light. Solitary non-reef-building corals, such as *Astrangia* off the coast of New York, do not possess zooxanthellae and are termed **ahermatypic**. Zooxanthellae is a general term for a variety of photosynthetic dinoflagellates (genus *Symbiodinium*) that are mutualistic with several types of invertebrate species. To date, the many species of *Symbiodinium* cluster into eight genetic clades. Unlike the more typical dinoflagellates, zooxanthellae lose their flagella and cellulose cell walls. They occur in concentrations of up to 1 million cells/cm² of coral surface and often provide most of the color seen in corals. In fact, corals that grow in bright sunlight are often creamy white, whereas those in deep shade are nearly black. This difference is due to variations in the cellular concentrations of photosynthetic pigments of the zooxanthellae rather than differences in the densities of their cells.

Zooxanthellae and corals derive several benefits from each other. Thus, this relationship usually is considered a mutualistic one. Corals provide the zooxanthellae with a stable, protected environment and an abundance of nutrients (CO_2 and nitrogenous and phosphate wastes from cellular respiration of the coral). In return, the host corals receive photosynthetic products (O_2 and energy-rich organic substances) from the symbiotic algae by stimulating or promoting their release with specific signal molecules that appear to alter the membrane permeability of the naked dinoflagellates. These zooxanthellae photosynthetically produce 10 to 100 times more carbon than is necessary for their own cellular needs, and almost all of this excess is transferred to the coral. Nearly all of the carbon that is transferred to the coral is respired and not used to build new coral tissue because it is low in nitrogen and phosphorus. This contribution by the zooxanthellae is sufficient to satisfy the daily energy needs of several species of corals. Soft corals are actually obligate symbionts, having lost the ability to capture and ingest plankton. The total contribution of symbiotic zooxanthellae to the energy budget of the reef is several times higher than phytoplankton production occurring in the waters above many reefs.

Hence, the answer to the mystery of how corals are able to construct enormous reefs in nutrient-poor waters is that they receive a significant supply of food from their algal associates, zooxanthellae. The coral animals also avoid the necessity of excreting some of their cellular wastes (which the algae absorb and use) and experience greater calcification rates than hermatypic corals that have been experimentally separated from their algal symbionts (**Figure 11.6 a** and **b**). Additional primary production on coral reefs is provided by several types of rather cryptic algae, cyanobacteria, and seagrasses. These include encrusting calcareous red algae, filamentous green algae that invade dead corals, a brown algal turf, photosynthetic symbionts in other reef invertebrates, macroalgae anchored in the sand, seagrass, and phytoplankton cells in the water column over the reef.

Despite the nutritional contribution of zooxanthellae, polyps of stony corals remain superbly equipped to prey on a

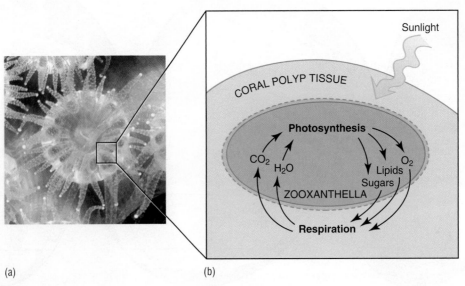

(a)   (b)

Figure 11.6 (a) Magnified view of coral polyps extending their tentacles to feed with algal symbionts visible. (b) Exchange of materials between zooxanthellae and their coral host.

(a) Courtesy of National Oceanic and Atmospheric Administration.

variety of external sources of food, and only soft corals depend solely on zooxanthellae. Corals with large polyps and tentacles, such as *Favia* or *Mussa*, feed exclusively on small fish and larger zooplankton, such as copepods, amphipods, and worms. Species with smaller polyps, such as *Porites* or *Siderastrea*, use ciliary currents to collect small plankton and detritus particles. Most coral polyps are capable of using **mesenterial filaments** to harvest particulate organic carbon from surrounding sediments. Corals also use their **mucus ciliary system** (analogous to the ciliated epithelium in the trachea of humans) to trap and ingest organic particles as small as suspended bacteria, bits of drifting fish slime, and even organic substances dissolved in passing seawater. Finally, the still controversial concept of **endo-upwelling** has been suggested as a possible source of additional dissolved nutrients, wherein geothermal heat deep within island reefs drives the upwelling of nutrient-rich water through the reef structure from depths of several hundred meters.

Corals are not the only animals on the reef that possess photosynthetic symbionts. Zooxanthellae also occur in other anthozoans, some medusae (such as the upside-down jellyfish, *Cassiopea*), sponges, and giant clams. In addition, it is well documented that sponges possess photosynthetic cyanobacteria. These photosynthetic symbionts are found in about 40% of sponge species from the Atlantic and Pacific Oceans, although their contribution to sponge ecology in the two oceans differs dramatically. On the Great Barrier Reef in the Pacific Ocean, 90% of the sponges on the outer reefs are **phototrophic** (they are flattened and obtain up to half of their energy from cyanobacteria), with 6 of 10 species studied producing three times as much oxygen as they consume. Very few of the sponges studied in the Caribbean Sea are phototrophic. This results in Caribbean sponges consuming 10 times more prey than their Pacific relatives. Perhaps this different reliance on energy from cyanobacteria is because primary productivity in the western Atlantic is higher than in the western Pacific.

Finally, nitrogen fixation, an activity that is light dependent, has recently been found to be associated with cyanobacteria living in the skeletons of various hermatypic corals. These nitrogen-fixing bacteria benefit from organic carbon excreted by the coral tissue. Corals also house other bacteria, archaea, and fungi, but little is known about the roles these microorganisms play. Perhaps these symbiotic associations are as important to corals as their mutualism with zooxanthellae, but more research is necessary.

The living richness of coral reefs stands in obvious contrast to the generally unproductive tropical oceans in which they live. The precise trophic relationships between producers and consumers on the reef are still largely unknown. Coral colonies seem to function as highly efficient trophic systems with their own photosynthetic, herbivorous, and carnivorous components. Crucial nutrients are rapidly recycled between the producer and consumer components of the coral colony. Because much of the nutrient cycling is accomplished within the coral tissues, little opportunity exists for the nutrients to escape from the coral production system. Coral colonies, therefore, are able to recycle their limited supply of nutrients rapidly between internal producer and consumer components and

keep productivity in coral reef communities relatively high (up to 5,000 gC/m² per year) compared with other regions of the ocean. Coral reefs are one example of an ecosystem adapted with high overall primary productivity despite low phytoplankton abundance.

Coral Reef Formation

Coral reefs occur in two general types: shelf reefs, which grow on continental margins, and oceanic reefs, which surround islands. Oceanic reefs may be divided into three general subtypes: **fringing reefs**, **barrier reefs**, and **atolls**. Most shelf reefs are fringing reefs, which form borders along shorelines. Some of the Hawaiian reefs and other relatively young oceanic reefs are also of this type. The longest fringing reef known extends throughout the Red Sea, extending about 400 km. Barrier reefs are further offshore and are separated from the shoreline by a lagoon. The Great Barrier Reef of Australia is by far the largest single biological feature on Earth, bordering about 2,000 km of Australia's northeast coast. Smaller barrier reefs occur in the Caribbean Sea. Atolls are generally ring-shaped reefs from which a few low islands project above the sea surface (**Figure 11.7**). The largest atoll known is Kwajalein Atoll in the Marshall Islands, which has a lagoon 100 km long and 55 m deep.

The famous evolutionary biologist Charles Darwin studied the morphology of coral reefs on several islands while serving as a naturalist aboard the H.M.S. *Beagle* during its voyage to circumnavigate the Earth from 1831 to 1836. His observations led him to propose that essentially all oceanic coral reefs were supported by volcanic mountains beneath their surfaces. Fringing reefs, barrier reefs, and atolls, he suggested, were sequential developmental stages in the life cycle of a single reef. Within the tropics, he argued, newly formed volcanic islands and submerged volcanoes that almost reach the sea surface are eventually populated by planktonic coral larvae from other nearby coral islands. The coral larvae settle and grow near the surface close to the shore, forming a fringing reef (**Figure 11.8**, left). The most rapid growth occurs on the outer sides of the reef where food and oxygen-rich waters are more abundant. Waves break loose pieces of the reef and move them down the slopes of the

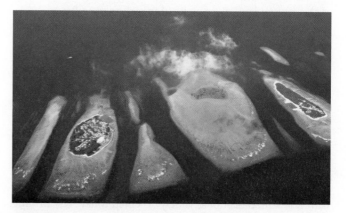

Figure 11.7 A satellite view of several of the hundreds of atolls that make up the Maldives.
© wilar/Shutterstock, Inc.

Types of coral reefs

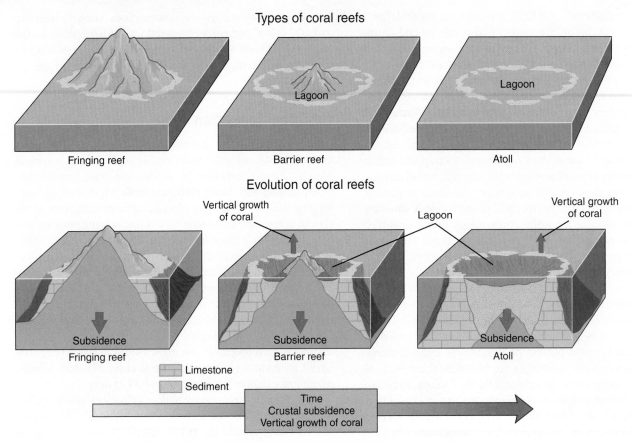

Fringing reef   Barrier reef   Atoll

Evolution of coral reefs

Vertical growth of coral

Lagoon

Vertical growth of coral

Subsidence   Subsidence   Subsidence
Fringing reef   Barrier reef   Atoll

Limestone
Sediment

Time
Crustal subsidence
Vertical growth of coral

Figure 11.8 The developmental sequence of coral reefs, from young fringing reefs (left), to barrier reefs (center), and finally to atolls (right).

volcano. More corals establish themselves on this debris and grow toward the surface. He reasoned that the weight of the expanding reef and the increasing density of the cooling volcano caused the island to sink slowly. If the upward growth of the reef keeps pace with the sinking island, the coral maintains its position in the sunlit surface waters. If the upward growth of the reef does not keep pace with the sinking island, the reef is pulled into the cold darkness below the photic zone and expires. Such a dead sunken reef, when associated with a flat-topped seamount, is called a **guyot** (pronounced "gee-oh").

As the island sinks away from the growing reef, the top of the reef widens. Eventually, this reef crest or flat becomes so wide that many of the corals on the quiet inner edge of the reef die because the water that reaches them is devoid of nutrients and oxygen and contains high concentrations of reef waste products. The dead corals are soon covered with reef debris and form a shallow lagoon. Delicate coral forms survive in the lagoon, protected from the waves by what is now a barrier reef (Figure 11.8, center). With further sinking, the volcanic core of the island may disappear completely beneath the surface of the lagoon and leave behind a ring of low-lying islands supported on a platform of coral debris, an atoll (Figure 11.8, right).

Remarkably, Darwin's concept of coral reef formation is, with a few modifications, widely accepted today. Test drilling on several atolls has revealed, as Darwin predicted, thick caps of carbonate reef material overlying submerged volcanoes. Two test holes drilled on Enewetak Atoll (the site of U.S. hydrogen

bomb tests in the 1950s) penetrated 1,268 m and 1,405 m into shallow-water reef deposits, respectively, before reaching the basalt rock of the volcano on which the reef had formed. For the past 60 million years, Enewetak apparently has been slowly subsiding as its surrounding reef grew around it. Because this transition from a fringing morphology through a barrier morphology to an atoll requires a great deal of time, and because the Atlantic Ocean is much younger than the Pacific Ocean, atolls are virtually nonexistent in the Atlantic.

Some anecdotal information reinforces the scientific data that support Darwin's hypothesis of coral reef development. For example, British explorer Captain James Cook discovered Hawaii in January 1779, during Makahika, a festival to honor the god Lono. The Hawaiian natives initially thought that Captain Cook was Lono, who was said to come from the sea. After realizing their mistake, they killed him. A monument was soon built in the surf to commemorate his arrival and death. Today, that monument can be found offshore at a depth of 20 m, yet the reef surrounding the island is still growing just under the surface of the sea.

For the past several hundred thousand years, the formation and melting of vast continental glaciers have produced extensive worldwide fluctuations in sea level. Darwin was aware of these fluctuations yet had no means of predicting their effects on coral reef development. Fifteen thousand years ago, during the last glacial maximum, the average sea level was about 150 m below its present level. As the ice melted, sea level gradually

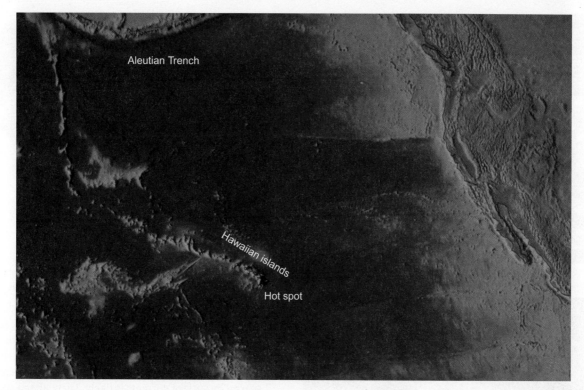

Figure 11.9 Chains of volcanoes along the Hawaiian Island–Emperor Seamount are carried, in a conveyer-belt fashion, north into deeper water by the movement of the Pacific Plate. Each volcano was formed over the "hot spot," a continuous source of new molten material presently under Hawaii, and is carried to its eventual destruction in the Aleutian Trench. Courtesy of NGDC/NESDIS/NOAA.

rose (about 1 cm/yr) until it reached its present level nearly 6,000 years ago. Many coral reefs did not grow upward quickly enough and perished. Those that did keep up with the rising sea are the living reefs we see today. Coral reefs in the Atlantic Ocean seem the most susceptible to glacier-induced changes in their morphology, and barrier reefs are most common in the Atlantic Ocean.

Coral reefs have also been subjected to the effects of global plate tectonics. The Hawaiian Islands and the reefs they support have been transported to the northwest by the movement of the Pacific Plate. Atolls at the northern end of the chain appear to have drowned as they reached the "Darwin Point," a threshold beyond which coral atoll growth cannot keep pace with recent changes in sea level (**Figure 11.9**). At the Darwin Point, only about 20% of the necessary $CaCO_3$ production is contributed by corals.

Reproduction in Corals

Corals reproduce in a variety of ways, both asexually and sexually. Most corals bud off new polyps along their margins asexually as they increase in diameter. Sometimes these new polyps sever the cenosarc and initiate a new colony that is a clone of their neighbor. Branching species, such as *Acropora*, are frequently broken by storms or ship anchors into clonal colonies by fragmentation, the production of new colonies from portions broken off from established colonies. Fragmentation decreases the risk of mortality of the genotype and avoids the

risk of high mortality of larvae and juveniles during sexual reproduction. In addition, fragmentation by species with high growth rates often results in that species dominating certain reef zones (such as the buttress zone discussed below), as well as rapid recolonization after a disturbance. Researchers also have observed "polyp bailout" in the laboratory, when polyps crawl out of their corallites and drift away. It is not known whether these polyps remain viable or whether this occurs naturally on coral reefs, or rather is an artifact of a laboratory setting.

Corals also reproduce sexually, either by brooding fertilized eggs internally or by spawning millions of gametes into the water column for external fertilization. In brooding species, the eggs remain in the gastrovascular cavities of the adults where they are fertilized by motile sperm cells. The developing zygotes and resultant larvae are brooded before they are released to settle nearby. Some evidence suggests that coral species with small polyps have low numbers of eggs combined with internal fertilization and brooding, whereas large-cupped species spawn huge quantities of eggs that are fertilized externally. In addition, the strategy of sexual reproduction used (brooding larvae vs. spawning gametes) is highly correlated with taxonomic affiliation at the family level. Members of the families Agariciidae, Dendrophylliidae, and Pocilloporidae commonly brood, whereas broadcast spawning is predominant in the Acroporidae, Caryophyllidae, Faviidae, and Rhizangidae. Family Poritidae includes both brooders and broadcasters.

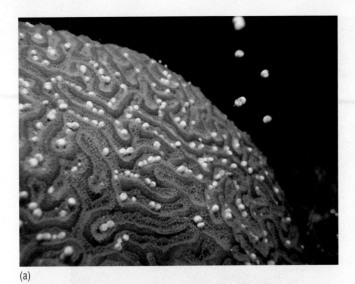

(a)

Figure 11.11 Micrograph of a planula larva of the coral *Pocillopora*.
© Valerie Hodgson/Visuals Unlimited.

(b)

Figure 11.10 Spawning corals. (a) Female brain coral, *Diploria*, releasing egg packets; (b) male star coral, *Montastraea*, releasing sperm.
(a) Courtesy of Emma Hickerson, Flower Garden Bank National Marine Sanctuary, National Oceanic and Atmospheric Administration; (b) Courtesy of Flower Garden Bank National Marine Sanctuary, National Oceanic and Atmospheric Administration.

Of nearly 200 species of corals studied on the Great Barrier Reef, 131 were hermaphroditic spawners, 37 were dioecious spawners, 11 were hermaphroditic brooders, and 7 were dioecious brooders. Hence, spawning by hermaphrodites seems to be the most common method of sexual reproduction among corals. Spawning is usually accomplished during a highly synchronous event known as *mass spawning*. On the Great Barrier Reef of Australia, mass spawning by corals is a spectacular sight. More than 100 of the 340 species of corals found there synchronously spawn on only one night each year, just a few days after the late spring full moon (**Figure 11.10**). A similar episode of mass spawning has been documented in the Gulf of Mexico in the evening 8 days after the full August moon. Mass spawning by corals seems to be induced by specific dark periods, and it can be delayed by experimentally extended light periods. Mass spawning also seems to be broadly influenced by temperature. Such highly seasonal spawning is surprising in the tropics, an area wherein reproduction throughout the year is said to be the norm because of relatively constant climatic conditions.

A few days after spawning, the fertilized eggs develop into ciliated **planula** larvae (**Figure 11.11**). These larvae, each already containing a supply of zooxanthellae, initially are positively phototactic; that is, they swim toward brighter light. This ensures that they remain near the sea surface where maximal dispersal by surface currents is likely. Then, after a specific time interval, they become negatively phototactic and attempt to settle on the seafloor. They thrive only if they encounter their preferred water and bottom conditions. From these planula larvae, new coral colonies develop and mature in about 7 to 10 years. Research has indicated that the larvae of *Pocillopora damicornis*, the commonly named cauliflower coral of the Indian and Pacific Oceans, are capable of reversible metamorphosis. In this species, the planula larva settles and begins to metamorphose into a juvenile. It forms a $CaCO_3$ exoskeleton, a mouth, and tentacles; however, if it is stressed within the first 3 days of settling, it will sever its attachments to its carbonate exoskeleton, revert back into a planula larva, and reenter the water column to search for an alternate settlement site. During their planktonic phase, coral larvae are capable of settling at new volcanic islands some distance from their island of origin. When they do, the form of the reef they eventually create depends on existing environmental conditions and the prior developmental history of reefs in the area.

It is unclear why some corals spawn synchronously and why this event occurs just several nights after the full moon. One advantage to mass spawning is that the chance of fertilization will increase greatly for one species. It is unclear, however, why mass spawnings are multispecies events, in that simultaneous spawning by many species may increase the risk of gamete loss via hybridization. Perhaps such an epidemic spawning event overwhelms (and satiates) active predators and filter feeders in the area, increasing the likelihood of gamete survival; however, these species also risk big losses by spawning on just a few nights each year. A sudden rain storm resulting in a drop in salinity of surface waters during a mass spawning event around Magnetic Island in November 1981 destroyed the entire reproductive effort of those corals for that year. Another hypothesis is that when environmental conditions necessary for development of gametes exist, all coral species in an area spawn to take advantage of these conditions. Some conditions that appear to be of importance for coral spawning are water temperature, day length, tidal height, and salinity. Mass spawning is not a universal behavior of reef corals; in the northern Red Sea, none of the major species of corals reproduces at the same time as any of the other major species.

The early life phases of all marine organisms are sensitive time periods, with survival rates very low for most species, especially those that use broadcast spawning methods. The first challenge is fertilization, involving clouds of sperm and eggs being released into the water with hopes that some sperm will find eggs of the same species. After fertilization takes place the zygote must remain in the water column in favorable water conditions for rapid growth. The larvae are subject to ocean water movements with little control over where they reside in the water column. Fish and other predators feed on newly fertilized eggs and larvae, further decreasing the chances of survival. If a larva lives long enough to reach settlement age, it must find a suitable area to settle to on the seafloor. If suitable habitat is available at the right time for settlement, timing that involves multiple factors, settlement will take place and the juvenile coral will grow rapidly. Recent work on *Acropora palmata*, a federally listed threatened species, indicates that all of the early life stages of this coral species are negatively impacted by decreased pH. Decreased pH from ocean acidification is occurring today and is predicted to become more extreme over time as more CO_2 enters the oceans. If early stages of corals are negatively impacted by ocean acidification, there may come a time when new adults will be rare.

Interestingly, calcareous green algae found in the Caribbean coral reef ecosystem also exhibit mass spawning. Nine Caribbean species in 5 genera participate in a predawn episode, with a total of 17 species of green algae exhibiting highly synchronous reproductive patterns. Unlike the coral phenomenon described previously, closely related algal species broadcast their gametes at different times, and the environmental or biological triggers of these events remain unknown. In all cases, gametes from both sexes remain motile for 40 to 60 minutes after release but sink quickly after combining to form a zygote.

Zonation on Coral Reefs

Environmental conditions that favor some coral reef inhabitants over others in a particular habitat depend a great deal on wave force, water depth, temperature, salinity, and a host of biological factors. These conditions vary greatly across a reef and provide for both horizontal and vertical zonation of the coral and algal species that form the reef. **Figure 11.12**, a cross section of an idealized Indo-Pacific atoll, includes the major features and zones of the reef.

The living base of a coral reef begins as deep as 150 m below sea level. Between 150 and 50 m on outer reef slopes, a few small, fragile species, such as *Leptoseris*, exist despite the low level of sunlight that penetrates to these depths. Above 50 m and extending up to the base of vigorous wave action (at a depth of approximately 20 m) is a transition zone between

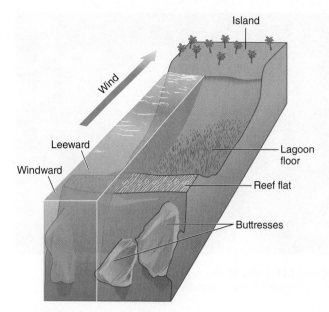

Figure 11.12 Cross-sectional zonation of an atoll.

deep- and shallow-water associations. In this zone, the corals and algae receive adequate sunlight, yet are sufficiently deep to avoid the adverse effects of surface waves. Several of the delicately branched species commonly found in the protected lagoon waters also occur in this relatively calm transition zone.

From a depth of about 20 m to just below the low-tide line is a rugged zone of spurs, or buttresses, radiating out from the reef. Interspersed between the buttresses are grooves that slope down the reef face. This windward profile of alternating buttresses and grooves is useful in dissipating some of the energy of waves that crash into the face of the reef, but damage to the reef and its inhabitants is inevitable. The grooves drain debris and sediment produced by wave impacts off the reef and into deeper water. Continual heavy surf makes it difficult to conduct detailed studies of the buttress zone, but it is known to be dominated by several species of encrusting coralline algae and by rapidly growing branching coral species (such as *Acropora*) that repair damage quickly and thrive when fragmented. Small fish seem to be in every hole and crevice on the reef, and many of the larger fishes of the reef—sharks, jacks, and barracudas—patrol the buttresses in search of food.

Most coral reefs are swept by the broad reaches of the trade winds. The waves generated by these winds crash as thundering breakers on the windward sides of reefs. Windward reefs are usually characterized by a low, jagged algal ridge that suffers the full fury of incoming waves. In this high-energy habitat, a few species of calcareous red algae, especially *Porolithon*, *Hydrolithon*, *Goniolithon*, and *Lithothamnion*, flourish and produce the ridge, creating new reef material as rapidly as the waves erode it. A few snails, limpets, and urchins (**Figure 11.13**) can also be found wedged into surface irregularities. Slicing across the algal ridge are surge channels that flush bits and fragments of reef material off the reef and down the seaward slope.

Figure 11.13 A sea urchin.
Courtesy of Dr. Dwayne Meadows, National Oceanic and Atmospheric Administration.

Figure 11.14 A giant clam, *Tridacna*, amid mixed corals. Note the blue mantle tissue that is brightly colored due to the presence of innumerable mutualistic zooxanthellae.
© Andy Lim/Shutterstock, Inc.

Extending behind the algal ridge to the island (or, if the island is absent, to the lagoon) is a reef flat, a nearly level surface barely covered by water at low tide in the Atlantic (Indo-Pacific reef flats are intertidal). The reef flat may be narrow or very wide, may consist of several subzones, and may have an immense variety of coral species and growth forms. In places where the water deepens to a meter or so, small raised microatolls occur. Microatolls are produced by a half dozen different genera of corals and, with other coral growth forms, provide the framework for the richest and most varied habitat on the reef. Burrowing sea urchins are common, and calcareous green algae and several species of large foraminiferans thrive and add their skeletons to the sand-sized deposits on the reef flat. The sand, in turn, provides shelter for other urchins, sea cucumbers, and burrowing worms and mollusks.

One of the most spectacular animals of the reef flat is the giant clam, *Tridacna*. The largest species of this genus occasionally exceeds a meter in length and weighs more than 100 kg. Some tridacnids sit exposed atop the reef platform; others rock slowly to work themselves into the growing coral structure beneath (**Figure 11.14**). Like corals and many other invertebrates, tridacnid clams house dense concentrations of zooxanthellae in specialized tissues, particularly the enlarged mantle that lines the edges of its shell. When the shell is open, the pigmented mantle tissues with their zooxanthellae are fully exposed to the energy of the tropical sun.

Tridacnid clams were long thought to "farm" their zooxanthellae in blood sinuses within the mantle and then transport them to the digestive glands, where they were then digested by single-celled amoebocytes. However, using elaborate staining and electron microscope techniques, scientists have demonstrated that the digestive amoebocytes of *Tridacna* selectively destroy only the old or degenerate zooxanthellae. Healthy zooxanthellae are maintained to provide photosynthetic products to their hosts in dissolved rather than cellular form. This selective capability of amoebocyte cells is a wonderful example of an advanced adaptation for survival in a sedentary marine organism.

The tranquil waters of the lagoon protect two general life zones: the lagoon reef and the lagoon floor. The lagoon reef is a leeward reef. It forms the shallow margin of the lagoon proper and is usually free of severe wave action. It lacks the challenging algal ridge characteristic of the windward reef and in its place has a more luxuriant stand of corals (**Figure 11.15**). Other algae, some specialized to burrow into coral, and uncountable species of crustaceans, echinoderms, mollusks, anemones, gorgonians, and representatives of many other animal phyla flourish in the lagoon reef. In this gentle, protected environment, single coral colonies of *Porites* and *Acropora* may achieve gigantic proportions. Branching bush- and treelike forms extend several meters from their bases. The plating, branching, and overtopping structures common in the protected lagoon are most likely

(a)

Figure 11.15 Variation in coral growth forms: (a) table coral, *Acropora*: (*Continued*)
(a) © Andy Lim/Shutterstock, Inc.

Figure 11.15 (*continued*) (b) brain coral, *Diploria*; and (c) staghorn coral, *Acropora*.
(b) © Lawrence Cruciana/Shutterstock, Inc.; (c) © Andy Lim/Shutterstock, Inc.

structural adaptations evolved in response to competition for particles of food and sunlight, two resources vital to the survival of reef-forming corals.

11.2   Coral Diversity and Catastrophic Mortality

The great diversity of species on coral reefs is legendary and rivals that of tropical rain forests. The classic explanation for this high diversity was that the uniform and predictable conditions on tropical coral reefs promoted high diversity by enabling species to become increasingly specialized relatively quickly. Recently this view has been challenged by an opposing argument that suggests that the high diversity of coral reefs is a nonequilibrium state in which diversity can persist only if it is disturbed. Like some rocky intertidal communities, coral reefs are subject to severe disturbances (e.g., hurricanes) often enough that equilibrium, or a climax stage, may never be reached, and high diversity is maintained by frequent catastrophic mortality. According to this view, some of the catastrophic mortality of corals and coral reef species that has been observed in the past 25 years can be viewed as natural perturbations of these communities rather than abnormal events. One common natural cause of catastrophic coral mortality is storm waves from hurricanes and typhoons. At Heron Island on the Great Barrier Reef, for example, the highest number of species of corals occurs on the crests and outer slopes that are constantly exposed to damaging waves. In fact, it has been reported that the most significant factor determining the spatial and temporal organization of Hawaiian coral reef communities is physical disturbance from waves. Nevertheless, there is still some cause for concern about coral loss, especially when new corals do not appear to be replacing those lost. The recent rate of loss of coral reefs worldwide may be higher than ever before, and it is likely that at least some of the loss is due to anthropogenic sources of pollution, disturbance, and changes to ocean chemistry due to ocean acidification. When coupled with natural causes of reef mortality, these human-induced mortality events may exceed a reef's ability to recover.

Coral reefs fringe about one sixth of the world's coastlines and are estimated to house about 25% of known marine species. Sadly, over half of those reefs are now threatened by human activities, with the Caribbean region being hardest hit. Moreover, the forecast for future mortality is even more dire. In 1994, the Global Coral Reef Monitoring Network (part of the International Coral Reef Initiative), a consortium of hundreds of coral reef scientists and managers from nearly 100 countries, published a *Call to Action* document concerning global coral reef health. Several updates have been made to this document that is now called the *Continuing Call to Action*, and in the latest edition published in 2013 it was confirmed that the world has effectively lost 19% of coral cover since 1950 and that over 60% of the world's reefs are under immediate and direct threat. The *Continuing Call to Action* document is an attempt to reach out to governments, individuals, or anyone with any influence on the health of our world ocean to keep the issue of coral reef health at the forefront of conservation efforts. The negative impacts humans are having on coral reefs worldwide are ongoing, will not be remedied quickly, and some may be irreversible.

In response to the declining trends in coral reefs in waters of the United States or U.S. territories, the National Marine Fisheries Service has approved requests to add numerous species to the threatened species list. As of 2016, 22 species have been listed, 7 from the Caribbean and 15 from the Indo-Pacific. Of the 22 total, 10 are in the genus *Acropora*, discussed throughout this chapter. Petitions to list other species have been submitted and rejected, and more are certainly being crafted as this

RESEARCH in Progress

The Quest to Preserve Coral Reefs in Times of a Changing Climate

It has been estimated that more than 50% of the world's coral reefs have perished in the last century. With climate change, ocean acidification, and other anthropogenic threats imminent, some scientists predict that coral reefs will be gone within 50 years. This prediction is grim, as coral reefs support countless other life forms in their vast ecosystem, including commercially important or rare fish and invertebrate species. From an economic standpoint, coral reefs support many economies worldwide as tourist attractions for diving, glass-bottom boats, fishing, and other ecotourism pursuits. A world ocean without corals would be an entirely different place than we experience today.

The threat of the great demise of coral reefs worldwide has inspired research on corals to investigate their basic life history characteristics and tolerance to stressors such as increased temperature and acidity. Surprisingly little research has been conducted on corals and their algal symbionts with respect to environmental stressors, which is the basic information necessary to predict the effects of environmental changes on corals. Most corals worldwide are limited by temperature due to their algal symbionts. Although temperature tolerances vary, for many corals temperatures above 32°C (90°F) can lead to coral bleaching events. Water temperature predictions are made through satellite monitoring and mapped to indicate potential coral bleaching events around the world based on our limited knowledge of coral heat tolerances (**Figure A**).

The prediction for climate change is an increase in water temperature worldwide by 2°F to 4°F (1°C) over the next century, potentially leading to many bleaching events.

Corals already living at the high end of their heat tolerance were once predicted to perish as climates warm several degrees, but research conducted by scientists at Stanford University suggests that some corals are currently capable of tolerating remarkably high temperatures, and some may actually be able to adapt to warmer temperatures if allowed time to acclimate. Some populations of *Acropora hyacinthus* (**Figure B**) in American Samoa live in waters with little mixing, leading to extremely high temperatures (35°C [95°F]). These temperatures are hotter than scientists believed corals should be able to survive in. Other nearby populations of the same species live in well-mixed waters with temperatures in what is considered the normal range for corals. Upon observing these corals surviving and thriving in very warm water, the following research questions were explored: (1) how is *Acropora hyacinthus* capable of surviving in 35°C (95°F) water? and (2) can *Acropora hyacinthus* currently living in cooler water adapt to warmer conditions, and if so, do they require an acclimation period?

To answer the question of how *Acropora hyacinthus* is adapting to warmer waters, researchers looked at the genes of corals. Genetic test results indicated that the corals living in high temperatures appear to be able to express genes that produce particular proteins that aid in resisting physiological damage from increased temperatures. Corals found residing in warmer water possess these genes and use them for protection and to thrive in water temperatures that should kill their algal symbionts and cause physiological stress to the corals. It was also discovered that the *Acropora hyacinthus* living in moderate temperatures possess the same genes, but do not express them while living in moderate temperatures. Other corals have been documented actually switching out

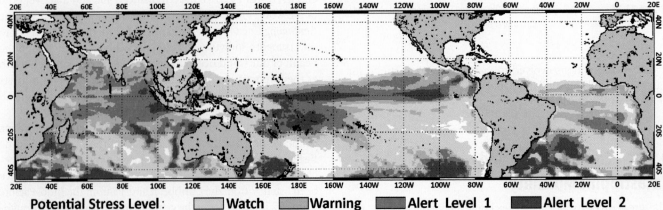

2015 Feb 10 NOAA Coral Reef Watch 60% Probability Coral Bleaching Thermal Stress for Feb-May 2015

Potential Stress Level :   Watch   Warning   Alert Level 1   Alert Level 2

Figure A Water temperature predictions for the entire ocean provide useful information to potentially predict coral bleaching events.
Courtesy of National Oceanic and Atmospheric Administration Reef Watch.

Figure B A healthy *Acropra hyacinthus* in the Maldives.
Courtesy of MDC Seamarc Maldives.

their algal symbionts for new ones that are more heat resistant when water temperatures rise.

Researchers set out to answer the question of the ability of *Acropora hyacinthus* populations living in moderate temperatures to adapt to much warmer waters by transplanting corals in their natural habitat from a cooler area to a warmer area. Corals transplanted from cooler waters to warmer ones bleached quickly. The same experiment was conducted in the winter, allowing transplanted corals to acclimate before temperatures rose significantly, and these transplanted corals survived. To verify these results, laboratory experiments were conducted in a more controlled setting. Corals that were allowed acclimation time, sometimes as little as 2 weeks, survived. These results confirmed the field experiment results and suggest that corals may be able to adapt to warmer waters relatively quickly, likely by turning on the genes that provide instructions for proteins that help protect the corals and their symbionts.

These study results are promising and indicate that some coral species may be more resilient than scientists initially thought. Although this is just one study, it sheds light on the adaptive mechanisms that corals possess for surviving in a changing environment. Corals have been in existence for millions of years and have survived wide swings in environmental conditions, so their ability to adapt is not surprising in some ways. The problem that still exists is that many anthropogenic environmental changes are taking place much more quickly than natural environmental fluctuations do. Whether corals can adapt and change quickly enough to keep up with increased temperature, acidity, and nutrients is still unknown, but the information gained from studies like this one are paramount to understanding the possibilities and to potentially intervening to save coral reefs worldwide.

Critical Thinking Questions

1. Is it surprising to you that some corals can resist physiological damage when water temperatures are extremely high? Explain your answer.

2. How can having an understanding of the potential adaptive capabilities of corals to changing climates help humans to preserve corals?

For Further Reading

Palumbi, S. R., D. J. Barshis, N. Traylor-Knowles, and R. A. Bay. 2014. Mechanisms of reef coral resistance to future climate change. *Science* 344:895–898.

Seneca, F. O., and S. R. Palumbi. 2014. The role of transcriptome resilience in resistance of corals to bleaching. *Molecular Ecology* 24:1467–1484.

book is written. Protecting corals from collection, pollution, and other harm due to humans will no doubt be instrumental in protecting and reviving this crucial ecosystem.

The main human activities that are implicated as causes for this unprecedented reef mortality include agricultural activities, deforestation, and coastal development, all of which introduce sediments, excessive nutrients, and assorted pollutants into coastal areas. A coating of sediment on a coral colony can smother it, clogging its feeding structures and increasing the colony's energy expenditure by causing its mucociliary system to work overtime to rid its surface of sediment particles. Sediments also decrease the photosynthetic output of zooxanthellae by shading them and reducing their light absorption.

Elevated nutrient levels occur when runoff from agricultural areas injects excess quantities of fertilizers in the waters that bathe coral reefs. Sewage runoff also supplies unnaturally high concentrations of nutrients to coastal areas. These increased concentrations of nitrogen and phosphorus enhance algal growth and enable macroalgae to dominate corals in their competition for space on the reef. They also result in phytoplankton blooms that cloud the water and further handicap zooxanthellae. These two common anthropogenic causes of coral mortality, increased sedimentation and nutrification, are thought to be responsible for the devastation of corals in the Florida Keys, in parts of Hawaii, and elsewhere that has occurred in recent decades.

Other reef herbivores seem to be equally important in helping corals maintain their dominance over rapidly growing seaweeds. A waterborne pathogen killed large numbers of a ubiquitous, long-spined, black sea urchin (*Diadema antillarum*) in the Caribbean Sea in 1983. It is estimated that 93% of the urchins in an area of 5 million square kilometers perished during what Knowlton called "the most extensive and severe mass mortality ever reported for a marine organism." The urchin mass mortality event is now rivaled by the sea star mass mortality seen in 2014 on rocky shorelines of the Pacific Coast. The rapid extermination of this urchin, an important grazer of algae, enabled algal populations to overgrow corals in their competition for space on reefs. Some Caribbean reefs still have not recovered and remain green, fuzzy remnants of their former beauty. Although the cause of this epidemic remains undetermined, some speculate that the Panama Canal enabled a virulent Pacific pathogen to make its way into the Caribbean Sea to cause these urchin deaths. If this is the case, humans are once again to blame for the resultant coral mortality.

Moreover, many reefs are badly overfished, and this removal of the majority of herbivorous teleosts from a reef by overfishing also enables macroalgae to overgrow corals quickly. Yet this is not the only impact that overfishing has on reefs. The methods used for removing desirable teleosts, either for food or for aquarium display, also result in reef destruction. Because the structural complexity of coral reefs provides countless homes and hiding places for reef fishes, traditional low-impact fishing methods (such as hook and line and netting) are inefficient ways to capture reef teleosts. Thus, fishers and shell collectors on reefs often turn to much more destructive methods, including dynamite, crowbars, and poisons, to obtain their catch. Dynamite and crowbars destroy the physical structure of the reef, and poisons (cyanide is used most commonly) can stun fish that then become available for the live fish trade (popular among Asian restaurants and aquarium hobbyists). It is estimated that the use of cyanide results in the unintentional deaths of about 50% of the fish exposed on the reef and the subsequent deaths of about 40% of captured fish during transport. This $1.2 billion industry centered in Southeast Asia, although lucrative, is very damaging and incredibly wasteful. It is estimated that only about 4% of Philippine reefs and less than 7% of Indonesian reefs remain unaffected by cyanide use.

Overfishing can also lead to outbreaks of coral predators, with massive mortality of corals being the obvious consequence. Over the past 50 years, several outbreaks of the coral-eating crown-of-thorns sea star, *Acanthaster planci* (**Figure 11.16a**), have occurred in the western Pacific Ocean. In some places, coral mortality adjacent to aggregations of the sea stars approached 100%. During the first outbreak in the 1960s, ecologists speculated that these sudden occurrences of large populations of this damaging sea star were the result of human activities, specifically the disappearance of its major predator, the Pacific triton, *Charonia tritonus* (**Figure 11.16b**). A large and beautiful snail, the Pacific triton had been nearly exterminated by shell collectors, but now populations appear to be recovering. Others suspect that the population increases are due to natural causes, such as unusually frequent storms.

Research has shown that more recent *Acanthaster* outbreaks occurred about 3 years after periods of unusually abundant rainfall. Perhaps abnormally high rainfall causes nutrient runoff, which leads to plankton blooms that feed *Acanthaster* larvae, resulting in subsequent increases in their successful settlement. Others suggest that outbreaks of *Acanthaster* are a recent phenomenon that is caused by the overfishing of prawns, which are major predators of juvenile *Acanthaster*. Although still poorly understood, outbreaks of *Acanthaster* and resultant reef mortality may have been augmented by the overfishing of tritons and prawns by humans. Humans with good intentions to remove *Acanthaster* from reefs actually made the problem worse. Some locals were collecting the sea stars, chopping them up into pieces and dumping the pieces back into the ocean. Little did they know that *Acanthaster* are capable of regeneration, so some of the pieces of sea star regrew arms and survived; thus, one sea star became several in some cases. Luckily, *Acanthaster* does not appear to be as good at regeneration as some sea star species capable of regenerating an entire body from a relatively small fragment.

(a)

(b)

Figure 11.16 (a) The predatory sea star, *Acanthaster*, and (b) its major predator, the endangered Pacific triton, *Charonia*.
(a) Courtesy of David Burdick/NOAA; (b) Courtesy of AIMS/NOAA.

Previously unknown coral diseases also have begun plaguing coral reefs on a global scale. A fourfold increase in whitepox disease was documented at 160 reef sites that were monitored in Florida since 1996. Data show that 37% of all coral species in Florida have died since the study began, and 85% of elkhorn coral have expired. Elkhorn coral, sometimes described as "the sequoia of the reef," is a beautiful, orange, 3-m-tall branching species that so dominates Caribbean reefs that the seaward-facing Palmata zone is named in recognition of its ubiquitous occurrence (its scientific name is *Acropora palmata*). *Serratia marcescens*, a bacterium common in human feces and sewage, is the cause of whitepox disease, the affliction causing the death of elkhorn coral in Florida. The disappearance of this majestic species is perhaps the greatest aesthetic loss suffered to date.

Of the many new diseases currently plaguing corals, most are named by the appearance of the affected coral tissue (as in whitepox); black band disease, white band disease, brown band disease, red band disease, yellow band disease, yellow blotch

Figure 11.17 Black band disease overgrowing a coral head.
Courtesy of Paige Gill, Florida Keys National Marine Sanctuary, National Oceanic and Atmospheric Administration.

disease, black necrosing disease, white plague, and bleaching are most common. **Black band disease**, first reported in the 1970s in Belize and Bermuda, is now causing high mortalities in susceptible corals worldwide. This disease is characterized by a band of blackened necrotic tissue that advances several millimeters per day around coral colonies (**Figure 11.17**). Black band disease is caused by *Phormidium corallyticum*, a sulfate-reducing cyanobacterium that invades corals, attacks their zooxanthellae, feeds on dying coral tissues, and grows as a densely interwoven mat that separates the cenosarc from the coral's skeleton. This tissue damage eventually results in death because of the invasion of a consortium of opportunistic pathogens, such as *Beggiatoa* and *Desulfovibrio*. This consortium of bacteria, several of which are known only from humans (and their sewage), creates a sulfide-rich environment that prevents photosynthesis by zooxanthellae. Black band disease is the only coral disease that can be successfully treated. The infected tissue can be removed and the area covered with putty to prevent the disease from spreading.

White band disease was first reported in the late 1970s in Caribbean species of *Acropora* (elkhorn and staghorn corals are the most well-known members of this genus). By 1989, 95% of the elkhorn coral in St. Croix had succumbed to this disease, which also appears to involve a suite of pathogenic agents. **White plague** (or just *plague*) is a disease that resembles white band disease, only it moves and kills much more quickly. It first appeared in the Florida Keys in the 1980s. Although this disease, like most tissue-sloughing coral ailments, is poorly understood, the 1995 plague in Florida seems to have been caused by *Sphingomonas*, a common bacterium that causes infections, septicemia, and peritonitis in humans. Once again, sewage transport of this pathogen is suspected.

It is possible for all of these negative impacts on coral reefs to be managed on a local scale. Unfortunately, coral death is also caused by global-scale environmental changes. For example, a team of scientists from the U.S. Geological Survey believe that none of the many hypotheses offered to explain coral deaths around the world is adequate to explain the vast distribution of coral diseases, nor coral's inability to recover after a die-off. They suggest that the hundreds of millions of tons of dust carried by winds to the Americas from Africa and Asia each year may transport viable pathogens, nutrients, trace metals, and other organic contaminants that could contribute to reef deaths worldwide. Another global issue is **ocean acidification**. Carbon dioxide entering the atmosphere worldwide affects water chemistry of the world ocean. Corals are susceptible to ocean acidification during all life stages, as early life stages appear to be sensitive to slight changes in pH, and adult forms rely on a steady supply of $CaCO_3$ for reef formation. The changes in ocean chemistry due to ocean acidification are just one more set of challenges for coral survival.

Perhaps the most well-studied cause of episodic coral mortalities is **bleaching**, a recently characterized phenomenon first observed in the mid-1980s. Bleaching occurs when physiologically stressed, pathogen-free corals expel their mutualistic zooxanthellae (**Figure 11.18**). This results in a whitening of the colony (due to the $CaCO_3$ skeleton of the coral being visible through its now pigment-free cenosarc), and perhaps its death. Bleaching events have been correlated with increased sea-surface temperatures, such as those that occur in the tropical eastern Pacific

(a)

(b)

Figure 11.18   (a) Widespread bleaching on a Pacific coral reef, and (b) Coral bleaching on an individual brain coral becoming overgrown by algae.

(a) Courtesy of David Burdick/NOAA; (b) Courtesy of National Oceanic and Atmospheric Administration.

Ocean during the El Niño–Southern Oscillation (ENSO). For example, nearly all the living coral in the Galapagos Islands bleached and died after the severe 1982–1983 ENSO episode. Some species completely disappeared during this event in the eastern Atlantic. Researchers reported that the even more intense ENSO event of 1997–1998 resulted in the deaths of fully one sixth of our planet's coral via bleaching. This event enabled the bacterium *Vibrio shiloi* to invade Mediterranean corals, and the Maldives were so badly impacted that virtually none of their corals survived. Some are now concerned that the 1,200 atolls that constitute the Maldives nation (see Figure 11.7) are no longer protected from erosion, which may lead to the complete disappearance of this archipelago over time.

The first Caribbean bleaching event occurred in 1987–1988 and affected all species living down to 30 m depth (only *Madracis* and *Acropora* seem minimally affected). A second event occurred in 1990. Unlike Pacific episodes that are usually attributed to increased water temperatures, both of the mass bleaching events in the Caribbean Sea were not readily explained by temperature alone. Recent studies suggest that decreased temperatures, increased levels of ultraviolet radiation, increased sediment loads, changes in salinity, or toxic chemicals may also play roles in Caribbean bleaching episodes during periods of calm clear water that occur during ENSO events.

However, the actual cause and mechanism of coral bleaching remain unknown. Do corals evict malfunctioning algae, thus hurting themselves in the process, or do the zooxanthellae voluntarily leave stressed coral polyps? Some have even suggested that latent viral infections are induced by the above coral stressors.

Incontrovertible evidence links greenhouse gases, climate change, and coral bleaching. Projected increases in atmospheric CO_2 and global temperatures during the next 50 years will rapidly exceed the conditions under which coral reefs have thrived for 500,000 years. Researchers have found that coral species with unusual algal symbionts are able to achieve increased thermal tolerance, so perhaps adaptive shifts between corals and new zooxanthellae clades will confer increased resistance to future climate change.

11.3   Coral Reef Fishes

Associated with the reef and lagoon but with the mobility to escape the limitations of a benthic existence are thousands of species of reef fishes (**Figure 11.19**). These fishes find protection on the reef; prey on the plants, algae, and animals living there; and sometimes nibble at the reef itself. These assemblages of shallow-water coral reef fishes are easily observed by divers and have been intensively studied for decades. Less well known are the fish of the deeper portions of coral reef communities (below 100 m). Submersible-based studies recently demonstrated that as one works down the reef face into deep water the same general assemblages are present, but individual numbers and species diversity both diminish. Coral reef fishes are extremely diverse in their form, coloration, and behaviors. Here we explore some representatives of the major groups of these organisms.

Figure 11.19 Some common reef fishes on a tropical Caribbean reef: (1) nurse shark (*Ginglymostoma*), (2) reef shark (*Carcharhinus*), (3) barracuda (*Sphyraena*), (4) surgeonfish (*Acanthurus*), (5) butterflyfish (*Chaetodon*), (6) angelfish (*Pomacanthus*), (7) hawkfish (*Amblycirrhitus*), (8) grouper (*Mycteroperca*), (9) moray eel (*Gymnothorax*), (10) stingray (*Dasyatis*), (11) grunt (*Haemulon*), (12) soldierfish (*Myripristis*), (13) porcupinefish (*Diodon*).

Coral Reef Sharks and Rays

Sharks are often described as large, voracious predators. Yet about 80% of known species are less than 2 m in length, half of all shark species are less than 60 cm long, and some species barely attain a length of 30 cm. Many of these smaller sharks are found only on coral reefs.

Without question, nurse sharks, carpet sharks, wobbegongs, and bamboosharks in the order Orectolobiformes dominate coral reefs. Some carcharhinid sharks, such as reef sharks, blacktips and whitetips, lemons, bulls, and tigers, also frequent coral reefs. All of these reef-dwelling sharks contradict the myth that sharks must constantly swim to breathe by coasting to a stop and resting on the seafloor for many hours at a time. Thanks to their inshore existence, they have developed the ability to flex their muscular gill slits and create the necessary flow of water over their gills even while stationary.

Most sharks that inhabit coral reefs also fail to fit the standard view of sharks as apex predators. Some consume large invertebrates, such as conchs, sea urchins, and clams from the seafloor. Caribbean nurse sharks suck sleeping wrasses from the sand under which they sleep. Many cryptic species are ambush predators, launching themselves from the reef when a prey species swims nearby. The numerous dermal flaps on the jaw margin of wobbegongs may function as lures to "bait"

prey near their mouths (**Figure 11.20**). Nurse sharks and some rays perch on extended pectoral fins, perhaps in an attempt to attract prey to the cavelike space that they create just under their chins. Some reef sharks are masters when it comes to extracting prey from reef crevices or using their snouts to flip coral rubble to reveal hidden crustaceans or annelids, and filter-feeding whale sharks and manta rays routinely visit reefs to consume the reproductive products of spawning corals and fish. The importance of sharks and other apex predators on reefs is highlighted when their numbers decline; apex predators help to maintain greater biodiversity in the ecosystem, and when their numbers decline, shifts in community structure occur. On coral reefs the shift is often toward algal-dominated reefs.

Figure 11.20 Dermal flaps around the mouth of a wobbegong, a benthic reef shark.
Courtesy of John Morrissey.

Case Study

The Lionfish Story: A Nonnative Invasion of the Sea Like Never Before

A recent invasion has taken place in the western Atlantic along the East Coast of the United States, the Caribbean, the Gulf of Mexico, and spanning the coasts of Central and South America. It is an alien invasion, but not of the extraterrestrial kind. Two very similar lionfish species, the devil firefish (*Pterois miles*) and the red lionfish (*Pterois volitans*), are reproducing rapidly with no end in sight, and these species are not native to the Atlantic Ocean. Lionfish are scorpionfish from the Indo-Pacific and are popular aquarium fish due to their beautiful and elaborate fins and coloration (**Figure A**). Unfortunately, and for unknown reasons, some aquarists release their lionfish into the ocean in areas that are not their natural homes, and populations of invasive lionfish are now thriving. The presence of these nonnative species is disrupting the natural flow of the food web and leading to fierce competition with native fish species.

Lionfish have modified dorsal and anal fins with elongated poisonous spines, making them virtually inedible by natural predators. Not only do the lionfish lack predators in their new home, but they are voracious carnivores, and they are not picky eaters. Gut content studies have revealed that lionfish choose to eat a large variety of juvenile or small adult fish and crustaceans, many of them commercially important species. In Belize, nearly half the stomach gut content of lionfish studied was a critically endangered wrasse species. Lionfish food preferences also overlap with native snappers and groupers; thus, they are directly competing with these commercially important, and in some cases low-abundance, fish species for food. In areas of heavy infestation it is estimated that lionfish have reduced their fish prey abundance by approximately 90%, which is unsustainable.

In response to the dramatic increases in abundance of nonnative lionfish, scientists set out to find out as much as possible about these invasive species so that efforts to eradicate them can be effective and efficient. They discovered that lionfish are sexually

Figure A A lionfish displaying long spines on the dorsal and anal fins glides over the seafloor. Courtesy of National Oceanic and Atmospheric Administration.

mature around the young age of 1 year and reproduce throughout the year every few days, which is why nonnative populations have exploded so quickly. They also estimate that lionfish can live around 15 years, which is a relatively long life span for a teleost fish, leading to a large reproductive potential for each individual fish. They appear to adapt well to all habitats, from the shoreline out to approximately 183 m (600 ft.). It was thought that lionfish required warm waters, but the presence of individuals along the central East Coast of the United States in waters barely above 13°C (56°F) has scientists questioning the presumed temperature requirements (**Figure B**). The rough abundance estimate as of 2015 is in the millions in the Gulf of Mexico region alone, and those numbers are predicted to keep climbing.

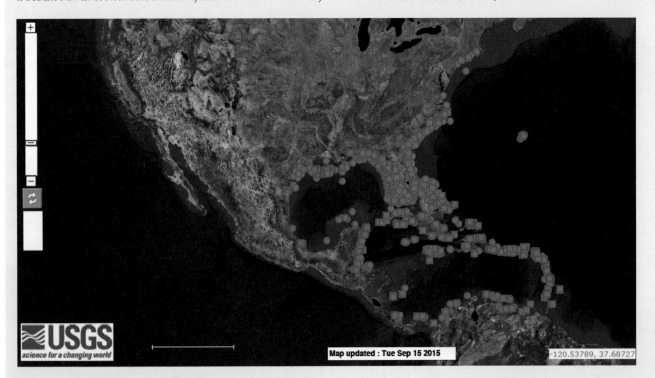

Figure B Distribution of documented established lionfish populations according to the U.S. Geological Survey as of September 2015. Undocumented invasions and those outside of the United States or its territories are not displayed on this map.
Courtesy of United States Geological Survey.

Case Study   *(Continued)*

Figure C   A research diver counting and observing lionfish.
Courtesy of National Oceanic and Atmospheric Administration.

It is clear that the lionfish introduction has created a remarkable problem in the Atlantic, Caribbean, and Gulf Coast regions. The question now is what can be done about it, if anything. Some scientists have predicted that lionfish are already too numerous to eradicate. Others suggest that great efforts should be made to remove as many as possible as quickly as possible. Although the invasive lionfish currently have no natural predators, it is possible that over time some larger fish or sharks may acquire a taste for them and figure out a way to eat them, despite their spiny fins. If humans make a valiant effort to remove lionfish and one or several predators start to consume them, the hope is that their impacts on the communities they reside in may be reduced. Fishers are highly encouraged to capture lionfish, and the most effective fishing methods are spearfishing and hand netting. These fishing methods are limiting in that they must occur at depths where humans can safely dive, and capturing large numbers of lionfish is challenging (**Figure C**). Communities of concerned citizens and scientists around Florida and the Caribbean have responded by organizing lionfish tournaments and competitions to encourage fishing. Another angle is to encourage the use of lionfish as food for humans. They are considered a delicacy in some parts

of the world, and their flesh is mild and flavorful. These are two species that fishers are encouraged to overfish! In all likelihood, these efforts will not be enough to wipe out lionfish populations, so innovative ways to remove these fish will be necessary for success. Scientists and citizens are working hard on these innovations as you read this case study, and one can only hope that they are successful.

Critical Thinking Questions

1.   Do you think it is possible to eradicate nonnative lionfish? Why or why not? If you do think it is possible, propose a novel way to remove them from the Atlantic.

2.   Consider the extreme appearance of lionfish, with their long, sharp spines. What selective pressures can you identify that may have driven lionfish species toward such extreme body designs?

Coral Reef Teleosts

Several groups of derived bony fishes are common to coral reefs worldwide (**Figure 11.21**). These include grunts, snappers, cardinalfish, moray eels, porcupinefish, butterflyfish, squirrelfish, groupers, triggerfish, gobies, blennies, wrasses, parrotfish, surgeonfish, and seahorses. Many of these fish are thought to be major importers of important limiting nutrients to local reef systems by foraging on pelagic prey during the day and then defecating at night while resting on the reef. Others feed in surrounding seagrass meadows at night and defecate on the reef while resting during the day. The results of this off-reef predation are converted through detritus food chains to dissolved nutrients usable by plants, phytoplankton, and the coral-based zooxanthellae.

Recent research on parrotfish has revealed an additional and crucial role of this group of fish to coral reef health. Caribbean coral reefs have declined dramatically in the past 50 years, and although there are several likely causes (as discussed earlier), one fact is clear: reefs with healthy populations of herbivorous parrotfish have healthier corals. Parrotfish graze heavily on the normally limited amount of macroalgae that grow on the reef. If these grazers are removed, macroalgae overgrow the reef and block the sun from reaching the symbiotic algae living within the corals. A direct result of macroalgae growth is death of corals and a conversion from healthy reef to what is termed **pavement**, the leftover dead coral skeleton with macroalgae growth on top. Some scientists claim that in the Caribbean immediately preserving the amount of grazers is more important than global warming and all of the other threats to coral reefs combined. These same scientists claim that Caribbean coral reefs will be gone in 20 years if populations of herbivores are not restored. Healthy reefs are more resilient and may be able to adapt to larger, gradual environmental changes or sudden and brief disturbances such as hurricanes, but unhealthy reefs have very little chance for survival with additional environmental stress, whether it is human-made or naturally occurring.

Figure 11.21 Numerous species of teleost fishes are associated with coral reefs. This photo was taken at the Pearl and Hermes Atoll, part of the northwestern Hawaiian Islands.
Courtesy of National Oceanic and Atmospheric Administration.

Figure 11.22 Two remoras with modified dorsal fins attached to a manta ray.
Courtesy of National Oceanic and Atmospheric Administration.

Symbiotic Relationships

Excellent examples of all the types of symbiosis can be found in many of the abundant animal groups of the coral reef. Our discussions are limited to some of the better-known symbiotic relationships involving coral reef fishes. These relationships span the entire spectrum of symbiosis, from very casual commensal associations to highly evolved parasitic relationships.

Remoras (**Figure 11.22**) associate with sharks, billfish, parrotfish, sea turtles, and even the occasional dolphin in a mutualistic symbiosis. The remora's first dorsal fin is modified as a sucking disc and is used to attach itself to its host. From its attached position, it feeds on scraps from the host and often cleans the host of external parasites. Thus, the remora gains food, a free ride, and protection via proximity, whereas the host rids itself of many ectoparasites. A similar association in the open ocean exists between sharks and pilotfish (*Naucrates*). The pilotfish swim below and in front of their hosts and scavenge bits of food from the shark's meal. It has been speculated that pilotfish may attract prey species to the shark.

It is common for smaller defenseless fish to live on or near better-defended species of reef invertebrates. For example, shrimpfish often hover vertically in a head-down position among the long, sharp spines of sea urchins in a commensal symbiosis (**Figure 11.23**). The shrimpfish acquire protection from the sea urchin without affecting it. Brightly colored clownfish and anemonefish find equally effective shelter by nesting among the stinging tentacles of several species of sea anemones (**Figure 11.24**). This relationship, somewhat more complex that those just described, is also probably a mutualistic one. In return for the protection they obtain, clownfish assume the role of "bait" and lure other fish within reach of the anemone. They occasionally collect morsels of food and, in at least one observed instance, catch other fish and feed them to the host anemone. Clownfish, however, are not immune to the venomous cnidocytes of all sea anemones. Although some clownfish are innately protected from some anemone species (i.e., their protection results from their normal development rather than from contact with chemical, visual, or mechanical stimuli from an anemone), researchers have demonstrated that some clownfish must acclimate to some anemone species. Moreover, they

Figure 11.23 An urchin clingfish, *Diademichthys,* nestled between urchin spines in Indonesia.
© WaterFrame/Alamy Stock Photo.

Figure 11.25 A nearly transparent cleaner shrimp, *Periclimenes,* on a Caribbean sponge.
Courtesy of Dr. Anthony R. Picciolo, NOAA/NODC.

also reported that other clownfish are unable to acclimate to certain species of anemones.

The increased popularity of skin diving and scuba diving has led to many more observations of coral reef fishes, and as a result has revealed some remarkable cleaning associations involving a surprising number of animals. Cleaning symbiosis is a form of mutualism; one partner picks external parasites and damaged tissue from the other. The first partner gets the parasites to eat; the other partner has an irritation removed.

The behavioral and structural adaptations of cleaners are well developed in a half dozen species of shrimps (**Figure 11.25**) and several groups of small fishes. Tropical cleaning fishes include juvenile butterflyfish, angelfish, and damselfish, but only some neon gobies in the Atlantic (*Elacatinus*) and cleaner wrasses in the Pacific (*Labroides*) are cleaning specialists throughout their lives (**Figure 11.26**). All tropical cleaning fish are brightly marked, are equipped with pointed, pincer-like snouts and beaks, and occupy a cleaning station around an obvious rock outcrop or coral head. Most are solitary; a few species, however, live in pairs or larger breeding groups.

Host fish approach cleaning stations, frequently queuing up and jockeying for position near the cleaner. Often, they assume unnatural and awkward poses similar to courtship displays. As the cleaner fish moves toward the host, it inspects the host's fins, skin, mouth, and gill chambers and then picks away parasites, slime, and infected tissue.

In the Bahamas, one study tested the cleaner's role in subduing parasites and the infections of other reef fishes. Two weeks after all known cleaner fish were removed from two small reefs, the areas were vacated by nearly all but territorial fish species. Those species that remained had an overall ratty appearance and showed signs of increased parasitism, frayed fins, and ulcerated skin. It was concluded that symbiotic cleaners were essential in maintaining healthy fish populations in this particular study area.

Others conducted similar studies on a Hawaiian reef. In this situation, the small cleaner wrasse, *Labroides phthirophagus* (the major cleaner on the reef), was excluded from the study site for more than 6 months. During that time, no increase in the level of parasite infestation was observed. This result suggests

Figure 11.24 A clownfish, *Amphiprion,* nestled within the protective tentacles of its host anemone.
© Russell swain/Shutterstock, Inc.

Figure 11.26 Neon gobies, *Elacatinus,* clean the head of a large green moray, *Gymnothorax.*
© Kelpfish/Shutterstock, Inc.

that for some cleaner–host associations the role of the cleaner is not crucial. The cleaner may be dependent on the host for food, but the host's need for the cleaner seems to be variable.

The fine line separating mutualistic cleaning of external parasites and actual parasitism of the host fish is occasionally crossed by cleaner fish. In addition to unwanted parasites and diseased tissue, some cleaners take a little extra healthy tissue or scales or graze on the skin mucus secreted by the host. Thus, the total range of associations displayed by cleaning fish encompasses mutualism, commensalism, and parasitism.

Because parasitism is such a widespread way of life in the sea, few fish avoid contact with parasites throughout their lives. The groups notorious for creating parasitic problems in humans—viruses, bacteria, flatworms, roundworms, and leeches—also plague marine fish. Despite the bewildering array of parasites that infest fish, very few fish become full-time parasites themselves. A remarkable exception is pearlfish. They find refuge in the intestinal tracts of sea cucumbers, the stomachs of certain sea stars, the body cavities of sea squirts, and the shells of clams. After this association is established, some pearlfish assume a parasitic existence, feeding on and seriously damaging the host's respiratory structures and gonads. When seeking a host sea cucumber, pearlfish detect a chemical substance from the cucumber and then orient themselves toward the respiratory current coming from the cucumber's cloaca. Sea cucumbers draw in and expel water through their cloacae for gas exchange. The fish enters the digestive tract tail first via the cloaca. The hosts are not willing participants in this relationship. They sometimes eject their digestive and respiratory organs in an attempt to rid themselves of the symbiont. In fact, sea cucumbers of the genus *Actinopyga* have evolved five teeth on their cloacal margin, perhaps as a pearlfish-exclusion mechanism.

Coloration

Against the colorful background of their coral environment, reef fishes have evolved equally brilliant hues and color patterns. The colors are derived from skin or internal pigments and from iridescent surface features (like those of a bird's feathers) with optical properties that produce color effects. Most fish form accurate visual color images of what they see, but like humans, they are susceptible to misleading visual images and camouflage.

Our interpretation of the adaptive significance of color in fish falls into three general categories: concealment, disguise, and advertisement. Some seemingly conspicuous fish resemble their coral environment so well that they are nearly invisible when in their natural setting. Extensive color changes often supplement their basic camouflage when they are moving to different surroundings. These rapid color changes are accomplished by expanding and contracting the colored granules of pigmented cells (**chromatophores**) in the skin and are governed by the direct action of light on the skin, by hormones, and by nerves connected to each chromatophore. As the chromatophore pigments disperse, the color changes become more obvious (**Figure 11.27**). When the granules are contracted, the pigment retreats to the center of the cell, and little of it is visible. Other cells, called **iridocytes**, contain reflecting crystals of

(a)

(b)

Figure 11.27 A well-camouflaged scorpionfish, *Scorpaena* (a), with magnified chromatophores from a section of skin (b). The multicolored pigments of some are expanded and diffused; others are densely concentrated in small spots.

(a) © Frank Boellmann/Shutterstock, Inc.; (b) © Rene Frederic/age fotostock.

guanine. Impressive to the observer, iridocytes can produce an entire spectrum of colors within a few seconds.

Several distinctive fish conceal themselves with color displays reminiscent of disruptive coloration, or *dazzle camouflage*. Bold contrasting lines, blotches, and bands tend to disrupt the fish's image and draw attention away from recognizable features such as eyes. Eyes are common targets for attack by predators, and thus a disguised eye is a protected eye. One common strategy masks the eye with a dark band across the black, staring pupil so that it appears to be continuous with some other part of the body (**Figure 11.28**). To carry the deception even further, masks hiding the real eyes are sometimes accompanied by fake eyespots on other parts of the body or fins. Eyespots, intended as visual attention-getters, are usually set off by concentric rings to form a bull's eye. As a result, predatory attacks are more likely to be directed away from the eyes and head and drawn to less vital parts of the body that may not lead to death upon attack.

The flashy color patterns of cleaning fish serve different functions. If the fish are to attract any business, they must be conspicuous. Thus, they advertise themselves and their location

Occasionally, a species capitalizes on the advertisement displays of another fish by closely mimicking its appearance. The cleaner wrasse (*Labroides*, the upper fish in **Figure 11.29**) is nearly immune to predation because of the cleaning role it performs for its potential predators. Over much of its range, *Labroides* live close to a small blenny (*Aspidontus*). The blenny so closely resembles *Labroides* in size, shape, and coloration (lower fish in Figure 11.29) that it fools many of the predatory fish that approach the wrasse's cleaning station. Not content to share the *Labroides* immunity to predation, the blenny also uses its disguise to prey on fish that mistakenly approach it for cleaning. This ability to disguise and thereby be protected is known as **mimicry**.

(a)

Figure 11.28 Disruptive coloration patterns of two species of butterflyfish, *Chaetodon*.
(a) © cbpix/Shutterstock, Inc.; (b) © Lawrence Cruciana/Shutterstock, Inc.

with bright, startling color combinations. These bold advertisement displays are also useful for sexual recognition. One or both sexes of certain species assume bright color patterns during the breeding period. The colors play a prominent role in the courtship displays, which lead to spawning. During this period, the positive value gained from sexual displays must offset the adverse impact of attracting hungry predators. Between breeding periods, these fish usually assume a drab, less conspicuous appearance.

Advertisement displays are also used to warn potential predators that their prey carry sharp or venomous spines, poisonous flesh, or other features that would be painful or dangerous if eaten. Predatory fish recognize the color patterns of unpalatable fish and learn to avoid them. This type of bright coloration to indicate harm and deter predators is seen throughout the animal kingdom.

(a)

(b)

Figure 11.29 A cleaner wrasse, *Labroides dimidiatus* (a), and its mimic, the bluestriped fangblenny, *Plagiotremus rhinorhynchos* (b).
(a) © vkilikov/Shutterstock; (b) © RGB Ventures/SuperStock/Alamy Stock Photo.

Only in the clear waters of the tropics and subtropics does color play such a significant role in the lives of shallow-water animals. In the more productive and turbid waters of temperate and colder latitudes, light does not penetrate as deeply nor is the range of colors available. In coastal waters and kelp beds, monotony and drabness of appearance, not brilliance, are the keys to camouflage. In the deep ocean, color is even less important. Without light to illuminate their pigments, it matters little whether deep-water organisms appear red, blue, black, or chartreuse when viewed at the surface. In the abyss, they would all assume the uniform blackness of their surroundings were it not for bioluminescence.

Spawning and Recruitment

Coral reef teleosts are a very diverse lot, yet most share a common life-history strategy: most adult reef teleosts are benthic fishes that spawn in the water column. Only about 20% to 30% of reef species (damselfish, gobies, and triggerfish) deposit 1-mm-long benthic eggs that stick to the substrate until they hatch after 1 to 4 days (longer in some species). Of these, damselfish are the most conspicuous, and the courtship and mating rituals of some damselfish are well known (**Figure 11.30**). For example, bicolor damselfish, *Stegastes partitus*, mate between full and new moons. The male builds a nest and then attempts to persuade females to deposit their eggs in his nest by performing a series of dips in the water column. Females typically choose the male that performs the most dips in a given time. Presumably, the female uses his dance to assess his health and fitness. Because the male must guard her eggs for 3 to 4 days (**Figure 11.31**), the female uses the rate of dipping during courtship to determine which male in her vicinity has the most energy stored as fat, which is energy that will be very useful while he guards her eggs relentlessly before their hatching.

Figure 11.30 Two bicolor damselfish mate inside a discarded PVC pipe on a Caribbean reef.
Courtesy of Dr. Michael P. Robinson, University of Miami.

Figure 11.31 A sergeant major (*Abudefduf*) guards its purple egg mass in the Caribbean Sea.
© David Fleetham/Alamy Images.

The majority of reef teleosts are pelagic spawners. In some cases, many species, 30 or more at any given time, will assemble around the same coral promontory to broadcast as many as 50,000 eggs/female and over 1,000,000 sperm/male into the water column (**Figure 11.32**) during the course of an hour or so. The coral pinnacle that is selected is not obviously different from neighboring outgrowths, but the fish seem to understand the difference. In fact, if one removes all female bluehead wrasses from a reef, the new set of replacement females that arrives will pick many (maybe completely) new spawning sites, and these sites will become the new "traditional" sites on that reef. This is one of the few examples of "culture" in fish. Wrasses are known to travel 1.5 km to a spawning site (a distance equal to 15,000 body lengths, or the equivalent of a 55-km round-trip for a human), and Nassau groupers, *Epinephelus striatus*, may travel up to 240 km to spawn.

After fertilization, these pelagic eggs drift away from the reef and disperse for a period of time that ranges from 1 day to a year or more. This time period, known as the **pelagic larval duration**, influences dispersal greatly and is of intense interest because of its influence on the potential efficacy of national marine sanctuaries.

In 1972, a century after the United States established the first national park at Yellowstone, legislation was passed to create the National Marine Sanctuary Program (NMSP). The intent of this legislation was to provide similar protection to selected coastal habitats as provided for land areas designated as national parks. The designation of an area as a marine sanctuary says to all that, like our national parks, this is a safe refuge where people can observe organisms in their natural environment, but tread lightly. Three decades later, only 13 marine sanctuaries and 1 marine national monument have been designated (**Figure 11.33**). In addition, the National Estuarine Research Reserve System has designated 28 national estuarine research reserves. A recent designation of additional protected areas has led to the umbrella term of *marine protected areas* (MPAs)

Figure 11.32 Dog snappers, *Lutjanus jocu*, return to the reef after a spawning run in the water column off Belize.

© Doug Perrine/Seapics.com.

for all marine areas that are protected for conservation. Over 400 MPAs with varying degrees of protection have been established in the United States and U.S. territories; some MPAs are fully protected no-take reserves, others allow fishing of particular species, and some are simply recognized as special areas and allow fishing for all species (except those protected by the Endangered Species Act).

The national system of MPAs should be viewed as a crucial part of new management practices in which whole communities, and not just individual species, are offered some degree of protection from habitat degradation. Nevertheless, because of the interactions between and inexplicable changes of the pelagic larval durations of reef invertebrates and teleosts and surface currents, one can never be certain that protection of a spawning population on one reef will ensure augmented recruitment of juveniles into that sanctuary. It seems just as likely that efforts at one location will result in huge benefits in terms of larval recruitment somewhere down current. Therefore, a great deal of research effort is also being directed to the study of **settlement**, or the passage from the pelagic existence of a larva to the benthic life of a reef-dwelling fish or invertebrate. Hypothetical recruitment factors that are being investigated include active mechanisms, such as larvae being attracted to reef sounds (such as waves breaking on the reef crest or the snapping of shrimp) or reef smells (in a mechanism analogous to the homing behavior of salmon), as well as passive mechanisms that may enable the settlement of postlarval juveniles (such as retention in gyres around some Hawaiian islands or the Florida Keys).

It is estimated that only about 1% of the eggs spawned survive to produce an individual that will settle on a reef, but successful settlement does not mean that the intense mortality is over. About 90% of those teleost larvae that settle on a reef are eaten during their first night. If a larva is lucky enough to

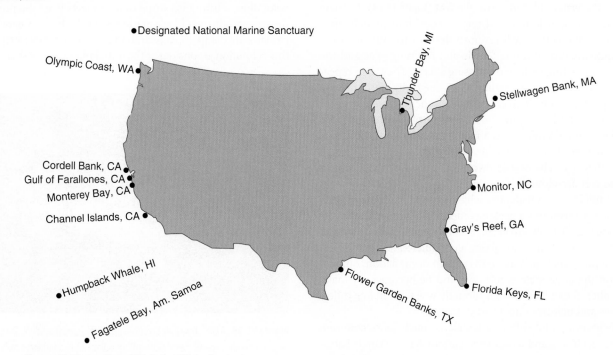

Figure 11.33 Locations of U.S. National Marine Sanctuaries.

survive its first night on the reef, it then must metamorphose into a benthic juvenile within a day or two. Only 50% of those that begin metamorphosis survive to complete the transformation. After they acquire the morphology, appearance, and behavior of a juvenile, mortality decreases greatly.

Sexual Systems in Reef Fishes

Reef teleosts display a great variety of sexual systems, from relatively straightforward **gonochorism**, with separate males and females (as seen in grunts, snappers, and most damselfish), to complex systems involving **hermaphroditic** individuals that play the roles of both genders during their adult lives. Simultaneous hermaphrodites function as males and females at the same time, whereas sequential hermaphrodites are born as one gender and then change sex during their life.

The best-known simultaneous, or synchronous, hermaphrodite on the reef is the hamlet (a relative of groupers in family Serranidae). All hamlets are monogamous, forming faithful pairs that often last throughout the breeding season, and perform synchronous mating rituals that involve choosing a sex for each event. During courtship, one member of the pair will "act male" and will release sperm during a stereotypical "clasping" in the water column above the reef (**Figure 11.34**). Its mate will play the female role and release eggs. Immediately after spawning, the couple will reverse roles and recourt, with the first individual now acting female and releasing eggs and the second fish acting male and releasing sperm. They then repeat this ritual 2 to 15 times, trading roles repeatedly, until they are both spent and unable to court with another fish.

Sequential hermaphrodites come in two varieties, those that begin their adult life as a male (**protandry**) and those that begin their adult life as a female (**protogyny**). Both types change sex as they age. The best-known protandrous hermaphrodites on the reef are clownfish (see Figure 11.24), which all begin their adult lives as a male. In each clownfish community (often a single sea anemone), a pecking order exists wherein the largest individual is an adult female, the second largest is an adult male, and the remaining individuals are all undifferentiated juveniles. If the alpha female is removed via death or predation, oppression on the large male is released, and he becomes a female. Soon after, the individual that formerly was third in line differentiates and matures into an adult male.

Beginning adult life as a female is much more common among reef fishes, and wrasses and parrotfish in the family Labridae are perhaps the best-known protogynous hermaphrodites.

Figure 11.34 Clasping hamlets above a reef.
© WaterFrame/Alamy Images.

The bucktooth parrotfish, *Sparisoma radians*, of Caribbean coral reefs has a particularly interesting sexual system that involves several phases of male and female fish that look and act quite different, but are all the same species. All *S. radians* begin life as females, so all males have changed sex at some point from an initial-phase female to a male. The most conspicuous phase is the large phase of the terminal-phase male, with bold and bright coloration, a turquoise stripe that connects from the mouth to the eye and distinct black markings on the body (**Figure 11.35**). Females are comparatively drab in color with no eye stripe, and this coloration is considered the initial phase. Some small males

Figure 11.35 A late-phase male bucktooth parrotfish, *Sparisoma radians*. The distinctive turquoise mouth to eye stripe and black markings indicate a fully transformed male.
Courtesy of Kirk Kilfoyle, Nova Southeastern University.

change only their behavior and physiology, but retain the look of an initial-phase female.

Spawning in *S. radians* occurs daily in the afternoon. Terminal-phase males are territorial and defend a **harem** of females that they spawn with in the same location each day. Courtship behavior begins a few minutes before spawning, as the male swims over to and brushes up against a female. Eventually the two fish join together in a vertical spiraling movement toward the surface where sperm and eggs are released. It is during this time that drab-colored initial-phase males take advantage of their cryptic coloration to blend into the environment and sneak up to release sperm into the spawning event. The spawning ritual is very costly in that leaving the safety of the reef or seagrass makes them extremely vulnerable to predation. It can be presumed that releasing gametes into the water column allows for better dispersal of gametes and soon after, fertilized eggs.

The various phases of *S. radians* are quite unique among fish and somewhat of a mystery to scientists. Living in the seagrass bordering the reef are small versions of males and females. These small individuals are not necessarily younger than the large-phase individuals, just smaller, and genetic tests reveal that they are the same species as the larger individuals.

The small-phase fish spawn with one another, and small initial- and terminal-phase males sometimes sneak in to spawn with larger females, too. Another unique feature of this species is the mechanism of sex change. In most sequentially hermaphroditic haremic fish, the largest female changes sex when the male is removed by predation. In most cases for *S. radians*, it is the second largest female that changes sex and takes over as the terminal-phase male. Sex change begins with morphological changes (color change) and is followed by behavioral and eventually physiological changes (**Figure 11.36**). About 20 days after the first signs of sex change, the newly transformed male looks, acts, and spawns as a male.

The tropical cleaner fish *Labroides* (also in the family Labridae; see Figure 11.29) is also protogynous. This inhabitant of the Great Barrier Reef of Australia occurs in small social groups of about 10 individuals. Each group consists of one dominant male and several females existing in a hierarchical harem. This type of social and breeding organization is termed **polygyny**. Only the dominant, most aggressive individual functions as the male and, by himself, contributes half the genetic information to be passed on to the next generation. In the event the dominant male of a *Labroides* population dies or is removed, the most dominant of the remaining females immediately assumes

Figure 11.36   A newly transformed male bucktooth parrotfish, *Sparisoma radians*. The turquoise mouth to eye stripe indicates that sex change from female to male has begun, but the gray coloring and lack of black on the body indicate that sex change is not yet complete.
Courtesy of Kirk Kilfoyle, Nova Southeastern University.

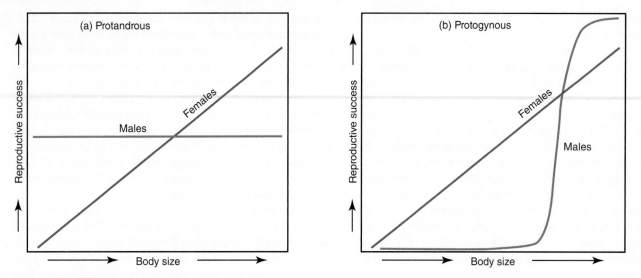

Figure 11.37 Relative reproductive success experienced by males and females of (a) protandrous fishes (e.g., clownfish) and (b) protogynous fishes (e.g., wrasses).

the behavioral role of the male. Within 2 weeks, the dominant individual's color patterns change, and the sex transformation to a male is complete. In this manner, males are produced only as they are needed, and then only from the most dominant of the remaining members of the population.

Before ending our discussion of hermaphroditic reef fishes, it is helpful to consider the benefits of such a sexual system. Why should an individual change its sex during its life, and under what circumstances should a reef fish begin life as a male (as opposed to a female)? Biologists hypothesize that changing sex is advantageous when reproductive success is closely tied to body size, and when being one sex (as opposed to the other) results in increased reproductive success at a given body size (this is known as the *size-advantage model*). **Figure 11.37a** shows the relative reproductive success of male and female clownfish from the time they are small adults to sometime later when they are older and larger. Body size does not influence the reproductive success of male clownfish. It remains the same as they grow because they produce countless sperm cells at all body sizes and they do not compete with other males for access to females; they simply mate with the adult female inhabiting their sea anemone. Yet female reproductive success among clownfish is greatly influenced by body size because eggs are very large and expensive to produce. Therefore, young (small) females do not produce nearly as many eggs as larger (older) females. Hence, it is advantageous for clownfish to be male while young (and small) and then change into females once older (and larger). Via this sexchanging strategy, they achieve maximal reproductive success at all sizes.

Figure 11.37b presents the relative reproductive success achieved by protogynous fishes, such as the bucktooth parrotfish described above. Female parrotfish and wrasses are very similar to female clownfish in that they only produce a large number of eggs when they possess a large body. Unlike male clownfish, small (young) male parrotfish experience very limited reproductive success because they are too small to compete for access to females and their eggs. Hence, their reproductive success is very low and unpredictable while young and small and then suddenly sky rockets after they become large enough to gain access to eggs in the water column. Hence, unlike the situation experienced by clownfish, it is better for a parrotfish to begin life as a small female (because small males do not reproduce much at all) and later change into a male after becoming large and competitive.

But the size-advantage model does not explain all changes of sex. The coral goby, *Paragobiodon echinocephalus*, changes sex both ways. Coral gobies are small, black fish with orange heads that live on the branching coral *Stylophora pistillata*. On each coral colony, only the two largest fish (one male and one female of similar sizes) breed monogamously. The reproductive success of the pair is positively correlated with size in both sexes; larger males are more successful at guarding benthic eggs, and larger females can produce more eggs. Hence, the size-advantage model does not explain the factors that induce sex change in this species. Instead, change in social rank determines the direction of sex change. When a coral goby loses its mate, it prefers to change sex in either direction to form a mating bond with the nearest adult goby, as opposed to traveling a great distance to locate a heterosexual mate.

STUDY GUIDE

TOPICS FOR DISCUSSION AND REVIEW

1. Summarize the physical and biological limitations to coral reef distribution, and then explain why coral reefs do not form at all latitudes and depths.

2. Tally the pros and cons experienced by each member of the symbiotic relationship between corals and zooxanthellae, and demonstrate that this relationship is mutualistic.

3. Why is hermaphroditic spawning the most common method of sexual reproduction in reef-building corals?

4. What can be done to slow or end the global destruction of coral reefs that we are experiencing currently?

5. What is a marine protected area (MPA)? A marine reserve? Are all MPAs "no take" reserves? How do MPAs benefit neighboring areas?

6. Describe the relationship between clownfish and sea anemones, listing the benefits and disadvantages experienced by each.

7. Generate a list of all potential cues that postlarval reef fishes could use to locate the coral reef on which they eventually settle.

8. Describe the main difference between protogynous hermaphroditism and protandrous hermaphroditism in coral reef fishes, and explain the advantages of each mating system.

9. What sex change feature is so different about the parrotfish, *Sparisoma radians*, compared to other hermaphroditic fishes?

KEY TERMS

ahermatypic 302
atoll 303
barrier reef 303
black band disease 313
bleaching 314
cenosarc 300
chromatophore 320
columella 300
corallite 300
endo-upwelling 303
fringing reef 303
gonochorism 324
guyot 304
harem 325
hermaphroditic 324
hermatypic 302
iridocyte 320
mesenterial filament 303
mimicry 321
mucusciliary system 303
ocean acidification 314
pavement 318
pelagic larval duration 322
phototrophic 303
planula 306

polygyny 325
polymorphic 300
protandry 324
protogyny 324
septa 300
settlement 323
white band disease 313
white plague 313
zooxanthellae 302

KEY *GENERA*

Abudefduf
Acanthaster
Acanthurus
Acropora
Actinopyga
Amblycirrhitus
Amphiprion
Aspidontus
Astrangia
Beggiatoa
Carcharhinus
Cassiopea
Chaetodon
Charonia
Dasyatis
Diademichthys
Desulfovibrio
Diadema
Diodon
Diploria
Elacatinus
Favia
Ginglymostoma
Goniolithon
Gymnothorax
Haemulon
Hydrolithon
Labroides
Leptoseris
Lithothamnion
Lutjanus
Madracis
Montastraea
Mussa
Mycteroperca
Myripristis
Naucrates
Orbicella
Paragobiodon
Periclimenes
Phormidium
Plagiotremus
Pocillopora
Pomacanthus
Porites
Porolithon
Scorpaena
Serratia
Siderastrea
Sparisoma
Sphingomonas
Sphyraena
Stylophora
Symbiodinium
Tridacna
Vibrio

REFERENCES

Adey, W. H. 1978. Coral reef morphogenesis: A multidimensional model. *Science* 202:831–837.
Aeby, G. S. 1991. Costs and benefits of parasitism in a coral reef system. *Pacific Science* 45:85–86.

Albright, R., B. Mason, M. Miller, and C. Langdon. 2010. Ocean acidification compromises recruitment success of the threatened Caribbean coral *Acropora palmata*. *Proceedings of the National Academy of Sciences USA* 107(47):20400–20404.

Arvedlund, M., and L. E. Nielsen. 1996. Do the anemonefish *Amphiprion ocellaris* (Pisces: Pomacentridae) imprint themselves to their host sea anemone *Heteractis magnifica* (Anthozoa: Actinidae)? *Ethology* 102:197–211.

Babcock, R. C., P. L. Bull, A. J. Heyward, J. K. Oliver, C. C. Wallace, and B. L. Willis. 1986. Synchronous spawnings of 105 scleractinian coral species on the Great Barrier Reef. *Marine Biology* 90:379–394.

Babcock, R. C., B. L. Willis, and C. J. Simpson. 1994. Mass spawning of corals on a high latitude reef. *Coral Reefs* 13:101–109.

Baker, A. C. 2001. Reef corals bleach to survive change. *Nature* 411:765–766.

Barber, C. V., and V. Pratt. 1998. Poison and profit: Cyanide fishing in the Indo-Pacific. *Environment* 40:5–34.

Barlow, G. W. 1972. The attitude of fish eye-lines in relation to body shape and to stripes and bars. *Copeia* 1:4–12.

Bellwood, D. R. 1995. Direct estimates of bioerosion by two parrot fish species, *Chlorurus gibbus* and *C. sordidus*, on the Great Barrier Reef, Australia. *Marine Biology* 121:419–429.

Birkeland, C. 1982. Terrestrial runoff as a cause of outbreaks of *Acanthaster planci* (Echinodermata: Asteroidea). *Marine Biology* 69:175–185.

Birkeland, C. 1989. The Faustian traits of the crown of thorns starfish. *American Scientist* 77:154–163.

Bjorndal, K. A. 1995. *Biology and Conservation of Sea Turtles*. Washington, DC: Smithsonian Institution Press.

Blair, S. M., T. L. McIntosh, and B. J. Mostkoff. 1994. Impacts of Hurricane Andrew on the offshore reef systems of the central and northern Dade County, Florida. *Bulletin of Marine Science* 54:961–973.

Bolden, S. K. 2000. Long-distance movement of a Nassau grouper (*Epinephelus striatus*) to a spawning aggregation in the central Bahamas. *Fishery Bulletin* 98:642–645.

Brown, B. E., R. P. Dunne, and H. Chansang. 1996. Coral bleaching relative to elevated seawater temperature in the Andaman Sea (Indian Ocean) over the last 50 years. *Coral Reefs* 15:151–152.

Brown, B. E., and J. C. Ogden. 1993. Coral bleaching. *Scientific American* 268:64–70.

Bruckner, A. W., and R. J. Bruckner. 1998. Treating coral disease. *Coastlines* 8(3):10–11.

Buddemeier, R. W., and Fautin, D. G. 1993. Coral bleaching as an adaptive mechanism. *BioScience* 43:320–326.

Chamberlain, J. A. 1978. Mechanical properties of coral skeleton: Compressive strength and its adaptive significance. *Paleobiology* 4:419–435.

Cheshire, A. C., C. R. Wilkinson, S. Seddon, and G. Westphalen. 1997. Bathymetric and seasonal changes in photosynthesis and respiration of the phototrophic sponge *Phyllospongia lamellose* in comparison with respiration by the heterotrophic sponge *Ianthella basta* on Davies Reef, Great Barrier Reef. *Marine and Freshwater Research* 48:589–599.

Clifton, K. E. 1997. Mass spawning by green algae on coral reefs. *Science* 275:1116–1118.

Connell, J. H. 1978. Diversity in tropical rain forests and coral reefs. *Science* 199:1302–1310.

Côté, I. M., S. J. Green, and M. A. Hixon. 2013. Predatory fish invaders: Insights from Indo-Pacific lionfish in the western Atlantic and Caribbean. *Biological Conservation* 164:50–61.

Dana, T. F. 1975. Development of contemporary Eastern Pacific coral reefs. *Marine Biology* 33:355–374.

Darwin, C. 1962. *The Structure and Distribution of Coral Reefs*. Berkeley, CA: University of California Press.

Dollar, S. J. 1982. Wave stress and coral community structure in Hawaii. *Coral Reefs* 1:71–81.

Dubinsky, Z., and P. L. Jokiel. 1994. Ratio of energy and nutrient fluxes regulates symbiosis between zooxanthellae and corals. *Pacific Science* 48:313–324.

Eakin, C. M. 1996. Where have all the carbonates gone? A model comparison of calcium carbonate budgets before and after the 1982–1983 El Niño at Uva Island in the eastern Pacific. *Coral Reefs* 15:109–119.

Earle, S. A. 1991. Sharks, squids, and horseshoe crabs—the significance of marine biodiversity. *BioScience* 41:506–509.

Edmunds, P. J. 2000. Recruitment of scleractinians onto skeletons of corals killed by black band disease. *Coral Reefs* 19:69–74.

Edmunds, P. J., and R. C. Carpenter. 2001. Recovery of *Diadema antillarum* reduces macroalgal cover and increases abundance of juvenile corals on a Caribbean reef. *Proceedings of the National Academy of Sciences USA* 98:5067–5071.

Fadlallah, Y. H. 1983. Sexual reproduction, development and larval biology in scleractinian corals: A review. *Coral Reefs* 2:129–150.

Falkowski, P. G., Z. Dubinsky, L. Muscatine, and L. McCloskey. 1993. Population control in symbiotic corals. *BioScience* 43:606–611.

Falkowski, P. G., Z. Dubinsky, L. Muscatine, and J. W. Porter. 1984. Light and the bioenergetics of a symbiotic coral. *BioScience* 34:705–709.

Fankboner, P. V. 1971. Intracellular digestion of symbiotic zooxanthellae by host amoebocytes in giant clams (Bivalvia: Tridachnidae), with a note on the nutritional role of the hypertrophied siphonal epidermis. *Biological Bulletin* 141:222–234.

Fonseca, M. S., J. C. Zieman, G. W. Thayer, and J. S. Fisher. 1983. The role of current velocity in structuring eelgrass (*Zostera marina* L.) meadows. *Estuarine and Coastal Shelf Science* 17:367–380.

Gardner, T. A., I. M. Côté, J. A. Gill, A. Grant, and A. R. Watkinson. 2003. Long-term region-wide declines in Caribbean corals. *Science* 301:958–960.

Garrison, V. H., E. A. Shinn, W. T. Foreman, D. W. Griffin, C. W. Holmes, C. A. Kellogg, M. S. Majewski, L. L. Richardson, K. B. Ritchie, and G. W. Smith. 2003. African and Asian dust: From desert soils to coral reefs. *BioScience* 53:469–480.

Gittings, S. R., G. S. Boland, K. Deslarzes, and T. J. Bright. 1992. Mass spawning and reproductive viability of reef corals at the East Flower Garden Bank, northwest Gulf of Mexico. *Bulletin of Marine Science* 51:420–428.

Gleason, D. F., and G. M. Wellington. 1993. Ultraviolet radiation and coral bleaching. *Science* 365:836–838.

Glynn, P. W. 1990. El Niño–Southern Oscillation 1982–1983: Nearshore population, community, and ecosystem responses. *Annual Review of Ecology and Systematics* 19:309–345.

Glynn, P. W., and L. D'Croz. 1990. Experimental evidence for high temperature stress as the cause of El Niño-coincident coral mortality. *Coral Reefs* 8:181–191.

Goreau, T. F. 1959. The ecology of Jamaican coral reefs. I. Species composition and zonation. *Ecology* 40:67–90.

Goreau, T. F. 1990. Coral bleaching in Jamaica. *Nature* 343:417.

Goreau, T. F., and N. I. Goreau. 1959. The physiology of skeleton formation in corals. II. Calcium deposition by hermatypic corals under various conditions in the reef. *Biological Bulletin* 117:239–250.

Goreau, T. F., N. I. Goreau, and C. M. Yonge. 1971. Reef corals: Autotrophs or heterotrophs? *Biological Bulletin* 141:247–260.

Grigg, R. W. 1982. Darwin Point: A threshold for atoll formation. *Coral Reefs* 1:29–34.

Halstead, B. W. 1988. *Poisonous and Venomous Marine Animals of the World*. Burbank, CA: Darwin Publications.

Halstead, B. W., P. S. Auerbach, and D. R. Campbell. 1990. *A Colour Atlas of Dangerous Marine Animals*. Boca Raton, FL: CRC Press.

Harrison, P. L., R. C. Babcock, G. D. Bull, J. K. Oliver, C. C. Wallace, and B. L. Willis. 1984. Mass spawning in tropical reef corals. *Science* 223:1186–1189.

Harvelt, C. D., K. Kim, J. M. Burkholder, R. R. Colwell, P. R. Epstein, D. J. Grimes, E. E. Hofmann, E. K. Lipp, A. D. M. E. Osterhaus, R. M. Overstreet, J. W. Porter, G. W. Smith, and G. R. Vasta. 1999. Emerging marine diseases—climate links and anthropogenic factors. *Science* 285:1505–1510.

Hatcher, B. G. 1990. Coral reef primary productivity: a hierarchy of pattern and process. *Trends in Ecology and Evolution* 5:149–155.

Hawkins, J. P., C. M. Roberts, and T. Adamson. 1991. Effects of a phosphate ship grounding on a Red Sea coral reef. *Marine Pollution Bulletin* 22:538–542.

Hemminga, M. A., and C. A. Duarte. 2001. *Seagrass Ecology*. Cambridge, UK: Cambridge University Press.

Hemminga, M. A., P. G. Harrison, and F. van Lent. 1991. The balance of nutrient losses and gains in seagrass meadows. *Marine Ecology Progress Series* 71:85–96.

Highsmith, R. C. 1982. Reproduction by fragmentation in corals. *Marine Ecology Progress Series* 7:207–226.

Hillis-Colinvaux, L. 1986. Historical perspectives on algae and reefs: Have reefs been misnamed? *Oceanus* 29:43–48.

Hughes, T. P. 1994. Catastrophes, phase shifts, and large-scale degradation of a Caribbean coral reef. *Science* 265:1547–1550.

Hughes, T. P., A. H. Baird, D. R. Bellwood, M. Card, S. R. Connolly, C. Folke, R. Grosberg, O. Hoegh-Guldberg, J. B. C. Jackson, J. Kleypas, J. M. Lough, P. Marshall, M. Nyström, S. R. Palumbi, J. M. Pandolfi, B. Rosen, J. Roughgarden. 2003. Climate change, human impacts, and the resilience of coral reefs. *Science* 301:929–933.

Hutchings, P. A. 1986. Biological destruction of coral reefs: A review. *Coral Reefs* 4:239–252.

Irlandi, E. A., and C. H. Peterson. 1991. Modification of animal habitat by large plants: Mechanisms by which sea grasses influence clam growth. *Oecologia* 87:307–318.

Jackson, J. B. C. 1991. Adaptation and diversity of reef corals. *BioScience* 41:475–482.

Jackson, J. B. C. 1997. Reefs Since Columbus. *Coral Reefs* 16:S23–S32.

Jackson, J. B. C., M. K. Donovan, K. L. Cramer, and V. V. Lam, eds. 2014. *Status and Trends of Caribbean Coral Reefs: 1970–2012*. Gland, Switzerland: Global Coral Reef Monitoring Network, IUCN.

Jackson, J. B. C., and T. P. Hughes. 1985. Adaptive strategies of coral-reef invertebrates. *American Scientist* 73:265–274.

Johannes, R. E., W. J. Wiebe, C. J. Crossland, D. W. Rimmer, and S. V. Smith. 1983. Latitudinal limits of coral reef growth. *Marine Ecology Progress Series* 11:105–111.

Jones, G. P. 1990. The importance of recruitment to the dynamics of a coral reef fish population. *Ecology* 71:1691–1698.

Jones, O. A., and R. Endean, eds. 1976. *Biology and Geology of Coral Reefs*. Vols. I and II. New York: Academic Press.

Kinzie, R. A., III. 1993. Effects of ambient levels of solar ultraviolet radiation on zooxanthellae and photosynthesis of the reef coral *Montipora verrucosa*. *Marine Biology* 116:319–327.

Klumpp, D. W., B. L. Bayne, and A. J. S. Hawkins. 1992. Nutrition of the giant clam *Tridacna gigas* (L.). I. Contribution of filter feeding and photosynthesis to respiration and growth. *Journal of Experimental Marine Biology and Ecology* 155:105–122.

Knowlton, N. 2001. Sea urchin recovery from mass mortality: New hope for coral reefs? *Proceedings of the National Academy of Sciences USA* 98:4822–4824.

Kramarsky-Winter, E., M. Fine, and Y. Loya. 1997. Coral polyp expulsion. *Nature* 387:137.

Lane, D. J. W. 1996. A crown-of-thorns outbreak in the eastern Indonesian Archipelago, February 1996. *Coral Reefs* 15:209–210.

Lema, K., B. L. Willis, and D. G. Bourne. 2012. Corals form characteristic associations with symbiotic nitrogen-fixing bacteria. *Applied Environmental Microbiology* 78(9):3136–3144.

Lessios, H. A. 1988. Mass mortality of *Diadema antillarum* in the Caribbean: What have we learned? *Annual Review of Ecology and Systematics* 19:371–393.

Losey, G. S., Jr. 1972. The ecological importance of cleaning symbiosis. *Copeia* 4:820–833.

Mariscal, R. N. 1972. Behavior of symbiotic fishes and sea anemones. In: H. E. Winn and B. L. Olla, eds. *Behavior of Marine Animals*. New York: Plenum.

Marshall, A. T. 1996. Calcification in hermatypic and ahermatypic corals. *Science* 271:637–639.

Miller, M. W., and M. E. Hay. 1998. Effects of fish predation and seaweed competition on the survival and growth of corals. *Oecologia* 113:231–238.

Morris, J. A., and P. Whitfield. 2009. Biology, Ecology, Control and Management of the Invasive Indo-Pacific Lionfish: An Updated Integrated Assessment. NOAA Technical Memorandum NOS NCCOS 99.

Munoz, R. C., and R. R. Warner. 2003. Alternative contexts of sex change with social control in the bucktooth parrotfish, *Sparisoma radians*. *Environmental Biology of Fishes* 68(3):307–319.

Muscatine, L. 1980. Productivity of zooxanthellae. In: P. G. Falkowski, ed. *Primary Productivity in the Sea*. New York: Plenum. pp. 381–402.

Muscatine, L., and J. W. Porter. 1977. Reef corals: Mutualistic symbiosis adapted to nutrient-poor environments. *BioScience* 27:454–460.

Odum, H. T., and E. P. Odum. 1955. Trophic structure and productivity of a windward coral reef community on Eniwetok Atoll. *Ecological Monographs* 25:291–320.

Ogden, J. C., J. W. Porter, N. P. Smith, A. Szmant, W. Jaap, and D. Forcucci. 1994. A long-term interdisciplinary study of the Florida Keys seascape. *Bulletin of Marine Science* 54:1059–1071.

Oliver, J., and R. Babcock. 1992. Aspects of the fertilization ecology of broadcast spawning corals: Sperm dilution effects and in situ measurements of fertilization. *Biological Bulletin* 183:409–417.

Pandolfi, J. M. 2003. Global trajectories of the long-term decline of coral reef ecosystems. *Science* 301:955–958.

Patterson, K. L., J. W. Porter, K. B. Ritchie, W. W. Polson, E. Mueller, E. C. Peters, D. L. Santavy, and G. W. Smith. 2002. The etiology of white pox, a lethal disease of the Caribbean elkhorn coral, *Acropora palmata*. *Proceedings of the National Academy of Sciences USA* 99:8725–8730.

Porter, J. W. 1976. Autotrophy, heterotrophy, and resource partitioning in Caribbean reef-building corals. *American Naturalist* 110:731–742.

Porter, J. W., P. Dustan, W. C. Jaap, K. L. Patterson, V. Kosmynin, O. W. Meier, M. E. Patterson, and M. Parsons. 2001. Patterns of spread of coral diseases in the Florida Keys. *Hydrobiologia* 460:1–24.

Porter, J. W., and O. W. Meier. 1992. Quantification of loss and change in Floridian reef coral populations. *American Zoologist* 32:625–640.

Poulin, R., and A. S. Grutter. 1996. Cleaning symbioses: Proximate and adaptive explanations. *BioScience* 46:512–516

Reaka-Kudla, M. L., J. S. Feingold, and P. W. Glynn. 1996. Experimental studies of rapid bioerosion of coral reefs in the Galápagos Islands. *Coral Reefs* 15:101–107.

Richardson, L. L, W. M. Goldberg, K. G. Kuta, R. B. Aronson, G. W. Smith, K. B. Ritchie, J. C. Halas, J. S. Feingold, and S. L. Miller. 1998. Florida's mystery coral-killer identified. *Nature* 392:557–558.

Richmond, R. H. 1985. Reversible metamorphosis in coral planula larvae. *Marine Ecology Progress Series* 22:181–185.

Riegl, B., C. Heine, and G. M. Branch. 1996. Function of funnel-shaped coral growth in a high-sedimentation environment. *Marine Ecology Progress Series* 145:87–93.

Robertson, D. R. 1972. Social control of sex reversal in a coral-reef fish. *Science* 177:1007–1009.

Rocha, L., C. R. Rocha, C. C. Baldwin, L. A. Weight, and M. McField. 2015. Invasive lionfish preying on critically endangered reef fish. *Coral Reefs* 34(3):803–806.

Rogers, C. S. 1983. Sublethal and lethal effects of sediments applied to common Caribbean reef corals in the field. *Marine Pollution Bulletin* 14:378–382.

Rothans, T. C., and A. C. Miller. 1991. A link between biologically imported particulate organic nutrients and the detritus food web in reef communities. *Marine Biology* 110:145–150.

Rougerie, F., and B. Wauthy. 1993. The endo-upwelling concept: From geothermal convection to reef construction. *Coral Reefs* 12:19–30.

Rowan, R. D. A. P. 1991. A molecular genetic classification of zooxanthellae and the evolution of animal–algal symbioses. *Science* 251:1348–1351.

Rützler, K., D. L. Santavy, and A. Antonius. 1983. The black band disease of Atlantic reef corals. III. Distribution, ecology, and development. *Marine Ecology* 4:329–358.

Sale, P. F. 1974. Mechanisms of coexistence in a guild of territorial reef fishes. *Marine Biology* 29:89–97.

Sale, P. F., ed. 2002. *Coral Reef Fishes: Dynamics and Diversity in a Complex Ecosystem*. San Diego: Academic Press.

Schlichter, D., H. Kampmann, and S. Conrady. 1997. Trophic potential and photoecology of endolithic algae living within coral skeletons. *Marine Ecology* 18:299–317.

Schuhmacher, H., and H. Zibrowius. 1985. What is hermatypic? *Coral Reefs* 4:1–9.

Scott, R. D., and H. R. Jitts. 1977. Photosynthesis of phytoplankton and zooxanthellae on a coral reef. *Marine Biology* 41:307–315.

Shapiro, D.Y. 1987. Differentiation and evolution of sex change in fishes. *BioScience* 37:490–497.

Shashar, N., Y. Cohen, Y. Loya, and N. Sar. 1994. Nitrogen fixation (acetylene reduction) in stony corals: Evidence for coral-bacteria interactions. *Marine Ecology Progress Series* 111:259–264.

Shlesinger, Y., and Y. Loya. 1985. Coral community reproductive patterns: Red Sea versus the Great Barrier Reef. *Science* 228:1333–1335.

Sorokin, Y. I. 1972. Bacteria as food for coral reef fauna. *Oceanology* 12:169–177.

Sponaugle, S., T. Lee, V. Kourafalou, and D. Pinkard. 2005. Florida Current frontal eddies and the settlement of coral reef fishes. *Limnology and Oceanography* 50:1033–1048.

Sponaugle, S., K. Grorud-Colvert, and D. Pinkard. 2006. Temperature-mediated variation in early life history traits and recruitment success of the coral reef fish *Thalassoma bifasciatum* in the Florida Keys. *Marine Ecology Progress Series* 308:1–15.

Stafford-Smith, M. G., and R. F. G. Ormond. 1992. Sediment rejection methods of 42 species of Australian scleractinian corals. *Australian Journal of Marine and Freshwater Research* 43:683–705.

Stat, M., E. Morris, and R. D. Gates. 2008. Functional diversity in coral–dinoflagellate symbiosis. *Publication of the National Academy of Sciences USA* 105:9256–9261.

Stoddart, D. R. 1973. Coral reefs: The last two million years. *Geography* 58:313–323.

Sweet, M. J., A. Croquer, and J. C. Bythell. 2014. Experimental antibiotic treatment identifies potential pathogens of white band disease in the endangered Caribbean coral *Acropora cervicornis*. *Proceedings of the Royal Society B* 281(1788).

Thayer, G. W., D. W. Engel, and K. A. Bjorndal. 1982. Evidence for short-circuiting of the detritus cycle of seagrass beds by the green turtle, *Chelonia mydas*. *Journal of Experimental Marine Biology and Ecology* 62:173–183.

Thresher, R. E. 1984. *Reproduction in Reef Fishes*. Neptune City, NJ: T.H.F. Publications.

Tribble, G. W., M. J. Atkinson, F. J. Sansone, and S. V. Smith. 1994. Reef metabolism and endo-upwelling in perspective. *Coral Reefs* 13:199–201.

Vernon, J. E. N. 1995. *Corals in Space and Time*. Sydney: University of New South Wales Press.

Walbran, P. D., R. A. Henderson, A. J. T. Jull, and M. J. Head. 1989. Evidence from sediments of long-term *Acanthaster planci* predation on corals of the Great Barrier Reef. *Science* 245:847–850.

Warner, R. R. 1984. Mating behavior and hermaphroditism in coral reef fishes. *American Scientist* 72:128–136.

Warner, R. R. 1990. Male versus female influences on mating-site determination in a coral reef fish. *Animal Behavior* 39:540–548.

Wilkinson, C. R. 1987. Interocean differences in size and nutrition of coral reef sponge populations. *Science* 236:1654–1657.

Wilson, J. R., and P. L. Harrison. 1998. Settlement-competency periods of larvae of three species of scleractinians. *Marine Biology* 131:339–345.

Yamamuro, M., H. Kayanne, and M. Minagawa. 1995. Carbon and nitrogen stable isotopes of primary producers in coral reef ecosystems. *Limnology and Oceanography* 40:617–621.

The Open Sea

STUDENT LEARNING OUTCOMES

1. Describe the various types of life forms that inhabit the open ocean.

2. Explain the different geographic patterns of distribution for pelagic organisms.

3. Summarize the unique adaptations exhibited by organisms found in the deeper zones of the ocean.

4. Explain why organisms of the open ocean migrate vertically in predictable patterns.

5. Describe the various methods pelagic organisms have for buoyancy.

6. Summarize the methods pelagic organisms use for orienting in the sea.

CHAPTER OUTLINE

A mixture of captured zooplankton, meroplankton, and holoplankton, collected in the Straits of Florida.
Courtesy of Cedric Guigand, Rosenstiel School of Marine & Atmospheric Science, University of Miami.

Life in the pelagic division of the marine environment exists in a vast three-dimensional medium with generally low nutrient availability. Microscopic bacteria and protists, the major groups of small herbivores, and many larger predators live in near-surface waters. The distribution of pelagic animals reflects their dependency on the primary producers of the sea for food. Away from continental shelves, reefs, and coastal upwelling, primary production rates tend to be low and phytoplankton blooms are patchy. As a result, pelagic animals congregate in or near the photic zone, and they typically are less abundant than animal populations along the coasts of temperate or tropical seas. At greater depths, population densities diminish rapidly, but animal life never completely disappears, even in the deepest parts of the ocean.

The pelagic division is a realm that presents few obvious ecological niches for its inhabitants, as conditions appear uniform over large areas. The small local variations that do exist in water temperature and chemical characteristics tend to be smoothed out by turbulent mixing and diffusion processes, and thus variations in light intensity, water temperature, and food availability change only on large horizontal scales. Despite the appearance of a mundane environment, some phenomenal animals are adapted to survive and thrive in the open sea during various stages of their life histories. In this chapter, we will explore some of these animals, their distributions along vertical and horizontal planes, and some of the critical adaptations they possess for orienting and maintaining positions in desired locations in the open sea.

12.1   Inhabitants of the Pelagic Division

The pelagic division of the sea is home to two major ecological groups of marine animals: zooplankton and nekton. Zooplankton are represented by **holoplankton** (**Figure 12.1**), the permanent planktonic forms, and temporary **meroplankton** (**Figure 12.2**). Meroplankton include larval stages of shallow-water broadcast-spawning invertebrates and fish, and many of these organisms will leave the open ocean once they reach settlement age or size. Most nektonic animals also begin their lives as meroplankton, and as they grow and improve their swimming capabilities they eventually graduate to the status of nekton, remaining in the open ocean. Meroplankton tend to be concentrated in nearshore neritic provinces over continental shelves and near shallow banks, reefs, and estuaries where the adults that produced them live. Meroplankton destined for a nektonic lifestyle upon adulthood may be found further from the coast or other physical structures; the adults are adapted for life in less productive areas or are highly mobile, swimming many kilometers between nearshore and offshore habitats. Their patterns of abundance are strongly related to

Figure 12.1 Mixed holoplankton sample including copepods and ostracods, among other crustaceans.
Courtesy of National Oceanic and Atmospheric Administration.

(a)

(b)

Figure 12.2 (a) An example of a meroplanktonic fish larva, a black grouper, *Mycteroperca bonaci*. (b) An example of a meroplanktonic slipper lobster, family Scyllaridae.
(a) Courtesy of Evan D'Alessandro, University of Miami, Rosenstiel School of Marine & Atmospheric Science; (b) Courtesy of Evan D'Alessandro, University of Miami, Rosenstiel School of Marine & Atmospheric Science.

Figure 12.3 A pelagic sea cucumber found swimming in approximately 3,200 m deep water.
Courtesy of National Oceanic and Atmospheric Administration *Okeanos Explorer* Program, INDEX-SATAL 2010.

the seasonal distribution and productivity cycles of local phytoplankton communities.

More than 7,000 species of holoplankton have been described. Prominent among these are members of all three nonparasitic protozoan phyla, as well as cnidarians, ctenophores, chaetognaths, crustacean arthropods, and invertebrate chordates. Interestingly, some groups known to be benthic, such as echinoderms, have holoplankton representatives (**Figure 12.3**).

Holoplankton use flotation and buoyancy techniques similar to those found in phytoplankton. Because most holoplankton are characteristically small, they increase their frictional resistance to the water by having high surface area-to-body volume ratios. Structures such as spines, hairs, wings, and other surface extensions also increase frictional resistance to sinking (**Figure 12.4**). Unicellular microscopic protists are so small that they have great difficulty in overcoming the viscous forces between water molecules by swimming. They experience almost no glide in their microscopic world; when they stop swimming, they instantly stop moving, and thus these microscopic cells must swim continuously if they are to move at all.

A large variety of gelatinous zooplankton exist, including numerous jellyfish medusae, siphonophores, pelagic mollusks (**Figure 12.5a**), ctenophores (**Figure 12.5b**), and tunicates. The tunicates include barrel-shaped salps, with life cycles alternating between solitary individuals and colonial clones (**Figure 12.5c**), as well as the smaller appendicularians. Because the bodies of all of these gelatinous species are isotonic to seawater, their body densities are very close to that of seawater, and buoyancy is not a problem. Their small proportion of organic material accounts for their low metabolic rates and for their potentially large body sizes, which range from about 1 mm to several meters. Their relatively large sizes in comparison with most other zooplankton and nearly transparent appearance in water offer some protection from larger pelagic predators, such as sea turtles.

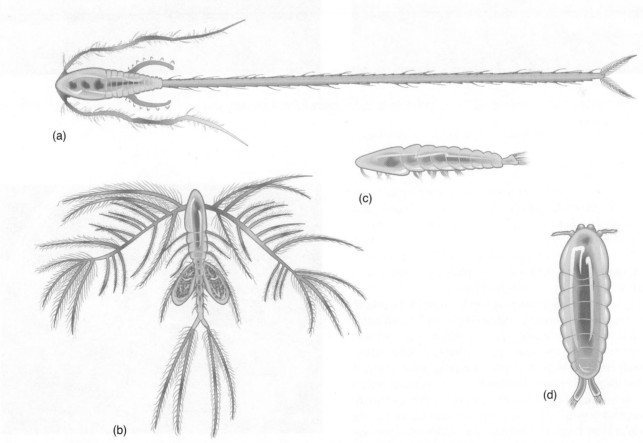

(a)

(b)

(c)

(d)

Figure 12.4 Some planktonic copepods exhibiting structural adaptions for floatation: (a) *Aegisthus*; (b) *Oithona*; (c) side and (d) top views of *Sapphirina*.
Data from H. U. Sverdrup. *The Oceans: Their Physics, Chemistry, and Biology*. Prentice Hall, 1970.

(a)

(c)

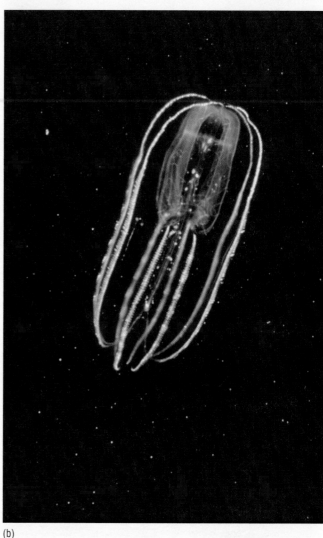

(b)

Figure 12.5 Some large gelatinous zooplankton. (a) A pelagic mollusk, *Corolla*. (b) A ctenophore, *Bolinopsis*, swimming with eight rows of ciliated combs. (c) A colony of salps (*Pegea*) cloned from a single parent.

(a) © David Wrobel/Visuals Unlimited; (b) Courtesy of OAR/National Undersea Research Program/NOAA; (c) © Eric Prine/age fotostock.

Crustaceans are the most numerous and widespread species of holoplankton. Copepods, euphausiids, amphipods, and decapods all contribute substantially to near-surface plankton communities. Calanoid copepods, such as *Calanus*, account for the bulk of herbivorous zooplankton in the 1- to 5-mm size range. Euphausiids are the giants of the planktonic crustaceans, yet they seldom exceed 5 cm (**Figure 12.6**).

The number of holoplanktonic species living in the pelagic province of the world ocean is matched by a nearly equal number of nektonic species roaming the same waters. These animals represent most of the taxonomic groups that have achieved the large body sizes and well-developed swimming powers needed to exploit the pelagic realm. Absolute body size is crucial; once a well-muscled animal exceeds a few centimeters in body length, the viscous forces of water that limit continuous swimming by zooplankton begin to diminish, and efficient swimming becomes possible. Kilometer-scale distances can be covered in minutes rather than hours or days, and horizontal migrations

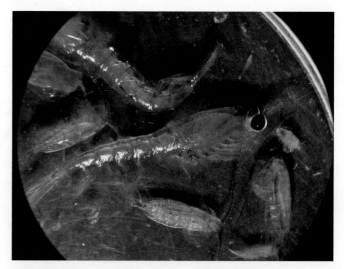

Figure 12.6 Calanoid copepods and euphausiids (krill).

Courtesy of National Oceanic and Atmospheric Administration.

to improve conditions for survival become possible and even feasible.

Most nekton are vertebrates, and most marine vertebrates are teleost fishes. Of the numerous groups of marine invertebrates that live in the sea, only squids and a few species of shrimps are truly nektonic. In some regions, vast numbers of small squids less than 1 m in length form important intermediate links in pelagic food webs (**Figure 12.7**). The giant squid,

Architeuthis, lives at greater depths. These large animals occasionally reach 18 m in length and weights of about 2 tons, with tentacles approximately 12 m long and eyes as large as soccer balls.

DID YOU KNOW?

Some nektonic fish transit thousands of kilometers in a year. Tagging studies of Atlantic bluefin tuna (*Thunnus thynnus*) revealed that this species makes transatlantic migrations annually to the Gulf of Mexico, the eastern Atlantic, or the Mediterranean Sea. Some individual fish even show site fidelity, returning to the same spawning areas each year. This type of very long distance migration was once thought to be limited to animals such as sea turtles and marine mammals, but now we know that some fish species can return to the same sites at distant locations repeatedly, too.

Figure 12.7   One of the common squids found off the coast of southern California, *Loligo opalescens*, the market squid.

Courtesy of National Oceanic and Atmospheric Administration.

Case Study

The Longest Fish in the Sea

Common name: Oarfish

Scientific name: *Regalecus glesne* (several species exist)

Distribution: *Regalecus glesne* is found around the globe from latitudes of 72° N to 52° S.

One theme that applies to many animals of the mesopelagic and deeper open-ocean realm is that we know very little about their biology, behavior, or life histories. In order to study an organism, scientists must encounter them, and many mesopelagic organisms are extremely difficult to find except for rare sightings and captures. The oarfish is no exception; with elusive behaviors and only rare occurrences in shallow or surface waters, we know very little about this fish. Several videos recorded by submersibles and information gleaned from specimens washed ashore have provided glimpses into the lives of oarfish, and they appear to live a very interesting existence with unique adaptations for survival.

Current evidence indicates that oarfish live in pelagic oceanic waters down to 1,000 m. Their overall body design is very odd, with a large head; slender and elongated body; and long, thin fins (**Figure A**). Their pectoral fins contain light-emitting structures, leading to a curious sight as they swim through the darkness. Individuals are known to reach lengths of up to 8 m, making them the longest fish species (**Figure B**). Several specimens washed ashore were missing a portion of the posterior portions of their bodies and also showed signs of regrowth of the tissue. Given that this state has been documented for numerous specimens, it appears that oarfish can self-amputate a portion of their body without suffering damage to any organs. Because oarfish are so long, it is likely that they will resort to such self-amputation in order to survive if a predator bites the tail region, losing a section of the body that is not vital for survival.

Figure A   An oarfish washed up on a southern California beach.
Courtesy of Katia Cao.

Evidence from oarfish specimens indicates that several of these self-amputation events may take place in a lifetime.

Adult specimens washed ashore appear to kill themselves by swimming out of the water onto shore. It is unclear why this occurs, but it may be due to illness, an inability to swim against strong currents that take them to shore, or simply old age. Between 2013 and 2015, several specimens washed up on beaches in southern California, providing scientists with material to examine and dissect (**Figure C**). These dissections provided new information about oarfish reproductive organs, which are also very long, and their

(continues)

Case Study (*continued*)

Figure B A giant oarfish collected near San Diego, California.
Courtesy of Wm. Leo Smith.

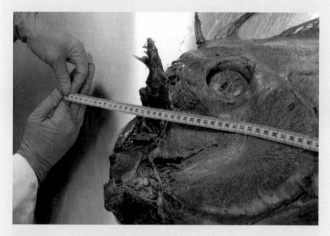

Figure C An oarfish head ready for dissection.
Courtesy of National Oceanic and Atmospheric Administration.

tissue was sent to labs around the world for genetic analysis. Examination of the head confirmed that oarfish lack teeth and must filter the water for food.

Videos of oarfish swimming provided a monumental surprise: oarfish orient their bodies vertically while using their long, slender dorsal fin to maintain their position in the water column. It was previously assumed that oarfish swam like most other fish, oriented horizontally, so footage of their odd vertical swimming behavior has given scientists something to ponder.

Although resembling a sea serpent at first glance, oarfish are actually highly evolved filter feeders that appear to live quite uneventful lives in the darkness of the mesopelagic zone. Their ecological role is unknown, and their morphology and swimming behaviors are unique. This is yet another example of how little we know about the deeper portions of the oceans, and how much more there is to discover.

Critical Thinking Questions

1. It has been hypothesized that oarfish lose a portion of their tails when attacked by a predator. Devise your own alternative hypothesis for this behavior.

2. Based on the information presented in the Case Study and upon observing images of oarfish, devise a research question for this species and provide a plan to test your hypothesis.

12.2   Geographic Patterns of Distribution

Away from the influences of continental borders, life in the upper 200 m of the world ocean drifts along in the large semienclosed current gyres found around the globe. This upper layer of the oceanic province, the epipelagic zone, is approximately coincident with the photic zone where phytoplankton can exist. In marked contrast to the numerous life zones available to animals on the sea bottom, the **epipelagic zone** is partitioned into only a few major habitats reflecting the major marine climatic zones. Each major epipelagic habitat is broadly defined by its own unique combination of water temperature and salinity characteristics and is occupied by a suite of species that, over long periods of time, have adapted to that set of environmental conditions.

These geographic patterns of distribution are nicely illustrated by six closely related species of planktonic euphausiids (**Figure 12.8**) yet are indicative of the general large-scale distribution of many other epipelagic animal species. Upon examining Figure 12.8 a few patterns are apparent. First, there is a tendency for each species to be distributed in a broad latitudinal band across one or more oceans. Well-defined patterns of tropical (*Euphausia diomedeae*), subtropical (*E. brevis*), and south polar (*E. superba*) distribution are evident. Some species, such as *E. diomedeae* and *E. brevis*, are broadly tolerant to their environmental regimes and occupy wide latitudinal bands. Other species (*E. longirostris*, *Thysanoessa gregaria*, and *E. superba*) occupy narrower latitudinal ranges. Similar regimes in both the Northern and Southern Hemispheres are frequently inhabited by the same species. The subtropical *E. brevis* and

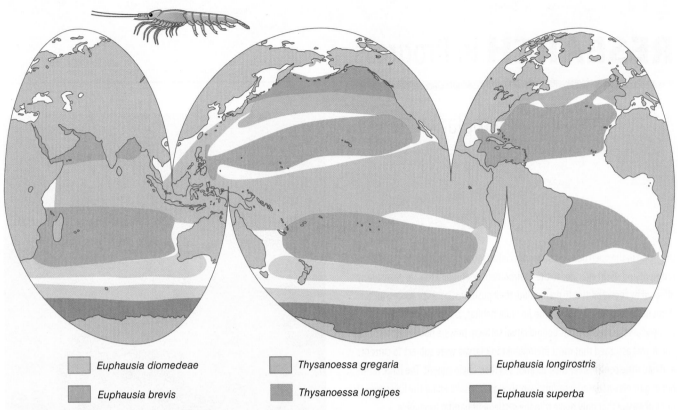

Figure 12.8 The global distribution of six species of epipelagic euphausiids.

the temperate-water *T. gregaria* exhibit this **antitropical distribution**. Often these lower-latitude species extend into all three major ocean basins (*E. brevis* and *T. gregaria*), but occasionally they do not (*E. diomedeae* is absent from the tropical Atlantic). High-latitude species in the Southern Hemisphere (*E. longirostris* and *E. superba*) also extend around the globe, aided by extensive oceanic connections between Antarctica and the other southern continents. Similar circumglobal distributions are more difficult to accomplish in the higher latitudes of the Northern Hemisphere (*T. longipes* is present only in the North Pacific).

The boundaries of these zones overlap slightly, but analyses of the distribution patterns of numerous other species of zooplankton confirm that these boundaries define, in a very real way, the major epipelagic habitats of the oceanic province; however, smaller-scale variations of environmental features do exist within these major habitats and do influence the structure of pelagic communities by contributing additional texture to the physical–chemical terrain, creating potential niches for occupation. Zooplankton, like phytoplankton, have patchy distributions at virtually all levels of sampling, from several kilometers down to microscopic distances. Some of this patchiness develops from the local effects of grazing or predation, some from responses to chemical or physical gradients, and some from the social responses of planktonic species to each other. At times, some zooplankton aggregate and some avoid one another. Zooplankton social behavior depends on a variety of factors, including whether a scenario calls for a "safety in numbers" approach to avoid predation or a more solitary existence while competing for a limited resource such as food.

The high primary production that occurs around the Antarctic continent during the short polar summer supports a trophic system quite unlike any found in northern latitudes. Essentially, all Antarctic communities depend directly or indirectly on one species of dominant pelagic herbivore, *E. superba* (**Figure 12.9**). This 6-cm-long, 1-g, superb euphausiid is the largest of all euphausiid species (collectively, these crustaceans

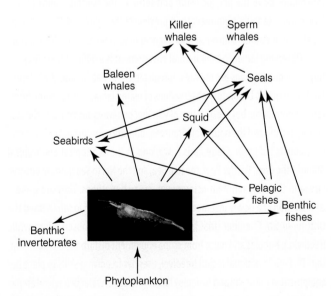

Figure 12.9 Antarctic food web with krill, *Euphausia superba*, occupying a central position on the second trophic level.

RESEARCH in Progress

A Day in the Life of Meroplankton

Orienting in the open ocean requires special adaptations for detecting the environment and slight changes to the environment. Meroplankton are particularly affected by their abilities to manipulate their positions, because successful settlement to their next habitat, the juvenile habitat, is crucial for survival. For many years the mechanisms behind larval settlement were poorly understood, and it was assumed that most meroplanktonic larvae were subject to currents and had little control over their locations in their environment. The large number of gametes released by adults was thought to make up for the low survival rates of larvae that only make it to the settlement habitat by chance.

Research efforts have been focused on larval settlement in an attempt to tease out cues for this dramatic shift in habitat use. Environmental cues such as lunar phase, tidal cycle, water temperature, scent, and salinity have now been correlated with the timing of settlement events for numerous species. Settlement often has an age and/or size requirement, and once one or both of these requirements have been met larvae use one or more of the environmental cues listed above to time their transition from the meroplankton to the benthos. Now that it has been established that a variety of environmental cues influence settlement, several questions remain: (1) how do larvae with limited swimming capabilities leave the pelagic realm and settle to the benthic realm? (2) are larvae more capable swimmers than previously thought? and (3) are larvae capable of orienting to increase their chances of settling to a suitable habitat?

Observing larvae in their natural environment is a difficult task. They are tiny, their distributions are mainly unpredictable and patchy, and their depths range drastically. Laboratory observations of reared larvae provide some interesting information, but the question of whether behaviors are similar in the laboratory setting and the natural environment introduces uncertainty. One group of scientists led by Dr. Claire Paris is working in the subtropical and tropical Atlantic, the Great Barrier Reef, the Red Sea, and the North Sea fjords to observe larvae in their environment using novel research methods. They are using a combination of a newly invented floating laboratory placed *in situ* (named the Drifting *In Situ* Chamber [DISC]; **Figure A**) and personal observations while freediving. Bubbles and noise from scuba equipment disturb the environment, thus Dr. Paris, a national record freediver, uses her freediving skills to place her equipment in the ocean and to deploy larvae into underwater behavioral arenas with little disturbance to the animals and the surrounding area (**Figure B**). The DISC allows for tracking of the orientation and behavior of fish larvae in their

Figure A   The DISC device invented by Dr. Claire Paris.

Courtesy of Dr. Claire Paris, Rosenstiel School of Marine & Atmospheric Science, University of Miami. Photo by Robin Faillettaz.

Figure B   Dr. Claire Paris and her field assistant (and husband), Ricardo Paris, recovered an oceanographic sensor by freediving.

Courtesy of Dr. Claire Paris, Rosenstiel School of Marine & Atmospheric Science, University of Miami. Photo by William Trubridge.

natural environment because it drifts with the currents and is made of transparent acrylic and mesh material, keeping it open to sight, sound, odor, and other natural environmental fluctuations that may exist. Environmental sensors and a camera that operates all hours of the day and night are mounted on the DISC to record environmental parameters and swimming behaviors (**Figure C**).

The results of the DISC floating laboratory experiments are intriguing thus far and appear to coincide with observations of larvae swimming freely outside of a DISC environment. In a study conducted at several sites around the Great Barrier Reef, hundreds of damselfish larvae clearly oriented themselves so that they were swimming in a southerly direction. Their swimming behavior was affected by the amount of sun available, with swimming behaviors more consistent under sunny skies and less so under cloudy skies. Time of day and angle of the sun (i.e., sun azimuth and elevation) also affected swimming behaviors. Results of observations of larvae in the DISC were very similar to those of freedivers making direct observations of free-swimming larvae. The DISC, however, allows cue manipulation to find the mechanisms for orientation. For example, olfactory cues propagating in the open ocean as turbulent ebb plumes help fish larvae to find their way back to isolated atolls.

DISC observations of lobster postlarvae in the Florida Straits revealed clear patterns of swimming direction in which they were keeping a significant bearing. In addition, the lobster postlarvae swam day and night, in contrast to results from a laboratory investigation that indicated swimming during the day only. Swimming orientation was generally against the prevailing northward-flowing Gulf Stream and was adjusted to veer toward the coast during ebb tides; these adjustments in swimming direction allowed them to remain on a shoreward trajectory. Lobster postlarvae also adjusted their swimming direction relative to wind direction, presumably using the wind to help them swim toward the coast.

The results of the DISC studies indicate that although larvae may not be strong swimmers given their small size, they are very capable of controlling their positions in the water to produce their desired end result: finding appropriate settlement habitat. Fish and invertebrate larvae appear to be very sensitive to small changes in their environment and use these changes as cues for orienting, not only for the final settlement act, but during the days or weeks prior to settlement to maintain a suitable location along their journey. This area of research is wide open for new discoveries, as the vast majority of marine animals have a pelagic larval phase, and additional cues for larval orientation certainly exist and have yet to be discovered.

Critical Thinking Questions

1. Do you think that use of the DISC and other similar devices allows for observations of the natural behavior of zooplankton? Why or why not?

2. It appears that some larvae orient better in sunny conditions than in cloudy conditions. Why do you think this is the case?

For Further Reading

Kough, S. A., C. B. Paris, and E. Staaterman. 2014. The *in situ* swimming and orientation behavior of spiny lobster (*Panulirus argus*) post larvae. *Marine Ecology Progress Series* 504:207–219.

Leis, J. M., C. B. Paris, J.-O. Irisson, M. N. Yerman, and U. E. Siebeck. 2014. Orientation of fish larvae *in situ* is consistent among locations, years and methods, but varies with time of day. *Marine Ecology Progress Series* 505:193–208.

Paris, C. B., J. Atema, J.-O. Irisson, M. Kingsford, G. Gerlach, and C. M. Guigand. 2013. Reef odor: A wake up call for navigation in reef fish larvae. *PLoS ONE* 8(8):e72808.

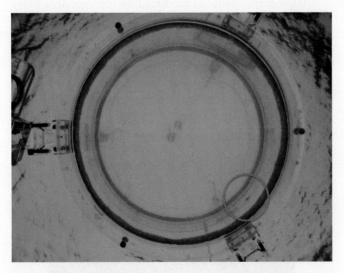

Figure C A snapper swimming and orienting within the DISC device, photographed by GoPro time lapse. The red circle indicates the location of the fish.

Courtesy of Dr. Claire Paris, Rosenstiel School of Marine & Atmospheric Science, University of Miami.

are commonly referred to as *krill*), and it occurs in enormous swarms around much of the Antarctic continent. This herbivore is adaptable in its taste for phytoplankton, shifting between filtering abundant pelagic diatoms and scraping ice algae from the undersides of pack-ice formations. These swarms of *E. superba* represent standing stocks of 200 to 250×10^6 metric tonnes, equivalent to about 2×10^{12} individual animals.

In its central trophic role in Antarctic food webs, *E. superba* is the staple prey for benthic invertebrates, pelagic fishes and squid, seabirds, and several species of marine mammals, including the largest carnivores on Earth, the blue and fin whales. One of the keys to the ecological success and dominance of *E. superba* is its ability to survive several long polar winter nights without food during its 5-year lifetime. Diurnal vertical migration is completely suppressed, and *E. superba* descend to their winter depths of 250 to 500 m and cease feeding. By reducing their metabolic rates during winter, they can survive for more than 7 months without food. During such prolonged periods of no food, these crustaceans metabolize some of their structural body proteins and stored lipids, causing a reduction in body size and eliminating the need to molt and produce new exoskeletal material.

12.3   Vertical Distribution of Pelagic Animals

Although the epipelagic zone accounts for less than 10% of the ocean's volume, most pelagic animals are found there. Most epipelagic nekton are carnivorous predators of the higher trophic levels of pelagic food webs. They are typically large in size when compared with zooplankton, are effective swimmers, and some accomplish impressive feats of long-distance swims to locate food or to improve their chances for successful reproduction. Their ability to move rapidly over appreciable distances tends to erase the sharper distributional boundaries exhibited by zooplankton.

Epipelagic animals of the open ocean seldom exhibit the bright coloration so common in coral reef fishes and invertebrates. Instead, **countershading** is a common pattern of coloration (**Figure 12.10**). Many abundant fish, whales, and squids have dark, often green or blue, pigmentation on their dorsal surfaces, with silvery or white pigmentation on their ventral surfaces. When viewed from above, the pigmented upper surfaces of countershaded fish blend with the darker background below. From beneath, the lighter undersides may be difficult to distinguish from the ambient light coming from the sea surface. From either view, these fish tend to blend visually into, rather than stand out against, their watery background. The functions of countershading may be many, from protecting prey species from visual detection by their predators to hiding predators from their prey for a sneak attack. Flashing of silvery bellies or white side stripes during abrupt turns may alert individuals in a school to the maneuvers of their immediate neighbors so they remain in the school. Finally, skin pigments may have additional structural properties or physiological functions completely independent of the visual appearance of the animal.

Figure 12.10   A manta ray clearly exhibiting countershading with a dark dorsal surface and light ventral surface.
© Petra Christen/Shutterstock, Inc.

Below the sunlit waters of the epipelagic zone lies the mesopelagic zone, a world where animals live in very dim light and depend on primary production from the photic zone above. The **mesopelagic zone** extends from the bottom of the epipelagic zone down to about 1,000 m. Most members of the mesopelagic zone rely totally on the flux of particles from above for food. This downward transport is accelerated by the conversion of nutrients in food to fecal pellets (**Figure 12.11**) and organic aggregates. The fecal pellets of calanoid copepods, for instance, sink about 10 times faster than do the individual phytoplankton cells constituting the pellet.

Macroscopic particles produced by incorporating living and dead material into irregularly shaped organic aggregates are known as **marine snow**. These aggregates, ranging from one to several millimeters in size, are composed of living and dead phytoplankton cells, abundant bacteria, exoskeletons shed by crustaceans, and other detrital material. Sometimes these aggregates also include fecal pellets, and thus the distinction between

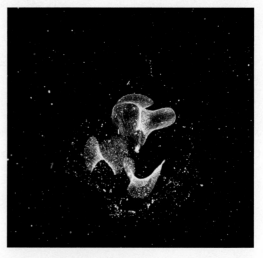

Figure 12.11   The gelatinous home of a larvacean sinks through the mesopelagic zone amid a storm of marine snow (tiny white dots).
Courtesy of Dr. Alice Alldredge, University of California, Santa Barbara.

the two types of particles blurs somewhat. Both types of particles serve as sites of additional aggregation by other members of the plankton community. Bacteria inoculate the particles and initiate processes of decomposition. Dinoflagellates exploit the nutrients released by the activity of the bacteria, and grazing herbivores are attracted by the concentration of energy-rich organic material, both living and dead, and move in for a meal.

Considerably less is known about the biology of mesopelagic animals than about epipelagic animals, and less still is known about animals living below 1,000 m. Fish living in the mesopelagic zone are typically much smaller than fish of the epipelagic zone. Mesopelagic fish seldom exceed 10 cm in length, and many are equipped with well-developed teeth and large mouths (**Figure 12.12a**). Only very dim light penetrates from above, and many species have evolved large eyes sensitive to low light intensities (**Figure 12.12b**) to detect prey and predators alike.

Correlated with large eyes is the presence of **photophores**, light-producing organs most commonly arranged on the ventral surface of the body (Figure 12.12 and **Figure 12.13**). The position and arrangement of photophores suggest several likely functions. The light produced by the ventral photophores usually approximates the intensity and spectral quality of the background light found at the normal daytime depths of these fish.

Figure 12.13 A midwater lanternfish, *Bolinichthys*. Small yellow dots are light-producing photophores.

The light from the photophores may disrupt the visual silhouette of the fish when observed from below by making the fish's silhouette visually blend with the faint background light from above in a manner similar to countershading in near-surface fish. Although the light from photophores is not very bright, it may provide a light source for the fish during the darkest hours of the day so it can orient or forage. Elaborate arrangements of photophores are species specific, suggesting that photophores may also be used for species identification. With little to be seen at these depths except the pattern of photophores, appropriate mate selection may depend on the existence of species-specific patterns of photophores.

Below the mesopelagic zone, light from the surface is so dim that it cannot be detected by the human eye, nor does it stimulate the visual systems of most deep-sea fishes. The light seen at depths below 1,000 m comes largely from photophores. At these depths, photophores are used as lures for prey, as species-recognition signals, and possibly even as lanterns to illuminate small patches of the surrounding blackness. Most fish found at these depths are not vertical migrators. Instead, some depend on the unpredictable sinking of food particles from the more heavily populated and productive waters above. These fish are typically small and have flabby, soft, nearly transparent flesh supported by very thin bones (**Figure 12.14**). Others feed on mesopelagic fish, often engulfing fish that are nearly their own size.

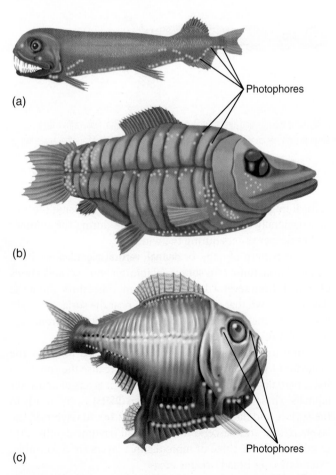

Figure 12.12 Some mesopelagic fishes: (a) loosejaw, *Aristostomias*; (b) barreleye, *Opisthoproctus*; and (c) hatchetfish, *Argyropelecus*. All are 5 to 20 cm in length.

DID YOU KNOW?

The marine snow that is not eaten by grazers, used by phytoplankton, or broken down by bacteria sinks to the bottom of the seafloor forming a carpet of organic material, collectively known as a muddy *ooze*. As much as three quarters of the seafloor around the world is covered by this blanket of ooze, and it is estimated that about 6 m of ooze collects every million years. Although made up of tiny individual particles, collectively marine snow contributes to massive amounts of nutrients in the water column and at the deepest depths of the seafloor.

12.4   Vertical Migration: Tying the Upper Zones Together

Zooplankton and small nekton, such as lanternfish (Figure 12.13), are poor long-distance travelers. They can, however, experience very different environmental conditions by vertically moving modest distances of a few tens of meters. Water

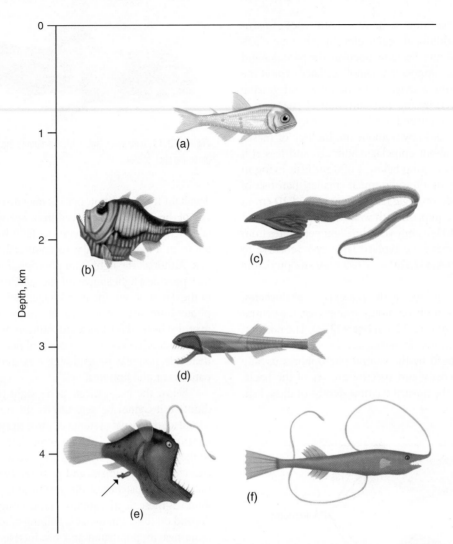

Figure 12.14 A few fishes of the deep sea, shown at their typical depths. Most have reduced bodies, large mouths, and lures to attract prey. (a) A lanternfish, *Bolinichthys*; (b) a hatchetfish, *Argyropelecus*; (c) a gulper eel, *Eurypharynx*; (d) a bristlemouth, *Cyclothone*; (e) a female anglerfish, *Melanocetus*, with an attached male (arrow); and (f) another anglerfish, *Gigantactis*.

temperature, light intensity, pressure, and food availability all change markedly as the distance from the sea surface increases.

The mesopelagic zone offers some distinct advantages when compared with life nearer to the sea surface. Prey species are more difficult to detect by their predators in the dim light. Decreased water temperatures at mid depths lower the metabolic rates and the food and oxygen needed to maintain those rates. The cold water, with its increased density and viscosity, also slows the sinking rates of food particles. The most abundant supply of food particles, however, still resides in the epipelagic zone just above.

To exploit the benefits of both zones better, large numbers of mesopelagic animals periodically migrate upward to feed in near-surface waters. The most common pattern of vertical migration occurs on a daily cycle. At dusk, these midwater animals ascend to the photic zone and feed throughout the night. Before daybreak, they begin migrating to deeper, darker waters to spend the day hidden in the dark. The following evening, the pattern is repeated. In Antarctic waters, the daily pattern of

vertical migration often breaks down out of necessity; the animals generally remain in the photic zone during the summer and in deeper waters during the winter.

This pattern of daily, or **diurnal**, **vertical migration** has been deduced from numerous sources of information. Net collections of animals from several depths at different times throughout the day have shown that more animals are near the surface at night than during the day (**Figure 12.15**). Direct observations from submersible vehicles also support these conclusions.

Ship-mounted sonar devices also are used to study the behavior of vertically migrating animals because the sonic signal is partially reflected by concentrations or layers of midwater animals. These **deep sound-scattering layers (DSSLs)** ascend nearly to the surface and disperse at dusk (**Figure 12.16**). At daybreak, the layers reform and descend to their usual daytime depths (200 to 600 m). Often, three or more distinct layers are discernible, extending over broad oceanic areas.

The species composition of the DSSL is still an unsettled question. Most inhabitants of the mesopelagic zone are too

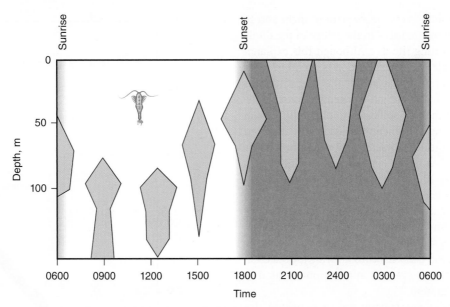

Figure 12.15 A generalized kite diagram of net collections of adult female copepods, *Calanus finmarchicus*, during a complete 1-day vertical migration cycle. The width of each part of each "kite" represents relative numbers of animals. Night hours are shaded.

small or too sparsely distributed to reflect sound signals strongly. Net tows and observations from manned submersibles suggest three groups of animals that cause the deep scattering layers: euphausiids, small fish (primarily lanternfish, Figure 12.13), and siphonophores (**Figure 12.17**). Euphausiids and small fish are often abundant members at depths where the DSSLs occur. The strong echoes of sound pulses may be due, in part, to the resonating qualities of the gas-filled swim bladders of lanternfish and gas-filled floats of siphonophores, known as **pneumatophores**.

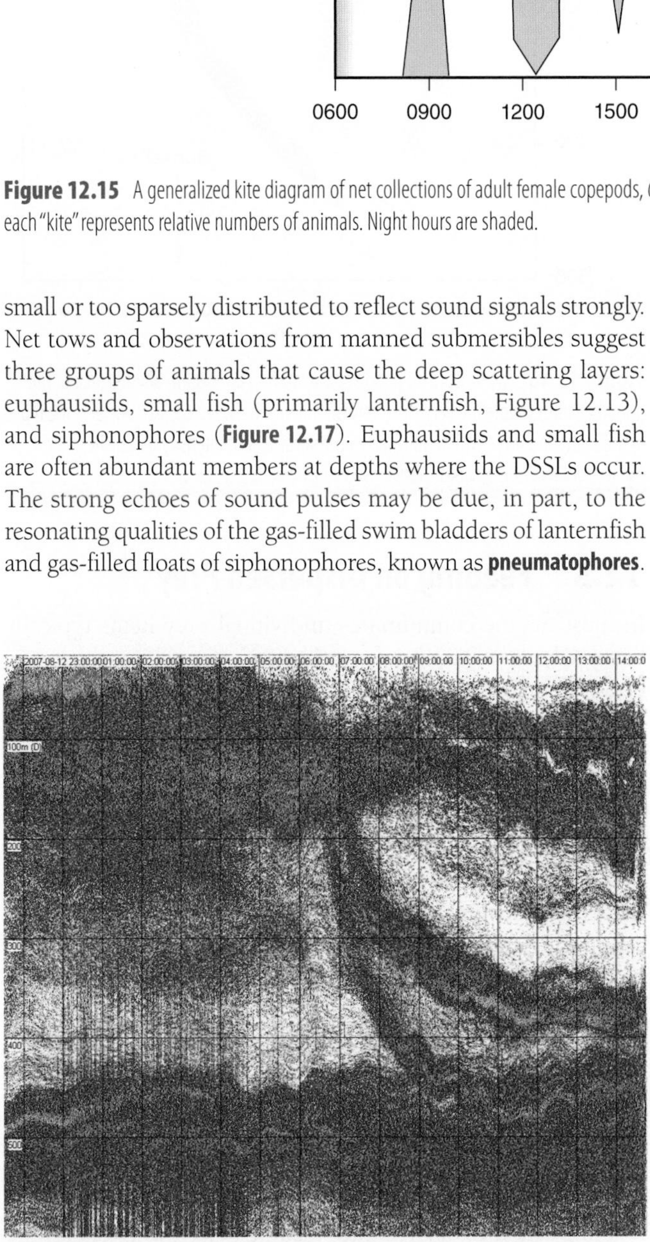

Relatively few swim bladders or pneumatophores are necessary to produce strong echoes at certain sound frequencies. Whatever the composition of the DSSLs, they are merely sound-reflecting indicators of much more extensive vertically migrating assemblages of animals not detected by sonar. Undoubtedly, members of the DSSLs graze on smaller vertical migrators and are, in turn, preyed on by larger fish and squids.

These vertical migrations, only a few hundred meters in extent, occur over short time periods. The copepod *Calanus*, for example, is only a few millimeters long, yet it can swim upward at 15 m/h and descend at 50 m/h. Larger 2-cm euphausiids swim in excess of 100 m/h. If diurnal vertical migrations are foraging trips from below into the productive photic zone, why do these animals descend after feeding? Why do they not remain in the photic zone? Many explanations for the adaptive value of vertical migration have been offered.

The most obvious explanation is that diurnal vertical migration enables animals to capitalize on the more abundant

Figure 12.16 A sonar record of diurnal vertical migration of a mesopelagic community. At night (left side of figure) the community is seen at a depth of about 100 m. At dawn (0700 hours in the center of the figure) the entire community descends to a depth of about 400 m, where it will remain until ascending once again at dusk.
Courtesy of Andrew Brierley, Ph.D., Gatty Marine Laboratory, University of St. Andrews.

Figure 12.17 A midwater siphonophore with brilliant red coloration.
Courtesy of National Oceanic and Atmospheric Administration *Okeanos Explorer* Program.

food resources of the photic zone in the dark of night and to escape visual detection by predators in the refuge of the dimly lit mesopelagic zone during the day. Although this explanation is certainly plausible, vertical migration is useful in other ways, too. Lower water temperatures at the deeper depths reduce an animal's metabolic rate and its energy requirements. The energy conserved may be sufficient to offset the lack of food during the day and the energy expenditures incurred during the actual migration. These energy expenditures are not insignificant. If a 3-cm-long copepod spends its days at 400 m and its nights at 100 m, then it swims a round trip of 600 m, or 20,000 body lengths. This is equivalent to a 6-ft.-tall human swimming 22.7 miles! (Or a 1.8-m tall person swimming 36 km.) Clearly, an energetic benefit must factor into an explanation for diurnal vertical migration, and each of these explanations offers plausible mechanisms favoring the selection of individuals possessing the genetic information required to accomplish vertical migration.

Vertical migrations of zooplankton and small nekton also occur on seasonal time scales. While in the late copepodite stage, *Calanus finmarchicus* spends the winter months of low primary productivity in the North Atlantic at depths near 1,000 m. When the spring diatom bloom develops, it molts to the adult form and begins to rise into the photic zone.

Daily or seasonal changes in light intensity seem to be the most likely stimulus for vertical migrations. Experiments with mixed coastal zooplankton populations have demonstrated that under constant light and temperature conditions some species of copepods maintained their diurnal migratory behavior as a circadian rhythm for several days without relying on external cues such as light intensity. Other species did not migrate and apparently require light or another external stimulus to initiate vertical migration behavior. Electric lights lowered into the water at night and bright moonlight can drive near-surface DSSLs downward. Even midday solar eclipses influence vertical migrators by causing them to move toward the surface for the duration of an eclipse. Collectively, these responses demonstrate a sensitivity to light intensity by natural populations of vertical migrators. Each DSSL follows an **isolume** (a constant light intensity) characteristic of the top of the layer at its normal daytime depth (**Figure 12.18**). As the sunlight intensity decreases in late afternoon, the isolume moves toward the sea surface, and the DSSL follows with a precision seldom seen in natural populations.

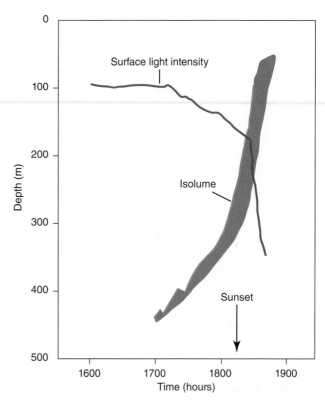

Figure 12.18 The upward migration of a scattering layer (blue-shaded area of the graph) at sunset. Note the very close correspondence between the isolume and the top of the scattering layer, with both rising as the surface light intensity diminishes. Modified from B. P. Boden and E. M. Kampa. *Symp Zool Soc Lond* 19(1967):15–26.

12.5 Feeding on Dispersed Prey

In most marine communities, individual prey items typically become larger in size but fewer in number and biomass at successively higher trophic levels. Two generalized food webs are shown in **Figure 12.19** to illustrate some fundamental differences in trophic linkages between oceanic regions of high and low productivity. Food webs of subtropical waters leading to tuna-sized predators that are low in biomass begin with very small phytoplankton and bacteria and include numerous trophic levels. The Antarctic upwelling system, in contrast, is characterized by relatively large (though still microscopic) primary producers, larger filter-feeding herbivores (the euphausiid krill described previously), extremely large carnivores, and fewer trophic levels.

In tropical and subtropical pelagic environments, small and dispersed prey items insufficient to support large numbers of large predators are the rule. To secure adequate food supplies in this environment, larger pelagic animals either must capture large but widely scattered prey items or be able to harvest very large numbers of smaller, more abundant prey efficiently. The foraging activities of large filter-feeding baleen whales, basking sharks, manta rays, and whale sharks are mostly limited to more productive coastal or high-latitude waters. Predators focused on one particular prey type per feeding event range in size from large tunas, billfish, and dolphins over a meter or two in length to centimeter-long voracious chaetognaths (**Figure 12.20**) and

DID YOU KNOW?

Although most vertically migrating organisms move up to shallower water during the nighttime to feed, there are some notable exceptions. Young Atlantic herring undergo a reverse diel vertical migration, heading into deeper waters at night. It appears that they do not feed at night, so they take advantage of the cooler temperatures found at depth, lowering their metabolic rates. These fish appear to migrate vertically to save energy, not to feed or evade predators.

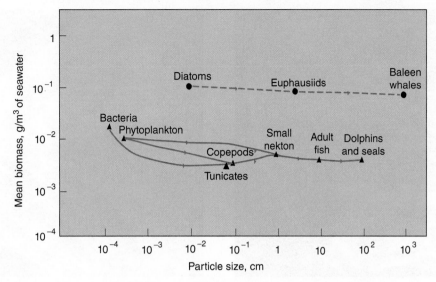

Figure 12.19 The relationship between food particle size and biomass in two pelagic food chains. Note that the biomass in the Antarctic (dashed lines) is about 10 times higher at all trophic levels than those in subtropical gyres (solid lines), and the biomass of each food chain decreases at higher trophic levels.
Redrawn from J. H. Steele. *Oceanus* 25(1980):3–8.

barely visible filter-feeding copepods. Chaetognaths are found throughout the world ocean in numbers sufficient to decimate whole broods of young fish. In addition to their significant role in pelagic food webs, some species of chaetognaths are well-known as biological indicators of distinctive types of surface ocean water.

Copepods are common prey of chaetognaths. All adult calanoid copepod species are similar in body form and general feeding behavior, suggesting the evolution of a very successful functional form. Copepods and other small, pelagic particle

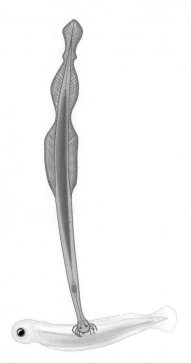

Figure 12.20 A small chaetognath, *Sagitta*, capturing and consuming a fish larva its own size.

grazers are typically exposed to a wide spectrum of food particle sizes. Food particles range from abundant minute bacteria through the common types of phytoplankton to organic aggregates and large centric diatoms. This size spectrum presents several opportunities for small, versatile particle grazers to adopt feeding strategies that select for optimal-sized food items.

Although copepods prey on large phytoplankton cells when they are available, calanoid species can capture the smaller, more abundant microplankton with a basket-like filtering mechanism derived from their complex, feathery feeding appendages (**Figure 12.21**). The hairlike setae on the appendages of *Calanus* are fine enough to retain nanoplankton-sized food particles larger than 10 μm. Laboratory observations have revealed that food particles are carried into the filter basket by currents generated by the feeding appendages and five pairs of more posterior thoracic swimming legs. Special long setae on the feeding appendages remove the trapped particles and direct them to the mouth. With this filtering mechanism, *Calanus* species are capable of exploiting a wide size range of food particles.

Even though *Calanus* and similar copepods can shift rapidly from one food particle size to the other, they prefer larger food particles to smaller ones. Studies using high-speed photomicrography techniques have shown that calanoid copepods efficiently capture sparsely scattered, single, large food items such as protozoans, small fish eggs, and large diatoms. As the copepods' feeding actions drive them forward in the water, their antennae extend laterally to function as an array of sensory receptors to detect minute disturbances surrounding larger food items (**Figure 12.22**). If the food particle is detected near the end of an antenna, the animal quickly adjusts its swimming direction to bring the particle within reach of an extended feeding appendage. The mouthparts then seize and manipulate the particles before eating them. Because filtering and large-particle

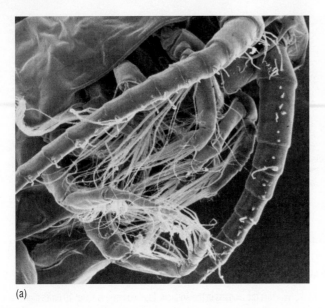

(a)

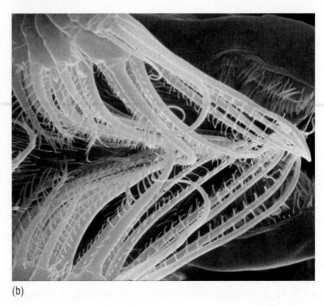

(b)

Figure 12.21 (a) A scanning electron micrograph of the thorax and filter-feeding mechanism of *Calanus*, shown in side view. (b) Higher-magnification ventral view of *Calanus*, showing the filtering basket formed by the second maxillae.
(a) Courtesy of M. M. Friedman and J. Rudi Strickler; (b) Courtesy of M. M. Friedman and J. Rudi Strickler.

seizure cannot operate simultaneously, this mode of feeding is interrupted when the copepods are filtering small particles.

Such particle-size selectivity by copepods is likely an important factor in stabilizing phytoplankton populations. When phytoplankton populations of a particular cell size become more abundant through growth and reproduction, they attract increased grazing pressures as more copepods shift feeding strategies to concentrate on them. It is unlikely, however, that the phytoplankton population would be grazed to extinction. Several copepod species exhibit ingestion rates that are dependent on the concentration of phytoplankton cells (**Figure 12.23**). For a certain food particle size, ingestion rates increase with increasing particle density to some crucial maximum. Beyond the maximum particle density, some aspect of the copepod's food-processing system appears to become saturated, and no further increase in

ingestion rates occurs. Conversely, ingestion rates decrease with decreasing phytoplankton densities; copepods most likely will shift to another more optimal concentration of food particles before the first population is exhausted completely.

In contrast to the rigid filter devices of crustaceans, some gelatinous herbivores rely on nets or webs of mucus to ensnare food particles. One unusual example is *Corolla* (Figure 12.5a), one of the few planktonic gastropod mollusks. When feeding, *Corolla* secretes a mucous web that often exceeds 2 m in diameter. The free-floating web spreads horizontally as it is produced, maintaining a single point of attachment at the animal's mouth. As the animal and its web slowly sink, bacteria and small phytoplankton become trapped in the mucus. The web, with its load of food, is formed into a mucous string and ingested. *Corolla* then swims upward to repeat the behavior.

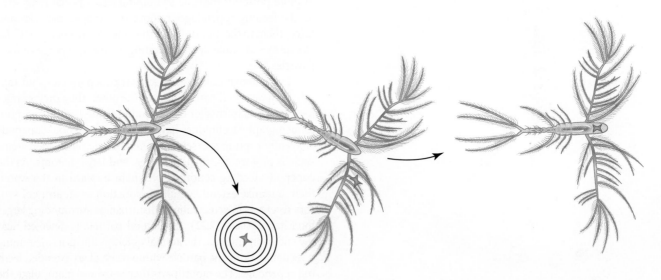

Figure 12.22 Copepod (*Oithona*) detection and capture behavior of individual diatoms (green). Sensory cells arranged in arrays on large antennae provide information to detect prey and guide the response until the capture is made (right).

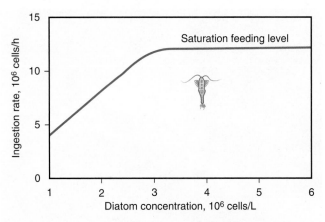

Figure 12.23 The ingestion rate of a copepod, *Calanus*, as a function of the concentration of its food (in this case, the diatom *Thalassiosira*). The ingestion rate peaks near 3,000 diatom cells/mL, and no further increase is seen even at much higher concentrations.

Modified from B. W. Frost. *Limnol Oceanogr* 17(1972):805–815.

Another elaborate mucous feeding system is found in appendicularians, small tadpole-shaped invertebrate chordates. Like *Oikopleura* (**Figure 12.24**), most appendicularians live enclosed within delicate, transparent mucous bubbles. Food-laden water, pumped by the tail beat of the occupant, enters the bubble through openings at one end. These openings are screened with fine-meshed grills to exclude large phytoplankton cells. Smaller cells enter the bubble and are trapped on a complex, internal mucous feeding screen. Every few seconds, the animal sucks the particles off the screen and into its mouth. When the grill in the bubble wall becomes clogged or the interior is fouled with feces, the entire bubble is abandoned and a new one is constructed, sometimes in as little as 10 minutes. With this feeding mechanism, even bacteria-sized particles can be harvested efficiently because these animals achieve filtering rates several times greater than those of calanoid copepods. The larger gelatinous salps are even more efficient feeders; a small chain of colonial salps (such as those shown in Figure 12.5c) is capable of filtering as much water as 3,000 copepods can in the same amount of time.

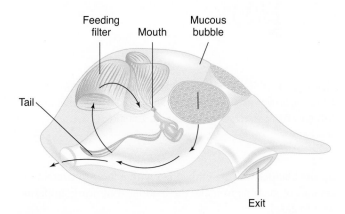

Figure 12.24 The appendicularian, *Oikopleura*, within its mucous bubble. Arrows indicate path of water flow.

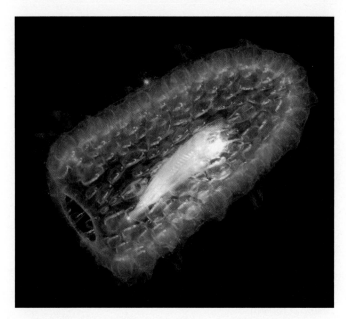

Figure 12.25 A colonial pelagic pyrosome tunicate with a fish inside it.

© Timothy Ewing/Getty Images.

The mechanisms used by zooplankton to glean small, diffuse food particles from the water reflect the crucial roles these animals play in pelagic food webs. Planktonic tunicates (**Figure 12.25**) are the largest herbivorous grazers to exploit the very small phytoplankton found in tropical and subtropical areas of low productivity. Because their filters clog quickly when they encounter the high phytoplankton densities of upwelling areas or temperate spring blooms, they are less successful competitors in these areas. It is in these regions of high productivity that planktonic crustaceans, particularly the calanoid copepods and euphausiids, thrive.

DID YOU KNOW?

Some copepods create feeding vortices around themselves to bring food particles toward their heads. They quickly move their posterior feeding limbs in a way that creates a small circular water current pattern around them. Particles are trapped in the circular current and then picked up by setae (small hairlike projections) on the anterior feeding appendages. Finally, the particles can be moved into the mouth for a meal.

12.6   Buoyancy

Living and moving in three dimensions well above the seafloor creates some buoyancy problems for pelagic animals because the bone and muscle tissues needed for locomotion in these animals are denser than seawater. Stored fats and oils, which are less dense than water, are common buoyancy devices used by some pelagic marine animals. Whales, seals, and penguins maintain thick blubber layers just under the skin. Many sharks and a variety of teleosts store large quantities of oils in their livers and muscle tissues. In fact, in some species of sharks the liver accounts for about one third of their body weight.

Fats and oils, however, are only slightly less dense than seawater. This fact poses a serious challenge for many small but active nektonic species. They cannot energetically afford to carry around a huge oily liver or a thick blubber layer, nor can they sacrifice muscle and bone to lighten their load. The solution for many marine animals is an internal gas-filled flotation organ. At sea level, air is only about 0.1% as dense as seawater, and a small volume of air can provide a large amount of buoyancy. The buoyancy derived from a volume of gas depends on the volume of seawater the gas displaces. Unlike fats or oils, gases are compressible; they occupy different volumes at different pressures and depths. At sea level, the pressure created by Earth's envelope of air is about 1 kg/cm^2, or 1 atm. Below the sea surface, the water pressure increases about 1 atm for each 10-m increase in depth. Thus, the total pressure experienced by a fish at 5,000 m is 501 atm (more than 3.5 tons/in.^{2}).

A few genera of colonial cnidarians maintain positive buoyancy by secreting gases into a float, or pneumatophore. *Velella* (sometimes called by-the-wind sailor; Jack-by-the-wind; or, simply, purple sail; **Figure 12.26a**) and the larger Portuguese man-of-war, *Physalia* (**Figure 12.26b**), have large pneumatophores and float at the sea surface. The pneumatophore acts as a sail to catch surface breezes and transport the colony long distances. Both *Velella* and *Physalia*, with only one species in each genus, have worldwide distributions.

Other siphonophores with gas floats (Figure 12.17) are neutrally buoyant and can easily change their vertical position in the water column by swimming. A gas gland within the pneumatophore secretes gas into the float. Excess gases are vented through a small pore that is opened and closed by a muscular valve, or sphincter, and the siphonophore can adjust its buoyancy, as needed.

Gas is also used for buoyancy by a small planktonic nudibranch, *Glaucus*. It produces and stores intestinal gases to offset the weight of its body. The planktonic snail *Janthina* forms a cluster of bubbles at the surface with its mouth and clings to it so that its thin, purple shell does not cause it to sink into the depths. These adaptations are apparently related to the preference of the nudibranch and snail for feeding on the soft parts of *Velella* (which also floats at the surface). These zooplankton adapted to live permanently at the sea surface are known as **neuston** and are relatively rare. Another example of a neustonic animal is the water strider.

Air in the lungs of mammals, reptiles, and birds also can provide some buoyancy; however, it is the teleost fish that have the most sophisticated system for using air to solve their buoyancy problems. Many teleosts, especially active species with extensive muscle and skeletal tissue, have body densities about 5% greater than that of seawater. To achieve neutral buoyancy, many of these fish have an internal swim bladder filled with gases (mostly N$_2$ and O$_2$, the most abundant gases in the atmosphere and also dissolved in seawater). The swim bladders of bony fish develop embryonically from an outpouching of the esophagus (**Figure 12.27**). The densely woven fibers that make

(a)

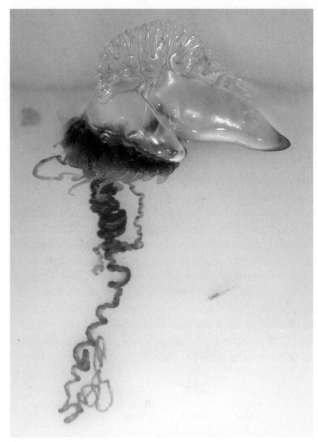

(b)

Figure 12.26 (a) Several purple sails, *Velella*, stranded on a beach in central California by the wind and (b) the Portuguese man-of-war, *Physalia*, floating at the sea surface. The trailing tentacles of *Physalia* may reach 50 m in length.

up the bladder wall are embedded with a layer of overlapping crystals of guanine to make the bladder wall nearly impermeable to gases.

The connection between the esophagus and swim bladder, called the **pneumatic duct**, is present during the larval or

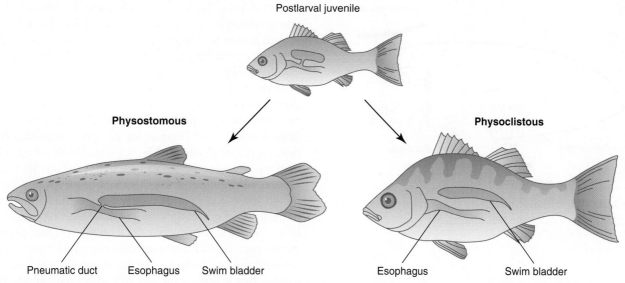

Postlarval juvenile

Physostomous   **Physoclistous**

Pneumatic duct   Esophagus   Swim bladder   Esophagus   Swim bladder

Figure 12.27 The two different types of swim bladders seen in fish that retain a swim bladder as adults.

juvenile stages of all teleosts. In some species, the pneumatic duct remains intact in the adult. This is the **physostomous swim bladder** condition. In other species, the duct disappears as the fish matures to create a **physoclistous swim bladder** (Figure 12.27). Nearly half of the species of teleosts, however, lose not only the pneumatic duct but also the entire swim bladder when they mature. Swim bladders are notably absent in benthic fish (flounders), highly active fish (tuna), and deep-sea fish (Figure 12.14). These fish lose their swim bladders because they do not need them as adults and in some cases have lifestyles that would be impeded by the presence of a gas-filled swim bladder.

Swim bladders are not rigid structures; the volume of water they displace is subject to changing water pressures at different depths. To maintain neutral buoyancy at different depths, the volume of a fish's swim bladder must remain constant. A fish that swims downward experiences greater external water pressure, which squeezes its swim bladder and reduces the bladder's volume and the fish's buoyancy. The fish must then increase the quantity of gas in the bladder to compensate for the volume change.

An ascending fish has the opposite problem: it must get rid of swim bladder gases as rapidly as they expand. Some shallow-water physostomous fish fill their swim bladders simply by gulping air at the sea surface. They also release excess gases through their pneumatic duct and eventually out the mouth or gills. Fish with physoclistous swim bladders, however, lack a pneumatic duct, and their ability to add or remove bladder gases rapidly to compensate for a rapid depth change is limited. If a deep-water fish with a physoclistous swim bladder ascends rapidly, the decreased water pressure enables the gases within the somewhat elastic swim bladder to expand and reduce the overall density of the fish. The density decrease may be so great that the fish is unable to descend for some time. This is well illustrated by the appearance of fish brought to the surface (unwillingly, of course) from depth on fishing lines or in trawls.

It is not unusual for the swim bladders of such fish to expand so much that severe internal organ damage occurs (**Figure 12.28**). During a slower natural ascent, excess gases are reabsorbed back into the bloodstream, but this takes some time. This is accomplished at a specialized region of the swim bladder, the oval body (**Figure 12.29**), which is richly supplied with blood vessels for reabsorption of gases. The oval body is isolated from the remainder of the swim bladder by a sphincter that controls the flow of bladder gases to the oval body.

Both types of fish swim bladders have gas glands that regulate the secretion of gas from the blood into the bladder when these fish are below the sea surface and have no access to air. Because the process of filling the swim bladder is the same in

Figure 12.28 A bocaccio rockfish, *Sebastes paucispinis*, with barotrauma is tagged to track its survival. Bulging eyes and a protruding stomach are common external signs of barotrauma.
National Oceanic and Atmospheric Administration.

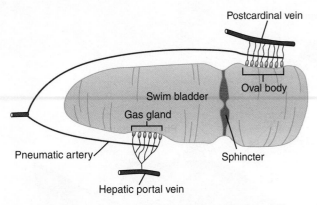

Figure 12.29 A physoclistous swim bladder and associated blood vessels. The area of the gas gland is diagrammed in greater detail in Figure 12.30.

DID YOU KNOW?

Fishers practicing catch-and-release techniques are encouraged to bring up their catch as slowly as possible. Allowing a captured fish to adjust to the changes in pressure and release gases from its swim bladder will give it the best chance of survival when thrown back into the water. This technique can prevent permanent damage to a fish and its organs, including the common and somewhat disturbing sight of a fish reeled in quickly with its eyes bulging and stomach protruding from its mouth!

---

both types, only the physoclistous swim bladder is described. Fish with gas-filled swim bladders have been collected from depths as great as 7,000 m. The gas pressure needed within the swim bladder to balance the water pressure at that depth is about 700 atm (5 tons/in.^{2}). Such extreme gas pressures are achieved by a dramatic increase in the O_2 concentration of the bladder gases. Oxygen commonly accounts for more than 50%, and occasionally exceeds 90%, of the gas mixture of the swim bladders of deep-sea fishes.

How are O_2 and N_2, which are dissolved in seawater at pressures no greater than 1 atm, concentrated in swim bladders at pressures as great as 700 atm? The answer includes several levels of mechanisms. Below the sea surface, fish must fill their swim bladders with gases absorbed from seawater by their gills. These gases are transported in the blood to the gas gland of the swim bladder and then are secreted into the bladder at pressures equal to external water pressures. When highly oxygenated hemoglobin reaches the gas gland of a swim bladder (**Figure 12.30**), the O_2 must be induced to leave the hemoglobin and diffuse into the swim bladder, often in the face of high O_2 pressures within the bladder. When a fish descends and its swim bladder volume must be increased, stretch receptors in the swim bladder wall stimulate the gas gland to produce lactic acid. The lactic acid diffuses into the blood vessels and lowers the pH of the blood. Lower pH conditions reduce the oxygen-carrying capacity of hemoglobin and induce it to unload a large

portion of its O_2. The free O_2, which has not yet left the blood, is now no longer associated with the hemoglobin. The total effect of lactic acid on hemoglobin is sufficient to produce about 2 atm of O_2 pressure at the gas gland of the swim bladder.

Eventually, the O_2 will diffuse into the swim bladder if the O_2 pressure there is not greater than 2 atm. This mechanism alone, however, is capable of producing swim bladder gas pressures useful only to relatively shallow fish. Fish with gas-filled swim bladders living below about 20 m rely on another structure, an extensive **rete mirabile**, associated with the gas gland (Figures 12.29 and 12.30) to achieve that extra swim bladder gas pressure. A typical rete system may contain a few hundred or as many as 200,000 tiny, capillary-sized blood vessels. Each tiny rete vessel approaches the gas gland carrying oxygen-rich hemoglobin and then doubles back on itself without penetrating the swim bladder. These complex rete systems form another countercurrent exchange system that operates on the same principle as that described for fish gills to concentrate O_2. Here, too, this simple system works by creating and manipulating O_2 diffusion gradients.

Until the O_2 unloaded from hemoglobin by lactic acid at the gas gland is concentrated sufficiently to match the pressure in the swim bladder, it will simply be carried through the turn of the rete capillary and start to leave the gas gland; however, its path is adjacent to and parallel with a capillary carrying blood toward the gas gland. As the concentration of the dissolved O_2 of the blood leaving the gas gland is higher than that of the incoming capillary, the O_2 will diffuse across the capillary walls and back into the incoming blood of adjacent capillaries. The O_2 forced off the hemoglobin at the gas gland is thus trapped in this countercurrent diffusion loop as it tries to leave the gas gland. Given sufficient time and enough cycles around this loop, the pressure of O_2 in the capillaries will surpass even very high pressures of the swim bladder, and O_2 will diffuse from the gas gland into the bladder.

As one might expect, a long rete is capable of concentrating more O_2 at the gas gland than is a short one. Still, the rete does not need to be unmanageably long. A rete only 1 cm long can secrete O_2 at pressures up to 2,000 atm, well in excess of the swim bladder pressures needed in the deepest parts of the sea. The rete mirabile concentrates N_2 as well as O_2; however, the lack of a specialized transport system for N_2 (as hemoglobin is for O_2) relegates N_2 to the role of a minor gas in swim bladders, especially at great depths.

As the pressure of gases inside swim bladders increases, so do their densities. At 7,000 m, the greatest depth at which gas-filled swim bladders have been found, the gas within a swim bladder is so compressed that its density is almost as great as that of fat. For some fish at great depths, the constant energy expenditures needed to maintain a full swim bladder become unrealistic, and swim bladder gases are replaced with fat. Fat-filled swim bladders provide almost as much buoyancy as gases do at great depths but are much simpler to maintain because fat will not compress at high pressures. Fat-filled swim bladders are also found in many vertically migrating fish species, such as lanternfish (Figure 12.13), that move through pressure changes of 10 to 40 atm twice each day.

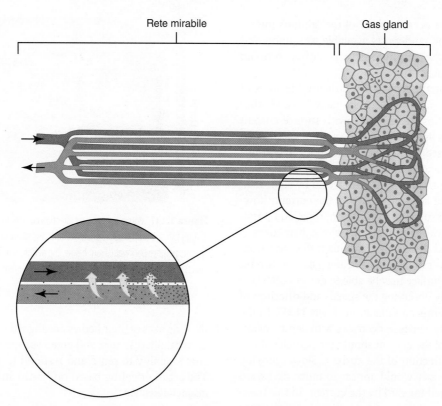

Rete mirabile   Gas gland

Figure 12.30 A simplified diagram of the rete mirabile and gas gland associated with the swim bladders of many bony fishes. The inset illustrates the countercurrent arrangement of blood flow (black arrows) and the diffusion of O_2 from outgoing to incoming blood vessels (white arrows).
Modified from W. W. Hoar. *General and Comparative Physiology*. Prentice Hall, 1983.

12.7 Orienting in the Sea

How do pelagic migratory species know where they are and where they are going? Before an animal can successfully accomplish a directed movement from one place to another, it must orient itself both in time and in space. From a human perspective, there are few obvious landmarks in the open ocean for organisms to follow, but the terrestrial habitats of humans are completely different from the watery pelagic habitats of sea creatures. Although landmarks in the open ocean are not obvious to us, pelagic organisms are extremely sensitive to very subtle changes in their environments and living conditions. Biological clocks operating on circadian and longer-period rhythms are important factors in the orientation process. A variety of environmental factors serve as cues to adjust or reset the timing of these rhythms. Well known among these timing factors is day length, which changes with predictable regularity through the seasons. Day length, water temperature, and food availability might serve as useful cues for following the passage of the seasons and as triggers for seasonal migrations of whales and other marine animals.

Day length is an environmental factor that varies predictably, but water temperature and food availability are factors that can change drastically in a much less predictable manner. When these factors change drastically and remain in a changed state for long periods, known animal distributions may change drastically, too. An example of such a scenario occurs during El Niño

years. As warmer-than-average sea surface temperatures remain in southern and central California, many species from warmer, southern waters along the U.S. West Coast venture northward into areas they do not normally inhabit. Sea turtles are found in central California, large game fish such as marlin venture into southern California from Mexican waters, and other tropical fish swim north. When water temperatures return to normal, animal distributions return to their known patterns.

Orientation in space is somewhat more complex than orientation in time, especially for animals migrating below the sea surface where directional information derived from the apparent position of the sun, moon, or stars is unavailable. It is known that eels, salmon, sharks, and many other fish, including fish larvae, have extremely keen olfactory senses. Since the 1970s, an impressive body of evidence has been gathered to support the idea that salmon use olfactory cues to guide them to their home stream. The sequential odor hypothesis of migrating salmon postulates that young salmon smolt are imprinted with a sequence of stream odors during their downstream trip to the ocean. When returning back to their home stream as adults, the remembered cues are played back in reverse to act as a sequence of sign stimuli that release a positive response to swim upstream. The chemical nature of the characteristic odors in stream water remains largely unidentified, but it appears that individual salmon can recognize not only their own species but also their own population by smell; however, these or similar odors are not likely to be concentrated sufficiently in the open ocean to guide the oceanic phase of the salmon's migration. What then are the guideposts available to salmon and other nekton migrating across huge expanses of open ocean well below the sea surface?

Currents are among the most stable structural features of the open ocean. The migratory patterns of many fish and other marine animals seem closely associated with surface current patterns. But how can a fish detect the direction or speed of an ocean current if it can see neither the surface nor the bottom? The sharp temperature and salinity gradients sometimes found at the edges of ocean currents might be detected by some fish, but only if they venture across the edge of the current; however, available evidence suggests that salmon, tuna, and possibly adult eels migrate within currents, not along their edges.

Current speeds and directions are difficult to detect from the surface if the observer is being carried along by the current, but animals below the surface may be able to detect ocean surface currents by visually observing the speeds and direction of horizontally moving debris and plankton (**Figure 12.31**). In the ocean, the water velocity generally decreases with depth. Swimmers near the bottom of the current should see particles above them moving in the direction of the current. Slower-moving particles below a swimmer would appear to move backward as the swimmer is carried forward by the current. In this manner, animals could determine the current direction and orient their swimming motions either in the same direction or directly against it.

When charged ions of seawater are moved by the ocean's currents through the magnetic field of Earth, a weak electrical potential is generated in a process that is similar to the operation of an electrical generator. These ocean current potentials have been measured with ship-towed electrodes and are used to compute current speeds. Some preliminary laboratory evidence suggests that at least the Atlantic eel, *Anguilla*, and the Atlantic salmon, *Salmo*, are sensitive to electrical potentials of the same magnitude as those generated by ocean currents. In addition, they are most sensitive to the electrical potential when

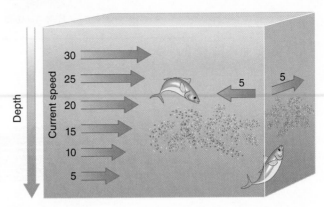

Figure 12.31 Possible speed and direction cues for fish in an ocean current. To a drifting fish above an accumulation of debris and plankton (shaded region), the debris appears to move backward. From below, the debris appears to be carried forward in the direction of the current. Numbers indicate relative current speeds.

the long axes of their bodies are aligned with the direction of the current. Sharks, rays, and green sea turtles also have a demonstrated ability to sense and respond to Earth's magnetic field. The act of orienting based on Earth's magnetic field is termed **magnetoreception**.

DID YOU KNOW?

Sharks are famous for their abilities to travel long distances and orient in their environments. Not only are sharks aware and in control of their own positions in the environment, they appear to be aware of the positions of scuba divers. In a study conducted on shark behavior in the presence of divers, sharks approached the divers from behind in over 80% of the interactions. Sharks are able to differentiate the front from the back side of humans, and it appears that they would rather remain unseen by humans. They are also known to attack prey from behind and below, but fortunately sharks seldom mistake humans for prey.

STUDY GUIDE

TOPICS FOR DISCUSSION AND REVIEW

1. Mesopelagic fishes differ from more familiar epipelagic and coastal fishes in many ways. Summarize these differences.

2. Summarize the proposed uses of photophores in marine animals.

3. What is the difference between holoplankton and meroplankton? List three well-known examples of each type of plankton.

4. Describe the common buoyancy structures used by pelagic marine animals.

5. Discuss the proposed advantages of vertical migration for mesopelagic species. Why do some mesopelagic species migrate to deeper waters at night?

6. Describe the various mechanisms used by zooplankton to collect diffuse food.

7. Identify the proposed cues used by marine animals to orient in space during their long migrations. Are larvae capable of orienting, or are they just subject to the current they are traveling in?

KEY TERMS

antitropical distribution 337	mesopelagic zone 340
countershading 340	neuston 348
deep sound-scattering layers (DSSLs) 342	photophores 341
diurnal 342	physoclistous swim bladder 349
epipelagic zone 336	physostomous swim bladder 349
holoplankton 332	pneumatic duct 348
isolume 344	pneumatophore 343
magnetoreception 352	rete mirabile 350
marine snow 340	vertical migration 342
meroplankton 332	

KEY *GENERA*

Aegisthus	Corolla
Architeuthis	Cyclothone
Argyropelecus	Euphausia
Aristostomias	Eurypharynx
Bolinichthys	Gigantactis
Bolinopsis	Glaucus
Calanus	Janthina
Loligo	Regalecus
Melanocetus	Sagitta
Oikopleura	Sapphirina
Oithona	Sebastes
Opisthoproctus	Thalassiosira
Pegea	Thysanoessa
Physalia	Velella

REFERENCES

Allan, J. D. 1976. Life history patterns in zooplankton. *American Naturalist* 110:165–180.

Alldredge, A. 1976. Appendicularians. *Scientific American* July:94–102.

Barham, E. G. 1966. Deep scattering layer migration and composition: Observations from a diving saucer. *Science* 151:1399–1403.

Bargu, S., C. L. Powell, S. L. Coale, M. Busman, G. J. Doucette, and M. W. Silver. 2002. Krill: A potential vector for domoic acid in marine food webs. *Marine Ecology Progress Series* 237:209–216.

Boden, B. P., and E. M. Kampa. 1967. The influence of natural light on the vertical migrations of an animal community in the sea. *Symposium of the Zoological Society of London* 19:15–26.

Boyd, C. M. 1976. Selection of particle sizes by filter-feeding copepods: A plea for reason. *Limnology and Oceanography* 21:175–179.

Bright, T., F. Ferrari, D. Martin, and G. A. Franceschini. 1972. Effects of a total solar eclipse on the vertical distribution of certain oceanic zooplankters. *Limnology and Oceanography* 17:296–301.

Brinton, E. 1962. The distribution of Pacific euphausiids. *Bulletin. Scripps Institution of Oceanography* 8:51–270.

Cushing, D. H. 1968. *Fisheries Biology*. Madison, WI: University of Wisconsin Press.

Denton, E. J., and J. P. Gilpin-Brown. 1973. Flotation mechanisms in modern and fossil cephalopods. *Advances in Marine Biology* 11:197–268.

Eastman, J. T., and A. L. DeVries. 1986. Antarctic fishes. *Scientific American* 255:106–114.

Gilmer, R. W. 1972. Free-floating mucus webs: A novel feeding adaptation for the open ocean. *Science* 176:1239–1240.

Giorgio, P. A., and Duarte, C. M. 2002. Respiration in the open ocean. *Nature* 420:379–384.

Hoar, W. W. 1983. *General and Comparative Physiology*. Englewood Cliffs, NJ: Prentice Hall.

Jackson, G. A. 1990. A model of the formation of marine algal flocs by physical coagulation processes. *Deep-Sea Research* 37:1197–1211.

Jensen, O. P., S. Hansson, T. Didrikas, J. D. Stockwell, T. R. Hrabik, T. Axenrot, and J. F. Kitchell. 2011. Foraging, bioenergetic, and predation constraints on diel vertical migration: Field observations and modeling of reverse migration by young-of-year herring *Clupea harengus*. *Journal of Fish Biology* 78:449–465.

Jumper, G. Y., Jr., and R. C. Baird. 1991. Location by olfaction: A model and application to the mating problem in the deep-sea hachetfish *Argyropelecus hemigymnus*. *American Naturalist* 138:1431–1458.

Kanwisher, J., and A. Ebling. 1957. Composition of swim bladder gas in bathypelagic fishes. *Deep-Sea Research* 4:211–217.

Keenleyside, M. H. A. 1979. *Diversity and Adaptation in Fish Behavior.* New York: Springer-Verlag.

Klimley, A. P. 1994. The predatory behavior of the white shark. *American Scientist* 82:122–133.

Koehn, R. K. 1972. Genetic variation in the eel, a critique. *Marine Biology (Berlin)* 14:179–181.

Lam, R. K., and B. W. Frost. 1976. Model of copepod filtering response to changes in size and concentration of food. *Limnology and Oceanography* 21:490–500.

Luschi, P. 2013. Long-distance animal migrations in the oceanic environment: Orientation and navigation correlates. *International Scholarly Research Notices Zoology* 2013:1–23.

Nicol, S., and W. de la Mare. 1993. Ecosystem management and the Antarctic krill. *American Scientist* 81:36–47.

O'Brien, W. J., H. I. Browman, and B. I. Evans. 1990. Search strategies in foraging animals. *American Scientist* 78:152–160.

Paris, C. B., J. Atema, J. Irisson, M. Kingsford, G. Gerlach, and C. M. Guigand. 2013. Reef odor: A wake up call for navigation in reef fish larvae. *PLoS ONE* 8(8):e72808. doi:10.1371/journal.pone.0072808.

Partridge, B. L. 1982. The structure and function of fish schools. *Scientific American* 246:114–123.

Pennisi, E. 1989. Much ado about eels. *BioScience* 39:594–598.

Perutz, M. F. 1978. Hemoglobin structure and respiratory transport. *Scientific American* 239:92–125.

Porter, K. G., and J. W. Porter. 1979. Bioluminescence in marine plankton: A coevolved antipredation system. *American Naturalist* 114:458–461.

Putman, N. F., M. M. Scanlan, E. J. Billman, J. P. O'Neil, R. B. Couture, T. P. Quinn, K. J. Lohmann, and D. L. G. Noakes. 2014. An inherited magnetic map guides ocean navigation in juvenile Pacific salmon. *Current Biology* 24(4):446–450.

Ritter, E., and R. Amin. 2013. Are Caribbean reef sharks, *Carcharhinus perezi*, able to perceive human body orientation? *Animal Cognition* 17:745–753.

Quetin, L. B., and R. M. Ross. 1991. Behavioral and physiological characteristics of the Antarctic krill, *Euphausia superba*. *American Zoologist* 31:49–63.

Richman, S., D. R. Heinle, and R. Huff. 1977. Grazing by adult estuarine calanoid copepods of the Chesapeake Bay. *Marine Biology* 42:69–84.

Rodin, E. Y., and S. Jacques. 1989. Countercurrent oxygen exchange in the swim bladders of deep-sea fish: A mathematical model. *Mathematical and Computer Modelling* 12:389–393.

Rommel, S. A., Jr., and J. D. McCleave. 1972. Oceanic electric fields: Perception by American eels? *Science* 176:1233–1235.

Royce, W., L. S. Smith, and A. C. Hartt. 1968. Models of oceanic migrations of Pacific salmon and comments on guidance mechanisms. *Fishery Bulletin* 66:441–462.

Rubenstein, D. I., and M. A. R. Koehl. 1977. The mechanisms of filter feeding: Some theoretical considerations. *American Naturalist* 111:981–994.

Scholander, P. F. 1957. The wonderful net. *Scientific American* 196:96–107.

Sheldon, R. W., A. Prakash, and W. H. Sutcliffe Jr. 1972. The size distribution of particles in the ocean. *Limnology and Oceanography* 17:327–340.

Silver, M. W., A. L. Shanks, and J. D. Trent. 1978. Marine snow: Microplankton habitat and source of small-scale patchiness in pelagic populations. *Science* 201:371–373.

Smith, D. L. 1996. *A Guide to Marine Coastal Plankton and Marine Invertebrate Larvae.* Dubuque, IA: Kendall/Hunt.

Smith, O. L., H. H. Shugart, R. V. O'Neill, R. S. Booth, and D. C. McNaught. 1971. Resource competition and an analytical model of zooplankton feeding on phytoplankton. *American Naturalist* 109:571–591.

Steele, J. H., ed. 1973. *Marine Food Chains.* Edinburgh: Oliver and Boyd.

Steele, J. H. 1976. Patchiness. In: D. H. Cushing and J. J. Walsh, eds. *Ecology of the Seas.* Oxford, UK: Blackwell Scientific.

Steele, J. H. 1980. Patterns in plankton. *Oceanus* 25:3–8.

Stoecker, D. K., and J. M. Capuzzo. 1990. Predation on protozoa: Its importance to zooplankton. *Journal of Plankton Research* 12:891–908.

Strickler, J. R. 1985. Feeding currents in calanoids: Two new hypotheses. *Symposium, Society of Experimental Biology* 39:459–485.

Sverdrup, H. U., M. W. Johnson, and R. H. Fleming. 1970. *The Oceans: Their Physics, Chemistry, and Biology.* Englewood Cliffs, NJ: Prentice Hall.

Tucker, D. W. 1959. A new solution to the Atlantic eel problem. *Nature (London)* 183:495–501.

Turner, J. T., P. A. Tester, and W. F. Hettler. 1985. Zooplankton feeding ecology. *Marine Biology* 90:1–8.

Vlymen, W. J. 1970. Energy expenditure of swimming copepods. *Limnology and Oceanography* 15:348–356.

Waickstead, J. H. 1976. *Marine Zooplankton.* London: E. Arnold.

Wilbur, R. J. 1987. Plastic in the North Atlantic. *Oceanus* 30:61–68.

Zaret, T. M., and J. S. Suffern. 1976. Vertical migration in zooplankton as a predator avoidance mechanism. *Limnology and Oceanography* 21:804–813.

The Deep-Sea Floor

STUDENT LEARNING OUTCOMES

1. Describe the living conditions in the deep sea and techniques used for sampling and observing this unique ecosystem.

2. Identify mechanisms involved in the transfer of dissolved gases and food particles from the shallow waters of the photic zone to the deep sea.

3. Describe the major life forms inhabiting the vast abyssal plains of the deep sea, and compare these forms to those inhabiting shallower waters.

4. Identify the diverse and unique life forms inhabiting hydrothermal vents and cold seeps.

5. Identify the primary producers supporting hydrothermal vent and cold seep communities, and compare the primary production strategies of chemosynthesis and photosynthesis.

CHAPTER OUTLINE

Active and inactive spires near the Kawio Barat submarine volcano at nearly 2,000 m depth. Image was captured by the *Little Hercules* remotely operated vehicle.
Courtesy of National Oceanic and Atmospheric Administration *Okeanos Explorer* Program, INDEX-SATAL 2010.

Beyond the continental shelves, the seafloor descends sharply down continental slopes to the perpetual dark and cold **abyssal zone** of the deep sea. Underlying the oceanic province of the world ocean are vast expanses of nearly flat and featureless abyssal plains and hills, plus a scattering of tectonically active ridges, rises, and trenches. Three fourths of the ocean bottom (the abyssal and **hadal zones**) lie at depths below 3,000 m. A large portion of the deep-ocean basin consists of broad, flat, sediment-covered abyssal plains. Abyssal plains typically extend seaward from the bases of continental slopes and oceanic ridge and rise systems at depths between 3 and 5 km below the sea surface. The long, narrow floors of marine trenches distributed around the margins of the deep-sea basins are generally deeper than 6 km. Trenches cover only 2% of the seafloor and are characterized by the most extreme pressure regimes experienced by living organisms anywhere on this planet.

Interrupting the continuity of abyssal plains are mountainous linear ridge and rise systems encircling the globe. The axes of these ridge and rise systems are the source of new oceanic crust, created as the global forces of plate tectonics rift apart the plates of Earth's crust. Growth of new crust is slow, about as fast as your fingernails grow, yet it causes about 60% of Earth's surface (or 85% of the seafloor) to be recycled every 200 million years. Several remarkable discoveries associated with ridge and rise–spreading centers have been made in the past several decades. These discoveries are radically changing our understanding of life in the deep sea and are clarifying how metal-rich mineral deposits are formed from chemical interactions between seawater and Earth's crust.

Our current understanding of the deep sea is in part due to the leaps and bounds made in marine survey technology over the past few decades. Humans can now explore the deep sea using specialized undersea vehicles and oceanographic sampling equipment. In this chapter, we explore the extreme physical characteristics and conditions of the deep sea, the unique living organisms that make the deep sea their home, and some of the technologies that have made these discoveries possible.

13.1   Living Conditions on the Deep-Sea Floor

It is not possible to establish a precise global depth boundary between the animals of the deep sea and the shallow-water fauna of the continental shelves. Generally, the boundary exists as a vague region of transition on the continental slopes bordering the deep-sea basin; however, animals of the "deep sea" commonly extend into shallower water in polar seas and, on occasion, even extend to the inner portions of high-latitude continental shelves. However we may define the deep sea, some

Figure 13.1 Fine-grained bottom sediments off the Oregon coast disturbed by the impact of a current-direction indicator.
Courtesy of James Sumich.

animals will defy our definitions and expectations when they venture between depth ranges.

Most of the sea bottom is covered with thick accumulations of fine sediment particles, skeletons of planktonic organisms, and other debris (**Figure 13.1**) that have settled from the surface. Marine sediments are derived from several sources. A few minerals precipitate from their dissolved state in seawater to produce irregular deposits on the seafloor. Manganese nodules (**Figure 13.2**) are a well-known example of this type of deposit. Most are flattened like a pancake, but during a survey of the deep Atlantic seafloor east of Barbados in 2015 a large area blanketed by manganese nodules as round and large as bowling balls was discovered. Such deposits have potential commercial importance as a source of minerals; however, they also may have an important influence on the structure of some benthic communities, and their removal in mining operations may cause catastrophic mortality in harvest areas. The idea of mining manganese nodules was dismissed by many due to the

Figure 13.2 Manganese nodules scattered on the surface of the seafloor in the Pacific Ocean.
© Institute of Oceanographic Sciences/NERC/Science Source.

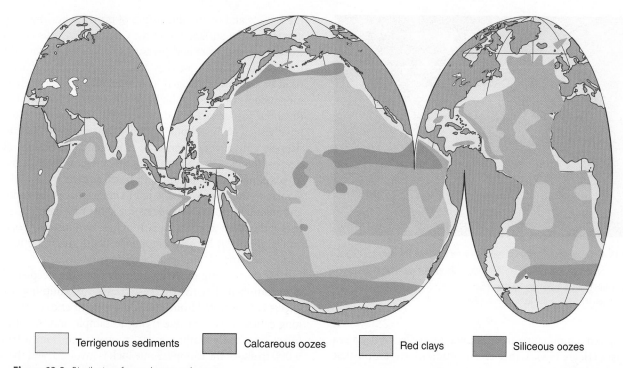

Terrigenous sediments   Calcareous oozes   Red clays   Siliceous oozes

Figure 13.3 Distribution of ocean-bottom sediments.
Data from R. I. Tait, and M. R. Howe. 1968. Some observations of thermohaline stratification in the deep ocean. *Deep-Sea Research* 15: 275–280, and H. U. Sverdrup, M. W. Johnson, and R. H. Fleming, 1942. *The Oceans, Their Physics, Chemistry and General Biology.* Prentice Hall Inc., New York.

great depths of occurrence and the vast effort that would be involved in their collection, but recent advances in technologies used for offshore oil and gas recovery, dredging, and telecommunications have led interested parties in the mining industry to rethink the potential for deep-sea mining of manganese nodules or other minerals.

Sedimentary materials found in the deep-ocean basins away from continental margins are composed of the mineralized skeletal remains of planktonic organisms. These deposits, known as **oozes**, are characterized by their chemical composition. Siliceous oozes contain cell walls of diatoms and the internal silicate skeletons of planktonic radiolarians. The skeletons of other planktonic protozoans, the foraminiferans, constitute most of the extensive calcareous oozes found on the ocean floor. These oceanic oozes accumulate very slowly, approximately 1 cm of new sediment every 1,000 years. In the deep sea, oceanic oozes exhibit distributional patterns that reflect the surface abundance of their biological sources. These patterns are shown in **Figure 13.3**.

DID YOU KNOW?

When siliceous oozes become very concentrated they form what is known as *diatomaceous earth*. This term may sound familiar, because diatomaceous earth is used by humans for a variety of purposes. Common uses are for pool filters and as a natural pesticide, and some farmers even use it in livestock feed as a sort of intestinal cleanser. Diatomaceous earth is mined for these uses, although not currently in the deepest seas where it is much less accessible.

To humans, the deep-sea bottom is one of the most uninviting and inaccessible environments on Earth. Below 3,000 m,

the water is cold, averaging 2°C and dipping slightly below 0°C in polar regions (**Table 13.1**). Water temperatures at these depths vary little on time scales of years to decades, creating an extremely constant thermal environment for the inhabitants of the deep sea. Pressures created by the overlying water are tremendous, ranging from 300 to 600 atmospheres (atm) on the abyssal seafloor and exceeding 1,000 atm in the deepest trenches. Laboratory studies have confirmed that metabolic rates of deep-sea bacteria are lower at pressures normally experienced on the seafloor than they are at sea-surface pressures. Less clear is the response of multicellular organisms to high pressures. Several studies have suggested that pressure-induced reductions in metabolic rates may lead to decreased growth rates, lowered reproductive rates, and increased life spans in

TABLE 13.1 Characteristics of a Typical Abyssal Plain Habitat at 3,000 m

Water pressure	300 atm
Water temperature	1°C to 2°C
Salinity	34.5% to 35%
Dissolved oxygen	5 ppm
Light	Bioluminescence only
Current speed	Slow, < 1 cm/s or 0.7 km/day
Sediment	
Type	Soft fine oozes or clay
Deposition rate	< 0.01 mm/yr
Organic content	0 to 0.5%

Figure 13.4 Gigantism is surprisingly common in the deep sea. The deep-sea isopod *Bathynomus giganteus* can exceed 76 cm in length, unlike its relatives living in shallower waters that only grow to maximum lengths of 5 cm.
Courtesy of Expedition to the Deep Slope 2006, National Oceanic and Atmospheric Administration.

the deep sea, culminating in occasional examples of deep-sea gigantism (**Figure 13.4**). Other recent studies have found that the depth-related decline in metabolic rates of crustaceans can be explained as metabolic adjustments to temperature declines with increasing depth and not to a separate depth or pressure effect. Because the effects of high pressures on growth rates and maximum sizes of deep-sea animals are not yet very well understood, this question will not be resolved without more study.

The effects of intense pressure are greatest in trenches, where water depths typically exceed 6,000 m. Individual trenches exist as "islands" of extreme-pressure habitats for animal life, isolated from each other by broad expanses of abyssal plains. Because of the enormous depths and pressures characterizing trenches, studies of deep-sea life have been fragmentary at best, inhibited by the high costs of sampling so far below

the surface and by the difficulties of bringing healthy deep-sea animals to the surface for study. Captured animals encounter extreme temperature and pressure changes when hauled from the bottom. By the time they reach the surface, they are often dead or seriously damaged. Consequently, we presently know more about features on the back side of Earth's moon than we do about vast areas of its deep-sea floor. In fact, our machines have visited Venus about 30 times, but we have been to the deepest part of the ocean only twice (and the bottom of the deepest trench is just 11 km from the surface). Recent advances in sampling techniques and land-based aquaria with conditions similar to the deep sea have led to some success in capturing and maintaining live organisms for short time periods. Scientists at the Monterey Bay Aquarium Research Institute collected an *Opisthoteuthis* flapjack octopus at depth, returned it to the surface unharmed, and it laid eggs in captivity.

Most sampling of living animals from the deep-sea floor is accomplished with dredges, trawls, or sampling scoops (**Figure 13.5**) operated from surface ships at the ends of extremely long cables. It is a bit like trying to sample animals of a rain forest by blindly trailing a scoop on a rope from a balloon 4,000 m above the canopy. Some luck is involved, and the sample returned to the ship is typically biased against animals fast enough to evade the sampler and those too fragile to survive the sampling experience and the trip to the surface. In the last quarter century, this kind of blind remote sampling has been supplemented with direct and video observations from manned submersibles and unmanned remotely operated vehicles (ROVs).

Despite the obvious difficulties with sampling the deep-sea floor, two general and contrasting pictures of animal life there are emerging. Abyssal plains are vast areas of soft mud habitats generally deeper than 3,000 m and extending over distances measured in hundreds or even thousands of kilometers. They are sparsely populated by animal communities with low biomass but with a surprisingly high number of different species.

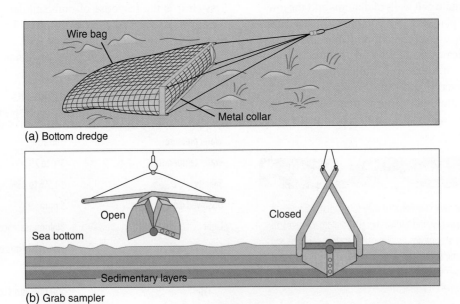

(a) Bottom dredge

(b) Grab sampler

Figure 13.5 Two types of seafloor samplers: (a) bottom dredge, which skims the surface of the sediment, and (b) grab sampler, which removes a quantitative "bite" of sediment and its inhabitants.

These animals are completely dependent on the rain of energy-rich material from the photic zone far above. In almost complete contrast to abyssal plains are deep-sea hydrothermal vent communities. They occupy small and localized rocky habitats on the axes of ridge and rise systems and are characterized by a relatively small number of species existing in an almost overwhelming amount of biomass. One feature of these vent communities that absolutely floored biologists at the time of their discovery was their complete nutritional independence from the primary productivity of the surface waters far above. Several other types of small vent or seep communities have since been discovered, also existing in similar states of nutritional independence from surface primary productivity. These unique and fascinating communities will be described later in this chapter.

13.2   Transfer of Oxygen and Energy from the Epipelagic Zone to the Deep Sea

The nonvent animal communities of the deep sea rely on phytoplankton in the photic zone far above for their oxygen and their food. Dissolved O_2 needed to support the metabolic activities of deep-sea animals must come from the surface, because O_2 simply is not produced in the deep sea. Diffusion and sinking of cold, dense water masses are the chief mechanisms of O_2 transport into the deep-ocean basin, with diffusion providing O_2 from the surface downward and sinking water masses filling the deeper ocean basins with water rich in O_2. As it diffuses downward from surface waters, dissolved O_2 is slowly diminished by animals and bacteria, leaving an oxygen minimum zone (OMZ) at intermediate depths (usually between 500 and 1,500 m). Below this zone, dissolved O_2 gradually increases to just above the sea bottom, as the bacteria that use up large amounts of O_2 in shallower waters are not present in high numbers below the OMZ. The presence of the OMZ reduces the abundance

and activity of mid-depth consumers. One study in the Pacific of a steep-sided volcanic seamount that penetrated the OMZ showed that the number and abundance of benthic species present were much greater just below the OMZ than within it.

DID YOU KNOW?

The oxygen minimum zone (OMZ) is home to organisms with some of the strangest adaptations seen in the oceans. To survive in a habitat with very little oxygen requires an unusual existence. Bacteria living in the OMZ are huge! In fact, the largest known bacterium, *Thiomargarita namibiensis*, is 0.75 mm, twice the size of the period at the end of this sentence, and was discovered in the OMZ. This bacterium oxidizes sulfur, as do many bacteria living in the deep sea. Animals in the OMZ have gills with huge surface areas and respiratory pigments with higher affinities for extracting impressive amounts of O_2 from the water. Although life would seem unlikely in areas with almost no oxygen, adaptations for survival allow bacteria and animals to thrive in these areas.

The other crucial and variable resource in the deep sea is food. The absence of sunlight means that there are no photosynthesizers. Food for deep-sea benthic communities, by necessity, comes from above. Only recently have studies determined the rate and condition of the food's arrival at the sea bottom. Several studies have demonstrated that, in contrast to earlier assumptions, a tight coupling exists between near-surface primary productivity and the eventual consumption of that production by inhabitants of the deep-sea floor (**Figure 13.6**). In most areas of the deep ocean, seasonal variations in the organic content of surface sediments reflect the seasonality of primary production occurring in the photic zone directly above. Such seasonal pulses of organic matter are often detectable several centimeters below the sediment surface as well, as infaunal animals rework this material through their burrowing and feeding activities.

(a)

(b)

Figure 13.6 Seafloor images showing the deposition of phytodetritus before (a) and 2 months after (b) a phytoplankton bloom in the photic zone above displays the dramatic differences in the seafloor after such an event.

(a, b) Courtesy of Dr. Richard Lampitt, National Oceanography Centre, Southampton.

Sinking rates of food-rich particles are largely a function of their size, with smaller particles sinking more slowly, but other factors, such as density, appear to affect sinking rates, too. In one study conducted near Antarctica, individual diatoms and radiolarians sank nearly as fast as the largest aggregated particles, likely due to their dense frustules and skeletons. As particles sink slowly through the photic zone, much of the near-surface phytoplankton is aggregated into larger gelatinous blobs or compacted into fecal pellets when eaten by zooplankton, which settle rapidly to the bottom as marine snow. This accumulation of microscopic particles into macroscopic fecal particles and gelatinous aggregates accelerates the transport of organic material to the abyss, falling from the surface waters in a few days rather than the weeks or months of settling time necessary for smaller phytoplankton particles. Thus, the time lag between the initial production of organic material in the photic zone and its arrival on the deep-sea floor may be a few days or as much as several weeks. On the way down, sinking particles of organic material are repeatedly consumed and defecated by pelagic scavengers and colonized by bacteria. Consequently, by passing through several digestive systems on their way to the abyss, sinking particles, especially the very small ones, lose much of their nutritive value by the time they settle to the bottom. The amount of photic zone productivity that actually reaches the sea bottom is typically only a few percent, and it is even less at greater depths. Despite the low amounts of productivity reaching the deep sea, it is this productivity that fuels most of the life found in this ecosystem.

13.3 Life on Abyssal Plains

Abyssal plains are flat and stable places occasionally interrupted by low abyssal hills. They are inhabited by a distinctive group of animals, and some exhibit structural adaptations that make them appear notably different from their shallow-water relatives. In addition, in deeper waters the dominant taxonomic groups shift. Echinoderms (especially sea cucumbers and crinoids), polychaete worms, pycnogonids, and isopod and amphipod crustaceans become abundant, whereas bivalves and other mollusks and echinoderms, such as sea stars, decline in number. Some taxonomic groups are virtually absent until relatively great depths are reached. Most species of pogonophorans, for instance, are found below 3,000 m, and 30% of known species are restricted to trenches below 5,000 m.

Low temperatures, high pressures, and a limited food supply led to an early and widespread belief that the rigorous and specialized climate of the deep sea would not support a highly diversified assemblage of animals. It was assumed that only a few highly adapted animals could succeed in the seemingly inhospitable abyss. Repeated sampling did reveal a marked decline in both density and biomass of organisms at greater depths, both presumed responses to the limiting effects of decreasing food and O_2 with depth; however, improvements in sampling equipment and sample analysis techniques in the past quarter century have shown that deep-sea assemblages contain a diversity of species comparable with or even exceeding the diversity of soft-bottom communities in shallow inshore waters. Nearly every new mission to sample the deep sea includes the discovery of previously unknown species. We are just scratching the surface of our knowledge of deep-sea life, and there is so much more to discover. Some classic sampling methods are described next, but a handful of scientists are working on more advanced methods as this book is being written.

Using an **epibenthic sledge**, a sampling device combining the best attributes of a bottom trawl and a dredge (**Figure 13.7**), researchers demonstrated an increase in species diversity with increasing depth along a transect from shallow New England coastal waters to deep-ocean basins near Bermuda nearly 5,000 m deep. Other researchers studying different deep-sea regions have not always found this pattern of increasing diversity with increasing depth; it seems to be strongly dependent on the choice of study location. Collectively, these studies dispel the basic hypothesis first supported by the *Challenger* expedition almost a century earlier that animal diversity decreases with increasing depth in the deep sea (**Table 13.2**). At the same

Figure 13.7 An epibenthic sledge used to sample epifauna and near-surface infauna.
© Courtesy of KC Denmark Research Equipment.

TABLE 13.2 Diversity of Major Animal Phyla (Indicated by Number of Species and Families) Collected from a Small Sampling Area at a Depth of 2,100 m off the New England Coast

Phylum	Number of Families	Number of Species
Annelida	49	385
Arthropoda	40	185
Mollusca	43	106
Echinodermata	13	39
Cnidaria	10	19
Pogonophora	5	13
Sipuncula	3	15
Echiura	2	4
Nemertea	1	22
Hemichordata	1	4
Priapulida	1	2
Brachiopoda	1	2
Ectoprocta	1	1
Chordata	1	1

Data from J. D. Gage and P. A. Tyler. *Deep-sea Biology: A Natural History of Organisms at the Deep-Sea Floor* (Cambridge University Press. 1991).

time, these studies confirm another *Challenger* observation: that the biomass of animal life in the sea becomes sparser at greater depths. Thus, although species diversity is surprisingly high at great depths, the average number of individuals per unit area and their average body sizes do tend to decrease consistently with depth. Yet so much of the deep-sea floor remains unsampled that we might yet find thousands more species if it is ever explored with the intensity that we currently direct to more accessible ecosystems, such as shallow coral reefs or tropical rain forests.

To explain the relatively high species diversity of abyssal plains (**Figure 13.8**), researchers have proposed several complementary explanations. First, the enormous extent of abyssal plains (about 200 million km²) provides large areas with few barriers to block dispersal, and thus species exhibit broad geographic ranges. When large areas of the deep sea are sampled at the same depth, approximately one additional species is added for each additional square kilometer sampled. When different depths are sampled, the rate at which additional species are collected is even greater. These trends suggest very high species diversity in the deep sea and also point out one of the basic difficulties in clarifying the pattern of that diversity: enormous areas must be sampled (at enormous costs) before we have an accurate picture of the number and distribution of species in the deep sea.

Second, within large areas of nearly constant deep-sea environments, small-scale variations in food availability and physical or biological disturbance create new opportunities for colonizing by additional species. Events such as food-falls of large animal carcasses or the sorting and modification of

sediment deposits when animals form tubes and compact sediments to produce fecal pellets and castings may serve to maintain some spatial diversity in what otherwise seems a very uniform environment.

Other biologists choose to view the surprising level of deep-sea biodiversity from a more global perspective. They ask, what do all highly diverse communities, such as the deep sea, coral reefs, and tropical rain forests, have in common? This different view of the question leads some to conclude that long-term environmental stability is what promotes exceptionally high biodiversity in these three communities. They suggest that chronically stable conditions, whether they are sunny and hot in a rain forest canopy or dark and icy on the abyssal plain, enable organisms to achieve extreme adaptational specializations such that competitive interactions between species are minimized. Proponents of this stability–time hypothesis maintain that the frequent physiological stress that accompanies less predictable environments will favor tolerance of a wide range of ambient conditions, thus preventing the increased biodiversity that results from divergence into separate niches. In short, environmental stability results in exceptional biodiversity that is maintained by highly refined resource partitioning.

Still others criticize this hypothesis by pointing out that biodiversity in the deep sea does not increase in a linear fashion with depth. The data show that there is a peak in biodiversity at a depth of 2,000 to 3,000 m on the continental slope for many taxa, including gastropod and bivalve mollusks, polychaete worms, and teleost fish. This "humpbacked" curve of biodiversity does not support the stability–time hypothesis because there is no evidence that environmental stability peaks on the continental slope. If anything, ambient conditions are more constant on the abyssal plains. What does peak at lower portions of the continental slope is the frequency of large-scale environmental disturbances, such as landslides and other geological phenomena. This disequilibrium explanation suggests that frequent environmental disturbances, such as hurricanes, cyclones, and typhoons on coral reefs and rain forests; storm surge and ice scour on rocky intertidal zones; or landslides along the continental slope, would decrease competitor abundance such that resources would rarely be limited. Under these disequilibrium conditions, where resources are plentiful and competitors are scarce, competitive interactions would be minimized, and many species could coexist while sharing the abundant resources. Additional biological disturbance on continental slopes comes in the form of specialized predators known as croppers (described later). Before we can determine why diversity appears high in certain areas of the deep sea it will be necessary to sample more of its area to make more robust comparisons between depths and geographic locations.

A portion of the smaller particles of organic material settling to the bottom is immediately claimed by suspension feeders. These are often epifauna clinging to rocky outcrops, shells, or manganese nodules. Other epifauna use stiltlike appendages to keep their bodies from sinking into the sediments. Most benthic animals in the deep sea, however, are infaunal **deposit feeders**. A variety of worms, mollusks, echinoderms, and crustaceans obtain their nourishment by ingesting accumulated detritus and

(a)

(b)

(c)

(d)

(e)

(f)

Figure 13.8 A variety of deep-sea organisms from various taxonomic groups: (a) slime star, *Hymenaster*; (b) a Venus flytrap anemone, *Actinoschphia*; (c) a deep-sea shrimp, *Glyphocrangon*, with relatively large eyes; (d) a purple sea cucumber, *Elasipod*; (e) a crown jellyfish, *Atolla*; and (f) a deep sea pycnogonid.
(a-f) Courtesy of Ocean Networks Canada.

digesting its organic material. These deposit feeders engulf sediments and process them through their digestive tracts. They extract nourishment from the organic material in the sediment in much the same manner as earthworms do. Analyses of their stomach contents indicate that many infauna indiscriminately engulf sediments containing smaller infauna and bacteria as well as dead organic material; others select organically rich substrates for consumption. It has been estimated that 30% to 40% of the organic material available at the bottom is first absorbed by benthic bacteria, which, in turn, is consumed by larger animals. An important component of the organic material is the chitinous exoskeletons shed by planktonic crustaceans; these cannot be digested by most animals but are decomposed and used by bacteria.

Large, rapidly sinking particles, including dead squids, fish, and an occasional whale or seal, may provide a significant, although unpredictable, supply of food to inhabitants of the deep-sea floor. When large food parcels do contribute significantly to the energy budget of the deep sea, they are rapidly consumed by various wide-ranging scavenger-predators.

Most scavenger-predators seem swift enough to evade bottom trawls and are best known from photographs taken by underwater cameras baited with food (**Figure 13.9**). They seem to be food generalists, capable of rapidly locating and dispatching large food items as they arrive on the bottom. Some of the food is eventually dispersed as detritus and fecal wastes to other benthic consumers.

Somewhat like saprobic fungi, several animal species are capable of extracting sufficient nourishment from sediments by conducting digestive processes outside their bodies. These animals absorb the products of digestion either through specialized organs or across the general body wall. Deep-sea pogonophorans, small worms related to the large tube worms recently discovered in association with deep-sea hot springs, depend on this type of **absorptive feeding**, as do numerous echinoderms. Sea stars are usually carnivorous, but a few species are quite opportunistic. When feeding on bottom sediments, some sea stars extrude their stomachs outside their bodies to digest and absorb organic matter from the sediments. The omnipresent bacteria also depend on extracellular digestion. As they absorb

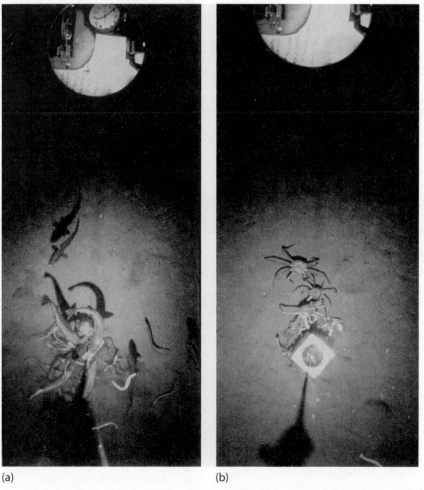

(a)   (b)

Figure 13.9 A bait can lowered to the seafloor at a depth of 1,390 m off the northern Baja California coast quickly attracts mobile scavenging fish (a). Several hours later (b), the same scene is dominated by slower invertebrates, including the tanner crabs seen here.

(a,b) © Scripps Institution of Oceanography Archives, UCSD.

Figure 13.10 Croppers, such as these foraging sea cucumbers known as "sea pigs," *Scotoplanes*, may keep deep-sea biodiversity high. This photo was taken at 2,300 m. Courtesy of Ocean Networks Canada.

nutrients and their populations grow, they, in turn, become a significant source of particulate food for deposit feeders.

The term **cropper** has been applied to deep-sea animals that have merged the roles of predator and deposit feeder (**Figure 13.10**). These croppers, by preying heavily on populations of smaller deposit feeders and bacteria, may be responsible for reducing competition for food and for encouraging coexistence between several species sharing the same food resource. Thus, although a deep-sea community as a whole is food limited, populations within the community need not be as long as they are heavily preyed on. With competition for food reduced by cropping, fewer species are pushed to extinction (on a local scale), and species diversity remains high. The disruptive effects of large croppers on smaller ones can be compared with predators, sea ice, or logs as disturbance mechanisms in rocky intertidal communities. By nonselectively reducing competition, they lessen the possibility of a species being excluded by resource competition.

In the physically stable environment of the deep sea, patterns of reproduction are thought to differ appreciably from the reproduction patterns of shallow-water benthic animals. Few deep-water benthic species produce planktonic larvae because the chances of larvae reaching the food-rich photic zone several kilometers above and successfully returning to the ocean floor for permanent settlement are extremely remote. To compensate for the absence of a dependable external food supply, fewer and larger (lecithotrophic) eggs are produced. The larvae are supplied with adequate yolks so that they can develop to fairly advanced stages before hatching. Brood pouches and other similar adaptations further enhance the chances of survival by protecting the eggs until they hatch. Consequently, the larvae have a reasonable chance of completing their development in one of the most severe environments on Earth.

DID YOU KNOW?

Although nearly invisible to predators most of the time due to a transparent body, when a jellyfish consumes a fish or other prey item it suddenly becomes visible due to the prey inside its gastrovascular cavity. In order to avoid being vulnerable to prey, some deep-sea jellyfish are entirely red, because red light is not visible at depth. Others are transparent except for their gastrovascular cavity where prey is digested. When a fish is consumed it is contained inside the red-colored cavity, allowing the jelly to remain inconspicuous. These adaptations reflect organisms taking advantage of the darkness of the deep sea for survival.

Case Study

Reproductive Biology of an Extreme Deep-Sea Octopus

Common name: Deep-sea octopus

Species name: *Graneledone boreopacifica*

The deep-sea octopus *Graneledone boreopacifica* is found in the North Pacific Ocean at great depths, generally around 2,000 m. This species may not look phenomenal upon initial inspection (**Figure A**), but it has some features not only unique to the world of octopuses, but unique to the animal kingdom. In general, octopuses are known to have just one reproductive period in their lifetime. Females lay and protect eggs (referred to as *brooding the eggs*), and shortly after the eggs hatch the mother dies. In the majority of shallow-water octopus species, the egg-tending period ranges from just 1 to 3 months. In deep-sea organisms, some biological processes, such as growth and reproduction, are hypothesized to take longer or be delayed in response to the extremely cold water temperatures. Metabolic rates decline in response to the cold, which likely leads to slow growth and even very long life spans of many organisms inhabiting great depths.

Individual *G. boreopacifica* females have been observed brooding eggs on several occasions by deep-sea submersibles and ROVs. In 2007, a study of a rock outcropping in Monterey Submarine Canyon in Monterey Bay, California, included an observation of an individual female near a rock where this species is known to breed, but no egg case was attached to the rock. The female was moving around the area, seemingly foraging. One month later the ROV returned to the same location and observed the same female, identifiable by very obvious scars on her body, tending to a clutch of eggs attached to the rocks. This observation led to a unique opportunity to track the amount of time this individual *G. boreopacifica* brooded her eggs. The ROV crew and scientists returned to the location 18 times over 4.5 years to document the progression of brooding and development of the octopus eggs. Each time they returned the same female octopus was guarding her eggs, and the eggs were growing in size (**Figure B**). After 53 months the octopus was gone and the egg cases were cracked open, indicating that the young had hatched (**Figure C**). This brooding period of over 4 years is the longest known for any animal on Earth!

In an earlier study of *G. boreopacifica*, the ROV *Tiburon* made an attempt to collect a female adult and her eggs. Upon being disturbed, the eggs suddenly hatched, and scientists were able to collect the newly hatched young, along with the adult. The specimens were studied back at the laboratory, and it was discovered that young of *G. boreopacifica* are the largest (5.5 cm) and most advanced in the octopus world. This was the first documented collection of live young, and because the hatching event

Case Study   (*Continued*)

Figure A An orange busy coral (*Trissopathes*) and deep-sea octopus
(*G. boreopacifica*) at nearly 2,000 m depth, photographed at the Davidson Seamount
in California.
Courtesy of National Oceanic and Atmospheric Administration and Monterey Bay Research Institute.

Figure C Broken egg capsules indicating the young
G. boreopacifica had hatched.
Courtesy of B. Robison, B. Seibel, and J. Drazen. *PLoS One*. 9(2014):e103437.

Figure B A female *G. boreopacifica* protecting her eggs that are nearly developed.
Notice the dark eyes visible through the egg capsule.
Courtesy of B. Robison, B. Seibel, and J. Drazen. *PLoS One*. 9(2014):e103437.

occurred upon disturbance, it can be assumed that the young may have been even larger
and more developed if left to hatch on their own.

In 2009, a clutch of a female *G. boreopacifica* was collected to determine paternity.
Because their reproductive methods likely include storing sperm, it has been assumed
that multiple paternity may be common in octopuses, but this had never before been
documented. Using microsatellite analysis scientists confirmed the suspicions of

multiple paternities, as at least two genetically distinct males had contributed sperm
to the hatchlings.

G. boreopacifica is not the only deep-sea octopus, but it has been relatively widely
studied for a deep-sea species. The discovery of so many extreme features in *G. boreopacifica* actually raises a variety of questions. If *G. boreopacifica* possesses extreme features, such as the longest brooding period of all animals and the largest and most
developed hatchlings of all octopus species, what about the other deep-sea octopuses?
G. boreopacifica has provided us with a small sampling of these unique features, but
many other extremes likely exist, and there are sure to be other surprising discoveries
in deep-sea mollusks.

Critical Thinking Questions

1. One of the studies described above confirmed that deep-sea octopus
   clutches have multiple fathers. What is the adaptive advantage for
   deep-sea organisms to use such a reproductive strategy?

2. What is one disadvantage of a really long brooding period?

3. What do you feel are the ethical concerns, if any, with collecting organisms from the deep sea where they likely cannot be safely returned?
   G. boreopacifica is not considered rare or in danger of extinction. Do
   you feel it is morally excusable to collect specimens in the name of
   science?

13.4   Vent and Seep Communities

Deep-sea vent and seep communities contain impressively
dense aggregations of animals, many living in symbiotic relationships with bacteria. Each vent or seep is like an oasis of
life in comparison to the patchy and dilute distribution of animals found on the seafloor of the vast abyssal plains at similar
depths. Vent communities are organized around hot-water and

other outflows vented from cracks in the seafloor. Most are
located on the axes of oceanic ridge or rise systems; a few others
are associated with oceanic island–arc systems. Seep communities tend to be more dispersed in areas where hydrocarbons,
particularly methane or other natural gases, are percolating up
through sediments to the surface of the seafloor (**Figure 13.11**).
Both communities were discovered relatively recently and have
led to extensive exploration and research efforts.

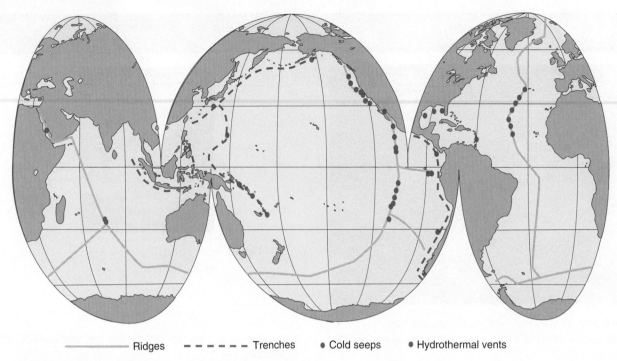

———— Ridges — — — — Trenches ● Cold seeps ● Hydrothermal vents

Figure 13.11 Approximate locations of confirmed hydrothermal vent communities (red dots) and cold seeps (purple dots). Most are associated with trenches or with actively spreading ridge systems.

Hydrothermal Vent Communities

The discovery of deep-sea animals intimately associated with hydrothermal vents has radically changed our understanding of life in the deep sea. The hydrothermal vents at the centers of these communities are the geological plumbing that channel hot, metal-enriched water from deeper within the crust to the surface of the seafloor. Because most of these vents are located on the axes of seafloor ridge or rise structures, they exist in very young rocks that form as crustal plates spreading apart. Rising molten magma cools, solidifies, and creates a new oceanic crust made of black volcanic basalt. If you could approach a hydrothermal vent from an adjacent abyssal plain, you would observe the fine sediment blanket that covers the basaltic ocean crust thinning until it disappears at the ridge or rise axis, exposing the rough surfaces of freshly cooled and solidified basalt. It is this zero-age oceanic crust, too young to have accumulated any sediment, that creates conditions for the development of hydrothermal vents and the unique animal and bacterial communities associated with them (**Figure 13.12**).

Not all hydrothermal vents are alike. Although most are dominated by freshly cooled basalt as described above, some are composed of massive boulders of sulfide that are cemented together by amorphous silica. These seafloor massive sulfide (SMS) deposits are equivalent to deposits of sulfide ore of volcanic origin that are mined on land. Extraction of SMS deposits for economic purposes, which are rich in lead, zinc, and copper sulfides, is hampered by the great depths at which these boulders form and the uncertainty surrounding the environmental impacts the harvest would cause (much like the manganese nodules described earlier). Hydrothermal vents in the

eastern Pacific are unlike those found in the North Atlantic Ocean. Although vents in both oceans are fractured perpendicularly into ridge segments, Pacific segments are separated by 10-km-long fracture zones, whereas Atlantic segments are separated by hundreds of kilometers. Moreover, vents in the eastern Pacific Ocean are approximately 200 m wide and about 10 m deep and are spreading relatively rapidly, whereas Atlantic vents are 1-km-wide, 2-km-deep cracks that spread very slowly.

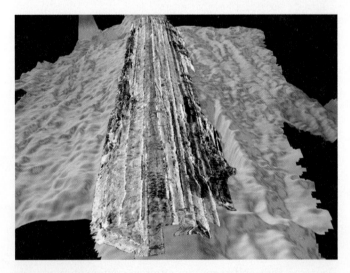

Figure 13.12 Sidescan sonar image (gray scale) overlaid onto multibeam bathymetry (color) for the East Pacific Rise (9° 25′ N to 9° 57′ N) at a depth of 2,500 m. The sidescan image covers an area of about 31 by 57 km and resolves seafloor features as small as 2 m.

Reprinted with permission from Woods Hole Oceanographic Institution; Courtesy of D. Fornari, M. Tivey, and H. Schouten. P. Johnson (U. Hawaii), who helped create the image. Funding provided by the National Science Foundation.

Figure 13.13 Red-plumed tube worms, *Riftia*, with sea anemones and a few other members of this unusual deep-sea community.
Courtesy of National Oceanic and Atmospheric Administration.

Deep-sea hydrothermal vent communities were first directly observed in 1977 when a team of geologists studying the Galápagos Rift Zone west of Ecuador discovered several remarkable assemblages of animals living in water nearly 3 km deep (**Figure 13.13**). Dense aggregations of large mussels, clams, giant tube worms, and crabs were found clustered in small areas of shimmering hot water pouring from the seafloor. Water temperature often exceeds 100°C at vent sites, in sharp contrast to the near-uniform 2°C abyssal water just a few meters away. Even though some of these vent animals are adapted to high water temperatures, it is the altered chemistry of the water, not the heat, that provides the basis of life in vent communities.

These hot-water plumes pouring from seafloor vents are the end products of seawater circulating through the many cracks and fissures of the new crust as it forms along the axes of the rise system (**Figure 13.14**). It has been estimated that a volume of water equivalent to the volume of all Earth's oceans circulates through these ridge crack systems approximately every 8 million years. As seawater percolates through these cracks, crustal rock temperatures above 350°C drive some complex chemical reactions to alter several properties of the circulating seawater. In addition to being heated well above the boiling point at sea level, vent water is devoid of dissolved O_2, is very acidic, and is enriched in metals such as manganese, iron, copper, zinc, and silicon. Sulfate (SO_4^{2-}), the third most abundant ion in seawater, reacts with water when heated to form hydrogen sulfide (H_2S). Despite being toxic to most animals, H_2S plays a central role in subsequent biological processes as the heated water emerges from the seafloor vent.

Some of the H_2S reacts with iron and other metal ions to form metal sulfides in concentrations several million times greater than those found in average seawater. At the vent opening, this super-enriched metal sulfide solution quickly mixes with cold surrounding seawater, and a dense, black "smoke" of iron sulfide particles is produced (**Figure 13.15**). The massive quantities of black particles created at chimney openings atop these vents are known as *black smokers*. These particles either settle immediately around the vent or are oxidized by oxygen from surrounding nonvent water to produce thick deposits of metal oxides in the vicinity of the vent.

Mixing of vent and nonvent waters in the proximity of vent chimneys creates strong gradients in dissolved O_2, pH, water temperature, and H_2S extending outward from the eye of the vent. Clouds of free-living bacteria, so abundant near vent openings that they make the water appear milky, are important primary producers in hydrothermal vent communities. This form of primary production, however, is a form of **chemosynthesis** rather than photosynthesis characteristic of the sunlit portions of the ocean. Like near-surface primary producers, these bacterial communities also experience large cyclic fluctuations in productivity rates. Rather than being seasonally driven, as are phytoplankton communities, vent bacteria periodically thrive and bloom in response to changes in flow rates of vent water caused by local shifts in crustal plates.

Vent bacteria use dissolved O_2 from the surrounding water to oxidize the abundant H_2S back to SO_4^{2-} to obtain energy to synthesize complex and energy-rich organic compounds

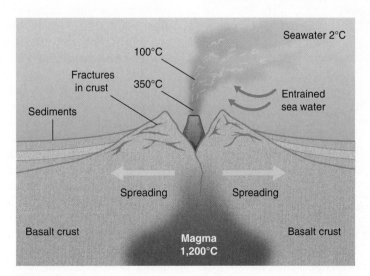

Figure 13.14 Cross section of a ridge axis and the plumbing connected to a vent chimney. Seawater flows through numerous fissures in the hot basalt crust of the seafloor.

Figure 13.15 Tubeworms thriving near the largest known black smoker vent, located off the Oregon coast and known as "Godzilla." Particles in the "smoke" are major sources of metal deposits around vent chimneys.
Courtesy of Ocean Networks Canada.

(**Figure 13.16**). In this manner, bacterial primary production occurs locally, and the vent community is independent of surface productivity for its food. Vent-based chemosynthesis is not totally independent of photosynthesis at the sea surface because the O_2 used in chemosynthesis originates from phytoplankton in the photic zone.

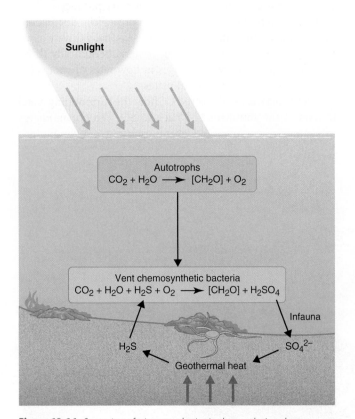

Figure 13.16 Comparison of primary production in photosynthetic and chemosynthetic systems. Transfer of O_2 from the surface to vent bacteria materially links these two marine primary-production systems.

Figure 13.17 Aggregations of large vent clams, *Calyptogena*, thrive near deep-sea hydrothermal vents.
Courtesy of Dr. Ana I. Dittel, University of Delaware.

Bacterial chemosynthesis is not a newly discovered form of primary productivity; it was first described in the late 19th century. Vent-based chemosynthesis, however, is unusual in that it is the principal energy source for the entire base of hydrothermal vent food webs. The unusual chemosynthetic bacteria of hydrothermal vent communities are the focus of intensive biochemical, physiological, and genetic studies. It is generally agreed that several species of bacteria tolerate extremely high water temperatures (100°C to 115°C) and that these "thermophiles" belong to the domain Archaea. The most abundant species of chemosynthetic bacteria have adapted to one of two markedly different vent community habitats: as suspended bacteria in and around black smoker fluids or as symbiotic bacteria living on or within tissues of vent-dwelling animals.

Suspended bacteria are consumed by dense aggregations of large clams and mussels found nowhere else in the sea (**Figure 13.17**). By trapping bacteria on their mucus-lined gills, as do their shallow-water suspension-feeding relatives, these large vent community bivalves may capitalize on an abundant, yet extremely small, supply of food particles. Some evidence also exists that these vent bivalves maintain endosymbiotic chemosynthetic bacteria within their gills and harvest them much as giant clams do in coral reef communities.

The large red-plumed vestimentiferid worm, *Riftia*, shown in Figure 13.13, completely lacks a digestive tract and is unable to harvest suspended bacteria. Instead, it maintains internal bacterial symbionts (endosymbionts) that oxidize H_2S within a specialized organ, the trophosome. Major blood vessels link the trophosome to the bright red plume, where gas exchange occurs (**Figure 13.18**). The blood of *Riftia* contains two proteins to deal with the high concentrations of dissolved H_2S and small amounts of dissolved O_2 in the deep-sea hot-spring environment. The first protein is a rare sulfide-binding protein used to transport large quantities of H_2S from the plume, where it is absorbed, to the trophosome, where it is used by bacterial symbionts. Hemoglobin, the other blood protein, is used to bind and transport O_2 within these animals. By binding O_2 and H_2S to different blood proteins, the high toxicity usually associated

(a)

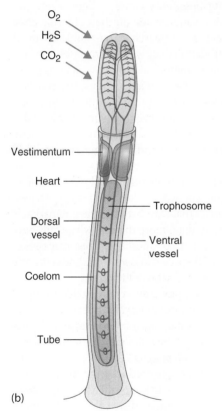

O₂
H₂S
CO₂

Vestimentum

Heart

Dorsal
vessel

Coelom

Tube

Trophosome

Ventral
vessel

(b)

Figure 13.18 External appearance (a) and internal anatomy (b) of the tube worm, *Riftia*.

(a) © Dr. Ken MacDonlad/SPL/Photo Researchers, Inc.

with H₂S is avoided, and the O₂ is prevented from prematurely oxidizing the H₂S before both are delivered to the endosymbionts in the trophosome.

In these deep-sea vent communities, bacterial use of O₂ from the water in the immediate vicinity of the vents is extensive. To compete effectively with bacteria for the limited available O₂ and to sequester O₂ away from H₂S, larger animals need some type of blood pigment with a strong affinity for O₂.

Figure 13.19 Aggregations of large vent crabs, *Bythograea*, with blue-pigmented blood visible through their carapaces.

Courtesy of Dr. Ana I. Dittel, University of Delaware.

Hemoglobin is abundant in *Riftia* and in large bivalve mollusks, including the large clam, *Calyptogena* (Figure 13.17). In the crabs most commonly found in vent communities (*Bythograea*, **Figure 13.19**), the blue pigment **hemocyanin** is used instead. The blood pigments of these vent animals are relatively insensitive to changing temperatures. Such thermal insensitivity is unusual and likely has evolved to accommodate the extremely wide range of water temperatures experienced over very short distances around vents. These pigments thus assist the larger deep-sea vent animals to extract the little O₂ left over by the abundant bacteria. With these adaptations, these populations of benthic animals have tapped into an unusual energy source—the heat and chemicals of Earth's crust—and have no need to rely on the fall-out products of the photic zone above.

DID YOU KNOW?

Most scientists agree with the hypothesis that life originated in the sea, and evidence indicates this occurred around 3.8 billion years ago. What is not clear is exactly how or where the first life forms evolved. Before life in deep-sea vents was discovered, it was assumed that life required sunlight, so naturally the first organisms evolved in waters of the photic zone. Now that we know sunlight is not a requirement and that life flourishes in seemingly toxic environments of deep-sea hydrothermal vents spewing hydrogen, sulfur, and CO₂, there are new hypotheses for where life evolved. The deep sea seems a likely candidate area, but evidence has yet to be found. What do you think?

Diversity of Vent Inhabitants

The assortment of organisms living around the vents in each ocean appears to vary dramatically, as the unique attributes of vents in different areas present different challenges for life. In fact, six major seafloor provinces are recognized. The benthic community that surrounds hydrothermal vents along the East Pacific Rise is dominated by *Riftia*, the spectacular tube worm described previously here, along with a mussel (*Bathymodiolus*),

Figure 13.20 Eyeless vent shrimp, *Rimicaris*, dominate deep hydrothermal vents in the North Atlantic.

Courtesy of National Oceanic and Atmospheric Administration *Okeanos Explorer* Program, Mid-Cayman Rise Expedition 2011.

yet they are not found in the Atlantic Ocean at all. Instead, deep hydrothermal vents (2,500 to 3,650 m) in the North Atlantic are dominated by the shrimp *Rimicaris exoculata* (**Figure 13.20**). These eyeless (hence their name), whitish shrimp swarm around North Atlantic vents by the hundreds of thousands of individuals, farming chemosynthetic bacteria on their highly modified gills that are contained in unusually large gill chambers, which cause the carapace of this shrimp to bulge laterally. Centered on their backs is a conspicuous white reflective patch, called the *dorsal organ*, which seems to function as a single eye, being most responsive to 500-nm light waves dorsal to the shrimp. Shallow vents in the North Atlantic (800 to 1,700 m) are colonized by far fewer shrimp but support a second species of *Bathymodiolus* mussel.

In the northeast Pacific, hydrothermal vents are dominated by bushes of skinny tube worms called *Ridgeia*. Although genetically identical, two growth forms of *Ridgeia* have been observed. The short, fat form is 30 cm long and 2 to 3 cm wide, whereas the long, skinny form can be 1 m long yet only as wide as a pencil.

The fifth biogeographic region recognized, vents of the western Pacific, includes limpets, mussels, barnacles, and "hairy" snails not seen elsewhere. Last, the central Indian Ocean is home to a vent community that seems to be intermediate to that of the Atlantic and Pacific assemblages described above, with Atlantic-type shrimp and Pacific-type snails and barnacles dominating the landscape. Biologists are very interested in the potential processes that may generate these distribution patterns and biogeographic limits, and most hypotheses center around larval dispersal of vent-dwelling taxa, the topic of the next section.

Larval Dispersal of Hydrothermal Vent Species

How are newly formed hydrothermal vents initially colonized when the nearest vent community with reproductively mature animals may be hundreds of miles away? Little is known about the spawning behaviors or early developmental stages of most vent animal species. *Riftia* is perhaps the best studied of these, and it is still not known if fertilization is internal or external in this tube worm. Free-swimming larvae of most vent community members remain elusive, and larval durations are not well known. Based on results of a laboratory study rearing *Riftia* larvae, it is estimated that *Riftia* have larval stages as long as 38 days, which is long enough for dispersal to new areas closer than about 100 km, depending on local currents. Reproductive strategies of members of these small and isolated vent communities must somehow balance the demands of retaining some of their progeny within their existing vent communities with the conflicting need to disperse larval forms to colonize newly formed vents.

Several "bridging" mechanisms to disperse larvae and explain the presence or absence of the same species at different hot springs have been proposed. Larvae may drift passively with bottom currents that flow parallel to ridge or rise axes, eventually reaching a new settlement site. Alternatively, non-swimming larvae may be transported as hitchhikers on other community members capable of walking (crabs) or swimming (fish) between vent sites.

Another intriguing possibility came to light as an outgrowth of a modest observation made off the California coast in 1987 from the deep-sea research submersible *Alvin*. A decomposing, but still identifiable, whale carcass on the seafloor was surrounded by numerous clams of two species previously seen only at vent sites. Dense mats of sulfide bacteria were also found. This suggests that hot-spring community members may "hop" from carcass to carcass, eventually moving long distances along the sea bottom. Such larval hops, if they occur, may be assisted by the vertical rise and eventual horizontal drift of warm plume water from an active vent. It is not yet known whether such **whale falls** occur frequently and closely enough to provide the larval-hopping network of food-rich material necessary to bridge the distances that exist between these remarkable deep-sea communities. To date, more than 20 whale falls have been found harboring ventlike communities. These carcasses of dead whales supported 10 species known from hot vents and 19 species normally associated with cold seeps.

Perhaps seafloor topography plays a role by decreasing or augmenting the dispersal abilities of larval forms. Others speculate that tectonic processes may open or close dispersal gateways. For example, the Logatchev site east of the Caribbean Sea is the only site where *Calyptogena* clams are found in the Atlantic, suggesting that they squeezed into the Atlantic from the eastern Pacific just before the Isthmus of Panama appeared.

The fact that vent communities discovered in the South Atlantic Ocean host distinct species that differ from mid-latitude vent communities suggests that strong currents may be responsible for helping or hindering larval dispersal; 1 million cubic meters of seawater flows through the Romanche and Chain Fracture Zones on the Atlantic equator each second, effectively preventing dispersal of larvae from the north. Entrance from the south may be prevented by equally powerful currents and huge chasms in the seafloor. The first deep-sea hydrothermal vent to be discovered in the Southern Ocean was documented in 2012 and hosts a variety of chemosynthetic organisms, including a new species of yeti crab (*Kiwa*), stalked barnacles, limpets, and anemones, to name a few.

Other researchers point to mechanisms that are suspected of influencing larval settlement in shallow waters, such as sea-floor composition or differences in seawater chemistry (in this case, because of differences in vent effluent in different areas). Some have speculated that life history characteristics of vent fauna may interact with vent geology in interesting and important ways. For example, rapidly spreading Pacific vents may favor rapid colonizers (i.e., *Riftia*), whereas slowly spreading vents in the Atlantic may favor populations of shrimp that grow more slowly.

Depth obviously plays a role in the North Atlantic, in that uncountable swarms of *Rimicaris* blanket deep Atlantic vent sites but are rare at more shallow sites. Finally, some biologists emphasize larval abilities and behaviors in their search for causes of the observed biogeographic differences. Some larvae swim upward upon hatching, whereas others swim downward and are carried in a different direction by a deeper current. Some larvae, such as the megalops larvae of the crab *Bythograea*, may be able to swim all the way to the sea surface to ride faster, wind-driven surface currents. In the laboratory, individuals of this species have survived surface pressures for 10 months and then metamorphosed into juvenile crabs successfully.

Clearly, a great deal of research still needs to be conducted on hydrothermal vent organisms. Obvious gaps in our biogeographic knowledge include the Arctic Ocean, which has never had deep-water connections to either the Atlantic or the Pacific, but hydrothermal vents are known from this area and likely contain unique and endemic vent communities. The Caribbean Sea is dominated by methane seeps that may act at stepping stones for larval dispersal. The Chile Rise in the southern Pacific Ocean is the union of three separate mid-oceanic ridges, and it may represent "Grand Central Station" for larval dispersal. Finally, other areas, such as the deepest spot on the planet, the Challenger Deep, and other locations near the Mariana Trench, are known to possess a full complement of vents and seeps, and their fauna have been observed briefly, but these areas remain largely unstudied.

Cold Seep Communities

Since their discovery in the 1980s, cold seeps have been reported from many regions, various tectonic settings, and different water depths, ranging from continental shelves to the deep sea. Like hot vents, they have been recognized as unique ecosystems that, particularly in the deep sea, represent oases (i.e., hotspots of biological production and biodiversity surrounded by vast, relatively barren abyssal plains). The resource fueling both seep and vent ecosystems is either methane or sulfide emanating from the seafloor. As in hot vents, primary producers are chemosynthetic bacteria using the enzymatic oxidation of reduced compounds as a basic metabolic energy source. This microbial consortium, consisting of mats of methane-oxidizing and sulfate-reducing bacteria, consumes sulfate and methane in a 1:1 molar ratio, producing sulfide and dissolved inorganic carbon according to the generalized net reaction:

$$SO_4^{-2} + CH_4 \rightarrow HS^- + HCO_3^- + H_2O$$

Just as in hydrothermal vents, a number of animals, such as various families of bivalves and large tube worms, live symbiotically with cold-seep bacteria. Studies using electron microscopy and isotopic analysis led to the conclusion that these animals are mainly relying, through their symbionts, on reduced sulfur produced in the sediment, with the exception of the mussels, which are probably able to use both sulfide produced in the sediment and/or methane in seep fluids.

Typical seep animals with chemosynthetic symbionts are restricted to depths below 370 m. A comparative analysis of the structure of seep communities with the neighboring fauna suggests that differences in predator abundance may determine this depth limitation. The abundance of predators, such as carnivorous crabs and sea stars, which can invade seeps from adjacent shallow habitats and efficiently prey on sessile bivalves living near the cold seeps, decreases greatly with depth. Hence, on the shelf and upper slope predation pressure may be high enough to prevent successful settlement of shallower seeps by organisms endemic to deeper seeps.

The seep community of the Florida Escarpment is fueled by the seepage of cold sulfide-, methane-, and ammonia-rich brine from the sediment in localized channels at the geological union of the limestone escarpment and the abyssal plain. Biomass at this cold seep is dominated by mussels similar to those found at hydrothermal vents (a species of *Bathymodiolus*) and vestimentiferan tube worms (*Escarpia* and *Lamellibrachia*) that are similar to *Riftia* and *Ridgeia,* described previously. Like their relatives around hydrothermal vents, both groups of animals provide hard substrate for communities of cold seep–associated species to colonize, such as a shrimp named after the submersible *Alvin, Alvinocaris*; squat lobsters (*Munidopsis*); and a zoarcid fish (*Pachycara*).

On the upper slope of the Gulf of Mexico, off Louisiana, cold methane seeps are usually colonized by two other tube worms, *Lamellibrachia* and *Seepiophila* (*Lamellibrachia* is usually larger and more abundant; **Figure 13.21**). Both species build

Figure 13.21 Deep-sea tube worms, *Lamellibrachia* grow in abundance near cold methane seeps in the seafloor.
© Ian R. MacDonald—AquaPix.

RESEARCH in Progress

Deep-Sea Submersibles for Seafloor Studies

For humans to explore the deeper parts of the seafloor, they must either use remotely operated vehicles (ROVs) or work from inside submersibles designed to protect human passengers from the enormous pressures and to provide oxygen, light, and other life support. Both ROVs and manned submersibles have their roles in deep-sea research. In this sense, there are close parallels between the vessels we use to explore the deep ocean and the ones used to explore the far reaches of space.

Both ROVs and submersibles have their disadvantages. ROVs have a tether attaching them to the boat, and this tether creates drag as the vehicle is moving, can become entangled in structures, and is subject to damage if stressed. A compromised tether often leads to an abandoned dive and hours of repair work on the deck of a boat before another dive attempt can be made. Manned submersibles are rather large, take much more effort to operate, and there are risks to the lives of the occupants if an operational malfunction occurs at great depths. Having access to both an ROV and submersible is ideal, although operating the two simultaneously takes great coordination efforts between the ship and vehicle pilots and crews.

The best-known manned deep submersible vehicle used for ocean research is the high-tech, but awkward-looking, 7-m-long mini-submarine dubbed *Alvin*, after Allyn Vine of Woods Hole Oceanographic Institution (WHOI), one of the early proponents for a deep submergence vehicle for seafloor studies. *Alvin* was delivered to Woods Hole in 1964. The submarine's first tender was crafted from a pair of surplus Navy pontoons and was named *Lulu* after Al Vine's mother. *Alvin* has room inside its 3-m-diameter titanium hull for a pilot and two researchers/passengers and for a myriad of sampling and recording instruments outside its hull (**Figure A**). Typical dives to *Alvin's* maximum depth limit take about 8 hours, and *Alvin* operates for 6 to 10 hours per dive. For the

Figure A *Alvin* has transported over 13,000 scientists on over 4,700 dives.
© Rod Catanach/MCT/Landov.

past 50 years, it has been the workhorse of manned deep-sea research efforts in the United States. *Alvin* has made more than 4,700 dives. A few highlights of its long and successful career are described here.

The first real test of the underwater capabilities of *Alvin* came in 1966 when a U.S. Air Force B-52 collided with an air tanker over Spain and dropped one of its H-bombs in the Mediterranean off the Spanish coast. The bomb was located and then lost during an attempt to attach lift lines. The bomb was relocated 2 weeks later and eventually recovered by a military submersible.

During the launch for Dive 308 in 1968, *Alvin's* support cables failed, and *Alvin* sank to the seafloor in 1,500 m of water. *Alvin* remained on the bottom until recovered in late 1969 by another submersible, the deep submergence vehicle *Aluminaut*. The sinking caused *Alvin* very little structural damage. Lunches left behind by the rapidly departing crew almost a year earlier were soggy but otherwise unchanged because of the near-freezing temperatures and absence of oxygen in the deep sea.

In 1977, *Lulu* carried *Alvin* through the Panama Canal and into the Pacific Ocean for the first time. Near the Galapagos Islands, *Alvin* discovered abundant, and previously unknown, animal life closely associated with hydrothermal vents 2,500 m below the sea surface. It was *Alvin* that allowed for this discovery, leading to entirely new areas of research in the marine sciences.

By 1980, *Alvin* had completed its 1,000th dive during another visit to the Galapagos Rift Zone. Three years later, *Alvin* was refitted to operate from a new support ship, the research vessel *Atlantis*. With its new mother ship, *Alvin* could operate in higher latitudes and in rougher weather. In 1986, *Alvin* made 12 dives to the wreck of RMS *Titanic* in the North Atlantic to test a new ROV called *Jason Jr.* and to document the wreck photographically.

By 1994, every piece of *Alvin* had been replaced, and her new titanium hull was capable of withstanding pressures to 4,500 m, which opened up an additional 20% of the seafloor. A new ship, also named *Atlantis*, which supports new ROVs to operate with *Alvin*, was launched in 1996 (**Figure B**). A major two-phase overhaul of *Alvin* was completed in 2013. A new and larger personnel sphere with five viewports was installed to increase viewing capabilities and enable scientists an overlapping field of view; new imaging systems with the latest high-definition technology were added; and the control system was replaced. The protocol for *Alvin* maintenance is to completely disassemble the sub every 3 to 5 years to inspect all of the components for damage and to replace parts.

Scientists using *Alvin* today have a variety of biological sampling tools available to them. Although observing and filming newly discovered species is quite exciting, actually collecting specimens is even more so. Push corers are soft sediment samplers and can be deployed using a manipulator arm. Scoop nets are used to literally scoop up specimens for capture. Slurp guns are used to vacuum up smaller organisms, bacterial mats, or other soft-bodied prizes. Collection boxes are used to hold specimens once they are collected by

Figure B *Alvin* on the deck of *Atlantis* after a research cruise. The crew is preparing the ship and submarine for another expedition.
Courtesy of Mark Spear, Woods Hole Oceanographic Institution.

a manipulator arm. As technology has improved over the years the scientific collecting equipment on *Alvin* has become quite impressive.

Jason II is one of a new generation of ROVs capable of routine operations to depths of 6,500 m and able to communicate data back to shore via the Internet. Other new ROVs include the towed imaging system *Argo II* and the DSL-120 side-scan sonar system. All three systems are part of the U.S. National Deep Submergence Facility operated by WHOI for the U.S. ocean sciences community. These are the ROVs the new *Atlantis* was designed for, enabling investigators to use the complete suite of manned and unmanned underwater tools from the same platform anywhere in the world.

In 2015, *Alvin* was officially certified to dive to 4,500 m. This new depth certification, along with the improvements described here, have produced a submersible with increased scientific and operational capabilities, including greatly improved access to most of the abyssal regions of the world's oceans. *Alvin* will continue to provide scientists with a view of the deep sea and the life that abounds that was unimaginable just half a century ago.

Critical Thinking Questions

1. Compare ROVs to submersibles. Which vessel type do you consider better to sample and survey the deep sea? Explain.

2. If you had the chance to go on a submersible dive on *Alvin*, would you go for it? If so, which part of the ocean would you be most interested in observing, and why?

semirigid tubes of chitin and protein during their lives, with a majority (76%) of their biomass being accounted for by this tube material. The tubes remain permanently anchored to the seafloor. Below the attachment point, rootlike, sulfide-permeable extensions of their tube and tissues are produced that take up sulfide at rates sufficient to fuel chemosynthesis by their bacterial symbionts. Hundreds, or even thousands, of individual worms of both species aggregate in roughly hemispherical, bushlike colonies that can reach over 2 m in height and several meters in diameter. Both species grow very slowly (relative to their hydrothermal vent relatives), and each can live for hundreds of years.

Predation pressure on these vibrant assemblages of the cold seep fauna can be inferred by differences in stable isotope characterization of potential predators. Carbon and sulfur stable isotope analyses show that rattail fish and synaphobranchid eels are significant seep predators, with giant isopods (*Bathynomus*), hagfish, and spider crabs (*Rochina*) playing a lesser role. Some carnivorous invertebrates, such as a sea star (*Sclerasterias*) and a snail (*Buccinum*), rely nearly 100% on cold-seep production. In addition to these resident predators, there is a high degree of movement in and out of the seep habitat by vagrant benthic predators that derive a substantial proportion of their nutritional needs from chemosynthetic production. With increasing size of the seep community and increasing depth (i.e., decreasing food supply from surface production), the significance of chemosynthetic cold-seep biomass as a food source for surrounding communities increases.

Surprisingly, tissue from the dominant cold-seep tube worms does not comprise a significant portion of the diet of the predators analyzed to date, and no significant damage has been observed in thousands of tube worms examined from cold seeps to date (and only nonlethal plume cropping has been reported from hydrothermal vent worms in the northeast Pacific). Clearly, direct predation on living tube worms represents a very rare event at these cold seeps.

The persistence of cold-seep communities depends on the continuous availability of the hydrocarbon seep, which is controlled mainly by tectonic processes, and on the protective isolation that prevents consumers from destroying the seep community. Since the 1930s, the use of natural gas has increased fivefold, and it now accounts for more than 25% of the world's energy consumption. With existing technology, the total supply of natural gas from terrestrial deposits will likely be used up within 60 years, yet offshore methane hydrates would provide the United States alone with an estimated potential natural gas reserve of 300,000 trillion cubic feet (tcf). Projections of hydrate gas reserves in the ocean south of Japan are 2,000 times that country's very small existing natural gas reserves. It is becoming clear that most of the world's gas hydrates are sequestered in the deep ocean. Because deep-sea hydrates are often associated with complex biological communities, as described previously, gas extraction and production from marine deposits would almost certainly disrupt or destroy these unique and fascinating communities.

STUDY GUIDE

TOPICS FOR DISCUSSION AND REVIEW

1. Summarize the ambient conditions present on the seafloor at a depth of 3,000 m.

2. What is the oxygen minimum zone? Where is it? How does it form?

3. Describe the technologies used for deep-sea exploration.

4. What are the different sources of food available to animals of the deep sea?

5. Why does the nutritional quality of surface particles decrease as they sink through the water column on their way to the abyssal plains?

6. Compare and contrast the benthic invertebrate fauna of the deep sea and the shallow subtidal zone.

7. If the deep sea is so inhospitable, why is the diversity of deep-sea animals similar to or even greater than that of shallow subtidal communities?

8. Why are infaunal deposit feeders so abundant in the deep sea? How do they feed?

9. Why is broadcast spawning less common in deep-sea animals than in shallow-water animals?

10. Where in the Atlantic Ocean would you predict that additional deep-sea hot springs and their associated chemosynthesis-powered communities might be found? Why?

KEY TERMS

absorptive feeding 363	epibenthic sledge 360
abyssal plain 360	hadal zone 356
abyssal zone 356	hemocyanin 369
chemosynthesis 367	ooze 357
cropper 364	whale fall 370
deposit feeder 361	

KEY *GENERA*

Actinoschphia	Buccinum
Alvinocaris	Bythograea
Atolla	Calyptogena
Bathymodiolus	Elasipod
Bathynomus	Escarpia
Glyphocrangon	Riftia
Graneledone	Rimicaris
Hymenaster	Rochina
Kiwa	Sclerasterias
Lamellibrachia	Scotoplanes
Munidopsis	Seepiophila
Opisthoteuthis	Thiomargarita
Pachycara	Trissopathes
Ridgeia	

REFERENCES

Arp, A. J., and J. J. Childress. 1983. Sulfide binding by the blood of the hydrothermal vent tube worm *Riftia pachyptila*. *Science* 219:295–297.

Baco, A. R., and C. R. Smith. 2003. High species richness in deep sea chemoautotrophic whale skeleton communities. *Marine Ecology Press Series* 260:109–114

Baker, E. T. 1996. Extensive distribution of hydrothermal plumes along the superfast-spreading East Pacific Rise, 13° 50′ –18° 40′ S. *Journal of Geophysics* 101:8685–8695.

Cary, S. C., et al. 1998. Worms bask in extreme temperatures. *Nature* 391:345–346.

Corliss, J. B., et al. 1979. Submarine thermal springs on the Galapagos Rift. *Science* 203:1073–1083.

Dayton, P. K., and R. R. Hessler. 1972. Role of biological disturbance in maintaining diversity in the deep sea. *Deep Sea Research* 19:199–208.

Distel, D. L. 1998. Evolution of chemoautotrophic endosymbiosis in bivalves. *BioScience* 48:277–286.

Fisher, C. R. 1996. Ecophysiology of primary production at deep-sea vents and seeps. *Biosystematics and Ecology Series* 11:313–336.

Gray, J. S. 1974. Animal–sediment relationships. *Oceanography and Marine Biology (Annual Review)* 12:223–261.

Haymon, R. M., and K. C. Macdonald. 1985. The geology of deep-sea hot springs. *American Scientist* 73:441–449.

Herring, P. 2002. *The Biology of the Deep Ocean*. Oxford, UK: Oxford University Press.

Hessler, R. R., J. D. Isaacs, and E. L. Mills. 1972. Giant amphipod from the abyssal Pacific Ocean. *Science* 175:636–637.

Isaacs, J. D., and R. A. Schwartzlose. 1975. Active animals of the deep-sea floor. *Scientific American* 233:84–91.

Jannasch, H. W. 1984. Chemosynthesis: The nutritional basis for life at deep-sea vents. *Oceanus* 27:73–78.

Kim, S. L., and L. S. Mullineaux. 1998. Distribution and nearbottom transport of larvae and other plankton at hydrothermal vents. *Deep Sea Research* 45:423–440.

Korunic, Z., and A. Mackay. 2000. Grain surface-layer treatment of diatomaceous earth for insect control. *Archives of Industrial Hygiene and Toxicology* 51(1):1–11.

Lampitt, R. S., B. J. Bett, K. Kiriakoulakis, E. E. Popova, O. Ragueneau, A. Vangriesheim, and G. A. Wolff. 2001. Material supply to the abyssal seafloor in the Northeast Atlantic. *Progress in Oceanography* 50:27–63.

Levin, L. A. 2002. Deep-ocean life where oxygen is scarce. *American Scientist* 90:436–444.

Levin, L. A. 2003. Oxygen minimum zone benthos: Adaptation and community response to hypoxia. *Oceanography and Marine Biology*, 41:1–45.

Lowry, J. K., and K. Dempsey. 2006. The giant deep-sea scavenger genus *Bathynomus* (Crustacea, Isopoda, Cirolanidae) in the Indo-West Pacific. In: B. Richer de Forges and J.-L. Justone, eds. *Résultats des Campagnes Musortom*, vol. 24. Paris: Mémoires du Muséum National d'Histoire Naturalle. pp. 163–192.

Lutz, R. A., T. M. Shank, and R. Evans. 2001. Life after death in the deep sea. *American Scientist* 89:422–431.

Malahoff, A. 1985. Hydrothermal vents and polymetallic sulfides of the Galapagos and Gorda/Juan De Fuca ridge systems and of submarine volcanoes. *Biological Society of Washington Bulletin* 6:19–41.

Margulis, L., D. Chase, and R. Guerrero. 1986. Microbial communities. *BioScience* 36:160–170.

Marsh, A. G., L. S. Mullineaux, C. M. Young, and D. T. Manahan. 2013. Larval dispersal potential of the tubeworm *Riftia pachyptila* at deep-sea hydrothermal vents. *Nature* 411:77-80.

Marshall, N. B. 1980. *Deep Sea Biology: Developments and Perspectives.* New York: Garland S.T.P.M.

McDonnell, A. M. P., and K. O. Buessler. 2010. Variability in the average sinking velocity of marine particles. *Limnology and Oceanography* 55(5):2085–2096.

Page, H. M., C. R. Fisher, and J. J. Childress. 1990. Role of filter feeding in the nutritional biology of a deep-sea mussel with methanotrophic symbionts. *Marine Biology* 104:251–257.

Pollard, R. T., M. I. Lucas, and J. F. Read. 2002. Physical controls on biogeochemical zonation in the Southern Ocean. *Deep Sea Research (Part II, Topical Studies in Oceanography)* 49:3289–3305.

Prieur, D. 1997. Microbiology of deep-sea hydrothermal vents. *Trends in Biotechnology* 15:242–244.

Rex, M. A. 1973. Deep-sea species diversity: Decreased gastropod diversity at abyssal depths. *Science* 181:1051–1053.

Robison, B., B. Seibel, and J. Drazen. 2014. Deep-sea octopus (*Graneledone boreopacifica*) conducts the longest-known egg-brooding period of any animal. *PLoS ONE* 9(7):e103437. doi:10.1371/journal.pone.0103437.

Rogers, A. D., P. A. Tyler, D. P. Connelly, J. T. Copley, R. James, and R. D. Larter. 2012. The discovery of new deep-sea hydrothermal vent communities in the Southern Ocean and implications for biogeography. *PLoS Biology* 10(1):e1001234. doi:10.1371/journal.pbio.1001234.

Rokop, F. J. 1974. Reproductive patterns in the deep-sea benthos. *Science* 186:743–745.

Sanders, N. K., and J. J. Childress. 1993. Plume surface area and blood volume in the hydrothermal vent worm *Riftia pachyptila. American Zoologist* 33:95A.

Secretariat of the Pacific Community. 2013. Deep sea minerals: Manganese nodules, a physical, biological, environmental, and technical review. Vol. 1B, SPC. Available from https://cld.bz/bookdata/h1Tu26r/basic-html/page2.html

Snelgrove, P. V. R. 1999. Getting to the bottom of marine biodiversity: Sedimentary habitats. *BioScience* 49:1129–1138.

Sokolova, M. N. 1970. Weight characteristics of meiobenthos in different regions of the deep-sea trophic areas of the Pacific Ocean. *Okeanologia* (in Russian) 10:348–356.

Tunnicliffe, V. R., R. W. Embley, J. Holden, D. Butterfield, G. J. Massoth, and S. Kim Juniper. 1997. Biological colonization of new hydrothermal vents following an eruption on Juan de Fuca Ridge. *Deep Sea Research* 44:1627–1644.

Voight, J. R., and J. C. Drazen. 2004. Hatchlings of the deep-sea octopus *Graneledone boreopacifica* are the largest and most advanced known. *Journal of Molluscan Studies* 70:406–408.

Voight, J. R., and K. A. Feldheim. 2009. Microsatellite inheritance and multiple paternity in the deep-sea octopus *Graneledone boreopacifica* (Mollusca: Cephalopoda). *Invertebrate Biology* 128(1):26–30.

Polar Seas

STUDENT LEARNING OUTCOMES

1. Compare and contrast environmental conditions and sea ice features of the Arctic and Antarctica.

2. Examine seasonal trends in primary productivity in the polar seas.

3. Explore the diversity of sea life inhabiting polar seas permanently and seasonally.

4. Investigate the major human threats to the health of polar seas and recent changes to the physical characteristics of the polar marine environment.

5. Analyze information presented on climate change, and examine evidence that humans are causing significant and potentially irreversible changes to our Earth.

CHAPTER OUTLINE

Ice and open water in the Beaufort Sea north of Alaska.
Courtesy of the National Oceanic and Atmospheric Administration.

olar seas represent some of the most extreme environments on Earth, with air and water temperatures that are inhospitable to most life forms. At both poles, the environments are extremely seasonal, with drastic changes in temperature, available habitat, salinity, and available nutrients. Despite these challenging conditions, many life forms are adapted for survival in the polar seas and thrive in the chilly waters either year-round or during seasonal migrations. Polar organisms have evolved spectacular adaptations for survival, many of which very likely still remain to be discovered. Migratory animals visiting polar seas are able to withstand the extreme conditions for short time periods, and they are often rewarded with an abundance of energy-rich food sources. Conditions in polar seas lead to seasonal spikes in phytoplankton when sunlight is available, and the entire food web takes advantage of these relatively short time periods to feed and grow. The health of polar seas has been a topic of fierce debate in recent years, as scientists warn that shifting sea ice patterns are problematic for many species and are likely due to climate change. In this chapter, we investigate the physical attributes of polar seas, examine a sampling of marine life inhabiting the ice and sea, and review recent information on how these phenomenal ecosystems are changing rapidly. Lastly, we will discuss ways that individuals can positively influence climate change.

14.1   Physical Characteristics

The two polar ends of Earth share several environmental characteristics that distinguish them from other marine environments. Both experience long winter nights without sunlight. Low levels of sunlight keep sea surface temperatures hovering around 0°C, even in summer. Large parts of both polar marine environments remain perpetually covered by permanent sea ice known as **fast ice**. Even larger areas freeze over in winter to form **pack ice** that thaws and disappears each summer. Polar seas are defined as those areas of the ocean characterized by a cover of either permanent fast ice or seasonal pack ice. The approximate geographic extent of fast and pack ice is shown in **Figure 14.1**, although in recent years the extent of **sea ice** at the North Pole

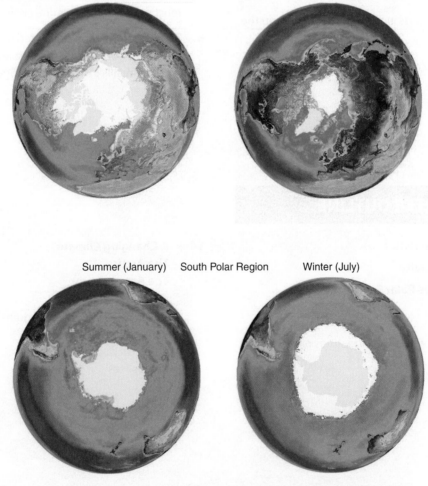

Figure 14.1 Approximate distribution of fast ice (summer) and pack ice (winter) in the north and south polar regions.
Courtesy of GeoEye and NASA SeaWiFS Project.

has been in decline, as discussed in Section 14.4. Sea ice is a solid physical structure encountered in no other marine ecosystem (**Figure 14.2**). It acts as a barrier to insulate seawater from continued chilling effects of the atmosphere in winter, and thus sea ice never more than a few meters thick forms in even the most extreme winter temperature regimes. Sea ice also provides a stable and nearly predator-free platform on which some birds and mammals can raise their young, and many rely on the sea ice for this function. Unfortunately for animals living above the sea ice, it also effectively inhibits many of them from moving easily from the ice surface to forage in the water below. To forage in the sea, they must either travel long distances to find a break in or edge of the ice or wait until some of the ice melts or breaks away.

Although there are some strong similarities in polar seas, there are also some strong contrasts between north and south polar marine environments. These contrasts lead to different primary productivity patterns, and as a result of food web dynamics a very different variety of plant and animal life occurs at each pole (discussed in Sections 14.2 and 14.3). The Arctic (North Pole) is a frozen ocean surrounded by continents; the Antarctic (South Pole) is a frozen continent surrounded by what is known as the Southern Ocean. This difference in the actual physical structures and amount of exposed ocean water found at each pole leads to other major variations in other features. Because Antarctica is a gigantic landmass, there is no seawater below it to provide heat on land. The Arctic is a relatively thin sheet of ice floating on top of seawater, and the seawater moderates temperatures above the ice.

A question one might have while reading this discussion is how really cold ocean water is able to warm the air. Although polar waters are quite chilly, the average temperature is around −1.1°C (30°F). This water temperature is relatively warm when compared to the land above the sea during most of the year, so heat is transferred from the sea to the air. Another reason the Antarctic is much colder than the Arctic is elevation. As elevation increases temperature decreases by approximately 6.5°C for each kilometer increase in elevation. The Arctic is at sea level, and the Antarctic has an average elevation of 2.3 km. The annual average temperature in the Arctic during summer is 0°C (32°F) and −40°C (−40°F) in winter. In Antarctica, the annual average temperature in summer is −28.2°C (−18°F) and

−60°C (−76°F) during winter. These values are averages, but recent studies indicate that water temperature, and as a result air temperature, are increasing at many locations at both poles. The effects of increased temperatures at the poles due to climate change will be discussed in Section 14.4. Some important physical features of the two polar regions are compared in **Table 14.1**.

TABLE 14.1   Comparison of Arctic and Antarctic Oceanographic Features

Feature	Arctic	Antarctic
Shelf	Broad, two narrow openings	Narrow, open to all oceans
River input	Several	None
Nutrients in photic zone	Seasonally depleted	High throughout year
Icebergs	Small, irregular, not in Arctic basin	Abundant, large, tabular
Pack ice (sea ice) averages[1]		
Maximum area	15×10^6 km^2	18×10^6 km^2
Minimum area	7×10^6 km^2	3×10^6 km^2
Age	Mostly multiyear	Mostly 1 year
Thickness	3.5 m	1.5 m
Trend in total amount (1979–2008)	4.1% decrease per decade	0.9% increase per decade
Snow thickness over ice	Relatively thin	Relatively thick

[1]Reported values are averages. In 2012 record lows were recorded for sea ice in the Arctic.
Data from the National Snow & Ice Data Center, University of Colorado, Boulder.

DID YOU KNOW?

One large reason for such cold temperatures at the poles is the angle of the sun hitting Earth at these extreme latitudes. At the equator, the sun hits Earth directly, at a 180° angle, so the sun's energy is concentrated over a small area at 0° latitude, and the equator is very hot. Moving away from the equator to the north and south, the angle of the sun hitting Earth changes, and the area over which the sun is shining on Earth increases. At the poles the sun is shining over a large area at extreme angles, so the magnitude of solar energy at these latitudes is the lowest of all areas on Earth. In addition, a large amount of the solar radiation that does reach the poles is reflected back because ice is very reflective (has a high **albedo**).

14.2   Primary Productivity

Light, or more correctly the lack of it, is the major limiting factor for plant or phytoplankton growth in polar seas. Sufficient light to sustain high phytoplankton growth rates lasts for only a few months during the summer. Even so, photosynthesis can continue around the clock during those few months to produce huge phytoplankton populations quickly. As the light intensity

Figure 14.2   Arctic sea ice in the summer. In recent years the amount of summer Arctic sea ice has declined dramatically.
Courtesy of National Oceanic and Atmospheric Administration.

and day length decline, the short summer diatom blooms also decline rapidly. Winter conditions in polar seas closely resemble those of temperate regions, except that the winter conditions last much longer. In polar seas, the complete cycle of production consists of a single short period of phytoplankton growth. This can be compared to a typical temperate spring bloom immediately followed by an autumn bloom and decline that alternates with an extended winter of reduced net production. In both the Arctic Ocean and around the Antarctic continent, the seasonal formation and melting of sea ice play a central role in shaping patterns of primary productivity.

As ice melts in the spring, the low salinity **meltwater** forms a low-density layer near the sea surface. This increases vertical stability, which encourages phytoplankton to grow near the sunlit surface. The melting ice also releases temporarily frozen phytoplankton cells, or **ice algae**, into the water to initiate the bloom (**Figure 14.3**). The individual phytoplankton cells produced in these summer blooms tend to be much larger than those in lower latitudes. These, in turn, feed relatively large planktonic copepods and benthic consumers, particularly amphipod crustaceans and bivalve mollusks. As the sea ice continues to melt toward the poles in early summer, the zone of high phytoplankton productivity follows, and the stage is then set for short food chains supporting a very productive, seasonal, migrating ice-edge community of diatoms, **krill**, birds, seals, fish, and whales. Animals that do exploit this production system must be prepared to endure long winter months of little primary production. As a result, many **homeotherms** have adjusted to this seasonal food supply with extended fasting periods and long migratory excursions to lower-latitude waters in winter where food is available.

Productivity patterns are different at the North and South Poles due to the amount of sunlight able to penetrate the ice. Historically, much of the Arctic Ocean was permanently covered by thick fast ice, and because so much of the year passed in darkness with almost no phytoplankton growth the annual average productivity rate there was low (about 25 gC/m² per yr). Phytoplankton blooms in the Bering Sea (**Figure 14.4**) were historically restricted to a narrow window of time during the summer. Recent trends in Arctic sea ice include higher rates of melting or thinning sea ice, so there is more exposed seawater and seawater under thinner ice that can receive enough sunlight for photosynthesis. As a result, the water column has less sea ice algae but more phytoplankton. The seasonal phytoplankton blooms have occurred earlier in recent years as well, which has extended the growing season and increased overall primary productivity. The estimated increase in primary productivity is approximately 20% so far, and it remains to be seen what primary productivity trends will persist if sea ice continues to decline. Around the Antarctic continent, upwelling of deep nutrient-rich water supports very high summertime net primary productivity (NPP) rates and annual average rates of about 150 gC/m² per year. In these regions, water that sank in the Northern Hemisphere returns to the surface in an uninterrupted zone of upwelling that extends around the entire continent to support a band of high phytoplankton productivity from the ice edge north to the **Antarctic Convergence** between 60° S and 70° S latitude, bringing with it a thousand-year accumulation of dissolved nutrients. The extraordinary fertility of the Antarctic seas stands in sharp contrast to the barrenness of the adjacent continent. It is because of these extremely productive seas that

Figure 14.4 A large phytoplankton bloom in the Bering Sea during summer.
Courtesy of NASA.

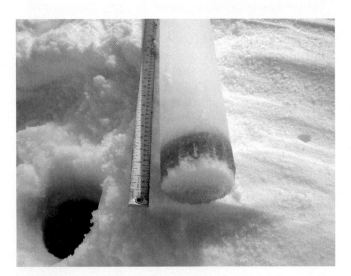

Figure 14.3 Ice algae frozen into the pack ice. When the pack ice melts, the algae are released and often initiate other free algae to bloom.
Courtesy of National Oceanic and Atmospheric Administration.

almost all Antarctic life forms, whether terrestrial or marine, depend on marine food webs supported by this massive upwelling. Large blooms of diatoms and other phytoplankton fuel the food web. Food web dynamics are complex in Antarctica, and recent research suggests that intricate relationships involving bacteria and the availability of certain vitamins also influence the physical condition of phytoplankton there. Comparable latitudes in the Arctic are interrupted by landmasses to form the small regional Bering, Baffin, Greenland, and Norwegian Seas.

DID YOU KNOW?

Several of the largest phytoplankton blooms ever observed have been located under Arctic sea ice. One such bloom extended at least 100 km horizontally under the sea ice! Researchers have been completely shocked by these observations, which challenge earlier views that the waters of the Arctic had overall low primary productivity and that phytoplankton were seriously light limited under the sea ice. The massive bloom was likely able to occur because the sea ice was thin enough to allow sufficient light to penetrate the water below. Whether these blooms occur regularly remains to be discovered, but if they do our current understanding of primary productivity in the Arctic will need to be revised.

14.3   Animal Life of the Polar Seas

Animals residing in polar regions face some of the harshest living conditions on Earth. Those living permanently at the poles are adapted for conditions few animals ever experience. Those that only visit seasonally typically take advantage of the mildest times when food is abundant and leave during the coldest, harshest seasons. Despite the challenges associated with life in a frozen environment, diversity at the poles is surprisingly high, and researchers have yet to explore many areas due to inhospitable conditions for human explorers and their scientific equipment. Humans visiting the poles during the coldest months can only venture outside for short periods of time and must wear survival suits to protect them against the elements. The recent loss of sea ice in the Arctic and changing conditions in Antarctica have prompted a wave of research at the poles to investigate the impacts of climate change on polar animals and food webs. In this section, we will explore just a small fraction of the very diverse assemblage of organisms that inhabit polar seas.

Invertebrates

Many invertebrates are very susceptible to changes in ocean chemistry due to their lack of adaptations for osmoregulation or thermoregulation. Therefore, invertebrates living in the frozen seas of the polar regions must have broad tolerances for the wide changes in salinity, light, and temperature that occur. They must also be capable of withstanding long periods of time with small amounts of food, as the polar seas are seasonal seas, with long periods of limited primary productivity. Many invertebrates discovered living in these regions are unique or even **endemic** to the region of discovery.

About 5,000 species of marine invertebrates have been identified in the Arctic, and approximately 90% live in the benthic realm. The majority of the known species are crustaceans; other notable groups include mollusks, annelids, and bryozoans. What was once thought to be a small proportion of Arctic marine invertebrates but likely includes many species are those found actually living in the sea ice, and some of these species are endemic. Sea ice provides an unusual habitat for many small animals that take advantage of algae trapped in the ice for food. Within the small mazelike structures embedded in Arctic sea ice lives an entire assemblage of small invertebrates, including worms and crustaceans. One commonly encountered example is the ice amphipod, *Gammarus wilkitzkii*, which is found in brine channels or pockets in the ice and reaches a body length of up to 5 cm (**Figure 14.5**). When the ice melts seasonally, algae (primarily diatoms) become available as a food source for amphipods and many other species and prompts blooms of additional algae in the water column nearby. Because these organisms are living in frozen areas and cannot **thermoregulate**, their body temperatures are the same as the surrounding ice. They must be able to withstand the harsh conditions and possess mechanisms to prevent the growth of ice crystals inside their bodies so they do not freeze from the inside out. Many of these animals harbor large organic molecules and lipids for energy storage when food is not available. Competition for food and space within the sea ice is low, because the number species capable of life in this environment is quite limited.

In contrast to the unique sea ice biota just described, some invertebrates living in the Arctic are found elsewhere and are widely distributed. Commonly seen in temperate seas, the blue mussel *Mytilus edulis* recently made a reoccurrence in the Arctic. This species was found in the Arctic region until around 1,000 years ago, but in 2004 it was discovered in subtidal locations around the Arctic. Some scientists hypothesize that the reoccurrence is due to increased water temperatures in the Arctic allowing for a more northern distribution. Research was conducted on the thermal tolerances of this species in an attempt

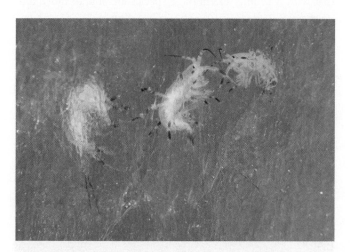

Figure 14.5 Ice amphipods, *Gammarus wilkitzkii*, which are found living in brine channels in sea ice.

Courtesy of Peter Leopold.

to explain why it appears to be moving northward in its distribution. It was discovered that blue mussels are able to physiologically adapt to a certain level of cold, and it is likely that constraints besides temperature limited the distribution of this species over the past 1,000 years.

Waters surrounding Antarctica appear to house even greater invertebrate species diversity than in the Arctic. It should be noted that historically more research has been conducted in Antarctica, so there have been more opportunities to catalog marine life in this region. Like Arctic sea ice, Antarctic sea ice also harbors algae, and in this case there is a very important grazer of sea ice algae, Antarctic krill, which picks off the algae from below the ice as it becomes available. Krill are relatively small crustaceans in the order Euphausiacea that inhabit all of the world's oceans. The name *krill* is a Norwegian word that means "whale food." Antarctic krill, *Euphausia superba*, are about 6 cm in length and can live up to 7 years (**Figure 14.6**). Compared to most other zooplankton, Antarctic krill are relatively large and energy dense, making them an ideal prey species for many fish and baleen whales. Antarctic krill are found in huge swarms of millions of individuals, sometimes blanketing hundreds of square kilometers of the ocean. Their high nutritional value and massive numbers make krill an integral part of the Antarctic food web and are the driving force for many seasonal species to migrate to the region. Although not very common in the United States, other countries harvest krill for human consumption, either in the whole form or for krill oil.

A group of invertebrates with low species diversity in Antarctic waters are decapods (e.g., crabs and lobsters). Decapods are very common worldwide, but they have a requirement of relatively warm water (greater than 0°C) due to their metabolism and inability to regulate magnesium at cooler temperatures. Only around 20 species of decapods have been identified in Antarctic waters, which is a very low number compared to all other areas of the world ocean. The recent discovery of large decapod crabs at several locations around Antarctica has led to concern that these species will intensively prey on many other organisms, potentially disrupting the food web. It is presumed that with warming waters these species are able to move into

areas that were previously uninhabitable due to temperature constraints. Some researchers have a different view and have hypothesized that large decapod crabs have been present in Antarctic waters for many decades or even longer, but have only recently been observed. There is much debate about the historic presence of decapods in Antarctica. Unfortunately crab and lobster fossils are not common in this area, so the debate will likely continue.

Fish

Polar fish display some of the most phenomenal adaptations to life in extremely cold environments. In general, as fish move water in through their mouths and over their gills their blood is cooled substantially as it is exposed to the cool seawater touching the gills. The cooling of blood is a huge disadvantage for fish living in the cold waters of the polar seas. They must offset the cooling of the blood by possessing specific physiological adaptations for minimizing their heat loss. Some of these adaptations will be described throughout this section as we investigate representative fish groups of both Arctic and Antarctic waters.

Waters of the Arctic and Antarctica house quite different fish species assemblages, with differing diversity and age patterns. Arctic waters are home to primarily **phylogenetically** young fish families, and many of the same species living in the nearby North Atlantic and North Pacific Oceans are also found in the Arctic. In the Antarctic, fish are distributed primarily based on ice cover, and there are more phylogenetically older families and a huge number of endemic species present. Although species numbers are moving targets for these regions under intensive research efforts, it is useful to compare the general species diversity of the seas. The Arctic **ichthyofauna** includes over 400 species in 96 families, whereas the Antarctic ichthyofauna includes around 300 species in 50 families. The larger degree of endemism and the older phylogenetic age of fish in the Antarctic is a reflection of the age of the ecosystem (22 million years for the Antarctic and less than 2 million years for the Arctic) and its isolation from other seas.

The majority of fish found in the Arctic are from six dominant groups: zoarcids (eelpouts), gadiforms (cod and its allies), cottids (a sculpin family), salmonids (salmon), pleuronectiforms (flatfish), and chondrichthyans (cartilaginous fish). Zoarcids (**Figure 14.7**) are found in relatively deep water and are also found in Antarctica. One common gadiform, the Arctic cod (*Arctogadus glacialis*), plays a crucial role in Arctic ecosystems as a prey item for such animals as seabirds, ringed seals, beluga whales, and narwhals. Compared to other gadiforms, this species is particularly rich in fat content, providing a high-energy meal for predators. Research is being conducted on the Arctic cod to assess their heat tolerance. It appears that this species is very sensitive to increases in temperature by just a few degrees, with decreased adult growth rates compared to other closely related species at a temperature of 2.5°C. Their growth and development at 0°C is rapid and healthy. Eggs appear to have an even narrower temperature range tolerance with a lethal limit of 5°C. These factors may seriously affect this species if water temperatures in Arctic seas continue to rise.

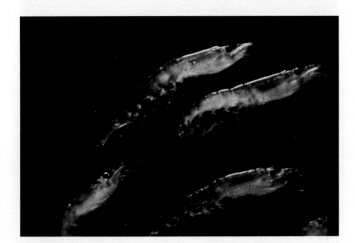

Figure 14.6 Several Antarctic krill swimming through the water column.
Courtesy of National Oceanic and Atmospheric Administration.

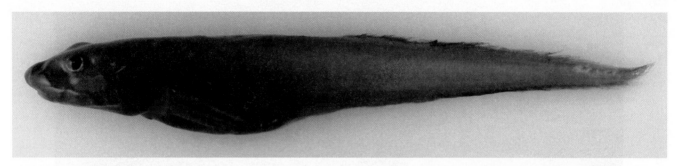

Figure 14.7 An eelpout collected from the southern Arctic.
Courtesy of C. W. Mecklenburg, National Oceanic and Atmospheric Administration.

Cottids, a family of sculpins, are benthic egg-laying fish with heavy spines and wide, thick bodies (**Figure 14.8**). Males of some species found in the Arctic region have a specialized structure for inserting sperm inside the female before she lays the fertilized eggs. Cottids are members of just one of three families of sculpins living in the Arctic, and this group of fish makes up the largest majority of Arctic ichthyofauna. Arctic salmonids are the main anadromous fish found in this region, including salmon, trout, chars, and whitefish. Recent northward shifts in salmon populations have occurred, and some scientists hypothesize that warmer water temperatures are allowing for these shifts in distributions. Pleuronectiforms are flatfish found inhabiting the seafloor, with the Arctic flounder as a classic example in this region.

The Arctic is home to over 20 species of chondrichthyans. Most Arctic chondrichthyans are skates and rays, but some shark species venture into Arctic waters, and one chimaera is known to inhabit the Arctic benthic realm. One large shark species, the Greenland shark (*Somniosus microcephalus*), is a bit of a mystery (**Figure 14.9**). It swims near the surface in the cold waters of the Arctic, where it is quite abundant. This species was thought to be restricted to the Arctic, but in recent years it has been sighted during deep-sea submersible dives in various other parts of the world. These sharks are large (over 7 m long) and swim slowly. Most of them appear to be almost blind due to a parasitic crustacean that latches onto their eye and damages the cornea. Despite their limited visual capabilities and slow swimming speeds, gut contents of Greenland sharks contain a huge variety of prey species, including fish, seals, and polar bears. Although Greenland sharks have not been observed feeding, it is clear that they will feed on **carrion**, dead and decaying flesh, since this is the most likely scenario that would lead to a shark eating a polar bear. In 2016 a study was published on Greenland shark aging and the results were astounding. It appears that this species can live to nearly 400 years, the oldest recorded age of any vertebrate.

Antarctica is dominated by one major fish group: the notothenioids, or ice fish. Notothenioids make up the large majority of the animal biomass and almost half of the fish species in Antarctica. It has been hypothesized that this group was able to slowly adapt to cooling water, and when other fish groups became extinct notothenioids survived and diversified greatly. Notothenoids are called "ice fish" because they contain specialized proteins that bind to ice crystals and inhibit their formation so that their blood does not freeze. The members of one family of notothenioids, Channichthyidae, have colorless blood that lacks **hemoglobin** and red blood cells, characteristics that are not seen in any other adult vertebrate group. These fish rely on O_2 dissolved in their plasma, which only transports about 10% as much O_2 as blood containing hemoglobin. To compensate for the lack of red blood cells and hemoglobin, they have larger hearts and blood vessels, an extremely high volume of blood, and dense capillary beds. Regardless of these adaptations, the only likely reason these fish are able to survive in their current state is that the Southern Ocean has high dissolved O_2 levels. Genetic tests have identified genes for making hemoglobin in these fish, but the genes are nonfunctional.

An investigation of sea ice communities in Antarctica led to the discovery of several fish and crustacean species living under 740-m-thick sea ice in the dark, with the nearest sunlight located 850 km away at the edge of the ice. A few different fish species were present, all with large eyes and one with a transparent body. The fish were not collected for identification, but scientists have suggested that they are likely notothenioids. Prior to this discovery, it was presumed that no fish or other **megafauna** would be able to survive in these areas lacking sunlight and an abundance of food. This kind of completely unpredicted discovery leads to the realization that there is still so much to learn about life in the polar seas.

Figure 14.8 A juvenile Arctic sculpin, *Myoxocephalus scorpioides*.
Courtesy of National Oceanic and Atmospheric Administration.

Figure 14.9 A rare sighting of a Greenland shark captured during a deep-sea submersible dive.
Courtesy of *Okeanos Explorer*, National Oceanic and Atmospheric Administration.

Although notothenioids are the dominant fish group in Antarctica, several other prominent groups exist, including the following: myctophids (lanternfish), liparids (snail fish), zoarcids (eelpouts; see Figure 14.7), and gadiforms (cods and their allies). Myctophids, liparids, and some zoarcids are found in relatively deep waters, and as a result possess adaptations for survival in cold temperatures, high pressures, and low light. Like the notothenoids, some gadiforms contain specialized chemicals in their blood that act as antifreeze, preventing their blood from freezing. Examples of gadiforms with specialized blood are the Antarctic toothfish and Patagonian toothfish, both commonly confused for one another and referred to as "Chilean sea bass" (**Figure 14.10**). These two species have been fished heavily in Antarctic and sub-Antarctic waters, and both grow slowly and mature late, all factors contributing to the concern for their population statuses. Countries purchasing fish sold as Chilean sea bass are encouraged to restrict their purchases to legally caught and documented fish.

Figure 14.10 Two Patagonian toothfish, also commonly known as Chilean sea bass, captured and measured for study.
Courtesy of Antarctic Marine Living Resources Program, National Oceanic and Atmospheric Administration.

Case Study

Narwhals, Odd Creatures of the Arctic

Narwhals (*Monodon monoceros*) are one of the most curious and mysterious animals inhabiting the sea. Numerous legends exist about the narwhal tusk, and there are many theories about what it is actually used for. In the Middle Ages, narwhal tusk was ground up and sold as unicorn horn, which physicians believed cured many ailments. Today, entire communities in the Arctic rely on narwhals and belugas for their very small local economies. Narwhals are hunted in these communities, although restrictions and catch limits have become strict in recent years due to concerns for the population status of this species. Narwhals are hunted using several methods, including being chased down and killed with a gun or, more traditionally, tracked along the ice and harpooned when an individual surfaces for air. When a narwhal is captured the meat is divided among families living in the small towns. Historically, the tusk was sold for a large profit, but the export of tusks has been banned, so some locals have started using tusks more traditionally as tools for survival.

An immediate question one asks upon observing a narwhal is, what do narwhals use their very odd-looking tusk for? A narwhal tusk is actually an enormous upper-left incisor, and it functions as an exceptionally sensitive sensory organ, with 10 million nerve endings that can detect salinity, temperature, and pressure. Only males regularly have tusks (about 3% of females do), and some males have been observed with two tusks. Because of the sexual dimorphism displayed by narwhals, one may conclude that the tusk is used for a mating ritual or maybe to fight for mates. Males have been observed swinging their tusks at one another in the presence of females, although the jousting matches do not seem to be very intense. Certainly the tusk is a sign of maleness, but beyond this we really do not know the function.

The narwhal is one of the few odontocete whales that inhabit the Arctic year-round, and it is also one of the most difficult organisms to study. Narwhals never leave Arctic waters and spend time in areas with high ice coverage. They use small ice-free areas for breathing and rarely surface for long (**Figure A**). They mate during the dark winter months between ice cracks, a time when observation by humans is nearly impossible due to inhospitable conditions in the Arctic. Because of their elusive existence, very few good photos of narwhals are available. For scientists to study them closely they must capture them with nets, which is difficult because they are so shy.

Studies in recent years have led to the capture and tagging of individuals that were released and tracked (**Figure B**). Because narwhals are deep divers, tagging studies

Figure B A research vessel used to study narwhals in the Arctic.
Courtesy of the National Oceanic and Atmospheric Administration/University of Washington.

provide an opportunity for scientists to observe environmental conditions in the water where they dive. The tags include temperature and depth sensors, thus providing data for areas that are very difficult to study. Areas of the polar seas that are under the ice are logistically difficult and very expensive to collect data from. Narwhals have proven to be excellent oceanographers, as they have recorded temperature data from deeper than 1,700 m in Baffin Bay off of Western Greenland, an area that lacked direct temperature data measurements since the early 2000s. In addition to narwhal data collections, data were collected by helicopter, which is a very expensive method of data collection (**Figure C**). The narwhal and helicopter-captured data indicate a rise in sea temperatures in this area, a fact that may threaten narwhals that rely heavily on ice for their lifestyles.

Data on narwhal population sizes are incomplete due to the lack of survey efforts, but there is concern for this species. They are specialist feeders, preferring fish such as halibut, and slight changes to the distribution of or access to their food sources would be fatal. Narwhals rely on sea ice for their feeding strategies, taking advantage of the ice to hide from predators and hunt in a predator-free area. They breed near the sea ice

Figure A A rare sighting of a group of narwhals surfacing, one with a tusk very apparent.
Courtesy of the University of Washington.

Figure C Preparation of a helicopter for collection of water temperature data to accompany narwhal-collected data.
Courtesy of the National Oceanic and Atmospheric Administration/University of Washington.

(continues)

Case Study (*continued*)

during very specific time periods and seem to rely on particular conditions for breeding behaviors. Narwhals have been identified as one of the species in the Arctic most vulnerable to a changing climate, and this is due to their reliance on sea ice. How and whether this strange animal is able to adapt to changing conditions in the Arctic remains to be seen. If serious declines in narwhal populations do occur, these declines would greatly alter food web dynamics in the Arctic, food webs that include humans that rely on narwhal meat for sustenance.

Critical Thinking Questions

1. Occasionally a female narwhal is observed with a tusk. What do you hypothesize might be the cause of this rare occurrence, and what could be the adaptive significance of females with tusks?

2. What can humans do to assist animals like narwhals if sea ice changes begin to drive them to extinction? Is there anything we can do to help?

Seabirds

Very few birds overwinter in the Arctic region, and fewer than 65 species breed in this region. Conditions are harsh, and nesting places are scarce. Dovekies (*Alle alle*), also known as little auks, are the most abundant seabird in the Atlantic Arctic, and possibly in the world. They nest on rocky slopes and swoop down to eat plankton from the sea. This species has been nesting earlier in recent years due to earlier availability of nesting sites when ice melts. Other Arctic birds are nesting later due to reduced food supply from warming Arctic waters and loss of sea ice.

The classic seabird inhabitants in waters around Antarctica are penguins. We will discuss two penguin species and will focus on the unique adaptations these birds have for life in the harsh Southern Ocean. The Research in Progress box in this chapter offers additional information on penguins and climate change.

Adélie penguins are moderate-sized birds, standing about 75 cm tall as adults (see Figure A in the Research in Progress box). During winter months, they are dispersed around the Antarctic continent north of the winter pack-ice edge where they feed on large, energy-rich Antarctic krill, *E. superba*. In early spring, adults leave the sea to return to their traditional nesting areas in rocky coastal areas blown free of drifting snow but where pack ice persists well into late spring. To reach their nesting areas, Adélie penguins must walk or toboggan over as much as 100 km of sea ice. Once at the nesting area, males and females establish pair bonds that may persist for several years, mate, and construct a crude nest of stones. Contrary to what scientists previously thought, Adélie penguins have been observed using alternate nesting locations when their traditional nesting site becomes too far to walk to due to sea ice shifts. This flexible nesting site choice is a behavior that will likely prolong the existence of this species as sea ice conditions change.

Typically, two eggs are laid a few days apart. After the second egg is laid, the female departs to resume feeding near the edge of the pack ice, which has migrated southward during the spring melt. Males remain on the nests to incubate the eggs until the females return several weeks later. By the time they resume feeding, males have been ashore without food for as many as 40 days. Through the summer, both parents are kept busy foraging on krill, filling their crops, and returning to the nest site

to feed their chicks by regurgitating partially digested krill. Seldom does the second chick to hatch in a nest survive, because it is unable to overcome the advantage in size and weight of its older sibling. By the end of summer, Adélie chicks have grown to the size of their parents, have developed adult plumage, and are ready to begin foraging for themselves near the ice edge. As winter pack ice forms, they move farther north and remain at sea until mature 3 years later.

Adélie penguins are obligate sea ice inhabitants, so shifting sea ice habitat greatly affects their livelihoods. The success of Adélie penguins requires an ideal amount of sea ice cover. Too much sea ice is a challenge because they do not walk well for long distances and need to reach water to forage for food to bring their young; too little sea ice does not provide sufficient area for them to raise their young.

Emperor penguins are the largest living penguins, exceeding 1.1 m in height. Emperor penguins have a reproductive pattern similar to that of Adélie penguins, with one crucial exception: they nest on the pack ice. For their chicks to grow to adult size before the pack ice melts out from under them, adults cannot nest near the ice edge. Instead, they walk far south of the edge of the pack ice in the dead of winter to start their breeding and nesting season in areas where the pack ice remains until they can complete their breeding season. A single egg is laid, and then the male takes over its incubation after the female very carefully transfers the egg to the male. Only at this time will the female break her long fast, walking across as much as 100 km of pack ice to forage in open water.

To attend an incubating egg in air temperatures as low as −70°C and isolate it from the cold ice on which they are standing, male emperor penguins carry their eggs on their feet and cover them with a flap of bare belly skin that is richly supplied with blood vessels to provide warmth for the egg. Here the egg develops for 64 days as the male continues to fast. During incubation, males huddle tightly together in large groups for additional protection against the biting winter winds. Over time, individuals near the edge of the huddle move toward the center so that everyone spends some time on the exposed edges of the huddle. The males must take extreme care while moving with the eggs on their feet. One tiny mistake leading to an egg slipping onto the ice will be the end of the developing young's life and a failed reproductive effort for both parents for the entire breeding season.

With the return of the spring sun, well-fed adult females also return to the nest site to relieve their starving mates, who have been fasting for more than 100 days. Males then face their own long march to open water before they can begin to feed. For the remainder of the summer, both parents forage for themselves and their chicks. In the water, male and female emperor penguins demonstrate exceptional diving abilities as they forgo krill to pursue midwater squids and fish. During feeding bouts, these animals make deep dives, sometimes exceeding 500 m while staying submerged for up to 15 minutes.

Marine Mammals

A variety of marine mammal species inhabit polar seas, many only seasonally, but some remain year-round. Baleen whales take advantage of the warmest months in polar seas when phytoplankton is most abundant and fueling the food web, leading to an enormous abundance of zooplankton prey. In Antarctic waters, krill is the primary zooplankton food source for several large whale species; in the Arctic, a *Calanus* copepod is the primary zooplankton food source. Although marine mammals are homeothermic and must maintain an internal body temperature that is much warmer than the surrounding waters of the polar seas, they are already adapted with a variety of mechanisms for maintaining body temperatures that are warmer than their environment. One example is **blubber**, located right underneath the lower epidermis and used to insulate internal body parts from environmental temperatures. Marine mammals that can survive in polar seas will have more extreme versions of adaptive mechanisms to remain warm in the very cold waters found at the poles.

In the Arctic, the true marine mammal residents are considered **ice-breeding**, because they rely on the ice for breeding and other critical activities for survival. Seven seal species, the walrus, narwhals, beluga and bowhead whales, and the polar bear are all considered organisms that rely on the ice for survival. Several large baleen whale species (e.g., gray and blue whales) visit the Arctic seasonally to feed, with marathon migrations that are some of the longest distances traveled by any animal. In Antarctic waters, the northern boundary is marked by the polar front, an area where water temperatures increase by as much as 10°C across the front. Animals living on the southern side of the front are considered Antarctic, and can survive the cool temperatures and harsh conditions experienced in this region. Marine mammals relying on these waters for survival include six seals, eight baleen whales, and seven odontocete whales. In the following discussion we will explore some of the most common marine mammal inhabitants of polar seas, highlighting their phenomenal adaptations for survival.

Polar Bears and Walruses

The polar bear, *Ursus maritimus,* is the animal icon of the Arctic (**Figure 14.11a**). Polar bears have completely white fur and slightly longer necks and larger body sizes than other bears. They are broadly distributed on ice-covered waters throughout the Arctic. In Hudson Bay and other regions that become

(a)

(b)

Figure 14.11 (a) A polar bear mother and her twin cubs at rest. (b) Hair-covered paws of a swimming polar bear.

(a) Courtesy of Captain Budd Christman, NOAA Corps; (b) © Emily Veinglory/Shutterstock, Inc.

completely ice free in summer, they are forced to spend several months on land fasting until the fall freeze-up.

Like most other terrestrial mammals, polar bears have few specialized adaptations for efficient swimming. They have large feet that form flat plates oriented perpendicular to the direction of motion, producing drag-based thrust to move the animal forward. Polar bears swim with a stroke like a crawl, pulling themselves through the water with their forelimbs while their hind legs trail behind. Despite their comparatively limited swimming capabilities, polar bears have been documented regularly swimming over 50 km (31 miles) in one trip. When traveling on land, their huge paws help distribute their body weight while their hair-covered foot pads increase friction between their feet and the ice, so land travel is less physically taxing and less energetically costly (**Figure 14.11b**). Activities on land also require less energy than swimming, because polar bears have several adaptations for thermoregulation that help them to take advantage of their skin exposed to the sun; black skin soaks up heat, and oily fur repels water so they dry more quickly.

Polar bears prey primarily on ice seals (ringed, bearded, ribbon, and sometimes harbor and harp) and occasionally hunt for walruses, beluga whales, and even birds. Bear predation of ice seals occurs primarily in unstable ice conditions where the bears can stalk seals resting out of water. Seal pups are especially vulnerable to bear predation. In the Canadian Arctic, ringed seals give birth to pups in snow caves above breathing holes in the ice. At weaning, these pups are 50% fat, and it is from feeding on these pups that polar bears secure most of their year's energy supply during the short period just after the ringed seal pups are weaned.

The extent of bear predation on walruses varies in different areas of the Arctic, with higher rates of predation in the Canadian Arctic than in the Bering Sea. Bears frequently fail to make a kill by stalking hauled-out walrus herds, yet they often manage to frighten herds into the water. In the ensuing stampede, walrus pups are sometimes injured or isolated, making them easy prey for the bears. It is not known whether bears commonly prey on walruses or seals underwater.

Like some other large marine mammals, polar bears endure prolonged seasonal fasts as the availability of food declines or when reproductive activities prevent foraging. When bears are forced to stay onshore in summer as the ice melts, they are forced to fast for up to 4 months until late autumn when the formation of fall ice permits them to regain access to the newly formed pack ice. Polar bears are one of the species that are negatively affected by the decline in sea ice and prolonged summer melting periods that are currently being experienced in the Arctic. Based on data from 1995 to 2015 collected by the U.S. Geological Survey, polar bears are experiencing a reduced survival rate of young and an overall population decline.

Polar bears breed in late spring when large males spend days tracking potential mates across large expanses of ice. Males compete aggressively for access to these females, because females typically breed only once every 3 to 4 years. Female polar bears are **induced ovulators**, so a female must mate repeatedly before she ovulates and can be fertilized. Once pregnant, delayed implantation blocks fetal development until fall, when females dig maternity dens to spend the winter and give birth. Usually,

Figure 14.12 Geographic distribution of the Atlantic (red) and Pacific (black) subspecies of walrus, *Odobenus rosmarus*.
Courtesy of NOAA.

only pregnant females occupy winter dens. Typically, two cubs are born about 2 months into the winter denning period. By March or April, the 3-month-old cubs have increased their birth weights 20-fold, and they are ready to leave their den with their mother. They remain with her until weaned 2 years later, when she will mate again and prepare for her next pregnancy.

Walruses are the only species in the pinniped family Odobenidae. They are found all around the Arctic basin where ice is sufficiently thin to break through to feed and where the sea bottom is not more than 80 m deep (**Figure 14.12**). Walruses are easily identified by their long paired tusks (**Figure 14.13**) that are present in both sexes but are shorter and more slender in females. Tusks

DID YOU KNOW?

Polar bears sometimes mate successfully with grizzly bears. Due to warmer temperatures, grizzly bears have expanded their range northward, and polar bears occasionally move further south than their typical range. Mating between grizzly and polar bears has produced "grolar bears" or "pizzly bears" that have also mated successfully to produce a second generation of young. If climates continue to shift we may see more hybridization of closely related marine mammals as species that did not previously interact begin to mingle.

Figure 14.13 Large male walrus displaying prominent tusks.
© Corbis/age footstock.

of males are larger, occasionally reaching 1 m in length. Male walruses use their tusks mainly in dominance displays directed toward other males; females use them to defend themselves and their young from aggressive males and from marauding polar bears. On occasion, both sexes use their tusks to pull themselves onto ice floes. Tusks are not usually used in feeding, although rarely they are used to kill seal prey by stabbing.

Although some walruses occasionally eat ringed and bearded seals and birds using their tongue as a vacuum pump to suck away the skin, blubber, and intestines, their typical prey are several species of benthic clams. Walruses forage for clams by swimming along the sea bottom in a head-down position, stirring up sediments with their snout and its very sensitive **vibrissae**. When a clam is located, it is either sucked into the mouth directly or is first excavated with jets of water squirted from the mouth. The soft tissue of large clams is sucked out of the shell by the tongue. Individual walruses consume large quantities of shallow-water benthic clams. They can harvest clams at an impressive rate of 40 to 60 clams per dive for a total of several thousand clams each day. During their foraging activities, walruses create long furrows and pits on the sea bottom and can be important agents of disturbance in shallow, soft-bottom Arctic benthic communities.

Polar Seals

Several species of seals, including ringed, harp, and hooded seals, breed on Arctic fast or pack ice. These species all show a reduced level or complete absence of the **polygyny** that is so strongly developed in elephant seals and other species of pinnipeds that breed in lower latitudes. This difference seems to be due to the difficulty males have in gaining and controlling access to multiple females in the physically unstable environment of pack ice. The extensive distribution of pack ice encourages breeding females to disperse widely while still providing the same protection from most predators that island rookeries do for elephant seals. Yet the temporary and unpredictable nature of the location, extent, and breakup time of typical pack ice restricts pupping to a short time period when the ice is most stable. Females of ice-pupping species cannot expect to return to the same "place" for pupping year after year, as elephant seal females do, so males are forced to search widely for dispersed females.

Mating in the water is common in these species, and this behavior provides males with even fewer opportunities to control the mating choices of females. To compensate somewhat, some ice-breeding seals have evolved complex vocalizations underwater to enhance their attractiveness and advertise their interest in mating. Mating in water is especially difficult for human observers to study, and it is much more so when it occurs under polar ice.

Like pack ice, the vast extent of Arctic fast ice also enables females breeding on fast ice to disperse widely, although access to cracks and holes in the ice usually promotes some clumping of individuals. In the Arctic, ringed seal females give birth to pups in isolated and hidden snow caves above breathing holes in the fast ice to avoid polar bears. Because females are isolated

Figure 14.14 A ringed seal resting near a convenient opening in the fast ice.
© jack stephens/Alamy Images.

and dispersed, individual male ringed seals typically have access to only one female and are considered **monogynous**. They have a broad Arctic distribution and are found wherever openings in fast ice occur, even as far north as the North Pole. Under the ice, they feed on small fish, krill, and planktonic amphipods. This small seal (**Figure 14.14**) is the most abundant pinniped in Arctic waters, although the Arctic subspecies was listed as threatened under the Endangered Species Act in 2012 and critical habitat was proposed for this species in 2014.

Harp seals are almost as common in the Arctic as ringed seals, although their distribution is more limited, extending eastward from Hudson Bay around Greenland to northern Siberia. A dorsal harp-like "V" and black hood is worn by adults. They exploit a broad range of prey species, concentrating on capelin and Arctic cod (also known as polar cod). This species is probably best known as the target species for human hunters of the white pelts of very young pups (**Figure 14.15**). These pups continue to be hunted each spring in the Canadian Atlantic

Figure 14.15 Harp seal pup, long a target of fur hunters, on Canadian pack ice.
© AbleStock.

before the breakup of the pack ice. The documented annual kill between 2010 and 2015 was 319,235 seals.

Hooded seals are unusual ice seals for many reasons. When females are ready to pup, they haul out on pack ice, loosely positioning themselves about 50 m apart. After their pup is born, their 4-day nursing period is the shortest of any pinniped, and probably the shortest of any large mammal. Soon after weaning, the female enters the water where she mates with an attendant male. The male soon returns to the ice to continue his search for other estrous females. Individual males may mate with as many as six to eight females in one breeding season. They are the only polygynous Arctic ice seal, and the appearance of adult males reflects the strong sexual dimorphism that is associated with polygyny. Male hooded seals sport an unusual two-part nasal "hood" ornament that is inflated during competitive displays. When inflated by closing the nostrils, one part, the hood, enlarges to cover the face and top of the head (**Figure 14.16a**). Males can also inflate a very elastic nasal septum to form a large membranous pink balloon that extends forward from one nostril (**Figure 14.16b**).

The most abundant seal on Earth is the crabeater seal; its total population probably exceeds 10 million animals. Despite their common and taxonomic names (*Lobodon carcinophagus* means "lobe-toothed crabeater"), crabeater seals forage almost exclusively on *E. superba*. This species has highly modified teeth (**Figure 14.17a**) that are effective as strainers to collect krill. Crabeater seals are circumpolar, spending the entire year on pack ice. They migrate on a seasonal schedule as the ice edge advances and retreats with the pattern of freezing and thawing.

Like Arctic pack-ice seals described earlier, crabeater seals are monogynous. Male crabeater seals have been observed to guard a female on the ice for 1 or 2 weeks after her pup is weaned, presumably after mating has occurred. Mate guarding by a male may enhance his prospects for paternity by preventing other males from access to that female after (and possibly even before) he has mated with her.

Eighty percent of crabeater seal pups die before their first birthday. This surprisingly high mortality rate seems to be due to predation by another ice seal, the leopard seal. For those crabeater seals that survive past their first birthday, most exhibit extensive scarring from previous leopard seal attacks. Leopard seals have teeth very much like those of crabeater seals, adapted

(a)

(b)

Figure 14.16 Displaying male hooded seals, (a) one with an inflated hood and (b) one with an inflated nasal balloon.

(a) © Biosphoto/Bruemmer Fred/Peter Arnold, Inc. (b) © Fred Bruemmer/Peter Arnold, Inc.

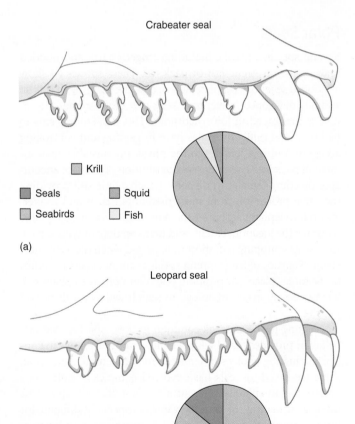

Crabeater seal

Krill

Seals   Squid

Seabirds   Fish

(a)

Leopard seal

(b)

Figure 14.17 Crabeater seal (a) and leopard seal (b) teeth specialized for filtering krill, with the proportion of krill in their respective diets indicated.

Figure 14.18 A Weddell seal lingers near its preferred breathing hole.
© Kim Westerskov/Alamy Images.

Figure 14.19 A beluga whale at the surface.
© Ryan Morgan/Shutterstock, Inc.

for straining krill from water. About half their total diet comes from krill (**Figure 14.17b**); however, these large and supple inhabitants of Antarctic pack-ice regions are also major predators of penguins and crabeater seal pups.

Weddell seals are found associated with cracks and leads in fast ice around the entire Antarctic continent (**Figure 14.18**). These openings in the ice must be kept ice free as breathing holes by constant abrasion by the occupant of the hole. Each animal has a preferred breathing hole and returns to it after each dive. This behavior has been used by a succession of researchers to equip free-swimming Weddell seals with electronic equipment to monitor physiological responses during dives without restraining them and still have a reasonable expectation that the seal will return to the same breathing hole with the expensive equipment package.

The major contributors to the diets of Weddell seals are Antarctic cod and other large bottom fish and squid. To harvest these, Weddell seals are extremely good divers. Some dives exceed 80 minutes in duration and 600 m in depth. These exceptional breath-holding abilities permit male Weddell seals to maintain under-ice breeding territories near breathing holes by excluding other males. Each male is limited to mating only with the females that congregate around his ice hole. Because males must display and mate underwater, their smaller size in comparison with females may make males more agile swimmers and possibly more attractive as potential mates. Males may enhance their attractiveness as mates with long and complex trilling vocalizations. Because Weddell seals mate underwater, it is unclear which gender is responsible for mate choice, but female choice is certainly suspected.

Odontocetes

Arctic waters are inhabited or at least visited by three species of toothed whales: belugas, narwhals, and killer whales. Belugas and the closely related narwhals are medium-sized whales intimately associated with pack-ice habitats. Belugas, also known as white whales, are actually born gray and gradually become creamy white as adults (**Figure 14.19**). Adaptations to ice-covered

waters include a very flexible neck, small flippers, and a thick layer of insulating blubber. In place of a dorsal fin, belugas have a stout dorsal ridge used to break through ice from below. They feed on seasonally and locally abundant fish and invertebrates, both in the water column and on the seafloor.

The most obvious feature of narwhals (**Figure 14.20**) is the long spiral tusk regularly possessed by males. Details of the biology of narwhals remain spotty because of their preference for remote polar areas over deep water covered with heavy pack ice in winter. Their general range extends over most of the Atlantic portion of the Arctic Ocean, with only occasional stragglers seen on the Pacific side. They feed mostly on Arctic cod and halibut, species of fishes often associated with the undersides of pack ice. Narwhals are excellent divers, sometimes to depths greater than 1,000 m, where they occasionally feed on bottom fish and midwater squid. For more information on narwhals, see the Case Study feature.

Despite the differences in body sizes and specializations in feeding behaviors and mechanisms displayed by rorquals (a family of baleen whales), all the rorqual species foraging in Antarctic waters exploit a single prey species, the enormously

Figure 14.20 A rare display of a narwhal from below. The tusked individual (foreground) is a male, and the individual lacking a tusk (background) is a female.
© David Fleetham/Alamy Images.

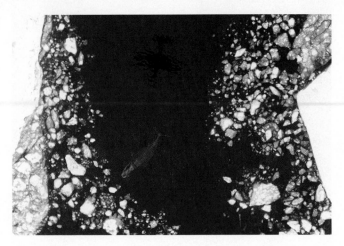

Figure 14.21 A bowhead whale making its way around and through Arctic ice.
Courtesy of Amelia Brower, National Oceanic and Atmospheric Administration.

abundant Antarctic krill, described earlier. Their summertime distribution in Antarctic waters closely mimics the distribution of krill concentrations. Before the advent of pelagic commercial whaling in Antarctic waters at the end of the 19th century, this plentiful food supply supported large populations of all rorqual species. Today, blue, fin, and humpback whale populations still remain depleted as a result of the intense commercial whaling during the first half of the 20th century. For example, 29,000 blue whales were harvested in a single year, 1929, and a total of 1.4 million whales (71 million tons of whale biomass) were removed from Antarctic waters by 1960. Today, a total of approximately 2,200 blue whales feed in Antarctic waters, but there is hope for this species. Populations in Antarctic waters appear to be rebounding, although slowly, and genetic studies indicate large diversity, which is favorable for survival of a rebounding population.

With the larger whale species so efficiently removed from Antarctic food webs, the numbers of smaller and commercially less attractive minke whales, Adélie penguins, crabeater and leopard seals, and fish that rely on *E. superba* have responded to the reduced competition with changes in their reproductive rates. The age at sexual maturity of minke whales, for instance, has declined from greater than 15 years in the 1930s to about 7 years at present, and females have shifted from reproducing every 2 years to every year. As a consequence, the Antarctic population of this small rorqual species has exploded.

Although several baleen whales visit the Arctic seasonally, the bowhead whale, *Balaena mysticetus*, is the only baleen whale that is a permanent resident of the Arctic and adjacent waters (**Figure 14.21**). Bowhead whales are characterized by the lack of a dorsal fin and a white-colored chin. They have a huge skull that makes their bodies look a bit out of proportion, but the skull is very strong and serves the purpose of plowing through thick ice with the head. One reason this species is able to survive year-round in and around the Arctic is that it has the thickest blubber of any whale, up to 50 cm thick. Like other baleen whales, bowheads feed primarily on zooplankton, as well as on fish and invertebrates.

14.4 A Changing Climate

Throughout this chapter there have been numerous references to climate change, particularly warming of the air and sea in the Arctic and Antarctic and changes in sea ice thickness and melting rates. Some changes to the polar seas are very apparent because we can actually visualize the changing environment. There is no doubt that these changes are occurring; they are being measured, counted, and observed by scientists using highly advanced equipment and many millions of dollars of research funds. The question is how much of the change is due to human influences? Although there has been fierce debate about this topic among scientists, politicians, and the average citizen, one thing is now clear: we know humans are contributing to increased carbon emissions, and an unnatural amount of carbon in the form of CO_2 and other greenhouse gases traps energy in the atmosphere and causes Earth to warm. Some of the drastic changes to our Earth have been and are being caused by humans.

Humans naturally tend to think in human time scales, but we are newborns compared to the age of Earth. Changing climates have been a theme for the approximately 4.3-billion-year life history of Earth, and some changes were necessary for life to exist on Earth. Throughout time, since the first life forms existed, organisms have adapted to changes or become extinct. Our Earth is a changing Earth, so the concern for us is not the fact that changes are occurring. The problem many scientists are concerned with is that humans are contributing to climate changes that may occur too quickly for many organisms to adapt. In the following discussion we will explore some of the most apparent effects of climate change on the polar seas, referring to the latest research available in this area and comparing recent trends to historical data. The temperature and ice data presented have been compiled from the National Snow and Ice Data Center (NSIDC), a government organization in the United States that has been collecting and compiling data on the polar seas since the 1950s, although its name has changed a few times since then. The NSIDC provides very recent data on conditions at the poles and historical data for comparisons.

The Arctic

Until recently, sea ice trends in the Arctic were relatively steady, with distinct seasons of sea ice melt and predictable thickness of the ice. Animals and plants living in the polar seas adapted to these fairly predictable seasonal conditions, taking advantage of times when food was abundant and preparing for long periods of fasting or reduced food availability. Observing trends from 1981 to 2010 and comparing average data from this time period to each of the past 5 years indicates that sea ice extent has decreased significantly. In 2012, the lowest summer sea ice values ever recorded were observed, and as of 2015 the sea ice has only slightly recovered, with 2015 measuring the fourth lowest values on record (**Figure 14.22**). During all years, the largest differences in snow ice coverage are during summer months. **Figure 14.23** displays the coverage of sea ice in the summer (right) and winter (left) during the lowest recorded sea ice extent in

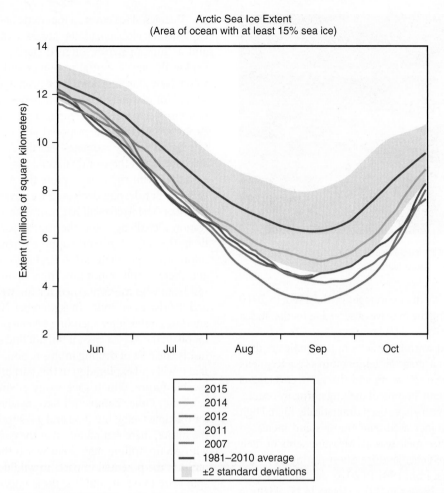

Figure 14.22 Average Arctic sea ice extent for 1981 to 2010 compared to the five lowest ice measurements recorded. Data are provided by the National Snow and Ice Data Center.
Courtesy of National Snow and Ice Data Center.

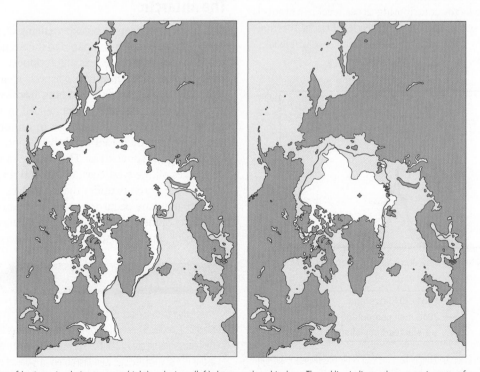

Figure 14.23 The extent of Arctic sea ice during summer (right) and winter (left) shown as the white layer. The red line indicates the average ice extent from 1981 to 2010.
Courtesy of National Snow and Ice Data Center.

Figure 14.24 Arctic sea ice at low levels during the summer.
Courtesy of Patrick Kelley, United States Geological Survey, United States Coast Guard.

2012 compared to the median coverage for the 1981 to 2010 time period, emphasizing the massive observed reduction in sea ice extent (**Figure 14.24**).

Sea ice decline and warming waters in the Arctic are due, in part, to warmer than average air temperatures. Sea ice melts at the surface when the air is warm and the snow coverage is low, and although a certain amount of melting normally occurs, melting rates in recent years have been dramatically high. High rates have been due to increased temperatures and increased length of the summer ice melt season. Measurements of melt percentage taken at the Greenland Ice Sheet during 2015 are compared to the average from 1981 to 2010, and the melt percentage is dramatically higher in 2015 (**Figure 14.25**). Melting rates have been increasing in recent years, shrinking the size and thickness of the ice sheet, and this excess melt water that should be ice is one factor contributing to sea level rise around the globe. Another prominent factor in sea level rise is thermal expansion caused by the warming of the ocean, because water expands as it warms.

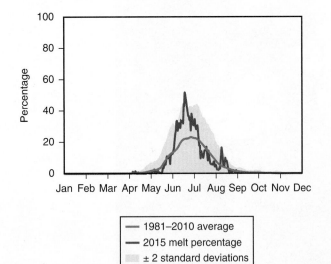

Figure 14.25 The extent of ice melt throughout the year at the Greenland ice sheet in 2015 compared to the average for 1981 to 2010.
Courtesy of National Snow and Ice Data Center/Thomas Mote, University of Georgia.

Decreased ice cover in the Arctic has led to more area to be exploited by global markets. Access to the Arctic has improved, and as human nature often dictates, people have begun to exploit the area's natural resources and anything of monetary value. New shipping lanes have been opened to increase accessibility for mariners, and new fisheries are being explored now that boats can access a larger area more frequently. Oil and gas exploration has been proposed and investigated, as the U.S. Geological Survey estimates that the Arctic houses a considerable amount of the world's natural gas and oil supplies. Several countries with rights to explore the Arctic have attempted to drill oil, but despite decreased ice coverage, drilling is a risky business. The Arctic still has many hazards that make safe navigation a challenge, and the nearby towns are small with little in the way of ground support when ships need repairs or unpredicted problems arise. Regulations for drilling are strict, and there is stiff opposition from environmentalists and local residents who are concerned for the well-being of marine life and the risk of oil spills. In September 2015, Royal Dutch Shell announced that after spending several billion dollars exploring for oil off the Alaska coast it did not find sufficient oil and gas to justify the cost of drilling in this region. The company claimed that it will end exploration in this part of the Arctic for the foreseeable future, which gave many environmentalists cause to celebrate. Other companies have reached similar conclusions for the time being, but if oil and gas supplies become limited in the future, there is no doubt that the costs and risks associated with Arctic drilling may seem worth the benefits. **Figure 14.26** outlines the potential impacts to wildlife in the Arctic if an oil spill were to occur, and it is these impacts that local residents and concerned citizens worldwide fear.

The Antarctic

The Antarctic appears to be experiencing very different changes due to climate change compared to the Arctic, which is not surprising given that these seas are at opposite ends of the globe and have very different physical aspects regarding ice coverage. Although some areas of the Antarctic have experienced declines in sea ice, other areas have experienced increases in ice coverage. It is hypothesized that changes in atmospheric weather patterns leading to increased winds have contributed to increased ice coverage in some areas. The decline in ice and collapse of ice shelves off the west coast of Antarctica appear to be due to warm ocean currents under the ice shelves. Although not uniform, changes are occurring to ice in Antarctica.

Some of the changes caused by global warming in Antarctica can be attributed to what is known as the **Antarctic ozone**

DID YOU KNOW?

The term *ozone hole* does not refer to an actual hole in the ozone layer, as no place is completely devoid of ozone. The term refers to an area with ozone concentrations below 220 Dobson units, a historical threshold set by scientists in the 1970s when declines were first observed. Satellite instruments are used to monitor the status of the ozone layer and report the size of the layer over Earth.

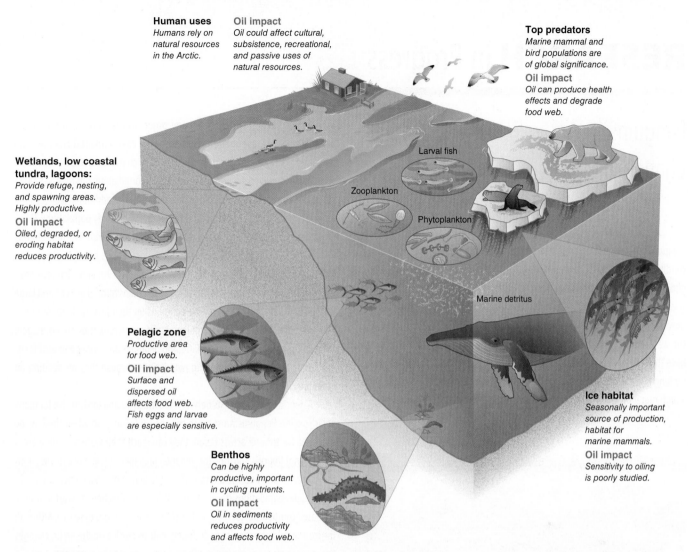

Human uses
Humans rely on natural resources in the Arctic.

Oil impact
Oil could affect cultural, subsistence, recreational, and passive uses of natural resources.

Top predators
Marine mammal and bird populations are of global significance.

Oil impact
Oil can produce health effects and degrade food web.

Larval fish

Zooplankton

Phytoplankton

Wetlands, low coastal tundra, lagoons:
Provide refuge, nesting, and spawning areas. Highly productive.

Oil impact
Oiled, degraded, or eroding habitat reduces productivity.

Pelagic zone
Productive area for food web.

Oil impact
Surface and dispersed oil affects food web. Fish eggs and larvae are especially sensitive.

Marine detritus

Ice habitat
Seasonally important source of production, habitat for marine mammals.

Oil impact
Sensitivity to oiling is poorly studied.

Benthos
Can be highly productive, important in cycling nutrients.

Oil impact
Oil in sediments reduces productivity and affects food web.

Figure 14.26 Potential impacts to the Arctic ecosystem and humans in the event of an oil spill in the region, as predicted by the National Oceanic and Atmospheric Administration.
Courtesy of Kate Sweeney, National Oceanic and Atmospheric Administration.

hole, a thinning of the ozone layer above Antarctica. An obvious impact from a thinner ozone layer is increased ultraviolet radiation, but less obvious impacts include changes to weather patterns that affect sea life and ice patterns, including winds

forcing warm air over the continent. These winds were one of the processes that led to the breakup of the 3,250 km² Larsen Ice Shelf in 2002, a dramatic event that brought attention to the effects of climate change on Antarctica (**Figure 14.27**). Waters of the Antarctic Circumpolar Current are warming more rapidly than the world ocean as a whole. In a 2015 study, models predicted that, moving forward in time, the Antarctic ozone hole will become smaller until it is a nonissue by the year 2040. Whether this actually happens remains to be seen.

What Can You Do?

Because we know that humans are creating changes to our global environment, it is up to us to find solutions to slow our negative impacts from carbon emissions. As an individual, the question is, what can be done? Whether you are a future scientist, concerned citizen, or a citizen with no concern, the changes to our climate will affect you, those you care about, and most certainly future generations. The most obvious activity that directly affects our climate is the use of oil for transportation,

Figure 14.27 The Larsen Ice Shelf after the ice separated from the main ice.
Courtesy of National Snow and Ice Data Center.

RESEARCH in Progress

Penguins in Antarctica and Climate Change

Penguins are great examples of organisms with extreme adaptations for severe conditions. Unlike most other birds, they are flightless and rely on walking and gliding over snow to move on land; in the water, they glide and swim with their wings (**Figure A**). Their familiar waddle on land may seem inefficient, but their movements are not generally meant to be quick, but rather very directed. In the water they can swim quickly to hunt for prey or to evade predators. They have thick, heavy bodies that allow them to store energy so that they can fast for long periods, unlike other bird species that must eat constantly to maintain a high metabolism and slender body form.

Figure A   A large group of Adélie penguins waddling around the snow covered ground.
Courtesy of the National Oceanic and Atmospheric Administration.

As our Earth has changed many times over its history, some robust penguin species have made it through large environmental changes and remain today. Three species in particular—the chinstrap, gentoo, and Adélie penguins—survived a warming event that occurred approximately 15,000 to 20,000 years ago. They were able to take advantage of additional ice-free land for nesting and came out of the climate change with steady populations. However, these species are not all faring as well with the current warming trends in Antarctica. Chinstrap and Adélie penguin populations have plummeted by more than 50% at some study locations. During the 19th and 20th centuries, population numbers of these species rose as humans hunted whales and large fish species that competed with the penguins for their favorite food, Antarctic krill. Now that whale and fish populations are recovering, they are once again competing with penguins for krill. Gentoo penguins are more generalist feeders, so they are likely maintaining populations because they are feeding on other prey items.

A delicate balance exists between too much ice and too little ice for many polar species. Penguins must reach the sea for feeding, so when there is too much sea ice to walk across before they can reach their watery feeding areas they will not flourish. Too little ice, and their prey species will move further from the continent and become unavailable to the penguins. Researchers are tracking penguin populations closely to determine their reactions to warming seas; increased competition for food; and, in some cases, changes to sea ice. Although the ice around Antarctica is not declining quite as much as in the Arctic, changes to the ice cover are occurring at different areas around the continent.

In one study on Adélie penguins, researchers used penguin droppings, or *guano*, to investigate population sizes over the past 700 years. Not only is guano a good indication of the presence of penguins, but the presence of guano determines the dominant vegetation in the area. When penguins and their guano are present, algae is dominant due to the increased nutrients available from the guano. When penguins and their guano are absent, lichens are dominant because they are sensitive to trampling by penguins. The presence of penguin populations varied at locations around Antarctica, but both extremely high and extremely low temperatures led to low penguin population sizes. The results of the study indicated a rise in penguin populations during a cold period, which is in contrast to other published work using DNA analysis indicating low populations during cold periods and rising numbers as temperatures increase. The contrasting results of the two different studies highlight the great variability of penguin presence at individual sites over time, as the study using guano evidence was conducted in the southern portion of Antarctica and the genetic study in the northern portion.

As climate changes occur in Antarctica and the surrounding seas, scientists will be closely monitoring penguin populations to observe their response. Although penguins seem to be sensitive to ice characteristics, it is often surprising and remarkable how resilient organisms are, and penguins are no exception.

Critical Thinking Questions

1. Besides the studies being conducted in different locations in Antarctica, devise a hypothesis explaining why two studies on penguin populations using different methods might lead to different conclusions.

2. It appears that the chinstrap and Adélie penguin populations are plummeting with the current warming trend in Antarctic waters. During past warming events these species survived. What might be different about the current warming event in comparison to the one that took place 15,000 to 20,000 years ago?

3. Are declines in populations that occur over a few years a major cause for concern, or is the concern an exaggeration and a case of humans thinking in terms of human time scales?

For Further Reading

Clucas, G. V., M. J. Dunn, G. Dyke, S. D. Emslie, H. Levy, R. Naveen, M. J. Polito, O. G. Pybus, A. D. Rodgers, and T. Hart. 2014. A reversal of fortunes: Climate change "winners" and "losers" in Antarctic Peninsula penguins. *Scientific Reports* 4:5024.

Hu, Q., L. Sun, Z. Xie, S. D. Emslie, and X. Liu. 2013. Increase in penguin populations during the Little Ice Age in the Ross Sea, Antarctica. *Scientific Reports* 3:2472.

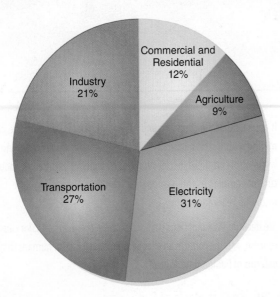

Figure 14.28 Human causes of greenhouse gas emissions in the United States during 2013.
Data courtesy of the United States Environmental Protection Agency.

which accounts for approximately 27% of **greenhouse gas emissions** (**Figure 14.28**). Unfortunately, many U.S. cities, and certainly suburbs, were planned with the use of a car in mind. Mass transit is not always an option for people for their everyday lives. If it is available, using mass transit is a great way to reduce your carbon emissions. If mass transit is unavailable, purchasing fuel-efficient vehicles and carpooling are additional ways to reduce carbon emissions from motor vehicles. Current technology has led to numerous advancements in motor vehicle efficiency, such as hybrid and electric cars, and although very expensive at the onset, these cars have already become more affordable for a larger demographic.

The majority of greenhouse gas emissions in the United States are a result of electricity production, which accounts for approximately 31% of emissions. Most of our electricity comes from burning fossil fuels, primarily coal and natural gas. Instead of decreasing our electricity production, we have increased emissions by 11% since 1990. Increases in population size and energy demand are the cause of this increase. Individuals can do some very simple things to decrease electricity use, such as using low-energy lightbulbs, turning off lights when not in use, and purchasing the most energy-efficient appliances. Although a serious monetary investment, adding solar panels to a house can allow residents to obtain most, if not all, of their energy from the sun. Many incentives exist for converting to solar power, and in some cases the monthly payment would be the same amount as the monthly electric bill. If we take this one step further, consider the scenario of converting a home to solar and purchasing an electric vehicle. This scenario would result in a near-zero-emissions transportation and electricity lifestyle.

A little less obvious and less direct solution to reducing carbon emissions is limiting our consumption of *stuff* as a whole and purchasing products made closer to home. Every product purchased was shipped to a store for purchase, and then consumers drive to the store to make the purchase. Limited natural resources (e.g., oil) are used in the production, selling, and purchasing processes. Purchasing reusable products is one solution to this problem. Purchasing products that last longer is another. The upfront cost may be more, but higher-quality products generally function better, almost always last longer, and may even be sourced sustainably and/or locally. Citizens of other first-world countries already live much more minimally and sustainably than Americans, and now it is our turn. It is up to individuals to act in an environmentally responsible manner, and the reward is a healthier planet for ourselves and for future generations of humans and all other inhabitants of this Earth. For more information on what we can do to reduce our carbon emissions, visit the Environmental Protection Agency's (EPA) website (https://www3.epa.gov).

STUDY GUIDE

TOPICS FOR DISCUSSION AND REVIEW

1. What are the main differences and similarities in the physical features of the Arctic and Antarctic regions?

2. Describe ice algae, and explain the current trends in ice algae in the Arctic.

3. Describe krill, and explain why it is such a valuable food source for animals in Antarctica.

4. What are some adaptations various fish species have evolved for living in frozen waters?

5. Why do most Arctic species of seals show a reduced level of polygyny and sexual dimorphism?

6. Explain how Antarctic filter-feeding rorquals engulf vast volumes of water to feed on krill.

7. Why have the numbers of minke whales, Adélie penguins, crabeater and leopard seals, and fish around Antarctica increased dramatically during the past 50 to 60 years?

8. How is climate change affecting the Arctic? Antarctica?

9. What are some of the main sources of carbon emissions in the United States?

10. What can you do to reduce your carbon emissions and help slow climate change?

KEY TERMS

albedo 379	ice-breeding 387
Antarctic Convergence 380	ichthyofauna 382
Antarctic ozone hole 394	induced ovulator 388
blubber 387	krill 380
carrion 383	megafauna 383
endemic 381	meltwater 380
fast ice 378	monogynous 389
greenhouse gas emissions 398	pack ice 378
hemoglobin 383	phylogenetic 382
homeotherm 380	polygyny 389
ice algae 380	sea ice 378
	thermoregulate 381
	vibrissae 389

KEY *GENERA*

Arctogadus	Euphausia
Balaena	Gammarus
Calanus	Lobodon
Monodon	Odobenus
Myoxocephalus	Somniosus
Mytilus	Ursus

REFERENCES

Arrigo, K. R. 2013. The changing Arctic Ocean. *Elementa Science of the Anthropocene.* doi:10.12952/journal.elementa.000010.

Arrigo, K. R., D. K. Perovich, R. S. Pickart, Z. W. Brown, and G. L. van Dijken. 2012. Massive phytoplankton blooms under Arctic sea ice. *Science* 336:1408.

Arrigo, K. R., and G. L. van Dijken. 2011. Secular trends in Arctic Ocean net primary production. *Journal of Geophysical Research* 116:C09011. doi:10.1029/2011JC007151.

Bertrand, E. M., J. P. McCrow, A. Moustafa, H. Zheng, J. B. McQuaid, T. O. Delmont, A. F. Post, R. E. Sipler, J. L. Spackeen, K. Xu, D. A. Bronk, D. A. Hutchins, and A. E. Allen. 2015. Phytoplankton–bacterial interactions mediate micronutrient colimitation at the coastal Antarctic sea ice edge. *Proceedings of the National Academy of Science USA* 112(32):9938–9943.

Clarke, A., and C. M. Harris. 2003 Polar marine ecosystems: Major threats and future change. *Review Environmental Conservation* 30:1–25.

Conservation of Arctic Flora and Fauna (CAFF). 2010. Arctic biodiversity trends 2010: Selected indicators of change. CAFF International Secretariat, Akureyri, Iceland.

Grebmeier, J. M., S. E. Moore, J. E. Overland, K. E. Frey, and R. R. Gradinger. 2010. Biological response to recent Pacific Arctic sea ice retreats. *EOS: Earth and Space Science News* 91:161–162.

Grémillet, D., J. Welcker, N. J. Karnovsky, W. Walkusz, M. E. Hall, J. Fort, Z. W. Brown, J. R. Speakman, and A. M. A. Harding. 2012. Little auks buffer the impact of current Arctic climate change. *Marine Ecology Progress Series* 454:197–206.

Griffiths, H. J., R. J. Whittle, S. J. Roberts, M. Belchier, and K. Linse. 2013. Antarctic crabs: Invasion or endurance? *PLoS ONE* 8(7):e66981. doi:10.1371/journal.pone.0066981.

Irvine, J. R., R. W. Macdonald, R. J. Brown, L. Goodbout, J. D. Reist, and E. C. Carmack. 2009. Salmon in the Arctic and how they avoid lethal low temperatures. *North Pacific Anadromous Fish Commission Bulletin* 5:39–50.

Kwok, R. 2015. Sea ice convergence along the Arctic coasts of Greenland and the Canadian Arctic Archipelago: Variability and extremes (1992–2014). *Geophysical Research Letters* (Accepted). doi:10.1002/2015GL065462.

Laidre, K. L., M. P. Heide-Jorgensen, W. Ermold, and M. Steele. 2010. Narwhals document continued warming of southern Baffin Bay. *Journal of Geophysical Research* 115(10):1–11.

Laidre, K. L., H. Stern, K. M. Kovacs, L. Lowry, S. E. Moore, E. V. Regehr, S. H. Ferguson, Ø. Wiig, P. Boveng, R. P. Angliss, E. W. Born, D. Litovka, L. Quakenbush, C. Lydersen, D. Vongraven, and F. Ugarte. 2015. Arctic marine mammal population status, sea ice habitat loss, and conservation recommendations for the 21st century. *Conservation Biology* 29(3):724–737.

McBride, M. M., P. Dalpadado, K. F. Drinkwater, O. R. Godø, A. J. Hobday, A. B. Hollowed, T. Kristiansen, E. J. Murphy, P. H. Ressler,

S. Subbey, E. E. Hofmann, and H. Loeng. 2014. Krill, climate, and contrasting future scenarios for Arctic and Antarctic fisheries. *ICES Journal of Marine Science* 71(7):1934–1955. doi:10.1093/icesjms/fsu002.

Michel, C., T. C. Nielsen, C. Nozais, and M. Gosselin. 2002. Significance of sedimentation and grazing by ice micro- and meiofauna for carbon cycling in annual sea ice (northern Baffin Bay). *Aquatic Microbial Ecology* 30:57–68.

Nielsen, J., R. B. Hedeholm, J. Heinemeier, P. G. Bushnell, J. S. Christiansen, J. Olsen, C. B. Ramsey, R. W. Brill, M. Simon, K. F. Steffensen, J. F. Steffensen. 2016. Eye lens radiocarbon reveals centuries of longevity in the Greenland shark (Somniosus microcephalus). *Science* 353(6300):702–704.

Pagano, A. M., G. M. Durner, S. C. Amstrup, K. S. Simac, and G. S. York. 2012. Long-distance swimming by polar bears (*Ursus maritimus*) of the southern Beaufort Sea during years of extensive open water. *Canadian Journal of Zoology* 90(5):663–676.

Perrin, W. F., B. Wursig, and J. G. M. Thewissen. 2009. *Encyclopedia of Marine Mammals.* New York: Elsevier.

Sremba, A. L., B. Hancock-Hanser, T. A. Branch, R. L. LeDuc, and C. S. Baker. 2012. Circumpolar diversity and geographic differentiation of mtDNA in the critically endangered Antarctic blue whale (*Balaenoptera musculus intermedia*). *PLoS ONE* 7(3):e32579. doi:10.1371/journal.pone.0032579.

Strahan, S. E., L. D. Oman, A. R. Douglass, and L. Coy. 2015. Modulation of Antarctic vortex composition by the quasi-biennial oscillation. *Geophysical Research Letters* 42(10):4216–4223.

Thyrring, J., S. Rysgaard, M. E. Blicher, and M. K. Sejr. 2014. Metabolic cold adaptation and aerobic performance of blue mussels (*Mytilus edulis*) along a temperature gradient into the High Arctic region. *Marine Biology* 162(1):235–243.

Wassmann, P. 2011. Arctic marine ecosystems in an era of rapid climate change. *Progress in Oceanography* 90:1–17.

Harvesting Living Marine Resources

STUDENT LEARNING OUTCOMES

1. Provide examples of the most frequently captured marine species.

2. Summarize information on the oceanographic conditions in different areas of the world ocean, and relate it to the world's major fishing areas.

3. Compare the harvesting of marine resources to the harvesting of terrestrial resources for food.

4. Summarize the problems with fishing down the food web.

5. Describe aquaculture, and identify the pros and cons associated with this method of increasing seafood production.

6. Discuss examples of overexploited fisheries; the problems that led to each population crash; and how these problems may be explained, in part, by the "tragedy of the commons."

7. Summarize the challenges associated with international regulation of fisheries, including sealing and whaling.

8. Discuss the marine ornamental fishing industry and the problems associated with take of these valuable organisms.

9. Identify ways individuals can positively impact the marine environment and populations of fished species.

CHAPTER OUTLINE

A Chilean purse seiner hauling in a catch of approximately 400 tons of jack mackerel, *Trachurus*. The man standing near the side of the boat provides a size reference and reveals how enormous these nets are.
Courtesy of National Oceanic and Atmospheric Administration.

Like many environmental problems, the current state of dwindling marine fisheries is largely due to human population expansion. For much of human history, the seas were isolated from the pressures created by the material needs of our land-based human population. Two thousand years ago, the human population was probably between 200 and 300 million people. The daily challenges of survival kept life expectancies short. High birth rates were balanced by high death rates, and the population grew slowly. Not until 1650, when the Industrial Revolution in Europe brought advances in medicine and technological improvements in food production, did human mortality rates decline and the population double to 500 million. Presently, the human population is more than 7 billion people, and it is projected to reach 9.6 billion by 2050.

The sheer magnitude of the current human population creates enormous demands for food, fiber, minerals, and other commodities that support our modern social fabric. Today, most of the world ocean is intensively exploited for fishing; recreation; military purposes; commercial shipping; dumping of waste materials; and extraction of gas, oil, and other mineral resources. Although the impact of these activities varies geographically, no portion of the world ocean is isolated from their effects. These effects often transcend national borders, as well as the boundaries of many ecological and taxonomic groups. Where the cumulative biological effects of these societal uses of the sea are large and measurable, they introduce a complex suite of social, political, economic, aesthetic, and biological questions that cannot be resolved solely by scientific means.

The demand for food to sustain the present human population is enormous and is increasing at a rapid rate. The United Nations Food and Agriculture Organization (UNFAO) and the World Bank estimate that over 1 billion people are seriously undernourished. Uncounted millions of people starve to death each year, and hundreds of millions more are deprived of good health and vigor because of inadequate diets with little protein. Many people, therefore, regard marine sources of food as a crucial part of the solution to present and future food problems.

Fishing is a multibillion-dollar global industry. For our purposes, the term *fishing* is broadly defined as the capture or collection of wild marine plants and animals for food production or other commercial activities. These other activities range from collecting coral reef animals for the ornamental fish trade to killing seal pups for their skins. According to the UNFAO, in 2011 the global import/export trade in fishery commodities was about $260 billion. Most seafood is sold for direct human consumption, but an appreciable portion is also used for livestock fodder, pet foods, and a variety of nonfood industrial products.

The impact of fishing practices on the stability, and even the continued existence, of some marine populations has been severe because a relatively small number of species make up the bulk of the world's fish harvest. These species have borne the brunt of the human population's demands for seafood. Some species have been completely eliminated from traditional fishing grounds, whereas others face the imminent possibility of biological extinction.

At the beginning of the 19th century, cod in the North Atlantic were so abundant that boats often hit these fish while lowering their anchors. Haddock was a popular food fish in Europe, and salmon swam in North American coastal waters in uncounted millions. In central California, a new and promising sardine fishery was just getting started. In the Southern Hemisphere, Antarctic waters teemed with large baleen whales in summer months, and very few people knew of pollock, hake, or menhaden. Now, as we begin the 21st century, the haddock populations are no more; they peaked in 1929. The cod fishery of the North Atlantic has collapsed, and we buy more farm-raised salmon than fish caught from wild populations. The sardine "cannery row" popularized by John Steinbeck now exists only as rows of shops and boutiques for tourists. Even the largest animal on Earth, the blue whale, currently numbers less than 10% of what it did a century earlier, and pollock, hake, and menhaden are among the most abundant marine fishes captured. In this chapter, we examine the characteristics of marine fisheries and some of the mechanisms behind these dramatic changes in dominant marine animal populations. We also explore solutions to fisheries problems and recommendations for living as a steward of the marine environment.

15.1   A Brief Survey of Marine Food Species

The raw material of the fishing industry includes numerous species of bony and cartilaginous finfish; many mollusks and crustaceans (collectively referred to as *shellfish* even if, like squid, they lack shells); a variety of other aquatic animals (from worms to whales); and even some marine plants. Each year the UNFAO compiles and publishes global fishery catch statistics that are the basis of most of the figures and tables in this chapter. The unit of measure used by the UNFAO (and used in this chapter) is the tonne, equivalent to 1,000 kg, and also known as a metric ton. **Figure 15.1** summarizes capture results and indicates capture trends at decade intervals since 1970. The annual catch size has varied somewhat; however, the relative ranking of most groups has remained reasonably constant for the past three decades.

Clupeoid fishes, including anchovies (**Figure 15.2**), anchovetas, herrings, sardines, pilchards, and menhaden, are very abundant and account for about one third of the total commercial catch. A single species, the Peruvian anchoveta (*Engraulis ringens*), provided more than 13 million ton, or nearly 19%,

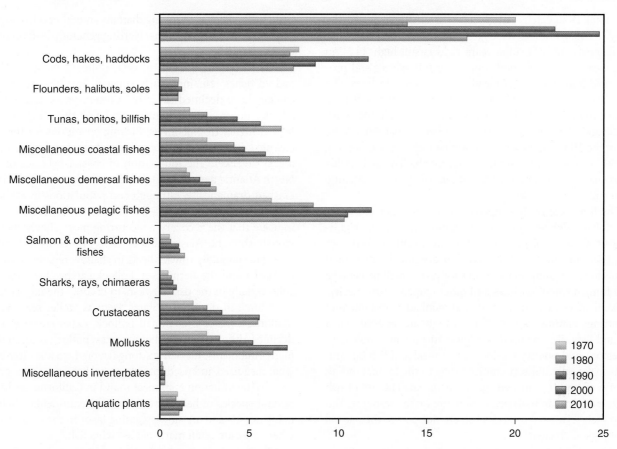

Figure 15.1   Global marine and estuarine harvest by major categories since 1970.
Data compiled from UNFAO statistics.

Figure 15.2   A compact school of filter-feeding anchovies.
© Ferenc Cegledi/Shutterstock, Inc.

of the total 1970 catch, but extensive overfishing caused the fishery to collapse a few years later. Since then the fishery has been inconsistent, but some years yield is still high. El Niño events clearly affect the food sources for this species and produce an additional factor that leads to population declines. The herring catch is an important part of the North Atlantic fishing industry. In the last decade, other herring fisheries in the South Atlantic and North Pacific oceans have been expanding. At its peak in the 1930s, the California sardine industry was landing over 500 thousand ton annually. The menhaden catch yields approximately 1 million ton annually, largely from the Atlantic Ocean and the Gulf of Mexico.

The huge size of the clupeoid fish catch is not indicative of its dollar value as an economic commodity nor its significance as a source of protein in human nutrition. Although canned sardines are common at any grocery store in the United States, direct consumption accounts for just a small percentage of the human use of sardines and other clupeoid fish species. Nearly all anchovies, anchovetas, and menhaden and much of the herrings, sardines, and pilchards caught are reduced to fish meal and fish oil to be used as a low-cost protein supplement in livestock and poultry fodder. As fish meal and fish oil, clupeoids make only indirect contributions to the human diet in the form of pork or chicken that humans eat. The use of fish to feed terrestrial animals places pressure on the resource, and due to their terrestrial lifestyles fish are not part of the natural diet of pigs or chickens.

Clupeoids are found in shallow coastal waters and in upwelling regions. Their schooling behavior (Figure 15.2) simplifies catching techniques and reduces harvesting expenses. Large purse seines (which may be 600 m long and 200 m deep, similar to the seine shown in **Figure 15.3**) are used to surround and trap entire schools. After encircled, the fish are ladled or pumped into the ship's hold. Most clupeoids are small, with an average adult length of between 15 and 25 cm. Their small size, however, is compensated for by other characteristics that enhance their economic usefulness. These fish have fine gill rakers that enable them to feed on small organisms close to the base of marine food chains. Mature Peruvian anchovetas feed

principally on chain-forming diatoms and other relatively large phytoplankton aggregations. Herring generally feed on herbivorous zooplankton.

The combined catch of cod, pollock, hake, and other gadoid fishes remained fairly constant for 30 years, but cod catches have declined severely in recent years. Gadoids usually live on or near the bottom, are larger than clupeoids, and feed at higher trophic levels. Fishing operations for these species are concentrated on continental shelves and other shallow areas. Cod were the foundation of coastal fisheries of many North Atlantic nations until the early 1990s, when populations crashed dramatically. Although cod populations in areas such as the Gulf of Maine appeared to be recovering, data from 2013 indicate that the recovery is occurring more slowly than previously thought. Alaska pollock, also known as walleye pollock, are caught by trawler fleets in shallow regions of the Gulf of Alaska and the Bering Sea. Pollock catches have increased substantially in the past 20 years to become the largest fishery by volume in the United States, even as Steller sea lion populations, a major predator of pollock, experienced a serious decline. Although a top fishery species, pollock is unknown to many Americans. It is fairly strong flavored, so it is often somewhat disguised and used for fish sticks, fish and chips, or even surimi, the imitation crab meat found in California sushi rolls. Several species of hakes abound on the continental shelves of many oceans and are slowly gaining favor in the United States, where they are often marketed as "whitefish."

Redfish, bass, sea perch, and other miscellaneous coastal rockfish live near the seafloor. No single species of this group dominates the catch statistics, yet the combined catch since 1980 consistently has amounted to more than 5 million ton. Sold fresh or frozen, most of these fish are popular fish market items. Miscellaneous pelagic fishes such as horse mackerel and jack mackerel generally feed on smaller anchovy-sized fish and squid. These mackerels resemble tuna in form, but they seldom exceed 30 to 40 cm in length, and their flavor is much stronger than tuna.

Tuna are individually the largest of the commercially exploited species discussed here. Tuna weighing more than 100 kg are not unusual, but most range from 5 to 20 kg (**Figure 15.4**). Yellowfin, bigeye, albacore, and skipjack tuna account for most tuna catches. Tuna are often the top carnivores in complex food chains and may be separated from the primary producers by seven or more trophic levels. These fish are active predators of smaller, more abundant animals, especially clupeoids and sauries and are captured in or near nutrient-rich waters where prey fish abound.

Figure 15.3 Tuna being ladled out of the purse seine shown in Figure 15.4.
Courtesy W. Perryman, National Marine Fisheries Service, NOAA.

DID YOU KNOW?

Because of pressure from consumers, large tuna-canning companies, such as Bumble Bee, have started providing detailed information about individual cans of tuna to consumers. Each can of tuna has an identifying number associated with it, and consumers can look up information such as catch location, harvesting method, and even a picture of the flag flown by the fishing vessel. When consumers demand information, companies are often forced to respond. Accountability is one step toward more sustainable fishing practices.

Figure 15.4 A modern tuna seiner with its purse seine surrounding a large school of tuna.

Courtesy W. Perryman, National Marine Fisheries Service, NOAA.

U.S. tuna fishers rely heavily on long-range purse seiners (Figure 15.4) to harvest schooling species of tuna, especially skipjack and yellowfin tuna. In the Eastern Tropical Pacific (ETP), yellowfin tuna exhibit a strong but poorly understood behavioral association with common (*Delphinus delphis*), spinner (*Stenella longirostris*), and pantropical spotted dolphins (*S. attenuata*). Because dolphins are easily visible at the sea surface, purse seining of yellowfin tuna is often simply a matter of setting the nets around a dolphin school with the assumption (usually correct) that a school of tuna swims just below. By the early 1970s, the global tropical tuna fishing fleet had killed several million dolphins unintentionally during capture efforts for tuna. Since then, tighter restrictions on both the design and operation of purse seining, imposed by the 1972 U.S. Marine Mammal Protection Act (MMPA), have reduced annual dolphin mortality due to purse seining to about 1,000 individuals each year for the U.S. tuna fleet. **Figure 15.5** provides a clear picture of dolphins trapped in a purse seine. Although they may appear healthy in the image, many dolphins are harmed as the net is retrieved or from encounters with other organisms caught in the net.

Figure 15.5 Dolphins surrounded by a tuna seining net.

Courtesy of National Oceanic and Atmospheric Administration.

Foreign nations are not bound by the MMPA, and their participation in the ETP tuna fishery was increasing each year. After facing embargoes on yellowfin tuna and tuna products and being pressured by consumer groups to provide "dolphin-safe" tuna, all nations exploiting the ETP tuna fishery successfully negotiated the Agreement on the International Dolphin Conservation Program (AIDCP) in February 1998. Since then, the ETP fishing countries have added fisheries observers on their fishing boats to provide more certainty that dolphin bycatch remains below the numbers allowed. Today, accidental dolphin mortality in the ETP has been reduced to less than 0.1% of the dolphin stock, which should have an insignificant impact. Unfortunately, dolphin populations in the ETP are not recovering at expected rates. Hypotheses to explain this lack of population growth include illegal and unreported dolphin mortality, habitat alterations in the area, or negative interactions with other species that might be delaying or preventing dolphin recovery.

To exploit the more widely dispersed bluefin and bigeye tuna, labor-intensive longline fishing methods are also used. These tuna aggregate in a narrow zone of relatively high primary productivity that straddles the equator in the eastern half of the Pacific Ocean. These large predators feed on sardines, lanternfish, other small fishes, and crustaceans, which, in turn, graze on the smaller zooplankton. The longline fisheries of the equatorial Pacific Ocean have developed into a valuable oceanic fishery, with Mexico and Japan each taking about 640,000 ton of yellowfin, bigeye, and bluefin tunas. The longline gear consists of floating mainlines that extend as far as 100 km along the surface. Hanging from each mainline are about 2,000 equally spaced vertical lines that terminate with baited hooks. After a longline set is in place, fishers move along the mainline, remove hooked tuna, and rebait and replace the hooks.

In the last decade, many fisheries have begun using gill nets. Designed to entangle fish in the net fabric, these large rectangular nets are either anchored to shore or allowed to drift. Although effective at entrapping fish, they are also invisible and indiscriminate killers of numerous nontarget species. These nets are fast becoming major causes of mortality for some species of sharks, sea turtles, and many seabird and marine mammal populations.

The world's commercial fishing fleet contributes more than 640,000 ton of accidentally lost nets, pots, traps, and setlines and deliberately discarded pieces of damaged fishing gear each year. The shift since 1940 from the use of natural fibers to virtually nonbiodegradable synthetic fibers for the construction of nets, lines, and other fishing gear has made commercial fisheries a major contributor to marine plastics pollution. The largest potential for lost and discarded fishing gear occurs in the North Pacific Ocean, where vessels of many nations operate under adverse climatic conditions. Over a million kilometers of gill net are now set annually by the squid fisheries, and the nations primarily responsible are Japan, Taiwan, and South and North Korea. These nets constitute a large potential source of derelict gear that when lost continues to drift for thousands of kilometers. Birds, mammals, and other animals that surface frequently are especially susceptible to entanglement in the floating net fabric or other plastic debris (**Figure 15.6**). When nets are left

Figure 15.6 Derelict fishing gear washed up on a beach. There is no way to know how many fish were trapped in this net and needlessly killed as it drifted around as a ghost net.
Courtesy of Dwayne Meadows, National Oceanic and Atmospheric Administration.

Case Study

The Wavering Identity of Steelhead Trout

Species name: *Oncorhynchus mykiss*

Common name: *Steelhead trout*

Endangered Species Act status: 1 distinct population segment (DPS) is endangered, 10 DPSs are threatened, and 1 DPS is a species of concern.

The steelhead trout, *Oncorhynchus mykiss*, is a somewhat mysterious fish, because its fate at birth is uncertain. All steelhead hatch in fast-flowing, well-oxygenated, gravel-bottom freshwater streams and rivers. Some steelhead, however, migrate from their natal streams or rivers to the ocean, living an anadromous existence. Others remain in freshwater their entire lives, and these are commonly known as *rainbow trout*. Steelhead that migrate to the sea have a slightly different appearance, with a slimmer profile, more silvery body color, and much larger size (up to 25 kg) than the rainbow trout (**Figure A**).

The mystery does not end with the differing freshwater and anadromous forms. Anadromous steelhead all spawn in their natal rivers, but they display one of two different methods of maturing before they spawn. Some forms mature completely while in the ocean, and then return to the natal stream just before spawning with fully developed gonads. These are referred to as *ocean-maturing* or *winter-run steelhead*. The other method of maturation occurs in the natal stream where fish return before their gonads are mature to spend several months to become fully mature and ready to spawn. These fish are referred to as *stream-maturing* or *summer-run steelhead*. In general, coastal streams are dominated by winter-run steelhead, and streams or rivers located more inland are dominated by summer-run steelhead. Steelhead can spawn more than one time, unlike other salmonids that spawn once and then die. For reasons unknown, some steelhead have been observed returning from the ocean to their natal rivers temporarily years before they are ready to spawn.

The variable lifestyles of this species have led to challenges for protection. Like many other coastal or estuarine species, steelhead are highly subject to pollution and

Case Study   *(Continued)*

Figure A   An adult steelhead trout swimming in a freshwater stream.
Courtesy of National Oceanic and Atmospheric Administration.

(b1)

(b2)

Figure B   An area with stream obstruction by a weir (1) was opened up to provide fish passage and to improve water quality (2).
Courtesy of National Oceanic and Atmospheric Administration.

other human-caused disturbances because they live in water bodies that are subject to habitat alteration or destruction, urban runoff, and other forms of pollution. Because steelhead have the requirement of returning to natal streams or rivers to spawn, access to these water bodies is essential for their successful reproductive efforts. The building of dams or weirs (low-lying dams) that are impassable to steelhead and other anadromous fishes such as various salmon species has severely limited their reproductive success. Slowly, dams are being removed (**Figure B**) or, more commonly, retrofitted with fish passage designs such as fish ladders to assist steelhead or salmon with safe passage upstream.

Dams are large and expensive structures, so convincing people that they need to be removed is no easy task. Searsville Dam is located on the property of Stanford University and is blocking the passage of steelhead in this area. The dam has generated much debate among scientists and the university, as conservationists and even some politicians are demanding its removal. The dam was used to divert water for irrigation of a large golf course and other campus landscaping, and although it is no longer used in this way, Stanford wants to preserve the dam and its rights to water usage. The diversion blocks fish passage and degrades fish habitat downstream of the dam. Instead of removing the dam, Stanford is considering cutting a hole in the bottom of it for fish passage. Many people feel this is not enough and favor complete removal. Stanford University is a highly ranked university, and its professors conduct impressive marine biology and coastal biology research. The fact that this institution will not comply with recommendations that will lead to the preservation of a threatened species is frustrating to say the least. The National Marine Fisheries Service (NMFS) is supposed to weigh in on this matter, but the agency has failed to provide requested documentation about steelhead populations and the effects of the dam. A district judge ordered the NMFS to provide a reason for its delay, and so for now the debate continues.

Other dam stories have more favorable endings for steelhead and other salmonids. A dam in central Oregon, the Stearns Dam, was completely removed from the landscape in the winter of 2014 to provide open fish passage into the Deschutes River Basin. The dam had been in place since 1911 and, along with several other dams in the area, was built to divert water to irrigate ranch lands. The process to remove the dam

took 10 years, due to the many legal and political issues that had to be resolved. Several other restoration projects in the Deschutes River area are ongoing, slowly providing more habitat and access for steelhead and salmon in this region.

Both steelhead and rainbow trout are popular aquaculture species, thus they are commonly eaten even though natural populations of steelhead are found in very low numbers. Farms exist all over the world and are expanding to meet market demands. In the summer of 2015, a major producer of seafood, Pacific Seafood Group, expanded its steelhead production by 50% at two sites in the Columbia River in Washington State. The group had permission for the expansion for years, but was waiting for market demand to go up and for access to additional research on steelhead to increase success in their fish production. With new research and aquaculture methods available and news that Chile will be decreasing its production of steelhead, 2015 seemed to be the right timing

(continues)

Case Study   (Continued)

to expand. Although aquaculture can be environmentally degrading, this particular area of the Columbia River was nutrient poor, so the fish farm appears to be adding valuable nutrients, for now. Pacific Seafood is the first U.S. steelhead facility to earn a Best Aquaculture Practices status, which is part of the Global Aquaculture Alliance program to recognize aquaculture facilities that "assure healthy foods produced by socially and environmentally responsible means."

The story of steelhead is a complicated one. Some populations are threatened, yet unlike almost all fished species the main source of population declines is not overfishing, but other human-caused environmental factors. Steelhead is a favorite food fish, so aquaculture is lucrative. The main issue preventing steelhead populations from recovering is the presence of dams, but dams are expensive and time-consuming to remove, so removal is not a simple matter. As populations disappear along the coast, the pressure to remove dams will increase. Whether natural steelhead populations have the opportunity to persist is in large part up to humans and our decisions.

Critical Thinking Questions

1. If steelhead and rainbow trout have such different lifestyles and habitat use, why do you think they are considered the same species? What type of biological information would lead to a designation of two separate species?

2. What problems might arise with raising an endangered fish species in captivity and selling it for human consumption?

3. Dams are expensive structures, so convincing cities or other entities to remove them is a difficult task. How do you place value on a fish species so that it can be compared to the value of something like a dam?

---

behind they continue to capture animals in an unnecessary and harmful practice termed *ghost fishing*.

15.2   Major Fishing Areas of the World Ocean

Areas of high primary productivity generally support robust fish populations. Shallow waters over continental shelves and near-surface banks encourage rapid regeneration of crucial nutrients in the photic zone, supporting healthy food webs. These nutrients are prevented from escaping into deeper water and accumulating below the upper mixed layer where return to the surface requires a much longer time. Fishing activities in neritic waters are effective because the resulting animal production is "crowded" into a water column usually less than 200 m deep. In contrast, organic material produced over deep-ocean basins is shared by the many consumers thinly dispersed within several thousand meters of water. These areas generally do not support thriving fisheries, although some pelagic species are caught over deep waters.

About 90% of the marine catch is taken from continental shelves and overlying neritic waters, a region representing less than 8% of the total oceanic area. Bottom fish and benthic invertebrates account for 10% to 15% of the total global catch. Halibut, flounder, sole, and other flatfish are benthic species caught in shallow waters using bottom trawls. High prices and stable markets for these fish have resulted in heavy fishing pressures for more than the past 100 years, and most populations are seriously overfished.

Other major fishing areas are centered in regions of upwelling where abundant supplies of crucial nutrients from deeper waters are returned to the photic zone. Upwelling may, depending on the locality, occur sporadically, on a seasonal basis, or continually throughout the year. The nutrient-rich waters surrounding the Antarctic continent sustain very high summertime primary production. Because of the challenges and costs associated with fishing in Antarctica, for decades the only fisheries were the near-defunct pelagic whaling industry and a small annual harvest of krill. Recently, other commercial fisheries have developed in Antarctic waters, including Patagonian toothfish, mackerel icefish, and Antarctic rock cod. Long distances to processing facilities and the absence of nearby population centers have had a restraining effect on the successful exploitation of this upwelling region to some extent, but fishers are beginning to move into this highly productive area.

Regions of coastal upwelling are most apparent along the west coasts of Africa and North and South America but occur to lesser degrees along other coastlines. High yields of commercial species result from increased rates of primary production over shallow bottoms. The greatest concentrations of clupeoids are found in regions of coastal upwelling: Peruvian anchoveta from the Peru Current upwelling area; pilchard from a similar area in the Benguela Current off the west coast of South Africa; and, before a drastic decline in the 1940s, sardines from the California Current. The Peruvian and Benguelan upwelling systems also support enormous populations of seabird and pinniped predators of these small fish.

DID YOU KNOW?

It doesn't always take much effort to create a fishery. With the need for fish to help feed the growing human population, new species to exploit are in high demand. In the 1980s, a famous chef, Julia Child, recommended monkfish tail on a cooking show, a fish that was not previously a target species in the United States. Since that time, monkfish has become a target species, and the fishery is now managed to avoid depletion. Most fishers chop off the tail to keep and throw back the rest of the body, as the delicate flavor of the tail has led to its nickname of "poor man's lobster."

15.3   A Perspective on Seafood Sources

Humans are omnivores by nature. We obtain nourishment from a large variety of plants and animals, yet the staples of the human diet can be narrowed to just a few types of plants and animals. The plant staples include cereals (rice, wheat, corn, and lesser amounts of other cereal grains), vegetables, fruits, nuts, and berries. Beef, poultry (plus the milk and egg products of these animals), pork, and fish provide the major share of the food to satisfy the carnivore in most of us. It is important to understand just how significant the present marine contribution of food is to the total human diet and how meaningful it may be in the future.

In **Table 15.1**, two categories of plant and animal foods from both terrestrial and oceanic production systems based on the state of technology used in the production of each food category are compared. Plant production is separated into gathering; the use of wild plants (**Figure 15.7**); and farming, the agricultural tending of domesticated plant species. Comparable categories for animals are used: hunting of wild animals and herding of genetically improved, controlled, domesticated animals. These terms are also applied to marine food items. Only a few marine algae, bivalve mollusks (**Figure 15.8a**), and fish, such as salmon (**Figure 15.8b**), are grown in abundance in controlled conditions collectively referred to as **aquaculture**. Table 15.1 lists the amount of food produced or captured in 2006 for land, freshwater, and ocean. Production statistics for wild plants and wild animals grown on land can only be approximated. The remaining statistics are provided by the UNFAO; the latest data set available for this analysis being from 2006. Since 2006, production from aquaculture has increased substantially, but the rate of increase is likely to stabilize soon as fish in culturing facilities are fed fish meal that requires take from the ocean.

The information presented in Table 15.1 forces some uncomfortable, yet undeniable, conclusions. Most seafoods harvested are animals three or four trophic levels above the primary producers. In great contrast, in terrestrial agricultural systems, more plants are harvested than animals, and losses to higher trophic levels are reduced. Furthermore, most of the marine harvest consists of wild stocks that are hunted rather than controlled and domesticated. Although time, energy, and

Figure 15.7 An artisanal fisher gathers kelp from the sea.
© Vespasian / Alamy Stock Photo.

a large amount of money have been invested in producing technologically advanced ships, nets, and fish-finding gear, very little of the monetary gains from increased catches are reinvested in these wild fish stocks. The role of aquaculture in the mix of seafood production is discussed in the next section.

Another negative trend in the use of living marine resources has developed in the past half century. In 1955, 86% of the world fish catch was consumed directly; the remainder was reduced to fish meal for use as a protein supplement for domestic livestock. By 1970, the fish meal fraction had increased to about 40% of the world catch, and it has remained at 30% to 35% since. In addition to feeding livestock, in recent years an increasing proportion of fish meal is being used to feed fish being farmed in aquaculture facilities or pens. Assuming the reduced fish meal in 2006 was fed to pigs and chickens with trophic efficiencies of 20%, only 6 million additional ton of edible pork and poultry were produced (Table 15.1). Thus, the actual amount of human food derived from the 2006 fish catch was not the total catch of 117 million ton. Instead, it was closer to 90 million ton; 84 million ton of edible fish and 6 million ton of livestock raised on fish meal derived from 33 million ton of fish, and projections for future harvests are similar. The

TABLE 15.1 Human Food from Land, Freshwater, and Ocean Production Systems for 2006

Food Source	Production Category	Food Production × 10⁶ tons		
		Land	**Freshwater**	**Ocean**
Plants	Gathering	160	No Data	1
	Farming	2,168	No Data	15
Animals	Hunting	129	10	81
	Herding	969	32	20
Totals		**3,426**	**42**	**117**

Used for fish meal: −33
For direct consumption: 84
Additional food from fish meal: +6
Total: direct consumption + additional food produced from fish meal: 90

(a)

(b)

Figure 15.8 (a) Oysters grown in aquaculture pens. (b) Rearing cages in Canadian nearshore salmon farm.

(a) © Doug Plummer/Photo Researchers, Inc.; (b) © Kevin Eaves/Shutterstock, Inc.

UNFAO reported that world fishery production in 2013 was 162 million ton.

Long food chains, unsophisticated production systems, and extensive industrial uses of marine organisms all severely limit the amount of marine food actually produced. The marine environment just does not produce very much food; for the past several decades, only 1% to 2% of the food consumed by the world human population came from the sea. The likelihood of greatly increasing that fraction of our marine diet depends on the magnitude of still unharvested fish stocks and how well we manage the future harvesting of these stocks.

How much fish can we continue to remove from the world ocean? The question is not easily answered, but there are several indications that we are already overharvesting most commercially important fish populations. The ultimate limit on marine sources of food is established by the rate of photosynthesis by primary producers. The marine production system begins at the first trophic level with a net production of 420 billion tonne of biomass annually (from 42 billion tonne of carbon). Neither markets nor techniques are available for economically harvesting phytoplankton, and thus the magnitude of the fish harvest is actually limited by the number of trophic levels in the food chain leading to harvestable fish and the efficiency with which animals at one trophic level use food derived from the previous trophic level.

The potential commercial value of a fish or other animal is, to a large degree, a function of its individual body size. These animals must be a certain minimum size before commercial exploitation is economically practical. In addition, the size of organisms at the base of marine food chains is an essential feature in establishing the number of steps in the food chain. Food chains based on smaller phytoplankton cells (such as those in subtropical gyres) generally consist of a greater number of trophic levels, as do food chains leading to large fish such as tuna.

In general, phytoplankton cells decrease in size from greater than 100 μm in coastal and upwelled waters to less than 25 μm in the open ocean. Several moderately sized zooplankton species such as *Euphausia pacifica*, which function as herbivores in coastal North Pacific waters, must move one step up the food chain and assume a carnivorous mode of feeding in offshore waters because the phytoplankton there are too small to be captured. Other forms of oceanic zooplankton occupying the third trophic level are no more than 1 or 2 mm in length. Virtually all species of herbivorous copepods in the open ocean are preyed on by chaetognaths that, in turn, become food for small fish.

Thus, in open-ocean environments, three or four trophic levels are required to produce animals only a few centimeters in length. Food chains leading to tuna, squid, and other commercially important open-ocean species consist of an average of five trophic levels, with an average number of three trophic levels for commercial species in coastal waters and one and one half for upwelling areas. The number of trophic levels for upwelling areas is low because many clupeoids taken from upwelling areas graze directly on phytoplankton without any intermediate trophic levels.

DID YOU KNOW?

A group of the world's leading ocean scholars and advocates have warned that humans should adjust their fishing efforts or be ready for the demise of fish populations as we know them. It has been estimated that populations of many of our favorite food fish have declined by up to 90% and that if we continue fishing at these rates many of the popular food fish species will be gone by the year 2050! The year 2050 is not very far away and is during many of our lifetimes. Its approach should cause many people to ask the question, "How can we prevent this?"

TABLE 15.2 Estimates of Harvestable Fish Production for Marine Production Provinces

Province	NAPP ($\times 10^9$ tons)[*]	Number of Trophic Levels	Trophic Efficiency (%)	Potential Fish Production ($\times 10^6$ tons)
Open ocean	324	5	10	3
Coastal	96	3	15	334
Upwelling	4	1.5	20	480
Total				**817**

[*]Net annual primary productivity (NAPP) is based on published values and multiplied by 10 to convert carbon weight to wet (live) weight.
Data from Longhurst et al., 1995; Pauly and Christiansen, 1995; Field et al., 1998; and Gregg et al., 2003.

Accurate estimates of ecological efficiencies in marine food chains also are difficult to achieve. Efficiency factors are based on the growth of organisms that is, in turn, a function of food assimilation minus waste and metabolic costs (such as respiration and locomotion). These factors vary widely between species, between individuals of the same species, and between various stages in the life cycles of a species. Young, growing individuals often exhibit efficiencies as high as 30%, but their ecological efficiencies decline to nearly 0% at maturity. Thus, efficiency estimates for populations composed of a variety of age groups can be little more than reasonably intelligent guesses. It is even more difficult to approximate the ecological efficiencies across entire trophic levels consisting of a diverse group of animal types, each with its own peculiar age structure and growth rate. With that caveat, current estimates of average ecological efficiencies of 10% for the oceanic province, 15% for coastal regions, and 20% for areas of upwelling are used here. With these estimates of net phytoplankton production (**Table 15.2**), average numbers of trophic levels, and efficiencies of exchange between the trophic levels for each marine production province, we can estimate the total potential production of fish from the sea (Table 15.2, last column) to be about 817 million ton annually.

A potential annual production of about 817 million ton of animals is not the same as an actual harvest of that magnitude. Large numbers of some commercially important species are consumed by seabirds, larger fish, and other predators. Furthermore, to maintain healthy populations available for continued exploitation, harvesting activities must allow a reasonably large fraction (generally one half to two thirds) of these populations to escape and reproduce so that harvesting can continue at that level.

Thus, using existing fishing methods to harvest presently exploited types of seafoods, the resource potential exists to double or, at best, triple our present global fish catch; however, a twofold or threefold increase of the 1% to 2% that seafood contributes to our present diet is still a very small portion of our total food needs, and continued increases in the human population will surely offset much of those gains in future fisheries production. In addition, unpredictable declines in fish populations are likely as fishing pressure increases, which will inhibit the growth potential of this resource. Other possibilities for increasing our future seafood harvests beyond the limits suggested by the numbers in Table 15.2 include expanding aquaculture and moving our harvesting efforts closer to the bases of marine food webs. In many parts of the world these measures have already been implemented.

15.4 Fishing Down the Food Web

In theory, if we fished one trophic level down the food web and harvested what currently harvested fish eat rather than harvesting the fish themselves, a 5- to 10-fold increase could be harvested because one trophic level and its associated energy loss would be eliminated. Rather than contemplating limits of 200 or 250 million ton of seafood each year, we might anticipate harvesting 1 or 2 billion ton instead. Serious problems, however, block this path to greatly increased marine harvests. Almost without exception, animals occupying lower trophic levels are smaller and more dispersed than the animals now harvested.

Although the technological developments necessary to harvest the zooplankton and smaller fish that comprise these lower trophic levels are not insurmountable, they may not be worthwhile. These smaller, more dispersed animals are more difficult to harvest; the extra energy needed to collect these small food items may exceed the energy gained from the additional harvest. In addition, many people favor the flavor and texture of the larger fish, and consumer demand drives fishery production. It is likely to continue to be more efficient and marketable for us to wait until larger animals have eaten the smaller ones before harvesting them.

To provide one example, a century ago the phytoplankton crop around the Antarctic continent supported a tremendous assemblage of zooplankton, particularly the krill *Euphausia superba*. In turn, krill fed large populations of blue, fin, and humpback whales. With these whale populations now seriously reduced, they no longer serve as intermediaries to harvest the krill for us. Fisheries scientists from several countries are test-harvesting krill with an eye to expanded future production. Preliminary estimates of an annual sustained harvest of 100 million and even 200 million ton have been suggested for this one species alone. The potential of such massive harvesting is a complex issue with several ramifications unrelated to food production.

A more serious consequence of fishing down the food web is the reverse of the one just described. When intensive commercial harvesting is focused on a trophic level near the base of a food web, larger predators occupying higher trophic levels are denied adequate prey, and their populations will suffer even though they are not directly targeted by harvesting activities.

Another problem inhibiting widespread use of krill is one common to any plan to use smaller marine animals: what is done with it after it is caught? There is little demand for fresh or frozen krill. It is nutritious, but few people will buy it in its natural form. The energy costs involved in catching krill, processing it to a palatable form, and transporting it to markets in the Northern Hemisphere are enormous. Thus, the likelihood of using krill and other nontraditional food species positioned on low trophic levels in marine food webs is presently marginal at best. Our most reasonable approach to satisfy our expanding food needs must continue to rely on improving techniques for terrestrial crop production using currently available techniques to grow well-known crops already in demand by society.

15.5   Aquaculture

Aquaculture is the application of agricultural techniques to grow, manage, and harvest aquatic animals and plants. It may vary from simple enhancement of natural populations by releasing hatchery-reared juvenile fish to intensive captive maintenance of species for their entire life span. Farming of aquatic plants and animals presently contributes nearly 50% to the total of aquatic food production, which is a great increase from just a decade ago. As a whole, aquaculture, which includes inland and marine culturing, is the fastest growing animal food-producing sector. Inland aquaculture accounts for the large majority of finfish farming, whereas **mariculture** (marine fish farming) accounts for the majority of mollusk and crustacean farming production. As of 2015, 575 species grown in freshwater, seawater, and brackish water have been catalogued as cultured species. Extensive research efforts have been made to increase production of farmed species to match humans' growing demands for seafood. Although seemingly a positive solution to the fish-production problem, aquaculture techniques are criticized for the use of antibiotics on fish and having negative environmental impacts in the form of pollution.

Salmon farming in Japan, Norway, Canada (see Figure 15.8b), Chile, and the United States has expanded rapidly in the past 30 years to about 2 million ton annually, exceeding the annual harvest of wild-caught salmon. Fish farming has been practiced for centuries in Southeast Asia, Japan, and China. Mullets and milkfish are grown in shallow estuarine ponds where they graze on algae, detritus, and small animals. Estuaries, salt marshes, and other productive coastal habitats are preferred for cultivating some fish species in closed pond systems; however, intensive aquaculture activities modify the nature of an estuary or salt marsh. In special circumstances, dissolved nutrients from sewer treatment plants or hot water from coastal power-generating stations can be used beneficially. Diverted into fish ponds, nutrients and the warm water could enhance the growth of primary producers to feed the fish, but these are rare circumstances. More commonly, aquaculture programs are, and will continue to be, restricted by adverse problems of coastal pollution and habitat destruction and by competing uses for the same land. For each fish pond installed, a portion of the native fish and shrimp populations already contributing to previously established local fisheries is displaced or denied access to these productive coastal waters.

The major expansion in global aquaculture has occurred with bivalve mollusks (clams, oysters, mussels, and scallops) and, to a much smaller extent, with shrimps, crabs, and other crustaceans (**Figure 15.9**). These animals also thrive in estuaries and other shallow coastal habitats. As consumers of detritus and bacteria close to the bases of their respective food webs, they are ideal species for aquaculture. In the United States, the expansion of finfish mariculture includes species such as black sea bass, sablefish, yellowtail, and pompano.

Aquaculture is not a glorious solution for all our marine food-production problems. When salmon are raised in crowded cage conditions, large quantities of antibiotics are used to suppress infections; the overall health of the fish is compromised, leading to questions about nutritional value; and animal wastes concentrated into small areas can create serious localized water-quality problems. As the aquaculture industry grows, environmental problems have become more apparent and are of great concern. A recent trend is the move toward sustainable aquaculture, producing fish without the use of hormones or antibiotics, similar to the recent movement toward organic terrestrial farming.

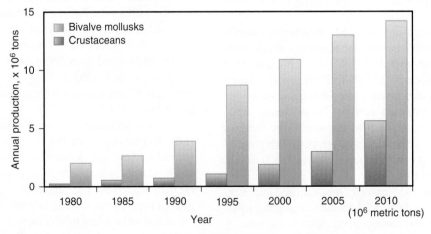

Figure 15.9 Growth of global maricultural production of marine crustaceans and bivalve mollusks.
Data compiled from UNFAO statistics.

15.6   The Problems of Overexploitation

Commercial and subsistence fishing represents a form of predation that has predictable effects on the prey species. When harvesting of a new population begins, initial catches are generally large and include a high proportion of large fish. Continued or increased fishing pressure tends to reduce the size of the population as well as the sizes of individual members of that population. If the fishing effort is adjusted to the growth and reproductive potential of the population, then a **maximum sustainable yield** of fish can be caught year after year without causing major upsets in the populations. Too often, however, the fishery is not regulated sufficiently, and fishing pressure becomes much greater than the population can withstand. When losses to fishing and natural predators together exceed replacement by young animals, populations decline, and these declines are quickly reflected in reduced catches.

Numerous examples of overfished stocks can be found in most segments of the fishing industry. Most of the popular species of halibut, plaice, cod, ocean perch, herring, and salmon of the North Atlantic and Pacific Oceans are already overexploited—so are many of the warm-water tuna stocks and most of the large whales. The two examples discussed here are presented to demonstrate some of the problems created by overfishing activities that seem in direct conflict with the fishing industry's own best interests.

North Atlantic Cod

Cod of the North Atlantic Ocean are the inheritors of the fishing pressure that used to be concentrated on haddock early in the 20th century. When the overfished haddock populations collapsed, fishing efforts shifted to the more abundant cod. The factory trawlers introduced in the 1950s revolutionized the practice of catching these fish, hauling them in by the ton rather than by handlining for individual fish. By 1970, over 1 million ton were being harvested annually (**Figure 15.10**). In retrospect, the eventual result seems entirely predictable. As the number of mature, reproducing individuals 7 years or older was reduced, the population diminished but the pressure to capture them did not. Existing fishing effort was increasingly concentrated on decreasing numbers of fish, and catches declined (Figure 15.10).

It is a sad irony that the pattern of decline, indicated by the right side of Figure 15.10, provides some of the best information about the actual sizes of cod populations. In other words, after you have eaten all the fish, you then know how many there used to be. After several years of blaming seal predation and deleterious oceanographic conditions, U.S. and Canadian regulatory agencies belatedly accepted the fact that the population had collapsed and closed nearly all cod fisheries in their waters in 2003. Several European nations followed suit for cod fishing areas on the east side of the North Atlantic. How quickly the cod populations might recover is unknown, as is the type and magnitude of response to the recovery by fishers and the governmental agencies responsible for that fishery. As of 2016, it appeared that North Atlantic cod populations were on the serious rebound, but whether populations will continue to rise and become stable remains to be seen.

The Peruvian Anchoveta

The Peruvian anchoveta (*E. ringens*) has become the classic textbook example of the consequences of overfishing. Unlike the Atlantic cod, which has been exploited for centuries, the commercial harvesting of Peruvian anchoveta has been sufficiently recent that reliable data exist on the initial growth and eventual

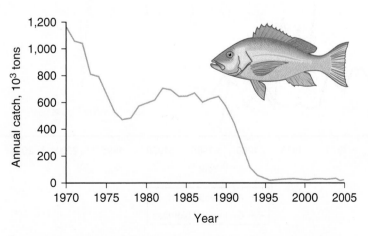

Figure 15.10 The dramatic decline in Northwestern Atlantic cod harvests, 1970 to 2005.
Data compiled from UNFAO statistics.

collapse of this fishery. This species is a typical **clupeoid** fish, almost identical to the anchovy shown in Figure 15.2. It is a small, fast-growing filter feeder that schools in the upwelling areas of the Peru Current. The first commercial use of the anchoveta was indirect. From the time of the Incas, guano deposits from the nesting colonies of seabirds that feed on anchoveta have been collected and used as a major source of fertilizer. These seabirds, primarily Peruvian boobies, brown pelicans, and guanay cormorants, annually convert about 4 million ton of anchoveta to an inexpensive fertilizer widely used by Peru's subsistence farmers.

Commercial exploitation of the Peruvian anchoveta for reduction to fish meal began in 1950. The next year, 7,000 ton were landed. After 1955, the growth of the fishery was explosive (**Figure 15.11**): more than 3 million ton were landed in 1960. By 1970, the catch of this one species had surpassed 13 million ton, almost one fifth of the entire world seafood harvest for that year. Nearly all the catch was taken by local fishers and reduced to fish meal and oil for export.

Accompanying the gigantic rise in commercial anchoveta catches was a drastic drop in the number of guano birds that depended on the anchoveta for food. From 28 million in 1956, the guano bird population was reduced to 6 million during the El Niño year of 1957. Phytoplankton populations were dramatically reduced. Anchoveta mortality increased, and the guano birds starved. After 4 years without an El Niño, the bird population had rebounded to 17 million, only to be hit by another El Niño in 1965. That time the bird population plummeted to 4 million, and since then recovery efforts have included marine reserves and protection of birds. Guano is harvested for fertilizer and has great economic value.

A substantive base of biological information was collected during the early development and growth of the Peruvian anchoveta fishery. This information base was supposed to help enact proper management procedures that would ensure a large and perpetual harvest from this immense population. The biological information suggested that a maximum sustainable yield of approximately 9.5 million ton annually was possible. At the time, the reduced bird populations were taking less than 1 million ton each year. Tonnage reports alone, however, are inadequate and sometimes misleading. As fishing pressure increased over the 1960s, the average size of fish being caught decreased, and the number needed to make a ton increased sharply. By 1970, small anchoveta were taken with such efficiency that 95% of the juvenile fish recruited into the population were captured before their first spawning.

A glance at Figure 15.11 shows that the catches of anchoveta for 1967 through 1971 exceeded the predicted maximum

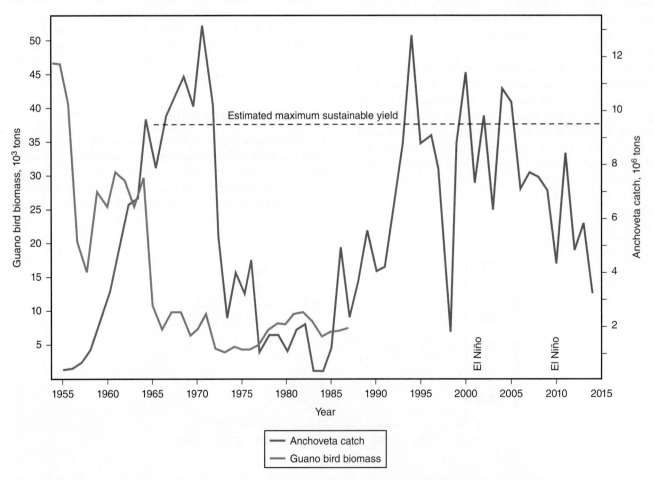

Figure 15.11 Changes in the anchoveta catch and the guano bird populations along the northwest coast of South America. El Niño years are indicated.

Data from Muck, P. and D. Pauly. 1987. Monthly anchoveta consumption of guano birds, 1953 to 1982. P. 219–233. In D. Pauly and I. Tsukayama, eds., *The Peruvian anchoveta and its upwelling ecosystem: three decades of change;* and UNFAO statistics.

sustainable yield by at least 1 million ton each year. The industry and the regulatory agencies responsible for managing the anchoveta stocks had ample warning of what was to come. In 1972, sampling surveys indicated that the anchoveta stocks had been severely depleted and recruitment of juvenile fish was poor. As expected, the 1972 catch dropped to little more than 3 million ton, less than half that of the previous year. Even worse was 1973, with an estimated catch of less than 3 million ton. The excessive fishing pressures of the previous decade were too much even for this tremendous fish population, and it finally collapsed. Since 1980, efforts to harvest anchoveta and the population's response to that fishing effort have been extremely variable. Catches have moved between a 1994 harvest that rivaled the record 1970 catch to less than 2 million ton 4 years later. In the 1990s, the huge bird populations that once relied on this species showed no signs of recovery to pre-1970 levels.

> **DID YOU KNOW?**
>
> Although some commercial fishers have a bad reputation for being inflexible and taking relentlessly from the sea, this is not the case for many. The fishing industry has a historic cultural heritage associated with it, and some fishers come from a long line of fishers making their hard-earned living from the sea. The majority of fishers are opposed to regulations because they do not want to lose their livelihoods, which is understandable. However, many are open to restrictions that allow fish populations to recover, such as seasonal closures during spawning seasons. Fishery closures and restrictions, although necessary to manage fish populations, greatly affect the lives of fishers and their families. Finding a balance between maintaining fish populations for future generations and allowing fishers to make a living is a complicated task.

15.7 The Tragedy of the Commons

Why have fishing enterprises and fishing nations repeatedly exploited the fish resources on which they depend to the point that returns on their fishing efforts decline and too often disappear? Fishing is too often considered a right without attendant responsibilities. Even with the maze of legal and economic considerations involved, incentives for these apparently self-defeating actions are not difficult to find. Cod, tuna, anchovetas, and other oceanic species are unowned resources belonging to no single nation or individual. Many exist outside the jurisdictional limits of all nations and are therefore open to access by any nation. Historically, the concept of open access to the high seas evolved in the 16th and 17th centuries when the right to navigate freely was more crucial than the freedom to fish, but as coastal fish stocks were depleted, fishers became increasingly dependent on more distant stocks in international waters. They eventually discovered that the freedom to fish on the high seas was fundamentally different from the freedom to navigate. Unlike navigation, fishing activities remove a valuable commodity from a common resource pool at the expense of everyone, including those who do not fish.

It is commonly assumed that oceanic resources are unowned and open to access for all people, but with today's advancing pace of technology, some nations have achieved the ability to exploit these resources much more quickly and effectively than others. If the fishing fleets of one nation fail to catch these unowned oceanic species, those of some other nation soon will. In species after species, such attitudes have led to inevitable and predictable results: increased competition for limited resources; duplication of effort; declining fishing efficiency; and, of course, overfishing.

Once in the net, a school of fish is no longer the property of all people; it belongs instead to those who set the net and haul the fish aboard. All people and all nations share the cost of losing the fish, the great whales, and the other marine animals that have nearly disappeared because of overfishing, yet the short-term profits derived from overfishing are not similarly shared. This is Hardin's concept of the "tragedy of the commons." The tragedy of this situation is that it best rewards those who most heavily exploit and abuse the unprotected living resources of the open ocean.

15.8 International Regulation of Fisheries

In the United States, a dedicated government agency, the National Marine Fisheries Service, is tasked with researching, assessing, and regulating marine fisheries and making additional recommendations as outlined by the Endangered Species Act. Fished species are monitored using periodic stock assessments that aim to provide population estimates that fisheries managers can use to make recommendations. Some other countries have similar agencies for managing fisheries resources within their jurisdictions, but many countries have no such monitoring, and international waters are not controlled by any one nation. Without controls to limit the access of fishers to fish populations in international waters, concerned nations have created a variety of multinational and international regulatory commissions for the purpose of governing the management and harvest of regional fish populations. Common strategies for management include setting quotas on the amount of fish harvested; establishing seasons and minimum size limits; defining acceptable gear, such as net mesh sizes; and limiting the number of boats that are allowed to participate in a fishery. The Northwest Atlantic Fisheries Organization, for instance, currently includes Canada, Denmark, Cuba, France, Iceland, Norway, Japan, Russia, Ukraine, the United States, and the European Union. The regulations established by this commission do not carry the weight of international law, but they are binding on member nations. Even so, they have failed to halt serious overfishing of cod and other valuable fish populations in the North Atlantic.

Other commissions, particularly those regulating the halibut and salmon fisheries in the North Pacific, have been more successful. Their very success in maintaining viable populations, however, is now creating new problems. Too many additional fishing boats from nations all over the world are attracted to these well-managed fisheries, magnifying management

RESEARCH in Progress

Using Technology to Reduce Bycatch

Bycatch is a very real determining factor for declining populations of the many species that are killed simply because they were in the wrong place at the wrong time. Bycatch not only includes the take of nontarget species or undersized target species, but often the bycatch specimens are thrown back overboard and completely wasted after being needlessly killed. Many types of fisheries include some degree of bycatch, with the most extreme being those that include the use of large nets, such as trawl or gill nets, because the nets typically capture anything located in their paths. As technology has allowed fishers to further exploit fisheries with more advanced boats, nets, and other equipment, even greater numbers of bycatch are taken each year.

In recent years, bycatch has become an issue of great concern, and many feel that although technology has allowed for increased mortality of bycatch, other technological advances are the answer to reducing bycatch and removing fishing pressure from nontarget species. The scientific community has invested time and money into the development of equipment that will reduce mortality from bycatch, and funds are available annually from a program funded by the National Oceanic and Atmospheric Administration, the Bycatch Reduction Engineering Program (BREP). BREP funds nongovernmental research on bycatch reduction projects using novel techniques and equipment (**Figure A**). Since the inception of the program in 2008, dozens of projects have been funded all around the United States, including the Hawaiian Islands and Guam, leading to technological advances that can be used in the fishing industry. The following are just a few examples of projects that have shown remarkable results: (1) development of a new type of circular fishing hook that reduced bluefin tuna bycatch by 56%; (2) the use of LED lights on trawl nets to reduce bycatch of Columbia River smelt in the pink shrimp fishery by 90%; and (3) the invention of turtle-excluder devices for trawl nets that allow sea turtles to swim out of the net instead of being brought up onto the boat (**Figure B**).

Once a nontarget species has been caught its fate is determined by how it is handled on the boat. With many commercial fishing practices it is time-consuming and cost-prohibitive to save bycatch and release it back to the sea unharmed, so unfortunately many fishers do not even try. Recreational fishers are much more likely to be capable of returning fish back to the sea with little harm, and as a result research is focusing on better handling practices for bycatch in recreational fisheries. The Atlantic cod has experienced an increase in recreational fishing pressure over the past decade, and discarded fish often do not survive. A study conducted in the Gulf of Maine examined postrelease mortality of cod for 30 days after release using transmitters and acoustic receivers. They were able to track the fish successfully and differentiate between living and nonliving specimens. These data can be used to help create more realistic catch limits and size restrictions for the recreational fishery as actual postrelease mortality rates become known. Data can also be used to help educate fishers about the most effective catch-and-release techniques for decreasing fish mortality.

Entanglement in nets and ropes is a cause of bycatch for many larger marine organisms, including whales, sharks, and sea turtles. North Pacific right whales are one of the most endangered species on Earth, and entanglement in fishing gear presents a very real threat to their existence. Researchers asked the question of whether different colored fishing gear and ropes would make the gear more visible to right whales, reducing the risk of entanglement. Laboratory studies were conducted to assess the visual capabilities of right whales, and the results of these studies were applied to field tests in the waters of Cape Cod Bay to ground-truth the laboratory results. In the field, a variety of rope colors were tested, and the use of red and orange ropes led to a significantly greater distance to detection than the black or green ropes currently used by fishers. These results are promising, because it appears that a simple change in color of the fishing gear will reduce entanglement events for this endangered species.

Many people do not think of sharks as vulnerable organisms, but sharks are just as likely to end up as bycatch as are many other marine animals. Sharks are voracious predators, so hook-and-line and longline fisheries often catch sharks as nontarget species. A group of scientists in Florida tested shark-repellent bait with longline equipment to attempt to deter sharks from being caught incidentally. A formula was devised that reduced shark bycatch by over

Figure A A close-up view of a composite panel fishing device, designed to reduce bycatch of a variety of fish and invertebrates in shrimp trawl nets.
Courtesy of National Oceanic and Atmospheric Administration.

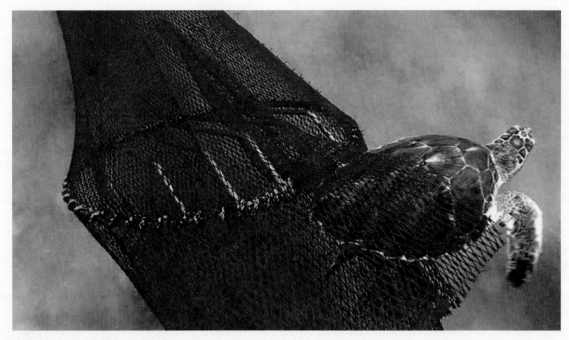

Figure B A turtle exclusion device (TED) in action with a turtle escaping out the back of the net.
Courtesy of National Oceanic and Atmospheric Administration.

70%, and further studies will confirm the best concentrations of the repellent. The key to this repellent is that it is long-lasting, which would allow fishers to apply it infrequently with little extra effort. This study provides fisheries managers with a product that can be recommended or even required by longline fishers to reduce shark mortality. In addition to saving sharks' lives, fishers will be able to catch more of their target species and reduce the risk of injuries that are possible when handling sharks on the boat.

Although fisheries bycatch is currently a source of unnecessary deaths of many marine organisms, the hope is that with new technologies and regulations mortality rates will decrease. With concerted efforts by regulatory agencies to provide funding for research on bycatch, reduction is possible. Targeted research on bycatch species, including physiological requirements and life history characteristics, is necessary to provide the fishing industry and fisheries managers with the information they need to make informed decisions concerning regulations.

Critical Thinking Questions

1. If an effective method to reduce bycatch is invented for a fishery, do you feel that fishers should be required to use it? Why or why not?

2. Think of a way fishers can be encouraged to harvest legal bycatch instead of throwing it overboard to die.

For Further Reading

Kraus, S., J. Fasick, T. Werner, and P. McFarron. 2014. Enhancing the visibility of fishing ropes to reduce right whale entanglements. *Bycatch Reduction Engineering Program 2014 Annual Report* 1:67–75.

Mandelman, J., C. Capizzano, W. Hoffman, M. Dean, D. Zemeckis, M. Stettner, and J. Sulikowski. 2014. Elucidating post-release mortality and "best capture and handling" methods in sublegal Atlantic cod discarded in Gulf of Maine recreational hook-and-line fisheries. *Bycatch Reduction Engineering Program 2014 Annual Report* 1:43–51.

Rice, P., B. DeSanti, and E. Stroud. 2014. Performance of a long lasting shark repellent bait for elasmobranch bycatch reduction during commercial pelagic longline fishing. *Bycatch Reduction Engineering Program 2014 Annual Report* 1:17–25.

problems, diminishing profits of individual fishers, and increasing the possibilities for eventual overexploitation.

To counter some of the difficulties of managing fisheries by multinational commissions, several individual nations have worked to develop individual systems to regulate fish populations that are nearby yet outside their own territorial or jurisdictional limits. In 1974, the U.S. government, under pressure from its own coastal fishing interests, ended its traditional policy of opposition to extended zones of control for coastal nations. In 1976, the United States passed the Fisheries Conservation and Management Act. Under this act, the United States assumed exclusive jurisdiction over fisheries management in a zone extending from shore 200 miles to sea. Entrance of foreign fishing vessels into this exclusive economic zone is allowed on a permit basis only, and then the fishers are allowed to harvest only fish stocks with sustained yields greater than the harvesting capacity of American fishers.

In 1982, the United Nations Conference on the Law of the Sea adopted a draft Law of the Sea (LOS) treaty that was to go into effect 12 months after ratification. Three decades later, the LOS treaty has been joined by 166 nations and the European Union. The LOS treaty includes many of the key features found in the U.S. Fisheries Conservation and Management Act, in particular a 200-mile-wide exclusive economic zone granting coastal nations sovereign rights with respect to natural resources (including fishing), scientific research, and environmental preservation. Additionally, the LOS treaty obliges nations to prevent or control marine pollution, to promote the development and transfer of marine technology to developing nations, and to settle peacefully disputes arising from competitive exploitation of marine resources.

DID YOU KNOW?

Fish don't care about the geographic boundaries set by humans, and this fact can make managing fisheries a particularly challenging task. Highly migratory species transit hundreds or thousands of kilometers annually, and along their routes they will spend time in multiple countries' jurisdictions, including international waters. Fish can be captured anywhere along their routes, and fishing restrictions set by one country may not be recognized by others. Regardless, some countries (like the United States) must set an example by creating and abiding by fishing quotas if there is any hope for these very valuable species to persist.

The imposition of 200-mile-wide exclusive economic zones by essentially all coastal nations of the world has dramatically changed the concept of open access for most of the world's continental shelves, coastal upwelling areas, and major fisheries lying within 200 miles of some nations' shorelines (**Figure 15.12**). This treaty is notably silent regarding the Antarctic upwelling area, because individual national territorial claims on the Antarctic continent are not recognized. Commercial ventures there, including either whale or krill harvesting, must be consistent with the goals of the 1982 Convention for the Conservation of Antarctic Marine Living Resources (CCAMLR), namely, the "maintenance of the ecological relationships between harvested, dependent, and related populations of Antarctic marine living resources and the restoration of depleted populations." This was the first in a slowly growing trend of national and international efforts to recognize the integrity of large marine ecosystems. Achievements of the CCAMLR over the past 30 years include establishing a Marine Protected Area in the Southern

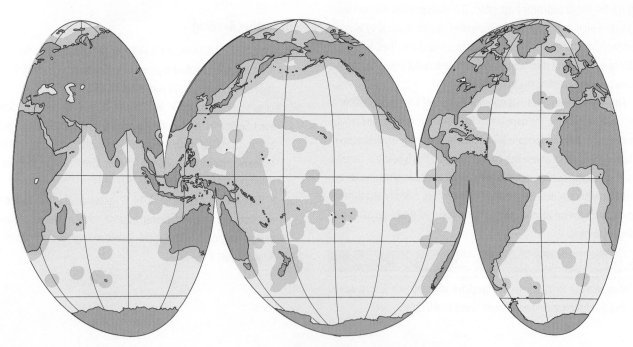

Figure 15.12 Worldwide extent of the 200-mile exclusive economic zones sanctioned by the United Nations LOS treaty.

Ocean to help manage vulnerable marine ecosystems; reducing seabird mortality; establishing a monitoring program; and attempting to address the problems of illegal, unreported, and unregulated fishing.

The waters around the Antarctic continent represent the last relatively unspoiled large marine ecosystem on Earth. The international cooperation demonstrated so far in protecting its living resources has been unusual in the long history of our attempts to protect marine resources. The waters around this remote continent can continue to serve both as a laboratory for improving understanding of living marine systems and as a model for creating approaches to preserving those resources as a common heritage of humankind.

15.9   Marine Ornamentals

It has been estimated that as many as 2 million people worldwide keep personal marine aquaria (600,000 households in the United States alone), supporting a trade in fish, corals, and other invertebrates that may be worth as much as $330 million annually. Most marine ornamental animals are collected and transported from Southeast Asia, although additional Indo-Pacific Islands are becoming more important as a source of desirable species. These collected species are usually shipped to the primary markets of the United States, Europe, and Japan (80% of stony corals and 50% of fish are exported to the United States).

This impressive trade could provide many jobs in coastal areas and significant economic incentive for coral reef conservation, because much of the Indo-Pacific region consists of low-income, developing island nations. In 2000, for example, 1 kg of Maldivian reef fishes destined for the aquarium trade was valued at nearly $500 (if used for food these fish would be worth just $6). In addition, 1 tonne of live coral has a value of $7,000 in the aquarium trade, but only $60 if used for limestone production, and 1 kg of live rock from Palau can earn as much as $4.40 if sent to aquarium hobbyists, but only 2 cents if used locally as construction material. Unfortunately, destructive collection techniques such as the use of dangerous chemicals or explosives, overexploitation of some species, and careless handling and transport all undermine potential benefits of the trade.

In 2000, the Global Marine Aquarium Database (GMAD) was created to obtain standardized species-specific trade data from importers and exporters as a first step toward long-term conservation and sustainable use of coral reefs and their inhabitants. Initially, the database was successful and gathered nearly 103,000 trade records covering about 2,400 species of fishes, corals, and other invertebrates involved in the hobby. The GMAD showed that 20 to 24 million individual fish in 1,471 species are traded worldwide each year. Nearly half of these are damselfish, with various angelfish, surgeonfish, wrasses, gobies, and butterflyfish comprising an additional 25% to 30% (and just 10 extremely popular fish species accounted for 36% of all fish traded between 1997 and 2002). GMAD also reports a total of 140 species of stony corals (11 to 12 million pieces),

61 species of soft corals (390,000 pieces), and 9 to 10 million individuals of 516 species of other invertebrates (mostly mollusks, cleaner shrimp, and sea anemones) were traded worldwide during the same time period. Entering data into the GMAD is completely voluntary, and unfortunately many marine ornamental collectors decide not to enter data, even though they have been encouraged to do so by scientists working in the ornamental fish field. Alternate forms of tracking marine ornamentals have been proposed, but there is not currently one global system tracking this trade.

Sadly, the popularity of marine ornamentals is often driven by their appearance rather than their ability to survive collection, handling, transport, or captivity. For example, although cleaner wrasses (*Labroides*), butterflyfish (*Chaetodon*), and the harlequin filefish (*Oxymonacanthus longirostris*) are characterized as "truly unsuitable" for home aquaria by the GMAD (mostly because of their highly restricted diets), they are still commonly traded. In addition, the attractive appearance of carnation or strawberry coral (*Dendronephthya*) makes it one of the most commonly traded soft corals, yet it typically dies within a few weeks of captivity because it lacks helpful zooxanthellae.

Opponents of the aquarium trade criticize the destructive methods used to collect marine ornamentals, the potential overharvesting of some species, the wasteful mortality caused by slipshod husbandry during shipping, and the introduction of exotic species that may disrupt local ecological relationships. Damaging harvesting techniques are diverse and common. Sodium cyanide and other harmful chemicals, such as rotenone and quinaldine, are used to stun fish so that they are easier to catch, yet overdoses are common, postcapture mortality of drugged fish is very high, and incidental poisoning and killing of adjacent reef creatures and corals is common. Collection for the aquarium trade damages the reef itself in other ways, including direct collection of corals for shipment, fragmentation of the reef to provide easier access to cryptic reef dwellers, and accidental damage to neighbors of targeted individuals.

Studies in Sri Lanka, Kenya, the Philippines, Indonesia, and Hawaii have revealed localized depletion of several target species. Supporters of the aquarium industry criticize these studies as preliminary and point out that a study in the Cook Islands suggests that aquarium fish there are being harvested sustainably (the catch per unit of effort was stable from 1990 to 1994). Nevertheless, the only comprehensive study to date assessed the impact of harvesting Hawaiian reef fishes for the aquarium trade. This study, conducted by Hawaii's Department of Land and Natural Resources, revealed that the abundance of 8 of the 10 most popular species was lower at harvest sites when compared to control sites where collecting was not permitted. It is reasonable to assume that corals targeted for the trade are even more vulnerable to overharvesting because of their extremely slow growth rates, but a great deal more data are needed before more definitive conclusions can be reached.

The death of reef fishes after their collection is caused by many factors, some of which lead to extreme rates of mortality. One report suggests that 75% of fish collected with narcotics, such as cyanide, die within hours of being taken from the reef.

Figure 15.13 A clownfish living symbiotically with a sea anemone. Clownfish are extremely popular in the aquarium trade and have been successfully cultured for many years.
© magnusdeepbelow/Shutterstock, Inc.

More typical mortality rates are caused by physical damage during capture, overcrowding and disease during shipping, and inattentive husbandry. A study of fish captured in Sri Lanka reported that 15% died during and immediately after capture, 10% died during shipping, and another 5% died before being sold to consumers in the United Kingdom, and the mortality did not end there. GMAD data show that corals in home aquaria experience 76% to 100% mortality within 18 months. Shockingly, the most "hardy" coral species, *Plerogyra* and *Catalaphyllia*, experience 54% and 60% mortality within 18 months of purchase, yet they remain popular in the industry.

DID YOU KNOW?

Clownfish (**Figure 15.13**) were some of the first marine fish to be cultured for the aquarium trade over 40 years ago. They are now bred widely, and breeders select for vibrant colors and various color patterns. Aquarium-raised fish often survive well in captivity and do not face the challenges of wild-caught fish. Successful breeding by aquarists lessens the need to capture fish from the wild and often leads to lower mortality rates and healthier fish, so consumers are happier, too.

Clearly, fisheries for marine ornamentals need to be managed to ensure sustainable use, to decrease conflict with other user groups (such as sport divers who prefer that no animal be removed from natural reefs), and to keep postcapture mortalities low. Although in their infancy, such efforts already include attempts to culture popular ornamentals for trade; as of 2014, 269 species were successfully bred by aquarists, although almost 90% of these are freshwater and only 10% marine. The implementation of typical management initiatives discussed elsewhere in this chapter, such as limiting entry into the fishery, establishing catch quotas and size limits, creating marine

reserves, and temporarily closing popular harvest sites, is beginning to take shape.

15.10   Sealing and Whaling

The large body sizes and high fat and oil content of pinnipeds and whales have long made them targets of commercial harvesting efforts. Sometimes the desired product was meat, but whale and seal oil and seal skins with dense insulating fur were also sought. In the United States, marine mammals have been protected by the Marine Mammal Protection Act since 1972. The act was amended in 1994 to include take guidelines for small subsistence harvesting activities for Alaskan Natives and for research and to include limits for take due to bycatch in commercial fishing efforts. Although marine mammals are highly protected in U.S. waters and in most parts of the world by other nations, a few nations are famous for illegal take of marine mammals, Japan being the most apparent, but not the only nation, to partake in illegal takes. The results of a 2011 study assessing worldwide marine mammal consumption since the year 1990 are stunning: people in at least 114 countries have consumed 1 or more of 87 marine mammal species, including threatened or endangered species. Although in many regions of the world marine mammal consumption is down due to strict laws and protection, other regions have increased consumption as marine mammals are caught incidentally in stronger fishing gear or in some cases are hunted directly. Marine mammal consumption appears to be directly related to poverty levels. Countries with high poverty levels resort to consuming marine mammals, especially when numbers of other marine food organisms such as large fish species are declining. Here we present some basic information about the past and present capture of pinnipeds and baleen whales.

Pinnipeds

The gregarious nature and relatively poor terrestrial locomotion of pinnipeds make them easy targets for sealers seeking skins and oil. Over the past 200 years, several pinniped species have been severely decimated by commercial harvests. The northern fur seal population numbered about 2.5 million when discovered by Russian sealers in the 18th century. By 1911, the population was reduced to about 100,000 animals. With international protection, 50,000 to 100,000 fur seals are killed each year in a limited hunt that has allowed a partial recovery to about half its pre-exploitation population size.

Nearly 100,000 ringed seal pups in the North Atlantic were killed for their skins, as were 250,000 to 370,000 harp seal pups, for each of the 5 years from 1996 to 2000. This 5-year period had the highest number of harp seals taken since the 1960s and was justified with the dubious assumption that harp seal predation was the major factor responsible for the recent decline in the cod harvest in the North Atlantic. Currently, Canada's Atlantic Seal Hunt Management Plan for 2015 stipulates Total Allowable Catches (TACs) for three phocid species. Licensed commercial sealers may land 400,000 harp seals, 8,200 hooded seals, and 60,000 gray seals. No TACs have been set for ringed or harbour seals (or any other seal that may stray into the area). Allowable take numbers have been increasing over the past 5 years despite a drop in value of seal products. In opposition to these practices, several countries, including the United States, have stopped trading in some or all products of commercial seal hunts, driving down the market value of seal pelts and seal oils.

Northern elephant seals were even more seriously decimated by sealers than were the smaller fur and harp seals. Once distributed from central California to the southern tip of Baja California, this species came under commercial hunting pressure in the early 19th century. A scant half century later, so few survived that they were not worth hunting. No elephant seals were sighted between 1884 and 1892. In 1892, C. Townsend, a collector for natural history museums, discovered eight animals on Isla de Guadalupe, 240 km off the coast of Baja California. Seven were promptly shot for museum specimens. Early census estimates suggest that in the 1890s as few as 25 individuals survived on a single inaccessible beach on Isla de Guadalupe. Beginning with protection afforded the northern elephant seal by Mexico and the United States in the early 20th century, the remnant population slowly recovered. Since then, the northern elephant seal has again spread throughout its former breeding range. The total population has swelled to more than 127,000 animals, and the future of this species seems secure.

Or does it? Is the present northern elephant seal population really as viable and hardy as the pre-exploitation population? Comparisons of blood proteins from 159 animals of the "recovered" population suggest that they are not. In marked contrast to proteins of other vertebrate species, no structural differences were demonstrated either between individuals or between separate groups of northern elephant seals breeding on different islands. The lack of structural differences in these proteins points to a complete absence of variation in the genes controlling the synthesis of these proteins. Researchers suggest that the absence of genetic variability in the existing northern elephant seal population is the result of a genetic bottleneck when the population was at its low point in the 1890s. It is quite conceivable that on that isolated beach on Isla de Guadalupe, a lone elephant seal male dominated the breeding of all the surviving sexually mature females for several years. If so, half the pool of genetic information possessed by the surviving representatives of this species was funneled through this single animal, and much of the genetic variability presumably inherent in the predecimation population was lost. The rapid recovery of the protected population indicates that genetic variability may not be essential to the short-term survival of this species; however, this species remains vulnerable to environmental changes occurring in an extended time frame because it may lack the genetic variability necessary to cope with such changing conditions.

For some species of pinnipeds, hunting is not the cause of their population decline. Rather, as predators near the tops of their food webs, they frequently encounter intense competition from an even more efficient predator: humans. Steller sea lions provide a dramatic example of the consequence of this competition for food resources. Steller sea lions in the

North Pacific numbered about 250,000 in 1960 and were important predators of pollock, salmon, herring, capelin, and rockfish. By 1985, the population had declined to 185,000 and by 1990 to 90,000 (**Figure 15.14**). The immediate cause of the decline was determined to be reduced pup and juvenile survival due to a scarcity of food. All of their usual prey

(a)

(b)

Figure 15.14 Steller sea lion rookery beach photographed on the same day of the year in (a) 1969 and (b) 1987.

(a,b) Courtesy of National Marine Mammal Laboratory, NMFS/NOAA.

species have experienced dramatic increases in commercial fishing pressure since 1975. Although it is difficult to establish a direct link between increased fishing and sea lion mortality, most biologists studying this predator–prey interaction agree that the reduction and geographic redistribution of the sea lions' prey species had a large negative effect on the survival of juvenile sea lions. There is a cause for optimism, however. A population census by the National Marine Fisheries Service in 2013 led to the removal of the eastern population from the list of threatened species.

DID YOU KNOW?

Although a staple in the diets of many people for centuries, the meat of whales and dolphins is now known to bioaccumulate high levels of toxins, decreasing its nutritional value dramatically. In 2000, whale meat was removed from Japanese store shelves in response to a study that found high levels of contaminants in the meat. Young people in Japan are generally more open minded and less traditional, and as they learn about the potential health impacts of contaminated meat and environmental issues surrounding whaling practices, many are choosing not to eat marine mammal meat.

Baleen Whales

The history of the global whaling industry has been a long and tragic one. Aboriginal hunting of coastal whales for food has occurred for several thousands of years. In the 18th and 19th centuries, however, whaling took a new and ominous turn. Pelagic whales became major items of commerce as demand for their oil increased and whaling as an industry grew into a profitable commercial enterprise. Ships from a dozen nations combed the oceans for whales that could be killed with hand harpoons and lances. Right, bowhead, and gray whales were favorite targets because they swam slowly and, once killed, floated conveniently at the sea surface. In his famed whaling novel, *Moby-Dick*, Herman Melville questioned the future of these great whales, faced, as they were, with "omniscient look-outs at the mastheads of the whale-ships, now penetrating even through Behring's straits, and into the remotest secret drawers and lockers of the world; and the thousand harpoons and lances darting along all the continental coasts; the moot point is, whether Leviathan can long endure so wide a chase, and so remorseless a havoc." By the end of the 19th century, the gray, right, and bowhead whales were all on the verge of extinction. Under strict international protection, the gray whale has since recovered, but the number of bowheads and right whales remains low.

The era of modern whaling was initiated in the late 19th century with the cannon-fired harpoon equipped with an explosive head. The explosive harpoon was so devastatingly effective that even the large rorquals, the blue and fin whales, were taken in large numbers. These whales had previously been ignored by whalers with hand harpoons because they were much too fast to be overtaken in sail- or oar-powered boats. The subsequent rapid decline of the whale stocks in the North Atlantic and Pacific Oceans forced ambitious whalers to seek new and untouched whaling areas. They found the Antarctic, the rich feeding grounds of the planet's largest whales. The discovery of the Antarctic whale populations began seven decades of slaughter unparalleled in the history of whaling.

Aided by pelagic factory ships fitted with stern ramps to haul the whale carcasses aboard for processing, the kill of large rorquals rose dramatically. From 176 blue whales in 1910, the annual take climbed to almost 30,000 in 1931 (**Figure 15.15**). After the peak year of 1931, blue whales became increasingly scarce. Blue whale catches declined steadily until they were commercially insignificant by the mid-1950s. In 1966, only 70 blue whales were killed in the entire world ocean. Only then, when substantial numbers could no longer be found to turn a profit, was the hunting of blue whales banned in the Southern Hemisphere.

The trend of increasing and then rapidly declining annual catches, shown by the blue whale curve in Figure 15.15, is distressingly similar to that of the Peruvian anchoveta fishery, but the ruthless exploitation of the great whales did not halt with the near extinction of the blue whale. As the blue whale populations gave out, whalers switched to the smaller, more numerous fin whales, catches of which skyrocketed to over 25,000 whales each year for most of the 1950s; however, by 1960, the fin whale catch also began to plummet, and whaling pressure was diverted to even smaller sei whales. The total sei whale population probably never exceeded 60,000. One third of the sei whale population was killed in 1965 alone. Whaling pressure quickly pushed the catches of this species far beyond its maximum sustainable yield. By the late 1960s, sei whales had followed their larger relatives to commercial extinction, and the whaling effort was shifted to the minke whale, an 8- to 9-m miniature version of a blue whale.

As early as 1940, whaling nations were faced with undeniable evidence that some populations of pelagic whales were seriously overexploited. In 1948, 20 whaling nations established the International Whaling Commission (IWC) to oversee the utilization and conservation of the world's whale resources. Unfortunately, the IWC had neither inspection nor enforcement powers. Only once during the decade from 1963 to 1973 did the whaling industry actually manage to catch the quotas established for it by the IWC. The quotas were, in effect, not quotas at all, because there were not enough whales to fill them. In 1974, the IWC, under pressure from several national governments and international conservation organizations, adopted a new set of management procedures for several geographically localized stocks of whales. Under these procedures, all species of baleen whales in the Antarctic except the minke are classified as protected stocks, and no harvesting is allowed. Four species, the blue, gray, humpback, and right whales, are protected in all oceans.

To some, the severe depletion of the large pristine stocks of krill-eating baleen whales led to the naive assumption that without these whales a large "surplus" of krill exists for our harvesting. Such an assumption ignores the evidence that a new equilibrium has developed in krill-based food chains. Minke whales, seabirds, and seals now play larger roles than they did a century ago. This is not a unique situation, because many fish stocks depleted by excessive harvesting pressures have been replaced with other (and, from a commercial point of view, often

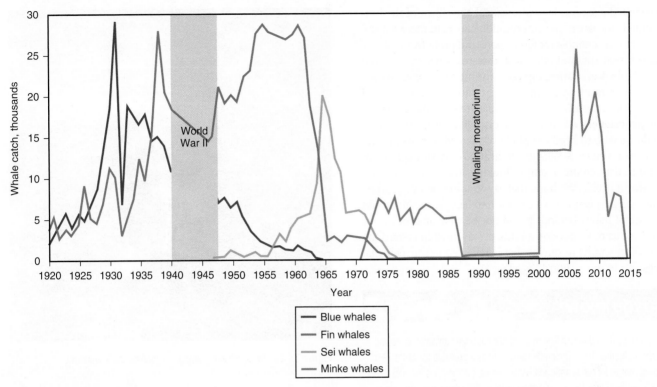

Figure 15.15 Harvests of blue, fin, sei, and minke whales in the Antarctic, 1920 to 2000. After the whaling moratorium was implemented in 1988, there were no reported catches of any whales except for minke whales.

Data compiled from UNFAO statistics.

less desirable) species. These new species assemblages, by their very existence in niches previously occupied by the exploited stocks, serve as a barrier to the recovery of those stocks.

The road to recovery for a large, slowly reproducing species such as the blue whale is fraught with unanswered questions. Can the few thousand remaining blue whales scattered over the world ocean encounter each other frequently enough to mate, reproduce, and add to their decimated numbers? Will additional blue whales overcome competition from other animals for food and other resources appropriated during their tragic decline? And, most important, when the population begins to increase, will humans refrain from exploiting it so that it can secure a more solid grip on survival? The reproductive resiliency of the great whales is largely unknown. They may bounce back. The gray whale did, but the right whale still has not.

In 1982, the IWC approved a 5-year moratorium on the commercial killing of large whales. This moratorium was seen as a crucial first step in assessing the impact of nearly a century of intense hunting of pelagic whale stocks. At the same time, it would give declining whale populations an opportunity to stabilize. Even as it went into effect, the moratorium was opposed by several nations with long whaling histories, especially Japan, Norway, and Iceland. These nations continued to take whales (particularly fin and minke whales) for so-called scientific research purposes during the moratorium.

When the second 5-year moratorium against whaling expired in 1986 it was extended, and it remains in place today, although the IWC still sets moratoriums for subsistence whaling. Some countries, especially those originally opposing the

moratorium, chose to resume limited whaling of several whale species. The IWC's 2013 report outlines whale catches for member nations and includes several illegal catch of minke whales by South Korea and several hundred catches under "special permit" of sei, Bryde's, minke, and sperm whales by Japan. The special permit exemption is a loophole Japan found in the IWC that cannot be closed without amending the treaty. Japan has also taken whales from inside the Southern Ocean Sanctuary in Antarctica, an act that was declared illegal by the IWC in 2014. In 2015, Japan took over 300 minke whales under a special research permit, which many other nations believe is a smokescreen for food harvesting. The battle to protect whales and prohibit whaling by countries with long whaling histories continues today and will likely continue for many years to come.

15.11 Concluding Thoughts: Developing a Sense of Stewardship

The world ocean is a great and forgiving ecosystem, for centuries providing food, transportation, and recreation while absorbing the pressures of human societies. Our continued use of the seas for these purposes depends on a better understanding of the consequences of our intrusion on the workings of marine ecosystems. We are the only species on Earth truly capable of understanding the motives and consequences of our past, present, and future actions, yet the history of our economic involvement with marine populations has repeatedly demonstrated a serious lack of practical awareness of the fundamental disturbances our intervention causes in these natural systems. In our scramble for food, sport, and profit from the sea, we have repeatedly, and often with disastrous results, violated existing ecological relationships and invented new ones. We are positioning ourselves with increasing frequency at the tops of heavily exploited marine food webs that are contaminated with the very substances we are afraid to dispose of near our homes. It would be wise to heed the plight of the cod, the anchoveta, and the great whales as sensitive indicators of what might be in store for us if we continue our unthinking and uncaring misuse of the world ocean. We hope that as we continue our journey through the 21st century we can learn to leave some of our old and wasteful habits behind. It will not be easy or simple, but each one of us must develop a sense of stewardship toward the world ocean and its resources that is reflected in our personal choices and in our political decisions.

DID YOU KNOW?

The Monterey Bay Aquarium Research Institute has been publishing a Seafood Watch Guide since 1999. The guide started out as a printable list of the best seafood choices for the Atlantic and Pacific Oceans, but now also includes recommendations for restaurants carrying sustainable seafood and aquaculture facilities farming fish or invertebrates using sustainable practices. Consumers and business owners can print a guide, search online by species, or download an application to their phone or computer providing them with a vast amount of searchable information to make informed decisions about seafood purchases. Seafood is rated as "Best Choice," "Good Alternative," or "Avoid" based on how well the fishery is managed, the current state of the population, and/or the impacts of the farming methods. Visit *www.seafoodwatch.org* to search for your favorite seafood and find out how it is rated.

On a brighter note, humans are now more than ever very capable of having a positive impact on our marine environment by making informed choices and acting responsibly. In recent decades, the amount of information available concerning the marine realm has increased dramatically, technology to assess and predict problems has improved, and new generations of humans are taught early that how we treat Earth and its inhabitants matters. Early warning systems are in place for meaningful environmental changes such as coral bleaching events and sea level changes. Various governmental agencies, nongovernmental organizations, and universities around the world are concentrating efforts on studying and educating people on the very real idea that the ocean is an invaluable resource that must be monitored, cared for, and protected.

Figure 15.16 Young children enjoying their local marine environment.
© Courtesy of Deanna Pinkard-Meier.

The great news is that as individuals we can do so much, whether we live near the ocean or not. For those frequenting the coast, be sure to dispose of trash properly and pick up any stray trash before leaving the beach. Purchase fewer plastic items, more reusable items, and recycle anything that can be recycled. Do not dump oil or other wastes into storm drains, and refrain from using chemical fertilizers on lawns and other plants. For those living near or far from the sea, donate to organizations that help to preserve oceans; teach children about the importance of environmental stewardship; reduce carbon emissions by driving less; and, when there is a choice, purchase more fuel-efficient vehicles. Make informed decisions about seafood purchases, which is incredibly easy to do with applications at the tips of our fingers, downloaded to a cell phone. Lastly, talk about what you care about. Inform others of your conscious decisions to protect and preserve the ocean. So many of our environmental problems stem from a simple misunderstanding of how our choices may affect the planet and the misguided idea that the actions of individuals do not matter. As a species, we have the tools, information, and capabilities to preserve our oceans in a state of good health for ourselves and for many future generations of fishers and ocean enthusiasts (**Figure 15.16**).

STUDY GUIDE

TOPICS FOR DISCUSSION AND REVIEW

1. Why is it that clupeoids are so easily harvested in vast numbers, such that they account for about one third of the world's total commercial catch? What is the primary use of these fish, once captured?

2. Regarding the trophic levels occupied by the plants and animals that we consume, how do seafood and terrestrial harvests differ and why?

3. The magnitude of our fish harvest is limited (hypothetically) by what two ecological factors?

4. Define *aquaculture*, and describe its relative advantages and disadvantages. Is aquaculture the answer to meeting the world's seafood needs?

5. Define *maximum sustainable yield*, and identify the biological information required to estimate its value correctly.

6. What is the tragedy of the commons with regard to marine fisheries?

7. What is an exclusive economic zone? Pretend that you and your classmates are the leaders of different nations, some technologically advanced (such as the United States) and others not (such as Indonesia). Then discuss the "fairness" of establishing an exclusive economic zone around your nation.

8. What criticisms are often made of the trade in marine ornamental species?

9. What two technological advances enabled whalers to harvest whales at previously unprecedented rates?

10. Discuss the personal choices that you can make to help ensure long-term sustainable use of marine resources.

KEY TERMS

aquaculture 409	mariculture 412
clupeoid 414	maximum sustainable yield 413

KEY *GENERA*

Catalaphyllia	Labroides
Chaetodon	Oxymonacanthus
Delphinus	Plerogyra
Dendronephthya	Stenella
Engraulis	Trachurus
Euphausia	

REFERENCES

Adey, W. H. 1987. Food production in low-nutrient seas. *BioScience* 37:340–348.

Alexander, L. M. 1993. Large marine ecosystems. *Marine Policy* 17:186–198.

Bardach, J. 1987. Aquaculture. *BioScience* 37:318–319.

Beddington, J. R., R. J. H. Beverton, and D. M. Lavigne. 1985. *Marine Mammals and Fisheries*. Boston: George Allen and Unwin.

Beddington, J. R., and R. M. May. 1982. The harvesting of interacting species in a natural ecosystem. *Scientific American* November:62–69.

Bell, F. W. 1978. *Food from the Sea: The Economics and Politics of Ocean Fisheries*. Denver: Westview Press.

Borgese, E. M. 1983. The law of the sea. *Scientific American* March:42–49.

Caddy, J. F., ed. 1988. *Marine Invertebrate Fisheries: Their Assessment and Management*. New York: Wiley-Interscience.

Castanza, R., F. Andrade, P. Antunes, M. van den Belt, D. Boersma, D. F. Boesch, F. Catarino, S. Hanna, K. Limburg, B. Low, M. Molitor, J. Gil Pereira, S. Rayner, R. Santos, J. Wilson, and M. Young. 1998. Principles for sustainable governance of the oceans. *Science* 281:198–199.

Cushing, D. H. 1981. *Fisheries Biology*. Madison: University of Wisconsin Press.

Cushing, D. H., and R. R. Dickson. 1976. The biological response in the sea to climatic changes. *Advances in Marine Biology* 14:1–122.

Dahlberg, K. A. 1991. Sustainable agriculture—fad or harbinger? *BioScience* 41:337–340.

Emery, K. O., and C. O. Iselin. 1967. Human food from ocean and land. *Science* 157:1279–1281.

Falkowski, P. G., ed. 1980. *Primary Productivity in the Sea*. New York: Plenum Press.

Field, C. B., M. J. Behrenfeld, J. T. Randerson, and P. Falkowski. 1998. Primary production of the biosphere: Integrating terrestrial and oceanic components. *Science* 281:237–240.

Food and Agricultural Organization of the United Nations. 1982. *Atlas of the Living Resources of the Seas*. Rome: Department of Fisheries (FAO).

Food and Agricultural Organization of the United Nations. 2002. *Yearbook of Fisheries Statistics, Catches, and Landings*. Rome: Department of Fisheries (FAO).

Food and Agricultural Organization of the United Nations. 2015. FAO Global Aquaculture Production database updated to 2013—Summary information. Available from http://www.fao.org/3/a-i4899e.pdf

Fye, P. M. 1982. The law of the sea. *Oceanus* 25:7–12.

Gulland, J. A. 1971. *The Fish Resources of the Ocean*. Surrey, England: Fishing News (Books).

Hammill, M. O., G. B. Stenson, T. Doniol-Valcroze, and A. Mosnier. 2011. Northwest Atlantic harp seals population trends, 1952–2012. DFO Canadian Science Advisory Secretariat Research Document.

Hardin, G. J. 1993. *Living Within Limits: Ecology, Economics, and Population Taboos*. New York: Oxford University Press.

Hempel, G., and K. Sherman, eds. 2003. *Large Marine Ecosystems of the World Change and Sustainability*. Amsterdam: Elsevier Science.

Heslinga, G. A., and W. K. Fitt. 1987. The domestication of reef-dwelling clams. *BioScience* 37:332–339.

Horn, M. H., and R. N. Gibson. 1988. Intertidal fisheries. *Scientific American* 258:64–70.

International Whaling Commission. 2013. Annual Report of the International Whaling Commission, August 2012–2013. Cambridge, UK.

Jarre, A., P. Muck, and D. Pauly. 1991. Two approaches for modelling fish stock interactions in the Peruvian upwelling ecosystem. *ICES Marine Science Symposium* 193:171–184.

Johansen, B. E. 2003. *The Dirty Dozen: Toxic Chemicals and the Earth's Future*. Westport, CT: Praeger Publishers.

Katz, S. L., R. Zabel, C. Harvey, T. Good, and P. Levin. 2003. Ecologically sustainable yield. *American Scientist* 91:150–157.

Koslow, J. A. 1997. Seamounts and the ecology of deep-sea fisheries. *American Scientist* 85:168–176.

Laws, R. M. 1985. The ecology of the southern ocean. *American Scientist* 73:26–40.

Levin, P. S., and M. H. Schiewe. 2001. Preserving salmon diversity. *American Scientist* 89:220–227.

Longhurst, A., S. Sathyendranath, T. Platt, and C. Caverhill. 1995. An estimate of global primary production in the ocean from satellite radiometer data. *Journal of Plankton Research* 17:1245–1271.

Marshall, N. B. 1980. *Deep-Sea Biology: Developments and Perspectives*. New York: Garland S.T.P.M. Press.

McCarthy, A. L., S. C. Steinessen, and D. Jones. 2015. Results of the acoustic-trawl surveys of walleye pollock (*Gadus chalcogrammus*) in the Gulf of Alaska, February–March 2014 (DY2014-01 and DY2014-03). AFSC Processed Rep. 2015-05, 85 p. Alaska Fish. Sci. Cent., NOAA, Natl. Mar. Fish. Serv., 7600 Sand Point Way NE, Seattle, WA 98115.

Muck, P. 1989. Major trends in the pelagic ecosystem off Peru and their implications for management. *PROCOPA Contribution* 90:386–403.

Murphy, R. C. 1925. *Bird Islands of Peru*. New York: G.P. Putnam and Sons.

Nicol, S., and W. de la Mare. 1993. Ecosystem management and the Antarctic krill. *American Scientist* 81:36–47.

Norse, E., ed. 1993. *Global Marine Biological Diversity: A Strategy for Building Conservation into Decision Making*. Washington, DC: Island Press.

Pauly, D., V. Christensen, R. Froese, and M. Palomares. 2000. Fishing down aquatic food webs. *American Scientist* 88:46–51.

Perrin, W. F. 2004. Chronological bibliography of the tuna–dolphin problem, 1941–2001. NOAA Tech. Memo. NMFS-SWFSC-356.

Radmer, R. J. 1996. Algal diversity and commercial algal products. *BioScience* 46:263–270.

Rhyne A. L., M. F. Tlusty, P. J. Schofield, L. Kaufman, J. A. Morris, Jr., A. W. Bruckner. 2012. Revealing the appetite of the marine aquarium fish trade: The volume and biodiversity of fish imported into the United States. *PLoS ONE* 7(5):e35808. doi:10.1371/journal.pone.0035808.

Robbards, M. D., and R. R. Reeves. 2011. The global extent and character of marine mammal consumption by humans: 1970–2009. *Biological Conservation* 144(2):2770–2786.

Robinson, M. A., and A. Crispoldi. 1975. Trends in world fisheries. *BioScience* 18:23–29.

Ross, R. M., and L. B. Quetin. 1986. How productive are Antarctic krill? *BioScience* 36:264–269.

Rothschild, B. J., ed. 1983. *Global Fisheries: Perspectives for the 1980s*. New York: Springer-Verlag.

Rudloe, J., and A. Rudloe. 1989. Shrimpers and sea turtles: A conservation impasse. *Smithsonian* 29:45–55.

Ryther, J. H. 1969. Photosynthesis and fish production in the sea. *Science* 166:72–76.

Ryther, J. H. 1981. Mariculture, ocean ranching, and other culture-based fisheries. *BioScience* 31:223–230.

Ryther, J. H., W. M. Dunstan, K. R. Tenore, and J. E. Huguenin. 1972. Controlled eutrophication—increasing food production from the sea by recycling human wastes. *BioScience* 22:144–152.

Scarff, J. E. 1980. Ethical issues in whale and small cetacean management. *Environmental Ethics* 3:241–279.

Schaefer, M. B. 1970. Men, birds, and anchovies in the Peru Current—dynamic interactions. *Transactions of the American Fisheries Institute* 99:461–467.

Shivji, M., S. Clarke, M. Pank, L. Natanson, N. Kohler, and M. Stanhope. 2002. Genetic identification of pelagic shark body parts for conservation and trade monitoring. *Conservation Biology* 16:1036–1047.

Sissenwine, M. P., and A. A. Rosenberg. 1993. U. S. fisheries: Status, long-term potential yields, and stock management ideas. *Oceanus* Summer:48–54.

Smith, W. O., Jr., and D. M. Nelson. 1986. Importance of ice edge phytoplankton production in the southern ocean. *BioScience* 36:251–257.

Stoecker, M. 2014. Habitat quality, rainbow trout occurrence, and steelhead recovery potential upstream of Searsville Dam. Report prepared by Stoecker Ecological, Santa Barbara, CA.

United Nations, Department of Economic and Social Affairs. *World Population Prospects: 2012 Revision*. June 2013. Available from http://esa.un.org/unpd/wpp/index.htm

Vitousek, P. M., H. A. Mooney, J. Lubchenco and J. M. Melillo. 1997. Human domination of Earth's ecosystems. *Science* 277:494–499.

Wade, P. R. 1995. Revised estimates of incidental kill of dolphins (Delphinidae) by the purse-seine tuna fishery in the eastern tropical Pacific, 1959–1972. *Fishery Bulletin* 93:345–354.

Walsh, J. J. 1984. The role of ocean biota in accelerated ecological cycles: A temporal view. *BioScience* 34:499–507.

Glossary

aboral Relating to side of the body opposite the mouth of radially symmetrical animals.

absorptive feeding A feeding method involving absorption of nutrients directly across the body wall or by using specialized organs.

abyssal plain The flat, sediment-covered floor of the ocean basin, usually 3,000 to 5,000 m deep.

abyssal zone Zone of the ocean at depths between 3,000 to 5,000 m.

accessory pigments Photosynthetic pigments found in marine plants that absorb light energy from the center of the visible light spectrum and transfer it to the green pigment chlorophyll *a*.

acoelomate Animal with three tissue layers but no body cavity, or coelom.

acrorhagi Nematocyst-armed defensive structures of sea anemones.

adenosine triphosphate (ATP) A complex organic compound composed of the molecule adenosine and three phosphate groups that serves in short-term energy storage and conversion in all organisms.

aerobic dive limit (ADL) The longest breath-hold dive time possible that does not lead to an increase in blood lactate concentration (i.e., a shift to anaerobic respiration).

aerobic respiration Cellular respiration occurring in the presence of oxygen.

ahermatypic Corals that do not build reefs.

albedo The amount of solar radiation reflected by a surface.

algal grazer An animal that consumes either macroalgae or thin algal films.

alveoli Minute air sacs in the lungs of vertebrates where gas exchange occurs across capillaries (singular = alveolus).

amniotic egg Eggs of reptiles, birds, and mammals containing an embryo that develops within an amniotic membrane.

ampulla of Lorenzini Specialized sensory organs forming a network of jelly-filled pores found primarily in cartilaginous fishes.

anadromous An animal (such as a salmon) that spends much of its life at sea and then returns to a freshwater stream or lake to spawn.

anaerobic respiration Cellular respiration occurring in the absence of oxygen.

ancestral character Any character that remains unchanged in the descendant from its condition or appearance in the ancestor.

androgen A male sexual hormone in vertebrates.

anoxic Without oxygen.

Antarctic Convergence An area encircling the Antarctic continent where cold, northward-flowing waters from the Antarctic meet the warmer waters from the subantarctic.

Antarctic ozone hole A thin layer of ozone over Antarctica that changes thickness seasonally.

antennae Elongated sensory structures projecting from the heads of arthropods (singular = antenna).

anterior The front end of a bilaterally symmetrical animal.

antitropical distribution A type of distribution in which similar climatic regimes in both the Northern and Southern Hemispheres are inhabited by the same species.

aorta Large artery (or arteries) carrying blood away from the heart to the body of vertebrates.

aphotic zone The portion of the water column, usually deeper than 1,000 m, where sunlight is absent.

apneustic breathing Breathing pattern exhibited by marine mammals in which several rapid breaths alternate with a prolonged breath-holding period.

aquaculture The captive breeding, or farming, of aquatic organisms.

aspect ratio Index of propulsive efficiency in fishes obtained by dividing the square of its caudal fin height by the caudal fin area.

atmosphere A layer of gases surrounding Earth that is held in place by gravity; also, a unit of pressure (atm) equal to 14.7 lbs./in.^{2} (or 760 mm Hg) and equivalent to the pressure created by a 1-inch-square, 10-meter-tall column of seawater.

atoll A ring-shaped chain of coral reefs from which a few low islands project above the sea surface, enclosing a central lagoon.

autosome A chromosome not designated as a sex chromosome.

autotrophic An organism that synthesizes its own organic nutrients from inorganic raw materials.

auxospore The naked cell of a diatom after the frustule has been shed.

avascular Not containing blood vessels.

baleen Rows of comblike material that hang from the outer edges of the upper jaws of filter-feeding whales.

bar-built estuary A type of estuary formed behind a coastal barrier or bar.

barrier island A long, narrow island of sand or mud running parallel to a coastal area.

barrier reef A coral reef separated from the shoreline by a lagoon.

bathyal zone Portion of the pelagic zone that extends from approximately 1,000 to 4,000 m; also known as the *bathypelagic zone*.

benthic division Pertaining to the seafloor and the organisms that live there.

benthos Marine organisms that live in or on the sea bottom.

bioaccumulation Increasingly concentrated accumulation of substances, especially pollutants, at successively higher trophic levels via food-chain amplification

biochemical oxygen demand (BOD) The quantity of dissolved O_2 needed to sustain a given community.

bioluminescence The natural production of light by a living organism.

biomass Biological material derived from living or recently living organisms.

biome Specific regions of the world with similar dominant vegetation types and animals present.

biosphere All of the available habitats on Earth where life may exist.

black band disease A disease of reef-forming corals caused by a sulfate-reducing cyanobacterium that results in a band of dead, blackened tissue surrounding a coral colony.

blade The flattened, usually broad, leaf-like structure of seaweeds.

blastocyst In mammals, a small and hollow ball of cells representing an early stage of embryonic development similar to the blastula of other animals.

blastula Early embryonic stage of many animals, consisting of a hollow ball of cells.

bleaching Expulsion of pigmented symbiotic zooxanthellae by reef-forming corals when stressed.

blubber The sheath of insulating fat and connective tissue of cetaceans and other marine mammals that aids in thermoregulation and has been used as a source of oil and food by humans.

bradycardia Dramatic slowing of the heart rate.

broadcast spawner Marine animals that reproduce by releasing eggs and sperm into the water.

bronchi Paired tubes of vertebrate respiratory systems that branch into each lung at the lower end of the trachea (singular = bronchus).

brooder An organism that holds developing young on or inside its body until it releases them for live birth.

bubble-net feeding A unique method of cooperative feeding used by some cetaceans in which one member of a group produces a curtain of ascending bubbles of exhaled air that cause prey to clump for easier capture.

buffer A chemical substance that tends to limit changes in pH levels.

byssal thread A strong elastic fiber produced by mussels to attach themselves to a solid substrate.

calorie A unit of heat energy equivalent to the amount of energy required to increase the temperature of 1 g of pure water 1°C.

Calvin cycle Also known as the *light-independent reactions*; in photosynthesis, a metabolic pathway found in the stroma of the chloroplast in which CO_2 is converted into sugars.

capillary wave The smallest wind waves that are displayed as ripples on the surface of the water.

carnivore An animal that obtains at least 90% of its caloric intake by eating other animals.

carpospore Diploid spores produced by the carposporophyte stage of red algae.

carposporophyte A diploid generation within the life cycle of some algae that produces carpospores.

carrion Dead, decaying animal flesh.

carrying capacity The maximum population size the environment can support, given available resources.

catadromous Fishes that migrate from freshwater to spawn in the ocean.

caudal peduncle The area where the caudal fin joins the rest of a fish's body.

cellular respiration The process of oxidizing sugars to CO_2 and water. Energy is released in the form of adenosine triphosphate (ATP) to fuel the cell's activities.

cenosarc A thin layer of flexible tissue that connects members of a coral colony.

centric A type of diatom with radial symmetry.

cephalization The evolutionary process of increasing specialization of the head, especially the brain and sense organs.

cerata Fingerlike branching external gills that project from the dorsal surface (mantle) of nudibranchs.

channel A narrow body of water between two land masses.

character Any countable or measurable attribute of an organism.

chemosynthesis Bacterial synthesis of organic material from inorganic substances using chemical energy.

chitin A flexible, impermeable fibrous substance consisting of polysaccharides found in the exoskeletons of arthropods and the cell walls of fungi.

chlorophyll The green photosynthetic pigment of plants, protists, and bacteria.

chloroplast A chlorophyll-containing organelle found in photosynthetic eukaryotes.

choanocytes Flagellated cells lining the spongocoel that extract food and oxygen from water passing through sponges.

chloride cells Cells in the gills of teleost fishes used for osmoregulation.

chromatophore Pigment cell that allows some marine organisms to change colors for camouflage.

chromosome A subcellular structure that contains the genetic information of the cell.

circadian clock (rhythm) A cycle of activity or behavior that recurs about once per day.

circalunadian rhythm A cycle of repeating activity each lunar day, or every 24.8 hours.

clade A group of organisms that includes a common ancestor and all of its descendants.

cladistics A biological classification system that uses shared derived characteristics to group organisms.

cladogram A branching diagram depicting hypothesized evolutionary relationships between groups of organisms.

clones Genetically identical group of individuals derived from a single individual.

closed circulatory system A circulatory system with blood confined in vessels, pumped by a muscular heart.

clupeoid In the suborder Clupeoidei, including herrings, anchovies, sardines, and similar small fishes.

cnidocyte Cell that contains the stinging organelle, or nematocyst, of cnidarians.

coastal lagoon A coastal body of water surrounded by rock, sand, mud, or coral that presents a partial barrier to the open sea.

coastal plain estuary An estuary formed as rising sea level floods a low-lying coastal river valley.

coccolith A small calcareous plate imbedded in the cell wall of coccolithophores, a type of phytoplankton.

coelomate An organism with a compartmentalized internal body cavity lined with mesoderm.

colony Many individuals of the same species living together in close association.

columella The central supporting structure of gastropod mollusk shells and of the corallites of corals.

commensalistic A symbiotic relationship in which the symbiont benefits without affecting the host in any measurable or demonstrable way.

community An assemblage of interacting populations living in a particular locale.

compensation depth The depth at which ambient light intensity is so diminished that the rate of photosynthesis (gross primary production) equals the rate of cellular respiration such that no net primary production occurs.

conduction The molecular transfer of heat through a medium.

connectivity The exchange of individuals among marine populations, often influenced by physical oceanographic factors.

consumer An organism that consumes and digests other organisms to satisfy its energy and material needs.

continental boundary current A surface ocean current flowing generally north or south along a continental edge.

continental drift The gradual movement of continents in response to seafloor spreading processes.

continental shelf The relatively smooth underwater extension of the edge of a continent that slopes gently seaward to a depth of 120 to 200 m.

continental slope The relatively steep portion of the sea bottom between the outer edge of the continental shelf (the shelf break) and the deep ocean basin or abyssal plain.

convection The transfer of energy via the flow or mixing within a volume of liquid or gas with regions of different temperature and different density.

convective mixing The vertical mixing of water masses driven by wind stresses or density changes at the sea surface.

convergent evolution Two different species with similar characteristics that did not arise from common ancestry, but from similar adaptive pressures.

coral bleaching An event that involves a coral polyp releasing its algal symbiont, resulting in a white coloration of the coral animal.

corallite The calcareous skeletal cup in which a coral polyp lives.

Coriolis effect The resulting change in direction of a moving object (counterclockwise in the Southern Hemisphere and clockwise in the Northern Hemisphere) due to the rotation of the Earth.

cost of transport (COT) The energy costs associated with an animal's locomotion.

countercurrent An ocean current that flows directly back into another current; also, in some animals, relating to paired blood vessels containing blood flowing in opposite directions or streams of blood and seawater flowing through a gill in opposite directions.

countershading The coloration pattern exhibited in pelagic animals for camouflage, with the upper surfaces darkly pigmented and the sides and ventral surfaces silvery or only lightly pigmented.

covalent bond A chemical bond between two atoms created by the sharing of electrons.

crest The tallest portion of a wave.

critical depth The depth to which a phytoplankton cell can be mixed yet still spend sufficient time above the compensation depth such that its daily respiratory needs are able to be met by its own photosynthetic production.

cropper A deep-sea organism that ingests living and nonliving matter to extract nutrients.

crossover speed The velocity at which it becomes more efficient for a swimming animal to leap above the surface, or porpoise, rather than to remain submerged.

ctene Bands of cilia found on the body surfaces of ctenophores.

cuticle A flexible and resilient exoskeleton that must be shed for growth.

cypris The final bivalved larval stage of a barnacle.

cytoplasm The material within a cell, excluding the nucleus.

dark reactions Also known as the *Calvin cycle* or *light-independent reactions*; in photosynthesis, a metabolic pathway found in the stroma of the chloroplast in which CO_2 is converted into sugars.

decomposer An organism that consumes and breaks down dead organic material.

deep sound-scattering layers (DSSLs) One or more layers of midwater marine animals that reflect and scatter the sound pulses of echo sounders.

delta A low-lying sediment deposit often found at the mouth of a river.

deposit feeder an animal that ingests sediments to digest food particles for nourishment.

derived character Any character whose condition or appearance is changed in the descendant or is different from that of its ancestor.

desiccation To dry out; lose water.

detritus Particulate dead organic matter, including excrement and other waste products, shed body parts (such as exoskeletons, skin, hair, or leaves), and minute dead organisms.

diffusion The transfer of substances along a gradient from regions of high concentrations to regions of low concentrations.

diploid Cells that contain two of each type of chromosome characteristic of its species.

diurnal Occurring daily or occurring during the day.

diurnal tide A tidal pattern with one high tide and one low tide each lunar day.

dorsal The back or upper surface of a bilaterally symmetrical animal.

echolocation A means of navigation and environmental assessment that includes producing high-frequency sounds and listening for reflected echoes as the sounds bounce off target objects.

ecological adaptation Short-term response expressed by an individual organism to its changing environment.

ecological succession The observed process of change in the species structure of an ecological community over time.

ecosystem All of the living and nonliving components of a given area.

ectotherm An animal whose body temperature is governed by external ambient temperatures.

eddy A relatively small loop of surface currents split off from a major current that may travel far for long time periods.

Ekman spiral A deflection of surface currents that is a consequence of the Coriolis effect wherein direction of flow rotates with increasing depth.

El Niño An ocean-atmosphere interaction leading to a cease in equatorial upwelling, warmer than normal water temperatures in the eastern Pacific, and a shift of rain patterns worldwide.

El Niño-Southern Oscillation (ENSO) An index used to predict and measure El Niño events, based on the Southern Oscillation.

electromagnetic radiation A form of radiant energy that is released by electromagnetic processes, including visible light.

elver A newly pigmented juvenile eel that swims upriver to take up residence in freshwater until sexual maturity.

emergent Rising out of fluid and only partially submerged.

endemic An organism native to a particular area and only found in that area.

endoplasmic reticulum A system of folded membranes within the cytosol of eukaryotic cells.

endotherm An animal whose body temperature is established by internal sources of metabolic heat.

endo-upwelling A process that carries Antarctic water into a reef network and drives it upward by the local geothermal gradient, providing nutrients at a reef crest.

epibenthic sledge A system for sampling benthic and benthopelagic fauna consisting of a metal frame with a net attached for containing specimens.

epibiont A species that lives on another species, using the host as a form of hard substrate.

epifauna Benthic animals that crawl about on the seafloor or are firmly attached to it.

epipelagic zone The upper 200 m of the oceanic province. See also *photic zone*.

epiphyte A plant that attaches itself to other organisms.

epitheca The larger portion of a diatom frustule.

estrogen A female sexual hormone in vertebrates.

eukaryotic Cells characterized by an organized nucleus and other membrane-bound intracellular organelles. Also refers to organisms containing one or more cells characterized by an organized nucleus and other membrane-bound intracellular organelles.

euryhaline The ability to withstand a wide range of environmental salinities.

eutrophication The introduction of chemical nutrients into an ecosystem.

evaporation For water, the conversion of liquid water to a gaseous state, also known as water vapor.

evolution Change in the heritable traits of biological populations over successive generations leading to species diversity.

evolutionary adaptation The changes occurring in a population of individuals over many generations by processes of natural selection.

exoskeleton An external supporting skeleton, commonly found in arthropods and mollusks.

exotherm A compound that gives off heat during formation and absorbs heat when it is broken down.

extant Still surviving or in existence; not extinct.

external auditory canal The sound-transmitting channel connecting the external and middle ears.

fast ice Sea ice that is connected or fastened to a coastline, to grounded icebergs, or to the seafloor along a shoal.

fecundity The number of offspring an organism can produce in a given time period.

feedback mechanism Control mechanisms in organisms and communities in which a change in a given factor either inhibits or stimulates processes controlling the production, release, or use of that factor.

fertilization The process wherein two haploid gametes unite to produce one diploid zygote.

fetch The extent of the ocean over which winds blow to create surface waves.

fineness ratio Ratio of an animal's body length to its maximum body diameter.

finlet Small median fin on the dorsal and ventral surfaces of the caudal peduncle of tuna and similar fishes.

fitness A measure of reproductive success.

fjord A deep coastal embayment formed by glacial erosion.

flagellum Whiplike structure used by cells for locomotion (plural = flagella).

flushing time The time required for all of the water of an estuary to be completely exchanged.

frictional drag The resistance created by an animal's body surface when it moves through a fluid medium.

fringing reef A large coral reef formation that closely borders the shoreline.

frustule The siliceous wall of a diatom; consists of two halves.

gamete An egg or sperm cell.

gametophyte A gamete-producing haploid generation of a plant.

gastrovascular cavity Blind digestive cavity of some lower invertebrates, with a single opening for both mouth and anus, that also plays a circulatory role.

gene A unit of inheritance.

gene flow The movement of genes from one population to another via immigration, emigration, seed dispersal, larval drift, etc.

gigantothermy The maintenance of a relatively high and constant body temperature by a large ectothermic animal that is the consequence of its low surface area-to-body volume ratio.

gill arch The skeletal supports inside the gill tissues of fishes.

globigerina ooze Thick blankets of foraminiferan tests that cover the seafloor.

gonads Animal reproductive organs.

gonochorism The state of having just one sex, as opposed to hermaphroditism.

grana Flattened saclike structures inside chloroplasts containing chlorophyll and other photosynthetic enzymes (singular = granum).

greenhouse gas emissions The release of greenhouse gases such as carbon dioxide, methane, nitrous oxide, and fluorinated gases into the atmosphere.

gross primary production The total amount of photosynthesis accomplished in a given period of time.

guyot A sunken volcanic island topped by a dead, usually flat coral reef.

gyre The large loop of interconnected surface currents within a single ocean basin, usually spanning 20° to 30° in latitude.

habitat The natural environment of an organism including its requirements for life.

hadal zone The deepest area of the oceans, typically around 6,000 m or deeper.

halocline A well-defined vertical salinity gradient in marine or estuarine water.

halophyte Flowering plants that are tolerant to seawater exposure, including complete submergence.

haploid Containing only one of each type of chromosome characteristic of its species.

haptera Short, sturdy rootlike structures that form the holdfast of seaweeds.

harem A group of female animals sharing a single male mate.

harmful algal bloom (HAB) A phytoplankton bloom including species that produce toxins, leading to fish kills and inedible shellfish.

heat capacity The measure of heat energy required to raise the temperature of 1 g of a substance 1°C.

hemocyanin A respiratory pigment used by many marine invertebrates.

hemoglobin A protein found in red blood cells that is an oxygen carrier.

herbivore An animal that obtains at least 90% of its caloric intake by eating plants.

hermaphrodite An animal that has the sex organs of both sexes at some point during its lifetime, sometimes simultaneously.

hermatypic A coral that contributes to the building of massive carbonate reefs and contains mutualistic dinoflagellates.

heterocercal A type of caudal fin found in sharks and rays, as well as some primitive bony fishes (such as sturgeons), wherein the vertebral column extends into the upper lobe of the tail.

heterotrophic An organism that cannot manufacture its own food, but must ingest food for nutrition.

high tide The highest level reached by the rising tide.

holdfast A structure that attaches seaweeds to the sea bottom or to other substrates.

holoplankton Species of zooplankton that are part of the plankton community throughout their lives.

homeostasis The condition in which the internal environment of an organism (e.g., salinity, temperature, blood sugar) remains relatively stable, within physiological limits.

homeotherm An animal, such as a bird or mammal, that maintains precisely controlled internal body temperatures using its own physiologically controlled heating and cooling mechanisms.

homocercal a type of caudal fin found in advanced bony fishes wherein the supporting skeletal rays are arranged symmetrically around the end of the vertebral column.

homologous Similar structures due to common ancestry.

Hox **genes** A group of related genes that control the major body plan features of an embryo.

hybrid An organism that is the result of two different species mating.

hydrogen bond (H-bond) A weak electrostatic bond formed by the attractive force between the charged ends of water molecules and other charged molecules or ions.

hydrostatic skeleton The body fluid contained within some animals, against which muscles work to provide shape changes.

hyperosmotic Relating to a water solution with a higher concentration of ions than that of an adjacent solution separated by a selectively permeable membrane.

hyphae Elongated and multinucleated threadlike structures making up the body, or mycelium, of a fungus.

hypoosmotic Relating to a water solution with a lower concentration of ions than that of an adjacent solution separated by a selectively permeable membrane.

hypotheca The smaller portion of a diatom frustule.

hypoxia A state of deficient oxygen within the tissues of an organism or the environment.

ice algae Various algal species found living in sea ice.

ice-breeding Animals that rely on the ice for crucial life processes such as breeding.

ichthyofauna The assemblage of fishes in an area.

induced drag The resistive force that is produced via the generation of lift.

induced ovulator A female animal that must be prompted to ovulate by the presence of sperm from a male.

infauna Animals that live within the sediments of the seafloor.

interstitial An animal that occupies the spaces (interstices) between sediment particles.

intertidal zone The vertical extent of the shoreline between the high and low tide lines.

ion an electrically charged atom or molecule formed by gaining or losing one or more electrons.

ionic bond In crystalline structures, an atomic bond formed between adjacent oppositely charged ions.

iridocyte A cell that occurs in the skin fish and some cephalopods that creates an iridescent appearance.

island An area of land, much smaller than a continent, that is completely surrounded by water.

isohaline Having the same salinity or a line on a chart or map connecting points of equal salinity.

isolume An intense effulgence of the sun creating a light variation for some marine organisms to follow.

isosmotic Relating to a water solution with the same concentration of ions (osmotic pressure) as an adjacent solution separated by a selectively permeable membrane.

junk A fat-filled organ of similar size to the spermaceti organ of sperm whales that is hypothesized to be homologous with the melon of smaller toothed whales.

kelp A group of large brown seaweeds.

keystone species A species with a greater impact on its community than its abundance would suggest.

king tide The very highest high and lowest low tides experienced in an area.

kleptoparasitic A feeding strategy where one animal takes prey from another that spent energy catching the prey.

krill Small crustaceans that provide food to many polar animals.

La Niña An ocean-atmosphere interaction leading to extreme equatorial upwelling, colder than normal water temperatures in the eastern Pacific, and a shift of rain patterns worldwide.

labyrinth organ One of a pair of equilibrium and balance organs in the inner ears of vertebrates that contains three fluid-filled semicircular canals (only one in hagfishes and two in lampreys).

lamella A thin layer of tissue in a fish gill.

Langmuir cells Parallel pairs of surface ocean convection cells driven by winds.

lanugo Soft, fine hair that covers the body of mammals, especially very young mammals.

last glacial maximum (LGM) The time at which the last major continental glacial advance in the Northern Hemisphere reached its maximum extent (about 18,000 years ago).

latent heat of fusion The heat that must be extracted from a liquid to freeze it to a solid at the same temperature; for water, it is 80 cal/g.

latent heat of vaporization The heat energy required to convert a liquid to a gas at the same temperature, for water, it is 540 cal/g.

latitude The angular distance north or south of the equator of a position on Earth's surface; measured in degrees (°).

leptocephalus larva Leaf-shaped, transparent larva of some eels, bonefish, and tarpon.

lichen An association between a fungus and a photosynthetic organism, usually a green alga or a cyanobacterium.

light-independent reactions Also known as the *Calvin cycle* or *dark reactions*; in photosynthesis, a metabolic pathway found in the stroma of the chloroplast in which CO_2 is converted into sugars.

light reactions The part of the photosynthetic process that, in the presence of light, captures energy to form ATP and NADPH to be used to synthesize complex organic molecules in the light-independent reactions.

limiting factor Any factor necessary for the growth or health of an organism that limits further growth if in insufficient supply.

lineage The line of descendants of an ancestor.

littoral The vertical extent of the shoreline between the high and low tide lines. Also known as the *intertidal zone*.

longitude The angular distance of a position on the Earth's surface east or west of the Greenwich Prime Meridian; measured in degrees (°).

lophophore Tentacle-bearing feeding structure of ectoprocts, brachiopods, and phoronids.

low tide The lowest level reached by a falling tide.

lunar day The time between one high tide and the same high tide on the next day, a period of about 24 hours and 50 minutes.

macroalgae Multicellular algae, often living attached to a substrate. Also called *seaweed*.

macrofauna Benthic infauna or epifauna larger than about 1 mm.

magnetoreception A sensory capability that involves the use of a magnetic field by organisms to orient in their environments and detect other organisms.

mangal A tropical community of mangrove plants and associated organisms.

mariculture The captive breeding, or farming, of marine organisms.

marine snow Detritus originating in the photic zone that sinks to the abyss, serving as the primary source of energy for deep-sea organisms.

matrilineal Used to describe the line of descent that follows the female side of the family.

maximum sustainable yield The optimal catch that may be taken from a fishing population each year without endangering its capacity to regenerate.

medusa Free-swimming, or jellyfish, generation of cnidarian life cycles.

megafauna The assemblage of macroscopic animals in an area.

meiofauna Benthic infauna that are 100 to 1,000 µm in size.

meiosis A process of cellular division used in sexual reproduction that reduces the chromosome number by half.

melon Fat-filled structure in the foreheads of toothed whales.

meltwater Water formed by the melting of ice and snow.

meristematic tissue Within some seaweeds and all plants, specific tissue sites where most cell division for growth occurs.

meroplankton Larval forms of benthic and nektonic adults that are temporary members of the plankton community.

mesenterial filaments Filamentous extensions of the mesentery tissue found in corals used to extract food particles from the water.

mucus ciliary system A method of feeding used by corals where food particles are trapped on mucus and moved toward the mouth by cilia.

mesoglea The acellular jellylike layer in the bodies of cnidarians.

mesopelagic zone The portion of the pelagic division that extends from the bottom of the epipelagic zone to a depth of 1,000 m.

metabolism The sum total of all biochemical processes occurring in a living organism.

metamere A repeated body unit, or segment, along the long axis of some bilateral animals.

microbe Microscopic bacteria and plankton.

microbial loop The portion of a planktonic food web composed of bacteria, phytoplankton, and microbial consumers and decomposers.

microfauna Benthic infauna that are less than 100 μm in size and colonize individual sediment particles.

mimicry The ability of an organism to disguise itself by assuming the behavior or appearance of another species.

mitochondria Organelles found in eukaryotic cells that are used for synthesis of ATP.

mixed semidiurnal tide Tidal pattern during a lunar day with two unequal high tides and two unequal low tides.

molt A process of punctuated growth in arthropods wherein they shed their old exoskeleton and replace it with a larger version.

monogynous The condition of having one mate at a time, sometimes for life.

mortality The rate at which individuals of a population die.

mudflat Coastal wetland that forms when mud is deposited by tides or rivers.

mutualism A type of symbiotic relationship in which both the symbiont and the host benefit from the association.

mycelium A network of hyphae that constitute the vegetative portion of a fungus.

mycophycobiosis Obligate mutualism between a marine fungus and a seaweed.

mycorrhiza Mutualism wherein a fungus colonizes the roots of a vascular plant.

myomere One of a series of muscle segments along the trunk of vertebrates, especially fishes.

nanoplankton Unicellular plankton with cells 5 to 20 μm in width.

natural selection The process wherein species that are better suited for their current, local environment survive longer and produce more offspring than other members of their population, which results in evolutionary change.

nauplius Microscopic free-swimming planktonic larval stage of barnacles and some other crustaceans (plural = nauplii).

nautical mile A measure of distance at sea equivalent to 1 minute of arc along a meridian of the Earth, or 1,852 m.

nekton Freely swimming pelagic animals.

nematocyst Organelle within the cnidocytes of cnidarians that is often venomous, used in prey capture and defense.

neritic province Pertaining to the portion of the marine environment that overlies the continental shelves.

net primary production The measure of primary production remaining after losses due to cellular respiration have been subtracted from gross primary production.

net primary productivity (NPP) The rate at which all of the plants and algae in an ecosystem produce net chemical energy.

neap tide a tide that recurs every two weeks (near quarter moons) that is characterized by a less-than-average tidal range.

neuromast A mechanosensory cell, similar to sensory hair cells, found in the lateral line system that detects water movements thereby aiding in prey capture, schooling behavior, and avoidance of obstacles and predators.

neuston Planktonic organisms living (usually floating) at or on the sea surface.

nitrification The biological breakdown of ammonia or ammonium to nitrite, followed by the breakdown of nitrite to nitrate.

nitrogen fixation Conversion by bacteria and cyanobacteria of atmospheric N_2 to other forms of nitrogen that can be directly used by eukaryotic plants.

node In phylogenetics, the point where two lineages diverge.

nonpoint source A source of pollution originating from a broad and vaguely defined source, such as fertilizers washed off fields into streams.

notochord An elongated, cartilaginous rod that forms the central skeletal support of chordate embryos.

nucleus The central organelle in a eukaryotic cell where chromosomes are contained.

nutrient regeneration Any of several mechanisms that transport nutrient-rich water from the seafloor up to the photic zone, such as upwelling and convective mixing.

ocean acidification A slight decrease in the pH of ocean water that leads to slightly more acidic waters and an imbalance of chemicals in the ocean.

oceanic province Pertaining to the portion of the marine environment that overlies the deep ocean basins, between continental shelf breaks.

olfaction the sense of smell, or the ability to detect and identify chemicals dissolved in air or water by using nasal sensory cells.

ooze The layer of fine sediment on the seafloor that is composed of mineralized skeletal remains of dead organisms settled from upper zones of the ocean.

oral Relating to the side of the body containing the mouth of radially symmetrical animals.

osculum The large aperture in a sponge's spongocoel through which water is expelled.

osmoregulator An organism that spends ATP to regulate the salt content of its internal fluid environment such that it remains relatively constant regardless of ambient salinity.

osmosis Diffusion of water from a more concentrated solution to an adjacent solution of lower concentration through a selectively permeable membrane.

osmotic conformer Organisms that tolerate large variations of internal ionic concentrations without serious damage by remaining isosmotic with the water around them.

osmotic pressure In hypoosmotic conditions, the internal fluid pressure that develops from the osmotic inflow of water.

oviparity The reproductive strategy of some vertebrates that reproduce via eggs that develop and hatch outside of the mother.

ovoviviparity An intermediate condition between viviparity and oviparity in which the eggs develop and hatch while still inside the mother, with subsequent live birth of the young.

oxygen minimum zone The layer of the water column below the photic zone where dissolved oxygen concentration is lowest.

ozone O_3, formed by the action of ultraviolet light on atmospheric O_2.

pack ice A large area of large ice pieces driven together.

Pangaea The supercontinent that consisted of all the present land masses prior to their breakup and subsequent drift to their present positions.

parasitism A symbiotic relationship where the host is harmed by the symbiont.

parthenogenesis A type of asexual reproduction that includes offspring developing from unfertilized eggs.

pavement Large areas of dead coral reef.

pelage The wool, fur, or hair of a mammal.

pelagic division Pertaining to the ocean's water column and the organisms that live there.

pelagic larval duration The length of time during which a particular larva remains a part of the plankton community prior to metamorphosis. See also *meroplankton*.

pennate Relating to diatoms that are elongated and display various degrees of bilateral symmetry.

pentamerous Five-sided radial body symmetry commonly displayed by echinoderms.

perennial A plant that lives for more than two growing seasons.

period In ocean waves, the time required for two successive waves to pass a reference point.

pheromone A chemical substance that is released from an animal that influences the behavior and development of other members of the same species.

phonic lips Enlarged folds of tissue associated with the nasal sacs of cetaceans that vibrate rapidly to produce the characteristic clicks and whistles of these animals.

photic zone The portion of the ocean where light intensity is sufficient to enable gross primary production (photosynthesis) to at least meet a cell's respiratory needs.

photophore Light-producing organs found in animals.

photosynthesis The process of converting solar energy into chemical energy.

phototrophic Obtaining energy from the sun to synthesize organic molecules for nutrition.

photosynthetically active radiation (PAR) The spectral range of solar radiation (400 to 700 nm) that photosynthetic organisms may use in the process of photosynthesis.

phylogenetic Based on natural evolutionary relationships.

phylogenetic relationship A relationship based on common ancestry.

phylogenetic tree A visual representation of natural evolutionary relationships.

physiology The study of the function of body components and whole organisms.

physoclistous swim bladder A swim bladder lacking an air passage to the esophagus. See also *pneumatic duct*.

physostomous swim bladder A swim bladder with an air passage or duct to the esophagus. See also *pneumatic duct*.

phytoplankton Photosynthetic members of the plankton community.

picoplankton Unicellular plankton less than 2 μm in width.

plankton A diverse ecological group of drifting organisms composed primarily of small phytoplankton, zooplankton, and larvae.

plankton bloom Rapid growth of phytoplankton within a water body.

planula The planktonic larval form of cnidarians.

plate tectonics The collective geological processes that move the crustal plates of the Earth and cause continental drifting and seafloor spreading.

pneumatic duct The connection between the esophagus and swim bladder of fishes.

pneumatophore Gas-filled float found in siphonophores.

pod A group of marine mammals living closely together, ranging in size from several individuals to several hundred individuals.

poikilotherm An animal that lacks the ability to control its body temperature physiologically.

point source A source of pollution originating at a definable point, such as a storm drain.

polar easterlies Winds that blow from east to west at very high latitudes.

pollen Microscopic grains discharged by male parts of flowers that contain the male gamete.

polygynous The condition of having more than one mate.

polygyny A type of social and breeding organization in which a male is dominant over and mates with several females.

polymorphic An organism that possesses more than one adult form.

polyp The attached, benthic generation of many types of cnidarians.

population A group of freely interbreeding organisms of the same species.

posterior The rear end of a bilaterally symmetrical animal.

pressure drag Hydrodynamic drag on an organism caused by its cross-sectional area.

primary producer Autotrophic organisms that synthesize organic compounds via photosynthesis or chemosynthesis.

primary production The synthesis of organic compounds from inorganic substances.

proboscis An extensible tubular organ used for feeding.

producer An organism, usually photosynthetic, that contributes to the production of organic compounds for a community.

prokaryotic Bacteria and cyanobacteria that lack the structural complexity and defined nucleus found in eukaryotes.

protandry A form of sequential hermaphroditism in which the individual matures first as a male, then transforms into an adult female later in life.

protogyny A form of sequential hermaphroditism in which the individual matures first as a female, then changes sex to an adult male later in life.

pseudocoelomate An animal with three tissue layers and a body cavity that is not compartmentalized or lined completely with mesoderm, as in a true coelom.

pycnocline The ocean layer, usually near the bottom of the photic zone, marked by a sharp change in density that separates the less dense surface waters from denser deep waters.

radula Rasping tonguelike structure of most mollusks.

raphe A groove in the frustules of pennate diatoms through which pseudopodia extend for locomotion.

red tide A bloom of toxic dinoflagellates that may cause serious mortality to other forms of marine life.

rete mirabile A complex of veins and arteries located very close together possessed by some vertebrates and used to store oxygen in some aquatic vertebrates.

retina The light-sensitive layer of nerve cells in the eyes of most vertebrates and some invertebrates.

rhizome The horizontal underground stem of seagrasses and some macroalgae.

ribosome A small particle consisting of proteins and ribosomal RNAs found inside the cell, providing the framework for protein synthesis.

Richter scale A logarithmic scale used to measure the magnitude of earthquakes.

ridge and rise system The interconnecting chain of seafloor mountains that trace the edges of crustal plates and the sites of production of new oceanic crust.

salinity A measure of the total amount of dissolved solutes in seawater, dominated by inorganic ions.

salt gland A specialized structure of osmoregulating vertebrates that excretes excess ions into seawater against the concentration gradient.

salt marsh A low-lying intertidal grass community found along shores of temperate estuaries.

salt-wedge estuary Also known as *highly stratified estuaries*, they are the least mixed estuaries, occurring when the mouth of a river flows directly into saltwater.

saprobe Organism that absorbs nutrients from detritus and other nonliving organic matter by secreting digestive enzymes externally and absorbing the resultant breakdown products.

school A well-defined social organization of marine animals, usually consisting of a single species, with all members of a similar size all swimming in the same direction at the same speed.

seafloor spreading A global process of oceanic crust moving away from ridge and rise systems where it formed.

sea ice Frozen ocean water.

seamount An isolated undersea volcano that rises well above the seafloor but falls well beneath the sea's surface.

seasonal delayed implantation Stalled gestation of some mammals wherein the blastocyst remains dormant in the female's uterus for some time before implantation on the uterine wall.

selectively permeable membrane A membrane that enables atoms, molecules, compounds, and ions with certain properties (such as small size in general, nonpolar, small polar, uncharged, and small charged) to pass through but is not permeable to larger atoms in general, large polar molecules, large compounds, and large ions.

semidiurnal tides Tidal patterns with two high tides and two low tides each lunar day.

sensory hair cell Specialized cells found in the inner ear of vertebrates that detect movements and vibrations resulting in the senses of equilibrium, balance, and hearing.

septa The partitions separating internal parts of organisms (singular = septum).

sequential hermaphrodite An organism changing sex from one sex to the other.

settlement The process wherein a meroplanktonic larva metamorphoses into a benthic juvenile and colonizes a substratum that it has selected using unidentified criteria.

sex chromosome One of a pair of chromosomes whose composition determines an organism's sex.

sexual dimorphism An obvious and often dramatic difference in the size, appearance, and/or behavior of adult males and females of the same species.

shelf break The point at which the outer edge of the continental shelf meets the upper edge of the continental slope, typically at a depth of 120 to 200 m.

simultaneous hermaphrodite An organism that displays both sexes at the same time.

siphon Tubelike structure of mollusks used to take in and expel water from the mantle cavity.

slightly stratified estuary Also known as *partially mixed estuaries*, freshwater and saltwater mix at all depths, but the lowest layer remains salty.

smolt A young salmon just before it migrates downstream and out to sea.

sonar Sound navigation and ranging system used to propagate sound underwater for navigation, communication, or detection of objects in the water or on the seafloor.

speciation event The origin of a new species via natural selection.

species A particular kind of organism; members share a common ancestry, similar morphology, and, if the species reproduces sexually, reproductive isolation from other species.

sperm competition A strategy used by copulating males competing with previous copulators by attempting to displace or dilute their sperm within a female's reproductive tract, thus increasing the probability of being the male to fertilize the female's gamete(s).

spermaceti organ A large organ in the forehead of sperm whales that is filled with a fine-quality liquid or waxy spermaceti oil.

spicules Small, mineralized skeletal structures of sponges.

spongin Fibrous material making up the flexible skeleton of many sponges.

spongocoel The internal cavity of sponges.

sporangia Structures that form and enclose spores.

sporophyte The multicellular diploid stage in the life cycle of an alga or plant.

spring tide A tide that recurs every 2 weeks (near new and full moons) that is characterized by a greater-than-average tidal range.

standing crop Total amount of plant or animal material in an area at any one time.

stenohaline Being able to tolerate exposure to only slight variations in salinity.

stigma The receptive portion of a flower that receives pollen.

stipe The stalk that supports the blades of an alga.

stroma The fluid of the chloroplast containing the enzymes for the light-independent reactions of photosynthesis.

surface tension The mutual attraction of water molecules at the surface of a water mass that creates a flexible molecular "skin" over the water surface.

swash A small marine invertebrate that can migrate up and down the seashore by riding the rising and falling tides, remaining moist but not fully submerged.

swim bladder Gas-filled buoyancy organ of bony fishes.

symbiont Either of two organisms that live within a symbiotic relationship.

symbiotic relationship A close relationship between two organisms of different species.

taxa Groups or categories in a taxonomic system of classification.

taxonomy The science of naming and grouping organisms according to their evolutionary relationships.

tectonic estuary An estuary that fills in a depression created by one or more faults in the nearshore crust.

temperature A relative intensity measure of the condition caused by heat.

thermocline The ocean layer, usually near the bottom of the photic zone, marked by a sharp change in temperature that separates the warmer surface waters from the colder deep waters.

thermohaline circulation Mixing of the world ocean on a global scale that is initiated by the sinking of very cold, very dense seawater at the poles.

thermoregulate The act of maintaining a constant body temperature.

tidal range Vertical distance between high and low tides.

tide A long-period wave noticeable as a periodic rise and fall of the sea surface along coastlines.

trace element an element needed for normal metabolism but available only in minute amounts from the environment.

trachea The windpipe of vertebrates.

trade winds Subtropical winds that blow from northeast to southwest in the Northern Hemisphere and from southeast to northwest in the Southern Hemisphere.

trench Deep area in the ocean floor, generally deeper than 6,000 m, that is the site of subduction of oceanic crust into the Earth's core.

trochophore Early free-swimming, ciliated larval stage of many marine mollusks, annelid worms, ectoprocts, and brachiopods.

trophic Pertaining to feeding or nutrition.

trophic level The position of an organism or species in a food chain or web.

trophonemata Long filaments extending from the uterine wall of some elasmobranchs that secrete a fluid that nourishes their embryos.

trough The lowest portion of a wave.

tsunami A series of waves of enormous size, speed, and wavelength caused by a large disturbance to the ocean, such as a large earthquake or landslide.

turnover rate The rate at which members of a population or community replace themselves.

tympanic bulla Bony case in the middle ear that encloses the sound-processing structures of mammals.

ultraplankton Unicellular plankton with cells 2 to 5 μm in width.

ultrasonic Frequencies of sound that are above the normal range of human hearing (usually greater than 18 to 20 kHz).

upwelling The process that carries nutrient-rich deep waters upward to the photic zone.

uric acid The main nitrogenous excretory product in birds, reptiles, some invertebrates, and insects.

vascular tissue A conducting tissue found in vascular plants that transports fluids and nutrients from throughout the plant.

vasoconstriction Narrowing of blood vessel diameter via contraction of smooth muscles in the vessel wall with resulting reduction in blood flow and increase in blood pressure.

veliger Late molluscan larval stage which develops from the trocophore larvae with large ciliated lobes (a velum).

ventral The underside of a bilaterally symmetrical animal.

ventrolateral Location toward the junction of the lateral and ventral sides.

vertebrae The series of articulated bones that make up the backbone of vertebrates (singular = vertebra).

vertical migration Daily or seasonal movement of marine animals between the photic zone and midwater depths.

vertically mixed estuary An estuary with low river flow and great mixing at all layers.

vibrissae Thick tactile hairs.

virus A tiny (20 to 300 nm) particle of protein-coated genetic material that infects a living organism intracellularly; considered nonliving.

viscosity The resistance of water molecules to external forces that would separate them.

visible light The portion of electromagnetic radiation to which the human eye is sensitive, usually represented as a spectrum of colors from shorter-wavelength violet to longer-wavelength red.

viviparity The reproductive strategy wherein embryos develop within the mother while receiving nutrients via a placenta with subsequent live birth.

water vapor Water droplets suspended in air.

wave A periodic, traveling undulation of the sea surface.

wave drag A force resistive to movement that is created by the production of surface waves when swimming at or near the sea surface.

wavelength The distance between two waves, measured at the crests or troughs.

westerlies Winds that blow primarily from the west in the mid-latitudes of both hemispheres.

whale fall The carcass of a whale that has fallen into the deep zones of the ocean and supports numerous species, providing food and habitat in an otherwise featureless environment.

white band disease An epidemic coral disease that most often infects elkhorn, staghorn, and other corals in genus *Acropora*.

white plague A wasting disease of corals caused by a bacterium common to untreated human sewage.

wind wave A periodic, traveling undulation of the sea surface caused by wind transferring energy from the air to the water.

wrack line An area of the beach where detached seaweeds and seagrasses wash up on shore and decompose.

zooplankton Heterotrophic members of the plankton community.

zooxanthellae Photosynthetic, mutualistic dinoflagellates found in corals, sea anemones, mollusks, and several other types of marine animals.

zygote A single diploid cell that is the product of the fusion of two haploid gametes.

Index

Page numbers followed by "*f*" and "*t*" indicate figures and tables respectively.